Discrete Mathematics and Symmetry

Discrete Mathematics and Symmetry

Special Issue Editor

Angel Garrido

MDPI • Basel • Beijing • Wuhan • Barcelona • Belgrade

Special Issue Editor
Angel Garrido
Department of Fundamental
Mathematics, Faculty of
Sciences, UNED
Spain

Editorial Office
MDPI
St. Alban-Anlage 66
4052 Basel, Switzerland

This is a reprint of articles from the Special Issue published online in the open access journal *Symmetry* (ISSN 2073-8994) from 2018 to 2020 (available at: https://www.mdpi.com/journal/symmetry/special_issues/Discrete_Mathematics_Symmetry).

For citation purposes, cite each article independently as indicated on the article page online and as indicated below:

LastName, A.A.; LastName, B.B.; LastName, C.C. Article Title. *Journal Name* **Year**, *Article Number*, Page Range.

ISBN 978-3-03928-190-9 (Pbk)
ISBN 978-3-03928-191-6 (PDF)

Contents

About the Special Issue Editor ix

Preface to "Discrete Mathematics and Symmetry" xi

Michal Staš
On the Crossing Numbers of the Joining of a Specific Graph on Six Vertices with the Discrete Graph
Reprinted from: *Symmetry* **2020**, *12*, 135, doi:10.3390/sym12010135 1

Abdollah Alhevaz, Maryam Baghipur, Hilal Ahmad Ganie and Yilun Shang
Bounds for the Generalized Distance Eigenvalues of a Graph
Reprinted from: *Symmetry* **2019**, *11*, 1529, doi:10.3390/sym11121529 13

Jelena Dakić, Sanja Jančić-Rašović and Irina Cristea
Weak Embeddable Hypernear-Rings
Reprinted from: *Symmetry* **2019**, *11*, 964, doi:10.3390/sym11080964 30

Ahmed A. Elsonbaty and Salama Nagy Daoud
Edge Even Graceful Labeling of Cylinder Grid Graph
Reprinted from: *Symmetry* **2019**, *11*, 584, doi:10.3390/sym11040584 40

Xiaohong Zhang and Xiaoying Wu
Involution Abel–Grassmann's Groups and Filter Theory of Abel–Grassmann's Groups
Reprinted from: *Symmetry* **2019**, *11*, 553, doi:10.3390/sym11040553 70

Yilun Shang
Isoperimetric Numbers of Randomly Perturbed Intersection Graphs
Reprinted from: *Symmetry* **2019**, *11*, 452, doi:10.3390/sym11040452 83

Shunyi Liu
Generalized Permanental Polynomials of Graphs
Reprinted from: *Symmetry* **2019**, *11*, 242, doi:10.3390/sym11020242 92

Sergei Sokolov, Anton Zhilenkov, Sergei Chernyi, Anatoliy Nyrkov and David Mamunts
Dynamics Models of Synchronized Piecewise Linear Discrete Chaotic Systems of High Order
Reprinted from: *Symmetry* **2019**, *11*, 236, doi:10.3390/sym11020236 104

Michal Staš
Determining Crossing Number of Join of the Discrete Graph with Two Symmetric Graphs of Order Five
Reprinted from: *Symmetry* **2019**, *11*, 123, doi:10.3390/sym11020123 116

Salama Nagy Daoud
Edge Even Graceful Labeling of Polar Grid Graphs
Reprinted from: *Symmetry* **2019**, *11*, 38, doi:10.3390/sym11010038 125

Xiujun Zhang, Xinling Wu, Shehnaz Akhter, Muhammad Kamran Jamil, Jia-Bao Liu, Mohammad Reza Farahani
Edge-Version Atom-Bond Connectivity and Geometric Arithmetic Indices of Generalized Bridge Molecular Graphs
Reprinted from: *Symmetry* **2018**, *10*, 751, doi:10.3390/sym10120751 151

Hu Zhao and Hong-Ying Zhang
Some Results on Multigranulation Neutrosophic Rough Sets on a Single Domain
Reprinted from: *Symmetry* **2018**, *10*, 417, doi:10.3390/sym10090417 . . . . **376**

Mobeen Munir, Asim Naseem, Akhtar Rasool, Muhammad Shoaib Saleem and Shin Min Kang
Fixed Points Results in Algebras of Split Quaternion and Octonion
Reprinted from: *Symmetry* **2018**, *10*, 405, doi:10.3390/sym10090405 . . . . **388**

Jianping Wu, Boliang Lin, Hui Wang, Xuhui Zhang, Zhongkai Wang and Jiaxi Wang
Optimizing the High-Level Maintenance Planning Problem of the Electric Multiple Unit Train Using a Modified Particle Swarm Optimization Algorithm
Reprinted from: *Symmetry* **2018**, *10*, 349, doi:10.3390/sym10080349 . . . . **406**

Jihyun Choi and Jae-Hyouk Lee
Binary Icosahedral Group and 600-Cell
Reprinted from: *Symmetry* **2018**, *10*, 326, doi:10.3390/sym10080326 . . . . **420**

Marija Maksimović
Enumeration of Strongly Regular Graphs on up to 50 Vertices Having S_3 as an Automorphism Group
Reprinted from: *Symmetry* **2018**, *10*, 212, doi:10.3390/sym10060212 . . . . **434**

About the Special Issue Editor

Angel Garrido, Ch. of Differential Geometry, Department of Fundamental Mathematics, Faculty of Sciences, UNED, Madrid, Spain. BIOGRAPHYFull Name: Ángel Laureano Garrido BullónCurrent position: Chairman of Mathematics. Permanent Professor Doctor of the Faculty of Sciences of the UNED. Department of Fundamental Mathematics.Professional Experience: Polytechnic University of Madrid; University of Manchester; UNED.Degree: Licenciado (Degree) in Exact Sciences in the Faculty of Sciences of the Complutense University of Madrid. Programmer and Analyst at IBM. Master in Artificial Intelligence by the UNED. Full PhD studies in Mathematics, Informatics and Philosophy. PhD in Philosophy (UNED), especially of Logic and Foundations, with Summa Cum Laude in Thesis, by unanimity, and First Extraordinary Prize.He is the author of 27 books, published in prestigious editorials. Editor-in Chief of Axioms journal. MDPI Foundation Verlag, Basel, Switzerland. 232 papers published. Editor of Symmetry and Education Sciences, MDPI, as well as Brain, Bacau University, amongst others. Co-Director of 100cias@uned.es, a publication of the Faculty of Sciences, UNED, Madrid. Gold Medal from the University of Bacau.First Birkhäuser Prize, ICM, 2016.

Preface to "Discrete Mathematics and Symmetry"

SYMMETRY AND GEOMETRY. One of the core concepts essential to understanding natural phenomena and the dynamics of social systems is the concept of "relation". Furthermore, scientists rely on relational structures with high levels of symmetry because of their optimal behavior and high performance. Human friendships, social and interconnection networks, traffic systems, chemical structures, etc., can be expressed as relational structures. A mathematical model capturing the essence of this situation is a combinatorial object exhibiting a high level of symmetry, and the underlying mathematical discipline is algebraic combinatorics—the most vivid expression of the concept of symmetry in discrete mathematics. The purpose of this Special Issue of the journal Symmetry is to present some recent developments as well as possible future directions in algebraic combinatorics. Special emphasis is given to the concept of symmetry in graphs, finite geometries, and designs. Of interest are solutions of long-standing open problems in algebraic combinatorics, as well as contributions opening up new research topics encompassing symmetry within the boundaries of discrete mathematics but with the possibility of transcending these boundaries. Prof. Dr. Angel Garrido, Guest Editor.

Angel Garrido
Special Issue Editor

symmetry

MDPI

Article

On the Crossing Numbers of the Joining of a Specific Graph on Six Vertices with the Discrete Graph

Michal Staš

Faculty of Electrical Engineering and Informatics, Technical University of Košice, 042 00 Košice, Slovakia; michal.stas@tuke.sk

Received: 19 December 2019; Accepted: 19 December 2019; Published: 9 January 2020

Abstract: In the paper, we extend known results concerning crossing numbers of join products of small graphs of order six with discrete graphs. The crossing number of the join product $G^* + D_n$ for the graph G^* on six vertices consists of one vertex which is adjacent with three non-consecutive vertices of the 5-cycle. The proofs were based on the idea of establishing minimum values of crossings between two different subgraphs that cross the edges of the graph G^* exactly once. These minimum symmetrical values are described in the individual symmetric tables.

Keywords: graph; good drawing; crossing number; join product; cyclic permutation

1. Introduction

An investigation on the crossing number of graphs is a classical and very difficult problem. Garey and Johnson [1] proved that this problem is NP-complete. Recall that the exact values of the crossing numbers are known for only a few families of graphs. The purpose of this article is to extend the known results concerning this topic. In this article, we use the definitions and notation of the crossing numbers of graphs presented by Klešč in [2]. Kulli and Muddebihal [3] described the characterization for all pairs of graphs which join product of a planar graph. In the paper, some parts of proofs are also based on Kleitman's result [4] on the crossing numbers for some complete bipartite graphs. More precisely, he showed that

$$\mathrm{cr}(K_{m,n}) = \left\lfloor \frac{m}{2} \right\rfloor \left\lfloor \frac{m-1}{2} \right\rfloor \left\lfloor \frac{n}{2} \right\rfloor \left\lfloor \frac{n-1}{2} \right\rfloor, \quad \text{for} \quad m \le 6.$$

Again, by Kleitman's result [4], the crossing numbers for the join of two different paths, the join of two different cycles, and also for the join of path and cycle, were established in [2]. Further, the exact values for crossing numbers of $G + D_n$ and of $G + P_n$ for all graphs G on less than five vertices were determined in [5]. At present, the crossing numbers of the graphs $G + D_n$ are known only for few graphs G of order six in [6–9]. In all these cases, the graph G is usually connected and includes at least one cycle.

The methods in the paper mostly use the combinatorial properties of cyclic permutations. For the first time, the idea of configurations is converted from the family of subgraphs which do not cross the edges of the graph G^* of order six onto the family of subgraphs whose edges cross the edges of G^* just once. According to this algebraic topological approach, we can extend known results for the crossing numbers of new graphs. Some of the ideas and methods were used for the first time in [10]. In [6,8,9], some parts of proofs were done with the help of software which is described in detail in [11]. It is important to recall that the methods presented in [5,7,12] do not suffice to determine the crossing number of the graph $G^* + D_n$. Also in this article, some parts of proofs can be simplified by utilizing the work of the software that generates all cyclic permutations in [11]. Its C++ version is located also on the website http://web.tuke.sk/fei-km/coga/, and the list with all short names of 120 cyclic permutations of six elements have already been collected in Table 1 of [8].

2. Cyclic Permutations and Corresponding Configurations of Subgraphs

Let G^* be the connected graph on six vertices consisting of one vertex which is adjacent with three non-consecutive vertices of the 5-cycle. We consider the join product of the graph G^* with the discrete graph D_n on n vertices. It is not difficult to see that the graph $G^* + D_n$ consists of just one copy of the graph G^* and of n vertices $t_1, \dots, t_n$, where any vertex t_j, $j = 1, \dots, n$, is adjacent to every vertex of the graph G^*. Let T^j, $j = 1, \dots, n$, denote the subgraph which is uniquely induced by the six edges incident with the fixed vertex t_j. This means that the graph $T^1 \cup \dots \cup T^n$ is isomorphic with $K_{6,n}$ and

$$G^* + D_n = G^* \cup K_{6,n} = G^* \cup \left(\bigcup_{j=1}^{n} T^j \right). \qquad (1)$$

In the paper, the definitions and notation of the cyclic permutations and of the corresponding configurations of subgraphs for a good drawing D of the graph $G^* + D_n$ presented in [8] are used. The *rotation* $\mathrm{rot}_D(t_j)$ of a vertex t_j in the drawing D is the cyclic permutation that records the (cyclic) counter-clockwise order in which the edges leave t_j, see [10]. We use the notation (123456) if the counterclockwise order of the edges incident with the vertex t_j is t_jv_1, t_jv_2, t_jv_3, t_jv_4, t_jv_5, and t_jv_6. Recall that a rotation is a cyclic permutation. Moreover, as we have already mentioned, we separate all subgraphs T^j, $j = 1, \dots, n$, of the graph $G^* + D_n$ into three mutually-disjoint families depending on how many times the edges of G^* are crossed by the edges of the considered subgraph T^j in D. This means, for $j = 1, \dots, n$, let $R_D = \{T^j : \mathrm{cr}_D(G^*, T^j) = 0\}$ and $S_D = \{T^j : \mathrm{cr}_D(G^*, T^j) = 1\}$. The edges of G^* are crossed by each other subgraph T^j at least twice in D. For $T^j \in R_D \cup S_D$, let F^j denote the subgraph $G^* \cup T^j$, $j \in \{1, 2, \dots, n\}$, of $G^* + D_n$, and let $D(F^j)$ be its subdrawing induced by D.

If we would like to obtain an optimal drawing D of $G^* + D_n$, then the set $R_D \cup S_D$ must be nonempty provided by the arguments in Theorem 1. Thus, we only consider drawings of the graph G^* for which there is a possibility of obtaining a subgraph $T^j \in R_D \cup S_D$. Since the graph G^* contains the 6-cycle as a subgraph (for brevity, we can write $C_6(G^*)$), we have to assume only crossings between possible subdrawings of the subgraph $C_6(G^*)$ and two remaining edges of G^*. Of course, the edges of the cycle $C_6(G^*)$ can cross themselves in the considered subdrawings. The vertex notation of G^* will be substantiated later in all drawings in Figure 1.

First, assume a good drawing D of $G^* + D_n$ in which the edges of G^* do not cross each other. In this case, without loss of generality, we can consider the drawing of G^* with the vertex notation like that in Figure 1a. Clearly, the set R_D is empty. Our aim is to list all possible rotations $\mathrm{rot}_D(t_j)$ which can appear in D if the edges of G^* are crossed by the edges of T^j just once. There is only one possible subdrawing of $F^j \setminus \{v_4\}$ represented by the rotation (16532), which yields that there are exactly five ways of obtaining the subdrawing of $G \cup T^j$ depending on which edge of the graph G^* can be crossed by the edge t_jv_4. We denote these five possibilities by $\mathcal{A}_k$, for $k = 1, \dots, 5$. For our considerations over the number of crossings of $G^* + D_n$, it does not play a role in which of the regions is unbounded. So we can assume the drawings shown in Figure 2. Thus, the configurations $\mathcal{A}_1$, $\mathcal{A}_2$, $\mathcal{A}_3$, $\mathcal{A}_4$, and $\mathcal{A}_5$ are represented by the cyclic permutations (165324), (165432), (146532), (165342), and (164532), respectively. Of course, in a fixed drawing of the graph $G^* + D_n$, some configurations from $\mathcal{M} = \{\mathcal{A}_1, \mathcal{A}_2, \mathcal{A}_3, \mathcal{A}_4, \mathcal{A}_5\}$ need not appear. We denote by $\mathcal{M}_D$ the set of all configurations that exist in the drawing D belonging to the set $\mathcal{M}$.

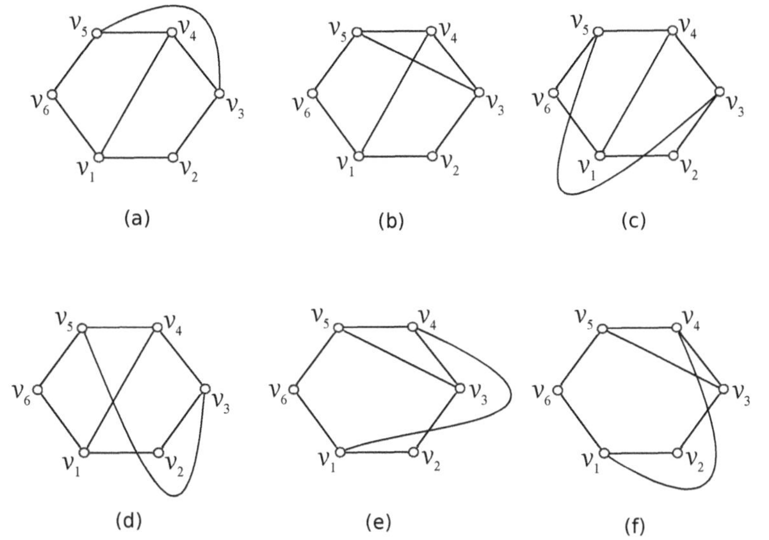

Figure 1. Six possible drawings of G^* with no crossing among edges of $C_6(G^*)$. **(a)**: the planar drawing of G^*; **(b)**: the drawing of G^* with $\mathrm{cr}_D(G^*) = 1$ and without crossing on edges of $C_6(G^*)$; **(c)**: the drawing of G^* only with two crossings on edges of $C_6(G^*)$; **(d)**: the drawing of G^* with $\mathrm{cr}_D(G^*) = 2$ and with one crossing on edges of $C_6(G^*)$; **(e)**: the drawing of G^* only with one crossing on edges of $C_6(G^*)$; **(f)**: the drawing of G^* with $\mathrm{cr}_D(G^*) = 2$ and with one crossing on edges of $C_6(G^*)$.

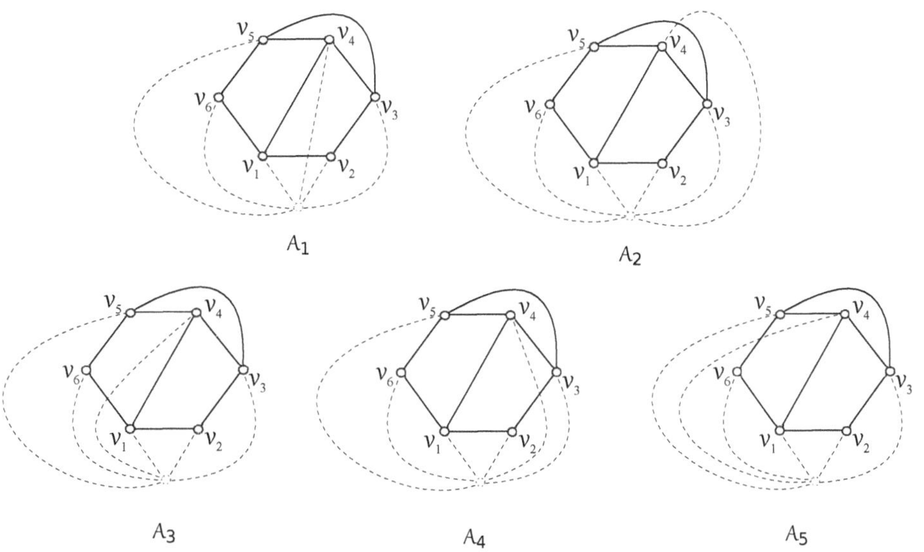

Figure 2. Drawings of five possible configurations from $\mathcal{M}$ of the subgraph F^j.

Recall that we are able to extend the idea of establishing minimum values of crossings between two different subgraphs onto the family of subgraphs which cross the edges of G^* exactly once. Let $\mathcal{X}$ and $\mathcal{Y}$ be the configurations from $\mathcal{M}_D$. We denote by $\mathrm{cr}_D(\mathcal{X}$ and $\mathcal{Y})$ the number of crossings in D between T^i and T^j for different $T^i, T^j \in S_D$ such that F^i and F^j have configurations $\mathcal{X}$ and $\mathcal{Y}$, respectively. Finally, let $\mathrm{cr}(\mathcal{X}, \mathcal{Y}) = \min\{\mathrm{cr}_D(\mathcal{X}, \mathcal{Y})\}$ over all possible good drawings of $G^* + D_n$ with $\mathcal{X}, \mathcal{Y} \in \mathcal{M}_D$.

Our aim is to determine $\mathrm{cr}(\mathcal{X},\mathcal{Y})$ for all such pairs $\mathcal{X},\mathcal{Y} \in \mathcal{M}$. In particular, the configurations $\mathcal{A}_1$ and $\mathcal{A}_2$ are represented by the cyclic permutations (165324) and (165432), respectively. Since the minimum number of interchanges of adjacent elements of (165324) required to produce cyclic permutation (165432) is two, we need at least four interchanges of adjacent elements of (165432) to produce cyclic permutation $\overline{(165324)} = (142356)$. (Let T^x and T^y be two different subgraphs represented by their $\mathrm{rot}(t_x)$ and $\mathrm{rot}(t_y)$ of length m, $m \geq 3$. If the minimum number of interchanges of adjacent elements of $\mathrm{rot}(t_x)$ required to produce $\mathrm{rot}(t_y)$ is at most z, then $\mathrm{cr}_D(T^x,T^y) \geq \lfloor \frac{m}{2} \rfloor \lfloor \frac{m-1}{2} \rfloor - z$. Details have been worked out by Woodall [13].) So any subgraph T^j with the configuration $\mathcal{A}_2$ of F^j crosses the edges of T^i with the configuration $\mathcal{A}_1$ of F^i at least four times; that is, $\mathrm{cr}(\mathcal{A}_1,\mathcal{A}_2) \geq 4$. The same reasoning gives $\mathrm{cr}(\mathcal{A}_1,\mathcal{A}_3) \geq 5$, $\mathrm{cr}(\mathcal{A}_1,\mathcal{A}_4) \geq 5$, $\mathrm{cr}(\mathcal{A}_1,\mathcal{A}_5) \geq 4$, $\mathrm{cr}(\mathcal{A}_2,\mathcal{A}_3) \geq 4$, $\mathrm{cr}(\mathcal{A}_2,\mathcal{A}_4) \geq 5$, $\mathrm{cr}(\mathcal{A}_2,\mathcal{A}_5) \geq 5$, $\mathrm{cr}(\mathcal{A}_3,\mathcal{A}_4) \geq 4$, $\mathrm{cr}(\mathcal{A}_3,\mathcal{A}_5) \geq 5$, and $\mathrm{cr}(\mathcal{A}_4,\mathcal{A}_5) \geq 4$. Clearly, also $\mathrm{cr}(\mathcal{A}_i,\mathcal{A}_i) \geq 6$ for any $i = 1,\ldots,5$. All resulting lower bounds for the number of crossings of two configurations from $\mathcal{M}$ are summarized in the symmetric Table 1 (here, $\mathcal{A}_k$ and $\mathcal{A}_l$ are configurations of the subgraphs F^i and F^j, where $k,l \in \{1,2,3,4,5\}$).

Table 1. The necessary number of crossings between T^i and T^j for the configurations $\mathcal{A}_k$, $\mathcal{A}_l$.

-	$\mathcal{A}_1$	$\mathcal{A}_2$	$\mathcal{A}_3$	$\mathcal{A}_4$	$\mathcal{A}_5$
$\mathcal{A}_1$	6	4	5	5	4
$\mathcal{A}_2$	4	6	4	5	5
$\mathcal{A}_3$	5	4	6	4	5
$\mathcal{A}_4$	5	5	4	6	4
$\mathcal{A}_5$	4	5	5	4	6

Assume a good drawing D of the graph $G^* + D_n$ with just one crossing among edges of the graph G^* (in which there is a possibility of obtaining of subgraph $T^j \in R_D \cup S_D$). At first, without loss of generality, we can consider the drawing of G^* with the vertex notation like that in Figure 1b. Of course, the set R_D can be nonempty, but our aim will be also to list all possible rotations $\mathrm{rot}_D(t_j)$ which can appear in D if the edges of G^* are crossed by the edges of T^j just once. Since the edges v_1v_2, v_2v_3, v_1v_6, and v_5v_6 of G^* can be crossed by the edges t_jv_3, t_jv_1, t_jv_5, and t_jv_1, respectively, these four ways under our consideration can be denoted by $\mathcal{B}_k$, for $k = 1,2,3,4$. Based on the aforementioned arguments, we assume the drawings shown in Figure 3.

Thus, the configurations $\mathcal{B}_1$, $\mathcal{B}_2$, $\mathcal{B}_3$, and $\mathcal{B}_4$ are uniquely represented by the cyclic permutations (165423), (126543), (156432), and (154326), respectively. Because some configurations from $\mathcal{N} = \{\mathcal{B}_1,\mathcal{B}_2,\mathcal{B}_3,\mathcal{B}_4\}$ may not appear in a fixed drawing of $G^* + D_n$, we denote by $\mathcal{N}_D$ the subset of $\mathcal{N}$ consisting of all configurations that exist in the drawing D. Further, due to the properties of the cyclic rotations, we can easily verify that $\mathrm{cr}(\mathcal{B}_i,\mathcal{B}_j) \geq 4$ for any $i,j \in \{1,2,3,4\}$, $i \neq j$. (Let us note that this idea was used for an establishing the values in Table 1)

In addition, without loss of generality, we can consider the drawing of G^* with the vertex notation like that in Figure 1e. In this case, the set R_D is also empty. Hence, our aim is to list again all possible rotations $\mathrm{rot}_D(t_j)$ which can appear in D if $T^j \in S_D$. Since there is only one subdrawing of $F^j \setminus \{v_3\}$ represented by the rotation (16542), there are four ways to obtain the subdrawing of F^j depending on which edge of G^* is crossed by the edge t_jv_3. These four possibilities under our consideration are denoted by $\mathcal{E}_k$, for $k = 1,2,3,4$. Again, based on the aforementioned arguments, we assume the drawings shown in Figure 4.

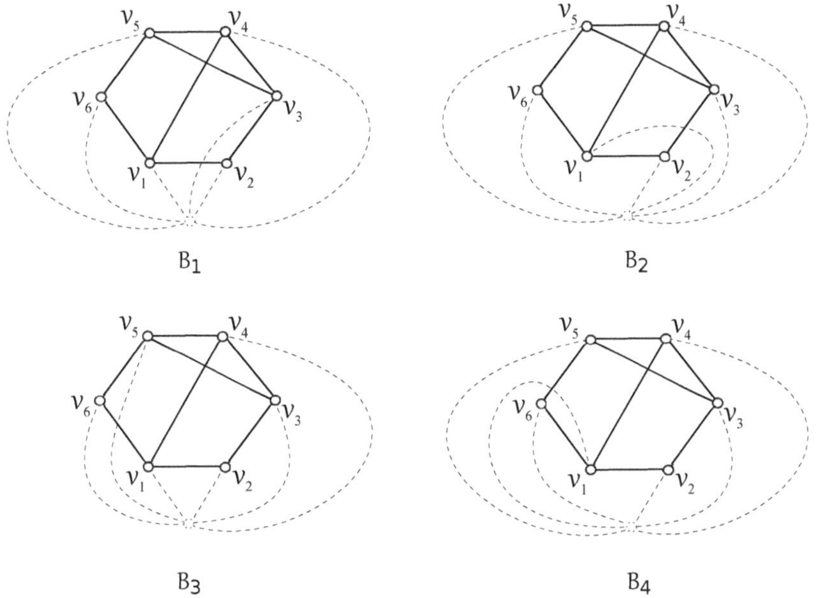

Figure 3. Drawings of four possible configurations from $\mathcal{N}$ of the subgraph F^j.

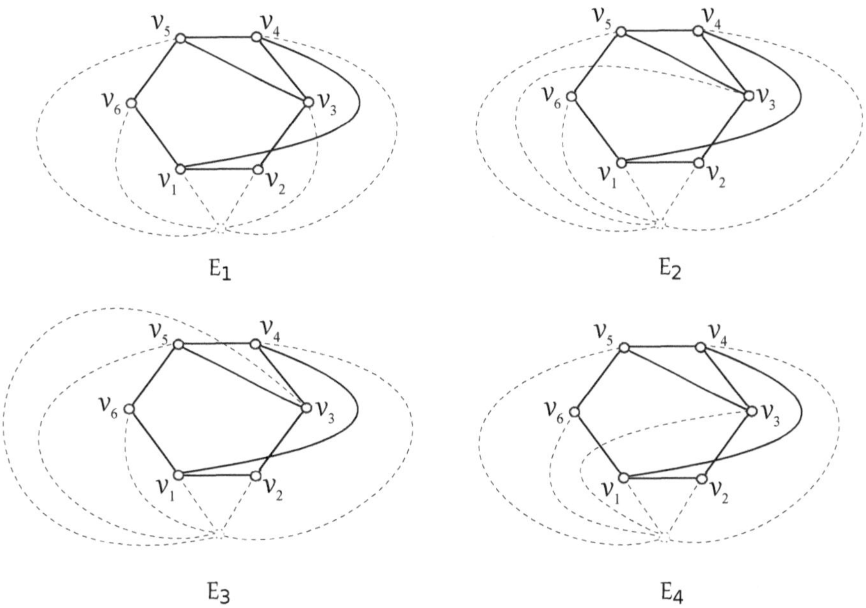

Figure 4. Drawings of four possible configurations from $\mathcal{O}$ of the subgraph F^j.

Thus, the configurations $\mathcal{E}_1$, $\mathcal{E}_2$, $\mathcal{E}_3$, and $\mathcal{E}_4$ are represented by the cyclic permutations (165432), (163542), (165342), and (136542), respectively. Again, we denote by $\mathcal{O}_D$ the subset of $\mathcal{O} = \{\mathcal{E}_1, \mathcal{E}_2, \mathcal{E}_3, \mathcal{E}_4\}$ consisting of all configurations that exist in the drawing D. Further, due to the properties of the cyclic rotations, all lower-bounds of number of crossings of two configurations from $\mathcal{O}$ can be summarized in the symmetric Table 2 (here, $\mathcal{E}_k$ and $\mathcal{E}_l$ are configurations of the subgraphs F^i and F^j, where $k, l \in \{1,2,3,4\}$).

Table 2. The necessary number of crossings between T^i and T^j for the configurations $\mathcal{E}_k$, $\mathcal{E}_l$.

-	$\mathcal{E}_1$	$\mathcal{E}_2$	$\mathcal{E}_3$	$\mathcal{E}_4$
$\mathcal{E}_1$	6	4	5	4
$\mathcal{E}_2$	4	6	5	5
$\mathcal{E}_3$	5	5	6	4
$\mathcal{E}_4$	4	5	4	6

Finally, without loss of generality, we can consider the drawing of G^* with the vertex notation like that in Figure 1f. In this case, the set R_D is also empty. So our aim will be to list again all possible rotations $\text{rot}_D(t_j)$ which can appear in D if $T^j \in S_D$. Since there is only one subdrawing of $F^j \setminus \{v_2\}$ represented by the rotation (16543), there are three ways to obtain the subdrawing of F^j depending on which edge of G^* is crossed by the edge $t_j v_2$. These three possibilities under our consideration are denoted by $\mathcal{F}_k$, for $k = 1, 2, 3$. Again, based on the aforementioned arguments, we assume the drawings shown in Figure 5.

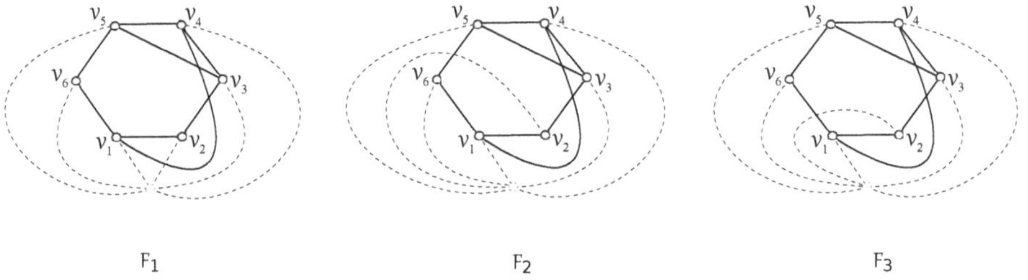

Figure 5. Drawings of three possible configurations from $\mathcal{P}$ of the subgraph F^j.

Thus, the configurations $\mathcal{F}_1$, $\mathcal{F}_2$, and $\mathcal{F}_3$ are represented by the cyclic permutations (165432), (162543), and (126543), respectively. Again, we denote by $\mathcal{P}_D$ the subset of $\mathcal{P} = \{\mathcal{F}_1, \mathcal{F}_2, \mathcal{F}_3\}$ consisting of all configurations that exist in the drawing D. Further, due to the properties of the cyclic rotations, all lower-bounds of number of crossings of two configurations from $\mathcal{P}$ can be summarized in the symmetric Table 3 (here, $\mathcal{F}_k$ and $\mathcal{F}_l$ are configurations of the subgraphs F^i and F^j, where $k, l \in \{1, 2, 3\}$).

Table 3. The necessary number of crossings between T^i and T^j for the configurations $\mathcal{F}_k$ and $\mathcal{F}_l$.

-	$\mathcal{F}_1$	$\mathcal{F}_2$	$\mathcal{F}_3$
$\mathcal{F}_1$	6	4	5
$\mathcal{F}_2$	4	6	5
$\mathcal{F}_3$	5	5	6

3. The Crossing Number of $G^* + D_n$

Recall that two vertices t_i and t_j of $G^* + D_n$ are *antipodal* in a drawing D of $G^* + D_n$ if the subgraphs T^i and T^j do not cross. A drawing is *antipodal-free* if it has no antipodal vertices. For easier and more accurate labeling in the proofs of assertions, let us define notation of regions in some subdrawings of $G^* + D_n$. The unique drawing of G^* as shown in Figure 1a contains four different regions. Let us denote these four regions by $\omega_{1,2,3,4}$, $\omega_{1,4,5,6}$, $\omega_{3,4,5}$, and $\omega_{1,2,3,5,6}$ depending on which of vertices are located on the boundary of the corresponding region.

Lemma 1. *Let D be a good and antipodal-free drawing of $G^* + D_n$, for $n > 3$, with the drawing of G^* with the vertex notation like that in Figure 1a. If $T^u, T^v, T^t \in S_D$ are three different subgraphs such that F^u, F^v, and F^t have three different configurations from the set $\{\mathcal{A}_i, \mathcal{A}_j, \mathcal{A}_k\} \subseteq \mathcal{M}_D$ with $i + 2 \equiv j + 1 \equiv k \pmod 5$, then*

$$\text{cr}_D(G^* \cup T^u \cup T^v \cup T^t, T^m) \geq 6 \qquad \text{for any } T^m \notin S_D.$$

Proof of Lemma 1. Let us assume the configurations $\mathcal{A}_1$ of F^u, $\mathcal{A}_2$ of F^v, and $\mathcal{A}_3$ of F^t. It is obvious that $\mathrm{cr}_D(T^u \cup T^v \cup T^t, T^m) \geq 3$ holds for any subgraph T^m, $m \neq u, v, t$. Further, if $\mathrm{cr}_D(G^*, T^m) > 2$, then we obtain the desired result $\mathrm{cr}_D(G^* \cup T^u \cup T^v \cup T^t, T^m) \geq 3 + 3 = 6$. To finish the proof, let us suppose that there is a subgraph $T^m \notin S_D$ such that T^m crosses exactly once the edges of each subgraph T^u, T^v, and T^t, and let also consider $\mathrm{cr}_D(G^*, T^m) = 2$. As $\mathrm{cr}_D(T^u, T^m) = 1$, the vertex t_m must be placed in the quadrangular region with four vertices of G^* on its boundary; that is, $t_m \in \omega_{1,4,5,6}$. Similarly, the assumption $\mathrm{cr}_D(T^t, T^m) = 1$ enforces that $t_m \in \omega_{1,2,3,4}$. Since the vertex t_m cannot be placed simultaneously in both regions, we obtain a contradiction. The proof proceeds in the similar way also for the remaining possible cases of the configurations of subgraphs F^u, F^v, and F^t, and the proof is done. □

Now we are able to prove the main result of the article. We can calculate the exact values of crossing numbers for small graphs using an algorithm located on a website http://crossings.uos.de/. It uses an ILP formulation based on Kuratowski subgraphs. The system also generates verifiable formal proofs like those described in [14]. Unfortunately, the capacity of this system is limited.

Lemma 2. $\mathrm{cr}(G^* + D_1) = 1$ *and* $\mathrm{cr}(G^* + D_2) = 3$.

Theorem 1. $\mathrm{cr}(G^* + D_n) = 6\left\lfloor \frac{n}{2} \right\rfloor \left\lfloor \frac{n-1}{2} \right\rfloor + n + \left\lfloor \frac{n}{2} \right\rfloor$ *for* $n \geq 1$.

Proof of Theorem 1. Figure 6 offers the drawing of $G^* + D_n$ with exactly $6\lfloor \frac{n}{2} \rfloor \lfloor \frac{n-1}{2} \rfloor + n + \lfloor \frac{n}{2} \rfloor$ crossings. Thus, $\mathrm{cr}(G^* + D_n) \leq 6\lfloor \frac{n}{2} \rfloor \lfloor \frac{n-1}{2} \rfloor + n + \lfloor \frac{n}{2} \rfloor$. We prove the reverse inequality by induction on n. By Lemma 2, the result is true for $n = 1$ and $n = 2$. Now suppose that, for some $n \geq 3$, there is a drawing D with

$$\mathrm{cr}_D(G^* + D_n) < 6\left\lfloor \frac{n}{2} \right\rfloor \left\lfloor \frac{n-1}{2} \right\rfloor + n + \left\lfloor \frac{n}{2} \right\rfloor \tag{2}$$

and that

$$\mathrm{cr}(G^* + D_m) \geq 6\left\lfloor \frac{m}{2} \right\rfloor \left\lfloor \frac{m-1}{2} \right\rfloor + m + \left\lfloor \frac{m}{2} \right\rfloor \qquad \text{for any integer } m < n. \tag{3}$$

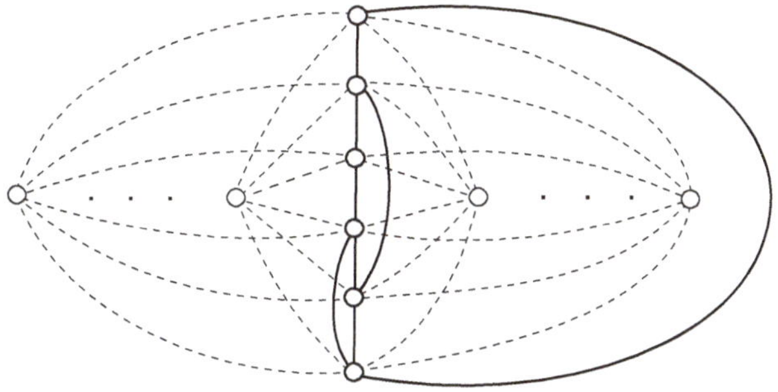

Figure 6. The good drawing of $G^* + D_n$ with $6\lfloor \frac{n}{2} \rfloor \lfloor \frac{n-1}{2} \rfloor + n + \lfloor \frac{n}{2} \rfloor$ crossings.

Let us first show that the considered drawing D must be antipodal-free. For a contradiction, suppose, without loss of generality, that $\mathrm{cr}_D(T^{n-1}, T^n) = 0$. If at least one of T^{n-1} and T^n, say T^n, does not cross G^*, it is not difficult to verify in Figure 1 that T^{n-1} must cross $G^* \cup T^n$ at least trice; that is, $\mathrm{cr}_D(G^*, T^{n-1} \cup T^n) \geq 3$. From [4], we already know that $\mathrm{cr}(K_{6,3}) = 6$, which yields that the edges of the subgraph $T^{n-1} \cup T^n$ are crossed by any T^k, $k = 1, 2, \ldots, n-2$, at least six times. So, for the number of crossings in D we have:

$$\mathrm{cr}_D(G^* + D_n) = \mathrm{cr}_D(G^* + D_{n-2}) + \mathrm{cr}_D(T^{n-1} \cup T^n) + \mathrm{cr}_D(K_{6,n-2}, T^{n-1} \cup T^n) + \mathrm{cr}_D(G^*, T^{n-1} \cup T^n)$$

$$\geq 6\left\lfloor\frac{n-2}{2}\right\rfloor\left\lfloor\frac{n-3}{2}\right\rfloor+n-2+\left\lfloor\frac{n-2}{2}\right\rfloor+6(n-2)+3=6\left\lfloor\frac{n}{2}\right\rfloor\left\lfloor\frac{n-1}{2}\right\rfloor+n+\left\lfloor\frac{n}{2}\right\rfloor.$$

This contradiction with the assumption (2) confirms that D is antipodal-free. Moreover, if $r=|R_D|$ and $s=|S_D|$, the assumption (3) together with $\mathrm{cr}(K_{6,n})=6\lfloor\frac{n}{2}\rfloor\lfloor\frac{n-1}{2}\rfloor$ imply that, in D, if $r=0$, then there are at least $\lceil\frac{n}{2}\rceil+1$ subgraphs T^j for which the edges of G^* are crossed just once by them. More precisely:

$$\mathrm{cr}_D(G^*)+\mathrm{cr}_D(G^*,K_{6,n})\leq \mathrm{cr}_D(G^*)+0r+1s+2(n-r-s)<n+\left\lfloor\frac{n}{2}\right\rfloor;$$

that is,

$$s+2(n-r-s)<n+\left\lfloor\frac{n}{2}\right\rfloor. \tag{4}$$

This enforces that $2r+s\geq n-\lfloor\frac{n}{2}\rfloor+1$, and if $r=0$, then $s\geq n-\lfloor\frac{n}{2}\rfloor+1=\lceil\frac{n}{2}\rceil+1$. Now, for $T^j\in R_D\cup S_D$, we discuss the existence of possible configurations of subgraphs $F^j=G^*\cup T^j$ in D.

Case 1: $\mathrm{cr}_D(G^*)=0$. Without loss of generality, we can consider the drawing of G^* with the vertex notation like that in Figure 1a. It is obvious that the set R_D is empty; that is, $r=0$. Thus, we deal with only the configurations belonging to the nonempty set $\mathcal{M}_D$ and we discuss over all cardinalities of the set $\mathcal{M}_D$ in the following subcases:

i. $|\mathcal{M}_D|\geq 3$. We consider two subcases. Let us first assume that $\{\mathcal{A}_i,\mathcal{A}_j,\mathcal{A}_k\}\subseteq\mathcal{M}_D$ with $i+2\equiv j+1\equiv k \pmod 5$. Without lost of generality, let us consider three different subgraphs T^{n-2}, T^{n-1}, $T^n\in S_D$ such that F^{n-2}, F^{n-1} and F^n have configurations $\mathcal{A}_i$, $\mathcal{A}_j$, and $\mathcal{A}_k$, respectively. Then, $\mathrm{cr}_D(T^{n-2}\cup T^{n-1}\cup T^n,T^m)\geq 14$ holds for any $T^m\in S_D$ with $m\neq n-2,n-1,n$ by summing the values in all columns in the considered three rows of Table 1. Moreover, $\mathrm{cr}_D(G^*\cup T^{n-2}\cup T^{n-1}\cup T^n,T^m)\geq 6$ is fulfilling for any subgraph $T^m\notin S_D$ by Lemma 1. $\mathrm{cr}_D(T^{n-2}\cup T^{n-1}\cup T^n)\geq 13$ holds by summing of three corresponding values of Table 1 between the considered configurations $\mathcal{A}_i$, $\mathcal{A}_j$, and $\mathcal{A}_k$, by fixing the subgraph $G^*\cup T^{n-2}\cup T^{n-1}\cup T^n$,

$$\mathrm{cr}_D(G^*+D_n)=\mathrm{cr}_D(K_{6,n-3})+\mathrm{cr}_D(K_{6,n-3},G^*\cup T^{n-2}\cup T^{n-1}\cup T^n)+\mathrm{cr}_D(G^*\cup T^{n-2}\cup T^{n-1}\cup T^n)$$

$$\geq 6\left\lfloor\frac{n-3}{2}\right\rfloor\left\lfloor\frac{n-4}{2}\right\rfloor+15(s-3)+6(n-s)+13+3=6\left\lfloor\frac{n-3}{2}\right\rfloor\left\lfloor\frac{n-4}{2}\right\rfloor+6n+9s-29$$

$$\geq 6\left\lfloor\frac{n-3}{2}\right\rfloor\left\lfloor\frac{n-4}{2}\right\rfloor+6n+9\left(\left\lceil\frac{n}{2}\right\rceil+1\right)-29\geq 6\left\lfloor\frac{n}{2}\right\rfloor\left\lfloor\frac{n-1}{2}\right\rfloor+n+\left\lfloor\frac{n}{2}\right\rfloor.$$

In addition, let us assume that $\mathcal{M}_D=\{\mathcal{A}_i,\mathcal{A}_j,\mathcal{A}_k\}$ with $i+1\equiv j \pmod 5$, $j+1\not\equiv k \pmod 5$, and $k+1\not\equiv i \pmod 5$. Without lost of generality, let us consider two different subgraphs T^{n-1}, $T^n\in S_D$ such that F^{n-1} and F^n have mentioned configurations $\mathcal{A}_i$ and $\mathcal{A}_j$, respectively. Then, $\mathrm{cr}_D(G^*\cup T^{n-1}\cup T^n,T^m)\geq 1+10=11$ holds for any $T^m\in S_D$ with $m\neq n-1,n$ also, by summing the values in Table 1. Hence, by fixing the subgraph $G^*\cup T^{n-1}\cup T^n$,

$$\mathrm{cr}_D(G^*+D_n)=\mathrm{cr}_D(K_{6,n-2})+\mathrm{cr}_D(K_{6,n-2},G^*\cup T^{n-1}\cup T^n)+\mathrm{cr}_D(G^*\cup T^{n-1}\cup T^n)$$

$$\geq 6\left\lfloor\frac{n-2}{2}\right\rfloor\left\lfloor\frac{n-3}{2}\right\rfloor+11(s-2)+4(n-s)+4+2=6\left\lfloor\frac{n-2}{2}\right\rfloor\left\lfloor\frac{n-3}{2}\right\rfloor+4n+7s-16$$

$$\geq 6\left\lfloor\frac{n-2}{2}\right\rfloor\left\lfloor\frac{n-3}{2}\right\rfloor+4n+7\left(\left\lceil\frac{n}{2}\right\rceil+1\right)-16\geq 6\left\lfloor\frac{n}{2}\right\rfloor\left\lfloor\frac{n-1}{2}\right\rfloor+n+\left\lfloor\frac{n}{2}\right\rfloor.$$

ii. $|\mathcal{M}_D|=2$; that is, $\mathcal{M}_D=\{\mathcal{A}_i,\mathcal{A}_j\}$ for some $i,j\in\{1,\dots,5\}$ with $i\neq j$. Without lost of generality, let us consider two different subgraphs T^{n-1}, $T^n\in S_D$ such that F^{n-1} and F^n have mentioned configurations $\mathcal{A}_i$ and $\mathcal{A}_j$, respectively. Then, $\mathrm{cr}_D(G^*\cup T^{n-1}\cup T^n,T^m)\geq 1+10=11$ holds for any $T^m\in S_D$ with $m\neq n-1,n$ also by Table 1. Thus, by fixing the subgraph $G^*\cup T^{n-1}\cup T^n$, we are able to use the same inequalities as in the previous subcase.

iii. $|\mathcal{M}_D| = 1$; that is, $\mathcal{M}_D = \{\mathcal{A}_j\}$ for only one $j \in \{1,\ldots,5\}$. Without lost of generality, let us assume that $T^n \in S_D$ with the configuration $\mathcal{A}_j \in \mathcal{M}_D$ of the subgraph F^n. As $\mathcal{M}_D = \{\mathcal{A}_j\}$, we have $\text{cr}_D(G^* \cup T^n, T^k) \geq 1 + 6 = 7$ for any $T^k \in S_D, k \neq n$ provided that $\text{rot}_D(t_n) = \text{rot}_D(t_k)$, for more see [13]. Hence, by fixing the subgraph $G^* \cup T^n$,

$$\text{cr}_D(G^* + D_n) = \text{cr}_D(K_{6,n-1}) + \text{cr}_D(K_{6,n-1}, G^* \cup T^n) + \text{cr}_D(G^* \cup T^n)$$

$$\geq 6\left\lfloor\frac{n-1}{2}\right\rfloor\left\lfloor\frac{n-2}{2}\right\rfloor + 7(s-1) + 3(n-s) + 1 = 6\left\lfloor\frac{n-1}{2}\right\rfloor\left\lfloor\frac{n-2}{2}\right\rfloor + 3n + 4s - 6$$

$$\geq 6\left\lfloor\frac{n-1}{2}\right\rfloor\left\lfloor\frac{n-2}{2}\right\rfloor + 3n + 4\left(\left\lceil\frac{n}{2}\right\rceil + 1\right) - 6 \geq 6\left\lfloor\frac{n}{2}\right\rfloor\left\lfloor\frac{n-1}{2}\right\rfloor + n + \left\lfloor\frac{n}{2}\right\rfloor.$$

Case 2: $\text{cr}_D(G^*) = 1$ with $\text{cr}_D(C_6(G^*)) = 0$. At first, without loss of generality, we can consider the drawing of G^* with the vertex notation like that in Figure 1b. Since the set R_D can be nonempty, two possible subcases may occur:

i. Let R_D be the nonempty set; that is, there is a subgraph $T^i \in R_D$. Now, for a $T^i \in R_D$, the reader can easily see that the subgraph $F^i = G^* \cup T^i$ is uniquely represented by $\text{rot}_D(t_i) = (165432)$, and $\text{cr}_D(T^i, T^j) \geq 6$ for any $T^j \in R_D$ with $j \neq i$ provided that $\text{rot}_D(t_i) = \text{rot}_D(t_j)$; for more see [13]. Moreover, it is not difficult to verify by a discussion over all possible drawings D that $\text{cr}_D(G^* \cup T^i, T^k) \geq 5$ holds for any subgraph $T^k \in S_D$, and $\text{cr}_D(G^* \cup T^i, T^k) \geq 4$ is also fulfilling for any subgraph $T^k \notin R_D \cup S_D$. Thus, by fixing the subgraph $G^* \cup T^i$,

$$\text{cr}_D(G^* + D_n) \geq 6\left\lfloor\frac{n-1}{2}\right\rfloor\left\lfloor\frac{n-2}{2}\right\rfloor + 6(r-1) + 5s + 4(n-r-s) + 1 = 6\left\lfloor\frac{n-1}{2}\right\rfloor\left\lfloor\frac{n-2}{2}\right\rfloor$$

$$+4n + (2r+s) - 5 \geq 6\left\lfloor\frac{n-1}{2}\right\rfloor\left\lfloor\frac{n-2}{2}\right\rfloor + 4n + \left(n - \left\lfloor\frac{n}{2}\right\rfloor + 1\right) - 5 \geq 6\left\lfloor\frac{n}{2}\right\rfloor\left\lfloor\frac{n-1}{2}\right\rfloor + n + \left\lfloor\frac{n}{2}\right\rfloor.$$

ii. Let R_D be the empty set; that is, each subgraph T^j crosses the edges of G^* at least once in D. Thus, we deal with the configurations belonging to the nonempty set $\mathcal{N}_D$. Let us consider a subgraph $T^j \in S_D$ with the configuration $\mathcal{B}_i \in \mathcal{N}_D$ of F^j, where $i \in \{1,2,3,4\}$. Then, the lower-bounds of number of crossings of two configurations from $\mathcal{N}$ confirm that $\text{cr}_D(G^* \cup T^j, T^k) \geq 1 + 4 = 5$ holds for any $T^k \in S_D, k \neq j$. Moreover, one can also easily verify over all possible drawings D that $\text{cr}_D(G^* \cup T^j, T^k) \geq 4$ is true for any subgraph $T^k \notin S_D$. Hence, by fixing the subgraph $G^* \cup T^j$,

$$\text{cr}_D(G^* + D_n) \geq 6\left\lfloor\frac{n-1}{2}\right\rfloor\left\lfloor\frac{n-2}{2}\right\rfloor + 5(s-1) + 4(n-s) + 1 + 1 = 6\left\lfloor\frac{n-1}{2}\right\rfloor\left\lfloor\frac{n-2}{2}\right\rfloor$$

$$+4n + s - 3 \geq 6\left\lfloor\frac{n-1}{2}\right\rfloor\left\lfloor\frac{n-2}{2}\right\rfloor + 4n + \left(\left\lceil\frac{n}{2}\right\rceil + 1\right) - 3 \geq 6\left\lfloor\frac{n}{2}\right\rfloor\left\lfloor\frac{n-1}{2}\right\rfloor + n + \left\lfloor\frac{n}{2}\right\rfloor.$$

In addition, without loss of generality, we can consider the drawing of G^* with the vertex notation like that in Figure 1e. It is obvious that the set R_D is empty; that is, the set S_D cannot be empty. Thus, we deal with the configurations belonging to the nonempty set $\mathcal{O}_D$. Note that the lower-bounds of number of crossings of two configurations from $\mathcal{O}$ were already established in Table 2. Since there is a possibility to find a subdrawing of $G^* \cup T^j \cup T^k$, in which $\text{cr}_D(G^* \cup T^j, T^k) = 3$ with $T^j \in S_D$ and $T^k \notin S_D$, we discuss four following subcases:

i. $\mathcal{E}_4 \in \mathcal{O}_D$. Without lost of generality, let us assume that $T^n \in S_D$ with the configuration $\mathcal{E}_4 \in \mathcal{O}_D$ of F^n. Only for this subcase, one can easily verify over all possible drawings D for which $\text{cr}_D(G^* \cup T^n, T^k) \geq 4$ is true for any subgraph $T^k \notin S_D$. Thus, by fixing the subgraph $G^* \cup T^n$,

$$\text{cr}_D(G^* + D_n) \geq 6\left\lfloor\frac{n-1}{2}\right\rfloor\left\lfloor\frac{n-2}{2}\right\rfloor + 5(s-1) + 4(n-s) + 1 + 1 = 6\left\lfloor\frac{n-1}{2}\right\rfloor\left\lfloor\frac{n-2}{2}\right\rfloor$$

$$+4n+s-3 \geq 6\left\lfloor\frac{n-1}{2}\right\rfloor\left\lfloor\frac{n-2}{2}\right\rfloor+4n+\left(\left\lceil\frac{n}{2}\right\rceil+1\right)-3 \geq 6\left\lfloor\frac{n}{2}\right\rfloor\left\lfloor\frac{n-1}{2}\right\rfloor+n+\left\lfloor\frac{n}{2}\right\rfloor.$$

ii. $\mathcal{E}_4 \notin \mathcal{O}_D$ and $\mathcal{E}_3 \in \mathcal{O}_D$. Without lost of generality, let us assume that $T^n \in S_D$ with the configuration $\mathcal{E}_3 \in \mathcal{O}_D$ of F^n. In this subcase, $\mathrm{cr}_D(G^* \cup T^n, T^k) \geq 1+5=6$ holds for any subgraph $T^k \in S_D$, $k \neq n$ by the remaining values in the third row of Table 2. Hence, by fixing the subgraph $G^* \cup T^n$,

$$\mathrm{cr}_D(G^*+D_n) \geq 6\left\lfloor\frac{n-1}{2}\right\rfloor\left\lfloor\frac{n-2}{2}\right\rfloor+6(s-1)+3(n-s)+1+1=6\left\lfloor\frac{n-1}{2}\right\rfloor\left\lfloor\frac{n-2}{2}\right\rfloor$$

$$+3n+3s-4 \geq 6\left\lfloor\frac{n-1}{2}\right\rfloor\left\lfloor\frac{n-2}{2}\right\rfloor+3n+3\left(\left\lceil\frac{n}{2}\right\rceil+1\right)-4 \geq 6\left\lfloor\frac{n}{2}\right\rfloor\left\lfloor\frac{n-1}{2}\right\rfloor+n+\left\lfloor\frac{n}{2}\right\rfloor.$$

iii. $\mathcal{O}_D = \{\mathcal{E}_1, \mathcal{E}_2\}$. Without lost of generality, let us consider two different subgraphs T^{n-1}, $T^n \in S_D$ such that F^{n-1} and F^n have mentioned configurations $\mathcal{E}_1$ and $\mathcal{E}_2$, respectively. Then, $\mathrm{cr}_D(G^* \cup T^{n-1} \cup T^n, T^k) \geq 1+10=11$ holds for any $T^k \in S_D$ with $k \neq n-1, n$ also by Table 2. Thus, by fixing the subgraph $G^* \cup T^{n-1} \cup T^n$,

$$\mathrm{cr}_D(G^*+D_n) \geq 6\left\lfloor\frac{n-2}{2}\right\rfloor\left\lfloor\frac{n-3}{2}\right\rfloor+11(s-2)+4(n-s)+4+2=6\left\lfloor\frac{n-2}{2}\right\rfloor\left\lfloor\frac{n-3}{2}\right\rfloor$$

$$+4n+7s-16 \geq 6\left\lfloor\frac{n-2}{2}\right\rfloor\left\lfloor\frac{n-3}{2}\right\rfloor+4n+7\left(\left\lceil\frac{n}{2}\right\rceil+1\right)-16 \geq 6\left\lfloor\frac{n}{2}\right\rfloor\left\lfloor\frac{n-1}{2}\right\rfloor+n+\left\lfloor\frac{n}{2}\right\rfloor.$$

iv. $\mathcal{O}_D = \{\mathcal{E}_i\}$ for only one $i \in \{1,2\}$. Without lost of generality, let us assume that $T^n \in S_D$ with the configuration $\mathcal{E}_1$ of F^n. In this subcase, $\mathrm{cr}_D(G^* \cup T^n, T^k) \geq 1+6=7$ holds for any $T^k \in S_D$, $k \neq n$ provided that $\mathrm{rot}_D(t_n) = \mathrm{rot}_D(t_k)$. Hence, by fixing the subgraph $G^* \cup T^n$,

$$\mathrm{cr}_D(G^*+D_n) \geq 6\left\lfloor\frac{n-1}{2}\right\rfloor\left\lfloor\frac{n-2}{2}\right\rfloor+7(s-1)+3(n-s)+1=6\left\lfloor\frac{n-1}{2}\right\rfloor\left\lfloor\frac{n-2}{2}\right\rfloor$$

$$+3n+4s-6 \geq 6\left\lfloor\frac{n-1}{2}\right\rfloor\left\lfloor\frac{n-2}{2}\right\rfloor+3n+4\left(\left\lceil\frac{n}{2}\right\rceil+1\right)-6 \geq 6\left\lfloor\frac{n}{2}\right\rfloor\left\lfloor\frac{n-1}{2}\right\rfloor+n+\left\lfloor\frac{n}{2}\right\rfloor.$$

Case 3: $\mathrm{cr}_D(G^*) = 2$ with $\mathrm{cr}_D(C_6(G^*)) = 0$. At first, without loss of generality, we can consider the drawing of G^* with the vertex notation like that in Figure 1c. It is obvious that the set R_D is empty, that is, the set S_D cannot be empty. Our aim is to list again all possible rotations $\mathrm{rot}_D(t_j)$ which can appear in D if a subgraph $T^j \in S_D$. Since there is only one subdrawing of $F^j \setminus \{v_1\}$ represented by the rotation (26543), there are three ways to obtain the subdrawing of F^j depending on which edge of G^* is crossed by the edge t_jv_1. These three possible ways under our consideration can be denoted by C_k, for $k = 1,2,3$. Based on the aforementioned arguments, we assume the drawings shown in Figure 7.

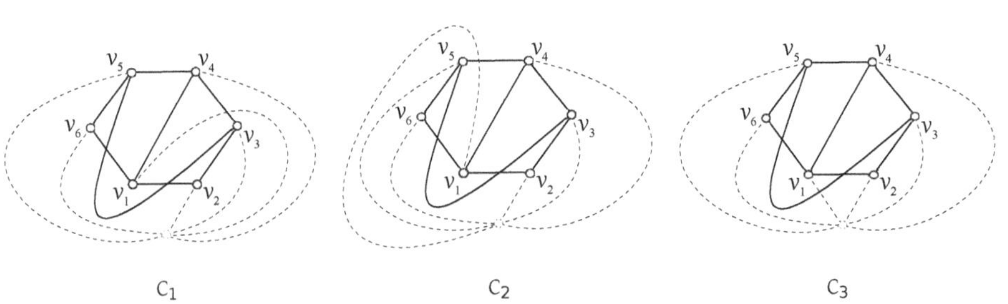

Figure 7. Drawings of three possible configurations of the subgraph F^j.

Thus the configurations $\mathcal{C}_1$, $\mathcal{C}_2$, and $\mathcal{C}_3$ are represented by the cyclic permutations (132654), (143265), and (165432), respectively. Further, due to the properties of the cyclic rotations we can easily verify that $\mathrm{cr}(\mathcal{C}_i, \mathcal{C}_j) \geq 4$ for any $i, j \in \{1,2,3\}$. Moreover, one can also easily verify over all possible drawings D that $\mathrm{cr}_D(G^* \cup T^j, T^k) \geq 4$ holds for any subgraph $T^k \notin S_D$, where $T^j \in S_D$ with some configuration $\mathcal{C}_i$ of F^j. As there is a $T^j \in S_D$, by fixing the subgraph $G^* \cup T^j$,

$$\mathrm{cr}_D(G^* + D_n) \geq 6\left\lfloor \frac{n-1}{2} \right\rfloor \left\lfloor \frac{n-2}{2} \right\rfloor + 5(s-1) + 4(n-s) + 2 + 1 = 6\left\lfloor \frac{n-1}{2} \right\rfloor \left\lfloor \frac{n-2}{2} \right\rfloor$$

$$+4n + s - 2 \geq 6\left\lfloor \frac{n-1}{2} \right\rfloor \left\lfloor \frac{n-2}{2} \right\rfloor + 4n + \left(\left\lceil \frac{n}{2} \right\rceil + 1\right) - 2 \geq 6\left\lfloor \frac{n}{2} \right\rfloor \left\lfloor \frac{n-1}{2} \right\rfloor + n + \left\lfloor \frac{n}{2} \right\rfloor.$$

In addition, without loss of generality, we can consider the drawing of G^* with the vertex notation like that in Figure 1d. In this case, by applying the same process, we obtain two possible forms of rotation $\mathrm{rot}_D(t_j)$ for $T^j \in S_D$. Namely, the rotations (165423) and (165432) if the edge t_jv_2 crosses either the edge v_3v_4 or the edge v_3v_5 of G^*, respectively. Further, they satisfy also the same properties like in the previous subcase, i.e., the same lower bounds of numbers of crossings on the edges of the subgraph $G^* \cup T^j$ by any T^k, $k \neq j$. Hence, we are able to use the same fixing of the subgraph $G^* \cup T^j$ for obtaining a contradiction with the number of crossings in D.

Finally, without loss of generality, we can consider the drawing of G^* with the vertex notation like that in Figure 1f. In this case, the set R_D is empty; that is, the set S_D cannot be empty. Thus, we can deal with the configurations belonging to the nonempty set $\mathcal{P}_D$. Recall that the lower-bounds of number of crossings of two configurations from $\mathcal{P}$ were already established in Table 3. Further, we can apply the same idea and also the same arguments as for the configurations $\mathcal{E}_i \in \mathcal{O}_D$, with $i = 1,2,3$, in the subcases ii.–iv. of Case 2.

Case 4: $\mathrm{cr}_D(G^*) \geq 1$ with $\mathrm{cr}_D(C_6(G^*)) \geq 1$. For all possible subdrawings of the graph G^* with at least one crossing among edges of $C_6(G^*)$, and also with the possibility of obtaining a subgraph T^j that crosses the edges of G^* at most once, one of the ideas of the previous subcases can be applied.

We have shown, in all cases, that there is no good drawing D of the graph $G^* + D_n$ with fewer than $6\lfloor \frac{n}{2} \rfloor \lfloor \frac{n-1}{2} \rfloor + n + \lfloor \frac{n}{2} \rfloor$ crossings. This completes the proof of the main theorem. □

4. Conclusions

Determining the crossing number of a graph $G + D_n$ is an essential step in establishing the so far unknown values of the numbers of crossings of graphs $G + P_n$ and $G + C_n$, where P_n and C_n are the path and the cycle on n vertices, respectively. Using the result in Theorem 1 and the optimal drawing of $G^* + D_n$ in Figure 6, we are able to postulate that $\mathrm{cr}(G^* + P_n)$ and $\mathrm{cr}(G^* + C_n)$ are at least one more than $\mathrm{cr}(G^* + D_n) = 6\lfloor \frac{n}{2} \rfloor \lfloor \frac{n-1}{2} \rfloor + n + \lfloor \frac{n}{2} \rfloor$.

Funding: This research received no external funding.

Acknowledgments: This work was supported by the internal faculty research project number FEI-2017-39.

Conflicts of Interest: The author declares no conflict of interest.

References

1. Garey, M.R.; Johnson, D.S. Crossing number is NP-complete. *SIAM J. Algebraic. Discrete Methods* **1983**, *4*, 312–316.
2. Klešč, M. The join of graphs and crossing numbers. *Electron. Notes Discret. Math.* **2007**, *28*, 349–355.
3. Kulli, V.R.; Muddebihal, M.H. Characterization of join graphs with crossing number zero. *Far East J. Appl. Math.* **2001**, *5*, 87–97.
4. Kleitman, D.J. The crossing number of $K_{5,n}$. *J. Comb. Theory* **1970**, *9*, 315–323.
5. Klešč, M.; Schrötter, Š. The crossing numbers of join products of paths with graphs of order four. *Discuss. Math. Graph Theory* **2011**, *31*, 312–331.

6. Berežný, Š.; Staš, M. Cyclic permutations and crossing numbers of join products of symmetric graph of order six. *Carpathian J. Math.* **2018**, *34*, 143–155.
7. Klešč, M. The crossing numbers of join of the special graph on six vertices with path and cycle. *Discret. Math.* **2010**, *310*, 1475–1481.
8. Staš, M. Cyclic permutations: Crossing numbers of the join products of graphs. In Proceedings of the Aplimat 2018: 17th Conference on Applied Mathematics, Bratislava, Slovakia, 6–8 February 2018; pp. 979–987.
9. Staš, M. Determining crossing numbers of graphs of order six using cyclic permutations. *Bull. Aust. Math. Soc.* **2018**, *98*, 353–362.
10. Hernández-Vélez, C.; Medina, C.; Salazar G. The optimal drawing of $K_{5,n}$. *Electron. J. Comb.* **2014**, *21*, 29.
11. Berežný, Š.; Buša J., Jr.; Staš, M. Software solution of the algorithm of the cyclic-order graph. *Acta Electrotech. Inform.* **2018**, *18*, 3–10.
12. Klešč, M.; Schrötter, Š. The crossing numbers of join of paths and cycles with two graphs of order five. In *Lecture Notes in Computer Science: Mathematical Modeling and Computational Science;* Springer: Berlin/Heidelberg, Germany, 2012; Volume 7125, pp. 160–167.
13. Woodall, D.R. Cyclic-order graphs and Zarankiewicz's crossing number conjecture. *J. Graph Theory* **1993**, *17*, 657–671.
14. Chimani, M.; Wiedera, T. An ILP-based proof system for the crossing number problem. In Proceedings of the 24th Annual European Symposium on Algorithms (ESA 2016), Aarhus, Denmark, 22–24 August 2016; Volume 29, pp. 1–13.

symmetry

MDPI

Article

Bounds for the Generalized Distance Eigenvalues of a Graph

Abdollah Alhevaz [1,*], Maryam Baghipur [1], Hilal Ahmad Ganie [2] and Yilun Shang [3]

1 Faculty of Mathematical Sciences, Shahrood University of Technology, Shahrood P.O. Box: 316-3619995161, Iran; maryamb8989@gmail.com
2 Department of Mathematics, University of Kashmir, Srinagar 190006, India; hilahmad1119kt@gmail.com
3 Department of Computer and Information Sciences, Northumbria University, Newcastle NE1 8ST, UK; yilun.shang@northumbria.ac.uk
* Correspondence: a.alhevaz@gmail.com or a.alhevaz@shahroodut.ac.ir

Received: 1 October 2019; Accepted: 13 December 2019; Published: 17 December 2019

Abstract: Let G be a simple undirected graph containing n vertices. Assume G is connected. Let $D(G)$ be the distance matrix, $D^L(G)$ be the distance Laplacian, $D^Q(G)$ be the distance signless Laplacian, and $Tr(G)$ be the diagonal matrix of the vertex transmissions, respectively. Furthermore, we denote by $D_\alpha(G)$ the generalized distance matrix, i.e., $D_\alpha(G) = \alpha Tr(G) + (1-\alpha)D(G)$, where $\alpha \in [0,1]$. In this paper, we establish some new sharp bounds for the generalized distance spectral radius of G, making use of some graph parameters like the order n, the diameter, the minimum degree, the second minimum degree, the transmission degree, the second transmission degree and the parameter α, improving some bounds recently given in the literature. We also characterize the extremal graphs attaining these bounds. As an special cases of our results, we will be able to cover some of the bounds recently given in the literature for the case of distance matrix and distance signless Laplacian matrix. We also obtain new bounds for the k-th generalized distance eigenvalue.

Keywords: distance matrix (spectrum); distance signlees Laplacian matrix (spectrum); (generalized) distance matrix; spectral radius; transmission regular graph

MSC: Primary: 05C50, 05C12; Secondary: 15A18

1. Introduction

We will consider simple finite graphs in this paper. A (simple) graph is denoted by $G = (V(G), E(G))$, where $V(G) = \{v_1, v_2, \ldots, v_n\}$ represents its vertex set and $E(G)$ represents its edge set. The *order* of G is the number of vertices represented by $n = |V(G)|$ and its *size* is the number of edges represented by $m = |E(G)|$. The *neighborhood* $N(v)$ of a vertex v consists of the set of vertices that are adjacent to it. The *degree* $d_G(v)$ or simply $d(v)$ is the number of vertices in $N(v)$. In a *regular* graph, all its vertices have the same degree. Let d_{uv} be the *distance* between two vertices $u, v \in V(G)$. It is defined as the length of a shortest path. $D(G) = (d_{uv})_{u,v \in V(G)}$ is called the *distance matrix* of G. $\overline{G}$ is the *complement* of the graph G. It has the same vertex set with G but its edge set consists of the edges not present in G. Moreover, the complete graph K_n, the complete bipartite graph $K_{s,t}$, the path P_n, and the cycle C_n are defined in the conventional way.

The *transmission* $Tr_G(v)$ of a vertex v is the sum of the distances from v to all other vertices in G, i.e., $Tr_G(v) = \sum\limits_{u \in V(G)} d_{uv}$. A graph G is said to be *k-transmission regular* if $Tr_G(v) = k$, for each $v \in V(G)$. The *transmission* (also called the *Wiener index*) of a graph G, denoted by $W(G)$, is the sum of distances between all unordered pairs of vertices in G. We have $W(G) = \frac{1}{2} \sum\limits_{v \in V(G)} Tr_G(v)$.

For a vertex $v_i \in V(G)$, $Tr_G(v_i)$ is also referred to as the *transmission degree*, or shortly Tr_i. The sequence of transmission degrees $\{Tr_1, Tr_2, \dots, Tr_n\}$ is the *transmission degree sequence* of the graph. $T_i = \sum_{j=1}^{n} d_{ij} Tr_j$ is called the second transmission degree of v_i.

Distance matrix and its spectrum has been studied extensively in the literature, see e.g., [6]. Compared to adjacency matrix, distance matrix encapsulates more information such as a wide range of walk-related parameters, which can be applicable in thermodynamic calculations and have some biological applications in terms of molecular characterization. It is known that embedding theory and molecular stability have to do with graph distance matrix.

Almost all results obtained for the distance matrix of trees were extended to the case of weighted trees by Bapat [12] and Bapat et al. [13]. Not only different classes of graphs but the definition of distance matrix has been extended. Indeed, Bapat et al. [14] generalized the concept of the distance matrix to that of q-analogue of the distance matrix. Let $Tr(G) = diag(Tr_1, Tr_2, \dots, Tr_n)$ be the diagonal matrix of vertex transmissions of G. The works [7–9] introduced the distance Laplacian and the distance signless Laplacian matrix for a connected graph G. The matrix $D^L(G) = Tr(G) - D(G)$ is referred to as the *distance Laplacian matrix* of G, while the matrix $D^Q(G) = Tr(G) + D(G)$ is the *distance signless Laplacian matrix* of G. Spectral properties of $D(G)$ and $D^Q(G)$ have been extensively studied since then.

Let A be the adjacency matrix and $\mathrm{Deg}(G) = \mathrm{diag}(d_1, d_2, \dots, d_n)$ be the degree matrix G. $Q(G) = \mathrm{Deg}(G) + A$ is the *signless Laplacian matrix* of G. This matrix has been put forth by Cvetkovic in [16] and since then studied extensively by many researchers. For detailed coverage of this research see [17–20] and the references therein. To digging out the contribution of these summands in $Q(G)$, Nikiforov in [33] proposed to study the *α-adjacency matrix* $A_\alpha(G)$ of a graph G given by $A_\alpha(G) = \alpha\,\mathrm{Deg}(G) + (1-\alpha)A$, where $\alpha \in [0,1]$. We see that $A_\alpha(G)$ is a convex combination of the matrices A and $\mathrm{Deg}(G)$. Since $A_0(G) = A$ and $2A_{1/2}(G) = Q(G)$, the matrix $A_\alpha(G)$ can underpin a unified theory of A and $Q(G)$. Motivated by [33], Cui et al. [15] introduced the convex combinations $D_\alpha(G)$ of $Tr(G)$ and $D(G)$. The matrix $D_\alpha(G) = \alpha Tr(G) + (1-\alpha)D(G)$, $0 \le \alpha \le 1$, is called *generalized distance matrix* of G. Therefore the generalized distance matrix can be applied to the study of other less general constructions. This not only gives new results for several matrices simultaneously, but also serves the unification of known theorems.

Since the matrix $D_\alpha(G)$ is real and symmetric, its eigenvalues can be arranged as: $\partial_1 \ge \partial_2 \ge \dots \ge \partial_n$, where ∂_1 is referred to as the *generalized distance spectral radius* of G. For simplicity, $\partial(G)$ is the shorthand for $\partial_1(G)$. By the Perron-Frobenius theorem, $\partial(G)$ is unique and it has a unique *generalized distance Perron vector*, X, which is positive. This is due to the fact that $D_\alpha(G)$ is non-negative and irreducible.

A column vector $X = (x_1, x_2, \dots, x_n)^T \in \mathbb{R}^n$ is a function defined on $V(G)$. We have $X(v_i) = x_i$ for all i. Moreover,

$$X^T D_\alpha(G) X = \alpha \sum_{i=1}^{n} Tr(v_i) x_i^2 + 2(1-\alpha) \sum_{1 \le i < j \le n} d(v_i, v_j) x_i x_j,$$

and λ has an eigenvector X if and only if $X \ne \mathbf{0}$ and

$$\lambda x_v = \alpha Tr(v_i) x_i + (1-\alpha) \sum_{j=1}^{n} d(v_i, v_j) x_j.$$

They are often referred to as the *(λ, x)-eigenequations* of G. If $X \in \mathbb{R}^n$ has at least one non-negative element and it is normalized, then in the light of the Rayleigh's principle, it can be seen that

$$\partial(G) \ge X^T D_\alpha(G) X,$$

where the equality holds if and only if X becomes the generalized distance Perron vector of G.

Spectral graph theory has been an active research field for the past decades, in which for example distance signless Laplacian spectrum has been intensively explored. The work [41] identified the graphs with minimum distance signless Laplacian spectral radius among some special classes of graphs. The unique graphs with minimum and second-minimum distance signless Laplacian spectral radii among all bicyclic graphs of the same order are identified in [40]. In [24], the authors show some bounding inequalities for distance signless Laplacian spectral radius by utilizing vertex transmissions. In [26], chromatic number is used to derive a lower bound for distance signless Laplacian spectral radius. The distance signless Laplacian spectrum has varies connections with other interesting graph topics such as chromatic number [10]; domination and independence numbers [21], Estrada indices [4,5,22,23,34–36,38], cospectrality [11,42], multiplicity of the distance (signless) Laplacian eigenvalues [25,29,30] and many more, see e.g., [1–3,27,28,32].

The rest of the paper is organized as follows. In Section 2, we obtain some bounds for the generalized distance spectral radius of graphs using the diameter, the order, the minimum degree, the second minimum degree, the transmission degree, the second transmission degree and the parameter α. We then characterize the extremal graphs. In Section 3, we are devoted to derive new upper and lower bounds for the k-th generalized distance eigenvalue of the graph G using signless Laplacian eigenvalues and the α-adjacency eigenvalues.

2. Bounds on Generalized Distance Spectral Radius

In this section, we obtain bounds for the generalized distance spectral radius, in terms of the diameter, the order, the minimum degree, the second minimum degree, the transmission degree, the second transmission degree and the parameter α.

The following lemma can be found in [31].

Lemma 1. *If A is an $n \times n$ non-negative matrix with the spectral radius $\lambda(A)$ and row sums $r_1, r_2, \ldots, r_n$, then*

$$\min_{1\le i\le n} r_i \le \lambda(A) \le \max_{1\le i\le n} r_i.$$

Moreover, if A is irreducible, then both of the equalities holds if and only if the row sums of A are all equal.

The following gives an upper bound for $\partial(G)$, in terms of the order n, the diameter d and the minimum degree δ of the graph G.

Theorem 1. *Let G be a connected graph of order n having diameter d and minimum degree δ. Then*

$$\partial(G) \le dn - \frac{d(d-1)}{2} - 1 - \delta(d-1), \tag{1}$$

with equality if and only if G is a regular graph with diameter ≤ 2.

Proof. First, it is easily seen that,

$$\begin{aligned} Tr_p = \sum_{j=1}^{n} d_{jp} &\le d_p + 2 + 3 + \cdots + (d-1) + d(n-1-d_p-(d-2)) \\ &= dn - \frac{d(d-1)}{2} - 1 - d_p(d-1), \quad \text{for all } p = 1, 2, \ldots, n. \end{aligned} \tag{2}$$

Let $Tr_{max} = \max\{Tr_G(v_i) : 1 \leq i \leq n\}$. For a matrix A denote $\lambda(A)$ its largest eigenvalue. We have

$$\begin{aligned} \partial(G) &= \lambda\Big(\alpha(Tr(G)) + (1-\alpha)D(G)\Big) \\ &\leq \alpha\lambda(Tr(G)) + (1-\alpha)\lambda(D(G)) \\ &\leq \alpha Tr_{max} + (1-\alpha)Tr_{max} = Tr_{max}. \end{aligned}$$

Applying Equation (2), the inequality follows.

Suppose that G is regular graph with diameter less than or equal to two, then all coordinates of the generalized distance Perron vector of G are equal. If $d = 1$, then $G \cong K_n$ and $\partial = n-1$. Thus equality in (1) holds. If $d = 2$, we get $\partial(G) = d_i + 2(n-1-d_i) = 2n-2-d_i$, and the equality in (1) holds. Note that the equality in (1) holds if and only if all coordinates the generalized distance Perron vector are equal, and hence $D_\alpha(G)$ has equal row sums.

Conversely, suppose that equality in (1) holds. This will force inequalities above to become equations. Then we get $Tr_1 = Tr_2 = \cdots = Tr_n = Tr_{\max}$, hence all the transmissions of the vertices are equal and so G is a transmission regular graph. If $d \geq 3$, then from the above argument, for every vertex v_i, there is exactly one vertex v_j with $d_G(v_i, v_j) = 2$, and thus $d = 3$, and for a vertex v_s of eccentricity 2,

$$\partial(G)x_s = d_s x_s + 2(n-1-d_s)x_s = \left(3n - \frac{3(3-1)}{2} - 1 - d_s(3-1)\right) x_s,$$

implying that $d_s = n-2$, giving that $G = P_4$. But the $D_\alpha(P_4)$ is not transmission regular graph. Therefore, G turns out to be regular and its diameter can not be greater than 2. □

Taking $\alpha = \frac{1}{2}$ in Theorem 1, we immediately get the following bound for the distance signless Laplacian spectral radius $\rho_1^Q(G)$, which was proved recently in [27].

Corollary 1. ([27], Theorem 2.6) *Let G be a connected graph of order $n \geq 3$, with minimum degree δ_1, second minimum degree δ_2 and diameter d. Then*

$$\rho_1^Q(G) \leq 2dn - d(d-1) - 2 - (\delta_1 + \delta_2)(d-1),$$

with equality if and only if G is (transmission) regular graph of diameter $d \leq 2$.

Proof. As $2D_{\frac{1}{2}}(G) = D^Q(G)$, letting $\delta = \delta_1$ in Theorem 1, we have

$$\rho_1^Q(G) = 2\partial(G) \leq 2dn - d(d-1) - 2 - 2\delta_1(d-1) \leq 2dn - d(d-1) - 2 - (\delta_1 + \delta_2)(d-1),$$

and the result follows. □

Next, the generalized distance spectral radius $\partial(G)$ of a connected graph and its complement is characterized in terms of a Nordhaus-Gaddum type inequality.

Corollary 2. *Let G be a graph of order n, such that both G and its complement $\overline{G}$ are connected. Let δ and Δ be the minimum degree and the maximum degree of G, respectively. Then*

$$\partial(G) + \partial(\overline{G}) \leq 2nk - (t-1)(t+n+\delta-\Delta-1) - 2,$$

where $k = \max\{d, \overline{d}\}, t = \min\{d, \overline{d}\}$ and $d, \overline{d}$ are the diameters of G and $\overline{G}$, respectively.

Proof. Let $\overline{\delta}$ denote the minimum degree of $\overline{G}$. Then $\overline{\delta} = n - 1 - \Delta$, and by Theorem 1, we have

$$\begin{aligned}\partial(G) + \partial(\overline{G}) &\leq dn - \frac{d(d-1)}{2} - 1 - \delta(d-1) + \bar{d}n - \frac{\bar{d}(\bar{d}-1)}{2} - 1 - \bar{\delta}(\bar{d}-1) \\ &= n(d+\bar{d}) - \frac{1}{2}(d(d-1) + \bar{d}(\bar{d}-1)) - 2 - \delta(d-1) - (n-1-\Delta)(\bar{d}-1) \\ &\leq 2nk - (t-1)(t+n+\delta-\Delta-1) - 2.\end{aligned}$$

□

The following gives an upper bound for $\partial(G)$, in terms of the order n, the minimum degree $\delta = \delta_1$ and the second minimum degree δ_2 of the graph G.

Theorem 2. *Let G be a connected graph of order n having minimum degree δ_1 and second minimum degree δ_2. Then for $s = \delta_1 + \delta_2$, we have*

$$\partial(G) \leq \frac{\alpha\Psi + \sqrt{\alpha^2\Psi^2 + 4(1-2\alpha)\Theta}}{2}, \tag{3}$$

where $\Theta = \left(dn - \frac{d(d-1)}{2} - 1 - \delta_1(d-1)\right)\left(dn - \frac{d(d-1)}{2} - 1 - \delta_2(d-1)\right)$ and $\Psi = 2dn - d(d-1) - 2 - s(d-1)$. Also equality holds if and only if G is a regular graph with diameter at most two.

Proof. Let $X = (x_1, x_2, \ldots, x_n)^T$ be the generalized distance Perron vector of graph G and let $x_i = \max\{x_k | k = 1, 2, \ldots, n\}$ and $x_j = \max_{k \neq i}\{x_k | k = 1, 2, \ldots, n\}$. From the ith equation of $D_\alpha(G)X = \partial(G)X$, we obtain

$$\partial x_i = \alpha Tr_i x_i + (1-\alpha)\sum_{k=1, k\neq i}^{n} d_{ik}x_k \leq \alpha Tr_i x_i + (1-\alpha)Tr_i x_j. \tag{4}$$

Similarly, from the jth equation of $D_\alpha(G)X = \partial(G)X$, we obtain

$$\partial x_j = \alpha Tr_j x_j + (1-\alpha)\sum_{k=1, k\neq j}^{n} d_{jk}x_k \leq \alpha Tr_j x_j + (1-\alpha)Tr_j x_i. \tag{5}$$

Now, by (2), we have,

$$\begin{aligned}&\left(\partial - \alpha\left(dn - \frac{d(d-1)}{2} - 1 - d_i(d-1)\right)\right)x_i \leq (1-\alpha)\left(dn - \frac{d(d-1)}{2} - 1 - d_i(d-1)\right)x_j \\ &\left(\partial - \alpha\left(dn - \frac{d(d-1)}{2} - 1 - d_j(d-1)\right)\right)x_j \leq (1-\alpha)\left(dn - \frac{d(d-1)}{2} - 1 - d_j(d-1)\right)x_i.\end{aligned}$$

Multiplying the corresponding sides of these inequalities and using the fact that $x_k > 0$ for all k, we obtain

$$\partial^2 - \alpha(2dn - d(d-1) - 2 - (d-1)(d_i + d_j))\partial - (1-2\alpha)\xi_i\xi_j \leq 0,$$

where $\xi_l = dn - \frac{d(d-1)}{2} - 1 - d_l(d-1), l = i, j$, which in turn gives

$$\partial(G) \leq \frac{\alpha(2dn - d(d-1) - 2 - s(d-1)) + \sqrt{\alpha^2(2dn - d(d-1) - 2 - s(d-1))^2 + 4(1-2\alpha)\Theta}}{2}.$$

Now, using $d_i + d_j \geq \delta_1 + \delta_2$, the result follows.

Suppose that equality occurs in (3), then equality occurs in each of the above inequalities. If equality occurs in (4) and (5), the we obtain $x_i = x_k$, for all $k = 1, 2, \ldots, n$ giving that G is a transmission regular graph. Also, equality in (2), similar to that of Theorem 1, gives that G is a graph of diameter at most two and equality in $d_i + d_j \geq \delta_1 + \delta_2$ gives that G is a regular graph. Combining all these it follows that equality occurs in (3) if G is a regular graph of diameter at most two.

Conversely, if G is a connected δ-regular graph of diameter at most two, then $\partial(G) = Tr_i = dn - \frac{d(d-1)}{2} - 1 - d_i(d-1)$. Also

$$\begin{aligned}
&\frac{\alpha(2dn - d(d-1) - 2 - s(d-1)) + \sqrt{\alpha^2(2dn - d(d-1) - 2 - s(d-1))^2 + 4(1-2\alpha)\Theta}}{2} \\
= &\frac{\alpha(2dn - d(d-1) - 2 - s(d-1)) + (2dn - d(d-1) - 2 - s(d-1))(1-\alpha)}{2} \\
= &\ dn - \frac{d(d-1)}{2} - 1 - \delta(d-1) = \partial(G).
\end{aligned}$$

That completes the proof. □

Remark 1. *For any connected graph G of order n having minimum degree δ, the upper bound given by Theorem 2 is better than the upper bound given by Theorem 1. As*

$$\begin{aligned}
&\frac{\alpha(2dn - d(d-1) - 2 - s(d-1)) + \sqrt{\alpha^2(2dn - d(d-1) - 2 - s(d-1))^2 + 4(1-2\alpha)\Theta}}{2}, \\
\leq &\frac{\alpha(2dn - d(d-1) - 2 - 2\delta(d-1)) + \sqrt{\alpha^2(2dn - d(d-1) - 2 - 2\delta(d-1))^2 + 4(1-2\alpha)\Phi}}{2}, \\
= &\frac{\alpha(2dn - d(d-1) - 2 - 2\delta(d-1)) + (2dn - d(d-1) - 2 - 2\delta(d-1))(1-\alpha)}{2} \\
= &\ dn - \frac{d(d-1)}{2} - 1 - \delta(d-1),
\end{aligned}$$

where $\Phi = (2dn - d(d-1) - 2 - 2\delta(d-1))^2$.

The following gives an upper bound for $\partial(G)$ by using quantities like transmission degrees as well as second transmission degrees.

Theorem 3. *If the transmission degree sequence and the second transmission degree sequence of G are $\{Tr_1, Tr_2, \ldots, Tr_n\}$ and $\{T_1, T_2, \ldots, T_n\}$, respectively, then*

$$\partial(G) \leq \max_{1 \leq i \leq n} \left\{ \frac{-\beta + \sqrt{\beta^2 + 4(\alpha Tr_i^2 + (1-\alpha)T_i + \beta Tr_i)}}{2} \right\}, \tag{6}$$

where $\beta \geq 0$ is an unknown parameter. Equality occurs if and only if G is a transmission regular graph.

Proof. Let $X = (x_1, \ldots, x_n)$ be the generalized distance Perron vector of G and $x_i = \max\{x_j | \ j = 1, 2, \ldots, n\}$. Since

$$\begin{aligned}
\partial(G)^2 X &= (D_\alpha(G))^2 X = (\alpha Tr + (1-\alpha)D)^2 X \\
&= \alpha^2 Tr^2 X + \alpha(1-\alpha) TrDX + \alpha(1-\alpha) DTrX + (1-\alpha)^2 D^2 X,
\end{aligned}$$

we have

$$\partial^2(G)x_i = \alpha^2 Tr_i^2 x_i + \alpha(1-\alpha)Tr_i \sum_{j=1}^{n} d_{ij}x_j + \alpha(1-\alpha) \sum_{j=1}^{n} d_{ij}Tr_j x_j + (1-\alpha)^2 \sum_{j=1}^{n}\sum_{k=1}^{n} d_{ij}d_{jk}x_k.$$

Now, we consider a simple quadratic function of $\partial(G)$:

$$\begin{aligned}(\partial^2(G)+\beta\partial(G))X &= (\alpha^2Tr^2X+\alpha(1-\alpha)TrDX+\alpha(1-\alpha)DTrX+(1-\alpha)^2D^2X)\\ &+ \beta(\alpha TrX+(1-\alpha)DX).\end{aligned}$$

Considering the ith equation, we have

$$\begin{aligned}(\partial^2(G)+\beta\partial(G))x_i &= \alpha^2Tr_i^2x_i+\alpha(1-\alpha)Tr_i\sum_{j=1}^{n}d_{ij}x_j+\alpha(1-\alpha)\sum_{j=1}^{n}d_{ij}Tr_jx_j\\ &+ (1-\alpha)^2\sum_{j=1}^{n}\sum_{k=1}^{n}d_{ij}d_{jk}x_k+\beta\left(\alpha Tr_ix_i+\alpha(1-\alpha)\sum_{j=1}^{n}d_{ij}x_j\right).\end{aligned}$$

It is easy to see that the inequalities below are true

$$\alpha(1-\alpha)Tr_i\sum_{j=1}^{n}d_{ij}x_j\le\alpha(1-\alpha)Tr_i^2x_i,\ \alpha(1-\alpha)\sum_{j=1}^{n}d_{ij}Tr_jx_j\le\alpha(1-\alpha)T_ix_i,$$

$$(1-\alpha)^2\sum_{j=1}^{n}\sum_{k=1}^{n}d_{jk}d_{ij}x_k\le(1-\alpha)^2T_ix_i,\ (1-\alpha)\sum_{j=1}^{n}d_{ij}x_j\le(1-\alpha)Tr_ix_i.$$

Hence, we have

$$\begin{aligned}&(\partial^2(G)+\beta\partial(G))x_i\le\alpha Tr_i^2x_i-\alpha T_ix_i+T_ix_i+\beta Tr_ix_i\\ &\Rightarrow\ \partial^2(G)+\beta\partial(G)-(\alpha Tr_i^2-(\alpha-1)T_i+\beta Tr_i)\le0\\ &\Rightarrow\ \partial(G)\le\frac{-\beta+\sqrt{\beta^2+4(\alpha Tr_i^2-(\alpha-1)T_i+\beta Tr_i)}}{2}.\end{aligned}$$

From this the result follows.

Now, suppose that equality occurs in (6), then each of the above inequalities in the above argument occur as equalities. Since each of the inequalities

$$\alpha(1-\alpha)Tr_i\sum_{j=1}^{n}d_{ij}x_j\le\alpha(1-\alpha)Tr_i^2x_i,\ \alpha(1-\alpha)\sum_{j=1}^{n}d_{ij}Tr_jx_j\le\alpha(1-\alpha)T_ix_i$$

and

$$(1-\alpha)^2\sum_{j=1}^{n}\sum_{k=1}^{n}d_{jk}d_{ij}x_k\le(1-\alpha)^2T_ix_i,\ (1-\alpha)\sum_{j=1}^{n}d_{ij}x_j\le(1-\alpha)Tr_ix_i,$$

occur as equalities if and only if G is a transmission regular graph. It follows that equality occurs in (6) if and only if G is a transmission regular graph. That completes the proof. □

The following upper bound for the generalized distance spectral radius $\partial(G)$ was obtained in [15]:

$$\partial(G)\le\max_{1\le i\le n}\left\{\sqrt{\alpha Tr_i^2+(1-\alpha)T_i}\right\},\tag{7}$$

with equality if and only if $\alpha Tr_i^2+(1-\alpha)T_i$ is same for i.

Remark 2. *For a connected graph G having transmission degree sequence* $\{Tr_1, Tr_2, \ldots, Tr_n\}$ *and the second transmission degree sequence* $\{T_1, T_2, \ldots, T_n\}$, *provided that* $T_i \leq Tr_i^2$ *for all i, we have*

$$\frac{-\beta + \sqrt{\beta^2 + 4\alpha Tr_i^2 + 4(1-\alpha)T_i + 4\beta Tr_i}}{2} \leq \sqrt{\alpha Tr_i^2 + (1-\alpha)T_i}.$$

Therefore, the upper bound given by Theorem 3 is better than the upper bound given by (7).

If, in particular we take the parameter β in Theorem 3 equal to the vertex covering number τ, the edge covering number, the clique number ω, the independence number, the domination number, the generalized distance rank, minimum transmission degree, maximum transmission degree, etc., then Theorem 3 gives an upper bound for $\partial(G)$, in terms of the vertex covering number τ, the edge covering number, the clique number ω, the independence number, the domination number, the generalized distance rank, minimum transmission degree, maximum transmission degree, etc.

Let $x_i = \min\{x_j \mid j = 1, 2, \ldots, n\}$ be the minimum among the entries of the generalized distance Perron vector $X = (x_1, \ldots, x_n)$ of the graph G. Proceeding similar to Theorem 3, we obtain the following lower bound for $\partial(G)$, in terms of the transmission degrees, the second transmission degrees and a parameter β.

Theorem 4. *If the transmission degree sequence and the second transmission degree sequence of G are* $\{Tr_1, Tr_2, \ldots, Tr_n\}$ *and* $\{T_1, T_2, \ldots, T_n\}$, *respectively, then*

$$\partial(G) \geq \min_{1 \leq i \leq n} \left\{ \frac{-\beta + \sqrt{\beta^2 + 4(\alpha Tr_i^2 + (1-\alpha)T_i + \beta Tr_i)}}{2} \right\},$$

where $\beta \geq 0$ *is an unknown parameter. Equality occurs if and only if G is a transmission regular graph.*

Proof. Similar to the proof of Theorem 3 and is omitted. □

The following lower bound for the generalized distance spectral radius was obtained in [15]:

$$\partial(G) \geq \min_{1 \leq i \leq n} \left\{ \sqrt{\alpha Tr_i^2 + (1-\alpha)T_i} \right\}, \tag{8}$$

with equality if and only if $\alpha Tr_i^2 + (1-\alpha)T_i$ is same for i.

Similar to Remark 2, it can be seen that the lower bound given by Theorem 4 is better than the lower bound given by (8) for all graphs G with $T_i \geq Tr_i^2$, for all i.

Again, if in particular we take the parameter β in Theorem 4 equal to the vertex covering number τ, the edge covering number, the clique number ω, the independence number, the domination number, the generalized distance rank, minimum transmission degree, maximum transmission degree, etc, then Theorem 4 gives a lower bound for $\partial(G)$, in terms of the vertex covering number τ, the edge covering number, the clique number ω, the independence number, the domination number, the generalized distance rank, minimum transmission degree, maximum transmission degree, etc.

$G_1 \nabla G_2$ is referred to as *join* of G_1 and G_2. It is defined by joining every vertex in G_1 to every vertex in G_2.

Example 1. (a) *Let* C_4 *be the cycle of order 4. One can easily see that* C_4 *is a 4-transmission regular graph and the generalized distance spectrum of* C_4 *is* $\{4, 4\alpha, 6\alpha - 2^{[2]}\}$. *Hence,* $\partial(C_4) = 4$. *Moreover, the transmission degree sequence and the second transmission degree sequence of* C_4 *are* $\{4, 4, 4, 4\}$ *and*

$\{16, 16, 16, 16\}$, *respectively. Now, putting* $\beta = Tr_{max} = 4$ *in the given bound of Theorem 3, we can see that the equality holds:*

$$\partial(C_4) \leq \frac{-4 + \sqrt{16 + 4(16\alpha + 16(1-\alpha) + 16)}}{2} = \frac{-4 + \sqrt{144}}{2} = 4.$$

(b) *Let* W_{n+1} *be the wheel graph of order* $n+1$. *It is well known that* $W_{n+1} = C_n \nabla K_1$. *The distance signless Laplacian matrix of* W_5 *is*

$$D^Q(W_5) = \begin{pmatrix} 5 & 1 & 2 & 1 & 1 \\ 1 & 5 & 1 & 2 & 1 \\ 2 & 1 & 5 & 1 & 1 \\ 1 & 2 & 1 & 5 & 1 \\ 1 & 1 & 1 & 1 & 4 \end{pmatrix}.$$

Hence the distance signless Laplacian spectrum of W_5 *is* $spec(W_5) = \left\{\frac{13+\sqrt{41}}{4}, \frac{13-\sqrt{41}}{4}, \frac{5}{2}, \frac{3}{2}^{[2]}\right\}$, *and then the distance signless Laplacian spectral radius is* $\rho_1^Q(W_5) = \frac{13+\sqrt{41}}{4}$. *Also, the transmission degree sequence and the second transmission degree sequence of* W_5 *are* $\{5, 5, 5, 5, 4\}$ *and* $\{24, 24, 24, 24, 20\}$, *respectively. As* $D_{\frac{1}{2}}(G) = \frac{1}{2}D^Q(G)$, *taking* $\alpha = \frac{1}{2}$ *and* $\beta = Tr_{max} = 5$ *in the given bound of Theorem 3, we immediately get the following upper bound for the distance signless Laplacian spectral radius* $\rho_1^Q(W_5)$:

$$\frac{1}{2}\rho_1^Q(W_5) \leq \frac{-5 + \sqrt{25 + 50 + 48 + 100}}{2} = \frac{-5 + \sqrt{223}}{2},$$

which implies that

$$\rho_1^Q(W_5) \leq -5 + \sqrt{223} \simeq 9.93.$$

3. Bounds for the k-th Generalized Distance Eigenvalue

In this section, we discuss the relationship between the generalized distance eigenvalues and the other graph parameters.

The following lemma can be found in [37].

Lemma 2. *Let X and Y be Hermitian matrices of order n such that $Z = X + Y$, and denote the eigenvalues of a matrix M by $\lambda_1 \geq \lambda_2 \geq \cdots \geq \lambda_n$.Then*

$$\lambda_k(Z) \leq \lambda_j(X) + \lambda_{k-j+1}(Y), \quad n \geq k \geq j \geq 1,$$
$$\lambda_k(Z) \geq \lambda_j(X) + \lambda_{k-j+n}(Y), \quad n \geq j \geq k \geq 1,$$

where $\lambda_i(M)$ is the ith largest eigenvalue of the matrix M. Any equality above holds if and only if a unit vector can be an eigenvector corresponding to each of the three eigenvalues.

The following gives a relation between the generalized distance eigenvalues of the graph G of diameter 2 and the signless Laplacain eigenvalues of the complement $\overline{G}$ of the graph G. It also gives a relation between generalized distance eigenvalues of the graph G of diameter greater than or equal to 3 with the α-adjacency eigenvalues of the complement $\overline{G}$ of the graph G.

Theorem 5. *Let G be a connected graph of order $n \geq 4$ having diameter d. Let $\overline{G}$ be the complement of G and let $\overline{q}_1 \geq \overline{q}_2 \geq \cdots \geq \overline{q}_n$ be the signless Laplacian eigenvalues of $\overline{G}$. If $d = 2$, then for all $k = 1, 2, \ldots, n$, we have*

$$(3\alpha - 1)n - 2\alpha + (1 - 2\alpha)d_k + (1 - \alpha)\overline{q}_k \leq \partial_k(G) \leq (2n - 2)\alpha + (1 - 2\alpha)d_k + (1 - \alpha)\overline{q}_k.$$

Equality occurs on the right if and only if $k = 1$ and G is a transmission regular graph and on the left if and only if $k \neq 1$ and G is a transmission regular graph.

If $d \geq 3$, then for all $k = 1, 2, \ldots, n$, we have

$$\alpha n - 1 + \lambda_k(A_\alpha(\overline{G})) + \lambda_n(M') \leq \partial_k(G) \leq n - 1 + \lambda_k(A_\alpha(\overline{G})) + \lambda_1(M'),$$

where $A_\alpha(\overline{G}) = \alpha \operatorname{Deg}(\overline{G}) + (1-\alpha)\overline{A}$ is the α-adjacency matrix of $\overline{G}$ and $M' = \alpha Tr'(G) + (1-\alpha)M$ with $M = (m_{ij})$ a symmetric matrix of order n having $m_{ij} = \max\{0, d_{ij} - 2\}$, d_{ij} is the distance between the vertices v_i, v_j and $Tr'(G) = \operatorname{diag}(Tr'_1, Tr'_2, \ldots, Tr'_n)$, $Tr'_i = \sum\limits_{d_{ij} \geq 3} (d_{ij} - 2)$.

Proof. Let G be a connected graph of order $n \geq 4$ having diameter d. Let $\operatorname{Deg}(\overline{G}) = diag(n - 1 - d_1, n - 1 - d_2, \ldots, n - 1 - d_n)$ be the diagonal matrix of vertex degrees of $\overline{G}$. Suppose that diameter d of G is two, then transmission degree $Tr_i = 2n - 2 - d_i$, for all i, then the distance matrix of G can be written as $D(G) = A + 2\overline{A}$, where A and $\overline{A}$ are the adjacency matrices of G and $\overline{G}$, respectively. We have

$$\begin{aligned} D_\alpha(G) &= \alpha Tr(G) + (1-\alpha)D(G) = \alpha(2n-2)I - \alpha \operatorname{Deg}(G) + (1-\alpha)(A + 2\overline{A}) \\ &= \alpha(2n-2)I - \alpha \operatorname{Deg}(G) + (1-\alpha)(A + \overline{A}) + (1-\alpha)\overline{A} \\ &= (3n\alpha - n - 2\alpha)I + (1-\alpha)J + (1-2\alpha)\operatorname{Deg}(G) + (1-\alpha)Q(\overline{G}), \end{aligned}$$

where I is the identity matrix and J is the all one matrix of order n. Taking $Y = (3n\alpha - n - 2\alpha)I + (1 - 2\alpha)\operatorname{Deg}(G) + (1-\alpha)Q(\overline{G})$, $X = (1-\alpha)J$, $j = 1$ in the first inequality of Lemma 2 and using the fact that $spec(J) = \{n, 0^{[n-1]}\}$, it follows that

$$\partial_k(G) \leq (2n-2)\alpha + (1-2\alpha)d_k + (1-\alpha)\overline{q}_k, \quad \text{for all} \quad k = 1, 2, \ldots, n. \tag{9}$$

Again, taking $Y = (3n\alpha - n - 2\alpha)I + (1-2\alpha)\operatorname{Deg}(G) + (1-\alpha)Q(\overline{G})$, $X = (1-\alpha)J$ and $j = n$ in the second inequality of Lemma 2, it follows that

$$\partial_k(G) \geq (3\alpha - 1)n - 2\alpha + (1-2\alpha)d_k + (1-\alpha)\overline{q}_k, \quad \text{for all} \quad k = 1, 2, \ldots, n. \tag{10}$$

Combining (9) and (10) the first inequality follows. Equality occurs in first inequality if and only if equality occurs in (9) and (10). Suppose that equality occurs in (9), then by Lemma 2, the eigenvalues ∂_k, $(3n-2)\alpha - n + (1-2\alpha)d_k + (1-\alpha)\overline{q}_k$ and $n(1-\alpha)$ of the matrices $D_\alpha(G)$, X and Y have the same unit eigenvector. Since $\mathbf{1} = \frac{1}{n}(1, 1, \ldots, 1)^T$ is the unit eigenvector of Y for the eigenvalue $n(1-\alpha)$, it follows that equality occurs in (9) if and only if $\mathbf{1}$ is the unit eigenvector for each of the matrices $D_\alpha(G)$, X and Y. This gives that G is a transmission regular graph and $\overline{G}$ is a regular graph. Since a graph of diameter 2 is regular if and only if it is transmission regular and complement of a regular graph is regular. Using the fact that for a connected graph G the unit vector $\mathbf{1}$ is an eigenvector for the eigenvalue ∂_1 if and only if G is transmission regular graph, it follows that equality occurs in first inequality if and only if $k = 1$ and G is a transmission regular graph.

Suppose that equality occurs in (10), then again by Lemma 2, the eigenvalues ∂_k, $(3n-2)\alpha - n + (1-2\alpha)d_k + (1-\alpha)\overline{q}_k$ and 0 of the matrices $D_\alpha(G)$, X and Y have the same unit eigenvector $\mathbf{x}$. Since $J\mathbf{x} = 0$, it follows that $\mathbf{x}^T\mathbf{1} = 0$. Using the fact that the matrix J is symmetric(so its normalized eigenvectors are orthogonal [43]), we conclude that the vector $\mathbf{1}$ belongs to the set of eigenvectors of the matrix J and so of the matrices $D_\alpha(G)$, X. Now, $\mathbf{1}$ is an eigenvector of the matrices $D_\alpha(G)$ and X, gives that G is a regular graph. Since for a regular graph of diameter 2 any eigenvector of $Q(\overline{G})$ and $D_\alpha(G)$ is orthogonal to $\mathbf{1}$, it follows that equality occurs in (10) if and only if $k \neq 1$ and G is a regular graph.

If $d \geq 3$, we define the matrix $M = (m_{ij})$ of order n, where $m_{ij} = \max\{0, d_{ij} - 2\}$, d_{ij} is the distance between the vertices v_i and v_j. The transmission of a vertex v_i can be written as $Tr_i = d_i + 2\overline{d}_i + Tr'_i$, where $Tr'_i = \sum\limits_{d_{ij} \geq 3} (d_{ij} - 2)$, is the contribution from the vertices which are at distance more than two from v_i. For $Tr'(G) = \text{diag}(Tr'_1, Tr'_2, \ldots, Tr'_n)$, we have

$$\begin{aligned} D_\alpha(G) &= \alpha Tr(G) + (1-\alpha)D(G) = \alpha \operatorname{Deg}(G) + 2\alpha \operatorname{Deg}(\overline{G}) + \alpha Tr'(G) + (1-\alpha)(A + 2\overline{A} + M) \\ &= \alpha(\operatorname{Deg}(G) + \operatorname{Deg}(\overline{G})) + (1-\alpha)(A + \overline{A}) + (\alpha \operatorname{Deg}(\overline{G}) + (1-\alpha)\overline{A}) + (\alpha Tr'(G) + (1-\alpha)M) \\ &= D_\alpha(K_n) + A_\alpha(\overline{G}) + M', \end{aligned}$$

where $A_\alpha(\overline{G})$ is the α-adjacency matrix of $\overline{G}$ and $M' = \alpha Tr'(G) + (1-\alpha)M$. Taking $X = D_\alpha(K_n)$, $Y = A_\alpha(\overline{G}) + M'$ and $j = 1$ in the first inequality of Lemma 2 and using the fact that $spec(D_\alpha(K_n)) = \{n-1, \alpha n - 1^{[n-1]}\}$, it follows that

$$\partial_k(G) \leq n - 1 + \lambda_k(A_\alpha(\overline{G}) + M'), \quad \text{for all} \quad k = 1, 2, \ldots, n.$$

Again, taking $Y = A_\alpha(\overline{G})$, $X = M'$ and $j = 1$ in the first inequality of Lemma 2, we obtain

$$\partial_k(G) \leq n - 1 + \lambda_k(A_\alpha(\overline{G})) + \lambda_1(M'), \quad \text{for all} \quad k = 1, 2, \ldots, n. \tag{11}$$

Similarly, taking $X = D_\alpha(K_n)$, $Y = A_\alpha(\overline{G}) + M'$ and $j = n$ and then $Y = A_\alpha(\overline{G})$, $X = M'$ and $j = n$ in the second inequality of Lemma 2, we obtain

$$\partial_k(G) \geq \alpha n - 1 + \lambda_k(A_\alpha(\overline{G})) + \lambda_n(M'), \quad \text{for all} \quad k = 1, 2, \ldots, n. \tag{12}$$

From (11) and (12) the second inequality follows. That completes the proof. □

It can be seen that the matrix M' defined in Theorem 5 is positive semi-definite for all $\frac{1}{2} \leq \alpha \leq 1$. Therefore, we have the following observation from Theorem 5.

Corollary 3. *Let G be a connected graph of order $n \geq 4$ having diameter $d \geq 3$. If $\frac{1}{2} \leq \alpha \leq 1$, then*

$$\partial_k(G) \geq \alpha n - 1 + \lambda_k(A_\alpha(\overline{G})), \quad \textit{for all} \quad k = 1, 2, \ldots, n,$$

where $A_\alpha(\overline{G}) = \alpha \operatorname{Deg}(\overline{G}) + (1-\alpha)\overline{A}$ is the α-adjacency matrix of $\overline{G}$.

It is clear from Corollary 3 that for $\frac{1}{2} \leq \alpha \leq 1$, any lower bound for the α-adjacency $\lambda_k(A_\alpha(\overline{G}))$ gives a lower bound for ∂_k and conversely any upper bound for ∂ gives an upper bound for $\lambda_k(A_\alpha(\overline{G}))$. We note that Theorem 5 generalizes one of the Theorems (namely Theorem 3.8) given in [8].

Example 2. (a) *Let C_n be a cycle of order n. It is well known (see [7]) that C_n is a k-transmission regular graph with $k = \frac{n^2}{4}$ if n is even and $k = \frac{n^2-1}{4}$ if n is odd. Let $n = 4$. It is clear that the distance spectrum of the graph C_4 is $\{4, 0, -2^{[2]}\}$. Also, since C_4 is a 4-transmission regular graph, then $Tr(C_4) = 4I_4$ and so $D_\alpha(C_4) = 4\alpha I_4 + (1-\alpha)D(C_4)$. Hence the generalized distance spectrum of C_4 is $\{4, 4\alpha, 6\alpha - 2^{[2]}\}$. Moreover, the signless Laplacian spectrum of $\overline{C_4}$ is $\{2^{[2]}, 0^{[2]}\}$. Since the diameter of C_4 is 2, hence, applying Theorem 5, for $k = 1$, we have,*

$$4\alpha = 4(3\alpha - 1) - 2\alpha + 2(1 - 2\alpha) + 2(1-\alpha) \leq \partial_1(C_4) = 4 \leq 6\alpha + 2(1 - 2\alpha) + 2(1-\alpha) = 4,$$

which shows that the equality occurs on right for $k = 1$ and transmission regular graph C_4.

Also, for $k = 2$, we have

$$4\alpha = 4(3\alpha - 1) - 2\alpha + 2(1 - 2\alpha) + 2(1-\alpha) \leq \partial_2(C_4) = 4\alpha \leq 6\alpha + 2(1 - 2\alpha) + 2(1-\alpha) = 4,$$

which shows that the equality occurs on left for $k = 2$ and transmission regular graph C_4.

(b) *Let C_6 be a cycle of order 6. It is clear that the distance spectrum of the graph C_6 is $\{9, 0^{[2]}, -1, -4^{[2]}\}$. Since C_6 is a 9-transmission regular graph, then $Tr(C_6) = 9I_6$ and so $D_\alpha(C_6) = 9\alpha I_6 + (1-\alpha)D(C_6)$. Hence, the generalized distance spectrum of C_6 is $\{9, 9\alpha^{[2]}, 10\alpha - 1, 13\alpha - 4^{[2]}\}$. Also, the α-adjacency spectrum of C_6 is $\{3, 2\alpha+1, 3\alpha^{[2]}, 5\alpha - 2^{[2]}\}$. Let M' be the matrix defined by the Theorem 5, hence the spectrum of M' is $\{1^{[3]}, 2\alpha - 1^{[3]}\}$. Since diameter of the graph C_6 is 3, hence, applying Theorem 5, for $k = 1$, we have*

$$8\alpha + 1 = 6\alpha - 1 + 3 + 2\alpha - 1 \leq \partial_1(C_6) = 9 \leq 5 + 3 + 1 = 9.$$

Also for $k = 2$, we have

$$10\alpha - 1 = 6\alpha - 1 + 2\alpha + 1 + 2\alpha - 1 \leq \partial_2(C_6) = 9\alpha \leq 5 + 2\alpha + 1 + 1 = 2\alpha + 7.$$

We need the following lemma proved by Hoffman and Wielandt [39].

Lemma 3. *Suppose we have $C = A + B$. Here, all these matrices are symmetric and have order n. Suppose they have the eigenvalues α_i, β_i, and γ_i, where $1 \leq i \leq n$, respectively arranged in non-increasing order. Therefore, $\sum_{i=1}^{n}(\gamma_i - \alpha_i)^2 \leq \sum_{i=1}^{n}\beta_i^2$.*

The following gives relation between generalized distance spectrum and distance spectrum for a simple connected graph G. We use $[n]$ to denote the set of $\{1, 2, \ldots, n\}$. For each subset S of $[n]$, we use S^c to denote $[n] - S$.

Theorem 6. *Let G be a connected graph of order n and let $\mu_1, \ldots, \mu_n$ be the eigenvalues of the distance matrix of G. Then for each non-empty subset $S = \{r_1, r_2, \ldots, r_k\}$ of $[n]$, we have the following inequalities:*

$$\begin{aligned} & \frac{2k\alpha W(G) - \sqrt{k(n-k)\left(n\sum_{i=1}^{n}\alpha^2 Tr_i^2 - 4\alpha^2 W^2(G)\right)}}{n} \\ \leq & \sum_{i\in S}(\partial_i + (\alpha-1)\mu_i) \\ \leq & \frac{2k\alpha W(G) + \sqrt{k(n-k)\left(n\sum_{i=1}^{n}\alpha^2 Tr_i^2 - 4\alpha^2 W^2(G)\right)}}{n}. \end{aligned}$$

Proof. Since $D_\alpha(G) = \alpha Tr(G) + (1-\alpha)D(G)$, then by the fact that $2\alpha W(G) = \sum_{i=1}^{n}(\partial_i + (\alpha-1)\mu_i)$, we get $2\alpha W(G) - \sum_{i\in S}(\partial_i + (\alpha-1)\mu_i) = \sum_{i\in S^C}(\partial_i + (\alpha-1)\mu_i)$. By Cauchy-Schwarz inequality, we further have that

$$\left(2\alpha W(G) - \sum_{i\in S}(\partial_i + (\alpha-1)\mu_i)\right)^2 \leq \sum_{i\in S^C} 1^2 \sum_{i\in S^C}(\partial_i + (\alpha-1)\mu_i)^2.$$

Therefore

$$\begin{aligned} & \left(2\alpha W(G) - \sum_{i\in S}(\partial_i + (\alpha-1)\mu_i)\right)^2 \\ & \leq (n-k)\left(\sum_{i=1}^{n}(\partial_i + (\alpha-1)\mu_i)^2 - \sum_{i\in S}(\partial_i + (\alpha-1)\mu_i)^2\right). \end{aligned}$$

By Lemma 3, we have that

$$\left(2\alpha W(G) - \sum_{i\in S}(\partial_i + (\alpha-1)\mu_i)\right)^2 + (n-k)\sum_{i\in S}(\partial_i + (\alpha-1)\mu_i)^2$$
$$\leq (n-k)\sum_{i=1}^{n}(\partial_i + (\alpha-1)\mu_i)^2 \leq (n-k)\sum_{i=1}^{n}\alpha^2 Tr_i^2.$$

Again by Cauchy-Schwarz inequality, we have that

$$\left(\frac{n-k}{k}\right)\left(\sum_{i\in S}(\partial_i + (\alpha-1)\mu_i)\right)^2 = \left(\sum_{i\in S}\sqrt{\frac{n-k}{k}}(\partial_i + (\alpha-1)\mu_i)\right)^2$$
$$\leq \sum_{i\in S}\left(\frac{n-k}{k}\right)\sum_{i\in S}(\partial_i + (\alpha-1)\mu_i)^2 = (n-k)\sum_{i\in S}(\partial_i + (\alpha-1)\mu_i)^2.$$

Therefore, we have the following inequality

$$\left(2\alpha W(G) - \sum_{i\in S}(\partial_i + (\alpha-1)\mu_i)\right)^2 + \left(\frac{n-k}{k}\right)\left(\sum_{i\in S}(\partial_i + (\alpha-1)\mu_i)\right)^2$$
$$\leq (n-k)\sum_{i=1}^{n}\alpha^2 Tr_i^2.$$

Solving the quadratic inequality for $\sum_{i\in S}(\partial_i + (\alpha-1)\mu_i)$, so we complete the proof. □

Notice that $\sum_{i=1}^{n}(\partial_i - \alpha Tr_i) = 0$ and by Lemma 3, we also have $\sum_{i=1}^{n}(\partial_i - \alpha Tr_i)^2 \leq (1-\alpha)^2\sum_{i=1}^{n}\mu_i^2 = 2(1-\alpha)^2\sum_{1\leq i<j\leq n}^{n} d_{ij}^2$. We can similarly prove the following theorem.

Theorem 7. *Let G be a connected graph of order n. Then for each non-empty subset $S = \{r_1, r_2, \ldots, r_k\}$ of $[n]$, we have:*

$$\left|\sum_{i\in S}(\partial_i - \alpha Tr_i)\right| \leq \sqrt{\frac{2k(n-k)(1-\alpha)^2\sum_{1\leq i<j\leq n} d_{ij}^2}{n}}.$$

We conclude by giving the following bounds for the k-th largest generalized distance eigenvalue of a graph.

Theorem 8. *Assume G is connected and is of order n. Suppose it has diameter d and δ is its minimum degree. Let*

$$\begin{aligned}\varphi(G) \;=\; \min\Bigg\{& n^2(n-1)\left(\frac{\alpha^2 n^2(n-1)}{4} + (1-\alpha)^2 d^2\right) - 4\alpha^2 W^2(G),\\ & n\left(\alpha^2\left(nd - \frac{d(d-1)}{2} - 1 - \delta(d-1)\right)^2 + (1-\alpha)^2 n(n-1)d^2\right) - 4\alpha^2 W^2(G)\Bigg\}.\end{aligned}$$

Then for $k = 1, \ldots, n$,

$$\frac{1}{n}\left\{2\alpha W(G) - \sqrt{\frac{k-1}{n-k+1}\varphi(G)}\right\} \leq \partial_k(G) \leq \frac{1}{n}\left\{2\alpha W(G) + \sqrt{\frac{n-k}{k}\varphi(G)}\right\}. \quad (13)$$

Proof. First we prove the upper bound. It is clear that

$$\text{trace}(D_\alpha^2(G)) = \sum_{i=1}^{k} \partial_i^2 + \sum_{i=k+1}^{n} \partial_i^2 \geq \frac{(\sum_{i=1}^{k} \partial_i)^2}{k} + \frac{(\sum_{i=k+1}^{n} \partial_i)^2}{n-k}.$$

Let $M_k = \sum_{i=1}^{k} \partial_i$. Then

$$\text{trace}(D_\alpha^2(G)) \geq \frac{M_k^2}{k} + \frac{(2\alpha W(G) - M_k)^2}{n-k},$$

which implies

$$\partial_k(G) \leq \frac{M_k}{k} \leq \frac{1}{n}\left\{2\alpha W(G) + \sqrt{\frac{n-k}{k}[n \cdot \text{trace}(D_\alpha^2(G)) - 4\alpha^2 W^2(G)]}\right\}.$$

We observe that

$$\begin{aligned} n \cdot \text{trace}(D_\alpha^2(G)) - 4\alpha^2 W^2(G) &= n\alpha^2 \sum_{i=1}^{n} Tr_i^2 + 2n(1-\alpha)^2 \sum_{1 \leq i < j \leq n} (d_{ij})^2 - 4\alpha^2 W^2(G) \\ &\leq n\alpha^2 \frac{n^3(n-1)^2}{4} + 2n(1-\alpha)^2 \frac{n(n-1)}{2} d^2 - 4\alpha^2 W^2(G) \\ &= n^2(n-1)\left(\frac{\alpha^2 n^2 (n-1)}{4} + (1-\alpha)^2 d^2\right) - 4\alpha^2 W^2(G), \end{aligned}$$

since $Tr_i \leq \frac{n(n-1)}{2}$, and

$$\begin{aligned} & n \cdot \text{trace}(D_\alpha^2(G)) - 4\alpha^2 W^2(G) \\ &= n\alpha^2 \sum_{i=1}^{n} Tr_i^2 + 2n(1-\alpha)^2 \sum_{1 \leq i < j \leq n} (d_{ij})^2 - 4\alpha^2 W^2(G) \\ &\leq n\alpha^2 \left(nd - \frac{d(d-1)}{2} - 1 - \delta(d-1)\right)^2 + 2n(1-\alpha)^2 \frac{n(n-1)}{2} d^2 - 4\alpha^2 W^2(G) \\ &= n\left(\alpha^2\left(nd - \frac{d(d-1)}{2} - 1 - \delta(d-1)\right)^2 + (1-\alpha)^2 n(n-1) d^2\right) - 4\alpha^2 W^2(G), \end{aligned}$$

since $Tr_i \leq nd - \frac{d(d-1)}{2} - 1 - d_i(d-1)$. Hence, we get the right-hand side of the inequality (13).

Now, we prove the lower bound. Let $N_k = \sum_{i=k}^{n} \partial_i$. Then we have

$$\begin{aligned} \text{trace}(D_\alpha^2(G)) = \sum_{i=1}^{k-1} \partial_i^2 + \sum_{i=k}^{n} \partial_i^2 &\geq \frac{\left(\sum_{i=1}^{k-1} \partial_i\right)^2}{k-1} + \frac{\left(\sum_{i=k}^{n} \partial_i\right)^2}{n-k+1} \\ &= \frac{(2\alpha W(G) - N_k)^2}{k-1} + \frac{N_k^2}{n-k+1}. \end{aligned}$$

Hence

$$\partial_k(G) \geq \frac{N_k}{n-k+1} \geq \frac{1}{n}\left\{2\alpha W(G) - \sqrt{\frac{k-1}{n-k+1}[n \cdot \text{trace}(D_\alpha^2(G)) - 4\alpha^2 W^2(G)]}\right\},$$

and we get the left-hand side of the inequality (13). □

By a *chemical tree*, we mean a tree which has all vertices of degree less than or equal to 4.

Example 3. *In Figure 1, we depicted a chemical tree of order $n = 5$.*

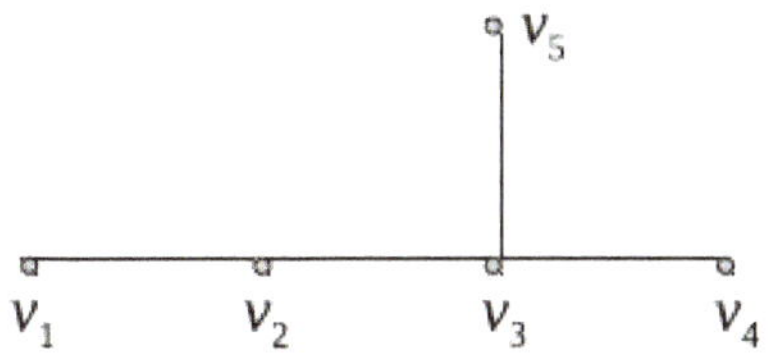

Figure 1. A chemical tree T.

The distance matrix of T is

$$D(T) = \begin{pmatrix} 0 & 1 & 2 & 3 & 3 \\ 1 & 0 & 1 & 2 & 2 \\ 2 & 1 & 0 & 1 & 1 \\ 3 & 2 & 1 & 0 & 2 \\ 3 & 2 & 1 & 2 & 0 \end{pmatrix}.$$

Let $\mu_1, \ldots, \mu_5$ be the distance eigenvalues of the tree T. Then one can easily see that $\mu_1 = 7.46$, $\mu_2 = -0.51$, $\mu_3 = -1.08$, $\mu_4 = -2$ and $\mu_5 = -3.86$. Note that, as $D_0(T) = D(T)$, taking $\alpha = 0$ in Theorem 8, then for $n = 5$ we get $-6\sqrt{\frac{k-1}{6-k}} \leq \mu_k \leq 6\sqrt{\frac{5-k}{k}}$, for any $1 \leq k \leq 5$. For example, $-6 \leq \mu_1 \leq 12$ and $-3 \leq \mu_2 \leq 7.3$.

4. Conclusions

Motivated by an article entitled "Merging the A- and Q-spectral theories" by V. Nikiforov [33], recently, Cui et al. [15] dealt with the integration of spectra of distance matrix and distance signless Laplacian through elegant convex combinations accommodating vertex transmissions as well as distance matrix. For $\alpha \in [0, 1]$, the generalized distance matrix is known as $D_\alpha(G) = \alpha Tr(G) + (1 - \alpha)D(G)$. Our results shed light on some properties of $D_\alpha(G)$ and contribute to establishing new inequalities (such as lower and upper bounds) connecting varied interesting graph invariants. We established some bounds for the generalized distance spectral radius for a connected graph using various identities like the number of vertices n, the diameter, the minimum degree, the second minimum degree, the transmission degree, the second transmission degree and the parameter α, improving some bounds recently given in the literature. We also characterized the extremal graphs attaining these bounds. Notice that the current work mainly focuses to determine some bounds for the spectral radius (largest eigenvalue) of the generalized distance matrix. It would be interesting to derive some bounds for other important eigenvalues such as the smallest eigenvalue as well as the second largest eigenvalue of this matrix.

Author Contributions: conceptualization, A.A., M.B. and H.A.G.; formal analysis, A.A., M.B., H.A.G. and Y.S.; writing—original draft preparation, A.A., M.B. and H.A.G.; writing—review and editing, A.A., M.B., H.A.G. and Y.S.; project administration, A.A.; funding acquisition, Y.S.

Funding: Y. Shang was supported by UoA Flexible Fund No. 201920A1001 from Northumbria University.

Acknowledgments: The authors would like to thank the academic editor and the four anonymous referees for their constructive comments that helped improve the quality of the paper.

Conflicts of Interest: The authors declare no conflict of interest.

References

1. Alhevaz, A.; Baghipur, M.; Ganie, H.A.; Pirzada, S. Brouwer type conjecture for the eigenvalues of distance signless Laplacian matrix of a graph. *Linear Multilinear Algebra* **2019**. [CrossRef]
2. Alhevaz, A.; Baghipur, M.; Hashemi, E. Further results on the distance signless Laplacian spectrum of graphs. *Asian-Eur. J. Math.* **2018**, *11*, 1850066. [CrossRef]

3. Alhevaz, A.; Baghipur, M.; Hashemi, E.; Ramane, H.S. On the distance signless Laplacian spectrum of graphs. *Bull. Malay. Math. Sci. Soc.* **2019**, *42*, 2603–2621. [CrossRef]
4. Alhevaz, A.; Baghipur, M.; Paul, S. On the distance signless Laplacian spectral radius and the distance signless Laplacian energy of graphs. *Discrete Math. Algorithms Appl.* **2018**, *10*, 1850035. [CrossRef]
5. Alhevaz, A.; Baghipur, M.; Pirzada, S. On distance signless Laplacian Estrada index and energy of graphs. *Kragujevac J. Math.* **2021**, *45*, 837–858.
6. Aouchiche, M.; Hansen, P. Distance spectra of graphs: A survey. *Linear Algebra Appl.* **2014**, *458*, 301–386. [CrossRef]
7. Aouchiche, M.; Hansen, P. Two Laplacians for the distance matrix of a graph. *Linear Algebra Appl.* **2013**, *439*, 21–33. [CrossRef]
8. Aouchiche, M.; Hansen, P. On the distance signless Laplacian of a graph. *Linear Multilinear Algebra* **2016**, *64*, 1113–1123. [CrossRef]
9. Aouchiche, M.; Hansen, P. Some properties of distance Laplacian spectra of a graph. *Czechoslovak Math. J.* **2014**, *64*, 751–761. [CrossRef]
10. Aouchiche, M.; Hansen, P. Distance Laplacian eigenvalues and chromatic number in graphs. *Filomat* **2017**, *31*, 2545–2555. [CrossRef]
11. Aouchiche, M.; Hansen, P. Cospectrality of graphs with respect to distance matrices. *Appl. Math. Comput.* **2018**, *325*, 309–321. [CrossRef]
12. Bapat, R.B. Determinant of the distance matrix of a tree with matrix weights. *Linear Algebra Appl.* **2006**, *416*, 2–7. [CrossRef]
13. Bapat, R.B.; Kirkland, S.J.; Neumann, M. On distance matrices and Laplacians. *Linear Algebra Appl.* **2005**, *401*, 193–209. [CrossRef]
14. Bapat, R.B.; Lal, A.K.; Pati, S. A q-analogue of the distance matrix of a tree. *Linear Algebra Appl.* **2006**, *416*, 799–814. [CrossRef]
15. Cui, S.Y.; He, J.X.; Tian, G.X. The generalized distance matrix. *Linear Algebra Appl.* **2019**, *563*, 1–23. [CrossRef]
16. Cvetković, D. Signless Laplacians and line graphs. *Bull. Acad. Serbe Sci. Arts Cl. Sci. Math. Natur. Sci. Math.* **2005**, *131*, 85–92. [CrossRef]
17. Cvetković, D. New theorems for signless Laplacians eigenvalues. *Bull. Acad. Serbe Sci. Arts Cl. Sci. Math. Natur. Sci. Math.* **2008**, *137*, 131–146.
18. Cvetković, D.; Simić, S.K. Towards a spectral theory of graphs based on the signless Laplacian I. *Publ. Inst. Math. (Beograd)* **2009**, *85*, 19–33. [CrossRef]
19. Cvetković, D.; Simić, S.K. Towards a spectral theory of graphs based on the signless Laplacian II. *Linear Algebra Appl.* **2010**, *432*, 2257–2272. [CrossRef]
20. Cvetković, D.; Simić, S.K. Towards a spectral theory of graphs based on the signless Laplacian III. *Appl. Anal. Discrete Math.* **2010**, *4*, 156–166.
21. Das, K.C.; Aouchiche, M.; Hansen, P. On distance Laplacian and distance signless Laplacian eigenvalues of graphs. *Linear Multilinear Algebra* **2019**, *67*, 2307–2324. [CrossRef]
22. Das, K.C.; Aouchiche, M.; Hansen, P. On (distance) Laplacian energy and (distance) signless Laplacian energy of graphs. *Discrete Appl. Math.* **2018**, *243*, 172–185. [CrossRef]
23. Diaz, R.C.; Rojo, O. Sharp upper bounds on the distance energies of a graph. *Linear Algebra Appl.* **2018**, *545*, 55–75. [CrossRef]
24. Duan, X.; Zhou, B. Sharp bounds on the spectral radius of a nonnegative matrix. *Linear Algebra Appl.* **2013**, *439*, 2961–2970. [CrossRef]
25. Fernandes, R.; de Freitas, M.A.A.; da Silva, C.M., Jr.; Del-Vecchio, R.R. Multiplicities of distance Laplacian eigenvalues and forbidden subgraphs. *Linear Algebra Appl.* **2018**, *541*, 81–93. [CrossRef]
26. Li, X.; Fan, Y.; Zha, S. A lower bound for the distance signless Laplacian spectral radius of graphs in terms of chromatic number. *J. Math. Res. Appl.* **2014**, *34*, 289–294.
27. Li, D.; Wang, G.; Meng, J. On the distance signless Laplacian spectral radius of graphs and digraphs. *Electr. J. Linear Algebra* **2017**, *32*, 438–446. [CrossRef]
28. Liu, S.; Shu, J. On the largest distance (signless Laplacian) eigenvalue of non-transmission-regular graphs. *Electr. J. Linear Algebra* **2018**, *34*, 459–471. [CrossRef]
29. Lu, L.; Huang, Q.; Huang, X. On graphs with distance Laplacian spectral radius of multiplicity $n-3$. *Linear Algebra Appl.* **2017**, *530*, 485–499. [CrossRef]

30. Lu, L.; Huang, Q.; Huang, X. On graphs whose smallest distance (signless Laplacian) eigenvalue has large multiplicity. *Linear Multilinear Algebra* **2018**, *66*, 2218–2231. [CrossRef]
31. Minć, H. *Nonnegative Matrices*; John Wiley & Sons: New York, NY, USA, 1988.
32. Lin, H.; Zhou, B. The distance spectral radius of trees. *Linear Multilinear Algebra* **2019**, *67*, 370–390. [CrossRef]
33. Nikiforov, V. Merging the *A*- and *Q*-spectral theories. *Appl. Anal. Discrete Math.* **2017**, *11*, 81–107. [CrossRef]
34. Shang, Y. Distance Estrada index of random graphs. *Linear Multilinear Algebra* **2015**, *63*, 466–471. [CrossRef]
35. Shang, Y. Estimating the distance Estrada index. *Kuwait J. Sci.* **2016**, *43*, 14–19.
36. Shang, Y. Bounds of distance Estrada index of graphs. *Ars Combin.* **2016**, *128*, 287–294.
37. So, W. Commutativity and spectra of Hermitian matrices. *Linear Algebra Appl.* **1994**, *212–213*, 121–129. [CrossRef]
38. Yang, J.; You, L.; Gutman, I. Bounds on the distance Laplacian energy of graphs. *Kragujevac J. Math.* **2013**, *37*, 245–255.
39. Wilkinson, J.H. *The Algebraic Eigenvalue Problem*; Oxford University Press: New York, NY, USA, 1965.
40. Xing, R.; Zhou, B. On the distance and distance signless Laplacian spectral radii of bicyclic graphs. *Linear Algebra Appl.* **2013**, *439*, 3955–3963. [CrossRef]
41. Xing, R.; Zhou, B.; Li, J. On the distance signless Laplacian spectral radius of graphs. *Linear Multilinear Algebra* **2014**, *62*, 1377–1387. [CrossRef]
42. Xue, J.; Liu, S.; Shu, J. The complements of path and cycle are determined by their distance (signless) Laplacian spectra. *Appl. Math. Comput.* **2018**, *328*, 137–143. [CrossRef]
43. Zhang, F. *Matrix Theory: Basic Results and Techniques*; Springer: New York, NY, USA, 1999.

symmetry

MDPI

Article

Weak Embeddable Hypernear-Rings

Jelena Dakić [1], Sanja Jančić-Rašović [1] and Irina Cristea [2,*]

[1] Department of Mathematics, Faculty of Natural Science and Mathematics, University of Montenegro, 81000 Podgorica, Montenegro
[2] Centre for Information Technologies and Applied Mathematics, University of Nova Gorica, 5000 Nova Gorica, Slovenia
* Correspondence: irina.cristea@ung.si or irinacri@yahoo.co.uk; Tel.: +386-0533-15-395

Received: 28 June 2019; Accepted: 18 July 2019; Published: 1 August 2019

Abstract: In this paper we extend one of the main problems of near-rings to the framework of algebraic hypercompositional structures. This problem states that every near-ring is isomorphic with a near-ring of the transformations of a group. First we endow the set of all multitransformations of a hypergroup (not necessarily abelian) with a general hypernear-ring structure, called the multitransformation general hypernear-ring associated with a hypergroup. Then we show that any hypernear-ring can be weakly embedded into a multitransformation general hypernear-ring, generalizing the similar classical theorem on near-rings. Several properties of hypernear-rings related with this property are discussed and illustrated also by examples.

Keywords: hypernear-ring; multitransformation; embedding

---

1. Introduction

Generally speaking, the embedding of an algebraic structure into another one requires the existence of an injective map between the two algebraic objects, that also preserves the structure, i.e., a monomorphism. The most natural, canonical and well-known embeddings are those of numbers: the natural numbers into integers, the integers into the rational numbers, the rational numbers into the real numbers and the real numbers into the complex numbers. One important type of rings is that one of the endomorphisms of an abelian group under function pointwise addition and composition of functions. It is well known that every ring is isomorphic with a subring of such a ring of endomorphisms. But this result holds only in the commutative case, since the set of the endomorphisms of a non-abelian group is no longer closed under addition. This aspect motivates the interest in studying near-rings, that appear to have applications also in characterizing transformations of a group. More exactly, the set of all transformations of a group G, i.e., $T(G) = \{f : G \to G\}$ can be endowed with a near-ring structure under pointwise addition and composition of mappings, such a near-ring being called the *transformation near-ring* of the group G.

In 1959 Berman and Silverman [1] claimed that every near-ring is isomorphic with a near-ring of transformations. At that time only some hints were presented, while a direct and clear proof of this result appeared in Malone and Heatherly [2] almost ten years later. Since $T(G)$ has an identity, it immediately follows that any near-ring can be embedded in a near-ring with identity. Moreover, in the same paper [2], it was proved that a group $(H, +)$ can be embedded in a group $(G, +)$ if and only if the near-ring $T_0(H)$, consisting of all transformations of H which multiplicatively commute with the zero transformation, can be embedded into the similar near-ring $T_0(G)$ on G under a kernel-preserving monomorphism of near-rings.

Similarly to near-rings, but in the framework of algebraic hyperstructures, Dašić [3] defined the hypernear-rings as hyperstructures with the additive part being a quasicanonical hypergroup [4,5] (called also a polygroup [6,7]), and the multiplicative part being a semigroup with a bilaterally absorbing element, such that the multiplication is distributive with respect to the hyperaddition on the left-hand side. Later on, this algebraic hyperstructure was called a *strongly distributive hypernear-ring*, or a *zero-symmetric hypernear-ring*, while in a hypernear-ring the distributivity property was replaced by the "inclusive distributivity" from the left (or right) side. Moreover, when the additive part is a hypergroup and all the other properties related to the multiplication are conserved, we talk about a *general hypernear-ring* [8]. The distributivity property is important also in other types of hyperstructures, see e.g., [9]. A detailed discussion about the terminology related to hypernear-rings is included in [10]. In the same paper, the authors defined on the set of all transformations of a quasicanonical hypergroup that preserves the zero element a hyperaddition and a multiplication (as the composition of functions) in such a way to obtain a hypernear-ring. More general, the set of all transformations of a hypergroup (not necessarily commutative) together with the same hyperaddition and multiplication is a strongly distributive hypernear-ring [3]. In this note we will extend the study to the set of all multimappings (or multitransformations) of a (non-abelian) hypergroup, defining first a structure of (left) general hypernear-ring, called the multitransformation general hypernear-ring associated with a hypergroup. Then we will show that any hypernear-ring can be weakly embedded into a multitransformation general hypernear-ring, generalizing the similar classical theorem on near-rings [2]. Besides, under same conditions, any additive hypernear-ring is weakly embeddable into the additive hypernear-ring of the transformations of a hypergroup with identity element that commute multiplicatively with the zero-function. The paper ends with some conclusive ideas and suggestions of future works on this topic.

2. Preliminaries

We start with some basic definitions and results in the framework of hypernear-rings and near-rings of group mappings. For further properties of these concepts we refer the reader to the papers [2,3,11,12] and the fundamental books [13–15]. For the consistence of our study, regarding hypernear-rings we keep the terminology established and explained in [8,16].

First we recall the definition introduced by Dašić in 1978.

Definition 1. *[12] A hypernear-ring is an algebraic system $(R, +, \cdot)$, where R is a non-empty set endowed with a hyperoperation $+ : R \times R \to P^*(R)$ and an operation $\cdot : R \times R \to R$, satisfying the following three axioms:*

1. *$(R, +)$ is a quasicanonical hypergroup (named also polygroup [6]), meaning that:*
   - *(a) $x + (y + z) = (x + y) + z$ for any $x, y, z \in R$,*
   - *(b) there exists $0 \in R$ such that, for any $x \in R$, $x + 0 = 0 + x = \{x\}$,*
   - *(c) for any $x \in R$ there exists a unique element $-x \in R$, such that $0 \in x + (-x) \cap (-x) + x$,*
   - *(d) for any $x, y, z \in R$, $z \in x + y$ implies that $x \in z + (-y)$, $y \in (-x) + z$.*
2. *$(R, \cdot)$ is a semigroup endowed with a two-sided absorbing element 0, i.e., for any $x \in R$, $x \cdot 0 = 0 \cdot x = 0$.*
3. *The operation "$\cdot$" is distributive with respect to the hyperoperation "$+$" from the left-hand side: for any $x, y, z \in R$, there is $x \cdot (y + z) = x \cdot y + x \cdot z$.*

This kind of hypernear-ring was called by Gontineac [11] a *zero-symmetric hypernear-ring*. In our previous works [10,16], regarding the distributivity, we kept the Vougiouklis' terminology [17], and therefore, we say that a hypernear-ring is a hyperstructure $(R, +, \cdot)$ satisfying the above mentioned axioms 1. and 2., and the new one:

3′. The operation "$\cdot$" is inclusively distributive with respect to the hyperoperation "$+$" from the left-hand side: for any $x, y, z \in R$, $x \cdot (y + z) \subseteq x \cdot y + x \cdot z$. Accordingly, the Dašić 's hypernear-ring (satisfying the axioms 1., 2., and 3.) is called *strongly distributive hypernear-ring*.

Furthermore, if the additive part is a hypergroup (and not a polygroup), then we talk about a more general type of hypernear-rings.

Definition 2. [8] *A general (left) hypernear-ring is an algebraic structure $(R, +, \cdot)$ such that $(R, +)$ is a hypergroup, $(R, \cdot)$ is a semihypergroup and the hyperoperation "$\cdot$" is inclusively distributive with respect to the hyperoperation "+" from the left-hand side, i.e., $x \cdot (y+z) \subseteq x \cdot y + x \cdot z$, for any $x, y, z \in R$. If in the third condition the equality is valid, then the structure $(R, +, \cdot)$ is called strongly distributive general (left) hypernear-ring. Besides, if the multiplicative part $(R, \cdot)$ is only a semigroup (instead of a semihypergroup), we get the notion of general (left) additive hypernear-ring.*

Definition 3. *Let $(R_1, +, \cdot)$ and $(R_2, +, \cdot)$ be two general hypernear-rings. A map $\rho : R_1 \to R_2$ is called an inclusion homomorphism if the following conditions are satisfied:*

1. $\rho(x+y) \subseteq \rho(x) + \rho(y)$
2. $\rho(x \cdot y) \subseteq \rho(x) \cdot \rho(y)$ *for all* $x, y \in R_1$.

A map ρ is called a good (strong) homomorphism if in the conditions 1. *and* 2. *the equality is valid.*

In the second part of this section we will briefly recall the fundamentals on *near-rings of group mappings*. A *left near-ring* $(N, +, \cdot)$ is a non-empty set endowed with two binary operations, the addition $+$ and the multiplication $\cdot$, such that $(N, +)$ is a group (not necessarily abelian) with the neutral element 0, $(N, \cdot)$ is a semigroup, and the multiplication is distributive with respect to the addition from the left-hand side. Similarly, we have a right near-ring. Several examples of near-rings are obtained on the set of "non-linear" mappings and here we will see two of them.

Let $(G, +)$ be a group (not necessarily commutative) and let $T(G)$ be the set of all functions from G to G. On $T(G)$ define two binary operations: "+" is the pointwise addition of functions, while the multiplication "$\cdot$" is the composition of functions. Then $(T(G), +, \cdot)$ is a (left) near-ring, called the *transformation near-ring* on the group G. Moreover, let $T_0(G)$ be the subnear-ring of $T(G)$ consisting of the functions of $T(G)$ that commute multiplicatively with the zero function, i.e., $T_0(G) = \{f \in T(G) \mid f(0) = 0\}$. These two near-rings, $T(G)$ and $T_0(G)$, have a fundamental role in embeddings. Already in 1959, it was claimed by Berman and Silverman [1] that every near-ring is isomorphic with a near-ring of transformations. One year later the proof was given by the same authors, but using an elaborate terminology and methodology. Here below we recall this result together with other related properties, as presented by Malone and Heatherly [2].

Theorem 1. [2] *Let $(R, +, \cdot)$ be a near-ring. If $(G, +)$ is any group containing $(R, +)$ as a proper subgroup, then $(R, +, \cdot)$ can be embedded in the transformation near-ring $T(G)$.*

Corollary 1. [2] *Every near-ring can be embedded in a near-ring with identity.*

Theorem 2. [2] *A group $(H, +)$ can be embedded in a group $(G, +)$ if and only if $T_0(H)$ can be embedded in $T_0(G)$ by a near-ring monomorphism which is kernel-preserving.*

Theorem 3. [2] *A group $(H, +)$ can be embedded in a group $(G, +)$ if and only if the near-ring $T(H)$ can be embedded in the near-ring $T(G)$.*

3. Weak Embeddable Hypernear-Rings

In this section we aim to extend the results related to embeddings of near-rings to the case of hypernear-rings. In this respect, instead of a group $(G, +)$ we will consider a hypergroup $(H, +)$ and then the set of all multimappings on H, which we endow with a structure of general hypernear-ring.

Theorem 4. *Let $(H, +)$ be a hypergroup (not necessarily abelian) and $T^*(H) = \{h : H \to P^*(H)\}$ the set of all multimappings of the hypergroup $(H, +)$. Define, for all $(f, g) \in T^*(H) \times T^*(H)$, the following hyperoperations:*

$$f \oplus g = \{h \in T^*(H) \mid (\forall x \in H)\ h(x) \subseteq f(x) + g(x)\}$$

$$f \odot g = \{h \in T^*(H) \mid (\forall x \in H)\ h(x) \subseteq g(f(x)) = \bigcup_{u \in f(x)} g(u)\}.$$

The structure $(T^*(H), \oplus, \odot)$ is a (left) general hypernear-ring.

Proof. For any $f, g \in T^*(H)$ it holds: $f \oplus g \neq \emptyset$. Indeed, for any $x \in H$, it holds $f(x) \neq \emptyset$ and $g(x) \neq \emptyset$ and thus, $f(x) + g(x) \neq \emptyset$. Therefore, for the map $h : H \to P^*(H)$ defined by: $h(x) = f(x) + g(x)$ for all $x \in H$, it holds $h \in f \oplus g$. Now, we prove that the hyperoperation $\oplus$ is associative. Let $f, g, h \in T^*(H)$ and set

$$L = (f \oplus g) \oplus h = \bigcup\{h' \oplus h \mid h' \in f \oplus g\} =$$

$$= \bigcup\{h' \oplus h \mid (\forall x \in H)\ h'(x) \subseteq f(x) + g(x)\}.$$

Thus, if $h'' \in L$, then, for all $x \in H$, it holds: $h''(x) \subseteq h'(x) + h(x) \subseteq (f(x) + g(x)) + h(x)$. Conversely, if h'' is an element of $T^*(H)$ such that: $h''(x) \subseteq (f(x) + g(x)) + h(x)$, for all $x \in H$, and if we choose h' such that $h'(x) = f(x) + g(x)$ for all $x \in H$, then $h' \in f \oplus g$ and $h'' \in h' \oplus h$ i.e., $h'' \in L$. So, $L = \{h'' \in T^*(H) | (\forall x \in H) h''(x) \subseteq (f(x) + g(x)) + h(x)\}$. On the other side, take $D = f \oplus (g \oplus h)$. Then, $D = \{h'' \in T^*(H) | (\forall x \in H) h''(x) \subseteq f(x) + (g(x) + h(x))\}$. By the associativity of the hyperoperation $+$ we obtain that $L = D$, meaning that the hyperoperation $\oplus$ is associative.

Let $f, g \in T^*(H)$. We prove that the equation $f \in g \oplus a$ has a solution $a \in T^*(H)$. If we set $a(x) = H$, for all $x \in H$, then $a \in T^*(H)$ and for all $x \in H$ it holds $g(x) + a(x) = H \supseteq f(x)$. So, $f \in g \oplus a$. Similarly, the equation $f \in a \oplus g$ has a solution in $T^*(H)$. Thus, $(T^*(H), \oplus)$ is a hypergroup.

Now, we show that $(T^*(H), \odot)$ is a semihypergroup. Let $f, g \in T^*(H)$. For all $x \in H$ it holds $g(x) \neq \emptyset$ and so $g(f(x)) \neq \emptyset$. Let $h : H \to P^*(H)$ be a multimapping defined by $h(x) = g(f(x))$, for all $x \in H$. Obviously, $h \in f \odot g$ and so $f \odot g \neq \emptyset$. Let us prove that $\odot$ is a associative. Let $f, g, h \in T^*(H)$. Set:

$$L = (f \odot g) \odot h = \bigcup\{h' \odot h \mid h' \in f \odot g\} = \{h' \odot h \mid (\forall x \in H)\ h'(x) \subseteq g(f(x))\} =$$

$$= \{h'' \mid (\forall x \in H)\ h''(x) \subseteq h(h'(x)) \wedge h'(x) \subseteq g(f(x))\}.$$

So, if $h'' \in L$, then $h''(x) \subseteq h(g(f(x)))$, for all $x \in H$. On the other side, if $h'' \in T^*(H)$ and $h''(x) \subseteq h(g(f(x)))$ for all $x \in H$, then we choose $h' \in T^*(H)$ such that $h'(x) = g(f(x))$ and consequently we obtain that $h'' \subseteq h(h'(x))$. Thus, $h'' \in L$. So, $L = \{h'' \in T^*(H) \mid (\forall x \in H)\ h''(x) \subseteq h(g(f(x)))\}$.

Similarly, $D = f \odot (g \odot h) = \{h'' \mid (\forall x \in H)\ h''(x) \subseteq h(g(f(x)))\}$. Thus, $L = D$.

It remains to prove that the hyperoperation $\oplus$ is inclusively distributive with respect to the hyperoperation $\odot$ on the left-hand side. Let $f, g, h \in T^*(H)$. Set $L = f \odot (g \oplus h) = \bigcup\{f \odot h' | h' \in g \oplus h\} = \bigcup\{f \odot h' | h' \in T^*(H) \wedge (\forall x) h'(x) \subseteq g(x) + h(x)\}$. So, if $k \in L$ then for all $x \in H$ it holds: $k(x) \subseteq h'(f(x)) \subseteq g(f(x)) + h(f(x))$.

On the other hand, $D = (f \odot g) \oplus (f \odot h) = \bigcup\{k_1 \oplus k_2 | k_1 \in f \odot g, k_2 \in f \odot h\}$. Let $k \in L$. Choose, $k_1, k_2 \in T^*(H)$ such that $k_1(x) = g(f(x))$ and $k_2(x) = h(f(x))$ for all $x \in H$. Then $k_1 \in f \odot g$ and $k_2 \in f \odot h$. Thus, $k(x) \subseteq k_1(x) + k_2(x)$ for all $x \in H$, i.e., $k \in k_1 \oplus k_2$ and $k_1 \in f \odot g, k_2 \in f \odot h$. So, $k \in D$. Therefore, $L \subseteq D$. □

Definition 4. *$T^*(H)$ is called the multitransformations general hypernear-ring on the hypergroup H.*

Remark 1. *Let $(G, +)$ be a group and $T(G)$ be the transformations near-ring on G. Obviously, $T(G) \subset T^*(G) = \{f : G \to P^*(G)\}$ and, for all $f, g \in T(G)$, it holds: $f \oplus g = f + g$, $f \odot g = f \cdot g$, meaning that the hyperoperations defined in Theorem 4 are the same as the operations in Theorem 1. It follows that $T(G)$ is a sub(hyper)near-ring of $(T^*(G), \oplus, \odot)$.*

Definition 5. *We say that the hypernear-ring $(R_1, +, \cdot)$ is weak embeddable (by short $W-$ embeddable) in the hypernear-ring $(R_2, +, \cdot)$ if there exists an injective inclusion homomorphism $\mu : R_1 \to R_2$.*

The next theorem is a generalization of Theorem 1 [5].

Theorem 5. *For every general hypernear-ring $(R, +, \cdot)$ there exists a hypergroup $(H, +)$ such that R is $W-$ embeddable in the associated hypernear-ring $T^*(H)$.*

Proof. Let $(R, +, \cdot)$ be a hypernear-ring and let $(H, +)$ be a hypergroup such that $(R, +)$ is a proper subhypergroup of $(H, +)$. For a fixed element $r \in R$ we define a multimapping $f_r : H \to P^*(H)$ as follows

$$f_r(g) = \begin{cases} g \cdot r, & \text{if } g \in R \\ r, & \text{if } g \in H \setminus R. \end{cases}$$

Let us define now the mapping $\mu : R \to T^*(H)$ as $\mu(r) = f_r$, which is an inclusion homomorphism. Indeed, if $a, b \in R$ then we have $\mu(a + b) = \{f_c \mid c \in a + b\}$ and $\mu(a) \oplus \mu(b) = f_a \oplus f_b = \{h \mid (\forall g \in H)\ h(g) \subseteq f_a(g) + f_b(g)\}$.

Consider $c \in a + b$ and $g \in H$. If $g \in R$, then $f_c(g) = g \cdot c \subseteq g \cdot (a + b) \subseteq g \cdot a + g \cdot b = f_a(g) + f_b(g)$. If $g \in H \setminus R$, then $f_c(g) = c \in a + b = f_a(g) + f_b(g)$. It follows that, for all $g \in H$, we have $f_c(g) \subseteq f_a(g) + f_b(g)$ and therefore $f_c \in \mu(a) \oplus \mu(b)$, meaning that $\mu(a + b) \subseteq \mu(a) \oplus \mu(b)$.

Similarly, there is $\mu(a \cdot b) = \{f_c \mid c \in a \cdot b\}$ and $\mu(a) \odot \mu(b) = f_a \odot f_b = \{h \in T^*(H) \mid (\forall g \in H)\ h(g) \subseteq f_b(f_a(g))\}$. Let $c \in a \cdot b$. Then, for $g \in R$, it holds: $f_c(g) = g \cdot c \subseteq g \cdot (a \cdot b) = (g \cdot a) \cdot b = f_b(f_a(g))$. If $g \in H \setminus R$, then there is $f_c(g) = c \in a \cdot b = f_b(a) = f_b(f_a(g))$. Thus, $f_c \in \mu(a) \odot \mu(b)$ and so $\mu(a \odot b) \subseteq \mu(a) \odot \mu(b)$.

Based on Definition 3, we conclude that μ is an inclusive homomorphism. It remains to show that μ is injective. If $\mu(a) = \mu(b)$, then for all $g \in H$, it holds $f_a(g) = f_b(g)$. So, if we choose $g \in H \setminus R$, then we get that $a = f_a(g) = f_b(g) = b$.

These all show that the general hypernear-ring R is W-embeddable in $T^*(H)$. □

Remark 2. *If $(R, +, \cdot)$ is a near-ring such that $(R, +)$ is a proper subgroup of a group $(G, +)$, then for a fixed $r \in R$ the multimapping f_r constructed in the proof of Theorem 5 is in fact a map from G to G, since in this case the multiplication $\cdot$ is an ordinary operation, i.e., $g \cdot r \in G$, for all $g \in R$. Thus $f_r : G \to G$ and thereby $\mu(R) \subseteq T(G)$. By consequence $\mu : R \to T(G)$ is an ordinary monomorphism. In other words, Theorem 5 is a generalization of Theorem 1.*

Example 1. *Let $(R, +, \cdot)$ be a left near-ring. Let P_1 and P_2 be non-empty subsets of R such that $R \cdot P_1 \subseteq P_1$ and $P_1 \subseteq Z(R)$, where $Z(R)$ is the center of R, i.e., $Z(R) = \{x \in R \mid (\forall y \in R) x + y = y + x\}$. For any $(x, y) \in R^2$ define:*

$$x \oplus_{P_1} y = x + y + P_1, \quad x \odot_{P_2} y = xP_2y.$$

Then the structure $(R, \oplus_{P_1}, \odot_{P_2})$ is a general left hypernear-ring [8,18]. Let $H = R \cup \{a\}$ and define on H the hyperoperation $\oplus'_P$ as follows:

$$x \oplus'_{P_1} y = \begin{cases} x \oplus_{P_1} y, & \text{if } x, y \in R \\ H, & \text{if } x = a \vee y = a. \end{cases}$$

It is clear that H is a hypergroup such that $(R, +)$ is a proper subhypergroup of $(H, +)$. Besides, based on Theorem 5, for every $r \in R$ the multimapping $f_r : H \to P^*(H)$ is defined as

$$f_r(g) = \begin{cases} g \odot_{P_2} r, & \text{if } g \in R \\ r, & \text{if } g = a \end{cases} = \begin{cases} gP_2r, & \text{if } g \in R \\ r, & \text{if } g = a. \end{cases}$$

Clearly it follows that $\mu : R \to P^*(H)$, defined by $\mu(r) = f_r$, is an inclusive homomorphism, so the general left hypernear-ring $(R, \oplus_{P_1}, \odot_{P_2})$ is W-embeddable in $T^*(H)$.

Example 2. *Consider the semigroup $(\mathbb{N}, \cdot)$ of natural numbers with the standard multiplication operation and the order "$\leq$". Define on it the hyperoperations $+_\leq$ and $\cdot_\leq$ as follows:*

$$x +_\leq y = \{z \mid x \leq z \vee y \leq z\}$$

$$x \cdot_\leq y = \{z \mid x \cdot y \leq z\}.$$

Then the structure $(\mathbb{N}, +_\leq, \cdot_\leq)$ is a strongly distributive general hypernear-ring (in fact it is a hyperring). This follows from Theorem 4.3 [19]. Furthermore, for any $a \notin \mathbb{N}$, it can be easily verified that $(\mathbb{N}, +_\leq)$ is a proper subhypergroup of $(\mathbb{N} \cup \{a\}, +'_\leq)$, where the hyperoperation $+'_\leq$ is defined by:

$$x +'_\leq y = \begin{cases} x +_\leq y, & \text{if } x, y \in \mathbb{N} \\ \mathbb{N} \cup \{a\}, & \text{if } x = a \vee y = a. \end{cases}$$

In this case, for a fixed $n \in \mathbb{N}$, we can define the multimapping $f_n : \mathbb{N} \cup \{a\} \to P^*(\mathbb{N} \cup \{a\})$ as follows:

$$f_n(g) = \begin{cases} g \cdot_\leq n, & \text{if } g \in \mathbb{N} \\ n, & \text{if } g = a \end{cases} = \begin{cases} \{k \in \mathbb{N} \mid g \cdot n \leq k\}, & \text{if } g \in \mathbb{N} \\ n, & \text{if } g = a \end{cases}$$

and therefore the mapping $\mu : \mathbb{N} \to P^*(\mathbb{N} \cup \{a\})$ is an inclusive homomorphism. Again this shows that the general hypernear-ring $(\mathbb{N}, +_\leq, \cdot_\leq)$ is W-embeddable in $T^*(\mathbb{N} \cup \{a\})$.

Example 3. *Let $R = \{0, 1, 2, 3\}$. Consider now the semigroup $(R, \cdot)$ defined by Table 1:*

Table 1. The Cayley table of the semigroup $(R, \cdot)$

$\cdot$	**0**	**1**	**2**	**3**
0	0	0	0	0
1	0	1	2	3
2	0	1	2	3
3	0	1	2	3

Define on R the hyperoperation $+_\leq$ as follows: $x +_\leq y = \{z \mid x \leq z \vee y \leq z\}$, so its Cayley table is described in Table 2:

Table 2. The Cayley table of the hypergroupoid $(R, +_\leq)$

$+_\leq$	**0**	**1**	**2**	**3**
0	R	R	R	R
1	R	{1,2,3}	{1,2,3}	{1,2,3}
2	R	{1,2,3}	{2,3}	{2,3}
3	R	{1,2,3}	{2,3}	{3}

Obviously, the relation $\leq$ is reflexive and transitive and, for all $x, y, z \in R$, it holds: $x \leq y \Rightarrow z \cdot x \leq z \cdot y$. Thus, $(R, +_\leq, \cdot)$ is an (additive) hypernear-ring. Let $H = R \cup \{4\}$ and define the hyperoperation $+_\leq$ as follows:

$$x +_\leq y = \begin{cases} x +_\leq y, & \text{if } x, y \in \{0, 1, 2, 3\} \\ H, & \text{if } x = 4 \vee y = 4 \end{cases}$$

It follows that $(R,+)$ is a proper subhypergroup of $(H,+)$ and for a fixed $r \in R$ it holds $f_r(x) = r$, for all $x \in H$. This implies that the mapping $\mu : H \to P^*(H)$, defined by $\mu(r) = f_r$ for any $r \in R$, is an inclusive homomorphism.

Now we will construct a left general additive hypernear-ring associated with an arbitrary hypergroup.

Theorem 6. *Let $(H,+)$ be a hypergroup and $T(H) = \{f : H \to H\}$. On the set $T(H)$ define the hyperoperation $\oplus_T$ and the operation $\odot_T$ as follows:*

$$f \oplus_T g = \{h \in T(H) \mid (\forall x \in H)\ h(x) \in f(x) + g(x)\},$$

$$(f \odot_T g)(x) = g(f(x)), \text{ for all } x \in H.$$

The obtained structure $(T(H), \oplus_T, \odot_T)$ is a (left) general additive hypernear-ring.

Proof. Let $f, g \in T(H)$. We prove that there exists $h \in T(H)$ such that $h(x) \in f(x) + g(x)$ for all $x \in H$. Let $x \in H$. Since $f(x) + g(x) \neq \emptyset$ we can choose $h_x \in f(x) + g(x)$ and define $h(x) = h_x$. Obviously, $h \in f \oplus_T g$. Now we prove that the hyperoperation $\oplus_T$ is associative. Let $f, g, h \in T(H)$. Set $L = (f \oplus_T g) \oplus_T h = \{h'' \mid (\forall x)\ h''(x) \in h'(x) + h(x) \wedge h'(x) \in f(x) + g(x)\}$ and $D = f \oplus_T (g \oplus_T h) = \{f'' \mid (\forall x)\ f''(x) \in f(x) + f'(x) \wedge f'(x) \in g(x) + h(x)\}$. Thus, if $h'' \in L$, then $h''(x) \in (f(x) + g(x)) + h(x) = f(x) + (g(x) + h(x))$. Thereby, for any $x \in H$, there exists $a_x \in g(x) + h(x)$ such that $h''(x) \in f(x) + a_x$. Define $f'(x) = a_x$. Then, $f' \in g \oplus_T h$ and for all $x \in H$ it holds $h''(x) \in f(x) + f'(x)$. Therefore, $h'' \in D$. So, $L \subseteq D$. Similarly, we obtain that $D \subseteq L$. Now, let $f, g \in T(H)$. We prove that the equation $f \in g \oplus_T h$ has a solution $h \in T(H)$. Since $(H,+)$ is a hypergroup, it follows that, for any $x \in H$, there exists $b_x \in H$ such that $f(x) \in g(x) + b_x$. Define $h : H \to H$ by $h(x) = b_x$. Then $h \in T(H)$ and $f \in g \oplus_T h$. Similarly, we obtain that the equation $f \in h \oplus_T g$ has a solution in $T(H)$. We may conclude that $(T(H), \oplus_T)$ is a hypergroup.

Obviously, $(T(H), \odot_T)$ is a semigroup, because the composition of functions is associative. Now we prove that the hyperoperation $\oplus_T$ is left inclusively distributive with respect to the operation $\odot_T$. Let $f, g, h \in T(H)$. Set $L = f \odot_T (g \oplus_T h) = \{f \odot_T k \mid k \in g \oplus_T h\}$ and $D = (f \odot_T g) \oplus_T (f \odot_T h) = \{h' \mid (\forall x \in H)\ h'(x) \in g(f(x)) + h(f(x))\}$. Let $k \in g \oplus h$. Then, for all $x \in H$, it holds $(f \odot k)(x) = k(f(x)) \subseteq g(f(x)) + h(f(x))$. Thus, $f \odot k \in D$, meaning that $L \subseteq D$. $\square$

For an arbitrary group G, Malone and Heatherly [2] denote by $T_0(G)$ the subset of $T(G)$ consisting of the functions which commute multiplicatively with the zero-function, i.e., $T_0(G) = \{f : G \to G \mid f(0) = 0\}$. Obviously, $T_0(G)$ is a sub-near-ring of $(T(G), +, \cdot)$. The next result extends this property to the case of hyperstructures.

Theorem 7. *Let $(H,+)$ be a hypergroup with the identity element 0 (i.e., for all $x \in H$, it holds $x \in x + 0 \cap 0 + x$), such that $0 + 0 = \{0\}$. Let $T_0(H) = \{f : H \to H \mid f(0) = 0\}$. Then, $T_0(H)$ is a subhypernear-ring of the general additive hypernear-ring $(T(H), \oplus_T, \odot_T)$.*

Proof. Let $f, g \in T_0(H)$. If $h \in f \oplus_T g$, then $h(0) \in f(0) + g(0) = 0 + 0 = \{0\}$, i.e., $h(0) = 0$. Thus, $h \in T_0(H)$. Let $f, g \in T_0(H)$. We prove now that the equation $f \in g \oplus a$ has a solution $a \in T_0(H)$. If we set $a(0) = 0$ and $a(x) = a_x$, where $f(x) \in g(x) + a_x$, for $x \neq 0$ and $a_x \in H$, then $a \in T_0(H)$ and $f \in g + a$. Similarly the equation $f \in a \oplus g$ has a solution $a \in T_0(H)$. Thus, $(T_0(H), \oplus_T)$ is a subhypergroup of $(T(H), \oplus_T)$. Obviously, if $f, g \in T_0(H)$, then it follows that $(f \odot_T g)(0) = g(f(0)) = g(0) = 0$, i.e., $f \odot_T g \in T_0(H)$. So, $(T_0(H), \odot_T)$ is a subsemihypergroup of $(T(H), \odot_T)$, implying that $T_0(H)$ is a subsemihypernear-ring of $T(H)$. $\square$

Theorem 8. *Let $(R,+,\cdot)$ be an additive hypernear-ring such that $(R,+)$ is a proper subhypergroup of the hypergroup $(H,+)$, having an identity element 0 satisfying the following properties:*

1. $0+0=\{0\}$ *and*
2. $0\cdot r=0$*, for all* $r\in R$.

Then the hypernear-ring $(R,+,\cdot)$ is $W-$embeddable in the additive hypernear-ring $T_0(H)$.

Proof. For a fixed $r\in R$, define a map $f:H\to H$ as follows

$$f_r(g)=\begin{cases} g\cdot r, & \text{if } g\in R \\ r, & \text{if } g\in H\setminus R. \end{cases}$$

Obviously, $f_r(0)=0\cdot r=0$. So, $f_r\in T_0(H)$ and, similarly as in the proof of Theorem 5, we obtain that the map $\rho:(R,+,\cdot)\to(T_0(H),\oplus_T,\odot_T)$ defined by $\rho(r)=f_r$ is an injective inclusion homomorphism. □

Example 4. *On the set $H=\{0,1,2,3,4,5,6\}$ define an additive hyperoperation and a multiplicative operation having the Cayley tables described in Tables 3 and 4, respectively:*

Table 3. The Cayley table of the hypergroupoid $(H,+)$

+	**0**	**1**	**2**	**3**	**4**	**5**	**6**
0	0	1	2	3	4	5	6
1	1	2	3	4	5	{0, 6}	1
2	2	3	4	5	{0, 6}	1	2
3	3	4	5	{0, 6}	1	2	3
4	4	5	{0, 6}	1	2	3	4
5	5	{0, 6}	1	2	3	4	5
6	6	1	2	3	4	5	0

Table 4. The Cayley table of the semigroup $(H,\cdot)$

·	**0**	**1**	**2**	**3**	**4**	**5**	**6**
0	0	0	0	0	0	0	0
1	0	5	4	3	2	1	0
2	0	1	2	3	4	5	0
3	0	0	0	0	0	0	0
4	0	5	4	3	2	1	0
5	0	1	2	3	4	5	0
6	0	0	0	0	0	0	0

The structure $(H,+,\cdot)$ is an (additive) hypernear-ring [16].

Let $R=\{0,3,6\}$. Then $(R,+,\cdot)$ is a hypernear-ring (in particular it is a subhypernear-ring of $(H,+,\cdot)$). Obviously, $(R,+)$ is a proper subhypergroup of the hypergroup $(H,+)$, which has the identity 0 such that $0+0=\{0\}$ and $0\cdot r=0$, for all $r\in R$. It follows that, for each $r\in\{0,3,6\}$, $f_r:H\to H$ is a map such that $f_0(g)=0$, for all $g\in H$,

$$f_3(g)=\begin{cases} g\cdot 3, & \text{if } g\in\{0,3,6\} \\ 3, & \text{if } g\in\{1,2,4\} \end{cases}=\begin{cases} 0, & \text{if } g\in\{0,3,6\} \\ 3, & \text{if } g\in\{1,2,4\}, \end{cases}$$

while

$$f_6(g)=\begin{cases} g\cdot 6, & \text{if } g\in\{0,3,6\} \\ 6, & \text{if } g\in\{1,2,4\} \end{cases}=\begin{cases} 0, & \text{if } g\in\{0,3,6\} \\ 6, & \text{if } g\in\{1,2,4\}. \end{cases}$$

Clearly, the map $\rho : (R, +, \cdot) \to (T_0(H), \oplus_T, \odot_T)$*, defined by* $\rho(r) = f_r$*, is an injective inclusion homomorphism, so the hypernear-ring* R *is W-embeddable in* $T_0(H)$*.*

Remark 3. *If* $(G, +)$ *is a group, then, for any* $f, g \in T(G) = \{f : G \to G\}$*, it holds* $f \oplus_T g = f + g$ *and* $f \odot_T g = f \cdot g$*, meaning that the transformation near-ring* $(T(G), +, \cdot)$ *of a group* G *is in fact the structure* $(T(G), \oplus_T, \odot_T)$*. Furthermore, if* $(R, +, \cdot)$ *is a zero-symmetric near-ring, i.e., a near-ring in which any element* x *satisfies the relation* $x \cdot 0 = 0 \cdot x = 0$*, then the map* ρ *constructed in the proof of Theorem 8 is the injective homomorphism* $\rho : R \to T_0(G)$*. Thus, according with Theorem 8, it follows that the zero-symmetric near-ring* $(R, +, \cdot)$ *is W-embeddable in the near-ring* $T_0(G)$*, where* $(G, +)$ *is any group containing* $(R, +)$ *as a proper subgroup.*

Remark 4. *If* $(G, +)$ *is a group, then the following inclusions hold:* $T_0(G) \subseteq T(G) \subseteq T^*(G)$*, where both* $T(G)$ *and* $T_0(G)$ *are sub-(hyper)near-rings of the hypernear-ring* $T^*(G)$*. Considering now* $(H, +)$ *a hypergroup, the same inclusions exist:* $T_0(H) \subseteq T(H) \subseteq T^*(H)$*, but generally* $T(H)$ *and* $T_0(H)$ *are not subhypernear-rings of* $T^*(H)$*.*

Proposition 1. *Let* $(H, +)$ *be a hypergroup with the identity element* 0 *(i.e., for all* $x \in H$ *it holds* $x \in x + 0 \cap 0 + x$*) such that* $0 + 0 = \{0\}$*. Let* $T_0^*(H) = \{f : H \to P^*(H) \mid f(0) = 0\}$*. Then,* $T_0^*(H)$ *is a subhypernear-ring of the general hypernear-ring* $(T^*(H), \oplus, \odot)$*.*

Proof. Let $f, g \in T_0^*(H)$. If $h \in f \oplus g$, then it holds $h(0) \subseteq f(0) + g(0) = 0 + 0 = \{0\}$. Since $h(0) \neq \emptyset$, it follows that $h(0) = \{0\}$. Thus, $h \in T_0^*(H)$. Let $f, g \in T_0^*(H)$. We prove that the equation $f \in g \oplus a$ has a solution $a \in T_0^*(H)$. If we set $a(0) = 0$ and $a(x) = H$, for all $x \neq 0$, then $a \in T_0^*(H)$ and, for all $x \neq 0$, it holds $g(x) + a(x) = H \supseteq f(x)$ and $g(0) + a(0) = \{0\} = f(0)$, meaning that $f \in g \oplus a$. Similarly, the equation $f \in a \oplus g$ has a solution in $T_0^*(H)$. So, $(T_0^*(H))$ is a subhypergroup of $(T^*(H), \oplus)$. Obviously, if $h \in f \odot g$, then $h(0) \subseteq g(f(0)) = \{0\}$. So, $h \in T_0^*(H)$. Thus $T_0^*(H)$ is a subsemihypergroup of $(T^*(H), \odot)$. Therefore, $T_0^*(H)$ is a subhypernear-ring of $(T^*(H), \oplus, \odot)$. □

4. Conclusions

Distributivity property plays a fundamental role in the ring-like structures, i.e., algebraic structures endowed with two operations, usually denoted by addition and multiplication, where the multiplication distributes over the addition. If this happens only from one-hand side, then we talk about near-rings. Similarly, in the framework of algebraic hypercompositional structures, a general hypernear-ring has the additive part an arbitrary hypergroup, the multiplicative part is a semihypergroup, and the multiplication hyperoperation inclusively distributes over the hyperaddition from the left or right-hand side, i.e., for three arbitrary elements x, y, z, there is $x \cdot (y + z) \subseteq x \cdot y + x \cdot z$ for the left-hand side, and respectively, $(y + z) \cdot x \subseteq y \cdot x + z \cdot x$ for the right-hand side. If the inclusion is substituted by equality, then the general hypernear-ring is called *strongly distributive*. We also recall here that there exist also hyperrings having the additive part a group, while the multiplicative one is a semihypergroup, being called *multiplicative hyperrings* [20].

The set of all transformations of a group G, i.e., $T(G) = \{g : G \to G\}$, can be endowed with a near-ring structure, while similarly, on the set of all multitransformations of a hypergroup H, i.e., $T^*(H) = \{h : H \to \mathcal{P}^*(H)\}$, can be defined a general hypernear-ring structure, called the *multitransformations general hypernear-ring* associated with the hypergroup H. We have shown that for every general hypernear-ring R there exists a hypergroup H such that R is weakly embeddable in the associated multitransformations general hypernear-ring $T^*(H)$ (see Theorem 5). Moreover, considering the set $T(H) = \{f : H \to H\}$ of all transformations of a hypergroup H, we have defined on it a hyperaddition and a multiplication such that $T(H)$ becomes a general additive hypernear-ring. We have determined conditions under which the set $T_0(H)$, formed with the transformations of H that multiplicatively commute with the zero function on H, is a subhypernear-ring of $T(H)$. Besides, an additive hypernear-ring satisfying certain conditions can be weakly embedded in the additive hypernear-ring $T_0(H)$ (see Theorem 8).

In our future work, we intend to introduce and study properties of $\Delta-$endomorphisms and $\Delta-$multiendomorphisms of hypernear-rings as generalizations of similar notions on near-rings.

Author Contributions: The authors contributed equally to this paper.

Funding: The third author acknowledges the financial support from the Slovenian Research Agency (research core funding No. P1-0285).

Conflicts of Interest: The authors declare no conflict of interest.

References

1. Berman, G.; Silverman, R.J. Near-rings. *Am. Math. Mon.* **1959**, *66*, 23–34. [CrossRef]
2. Malone, J.J.; Heatherly, H.E., Jr. Some Near-Ring Embeddings. *Quart. J. Math. Oxf. Ser.* **1969**, *20*, 81–85. [CrossRef]
3. Dašić, V. Hypernear-rings. In *Algebraic Hyperstructures and Applications (Xanthi, 1990)*; World Scientific Publishing: Teaneck, NJ, USA, 1991; pp. 75–85.
4. Bonansinga, P. Sugli ipergruppi quasicanonici. *Atti Soc. Peloritana Sci. Fis. Mat. Natur* **1981**, *27*, 9–17.
5. Massouros, C.G. Quasicanonical hypergroups. In *Algebraic Hyperstructures and Applications (Xanthi, 1990)*; World Scientific Publishing: Teaneck, NJ, USA, 1991; pp. 129–136.
6. Comer, S.D. Polygroups derived from cogroups. *J. Algebra* **1984**, *89*, 387–405. [CrossRef]
7. Davvaz, B. *Polygroup Theory and Related Systems*; World Scientific Publishing, Co. Pte. Ltd.; Hackensack, NJ, USA, 2013.
8. Jančić-Rašović, S.; Cristea, I. A note on near-rings and hypernear-rings with a defect of distributivity. *AIP Conf. Proc.* **1978**, *1978*, 34007.
9. Ameri, R.; Amiri-Bideshki, M.; Hoskova-Mayerova, S.; Saeid, A.B. Distributive and Dual Distributive Elements in Hyperlattices. *Ann. Univ. Ovidius Constanta Ser. Mat.* **2017**, *25*, 25–36. [CrossRef]
10. Jančić-Rašović, S.; Cristea, I. Division hypernear-rings. *Ann. Univ. Ovidius Constanta Ser. Mat.* **2018**, *26*, 109–126. [CrossRef]
11. Gontineac, M. On Hypernear-ring and H-hypergroups. In *Algebraic Hyperstructures and Applications (Lasi, 1993)*; Hadronic Press: Palm Harbor, FL, USA, 1994; pp. 171–179.
12. Dašić, V. A defect of distributivity of the near-rings. *Math. Balk.* **1978**, *8*, 63–75.
13. Clay, J. *Nearrings: Geneses and Application*; Oxford University Press: Oxford, UK, 1992.
14. Meldrum, J. *Near-Rings and Their Links with Groups*; Pitman: London, UK, 1985.
15. Pilz, G. *Near-Rings: The theory and Its Applications*; North-Holland Publication Co.: New York, NY, USA, 1983.
16. Jančić-Rašović, S.; Cristea, I. Hypernear-rings with a defect of distributivity. *Filomat* **2018**, *32*, 1133–1149. [CrossRef]
17. Vougiouklis, T. *Hyperstructures and Their Representations*; Hadronic Press: Palm Harbor, FL, USA, 1994.
18. Jančić-Rašović, S. On a class of $P_1 - P_2$ hyperrings and hypernear-rings. *Set-Val. Math. Appl.* **2008**, *1*, 25–37.
19. Jančić-Rašović, S.; Dasic, V. Some new classes of (m, n)-hyperrings. *Filomat* **2012**, *26*, 585–596. [CrossRef]
20. Ameri, R.; Kordi, A.; Hoškova-Mayerova, S. Multiplicative hyperring of fractions and coprime hyperideals. *Ann. Univ. Ovidius Constanta Ser. Mat.* **2017**, *25*, 5–23. [CrossRef]

symmetry

MDPI

Article

Edge Even Graceful Labeling of Cylinder Grid Graph

Ahmed A. Elsonbaty [1,2] and Salama Nagy Daoud [1,3,*]

1 Department of Mathematics, Faculty of Science, Taibah University, Al-Madinah 41411, Saudi Arabia; ahmad_elsonbaty@hotmail.com

2 Department of Mathematics, Faculty of Science, Ain Shams University, Cairo 11566, Egypt

3 Department of Mathematics and Compuer Science, Faculty of Science, Menoufia University, Shebin El Kom 32511, Egypt

* Correspondence: sdaoud@taibahu.edu.sa

Received: 26 March 2019; Accepted: 14 April 2019; Published: 22 April 2019

Abstract: Edge even graceful labeling (e.e.g., l.) of graphs is a modular technique of edge labeling of graphs, introduced in 2017. An e.e.g., l. of simple finite undirected graph $G = (V(G), E(G))$ of order $P = |(V(G)|$ and size $q = |E(G)|$ is a bijection $f : E(G) \to \{2, 4, \dots, 2q\}$, such that when each vertex $v \in V(G)$ is assigned the modular sum of the labels (images of f) of the edges incident to v, the resulting vertex labels are distinct mod$2r$, where $r = \max(p, q)$. In this work, the family of cylinder grid graphs are studied. Explicit formulas of e.e.g., l. for all of the cases of each member of this family have been proven.

Keywords: graceful labeling; edge even graceful labeling; cylinder grid graph

1. Introduction

The field of graph theory plays an important role in various areas of pure and applied sciences. One of the important areas in graph theory is graph labeling of a graph G which is an assignment of integers either to the vertices or edges or both subject to certain conditions. Graph labeling began nearly 50 years ago. Over these decades, more than 200 methods of labeling techniques were invented, and more than 2500 papers were published. In spite of this huge literature, just few general results were discovered. Nowadays, graph labeling has much attention from different brilliant researchers in graph theory, which has rigorous applications in many disciplines, e.g., communication networks, coding theory, X-ray crystallography, radar, astronomy, circuit design, communication network addressing, database management, and graph decomposition problems. More interesting applications of graph labeling can be found in References [1–11]. A function f is called a graceful labeling of a graph G if $f : V(G) \to \{0, 1, 2, \dots, q\}$ is injective and the induced function $f^* : E(G) \to \{1, 2, \dots, q\}$, defined as $f^*(e = uv) = |f(u) - f(v)|$, is bijective. This type of graph labeling was first introduced by Rosa in 1967 [12] as a $\beta-$ valuation, and later, Solomon W. Golomb [13] termed it as graceful labeling. A function f is called an odd graceful labeling of a graph G if $f : V(G) \to \{0, 1, 2, \dots, 2q-1\}$ is injective and the induced function $f^* : E(G) \to \{1, 3, \dots, 2q-1\}$, defined as $f^*(e = uv) = |f(u) - f(v)|$, is bijective. This type of graph labeling first introduced by Gnanajothi in 1991 [14]. For more results on this type of labeling, see References [15,16]. A function f is called an edge graceful labeling of a graph G if $f : E(G) \to \{1, 2, \dots, q\}$ is bijective and the induced function $f^* : V(G) \to \{0, 1, 2, \dots, p-1\}$, defined as $f^*(u) = \sum_{e=uv\in E(G)} f(e)(\text{mod}p)$, is bijective. This type of graph labeling was first introduced by Lo in 1985 [17]. For more results on this labeling see [18,19]. A function f is called an edge odd graceful labeling of a graph G if $f : E(G) \to \{1, 3, \dots, 2q-1\}$ is bijective and the induced function $f^* : V(G) \to \{0, 1, 2, \dots, 2q-1\}$ defined as $f^*(u) = \sum_{e=uv\subset E(G)} f(e)(\text{mod}2q)$ is injective. This type of graph labeling was first introduced by Solairaju and Chithra in 2009 [20]. For more results on this labeling, see References [21–23]. A function f is called an edge even graceful labeling of a graph G if $f: E(G) \to \{2,4,\ldots,2q-2\}$ is bijective and the induced function $f^*: V(G) \to \{0,2,4,\ldots,2q-2\}$, defined as $f^*(u) = \sum_{e=uv\in E(G)} f(e)(\text{mod}2r)$ where $r = \max\{p,q\}$, is injective. This type of graph labeling was first introduced by Elsonbaty and Daoud in 2017 [24,25]. For a summary of the results on these five types of graceful labels as well as all known labeling techniques, see Reference [26].

2. Cylinder Grid Graph

The Cartesian product $G_1 \times G_2$ of two graphs G_1 and G_2, is the graph with vertex set $V(G_1) \times V(G_2)$, and any two vertices (u_1, v_1) and (u_2, v_2) are adjacent in $G_1 \times G_2$ whenever $u_1 = u_2$ and $v_1v_2 \in E(G_2)$ or $v_1 = v_2$ and $u_1u_2 \in E(G_1)$. The cylinder grid graph $C_{m,n}$ is the graph formed from the Cartesian product $P_m \times C_n$ of the path graph P_m and the cycle graph C_n. That is, the cylinder grid graph consists of m copies of C_n represented by circles, and will be numbered from the innermost circle to the outer circle as $C_n^{(1)}, C_n^{(2)}, C_n^{(3)}, \ldots, C_n^{(m-1)}, C_n^{(m)}$ and we call them simply circles; n copies of P_m represented by paths transverse the m circles and will be numbered clockwise as $P_m^{(1)}, P_m^{(2)}, P_m^{(3)}, \ldots, P_m^{(n-1)}, P_m^{(n)}$ and we call them paths (see Figure 1).

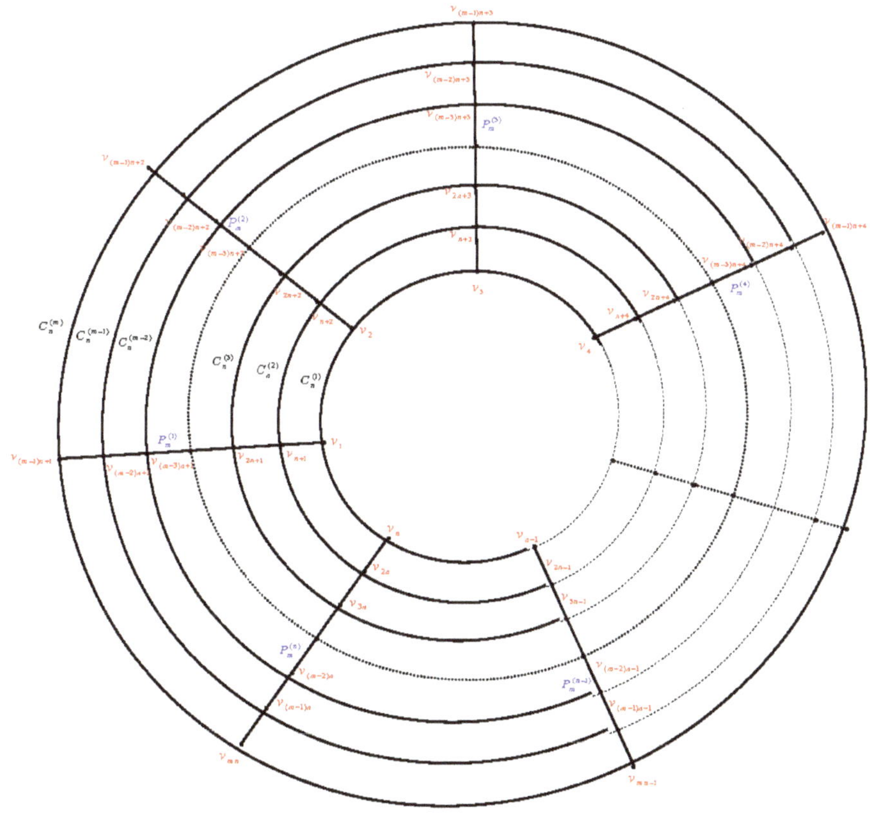

Figure 1. Cylinder grid graph $C_{m,n}$.

Theorem 1. *If m is an even positive integer greater than or equal 2 and $n \geq 2$, then the cylinder grid graph $C_{m,n}$, is an edge even graceful graph.*

Proof. Using standard notation $p = |V(C_{m,n})| = mn$, $q = |E(C_{m,n})| = 2mn - n$ and $r = \max(p,q) = 2mn - n$ and $f : E(C_{m,n}) \to \{2, 4, 6, \ldots, 4mn - 2n - 2\}$. Let the cylinder grid graph $C_{m,n}$ be as in Figure 2. □

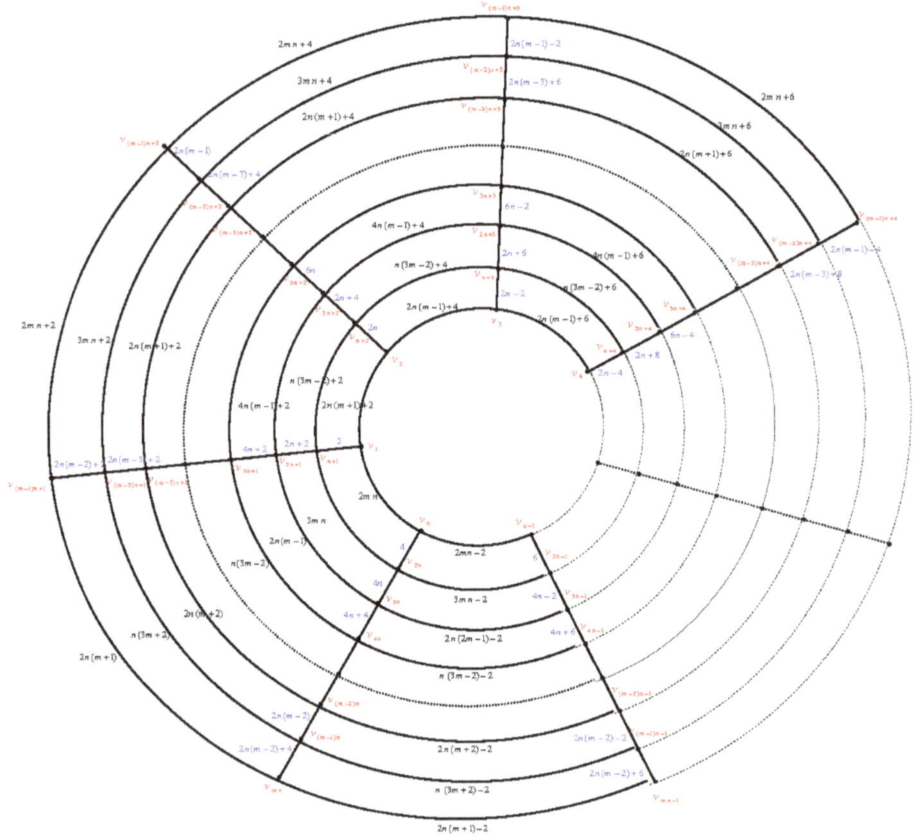

Figure 2. The cylinder grid graph $C_{m,n}$, m is even and $n \geq 2$.

First, we label the edges of the paths $P_m^{(k)}, 1 \leq k \leq n$ beginning with the edges of the path $P_m^{(1)}$ as follows: Move anticlockwise to label the edges $v_1v_{n+1}, v_nv_{2n}, v_{n-1}v_{2n-1}, \ldots, v_3v_{n+3}, v_2v_{n+2}$ by $2, 4, 6, \ldots, 2n-2, 2n$, then move clockwise to label the edges $v_{n+1}v_{2n+1}, v_{n+2}v_{2n+2}, v_{n+3}v_{2n+3}, \ldots, v_{2n-1}v_{3n-1}, v_{2n}v_{3n}$ by $2n+2, 2n+4, 2n+6, \ldots, 4n-2, 4n$, then move anticlockwise to label the edges $v_{2n+1}v_{3n+1}, v_{3n}v_{4n}, v_{3n-1}v_{4n-1}, \ldots, v_{2n+3}v_{3n+3}, v_{2n+2}v_{3n+2}$ by $4n+2, 4n+4, 4n+6, \ldots, 6n-2, 6n$ and so on. Finally, move anticlockwise to label the edges $v_{(m-2)n+1}v_{(m-1)n+1}, v_{(m-1)n}v_{mn}, v_{(m-1)n-1}v_{mn-1}, \ldots, v_{(m-2)n+3}v_{(n-1)m+3}, v_{(m-2)n+2}v_{(m-1)n+2}$ by $2n(m-1)+2, 2n(m-2)+4, 2n(m-2)+6, 2n(m-2)+8, \ldots, 2n(m-1)-2, 2n(m-1)$.

Secondly, we label the edges of the circles $C_n^{(k)}, 1 \leq k \leq m$ beginning with the edges of the innermost circle $C_n^{(1)}$ then the edges of outer circle $C_n^{(m)}$, then the edges of the circles $C_n^{(m-2)}, C_n^{(m-4)}, \ldots, C_n^{(2)}$.

Finally, we label the edges of the circles $C_n^{(m-1)}, C_n^{(m-3)}, \ldots, C_n^{(3)}$ as follows: $f(v_iv_{i+1}) = 2n(m-1) + 2i, 1 \leq i \leq n-1, f(v_nv_1) = 2mn; f(v_{(m-1)n+i}v_{(m-1)n+i+1}) = 2mn + 2i, 1 \leq i \leq n-1, f(v_{mn}v_{(m-1)n+1}) = 2n(m+1); f(v_{(k-1)n+i}v_{(k-1)n+i+1}) = n(3m-k) + 2i, 1 \leq i \leq n-1, f(v_{kn}v_{(k-1)n+1}) = n(3m-k+2), 2 \leq k \leq m-2; f(v_{(k-1)n+i}v_{(k-1)n+i+1}) = n(4m-k-1) + 2i, 1 \leq i \leq n-1, f(v_{kn}v_{(k-1)n+1}) = n(4m-k+1)$, $3 \leq k \leq m-1, k$ is odd.

Thus, the labels of corresponding vertices $\mathrm{mod}(4mn-2n)$ will be: $f^*(v_i) \equiv 2i+2; f^*(v_{n+i}) \equiv 2mn+2n+4i+2; f^*(v_{2n+i}) \equiv 4n+4i+2; f^*(v_{3n+i}) \equiv 2mn+6n+4i+2; \ldots; f^*(v_{(m-3)n+i}) \equiv 4mn-6n+4i+2; f^*(v_{(m-2)n+i}) \equiv 2mn-4n+4i+2; f^*(v_{(m-1)n+i}) \equiv 2mn+2i+2, 1 \le i \le n$.

Illustration: An e.e.g., l, of the cylinder grid graphs $C_{8,11}$ and $C_{8,12}$ are shown in Figure 3.

Theorem 2. *If $m = 3$ and n is an odd positive integer greater than 3, then the cylinder grid graph $C_{3,n}$, is an edge even graceful graph.*

Proof. Using standard notation $p = |V(C_{3,n})| = 3n$, $q = |E(C_{3,n})| = 5n$, $r = \max(p,q) = 5n$, and $f: E(C_{3,n}) \to \{2,4,6,\ldots,10n-2\}$. There are three cases:

Case (1): If $n \equiv 1 \mathrm{mod} 6$, let the cylinder grid graph $C_{3,n}$ be as in Figure 4.

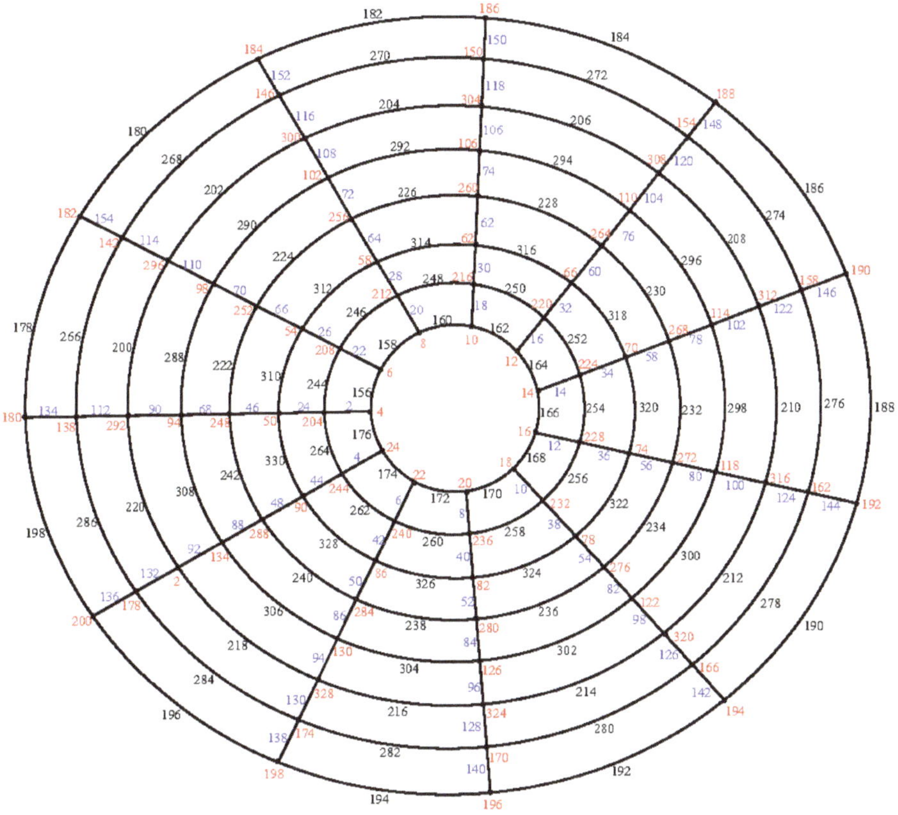

(a) $C_{8,11}$

Figure 3. *Cont.*

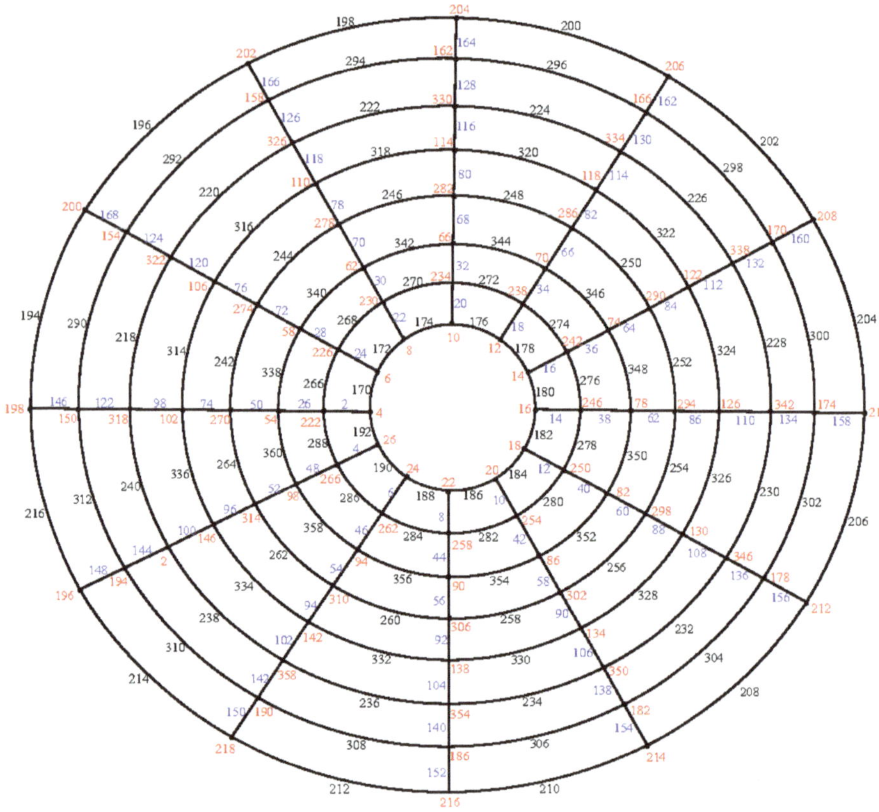

(b) $C_{8,12}$

Figure 3. An edge even graceful labeling (e.e.g., l.) of the cylinder grid graphs $C_{8,11}$ and $C_{8,12}$.

First, we label the edges of the paths $P_3^{(k)}, 1 \leq k \leq n$ beginning with the edges of the path $P_3^{(1)}$ as follows: Move clockwise to label the edges $f(v_1v_{n+1}) = 2, f(v_2v_{n+2}) = 6, f(v_iv_{n+i}) = 2i+2, 3 \leq i \leq n$. Then, move anticlockwise to label the edges $f(v_{n+1}v_{2n+1}) = 2n+4, f(v_{2n}v_{3n}) = 2n+6, f(v_{2n-1}v_{3n-1}) = 2n+8, f(v_{2n-2}v_{3n-2}) = 2n+10, \ldots, f(v_{n+3}v_{2n+3}) = 4n, f(v_{n+2}v_{2n+2}) = 4n+2$.

Secondly, we label the edges of the circles $C_n^{(k)}, 1 \leq k \leq 3$ beginning with the edges of the innermost circle $C_n^{(1)}$, then the edges of outer circle $C_n^{(3)}$, and then the edges of the circle $C_n^{(2)}$. Label the edges of the circle $C_n^{(1)}$ as follows: $f(v_1v_2) = 4n+4, f(v_2v_3) = 4n+6, \ldots, f(v_{\frac{n-1}{3}}v_{\frac{n+2}{3}}) = \frac{14n+10}{3}, f(v_{\frac{n+2}{3}}v_{\frac{n+5}{3}}) = \frac{14n+4}{3}, f(v_{\frac{n+5}{3}}v_{\frac{n+8}{3}}) = \frac{14n+16}{3}, f(v_{\frac{n+8}{3}}v_{\frac{n+11}{3}}) = \frac{14n+22}{3}, f(v_{\frac{n+11}{3}}v_{\frac{n+14}{3}}) = \frac{14n+34}{3}, f(v_{\frac{n+14}{3}}v_{\frac{n+17}{3}}) = \frac{14n+28}{3}, f(v_{\frac{n+17}{3}}v_{\frac{n+20}{3}}) = \frac{14n+40}{3}, f(v_{\frac{n+20}{3}}v_{\frac{n+23}{3}}) = \frac{14n+46}{3}, f(v_{\frac{n+23}{3}}v_{\frac{n+26}{3}}) = \frac{14n+58}{3}, f(v_{\frac{n+26}{3}}v_{\frac{n+29}{3}}) = \frac{14n+52}{3}, f(v_{\frac{n+29}{3}}v_{\frac{n+32}{3}}) = \frac{14n+64}{3}, f(v_{\frac{n+32}{3}}v_{\frac{n+35}{3}}) = \frac{14n+70}{3}, f(v_{\frac{n+35}{3}}v_{\frac{n+38}{3}}) = \frac{14n+82}{3}, f(v_{\frac{n+38}{3}}v_{\frac{n+41}{3}}) = \frac{14n+76}{3}, f(v_{\frac{n+41}{3}}v_{\frac{n+44}{3}}) = \frac{14n+88}{3}, f(v_{\frac{n+44}{3}}v_{\frac{n+47}{3}}) = \frac{14n+94}{3}, \ldots, f(v_{n-13}v_{n-12}) = 6n-22, f(v_{n-12}v_{n-11}) = 6n-24, f(v_{n-11}v_{n-10}) = 6n-20, f(v_{n-10}v_{n-9}) = 6n-18, f(v_{n-9}v_{n-8}) = 6n-14, f(v_{n-8}v_{n-7}) = 6n-16, f(v_{n-7}v_{n-6}) = 6n-12, f(v_{n-6}v_{n-5}) = 6n-10, f(v_{n-5}v_{n-4}) = 6n-6, f(v_{n-4}v_{n-3}) = 6n-8, f(v_{n-3}v_{n-2}) = 6n-4, f(v_{n-2}v_{n-1}) = 6n-2, f(v_{n-1}v_n) = 6n+2, f(v_nv_1) = 6n$. □

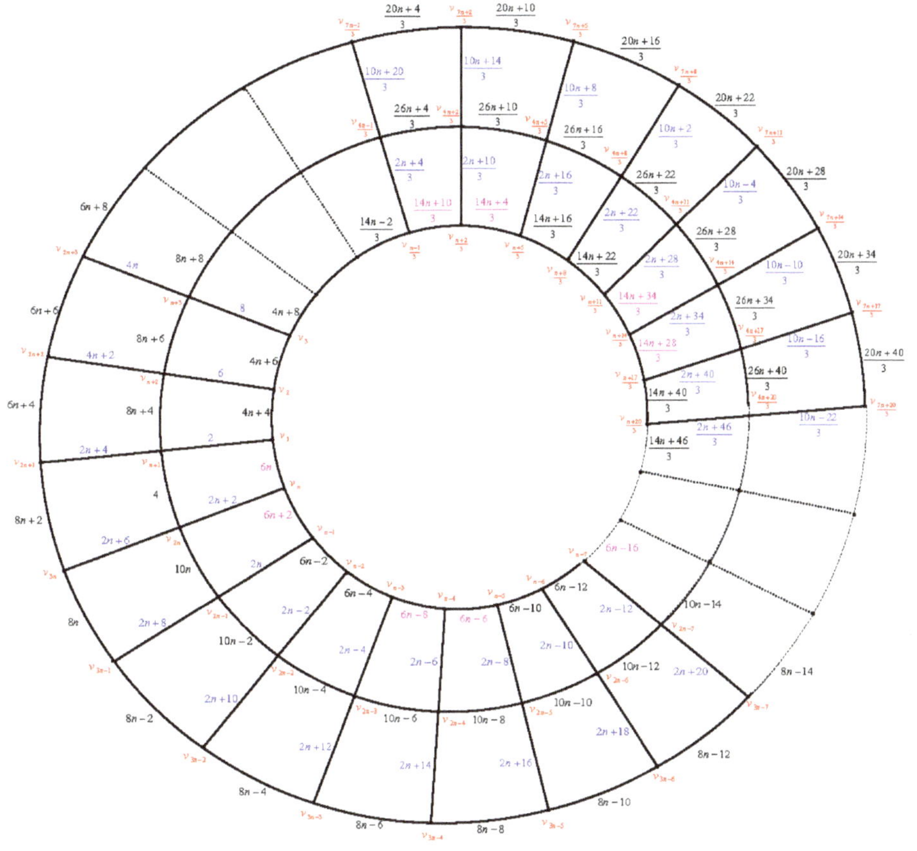

Figure 4. The cylinder grid graph $C_{3,n}$, $n \equiv 1\text{mod}6$.

Label the edges of the circle $C_n^{(2)}$ as follows: $f(v_{n+i}v_{n+i+1}) = 8n + 2i + 2, 1 \le i \le n-1, f(v_{2n}v_{n+1}) = 4$.

Label the edges of the circle $C_n^{(3)}$ as follows: $f(v_{2n+i}v_{2n+i+1}) = 6n + 2i + 2, 1 \le i \le n$.

The labels of corresponding of vertices mod$10n$ are as follows:

The labels of vertices of the circle $C_n^{(1)}$ are as follows: $f^*(v_1) \equiv 6, f^*(v_2) \equiv 8n+16, f^*(v_3) \equiv 8n+22, \ldots, f^*(v_{\frac{n-1}{3}}) \equiv 4, f^*(v_{\frac{n+2}{3}}) \equiv 8, f^*(v_{\frac{n+5}{3}}) \equiv 12, f^*(v_{\frac{n+8}{3}}) \equiv 20, f^*(v_{\frac{n+11}{3}}) \equiv 28,$ $f^*(v_{\frac{n+14}{3}}) \equiv 32, f^*(v_{\frac{n+17}{3}}) \equiv 36, f^*(v_{\frac{n+20}{3}}) \equiv 44, f^*(v_{\frac{n+23}{3}}) \equiv 52, f^*(v_{\frac{n+26}{3}}) \equiv 56, f^*(v_{\frac{n+29}{3}}) \equiv 60, f^*(v_{\frac{n+32}{3}}) \equiv 68,$ $f^*(v_{\frac{n+35}{3}}) \equiv 76, f^*(v_{\frac{n+38}{3}}) \equiv 80, f^*(v_{\frac{n+41}{3}}) \equiv 84, f^*(v_{\frac{n+44}{3}}) \equiv 92, f^*(v_{\frac{n+47}{3}}) \equiv 100, \ldots, f^*(v_{n-12}) \equiv 4n-68, f^*(v_{n-11}) \equiv 4n-64,$ $f^*(v_{n-10}) \equiv 4n-56, f^*(v_{n-9}) \equiv 4n-48, f^*(v_{n-8}) \equiv 4n-44, f^*(v_{n-7}) \equiv 4n-40, f^*(v_{n-6}) \equiv 4n-32, f^*(v_{n-5}) \equiv 4n-24,$ $f^*(v_{n-4}) \equiv 4n-20, f^*(v_{n-3}) \equiv 4n-16, f^*(v_{n-2}) \equiv 4n-8, f^*(v_{n-1}) \equiv 4n, f^*(v_n) \equiv 4n+4$.

The labels of vertices of the circle $C_n^{(2)}$ are $f^*(v_{i+1}) = 4i + 10, 1 \le i \le n-1, f^*(v_{2n}) = 4n + 12$.

The labels of vertices of the circle $C_n^{(3)}$ are $f^*(v_{2i+1}) = 6n + 2i + 8, 1 \le i \le n$.

Case (2): If $n \equiv 3\text{mod}6$, let the cylinder grid graph $C_{3,n}$ be as in Figure 5.

First, we label the edges of the paths $P_3^{(k)}, 1 \le k \le n$ beginning with the edges of the path $P_3^{(1)}$ as the same in case (1).

Secondly, we label the edges of the circles $C_n^{(k)}, 1 \le k \le 3$ beginning with the edges of the innermost circle $C_n^{(1)}$, then the edges of outer circle $C_n^{(3)}$, and then the edges of the circle $C_n^{(2)}$.

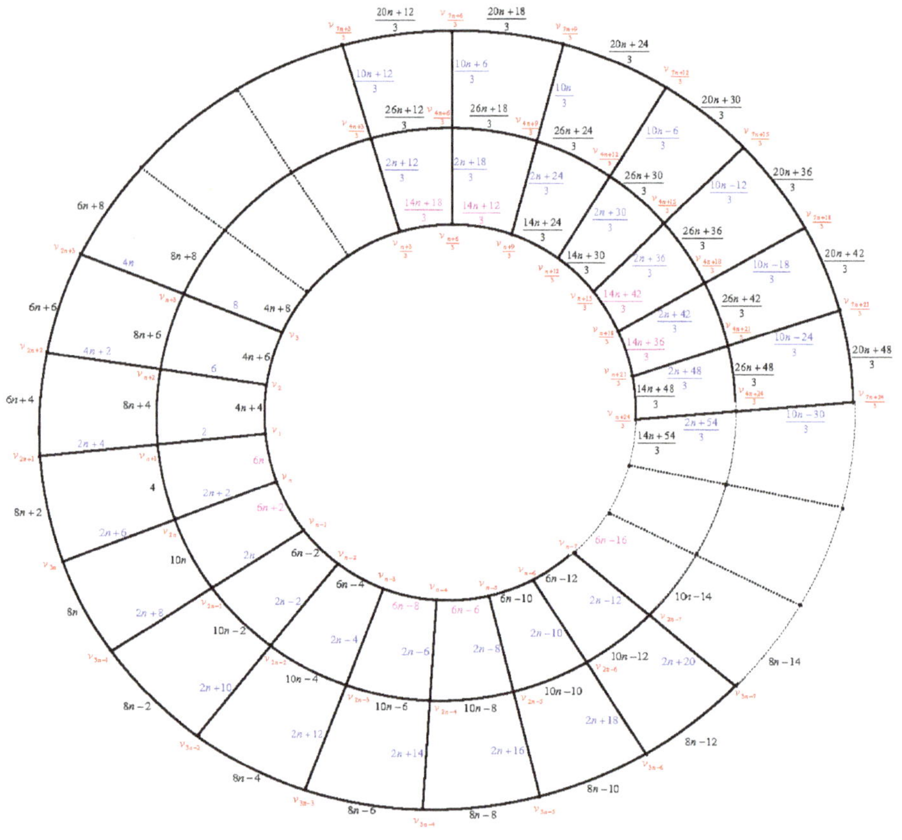

Figure 5. The cylinder grid graph $C_{3,n}$, $n \equiv 3 \bmod 6$.

Label the edges of the circle $C_n^{(1)}$ as follows: $f(v_1v_2)=4n+4, f(v_2v_3)=4n+6,\ldots, f(v_{\frac{n+3}{3}}v_{\frac{n+6}{3}})=\frac{14n+18}{3}, f(v_{\frac{n+6}{3}}v_{\frac{n+9}{3}})=\frac{14n+12}{3}, f(v_{\frac{n+9}{3}}v_{\frac{n+12}{3}})=\frac{14n+24}{3}, f(v_{\frac{n+12}{3}}v_{\frac{n+15}{3}})=\frac{14n+30}{3}, f(v_{\frac{n+15}{3}}v_{\frac{n+18}{3}})=\frac{14n+42}{3}, f(v_{\frac{n+18}{3}}v_{\frac{n+21}{3}})=\frac{14n+36}{3}, f(v_{\frac{n+21}{3}}v_{\frac{n+24}{3}})=\frac{14n+48}{3}, f(v_{\frac{n+24}{3}}v_{\frac{n+27}{3}})=\frac{14n+54}{3}, f(v_{\frac{n+27}{3}}v_{\frac{n+30}{3}})=\frac{14n+66}{3}, f(v_{\frac{n+30}{3}}v_{\frac{n+33}{3}})=\frac{14n+60}{3}, f(v_{\frac{n+33}{3}}v_{\frac{n+36}{3}})=\frac{14n+72}{3}, f(v_{\frac{n+36}{3}}v_{\frac{n+39}{3}})=\frac{14n+78}{3}, f(v_{\frac{n+39}{3}}v_{\frac{n+42}{3}})=\frac{14n+90}{3}, f(v_{\frac{n+42}{3}}v_{\frac{n+45}{3}})=\frac{14n+84}{3}, f(v_{\frac{n+45}{3}}v_{\frac{n+48}{3}})=\frac{14n+96}{3}, f(v_{\frac{n+48}{3}}v_{\frac{n+51}{3}})=\frac{14n+102}{3},\ldots, f(v_{n-13}v_{n-12})=6n-22, f(v_{n-12}v_{n-11})=6n-24, f(v_{n-11}v_{n-10})f(v_{n-9}v_{n-8})=6n-14, f(v_{n-8}v_{n-7})=6n-16, f(v_{n-7}v_{n-6})=6n-12, f(v_{n-6}v_{n-5})=6n-10, f(v_{n-5}v_{n-4})=6n-6, f(v_{n-4}v_{n-3})=6n-8, f(v_{n-3}v_{n-2})=6n-4, f(v_{n-2}v_{n-1})=6n-2, f(v_{n-1}v_n)=6n+2, f(v_nv_1)=6n$.

The labels of corresponding vertices mod$10n$ are as follows: The label of vertices of the circle $C_n^{(1)}$ are $f^*(v_1)\equiv 6, f^*(v_2)\equiv 8n+16, f^*(v_3)\equiv 8n+22,\ldots, f^*(v_{\frac{n}{3}-1})\equiv 10n-2, f^*(v_{\frac{n}{3}})\equiv 4, f^*(v_{\frac{n}{3}+1})\equiv 12, f^*(v_{\frac{n}{3}+2})\equiv 16, f^*(v_{\frac{n}{3}+3})\equiv 20, f^*(v_{\frac{n}{3}+4})\equiv 28, f^*(v_{\frac{n}{3}+5})\equiv 36, f^*(v_{\frac{n}{3}+6})\equiv 40, f^*(v_{\frac{n}{3}+7})\equiv 44, f^*(v_{\frac{n}{3}+8})\equiv 52, f^*(v_{\frac{n}{3}+9})\equiv 60, f^*(v_{\frac{n}{3}+10})\equiv 64, f^*(v_{\frac{n}{3}+11})\equiv 68, f^*(v_{\frac{n}{3}+12})\equiv 76, f^*(v_{\frac{n}{3}+13})\equiv 84, f^*(v_{\frac{n}{3}+14})\equiv 88, f^*(v_{\frac{n}{3}+15})\equiv 92, f^*(v_{\frac{n}{3}+16})\equiv 100,\ldots, f^*(v_{n-12})\equiv 4n-68, f^*(v_{n-11})\equiv 4n-64, f^*(v_{n-10})\equiv 4n-56, f^*(v_{n-9})\equiv 4n-48, f^*(v_{n-8})\equiv 4n-44, f^*(v_{n-7})\equiv 4n-40, f^*(v_{n-6})\equiv 4n-32, f^*(v_{n-5})\equiv 4n-24, f^*(v_{n-4})\equiv 4n-20, f^*(v_{n-3})\equiv 4n-16, f^*(v_{n-2})\equiv 4n-8, f^*(v_{n-1})\equiv 4n, f^*(v_n)\equiv 4n+4$.

The labels of vertices of the circles $C_n^{(2)}$ and $C_n^{(3)}$ are the same as in case (1).

Case (3): If $n \equiv 5 \bmod 6$, let the cylinder grid graph $C_{3,n}$ be as in Figure 6.

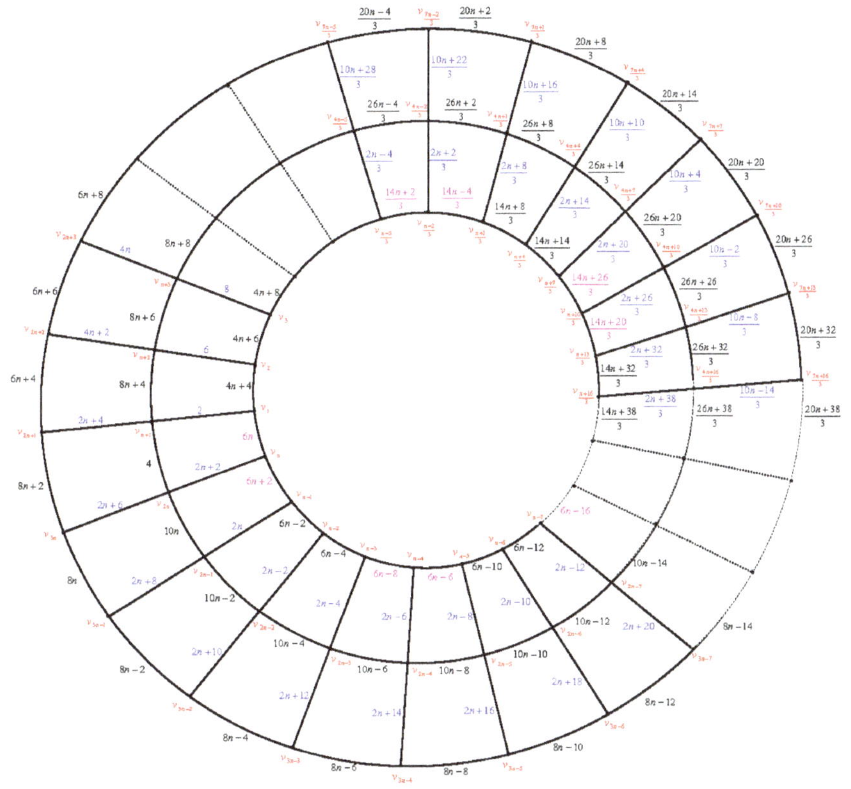

Figure 6. The cylinder grid graph $C_{3,n}$, $n \equiv 5 \bmod 6$.

First, we label the edges of the paths $P_3^{(k)}, 1 \le k \le 2$ beginning with the edges of the path $P_3^{(1)}$ as the same in case (1). Second, we label the edges of the circles $C_n^{(k)}, 1 \le k \le 3$ beginning with the edges of the innermost circle $C_n^{(1)}$, then the edges of outer circle $C_n^{(3)}$, and then the edges of the circle $C_n^{(2)}$.

Label the edges of the circle $C_n^{(1)}$ as follows: $f(v_1v_2) = 4n+4, f(v_2v_3) = 4n+6, \ldots, f(v_{\frac{n-5}{3}}v_{\frac{n-2}{3}}) = \frac{14n+2}{3}, f(v_{\frac{n-2}{3}}v_{\frac{n+1}{3}}) = \frac{14n-4}{3}, f(v_{\frac{n+1}{3}}v_{\frac{n+4}{3}}) = \frac{14n+8}{3}, f(v_{\frac{n+4}{3}}v_{\frac{n+7}{3}}) = \frac{14n+14}{3}, f(v_{\frac{n+7}{3}}v_{\frac{n+10}{3}}) = \frac{14n+26}{3}, f(v_{\frac{n+10}{3}}v_{\frac{n+13}{3}}) = \frac{14n+20}{3}, f(v_{\frac{n+13}{3}}v_{\frac{n+16}{3}}) = \frac{14n+32}{3}, f(v_{\frac{n+16}{3}}v_{\frac{n+19}{3}}) = \frac{14n+38}{3}, f(v_{\frac{n+19}{3}}v_{\frac{n+22}{3}}) = \frac{14n+50}{3}, f(v_{\frac{n+22}{3}}v_{\frac{n+25}{3}}) = \frac{14n+44}{3}, f(v_{\frac{n+25}{3}}v_{\frac{n+28}{3}}) = \frac{14n+56}{3}, f(v_{\frac{n+28}{3}}v_{\frac{n+31}{3}}) = \frac{14n+62}{3}, f(v_{\frac{n+31}{3}}v_{\frac{n+34}{3}}) = \frac{14n+74}{3}, f(v_{\frac{n+34}{3}}v_{\frac{n+37}{3}}) = \frac{14n+68}{3}, f(v_{\frac{n+37}{3}}v_{\frac{n+40}{3}}) = \frac{14n+80}{3}, f(v_{\frac{n+40}{3}}v_{\frac{n+43}{3}}) = \frac{14n+86}{3}, \ldots, f(v_{n-13}v_{n-12}) = 6n-22, f(v_{n-12}v_{n-11}) = 6n-24, f(v_{n-11}v_{n-10}) = 6n-20, f(v_{n-10}v_{n-9}) = 6n-18, f(v_{n-9}v_{n-8}) = 6n-14, f(v_{n-8}v_{n-7}) = 6n-16, f(v_{n-7}v_{n-6}) = 6n-12, f(v_{n-6}v_{n-5}) = 6n-10, f(v_{n-5}v_{n-4}) = 6n-6, f(v_{n-4}v_{n-3}) = 6n-8, f(v_{n-3}v_{n-2}) = 6n-4, f(v_{n-2}v_{n-1}) = 6n-2, f(v_{n-1}v_n) = 6n+2, f(v_nv_1) = 6n$.

The labels of corresponding vertices mod$10n$ are as follows: The labels of vertices of the circle $C_n^{(1)}$: $f^*(v_1) \equiv 6, f^*(v_2) \equiv 8n+16, f^*(v_3) \equiv 8n+22, \ldots, f^*(v_{\frac{n}{3}-5}) \equiv 10n-4, f^*(v_{\frac{n-2}{3}}) \equiv 0, f^*(v_{\frac{n+1}{3}}) \equiv 4, f^*(v_{\frac{n+4}{3}}) \equiv 12, f^*(v_{\frac{n+7}{3}}) \equiv 20, f^*(v_{\frac{n+10}{3}}) \equiv 24, f^*(v_{\frac{n+13}{3}}) \equiv 28, f^*(v_{\frac{n+16}{3}}) \equiv 36, f^*(v_{\frac{n}{3}+7}) \equiv 44, f^*(v_{\frac{n+19}{3}}) \equiv 44, f^*(v_{\frac{n+22}{3}}) \equiv 48, f^*(v_{\frac{n+25}{3}}) \equiv 52, f^*(v_{\frac{n+28}{3}}) \equiv 60, f^*(v_{\frac{n+31}{3}}) \equiv 68, f^*(v_{\frac{n+34}{3}}) \equiv 72, f^*(v_{\frac{n+37}{3}}) \equiv 76, f^*(v_{\frac{n+40}{3}}) \equiv 84, f^*(v_{\frac{n+43}{3}}) \equiv 92, f^*(v_{\frac{n+46}{3}}) \equiv 96, f^*(v_{\frac{n+49}{3}}) \equiv 100, \ldots, f^*(v_{n-12}) \equiv 4n-68, f^*(v_{n-11}) \equiv 4n-64, f^*(v_{n-10}) \equiv 4n-56, f^*(v_{n-9}) \equiv 4n-48, f^*(v_{n-8}) \equiv 4n-44, f^*(v_{n-7}) \equiv 4n-40, f^*(v_{n-6}) \equiv 4n-32, f^*(v_{n-5}) \equiv 4n-24, f^*(v_{n-4}) \equiv 4n-20, f^*(v_{n-3}) \equiv 4n-16, f^*(v_{n-2}) \equiv 4n-8, f^*(v_{n-1}) \equiv 4n, f^*(v_n) \equiv 4n+4$.

The labels of vertices of the circles $C_n^{(2)}$ and $C_n^{(3)}$ are the same as in case (1).

Illustration: An e.e.g., l. of the cylinder grid graphs $C_{3,25}, C_{3,27}$ and $C_{3,29}$ are shown in Figure 7.

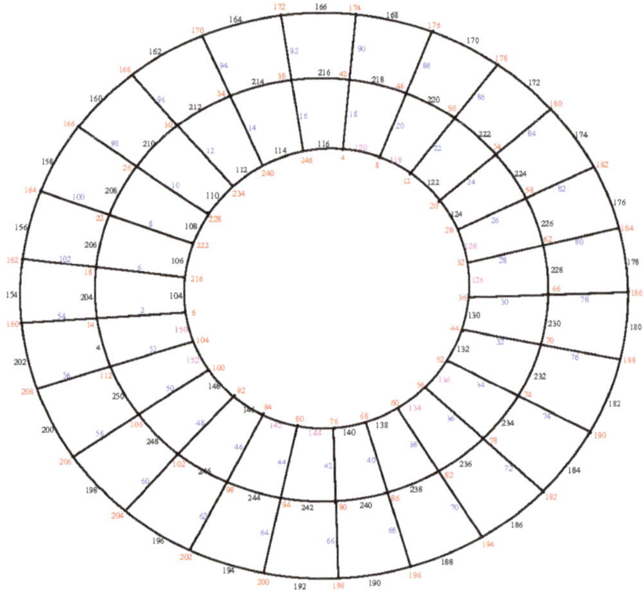

(**a**) $C_{3,25}$

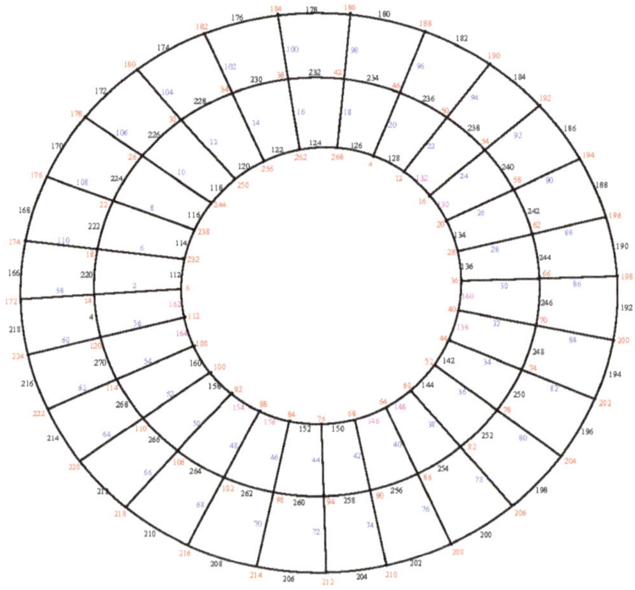

(**b**) $C_{3,27}$

Figure 7. *Cont.*

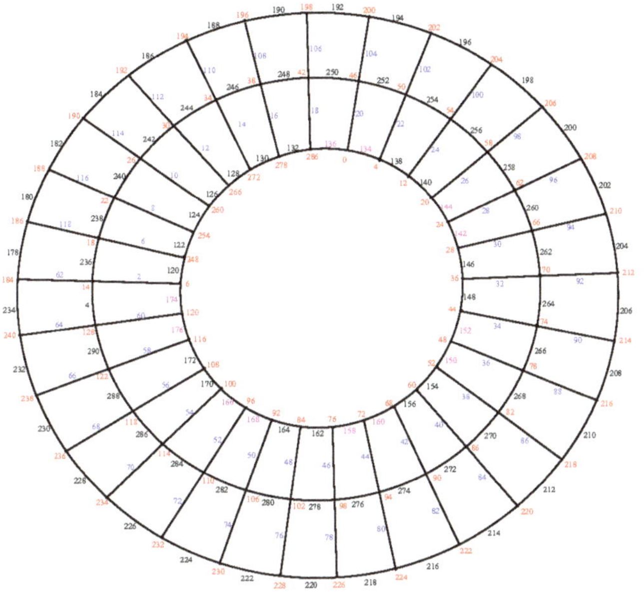

(**c**) $C_{3,29}$

Figure 7. An e.e.g., l. of the cylinder grid graphs $C_{3,25}, C_{3,27}$ and $C_{3,29}$.

Remark 1. *Note that $C_{3,5}$ is an edge even graceful graph but it does not follow the pervious rule (see Figure 8).*

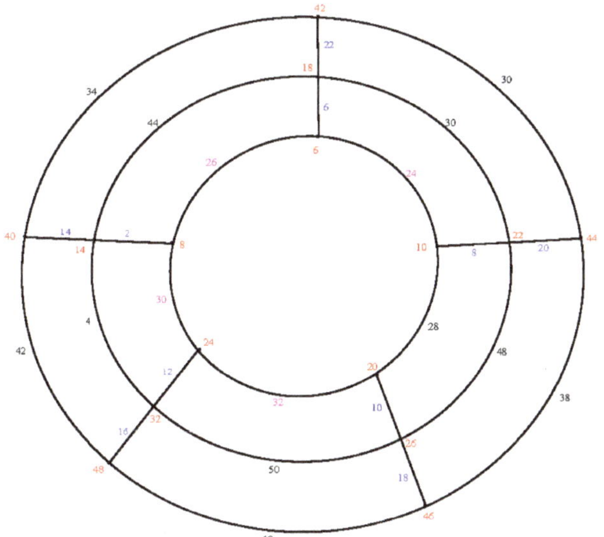

Figure 8. An e.e.g., l. of the cylinder grid graph $C_{3,5}$.

Theorem 3. *If m is an odd positive integer greater than 3 and n is an even positive integer, $n \geq 2$, then the cylinder grid graph $C_{m,n}$, is an edge even graceful graph.*

Proof. Using standard notation $p = |V(C_{m,n})| = mn$, $q = |E(C_{m,n})| = 2mn - n$ and $r = \max(p,q) = 2mn - n$ and $f : E(C_{m,n}) \rightarrow \{2, 4, 6, \ldots, 4mn - 2n - 2\}$. □

Let the cylinder grid graph $C_{m,n}$ be as in Figure 9. There are six cases:

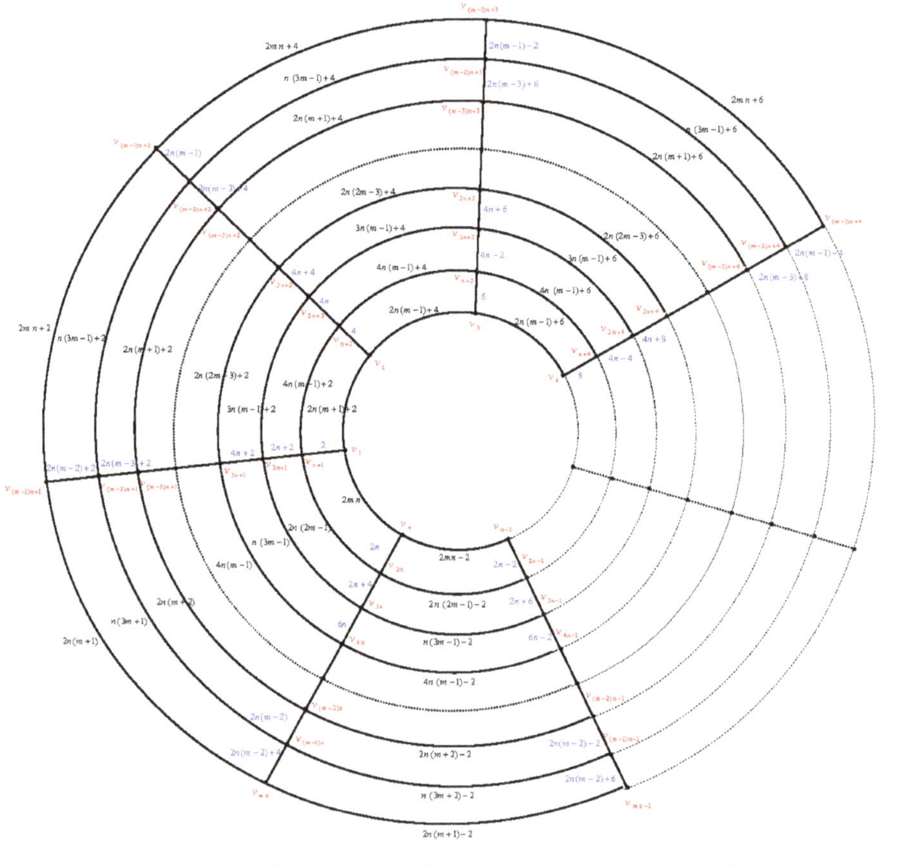

Figure 9. The cylinder grid graph $C_{m,n}$,m is odd greater than 3 and $n \geq 2$.

Case (1): $n \equiv 0 \bmod 12$. First, we label the edges of the paths $P_m^{(k)}, 1 \leq k \leq n$ beginning with the edges of the path $P_m^{(1)}$ as follows: Move clockwise to label the edges $v_1v_{n+1}, v_2v_{n+2}, v_3v_{n+3}, \ldots, v_{n-1}v_{2n-1}, v_nv_{2n}$ by $2, 4, 6, \ldots, 2n-2, 2n$, then move anticlockwise to label the edges $v_{n+1}v_{n+2}, v_{2n}v_{3n}, v_{2n-1}v_{3n-1}, \ldots, v_{n+3}v_{2n+3}, v_{n+2}v_{2n+2}$ by $2n+2, 2n+4, 2n+6, \ldots, 4n-2, 4n$, then move clockwise to label the edges $v_{2n+1}v_{3n+1}, v_{2n+2}v_{3n+2}, v_{2n+3}v_{3n+3}, \ldots, v_{3n-1}v_{4n-1}, v_{3n}v_{4n}$ by $4n+2, 4n+4, 4n+6, \ldots, 6n-2, 6n$ and so on.

Finally, move anticlockwise to label the edges $v_{(m-2)n+1}v_{(m-1)n+1}, v_{(m-1)n}v_{mn}, v_{(m-1)n-1}v_{mn-1}, \ldots, v_{(m-2)n+3}v_{(m-1)n+3}, v_{(m-2)n+2}v_{m(n-1)+2}$ by $2n(m-2)+2, 2n(m-2)+4, 2n(m-2)+6, \ldots, 2n(m-1)-2, 2n(m-1)$.

Secondly, we label the edges of the circles $C_n^{(k)}, 1 \le k \le m$ beginning with the edges of the inner most circle $C_n^{(1)}$, then the edges of outer circle $C_n^{(m)}$, then the edges of the circles $C_m^{(m-1)}, C_m^{(m-3)}, \ldots, C_m^{(2)}$. Finally, we label the edges of the circles $C_m^{(m-1)}, C_m^{(m-3)}, \ldots, C_m^{(2)}$.

Label the edges of the circle $C_n^{(1)}$ as follows: $f(v_1v_2) = 2n(m-1)+2, f(v_2v_3) = 2n(m-1)+6, f(v_3v_4) = 2n(m-1)+4, f(v_4v_5) = 2n(m-1)+8, f(v_5v_6) = 2n(m-1)+10, f(v_6v_7) = 2n(m-1)+14, f(v_7v_8) = 2n(m-1)+12, f(v_8v_9) = 2n(m-1)+16, f(v_9v_{10}) = 2n(m-1)+18, f(v_{10}v_{11}) = 2n(m-1)+22, f(v_{11}v_{12}) = 2n(m-1)+20, f(v_{12}v_{13}) = 2n(m-1)+24, \ldots, f(v_{n-7}v_{n-6}) = 2mn-14, f(v_{n-6}v_{n-5}) = 2mn-10, f(v_{n-5}v_{n-4}) = 2mn-12, f(v_{n-4}v_{n-3}) = 2mn-8, f(v_{n-3}v_{n-2}) = 2mn-2, f(v_{n-2}v_{n-1}) = 2mn-6, f(v_{n-1}v_n) = 2mn, f(v_nv_1) = 2mn-4$.

Label the edges of the circle $C_n^{(m)}$ as follows: $f(v_{(m-1)n+i}v_{(m-1)n+i+1}) = 2mn+2i, f(v_{mn}v_{(m-1)n+1}) = 2n(m+1), 1 \le i \le n-1$.

Label the edges of the circle $C_n^{(m-2)}$ as follows: $f(v_{(m-3)n+i}v_{(m-3)n+i+1}) = 2n(m+1)+2i, f(v_{(m-2)n}v_{(m-3)n+1}) = 2n(m+2), 1 \le i \le n-1$.

Label the edges of the circle $C_n^{(m-4)}$ as follows: $f(v_{(m-5)n+i}v_{(m-5)n+i+1}) = 2n(m+2)+2i, f(v_{(m-4)n}v_{(m-5)n+1}) = 2n(m+3), 1 \le i \le n-1$, and so on.

Label the edges of the circle $C_n^{(3)}$ as follows: $f(v_{2n+i}v_{2n+i+1}) = 3n(m-1)+2i, f(v_{3n}v_{2n+1}) = n(3m-1), 1 \le i \le n-1$,

Label the edges of the circle $C_n^{(m-1)}$ as follows: $f(v_{(m-2)n+i}v_{(m-2)n+i+1}) = n(3m-1)+2i, f(v_{(m-1)n}v_{(m-1)n+1}) = n(3m+1)-1, 1 \le i \le n-1$,

Label the edges of the circle $C_n^{(m-3)}$ as follows: $f(v_{(m-4)n+i}v_{(m-4)n+i+1}) = n(3m+2)+2i, f(v_{(m-3)n}v_{(m-4)n+1}) = 3n(m+1), 1 \le i \le n-1, \ldots$, and so on.

Label the edges of the circle $C_n^{(4)}$ as follows: $f(v_{3n+i}v_{3n+i+1}) = 2n(2m-3)+2i, f(v_{4n}v_{3n+1}) = 4n(m-1), 1 \le i \le n-1$,

Label the edges of $C_n^{(2)}$ as follows: $f(v_{n+i}v_{n+i+1}) = 4n(m-1)+2i, f(v_{2n}v_{2n+1}) = 2n(m-1), 1 \le i \le n-1$,

Thus, the labels of corresponding vertices mod$(4mn-2n)$ will be:

The label the vertices of $C_n^{(1)}$ are: $f^*(v_1) \equiv 0; f^*(v_2) \equiv 4mn-4n+12; f^*(v_3) \equiv 4mn-4n+16; f^*(v_4) \equiv 4mn-4n+20; f^*(v_5) \equiv 4mn-4n+28; f^*(v_6) \equiv 4mn-4n+36; f^*(v_7) \equiv 4mn-4n+40; f^*(v_8) \equiv 4mn-4n+44; f^*(v_9) \equiv 4mn-4n+52; f^*(v_{10}) \equiv 4mn-4n+60; f^*(v_{11}) \equiv 4mn-4n+64; f^*(v_{12}) \equiv 4mn-4n+68; \ldots; f^*(v_{n-6}) \equiv 4n-36; f^*(v_{n-5}) \equiv 4n-32; f^*(v_{n-4}) \equiv 4n-28; f^*(v_{n-3}) \equiv 4n-16; f^*(v_{n-2}) \equiv 4n-12; f^*(v_{n-1}) \equiv 4n-8; f^*(v_n) \equiv 4n-4$.

The label the vertices of $C_n^{(2)}, C_n^{(3)}, C_n^{(4)}, \ldots, C_n^{(m-2)}, C_n^{(m-1)}, C_n^{(m)}$ respectively are: $f^*(v_{n+i}) \equiv 4i+2; f^*(v_{2n+i}) \equiv 2mn+4n+4i+2; f^*(v_{3n+i}) \equiv 4n+4i+2; \ldots; f^*(v_{(m-3)n+i}) \equiv 4mn-6n+4i+2; f^*(v_{(m-2)n+i}) \equiv 2mn-6n+4i+2; f^*(v_{(m-1)n+i}) \equiv 2mn+2i+2, 1 \le i \le n$.

Case (2): $n \equiv 2 \bmod 12, n \neq 2$.

First, we label the edges of the paths $P_m^{(k)}, 1 \le k \le n$ begin with the edges of the path $P_m^{(1)}$ as the same in case (1).

Secondly, we label the edges of the circles $C_n^{(k)}, 1 \le k \le m$ begin with the edges of the inner most circle $C_n^{(1)}$, then the edges of outer circle $C_n^{(m)}$, then the edges of the circles $C_n^{(m-2)}, C_n^{(m-4)}, \ldots, C_n^{(3)}$.

Finally, we label the edges of the circles $C_m^{(m-1)}, C_m^{(m-3)}, \ldots, C_m^{(2)}$.

Label the edges of the circle $C_n^{(1)}$ as follows: $f(v_1v_2) = 2n(m-1)+2, f(v_2v_3) = 2n(m-1)+6, f(v_3v_4) = 2n(m-1)+4, f(v_4v_5) = 2n(m-1)+8, f(v_5v_6) = 2n(m-1)+10, f(v_6v_7) = 2n(m-1)+14, f(v_7v_8) = 2n(m-1)+12, f(v_8v_9) = 2n(m-1)+16, f(v_9v_{10}) = 2n(m-1)+18, f(v_{10}v_{11}) = 2n(m-1)+22, f(v_{11}v_{12}) = 2n(m-1)+20, f(v_{12}v_{13}) = 2n(m-1)+24, \ldots, f(v_{n-9}v_{n-8}) = 2mn-18, f(v_{n-8}v_{n-7}) = 2mn-14, f(v_{n-7}v_{n-6}) = 2mn-16, f(v_{n-6}v_{n-5}) = 2mn-12, f(v_{n-5}v_{n-4}) = 2mn-10, f(v_{n-4}v_{n-3}) = 2mn-6, f(v_{n-3}v_{n-2}) = 2mn-8, f(v_{n-2}v_{n-1}) = 2mn-4, f(v_{n-1}v_n) = 2mn-2, f(v_nv_1) = 2mn$.

Label the edges of the circle $C_n^{(2)}$ as follows: $f(v_{n+1}v_{n+2}) = 4n(m-1)+4, f(v_{n+2}v_{n+3}) = 4n(m-1)+2, f(v_{n+3}v_{n+4}) = 4n(m-1)+8, f(v_{n+4}v_{n+5}) = 4n(m-1)+6, f(v_{n+i}v_{n+i+1}) = 4n(m-1)+2i, 6 \leq i \leq n-2, f(v_{2n-1}v_{2n}) = 2n(2m-1), f(v_{2n}v_{n+1}) = 2n(2m-1)-2$. Label the edges of $C_n^{(m)}, C_n^{(m-2)}, C_n^{(m-4)}, \ldots, C_n^{(3)}$ and $C_n^{(m-1)}, C_n^{(m-3)}, C_n^{(m-5)}, \ldots, C_n^{(4)}$ as in case (1).

Thus, the labels of corresponding vertices $\bmod(4mn-2n)$ will be:

The label the vertices of $C_n^{(1)}$ are: $f^*(v_1) \equiv 4, f^*(v_2) \equiv 4mn-4n+12, f^*(v_3) \equiv 4mn-4n+16, f^*(v_4) \equiv 4mn-4n+20, f^*(v_5) \equiv 4mn-4n+28, f^*(v_6) \equiv 4mn-4n+36, f^*(v_7) \equiv 4mn-4n+40, f^*(v_8) \equiv 4mn-4n+44, f^*(v_9) \equiv 4mn-4n+52, f^*(v_{10}) \equiv 4mn-4n+60, f^*(v_{11}) \equiv 4mn-4n+64, f^*(v_{12}) \equiv 4mn-4n+68, f^*(v_{13}) \equiv 4mn-4n+76, \ldots, f^*(v_{n-8}) \equiv 4n-48, f^*(v_{n-7}) \equiv 4n-44, f^*(v_{n-6}) \equiv 4n-40, f^*(v_{n-5}) \equiv 4n-32, f^*(v_{n-4}) \equiv 4n-24, f^*(v_{n-3}) \equiv 4n-20, f^*(v_{n-2}) \equiv 4n-16, f^*(v_{n-1}) \equiv 4n-8, f^*(v_n) \equiv 4n-2$.

The label the vertices of the circle $C_n^{(2)}$ are: $f^*(v_{n+1}) \equiv 6, f^*(v_{n+2}) \equiv 10, f^*(v_{n+3}) \equiv 14, f^*(v_{n+4}) \equiv 18, f^*(v_{n+5}) \equiv 20, f^*(v_{n+i}) \equiv 4i+2, 6 \leq i \leq n-2, f^*(v_{2n-1}) \equiv 4n, f^*(v_{2n}) \equiv 4n+2$.

The label the vertices of $C_n^{(3)}, C_n^{(4)}, \ldots, C_n^{(m-2)}, C_n^{(m-1)}, C_n^{(m)}$ respectively are as the same as in case (1).

Remark 2. *In case $n = 2$. Let the edges of the cylinder grid graph $C_{m,2}$ are labeled as shown in Figure 10. The corresponding labels of vertices $\bmod(8m-4)$ are as follows: $f^*(v_1) \equiv 8, f^*(v_{2i+1}) \equiv 4m+8i+4, 1 \leq i \leq \frac{m-3}{2}, f^*(v_{2i}) \equiv 8i+6, 1 \leq i \leq \frac{m-1}{2}$; $f^*(v'_1) \equiv 12, f^*(v'_2) \equiv 20, f^*(v'_{2i+1}) \equiv 4m+8i+18, 1 \leq i \leq \frac{m-3}{2}, f^*(v'_{2i}) \equiv 8i+10, 2 \leq i \leq \frac{m-1}{2}$.*

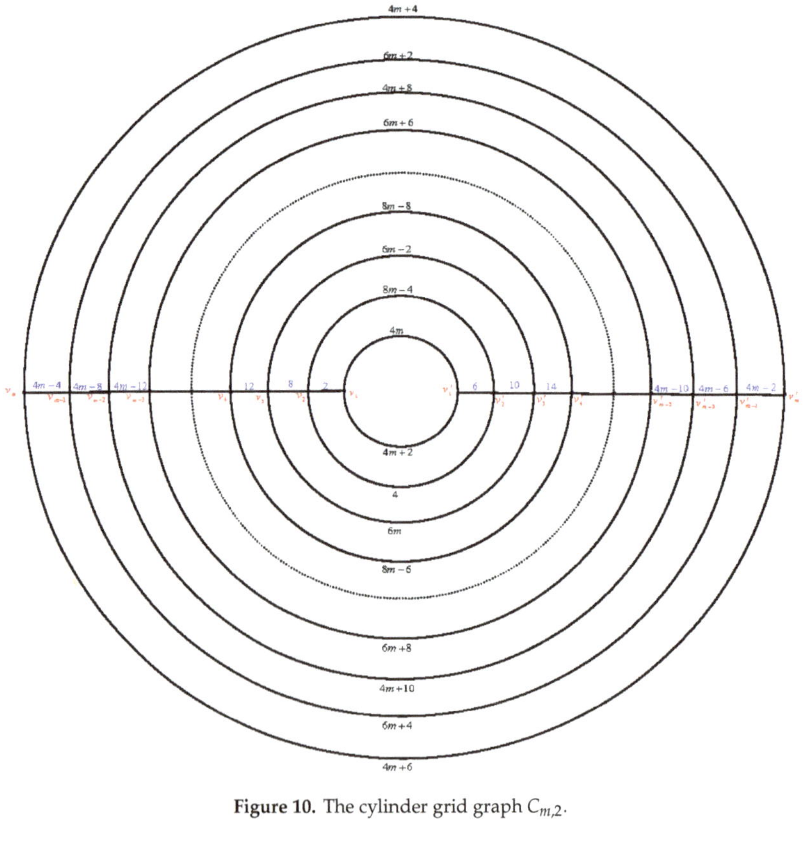

Figure 10. The cylinder grid graph $C_{m,2}$.

Case (3): $n \equiv 4 \bmod 12$.

First we label the edges of the paths $P_m^{(k)}, 1 \le k \le n$ begin with the edges of the path $P_m^{(1)}$ as the same in case (1).

Second we label the edges of the circles $C_n^{(k)}, 1 \le k \le m$ begin with the edges of the inner most circle $C_n^{(1)}$, then the edges of outer circle $C_n^{(m)}$, then the edges of the circles $C_n^{(m-2)}, C_n^{(m-4)}, \ldots, C_n^{(3)}$.

Finally we label the edges of the circles $C_m^{(m-1)}, C_m^{(m-3)}, \ldots, C_m^{(2)}$.

Label the edges of the circle $C_n^{(1)}$ as follows: $f(v_1v_2) = 2n(m-1)+2, f(v_2v_3) = 2n(m-1)+6, f(v_3v_4) = 2n(m-1)+4, f(v_4v_5) = 2n(m-1)+8, f(v_5v_6) = 2n(m-1)+10, f(v_6v_7) = 2n(m-1)+14, f(v_7v_8) = 2n(m-1)+12, f(v_8v_9) = 2n(m-1)+16, f(v_9v_{10}) = 2n(m-1)+18, f(v_{10}v_{11}) = 2n(m-1)+22, f(v_{11}v_{12}) = 2n(m-1)+20, f(v_{12}v_{13}) = 2n(m-1)+24, \ldots, f(v_{n-8}v_{n-7}) = 2mn-16, f(v_{n-7}v_{n-6}) = 2mn-14, f(v_{n-6}v_{n-5}) = 2mn-10, f(v_{n-5}v_{n-4}) = 2mn-12, f(v_{n-4}v_{n-3}) = 2mn-8, f(v_{n-3}v_{n-2}) = 2mn-2, f(v_{n-2}v_{n-1}) = 2mn-6, f(v_{n-1}v_n) = 2mn$, and $f(v_nv_1) = 2mn-4$.

Label the edges of $C_n^{(m)}, C_n^{(m-2)}, C_n^{(m-4)}, \ldots, C_n^{(3)}$ and $C_n^{(m-1)}, C_n^{(m-3)}, C_n^{(m-5)}, \ldots, C_n^{(4)}, C_n^{(2)}$ as in case (1).

Thus we have the labels of corresponding vertices of the circle $C_n^{(1)}$ $\bmod(4mn-2n)$ will be: $f^*(v_1) \equiv 0, f^*(v_2) \equiv 4mn-4n+12, f^*(v_3) \equiv 4mn-4n+16, f^*(v_4) \equiv 4mn-4n+20, f^*(v_5) \equiv 4mn-4n+28, f^*(v_6) \equiv 4mn-4n+36, f^*(v_7) \equiv 4mn-4n+40, f^*(v_8) \equiv 4mn-4n+44, f^*(v_9) \equiv 4mn-4n+52, f^*(v_{10}) \equiv 4mn-4n+60, f^*(v_{11}) \equiv 4mn-4n+64, f^*(v_{12}) \equiv 4mn-4n+68, f^*(v_{13}) \equiv 4mn-4n+76, \ldots, f^*(v_{n-7}) \equiv 4n-44, f^*(v_{n-6}) \equiv 4n-36, f^*(v_{n-5}) \equiv 4n-32, f^*(v_{n-4}) \equiv 4n-28, f^*(v_{n-3}) \equiv 4n-16, f^*(v_{n-2}) \equiv 4n-12, f^*(v_{n-1}) \equiv 4n-8, f^*(v_n) \equiv 4n-4$.

The label the vertices of $C_n^{(2)}, C_n^{(3)}, C_n^{(4)}, \ldots, C_n^{(m-2)}, C_n^{(m-1)}, C_n^{(m)}$ respectively are as same in case (1)

Remark 3. *In case $n = 4$. Let the the edges of the cylinder grid graph $C_{m,4}$ are labeled as shown in Figure 11. The corresponding labels of vertices $\bmod(16m-8)$ are as follows: $f^*(v_1) \equiv 6, f^*(v_2) \equiv 8, f^*(v_3) \equiv 16, f^*(v_4) \equiv 20; f^*(v_{4i+1}) \equiv 4i+10, 1 \le i \le 3, f^*(v_8) \equiv 28; f^*(v_{8k+i}) \equiv 8m+4i+16k-10, 1 \le i \le 4, 1 \le k \le \frac{m-5}{2}; f^*(v_{4m-11}) \equiv 0, f^*(v_{4m-10}) \equiv 2, f^*(v_{4m-9}) \equiv 4, f^*(v_{4m-8}) \equiv 10, f^*(v_{8k+4+i}) \equiv 4i+16k+10, 1 \le i \le 4, 1 \le k \le \frac{m-3}{2}$.*

Case (4): $n \equiv 6 \bmod 12$.

First, we label the edges of the paths $P_m^{(k)}, 1 \le k \le n$ begin with the edges of the path $P_m^{(1)}$ as the same in case (1).

Secondly, we label the edges of the circles $C_n^{(k)}, 1 \le k \le m$ begin with the edges of the inner most circle $C_n^{(1)}$, then the edges of outer circle $C_n^{(m)}$, then the edges of the circles $C_n^{(m-2)}, C_n^{(m-4)}, \ldots, C_n^{(3)}$.

Finally, we label the edges of the circles $C_m^{(m-1)}, C_m^{(m-3)}, \ldots, C_m^{(2)}$.

Label the edges of the circle $C_n^{(1)}$ as follows: $f(v_1v_2) = 2n(m-1)+2, f(v_2v_3) = 2n(m-1)+6, f(v_3v_4) = 2n(m-1)+4, f(v_4v_5) = 2n(m-1)+8, f(v_5v_6) = 2n(m-1)+10, f(v_6v_7) = 2n(m-1)+14, f(v_7v_8) = 2n(m-1)+12, f(v_8v_9) = 2n(m-1)+16, f(v_9v_{10}) = 2n(m-1)+18, f(v_{10}v_{11}) = 2n(m-1)+22, f(v_{11}v_{12}) = 2n(m-1)+20, f(v_{12}v_{13}) = 2n(m-1)+24, \ldots, f(v_{n-9}v_{n-8}) = 2mn-18, f(v_{n-8}v_{n-7}) = 2mn-14, f(v_{n-7}v_{n-6}) = 2mn-16, f(v_{n-6}v_{n-5}) = 2mn-12, f(v_{n-5}v_{n-4}) = 2mn-10, f(v_{n-4}v_{n-3}) = 2mn-6, f(v_{n-3}v_{n-2}) = 2mn-8, f(v_{n-2}v_{n-1}) = 2mn-4, f(v_{n-1}v_n) = 2mn+2, f(v_nv_1) = 2mn-2$.

Label the edges of $C_n^{(m-4)}, \ldots, C_n^{(3)}$ and $C_n^{(m-1)}, C_n^{(m-3)}, C_n^{(m-5)}, \ldots, C_n^{(4)}, C_n^{(2)}$ as in case (1).

Label the edges of the circle $C_n^{(m-2)}$ as follows: $f(v_{(m-3)n+1}v_{(m-3)n+2}) = 2n(m+2), f(v_{(m-3)n+i}v_{(m-3)n+i+1}) = 2n(m+1)+2i, 2 \le i \le n-1, f(v_{(m-2)n}v_{(m-3)n+1}) = 2n(m+2)+2$.

Label the edges of the circle $C_n^{(m)}$ as follows: $f(v_{(m-1)n+1}v_{(m-1)n+2}) = 2mn, f(v_{(m-1)n+i}v_{(m-1)n+i+1}) = 2mn+2i, 2 \le i \le n-1, f(v_{(m-2)n}v_{(m-3)n+1}) = 2n(m+2)$.

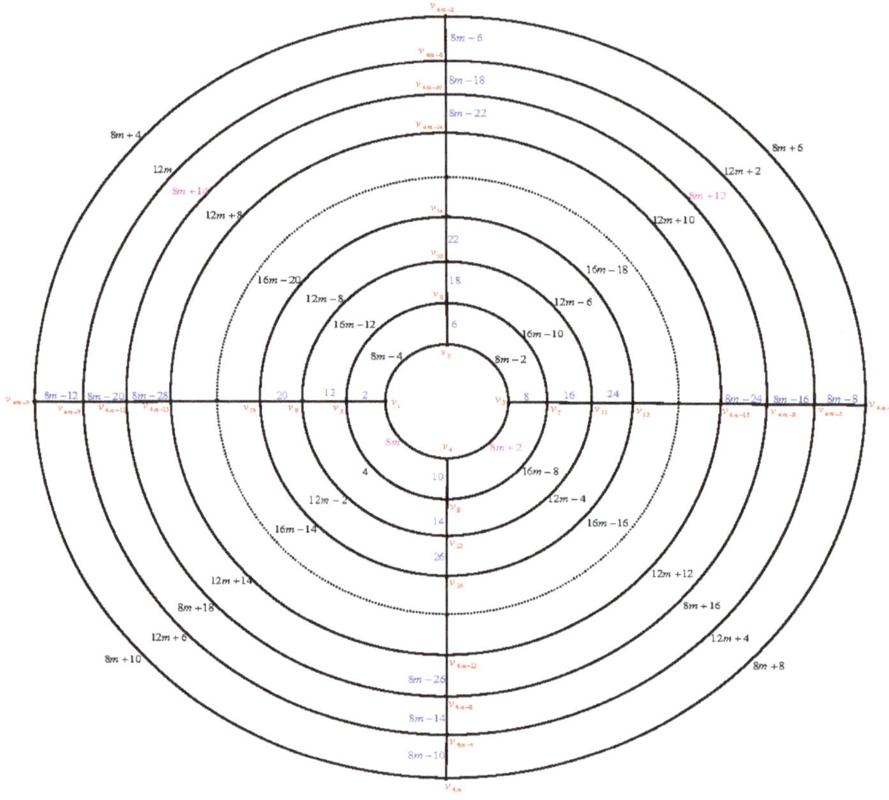

Figure 11. An e.e.g., l. of the cylinder grid graph $C_{m,4}$.

Thus we have the labels of corresponding vertices mod$(4mn - 2n)$ will be:

The labels the vertices of the circle $C_n^{(1)}$ are: $f^*(v_1) \equiv 2, f^*(v_2) \equiv 4mn - 4n + 12, f^*(v_3) \equiv 4mn - 4n + 16, f^*(v_4) \equiv 4mn - 4n + 20, f^*(v_5) \equiv 4mn - 4n + 28, f^*(v_6) \equiv 4mn - 4n + 36, f^*(v_7) \equiv 4mn - 4n + 40, f^*(v_8) \equiv 4mn - 4n + 44, f^*(v_9) \equiv 4mn - 4n + 52, f^*(v_{10}) \equiv 4mn - 4n + 60, f^*(v_{11}) \equiv 4mn - 4n + 64, f^*(v_{12}) \equiv 4mn - 4n + 68, \ldots, f^*(v_{n-8}) \equiv 4n - 48, f^*(v_{n-7}) \equiv 4n - 44, f^*(v_{n-6}) \equiv 4n - 40, f^*(v_{n-5}) \equiv 4n - 32, f^*(v_{n-4}) \equiv 4n - 24, f^*(v_{n-3}) \equiv 4n - 20, f^*(v_{n-2}) \equiv 4n - 16, f^*(v_{n-1}) \equiv 4n - 4; f^*(v_n) \equiv 4n$.

The labels the vertices of the circle $C_n^{(m-2)}$ are: $f^*(v_{(m-3)n+1}) \equiv 4mn - 6n + 6; f^*(v_{(m-3)n+2}) \equiv 4mn - 6n + 8; f^*(v_{(m-3)n+i}) \equiv 4mn - 6n + 4i + 2, 3 \le i \le n - 1, f^*(v_{(m-2)n}) \equiv 4mn - 4n + 4$.

The labels the vertices of $C_n^{(2)}, C_n^{(3)}, C_n^{(4)}, \ldots, C_n^{(m-1)}, C_n^{(m-3)}$ respectively are the same as in case (1).

The labels the vertices of $C_n^{(m)}$ are: $f^*(v_{(m-1)n+1}) \equiv 2mn + 2, f^*(v_{(m-1)n+2}) \equiv 2mn + 4, f^*(v_{(m-1)n+i}) \equiv 2mn + 2i + 2, 3 \le i \le n$. Case (5): $n \equiv 8 \bmod 12$.

First, we label the edges of the paths $P_m^{(k)}, 1 \le k \le n$ begin with the edges of the path $P_m^{(1)}$ as the same in case (1).

Secondly, we label the edges of the circles $C_n^{(k)}, 1 \le k \le m$ begin with the edges of the inner most circle $C_n^{(1)}$, then the edges of outer circle $C_n^{(m)}$, then the edges of the circles $C_n^{(m-2)}, C_n^{(m-4)}, \ldots, C_n^{(3)}$.

Finally we label the edges of the circles $C_m^{(m-1)}, C_m^{(m-3)}, \ldots, C_m^{(2)}$

Label the edges of the circle $C_n^{(1)}$ as follows: $f(v_1v_2) = 2n(m-1) + 2, f(v_2v_3) = 2n(m-1) + 6, f(v_3v_4) = 2n(m-1) + 4, f(v_4v_5) = 2n(m-1) + 8, f(v_5v_6) = 2n(m-1) + 10, f(v_6v_7) = 2n(m-1) + 14, f(v_7v_8) = 2n(m-1) + 12, f(v_8v_9) = 2n(m-1) + 16, f(v_9v_{10}) = 2n(m-1) + 18, f(v_{10}v_{11}) =$

$2n(m-1)+22, f(v_{11}v_{12}) = 2n(m-1)+20, f(v_{12}v_{13}) = 2n(m-1)+24, f(v_{13}v_{14}) = 2n(m-1)+26, \ldots, f(v_{n-9}v_{n-8}) = 2mn-20, f(v_{n-8}v_{n-7}) = 2mn-16, f(v_{n-7}v_{n-6}) = 2mn-14, f(v_{n-6}v_{n-5}) = 2mn-10, f(v_{n-5}v_{n-4}) = 2mn-12, f(v_{n-4}v_{n-3}) = 2mn-8, f(v_{n-3}v_{n-2}) = 2mn-6, f(v_{n-2}v_{n-1}) = 2mn-2, f(v_{n-1}v_n) = 2mn-4, f(v_nv_1) = 2mn+4$.

Label the edges of the circle $C_n^{(m)}$ as follows $f(v_{(m-1)n+1}v_{(m-1)n+2}) = 2mn+2, f(v_{(m-1)n+2}v_{(m-1)n+3}) = 2mn+6, f(v_{(m-1)n+i}v_{(m-1)n+i+1}) = 2mn+2i+2,\ 3 \le i \le n-1, f(v_{mn}v_{(m-1)n+1}) = 2mn$.

Label the edges of $C_n^{(m-2)}, C_n^{(m-4)}, \ldots, C_n^{(3)}$ and $C_n^{(m-1)}, C_n^{(m-3)}, C_n^{(m-5)}, \ldots, C_n^{(4)}, C_n^{(2)}$ as the same in case (1).

Thus we labels of corresponding vertices of the circle $C_n^{(1)}$ $\bmod(4mn-2n)$ will be: $f^*(v_1) \equiv 8, f^*(v_2) \equiv 4mn-4n+12, f^*(v_3) \equiv 4mn-4n+16, f^*(v_4) \equiv 4mn-4n+20, f^*(v_5) \equiv 4mn-4n+28, f^*(v_6) \equiv 4mn-4n+36, f^*(v_7) \equiv 4mn-4n+40, f^*(v_8) \equiv 4mn-4n+44, f^*(v_9) \equiv 4mn-4n+52, f^*(v_{10}) \equiv 4mn-4n+60, f^*(v_{11}) \equiv 4mn-4n+64, f^*(v_{12}) \equiv 4mn-4n+68, f^*(v_{13}) \equiv 4mn-4n+76, \ldots, f^*(v_{n-8}) \equiv 4n-52, f^*(v_{n-7}) \equiv 4n-44, f^*(v_{n-6}) \equiv 4n-36, f^*(v_{n-5}) \equiv 4n-32, f^*(v_{n-4}) \equiv 4n-28, f^*(v_{n-3}) \equiv 4n-20, f^*(v_{n-2}) \equiv 4n-12, f^*(v_{n-1}) \equiv 4n-8, f^*(v_n) \equiv 4n$.

The labels the vertices of the circle $C_n^{(m)}$ are: $f^*(v_{(m-1)n+1}) \equiv 2mn-4n+14, f^*(v_{(m-1)n+2}) \equiv 2mn+8, f^*(v_{(m-1)n+i}) \equiv 2mn+2i+6, 3 \le i \le n-1, f^*(v_{mn}) \equiv 2mn+4$. The labels the vertices of $C_n^{(2)}, C_n^{(3)}, C_n^{(4)}, \ldots, C_n^{(m-2)}, C_n^{(m-1)}$ respectively are as the same in case (1).

Case (6): $n \equiv 10 \bmod 12$. First we label the edges of the paths $P_m^{(k)}, 1 \le k \le n$ begin with the edges of the path $P_m^{(1)}$ as the same as in case (1).

Second we label the edges of the circles $C_n^{(k)}, 1 \le k \le m$ begin with the edges of the inner most circle $C_n^{(1)}$, then the edges of outer circle $C_n^{(m)}$, then the edges of the circles $C_n^{(m-2)}, C_n^{(m-4)}, \ldots, C_n^{(3)}$.

Finally we label the edges of the circles $C_m^{(m-1)}, C_m^{(m-3)}, \ldots, C_m^{(2)}$.

Label the edges of the circle $C_n^{(1)}$ as follows: $f(v_1v_2) = 2n(m-1)+2, f(v_2v_3) = 2n(m-1)+6, f(v_3v_4) = 2n(m-1)+4, f(v_4v_5) = 2n(m-1)+8, f(v_5v_6) = 2n(m-1)+10, f(v_6v_7) = 2n(m-1)+14, f(v_7v_8) = 2n(m-1)+12, f(v_8v_9) = 2n(m-1)+16, f(v_9v_{10}) = 2n(m-1)+18, f(v_{10}v_{11}) = 2n(m-1)+22, f(v_{11}v_{12}) = 2n(m-1)+20, f(v_{12}v_{13}) = 2n(m-1)+24, f(v_{13}v_{14}) = 2n(m-1)+26, \ldots, f(v_{n-9}v_{n-8}) = 2mn-18, f(v_{n-8}v_{n-7}) = 2mn-14, f(v_{n-7}v_{n-6}) = 2mn-16, f(v_{n-6}v_{n-5}) = 2mn-12, f(v_{n-5}v_{n-4}) = 2mn-10, f(v_{n-4}v_{n-3}) = 2mn-6, f(v_{n-3}v_{n-2}) = 2mn-8, f(v_{n-2}v_{n-1}) = 2mn-4, f(v_{n-1}v_n) = 2mn-2, f(v_nv_1) = 2mn$.

Label the edges of the circle $C_n^{(2)}$ as follows: $f(v_{n+1}v_{n+2}) = 4n(m-1)+4, f(v_{n+2}v_{n+3}) = 4n(m-1)+2, f(v_{n+i}v_{n+i+1}) = 4n(m-1)+2i, 3 \le i \le n-2, f(v_{2n-1}v_{2n}) = 2n(2m-1), f(v_{2n}v_{2n+1}) = 2n(2m-1)-2$.

Label the edges of $C_n^{(m)}, C_n^{(m-2)}, C_n^{(m-4)}, \ldots, C_n^{(3)}$ and $C_n^{(m-1)}, C_n^{(m-3)}, C_n^{(m-5)}, \ldots, C_n^{(4)}$ as in case (1).

Thus we have the labels of corresponding vertices $\bmod(4mn-2n)$ will be:

The labels the vertices of the circle $C_n^{(1)}$ are as follows: $f^*(v_1) \equiv 4, f^*(v_2) \equiv 4mn-4n+12, f^*(v_3) \equiv 4mn-4n+16, f^*(v_4) \equiv 4mn-4n+20, f^*(v_5) \equiv 4mn-4n+28, f^*(v_6) \equiv 4mn-4n+36, f^*(v_7) \equiv 4mn-4n+40, f^*(v_8) \equiv 4mn-4n+44, f^*(v_9) \equiv 4mn-4n+52, f^*(v_{10}) \equiv 4mn-4n+60, f^*(v_{11}) \equiv 4mn-4n+64, f^*(v_{12}) \equiv 4mn-4n+68, f^*(v_{13}) \equiv 4mn-4n+76, f^*(v_{14}) \equiv 4mn-4n+84, \ldots, f^*(v_{n-8}) \equiv 4n-48, f^*(v_{n-7}) \equiv 4n-44, f^*(v_{n-6}) \equiv 4n-40, f^*(v_{n-5}) \equiv 4n-32, f^*(v_{n-4}) \equiv 4n-24, f^*(v_{n-3}) \equiv 4n-20, f^*(v_{n-2}) \equiv 4n-16, f^*(v_{n-1}) \equiv 4n-8, f^*(v_n) \equiv 4n-2$.

The labels the vertices of the circle $C_n^{(2)}$ are as follows: $f^*(v_{n+1}) \equiv 6, f^*(v_{n+2}) \equiv 10, f^*(v_{n+3}) \equiv 12, f^*(v_{n+i}) \equiv 4i+2, 4 \le i \le n-2, f^*(v_{2n-1}) \equiv 4n, f^*(v_{2n}) \equiv 4n+2$. Label the vertices of $C_n^{(3)}, C_n^{(4)}, \ldots, C_n^{(m-2)}, C_n^{(m-1)}, C_n^{(m)}$ respectively are as the same as in case (1).

Illustration: The edge even graceful labeling of the cylinder grid graphs $C_{9,2}, C_{9,4}, C_{7,10}, C_{7,12}, C_{7,14}C_{7,16}C_{7,18}$ and $C_{7,20}$ are shown in Figure 12.

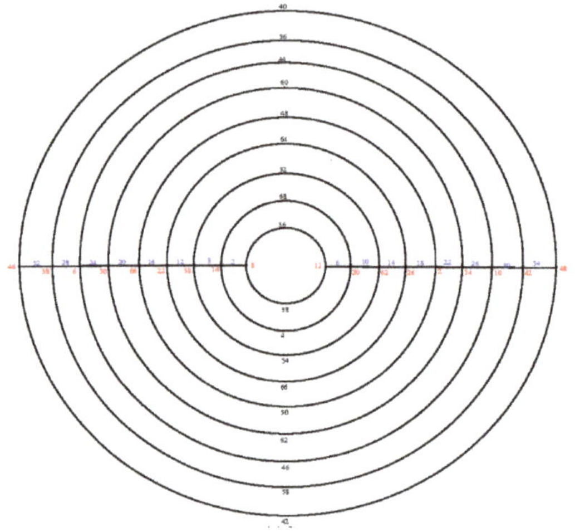

(a) $C_{9,2}$

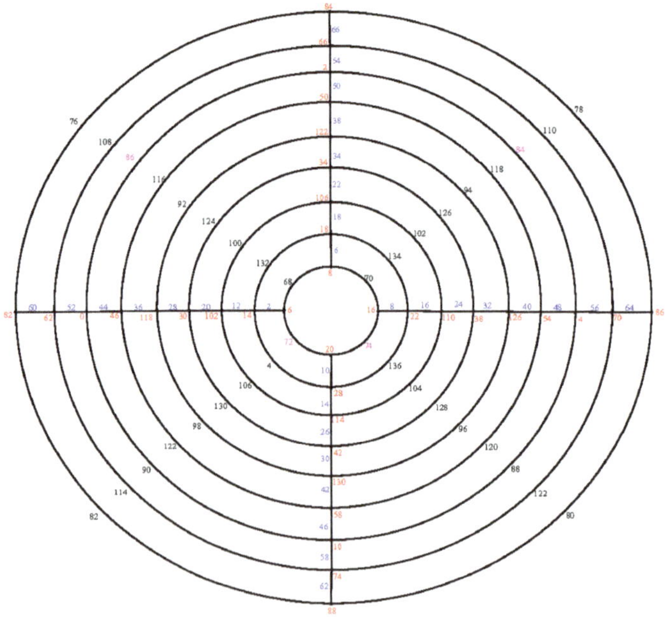

(b) $C_{9,4}$

Figure 12. *Cont.*

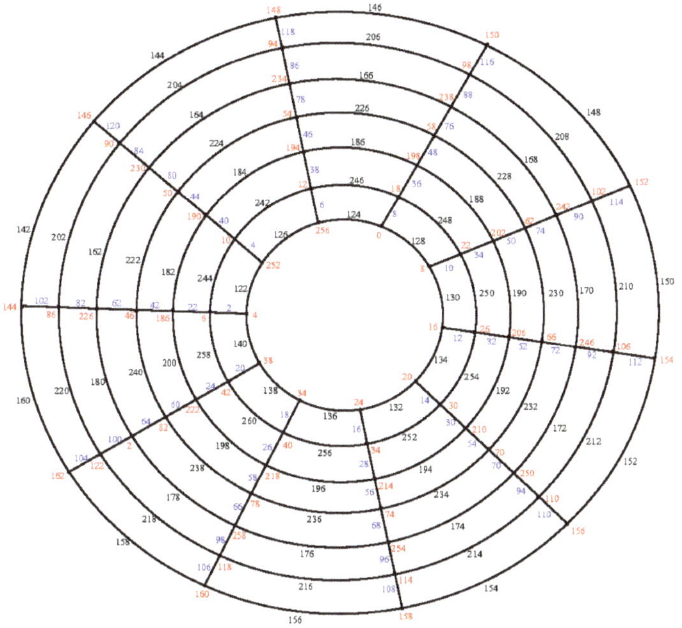

(c) $C_{7,10}$

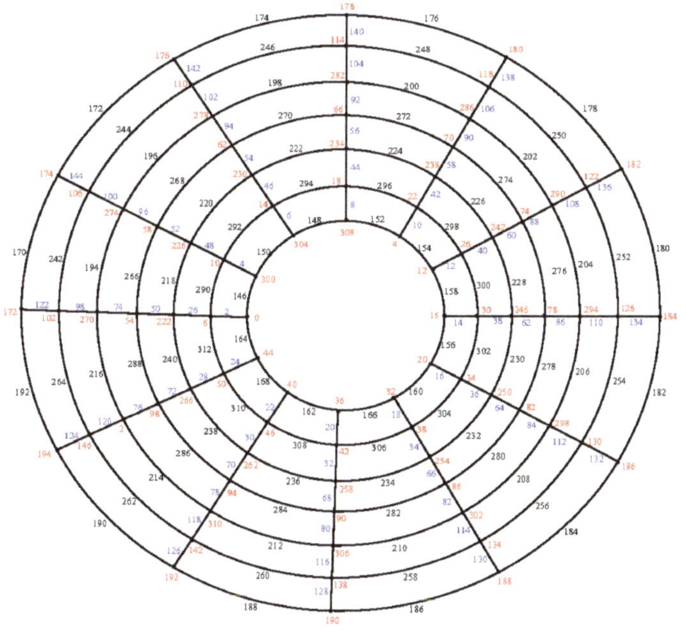

(d) $C_{7,12}$

Figure 12. *Cont.*

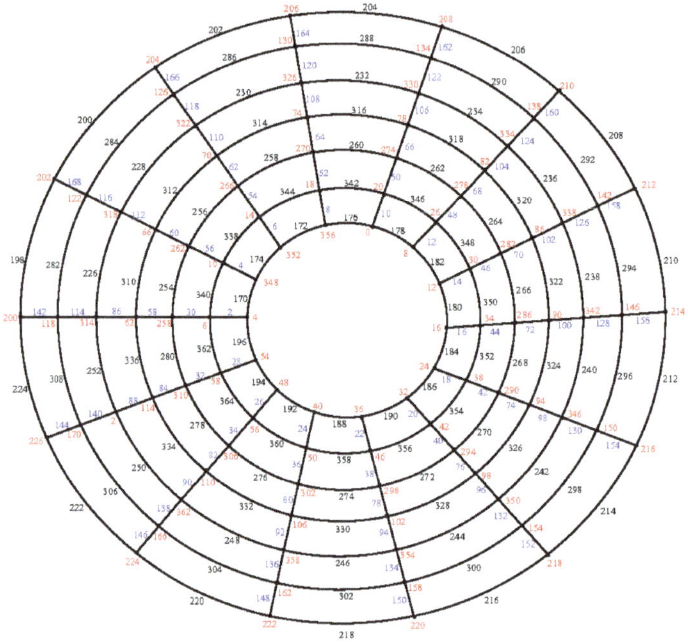

(e) $C_{7,14}$

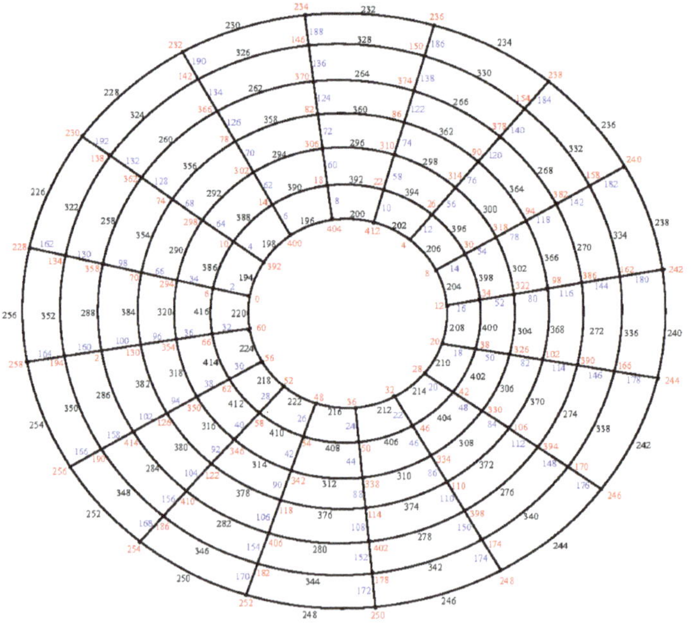

(f) $C_{7,16}$

Figure 12. *Cont.*

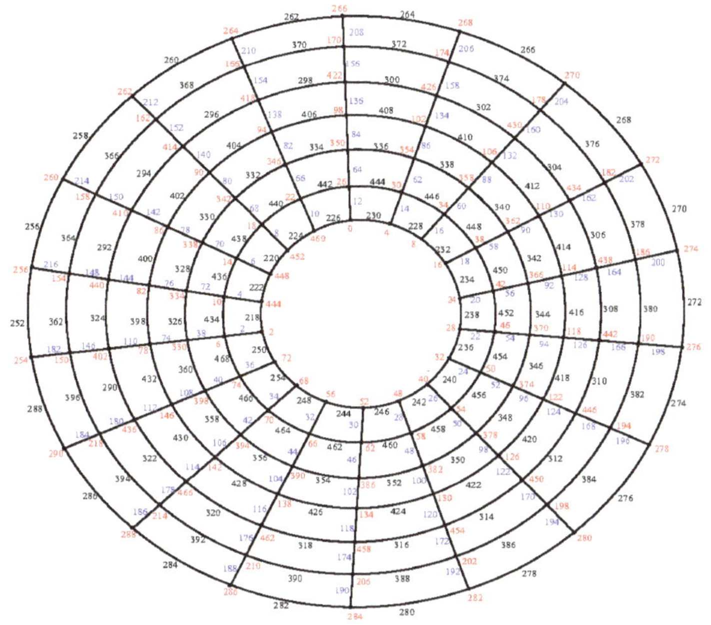

(g) $C_{7,18}$

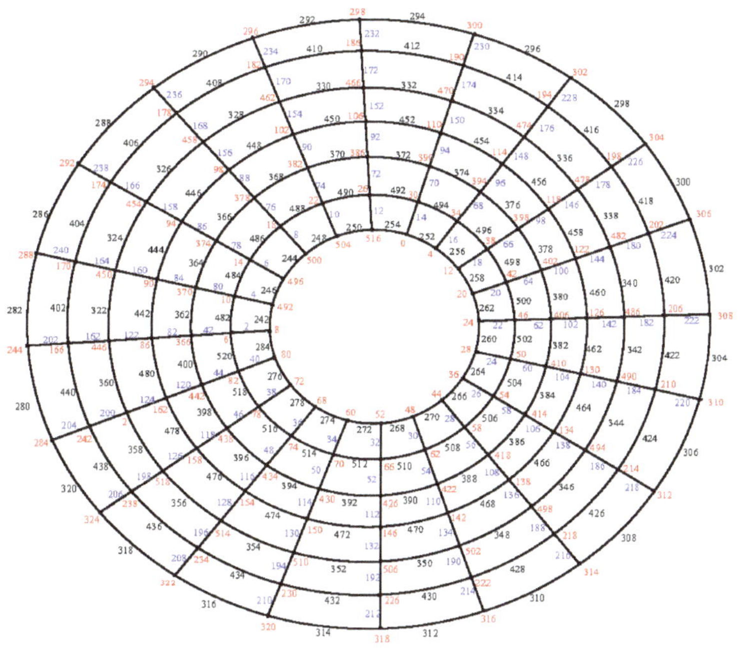

(h) $C_{7,20}$

Figure 12. An e.e.g., l. of the cylinder grid graphs $C_{9,2}$, $C_{9,4}$, $C_{7,10}$, $C_{7,12}$, $C_{7,14}$, $C_{7,16}$, $C_{7,18}$, and $C_{7,20}$.

Theorem 4. *If m is an odd positive integer greater than 3 and n is an odd positive integer, $n \geq 3$, then the cylinder grid graph $C_{m,n}$, is an edge even graceful graph.*

Proof. Using standard notation $p = |V(C_{m,n})| = mn$, $q = |E(C_{m,n})| = 2mn - n$, $r = \max(p,q) = 2mn - n$, and $f : E(C_{m,n}) \to \{2, 4, 6, \ldots, 4mn - 2n - 2\}$. □

Let the cylinder grid graph $C_{m,n}$ be as in Figure 9. There are six cases:

Case (1): $n \equiv 1 \bmod 12$.

First, we label the edges of the paths $P_m^{(k)}, 1 \leq k \leq n$ beginning with the edges of the path $P_m^{(1)}$ as follows: Move clockwise to label the edges $v_1v_{n+1}, v_2v_{n+2}, v_3v_{n+3}, \ldots, v_{n-1}v_{2n-1}, v_nv_{2n}$ by $2, 4, 6, \ldots, 2n-2, 2n$, then move anticlockwise to label the edges $v_{n+1}v_{n+2}, v_{2n}v_{3n}, v_{2n-1}v_{3n-1}, \ldots, v_{n+3}v_{2n+3}, v_{n+2}v_{2n+2}$ by $2n+2, 2n+4, 2n+6, \ldots, 4n-2, 4n$, then move clockwise to label the edges $v_{2n+1}v_{3n+1}, v_{2n+2}v_{3n+2}, v_{2n+3}v_{3n+3}, \ldots, v_{3n-1}v_{4n-1}, v_{3n}v_{4n}$ by $4n+2, 4n+4, 4n+6, \ldots, 6n-2, 6n$, and so on.

Finally, move anticlockwise to label the edges $v_{(m-2)n+1}v_{(m-1)n+1}, v_{(m-1)n}v_{mn}, v_{(m-1)n-1}v_{mn-1}, \ldots, v_{(m-2)n+3}v_{(m-1)n+3}, v_{(m-2)n+2}v_{m(n-1)+2}$ by $2n(m-2)+2, 2n(m-2)+4, 2n(m-2)+6, \ldots, 2n(m-1)-2, 2n(m-1)$.

Second, we label the edges of the circles $C_n^{(k)}, 1 \leq k \leq m$ beginning with the edges of the innermost circle $C_n^{(1)}$, then the edges of outer circle $C_n^{(m)}$, and then the edges of the circles $C_n^{(m-2)}, C_n^{(m-4)}, \ldots, C_n^{(3)}$.

Finally, we label the edges of the circles $C_m^{(m-1)}, C_m^{(m-3)}, \ldots, C_m^{(2)}$.

Label the edges of $C_n^{(1)}$ as follows: $f(v_1v_2) = 2n(m-1)+2, f(v_2v_3) = 2n(m-1)+4, f(v_3v_4) = 2n(m-1)+8, f(v_4v_5) = 2n(m-1)+6, f(v_5v_6) = 2n(m-1)+10, f(v_6v_7) = 2n(m-1)+12, f(v_7v_8) = 2n(m-1)+16, f(v_8v_9) = 2n(m-1)+14, f(v_9v_{10}) = 2n(m-1)+18, f(v_{10}v_{11}) = 2n(m-1)+20, f(v_{11}v_{12}) = 2n(m-1)+24, f(v_{12}v_{13}) = 2n(m-1)+22, f(v_{13}v_{14}) = 2n(m-1)+26, f(v_{14}v_{15}) = 2n(m-1)+28, \ldots, f(v_{n-7}v_{n-6}) = 2mn-14, f(v_{n-6}v_{n-5}) = 2mn-10, f(v_{n-5}v_{n-4}) = 2mn-12, f(v_{n-4}v_{n-3}) = 2mn-8, f(v_{n-3}v_{n-2}) = 2mn-6, f(v_{n-2}v_{n-1}) = 2mn-2, f(v_{n-1}v_n) = 2mn-4, f(v_nv_1) = 2mn$.

Then, label the edges of $C_n^{(m)}, C_n^{(m-2)}, C_n^{(m-4)}, \ldots, C_n^{(3)}$ and $C_n^{(m-1)}, C_n^{(m-3)}, C_n^{(m-5)}, \ldots, C_n^{(4)}, C_n^{(2)}$ as follows:

Label the edges of the circle $C_n^{(m)}$ as follows: $f(v_{(m-1)n+i}v_{(m-1)n+i+1}) = 2mn + 2i, f(v_{mn}v_{(m-1)n+1}) = 2n(m+1), 1 \leq i \leq n-1$.

Label the edges of the circle $C_n^{(m-2)}$ as follows: $f(v_{(m-3)n+i}v_{(m-3)n+i+1}) = 2n(m+1) + 2i, f(v_{(m-2)n}v_{(m-3)n+1}) = 2n(m+2), 1 \leq i \leq n-1$.

Label the edges of the circle $C_n^{(m-4)}$ as follows: $f(v_{(m-5)n+i}v_{(m-5)n+i+1}) = 2n(m+2) + 2i, f(v_{(m-4)n}v_{(m-5)n+1}) = 2n(m+3), 1 \leq i \leq n-1$, and so on.

Label the edges of the circle $C_n^{(3)}$ as follows: $f(v_{2n+i}v_{2n+i+1}) = 3n(m-1) + 2i, f(v_{3n}v_{2n+1}) = n(3m-1), 1 \leq i \leq n-1$,

Label the edges of the circle $C_n^{(m-1)}$ as follows: $f(v_{(m-2)n+i}v_{(m-2)n+i+1}) = n(3m-1) + 2i, f(v_{(m-1)n}v_{(m-1)n+1}) = n(3m+1), 1 \leq i \leq n-1$,

Label the edges of the circle $C_n^{(m-3)}$ as follows: $f(v_{(m-4)n+i}v_{(m-4)n+i+1}) = n(3m+2) + 2i, f(v_{(m-3)n}v_{(m-4)n+1}) = 3n(m+1), 1 \leq i \leq n-1, \ldots$, and so on.

Label the edges of the circle $C_n^{(4)}$ as follows: $f(v_{3n+i}v_{3n+i+1}) = 2n(2m-3) + 2i, f(v_{4n}v_{3n+1}) = 4n(m-1), 1 \leq i \leq n-1$,

Label the edges of $C_n^{(2)}$ as follows: $f(v_{n+i}v_{n+i+1}) = 4n(m-1) + 2i, f(v_{2n}v_{2n+1}) = 2n(m-1), 1 \leq i \leq n-1$,

Thus, the labels of corresponding vertices of the circle $C_n^{(1)} \bmod (4mn-2n)$ will be: $f^*(v_1) \equiv 4, f^*(v_2) \equiv 4mn - 4n + 10, f^*(v_3) \equiv 4mn - 4n + 18, f^*(v_4) \equiv 4mn - 4n + 22, f^*(v_5) \equiv 4mn - 4n + 26, f^*(v_6) \equiv 4mn - 4n + 34, f^*(v_7) \equiv 4mn - 4n + 42, f^*(v_8) \equiv 4mn - 4n + 46, f^*(v_9) = 4mn - 4n + 50, f^*(v_{10}) \equiv 4mn -$

$4n+58, f^*(v_{11}) \equiv 4mn-4n+66, f^*(v_{12}) \equiv 4mn-4n+70, f^*(v_{13}) \equiv 4mn-4n+74, f^*(v_{14}) \equiv 4mn-4n+82, \ldots, f^*(v_{n-7}) \equiv 4n-44,\ f^*(v_{n-6}) \equiv 4n-36, f^*(v_{n-5}) \equiv 4n-32, f^*(v_{n-4}) \equiv 4n-28, f^*(v_{n-3}) \equiv 4n-20, f^*(v_{n-2}) \equiv 4n-12, f^*(v_{n-1}) \equiv 4n-8, f^*(v_n) \equiv 4n-4$.

The labels of the vertices of $C_n^{(2)}, C_n^{(3)}, C_n^{(4)}, \ldots, C_n^{(m-2)}, C_n^{(m-1)}, C_n^{(m)}$, respectively, are as follows: $f^*(v_{n+i}) \equiv 4i+2; f^*(v_{2n+i}) \equiv 2mn+4n+4i+2; f^*(v_{3n+i}) \equiv 4n+4i+2; \ldots; f^*(v_{(m-3)n+i}) \equiv 4mn-6n+4i+2; f^*(v_{(m-2)n+i}) \equiv 2mn-6n+4i+2; f^*(v_{(m-1)n+i}) \equiv 2mn+2i+2, 1 \le i \le n$.

Case (2): $n \equiv 3 \bmod 12$.

First, we label the edges of the paths $P_m^{(k)}, 1 \le k \le n$ beginning with the edges of the path $P_m^{(1)}$ as the same in case (1).

Second, we label the edges of the circles $C_n^{(k)}, 1 \le k \le m$ beginning with the edges of the innermost circle $C_n^{(1)}$, then the edges of outer circle $C_n^{(m)}$, and then the edges of the circles $C_n^{(m-2)}, C_n^{(m-4)}, \ldots, C_n^{(3)}$.

Finally, we label the edges of the circles $C_m^{(m-1)}, C_m^{(m-3)}, \ldots, C_m^{(2)}$.

Label the edges of the circle $C_n^{(1)}$ as follows: $f(v_1v_2) = 2n(m-1)+2, f(v_2v_3) = 2n(m-1)+4, f(v_3v_4) = 2n(m-1)+8, f(v_4v_5) = 2n(m-1)+6, f(v_5v_6) = 2n(m-1)+10,\ f(v_6v_7) = 2n(m-1)+12, f(v_7v_8) = 2n(m-1)+16, f(v_8v_9) = 2n(m-1)+14, f(v_9v_{10}) = 2n(m-1)+18,\ f(v_{10}v_{11}) = 2n(m-1)+20, f(v_{11}v_{12}) = 2n(m-1)+24, f(v_{12}v_{13}) = 2n(m-1)+22, f(v_{13}v_{14}) = 2n(m-1)+26, \ldots,\ f(v_{n-9}v_{n-8}) = 2mn-18, f(v_{n-8}v_{n-7}) = 2mn-14, f(v_{n-7}v_{n-6}) = 2mn-16, f(v_{n-6}v_{n-5}) = 2mn-12,\ f(v_{n-5}v_{n-4}) = 2mn-10, f(v_{n-4}v_{n-3}) = 2mn-6, f(v_{n-3}v_{n-2}) = 2mn-8, f(v_{n-2}v_{n-1}) = 2mn, f(v_{n-1}v_n) = 2mn-2, f(v_nv_1) = 2mn-4$.

Label the edges of the circle $C_n^{(2)}$ as follows: $f(v_{n+i}v_{n+i+1}) = 4n(m-1)+2i, 1 \le i \le n-9, f(v_{2n-9}v_{2n-8}) = 2n(2m-1)-18, f(v_{2n-8}v_{2n-7}) = 2n(2m-1)-14, f(v_{2n-7}v_{2n-6}) = 2n(2m-1)-16,\ f(v_{2n-6}v_{2n-5}) = 2n(2m-1)-10, f(v_{2n-5}v_{2n-4}) = 2n(2m-1)-12, f(v_{2n-4}v_{2n-3}) = 2n(2m-1)-6,\ f(v_{2n-3}v_{2n-2}) = 2n(2m-1)-8, f(v_{2n-2}v_{2n-1}) = 2n(2m-1)-4, f(v_{2n-1}v_{2n}) = 2n(2m-1)-2, f(v_{2n}v_{n+1}) = 2n(2m-1)$.

Label the edges of $C_n^{(m)}, C_n^{(m-2)}, C_n^{(m-4)}, \ldots, C_n^{(3)}$ and $C_n^{(m-1)}, C_n^{(m-3)}, C_n^{(m-5)}, \ldots, C_n^{(4)}$ as in case (1).

Thus, the labels of corresponding vertices $\bmod (4mn-2n)$ will be:

The labels of the vertices of $C_n^{(1)}$ are as follows: $f^*(v_1) \equiv 0, f^*(v_2) \equiv 4mn-4n+10, f^*(v_3) \equiv 4mn-4n+18, f^*(v_4) \equiv 4mn-4n+22, f^*(v_5) \equiv 4mn-4n+26, f^*(v_6) \equiv 4mn-4n+34,\ f^*(v_7) \equiv 4mn-4n+42, f^*(v_8) \equiv 4mn-4n+46, f^*(v_9) \equiv 4mn-4n+50, f^*(v_{10}) \equiv 4mn-4n+58, f^*(v_{11}) \equiv 4mn-4n+66,\ f^*(v_{12}) \equiv 4mn-4n+70, f^*(v_{13}) \equiv 4mn-4n+74, f^*(v_{14}) \equiv 4mn-4n+82, \ldots, f^*(v_{n-8}) \equiv 4n-48, f^*(v_{n-7}) \equiv 4n-44,\ f^*(v_{n-6}) \equiv 4n-40, f^*(v_{n-5}) \equiv 4n-32, f^*(v_{n-4}) \equiv 4n-24, f^*(v_{n-3}) \equiv 4n-20, f^*(v_{n-2}) \equiv 4n-12, f^*(v_{n-1}) \equiv 4n-4, f^*(v_n) \equiv 4n-6$.

The labels of the vertices of the circle $C_n^{(2)}$ are as follows: $f^*(v_{n+i}) \equiv 4i+2, 1 \le i \le n-9, f^*(v_{2n-8}) \equiv 4n-28, f^*(v_{2n-7}) \equiv 4n-26, f^*(v_{2n-6}) \equiv 4n-22, f^*(v_{2n-5}) \equiv 4n-18, f^*(v_{2n-4}) \equiv 4n-14,\ f^*(v_{2n-3}) \equiv 4n-10, f^*(v_{2n-2}) \equiv 4n-8, f^*(v_{2n-1}) \equiv 4n-2, f^*(v_{2n}) \equiv 4n+2$.

The labels of the vertices of $C_n^{(3)}, C_n^{(4)}, \ldots, C_n^{(m-2)}, C_n^{(m-1)}, C_n^{(m)}$, respectively, are the same as in case (1).

Remark 4. *In case $n = 3$ and m is odd, $m \ge 3$.*

Let the label of edges of the cylinder grid graph $C_{m,3}$ be as in Figure 13.

Thus, the labels of corresponding vertices $\bmod (12m-6)$ are as follows:

The labels of the vertices of the circle $C_3^{(1)}$ are $f^*(v_1) \equiv 8, f^*(v_2) \equiv 12, f^*(v_3) \equiv 16$.

The labels of the vertices of the circle $C_3^{(3)}$ are $f^*(v_{3m-2}) \equiv 6m+10, f^*(v_{3m-1}) \equiv 6m+12, f^*(v_{3m}) \equiv 6m+14$.

The labels of the vertices of the circles $C_3^{(2)}, C_3^{(4)}, \ldots, C_3^{(m-1)}$ are $f^*(v_{3k+i}) \equiv 4i+6k+4, 1 \le i \le 3, 1 \le k \le m-2, k$ is odd.

The labels of the vertices of the circles $C_3^{(3)}, C_3^{(5)}, \ldots, C_3^{(m-2)}$ are $f^*(v_{3k+i}) \equiv 6m+4i+6k+10, 1 \le i \le 3, 2 \le k \le m-3, k$ is even.

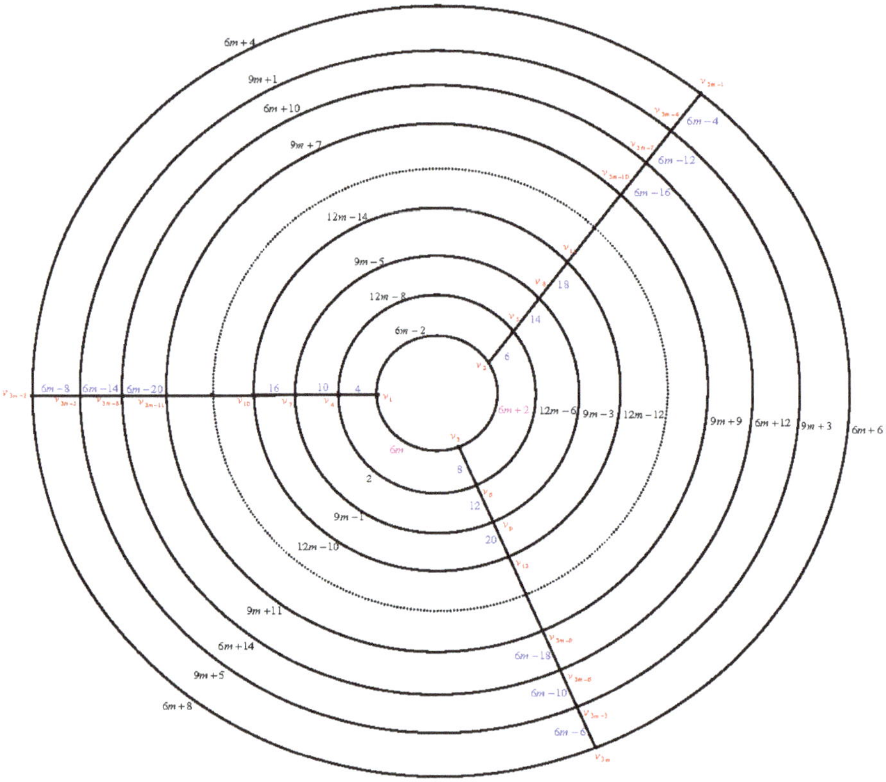

Figure 13. The cylinder grid graph $C_{m,3}$, m is odd, $m \geq 3$.

Case (3): $n \equiv 5 \bmod 12$.

First, we label the edges of the paths $P_m^{(k)}, 1 \leq k \leq n$ beginning with the edges of the path $P_m^{(1)}$ the same as in case (1).

Second, we label the edges of the circles $C_n^{(k)}, 1 \leq k \leq m$ beginning with the edges of the innermost circle $C_n^{(1)}$, then the edges of outer circle $C_n^{(m)}$, and then the edges of the circles $C_n^{(m-2)}, C_n^{(m-4)}, \ldots, C_n^{(3)}$.

Finally, we label the edges of the circles $C_m^{(m-1)}, C_m^{(m-3)}, \ldots, C_m^{(2)}$.

Label the edges of the circle $C_n^{(1)}$ as follows: $f(v_1v_2) = 2n(m-1)+2, f(v_2v_3) = 2n(m-1)+4, f(v_3v_4) = 2n(m-1)+8, f(v_4v_5) = 2n(m-1)+6$, $f(v_5v_6) = 2n(m-1)+10, f(v_6v_7) = 2n(m-1)+12, f(v_7v_8) = 2n(m-1)+16, f(v_8v_9) = 2n(m-1)+14$, $f(v_9v_{10}) = 2n(m-1)+18, f(v_{10}v_{11}) = 2n(m-1)+20, f(v_{11}v_{12}) = 2n(m-1)+24, f(v_{12}v_{13}) = 2n(m-1)+22$, $f(v_{13}v_{14}) = 2n(m-1)+26, \ldots, f(v_{n-8}v_{n-7}) = 2mn-16, f(v_{n-7}v_{n-6}) = 2mn-14, f(v_{n-6}v_{n-5}) = 2mn-10$, $f(v_{n-5}v_{n-4}) = 2mn-12, f(v_{n-4}v_{n-3}) = 2mn-8, f(v_{n-3}v_{n-2}) = 2mn-6, f(v_{n-2}v_{n-1}) = 2mn-2, f(v_{n-1}v_n) = 2mn-4$, $f(v_nv_1) = 2mn$. Label the edges of $C_n^{(m)}, C_n^{(m-2)}, C_n^{(m-4)}, \ldots, C_n^{(3)}$ and $C_n^{(m-1)}, C_n^{(m-3)}, C_n^{(m-5)}, \ldots, C_n^{(4)}, C_n^{(2)}$ as in case (1).

Thus, the labels of corresponding vertices of the circle $C_n^{(1)}$ $\bmod(4mn-2n)$ will be: $f^*(v_1) \equiv 4, f^*(v_2) \equiv 4mn-4n+10, f^*(v_3) \equiv 4mn-4n+18, f^*(v_4) \equiv 4mn-4n+22, f^*(v_5) \equiv 4mn-4n+26$, $f^*(v_6) \equiv 4mn-4n+34, f^*(v_7) \equiv 4mn-4n+42, f^*(v_8) \equiv 4mn-4n+46, f^*(v_9) \equiv 4mn-4n+50$, $f^*(v_{10}) \equiv 4mn-4n+58, f^*(v_{11}) \equiv 4mn-4n+66, f^*(v_{12}) \equiv 4mn-4n+70, f^*(v_{13}) \equiv 4mn-4n+74$, $f^*(v_{14}) \equiv 4mn-4n+82, \ldots, f^*(v_{n-7}) \equiv 4n-44, f^*(v_{n-6}) \equiv 4n-36, f^*(v_{n-5}) \equiv 4n-32, f^*(v_{n-4}) \equiv 4n-28, f^*(v_{n-3}) \equiv 4n-20, f^*(v_{n-2}) \equiv 4n-12, f^*(v_{n-1}) \equiv 4n-8, f^*(v_n) \equiv 4n-4$.

The labels of the vertices of $C_n^{(2)}, C_n^{(3)}, C_n^{(4)}, \ldots, C_n^{(m-2)}, C_n^{(m-1)}, C_n^{(m)}$, respectively, are the same as in case (1).

Case (4): $n \equiv 7 \bmod 12$.

First, we label the edges of the paths $P_m^{(k)}, 1 \le k \le n$ beginning with the edges of the path $P_m^{(1)}$ the same as in case (1).

Second, we label the edges of the circles $C_n^{(k)}, 1 \le k \le m$ beginning with the edges of the innermost circle $C_n^{(1)}$, then the edges of outer circle $C_n^{(m)}$, and then the edges of the circles $C_n^{(m-2)}, C_n^{(m-4)}, \ldots, C_n^{(3)}$.

Finally, we label the edges of the circles $C_m^{(m-1)}, C_m^{(m-3)}, \ldots, C_m^{(2)}$.

Label the edges of the circle $C_n^{(1)}$ as follows: $f(v_1v_2) = 2n(m-1)+2, f(v_2v_3) = 2n(m-1)+4, f(v_3v_4) = 2n(m-1)+8, f(v_4v_5) = 2n(m-1)+6$, $f(v_5v_6) = 2n(m-1)+10, f(v_6v_7) = 2n(m-1)+12, f(v_7v_8) = 2n(m-1)+16, f(v_8v_9) = 2n(m-1)+14$, $f(v_9v_{10}) = 2n(m-1)+18, f(v_{10}v_{11}) = 2n(m-1)+20, f(v_{11}v_{12}) = 2n(m-1)+24, f(v_{12}v_{13}) = 2n(m-1)+22$, $f(v_{13}v_{14}) = 2n(m-1)+26, \ldots, f(v_{n-10}v_{n-9}) = 2mn-20, f(v_{n-9}v_{n-8}) = 2mn-18, f(v_{n-8}v_{n-7}) = 2mn-14$, $f(v_{n-7}v_{n-6}) = 2mn-16, f(v_{n-6}v_{n-5}) = 2mn-12, f(v_{n-5}v_{n-4}) = 2mn-10, f(v_{n-4}v_{n-3}) = 2mn-6, f(v_{n-3}v_{n-2}) = 2mn-8$, $f(v_{n-2}v_{n-1}) = 2mn, f(v_{n-1}v_n) = 2mn-2, f(v_nv_1) = 2mn-4$.

Label the edges of the circle $C_n^{(2)}$ as follows: $f(v_{n+i}v_{n+i+1}) = 4n(m-1)+2i, 1 \le i \le n-9, f(v_{2n-9}v_{2n-8}) = 2n(2m-1)-18, f(v_{2n-8}v_{2n-7}) = 2n(2m-1)-14$, $f(v_{2n-7}v_{2n-6}) = 2n(2m-1)-16, f(v_{2n-6}v_{2n-5}) = 2n(2m-1)-10, f(v_{2n-5}v_{2n-4}) = 2n(2m-1)-12$, $f(v_{2n-4}v_{2n-3}) = 2n(2m-1)-6, f(v_{2n-3}v_{2n-2}) = 2n(2m-1)-8, f(v_{2n-2}v_{2n-1}) = 2n(2m-1)-4$, $f(v_{2n-1}v_{2n}) = 2n(2m-1)-2, f(v_{2n}v_{n+1}) = 2n(2m-1)$.

Label the edges of $C_n^{(m)}, C_n^{(m-2)}, C_n^{(m-4)}, \ldots, C_n^{(3)}$ and $C_n^{(m-1)}, C_n^{(m-3)}, C_n^{(m-5)}, \ldots, C_n^{(4)}$ as in case (1).

Thus, the labels of corresponding vertices $\bmod(4mn-2n)$ will be:

The labels of the vertices of the circle $C_n^{(1)}$ are as follows: $f^*(v_1) \equiv 0, f^*(v_2) \equiv 4mn-4n+10, f^*(v_3) \equiv 4mn-4n+18, f^*(v_4) \equiv 4mn-4n+22, f^*(v_5) \equiv 4mn-4n+26$, $f^*(v_6) \equiv 4mn-4n+34, f^*(v_7) \equiv 4mn-4n+42, f^*(v_8) \equiv 4mn-4n+46, f^*(v_9) \equiv 4mn-4n+50$, $f^*(v_{10}) \equiv 4mn-4n+58, f^*(v_{11}) \equiv 4mn-4n+66, f^*(v_{12}) \equiv 4mn-4n+70, f^*(v_{13}) \equiv 4mn-4n+74$, $f^*(v_{14}) \equiv 4mn-4n+82, \ldots, f^*(v_{n-9}) \equiv 4n-56, f^*(v_{n-8}) \equiv 4n-48, f^*(v_{n-7}) \equiv 4n-44, f^*(v_{n-6}) \equiv 4n-40$, $f^*(v_{n-5}) \equiv 4n-32, f^*(v_{n-4}) \equiv 4n-24, f^*(v_{n-3}) \equiv 4n-20, f^*(v_{n-2}) \equiv 4n-12, f^*(v_{n-1}) \equiv 4n-4, f^*(v_n) \equiv 4n-6$.

The labels of the vertices of the circle $C_n^{(2)}$ are as follows: $f^*(v_{n+i}) \equiv 4i+2, 1 \le i \le n-9, f^*(v_{2n-8}) \equiv 4n-28, f^*(v_{2n-7}) \equiv 4n-26, f^*(v_{2n-6}) \equiv 4n-22, f^*(v_{2n-5}) \equiv 4n-18$, $f^*(v_{2n-4}) \equiv 4n-14, f^*(v_{2n-3}) \equiv 4n-10, f^*(v_{2n-2}) \equiv 4n-8, f^*(v_{2n-1}) \equiv 4n-2, f^*(v_{2n}) \equiv 4n+2$.

The labels of the vertices of $C_n^{(3)}, C_n^{(4)}, \ldots, C_n^{(m-2)}, C_n^{(m-1)}, C_n^{(m)}$, respectively, are as in case (1).

Case (5): $n \equiv 9 \bmod 12$.

First, we label the edges of the paths $P_m^{(k)}, 1 \le k \le n$ beginning with the edges of the path $P_m^{(1)}$ the same as in case (1).

Second, we label the edges of the circles $C_n^{(k)}, 1 \le k \le m$ beginning with the edges of the innermost circle $C_n^{(1)}$, then the edges of outer circle $C_n^{(m)}$, and then the edges of the circles $C_n^{(m-2)}, C_n^{(m-4)}, \ldots, C_n^{(3)}$.

Finally, we label the edges of the circles $C_m^{(m-1)}, C_m^{(m-3)}, \ldots, C_m^{(2)}$.

Label the edges of the circle $C_n^{(1)}$ as follows: $f(v_1v_2) = 2n(m-1)+2, f(v_2v_3) = 2n(m-1)+4, f(v_3v_4) = 2n(m-1)+8, f(v_4v_5) = 2n(m-1)+6$, $f(v_5v_6) = 2n(m-1)+10, f(v_6v_7) = 2n(m-1)+12, f(v_7v_8) = 2n(m-1)+16, f(v_8v_9) = 2n(m-1)+14$, $f(v_9v_{10}) = 2n(m-1)+18, f(v_{10}v_{11}) = 2n(m-1)+20, f(v_{11}v_{12}) = 2n(m-1)+24, f(v_{12}v_{13}) = 2n(m-1)+22$, $f(v_{13}v_{14}) = 2n(m-1)+26, \ldots, f(v_{n-11}v_{n-10}) = 2mn-22, f(v_{n-10}v_{n-9}) = 2mn-18, f(v_{n-9}v_{n-8}) = 2mn-20$, $f(v_{n-8}v_{n-7}) = 2mn-16, f(v_{n-7}v_{n-6}) = 2mn-14, f(v_{n-6}v_{n-5}) = 2mn-10, f(v_{n-5}v_{n-4}) = 2mn-12$, $f(v_{n-4}v_{n-3}) = 2mn-8, f(v_{n-3}v_{n-2}) = 2mn-6, f(v_{n-2}v_{n-1}) = 2mn, f(v_{n-1}v_n) = 2mn-2, f(v_nv_1) = 2mn-4$.

Label the edges of the circle $C_n^{(2)}$ as follows: $f(v_{n+i}v_{n+i+1}) = 4n(m-1)+2i, 1 \le i \le n-8, f(v_{2n-7}v_{2n-6}) = 2n(2m-1)-12, f(v_{2n-6}v_{2n-5}) = 2n(2m-1)-14$, $f(v_{2n-5}v_{2n-4}) = 2n(2m-1)-6, f(v_{2n-4}v_{2n-3}) = 2n(2m-1)-10, f(v_{2n-3}v_{2n-2}) = 2n(2m-1)-8$, $f(v_{2n-2}v_{2n-1}) = 2n(2m-1)-4, f(v_{2n}v_{n+1}) = 2n(2m-1)$.

Label the edges of $C_n^{(m)}, C_n^{(m-2)}, C_n^{(m-4)}, \ldots, C_n^{(3)}$ and $C_n^{(m-1)}, C_n^{(m-3)}, C_n^{(m-5)}, \ldots, C_n^{(4)}$ as in case (1).

Thus, the labels of corresponding vertices mod$(4mn-2n)$ will be:

The labels of the vertices of the circle $C_n^{(1)}$ are as follows: $f^*(v_1) \equiv 0, f^*(v_2) \equiv 4mn-4n+10, f^*(v_3) \equiv 4mn-4n+18, f^*(v_4) \equiv 4mn-4n+22, f^*(v_5) \equiv 4mn-4n+26, f^*(v_6) \equiv 4mn-4n+34, f^*(v_7) \equiv 4mn-4n+42, f^*(v_8) \equiv 4mn-4n+46, f^*(v_9) \equiv 4mn-4n+50, f^*(v_{10}) \equiv 4mn-4n+58, f^*(v_{11}) \equiv 4mn-4n+66, f^*(v_{12}) \equiv 4mn-4n+70, f^*(v_{13}) \equiv 4mn-4n+74, f^*(v_{14}) \equiv 4mn-4n+82, \ldots, f^*(v_{n-10}) \equiv 4n-60, f^*(v_{n-9}) \equiv 4n-56, f^*(v_{n-8}) \equiv 4n-52, f^*(v_{n-7}) \equiv 4n-44, f^*(v_{n-6}) \equiv 4n-36, f^*(v_{n-5}) \equiv 4n-32, f^*(v_{n-4}) \equiv 4n-28, f^*(v_{n-3}) \equiv 4n-20, f^*(v_{n-2}) \equiv 4n-10, f^*(v_{n-1}) \equiv 4n-4, f^*(v_n) \equiv 4n-6$.

The labels of the vertices of the circle $C_n^{(2)}$ are as follows: $f^*(v_{n+i}) \equiv 4i+2, 1 \le i \le n-8, f^*(v_{2n-7}) \equiv 4n-24, f^*(v_{2n-6}) \equiv 4n-22, f^*(v_{2n-5}) \equiv 4n-16, f^*(v_{2n-4}) \equiv 4n-12, f^*(v_{2n-3}) \equiv 4n-14, f^*(v_{2n-2}) \equiv 4n-8, f^*(v_{2n-1}) \equiv 4n-2, f^*(v_{2n}) \equiv 4n+2$.

The labels of the vertices of $C_n^{(3)}, C_n^{(4)}, \ldots, C_n^{(m-2)}, C_n^{(m-1)}, C_n^{(m)}$, respectively, are as in case (1).

Case (6): $n \equiv 11 \bmod 12$.

First, we label the edges of the paths $P_m^{(k)}, 1 \le k \le n$ beginning with the edges of the path $P_m^{(1)}$ the same as in case (1).

Second, we label the edges of the circles $C_n^{(k)}, 1 \le k \le m$ beginning with the edges of the innermost circle $C_n^{(1)}$, then the edges of outer circle $C_n^{(m)}$, and then the edges of the circles $C_n^{(m-2)}, C_n^{(m-4)}, \ldots, C_n^{(3)}$.

Finally, we label the edges of the circles $C_m^{(m-1)}, C_m^{(m-3)}, \ldots, C_m^{(2)}$.

Label the edges of the circle $C_n^{(1)}$ as follows: $f(v_1v_2) = 2n(m-1)+4, f(v_2v_3) = 2n(m-1)+2, f(v_3v_4) = 2n(m-1)+6, f(v_4v_5) = 2n(m-1)+8, f(v_5v_6) = 2n(m-1)+12, f(v_6v_7) = 2n(m-1)+10, f(v_7v_8) = 2n(m-1)+14, f(v_8v_9) = 2n(m-1)+16, f(v_9v_{10}) = 2n(m-1)+20, f(v_{10}v_{11}) = 2n(m-1)+18, f(v_{11}v_{12}) = 2n(m-1)+22, f(v_{12}v_{13}) = 2n(m-1)+24, f(v_{13}v_{14}) = 2n(m-1)+28, \ldots, f(v_{n-8}v_{n-7}) = 2mn-16, f(v_{n-7}v_{n-6}) = 2mn-14, f(v_{n-6}v_{n-5}) = 2mn-10, f(v_{n-5}v_{n-4}) = 2mn-12, f(v_{n-4}v_{n-3}) = 2mn-8, f(v_{n-3}v_{n-2}) = 2mn-6, f(v_{n-2}v_{n-1}) = 2mn-2, f(v_{n-1}v_n) = 2mn-4, f(v_nv_1) = 2mn$.

Label the edges of the circle $C_n^{(2)}$ as follows: $f(v_{n+i}v_{n+i+1}) = 4n(m-1)+2i, 1 \le i \le n-2, f(v_{2n-1}v_{2n}) = 4mn, f(v_{2n}v_{n+1}) = 4mn-2$.

Label the edges of $C_n^{(m)}, C_n^{(m-2)}, C_n^{(m-4)}, \ldots, C_n^{(3)}$ and $C_n^{(m-1)}, C_n^{(m-3)}, C_n^{(m-5)}, \ldots, C_n^{(4)}$ as in case (1).

Thus, the labels of corresponding vertices mod$(4mn-2n)$ will be:

The labels of the vertices of the circle $C_n^{(1)}$ are as follows: $f^*(v_1) \equiv 6, f^*(v_2) \equiv 4mn-4n+10, f^*(v_3) \equiv 4mn-4n+14, f^*(v_4) \equiv 4mn-4n+22, f^*(v_5) \equiv 4mn-4n+30, f^*(v_6) \equiv 4mn-4n+34, f^*(v_7) \equiv 4mn-4n+38, f^*(v_8) \equiv 4mn-4n+46, f^*(v_9) \equiv 4mn-4n+54, f^*(v_{10}) \equiv 4mn-4n+58, f^*(v_{11}) \equiv 4mn-4n+62, f^*(v_{12}) \equiv 4mn-4n+70, \ldots, f^*(v_{n-7}) \equiv 4n-44, f^*(v_{n-6}) \equiv 4n-36, f^*(v_{n-5}) \equiv 4n-32, f^*(v_{n-4}) \equiv 4n-28, f^*(v_{n-3}) \equiv 4n-20, f^*(v_{n-2}) \equiv 4n-12, f^*(v_{n-1}) \equiv 4n-8, f^*(v_n) \equiv 4n-4$.

The labels of the vertices of the circle $C_n^{(2)}$ are as follows: $f^*(v_{n+1}) \equiv 4, f^*(v_{n+i}) \equiv 4i+2, 2 \le i \le n-2, f^*(v_{2n-1}) \equiv 4n, f^*(v_{2n}) \equiv 4n+2$.

The labels of the vertices of $C_n^{(3)}, C_n^{(4)}, \ldots, C_n^{(m-2)}, C_n^{(m-1)}, C_n^{(m)}$, respectively, are as the same as in case (1).

Illustration: An e.e.g.l. of the cylinder grid graphs $C_{9,3}, C_{7,9}, C_{7,11}, C_{7,13}, C_{7,15}, C_{7,17}$ and $C_{7,19}$ is shown in Figure 14.

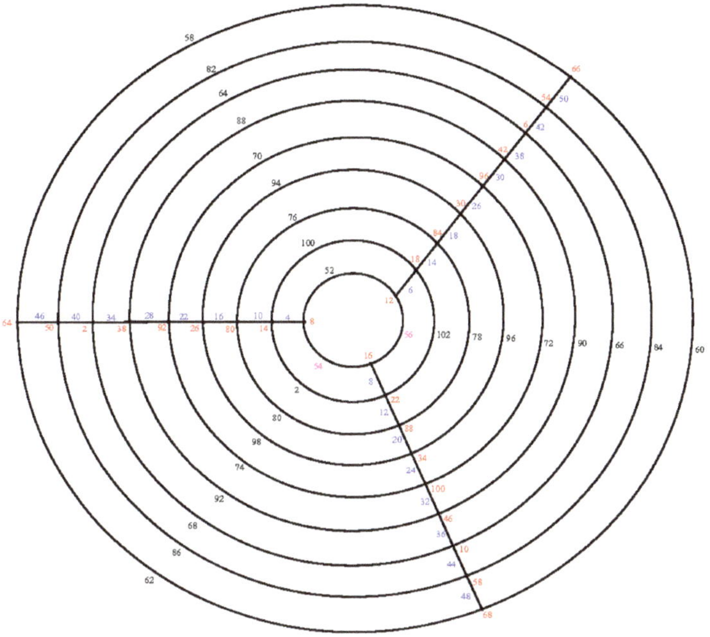

(**a**) $C_{9,3}$

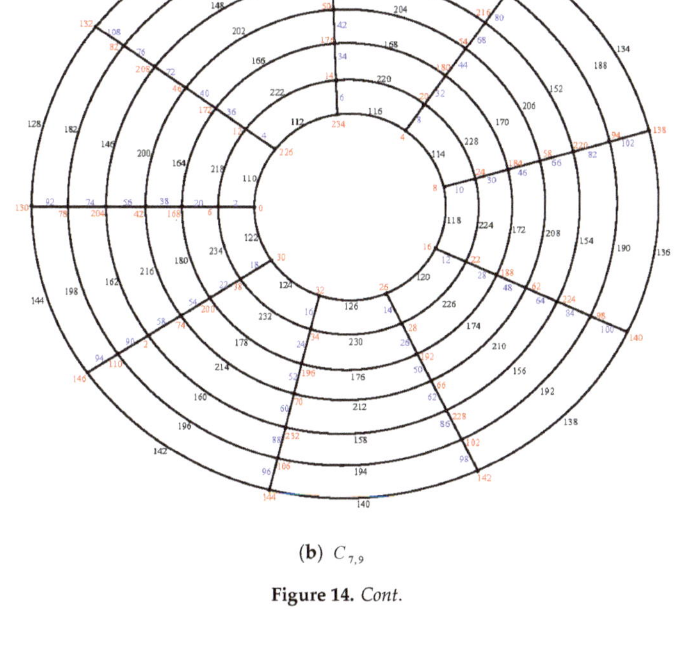

(**b**) $C_{7,9}$

Figure 14. *Cont.*

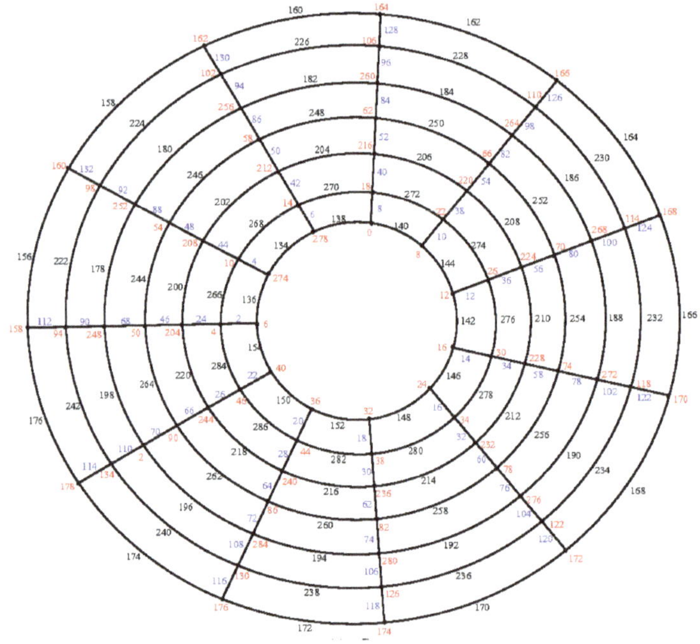

(**c**) $C_{7,11}$

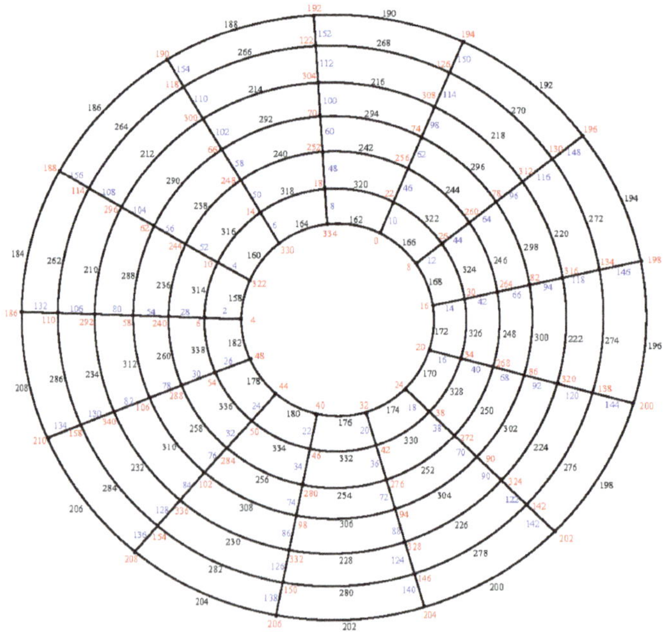

(**d**) $C_{7,13}$

Figure 14. *Cont.*

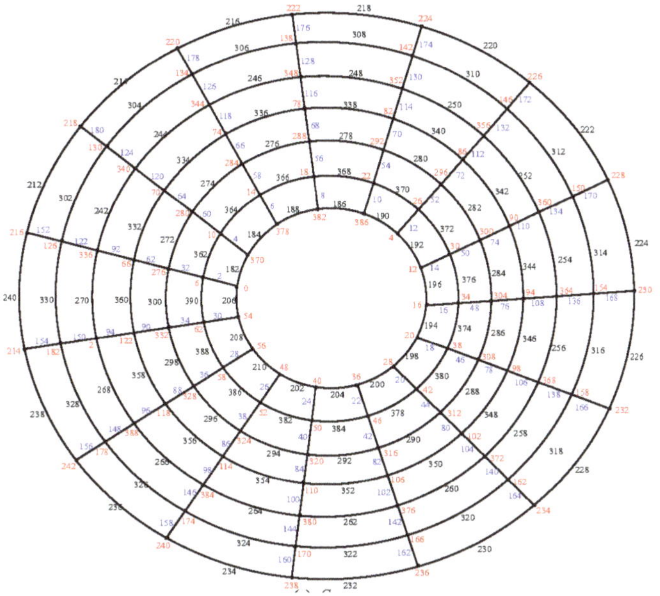

(e) $C_{7,15}$

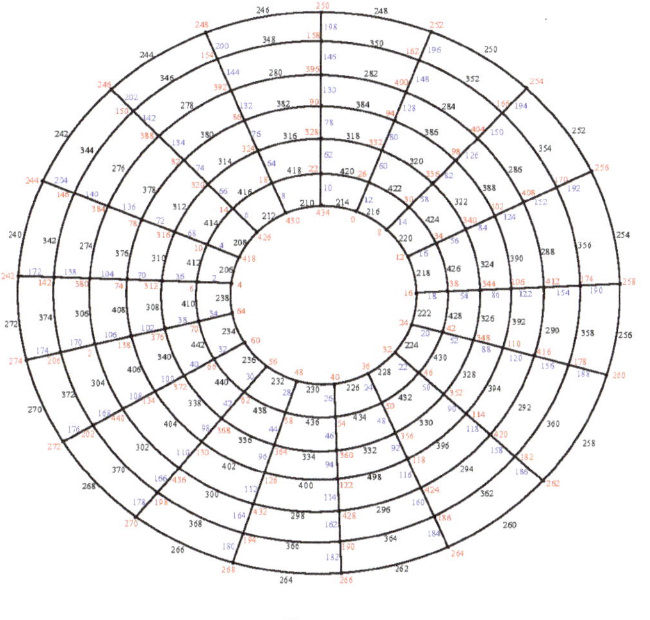

(f) $C_{7,17}$

Figure 14. *Cont.*

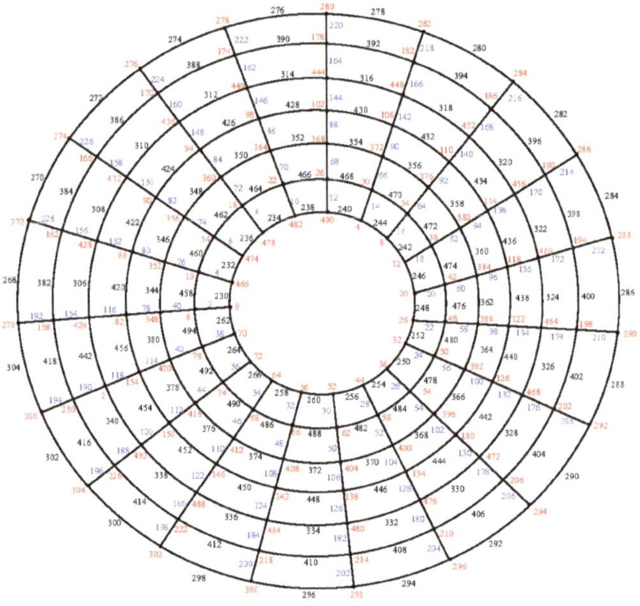

(g) $C_{7,19}$

Figure 14. An e.e.g.l. of the cylinder grid graphs $C_{9,3}, C_{7,9}, C_{7,11}, C_{7,13}, C_{7,15}, C_{7,17}$ and $C_{7,19}$.

3. Conclusions

In this paper, using the connection of labeling of graphs with modular arithmetic and theory of numbers in general, we give a detailed study for e.e.g., l. of all cases of members of the cylinder grid graphs. The study of necessary and sufficient conditions for e.e.g., l. of other important families including torus $C_m \times C_n$ and rectangular $P_m \times P_n$ grid graphs should be taken into consideration in future studies of e.e.g., l.

Author Contributions: All authors contributed equally to this work.

Funding: This work was supported by the deanship of Scientific Research, Taibah University, Al-Madinah Al-Munawwarah, Saudi Arabia.

Acknowledgments: The authors are grateful to the anonymous reviewers for their helpful comments and suggestions for improving the original version of the paper.

Conflicts of Interest: The authors declare that there are no conflicts of interest regarding the publication of this paper.

References

1. Acharya, B.D.; Arumugam, S.; Rosa, A. *Labeling of Discrete Structures and Applications;* Narosa Publishing House: New Delhi, India, 2008; pp. 1–14.
2. Bloom, G.S. Numbered Undirected Graphs and Their Uses, a Survey of a Unifying Scientific and Engineering Concept and Its Use in Developing a Theory of Non-Redundant Homometric Sets Relating to Some Ambiguities in X-ray Diffraction Analysis. Ph.D. Thesis, University of Southern California, Los Angeles, CA, USA, 1975.
3. Bloom, G.S.; Golomb, S.W. Numbered complete graphs, unusual rulers, and assorted applications. In *Theory and Applications of Graphs, Lecture Notes in Math, 642;* Springer: New York, NY, USA, 1978; pp. 53–65.
4. Bloom, G.S.; Golomb, S.W. Applications of numbered undirected graphs. *Proc. IEEE* **1977**, *65*, 562–570. [CrossRef]

5. Bloom, G.S.; Hsu, D.F. On graceful digraphs and a problem in network addressing. *Congr. Numer.* **1982**, *35*, 91–103.
6. Graham, R.L.; Pollak, H.O. On the addressing problem for loop switching. *Bell Syst. Tech. J.* **1971**, *50*, 2495–2519. [CrossRef]
7. Sutton, M.; Labellings, S.G. Summable Graphs Labellings and Their Applications. Ph.D. Thesis, the University of Newcastle, New South Wales, Australia, 2001.
8. Shang, Y. More on the normalized Laplacian Estrada index. *Appl. Anal. Discret. Math.* **2014**, *8*, 346–357. [CrossRef]
9. Shang, Y. Geometric assortative growth model for small-world networks. *Sci. World J.* **2014**, *2014*, 1–8. [CrossRef] [PubMed]
10. Shang, Y. Deffuant model of opinion formation in one-dimensional multiplex networks. *J. Phys. A Math. Theor.* **2015**, *48*, 395101. [CrossRef]
11. Gross, J.; Yellen, J. *Graph Theory and Its Applications*; CRC Press: Boca Raton, FL, USA, 1999.
12. Rosa, A. On certain valuations of the vertices of a graph. In *Theory of Graphs, Proceedings of the International Symposium, Rome, Italy, July 1966*; Gordan and Breach, Dunod: New York, NY, USA, 1967; pp. 349–355.
13. Golomb, S.W. How to Number a Graph. In *Graph Theory and Computing*; Read, R.C., Ed.; Cademic Press: New York, NY, USA, 1972; pp. 23–37.
14. Gnanajothi, R.B. Topics in Graph Theory. Ph.D. Thesis, Madurai Kamaraj University, Tamil Nadu, India, 1991.
15. Seoud, M.A.; Abdel-Aal, M.E. On odd graceful graphs. *Ars Comb.* **2013**, *108*, 161–185.
16. Gao, Z. Odd graceful labelings of some union graphs. *J. Nat. Sci. Heilongjiang Univ.* **2007**, *24*, 35–39.
17. Lo, S.P. On edge-graceful1abelings of graphs. *Congr. Numer.* **1985**, *50*, 231–241.
18. Kuan, Q.; Lee, S.; Mitchem, J.; Wang, A. On edge-graceful unicyclic graphs. *Congr. Numer.* **1988**, *61*, 65–74.
19. Lee, L.; Lee, S.; Murty, G. On edge-graceful labelings of complete graphs: Solutions of Lo's conjecture. *Congr. Numer.* **1988**, *62*, 225–233.
20. Solairaju, A.; Chithra, K. Edge-odd graceful graphs. *Electron. Notes Discret. Math.* **2009**, *33*, 15–20. [CrossRef]
21. Daoud, S.N. Edge odd graceful labeling of some path and cycle related graphs. *AKCE Int. J. Graphs Comb.* **2017**, *14*, 178–203. [CrossRef]
22. Daoud, S.N. Edge odd graceful labeling cylinder grid and torus grid graphs. *IEEE Access* **2019**, *7*, 10568–10592. [CrossRef]
23. Daoud, S.N. Vertex odd graceful labeling. *Ars Comb.* **2019**, *142*, 65–87.
24. Elsonbaty, A.; Daoud, S.N. Edge even graceful labeling of some path and cycle related graphs. *Ars Comb.* **2017**, *130*, 79–96.
25. Daoud, S.N. Edge even graceful labeling polar grid graph. *Symmetry* **2019**, *11*, 38. [CrossRef]
26. Gallian, J.A. A dynamic survey of graph labeling. *Electron. J. Comb.* **2017**, *22*, #DS6.

symmetry

MDPI

Article

Involution Abel–Grassmann's Groups and Filter Theory of Abel–Grassmann's Groups

Xiaohong Zhang * and Xiaoying Wu

Department of Mathematics, Shaanxi University of Science & Technology, Xi'an 710021, China; 46018@sust.edu.cn

* Correspondence: zhangxiaohong@sust.edu.cn

Received: 12 March 2019; Accepted: 8 April 2019; Published: 17 April 2019

Abstract: In this paper, some basic properties and structure characterizations of AG-groups are further studied. First, some examples of infinite AG-groups are given, and weak commutative, alternative and quasi-cancellative AG-groups are discussed. Second, two new concepts of involution AG-group and generalized involution AG-group are proposed, the relationships among (generalized) involution AG-groups, commutative groups and AG-groups are investigated, and the structure theorems of (generalized) involution AG-groups are proved. Third, the notion of filter of an AG-group is introduced, the congruence relation is constructed from arbitrary filter, and the corresponding quotient structure and homomorphism theorems are established.

Keywords: Abel–Grassmann's groupoid (AG-groupoid); Abel–Grassmann's group (AG-group); involution AG-group; commutative group; filter

---

1. Introduction

Nowadays, the theories of groups and semigroups [1–5] are attracting increasing attention, which can be used to express various symmetries and generalized symmetries in the real world. Every group or semigroup has a binary operation that satisfies the associative law. On the other hand, non-associative algebraic structures have great research value. Euclidean space $\mathbf{R}^3$ with multiplication given by the vector cross product is an example of an algebra that is not associative, at the same time; Jordan algebra and Lie algebra are non-associative.

For the generalization of commutative semigroup, the notion of an AG-groupoid (Abel–Grassmann's groupoid) is introduced in [6], which is also said to be a left almost semigroup (LA-semigroup). Moreover, a class of non-associative ring with condition $x(yz) = z(yx)$ is investigated in [7]; in fact, the condition $x(yz) = z(yx)$ is a dual distortion of the operation law in AG-groupoids.

An AG-groupoid is a non-associative algebraic structure, but it is a groupoid $(N, *)$ satisfying the left invertive law:

$$(a * b) * c = (c * b) * a, \text{ for any } a, b, c \in N.$$

Now, many characterizations of AG-groupoids and various special subclasses are investigated in [8–13]. As a generalization of commutative group (Abelian group) and a special case of quasigroup, Kamran extended the concept of AG-groupoid to AG-group in [14]. An AG-groupoid is called AG-group if there exists left identity and inverse, and its many properties (similar to the properties of groups) have been revealed successively in [15,16].

In this paper, we further analyze and study the structural characteristics of AG-groups, reveal the relationship between AG-groups and commutative groups, and establish filter and quotient algebra theories of AG-groups. The paper is organized as follows. Section 2 presents several basic concepts and results. Some new properties of AG-groups are investigated in Section 3, especially some examples of infinite AG-groups, and the authors prove that every weak commutative or alternative AG-group is a commutative group (Abelian group) and every AG-group is quasi-cancellative. In Section 4, two special classes of AG-groups are studied and the structure theorems are proved. In Section 5, the filter theory of AG-groups is established, the quotient structures induced by filters are constructed, and some homomorphism theorems are proved. Finally, the main results of this paper are systematically summarized via a schematic figure.

2. Preliminaries

First, we present some basic notions and properties.

A groupoid $(N, *)$ is called an AG-groupoid (Abel–Grassmann's groupoid), if for any $a, b, c \in N$, $(a*b)*c = (c*b)*a$. It is easy to verify that in an AG-groupoid $(N, *)$, the medial law holds:

$$(a*b)*(c*d) = (a*c)*(b*d), \text{ for any } a, b, c, d \in N.$$

Let $(N, *)$ be an AG-groupoid with left identity e, we have

$$a*(b*c) = b*(a*c), \text{ for any } a, b, c \in N;$$

$$(a*b)*(c*d) = (d*b)*(c*a), \text{ for any } a, b, c, d \in N.$$

$$NN = N, N*e = N = e*N.$$

An AG-groupoid $(N, *)$ is called a locally associative AG-groupoid, if it satisfies $a*(a*a) = (a*a)*a$, $\forall a \in N$.

An AG-groupoid $(N, *)$ is called an AG-band, if it satisfies $a*a = a$ $(\forall a \in N)$.

Definition 1. *([9,10]) Let $(N, *)$ be an AG-groupoid. Then, N is called to be quasi-cancellative if for any $a, b \in N$,*

$$a = a*b \text{ and } b^2 = b*a \text{ imply that } a = b; \text{ and} \tag{1}$$

$$a = b*a \text{ and } b^2 = a*b \text{ imply that } a = b. \tag{2}$$

Proposition 1. *([9,10]) Every AG-band is quasi-cancellative.*

Definition 2. *([14,15]) An AG-groupoid $(N, *)$ is called an AG-group or a left almost group (LA-group), if there exists left identity $e \in N$ (that is $e*a = a$, for all $a \in N$), and there exists $a^{-1} \in N$ such that $a^{-1}*a = a*a^{-1} = e$ $(\forall a \in N)$.*

Proposition 2. *([15]) Assume that $(N, *)$ is an AG-group. We get that $(N, *)$ is a commutative Abel–Grassmann's Group if and only if it is an associative AG-Group.*

Proposition 3. *([15]) Let $(N, *)$ be an AG-group with right identity e. Then, $(N, *)$ is an Abelian group.*

Proposition 4. *([15]) Let $(N, *)$ be an AG-group. Then, $(N, *)$ has exactly one idempotent element, which is the left identity.*

Proposition 5. *([11]) Let $(N, *)$ be an AG-groupoid with a left identity e. Then, the following conditions are equivalent,*

(1) *N is an AG-group.*
(2) *Every element of N has a right inverse.*
(3) *Every element a of N has a unique inverse a^{-1}.*
(4) *The equation $x*a = b$ has a unique solution for all $a, b \in N$.*

Proposition 6. *([16]) Let (N, *) be an AG-group. Define a binary operation ∘ as follows:*

$$x \circ y = (x * e) * y, \text{ for any } x, y \in N.$$

Then, (N, ∘) is an Abelian group, denote it by ret(N, *) = (N, ∘).

3. Some Examples and New Results of AG-Groups

In this section, we give some examples of AG-groups (including some infinite examples), and investigate the characterizations of weak commutative AG-groups, alternative AG-groups and quasi-cancellative AG-groups. Moreover, we obtain two subalgebras from arbitrary AG-group.

Example 1. *Let us consider the rotation transformations of a square. A square is rotated 90°, 180° and 270° to the right (clockwise) and they are denoted by φ_a, φ_b and φ_c, respectively (see Figure 1). There is of course the movement that does nothing, which is denoted by φ_e. The following figure gives an intuitive description of these transformations. Denote N = $\{\varphi_e, \varphi_a, \varphi_b, \varphi_c\}$.*

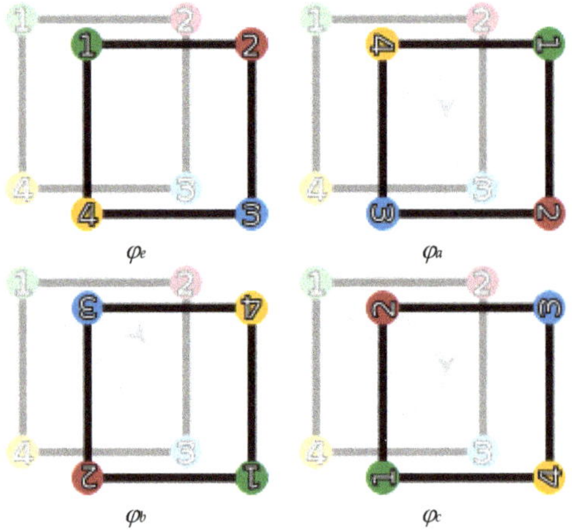

Figure 1. The rotation transformations of a square.

*Obviously, two consecutive rotations have the following results: $\varphi_e\varphi_e = \varphi_e$, $\varphi_a\varphi_c = \varphi_c\varphi_a = \varphi_e$, $\varphi_b\varphi_b = \varphi_e$. That is, ${\varphi_e}^{-1} = \varphi_e$, ${\varphi_a}^{-1} = \varphi_c$, ${\varphi_b}^{-1} = \varphi_b$, ${\varphi_c}^{-1} = \varphi_a$. Now, we define operations * on N as follows:*

$$\varphi_x * \varphi_y = {\varphi_x}^{-1}\varphi_y, \forall x, y \in \{e, a, b, c\}.$$

*Then, (N, *) satisfies the left invertive law, and the operation * is as follows in Table 1. We can verify that (N, *) is an AG-Group.*

Table 1. AG-group generated by rotation transformations of a square.

*	φ_e	φ_a	φ_b	φ_c
φ_e	φ_e	φ_a	φ_b	φ_c
φ_a	φ_c	φ_e	φ_a	φ_b
φ_b	φ_b	φ_c	φ_e	φ_a
φ_c	φ_a	φ_b	φ_c	φ_e

Example 2. *Let* $X = \{(a, b)|a, b \in \mathbf{R}-\{0\}\}$, *where* **R** *represents the set of all real numbers. Define binary operation* * *as follows:*

$$(a, b) * (c, d) = (ac, d/b), \text{ for any } (a, b), (c, d) \in X.$$

Then,

$$[(a, b) * (c, d)] * (e, f) = (ac, d/b) * (e, f) = (ace, fb/d);$$

$$[(e, f) * (c, d)] * (a, b) = (ec, d/f) * (a, b) = (ace, fb/d).$$

Therefore, $[(a, b) * (c, d)] * (e, f) = [(e, f) * (c, d)] * (a, b)$, *that is, the operation* * *satisfies left invertive law. For any* $(a, b) \in X$, $(1, 1)$ *is the left identity of* (a, b) *and* $(1/a, b)$ *is the left inverse of* (a, b)*:*

$$(1,1) * (a, b) = (a, b); (1/a, b) * (a, b) = (1, 1).$$

Therefore, $(X, *)$ *is an AG-Group.*

Example 3. *Let* $Y = \{(a, b)|a \in \mathbf{R}, b = 1 \text{ or } -1\}$, *where* **R** *represents the set of all real numbers. Define binary operation* * *as follows:*

$$(a, b) * (c, d) = (ac, b/d), \text{ for any } (a, b), (c, d) \in Y.$$

Then,

$$[(a, b) * (c, d)] * (e, f) = (ac, b/d) * (e, f) = (ace, b/df);$$

$$[(e, f) * (c, d)] * (a, b) = (ec, f/d) * (a, b) = (ace, f/bd).$$

Because $b, f \in \{1, -1\}$, $b^2 = f^2$, *and* $b/f = f/b$. *We can get* $b/df = f/bd$. *Therefore,* $[(a, b) * (c, d)] * (e, f) = [(e, f) * (c, d)] * (a, b)$, *that is, the operation* * *satisfies left invertive law. Moreover, we can verify that* $(1, 1)$ *is the left identity and* $(1/a, \pm 1)$ *is the left inverse of* $(a, \pm 1)$, *since*

$$(1, 1) * (a, b) = (a, 1/b) = (a, b); \text{ (because } b=1 \text{ or } -1)$$

$$(1/a, 1) * (a, 1) = (1, 1) \text{ and } (1/a, -1) * (a, -1) = (1, 1).$$

Therefore, $(Y, *)$ *is an AG-group.*

Example 4. *Let* $Z = \{(a, b)|a \in \mathbf{R}, b = 1, -1, i, \text{ or } -i\}$, *where* **R** *represents the set of all real numbers and I represents the imaginary unit. Define binary operation* * *as follows:*

$$(a, b) * (c, d) = (ac, b/d), \text{ for any } (a, b), (c, d) \in Z$$

Then,

$$[(a, b) * (c, d)] * (e, f) = (ac, b/d) * (e, f) = (ace, b/df);$$

$$[(e, f) * (c, d)] * (a, b) = (ec, f/d) * (a, b) = (ace, f/bd).$$

Because $b, f \in \{1, -1, i, -i\}$, *hence* $b^2 = f^2$, *and* $b/f = f/b$. *We can get* $b/df = f/bd$. *Therefore,* $[(a, b) * (c, d)] * (e, f) = [(e, f) * (c, d)] * (a, b)$, *that is, the operation* * *satisfies left invertive law. Therefore,* $(Z, *)$ *is an AG-groupoid. However, it is not an AG-group, since*

$$(1, 1) * (a, 1) = (a, 1), (1, 1) * (a, -1) = (a, -1);$$

$$(1, -1) * (a, i) = (a, i), (1, -1) * (a, -i) = (a, -i).$$

That is, $(1, 1)$ *and* $(1, -1)$ *are locally identity, not an identity.*

Definition 3. *Assume that* $(N, *)$ *is an AG-group.* $(N, *)$ *is said to be a weak commutative Abel– Grassmann's group (AG-group), if one of the following conditions holds:*

(1) $e^*x^{-1} = x^{-1}*e$, *for all x in N;*
(2) $e^*x = x^*e$, *for all x in N; or*
(3) $x^{-1}*y^{-1} = y^{-1}*x^{-1}$, *for all x, y in N.*

Theorem 1. *Let (N, *) be an AG-group. We can get that N is a weak commutative AG-group if and only if it is an Abelian group.*

Proof. First, we prove that the Conditions (1)–(3) in Definition 3 are equivalent for an AG-group (N, *).

(1)→(2): Suppose thatCondition (1) holds in the AG-group (N, *). For all x in N, by $(x^{-1})^{-1} = x$, we have $e^*(x^{-1})^{-1} = (x^{-1})^{-1}*e$, that is, $e^*x = x^*e$.

(2)→(3): Suppose that Condition (2) holds in the AG-group (N, *). For all x, y in N, by Proposition 3, we know that N is an Abelian group, that is, $x^*y = y^*x$, it follows that $x^{-1}*y^{-1} = y^{-1}*x^{-1}$.

(3)→(1): Suppose that Condition (3) holds in the AG-group (N, *). Then, for all x in N, we have $(e^{-1})^{-1}*x^{-1} = x^{-1}*(e^{-1})^{-1}$, that is, $e^*x^{-1} = x^{-1}*e$.

Now, we prove that an AG-group (N, *) satisfying Condition (2) in Definition 3 is an Abelian group. Through Condition (2), $e^*a = a^*e$ for any $a \in N$. Then, $a^*e = e^*a = a$, which means that e is right identity. Applying Proposition 3, we get that $(N, *)$ is an Abelian group. Moreover, obviously, every Abelian group is a weak commutative AG-group. Therefore, the proof is completed. □

Theorem 2. *Assume that (N, *) is an AG-group, we have that (N, *) is quasi-cancellative AG-groupoid, that is, if it satisfies the following conditions, for any x, y∈N,*

(1) $x = x * y$ *and* $y^2 = y^*x$ *imply that* $x = y$; *and*
(2) $x = y * x$ *and* $y^2 = x * y$ *imply that* $x = y$.

Proof. (1) Suppose that $x = x^*y$ and $y^2 = y^*x$, where $x, y \in N$. Then,

$$x = x * y = (e * x) * y = (y * x) * e = y^2 * e = (e * y) * y = y^2. \quad \text{(a)}$$

That is, $x = y^2$; it follows that $x^*y = y^*x$. Moreover, we have

$$y * e = y * (x^{-1} * x) = (e * y) * (x^{-1} * x) = (e * x^{-1}) * (y * x) = x^{-1} * (y * x) = x^{-1} * x = e. \quad \text{(b)}$$

$$\begin{aligned} x * e &= (x * y) * e = (x * y) * (x^{-1} * x) = (x * x^{-1}) * (y * x) = (x * x^{-1}) * y^2 = (x * x^{-1}) * (y * y) \\ &= (x * y) * (x^{-1} * y) = x * (x^{-1} * y) = (e * x) * (x^{-1} * y) = (e * x^{-1}) * (x * y) \\ &= x^{-1} * (x * y) = x^{-1} * x = e. \end{aligned} \quad \text{(c)}$$

Combining Equations (b) and (c), we can get

$$x = e^*x = (y^*e)^*x = (x^*e)^*y = e^*y = y.$$

(2) Suppose that $x = y^*x$ and $y^2 = x^*y$, where $x, y \in N$. Then,

$$x = y * x = (e * y) * (x * y) * e = y^2 * e = (e * y) * y = y^2 * e = (e * y) * y = y^2. \quad \text{(d)}$$

That is, $x = y^2$; it follows that $x^*y = y^*x$. Then, we have

$$y * e = y * (x^{-1} * x) = (e * y) * (x^{-1} * x) = (e * x^{-1}) * (y * x) = (e * x^{-1}) * x = e. \quad \text{(e)}$$

$$x * e = x * (y^{-1} * y) = (e * x) * (y^{-1} * y) = y^{-1} * (x * y) = y^{-1} * y^2 = y^{-1} * (y * y) = y * (y^{-1} * y) = y * e = e. \quad \text{(f)}$$

Combining Equations (e) and (f), we can get

$$x = e^*x = (y^*e)^*x = (x^*e)^*y = e^*y = y.$$

Hence, $(N, *)$ is quasi-cancellative AG-groupoid. □

Definition 4. *Let $(N, *)$ be an AG-group. Then, $(N, *)$ is called to be alternative, if it satisfies one of the following conditions,*

(1) $a*(a*b) = (a*a)*b, \forall a, b \in N$; *or*
(2) $a*(b*b) = (a*b)*b, \forall a, b \in N$.

Theorem 3. *Let $(N, *)$ be an AG-group. Then, $(N, *)$ is alternative if and only if it is an Abelian group.*

Proof. (1) Suppose that $(N, *)$ is an alternative AG-group, then Condition (2) in Definition 4 holds. Then, for any $a, b \in N$, $a*(b*b) = (a*b)*b$. Putting $b = e$ and applying left invertive law, we get that $a*e = a*(e*e) = (a*e)*e = (e*e)*a = e*a = a$; by Proposition 3, we know that $(N, *)$ is an Abelian group.

(2) Suppose that $(N, *)$ is an alternative AG-group, then Condition (1) in Definition 4 holds. For any $a, b \in N$, it satisfies $a*(a*b) = (a*a)*b$. Putting $b = e$, we have $(a*a)*e = a*(a*e)$. According to the arbitrariness of a, we can get that

$$((a*e)*(a*e))*e = (a*e)*((a*e)*e).$$

Then,

$$a*a = (e*a)*a = (a*a)*e = ((a*a)*(e*e))*e = ((a*e)*(a*e))*e = (a*e)*((a*e)*e) = (a*e)*a.$$

Let $b*a = e$, using Condition (1) in Definition 4, $(a*a)*b = a*(a*b)$. It follows that $(a*a)*b = ((a*e)*a)*b$. Thus,

$$a = e*a = (b*a)*a = (a*a)*b = ((a*e)*a)*b = (b*a)*(a*e) = e*(a*e) = a*e.$$

Applying Proposition (3), we know that $(N, *)$ is an Abelian group.

Conversely, it is obvious that every Abelian group is an alternative AG-group. Therefore, the proof is completed. □

Theorem 4. *Let $(N, *)$ be an AG-group. Denote*

$$U(N) = \{x \in N | \ x = x*e\}.$$

Then,

(1) *$U(N)$ is sub-algebra of N.*
(2) *$U(N)$ is maximal subgroup of N with identity e.*

Proof. (1) Obviously, $e \in U(N)$, that is, $U(N)$ is not empty. Suppose $x, y \in U(N)$, then $x*e = x$ and $y*e = y$. Thus, $x*y = (x*e)*(y*e) = (x*y)*e \in U(N)$. This means that $U(N)$ is a subalgebra of N.

(2) For any $x \in U(N)$, that is, $x*e = x$. Assume that y is the left inverse of x in N, then $y*x = e$. Thus,

$$x*y = (e*x)*y = ((y*x)*x)*y = (y*x)*(y*x) = e*e = e,$$

$$y = e*y = (x*y)*y = ((x*e)*y)*y = ((y*e)*x)*y = (y*x)*(y*e) = e*(y*e) = y*e.$$

It follows that $y \in U(N)$. Therefore, $U(N)$ is a group, and it is a subgroup of N with identity e. If M is a subgroup of N with identity e, and $U(N) \subseteq M$, then M is an Abelian group (by Proposition (3)) and satisfies $x*e = e*x = x$, for any $x \in M$. Thus, $M \subseteq U(N)$, it follows that $M = U(N)$. Therefore, $U(N)$ is maximal subgroup of N with identity e. □

Theorem 5. *Let (N, *) be an AG-group. Denote* $P(N) = \{x \in N | \exists a \in N, s.t\ x = a^*a\}$. *Then*

(1) *P(N) is the subalgebra of N;*
(2) *f is a homomorphism mapping from N to P(N), where* $f\colon N \to P(N), f(x)=x^*x \in P(N)$.

Proof. (1) Obviously, $e \in P(N)$, that is, $P(N)$ is not empty. Suppose $x, y \in P(N)$ and $a, b \in N$. Then, $a^*a = x$ and $b^*b = y$. Thus, $x^*y = (a^*a)^*(b^*b) = (a^*b)^*(a^*b) \in P(N)$. This means that $P(N)$ is a subalgebra of N.

(2) For any $x, y \in N$, we have

$$f(x^*y) = (x^*y)^*(x^*y) = (x^*x)^*(y^*y) = f(x)^*f(y).$$

Therefore, f is a homomorphism mapping from N to $P(N)$. □

4. Involution AG-Groups and Generalized Involution AG-Groups

In this section, we discuss two special classes of AG-groups, that is, involution AG-groups and generalized involution AG-groups. Some research into the involutivity in AG-groupoids is presented in [16,17] as the foundation, and further results are given in this section, especially the close relationship between these algebraic structures and commutative groups (Abelian groups), and their structural characteristics.

Definition 5. *Let (N, *) be an AG-group. If (N, *) satisfies* $a^*a = e$, *for any* $a \in N$, *then (N, *) is called an involution AG-Group.*

*We can verify that (N, *) in Example 1 is an involution AG-Group.*

Example 5. *Denote N = {a, b, c, d}, define operations * on N as shown in Table 1. We can verify that (N, *) is an involution AG-group (Table 2).*

Table 2. Involution AG-group (N, *).

*	a	b	c	d
a	a	b	c	d
b	b	a	d	c
c	d	c	a	b
d	c	d	b	a

Example 6. *Let (G, +) be an Abelian group. Define operations * on G as follows:*

$$x^*y = (-x) + y, \forall x, y \in G$$

where $(-x)$ *is the inverse of x in G. Then, (G, *) is an involution AG-group. Denote (G, *) by der (G, +) (see [15]), and call it derived AG-group by Abelian group (G, +).*

Theorem 6. *Let (N, *) be an AG-group. Then, (N, *) is an involution AG-Group if and only if it satisfies one of the following conditions:*

(1) $P(N) = \{e\}$, *where P(N) is defined as Theorem 5.*
(2) $(x^*x)^*x = x$ *for any* $x \in N$.

Proof. Obviously, (N, *) is an involution AG-group if and only if $P(N) = \{e\}$.

If (N, *) is an involution AG-group, then apply Definition 5, for any $x \in N$,

$$(x^*x)^*x = e^*x = x.$$

Conversely, if $(N, *)$ satisfies the Condition (2), then for any $x \in N$,

$$(x*x)*(x*x) = ((x*x)*x)*x = x*x.$$

This means that $(x*x)$ is an idempotent element. Using Proposition 4, we have $x*x = e$. Thus, $(N, *)$ is an involution AG-group. □

Theorem 7. *Let $(N, *)$ be an involution AG-group. Then, $(N, \circ) = ret\ (N, *)$ defined in Proposition 6 is an Abelian group, and the derived AG-group der $(N, \circ)$ by ret $(N, *)$ (see Example 5) is equal to $(N, *)$, that is,*

$$der(ret(N, *)) = (N, *).$$

Proof. (1) By Proposition 6 and Definition 5, $\forall x, y, z \in N$, we can get that

$$x \circ y = y \circ x;\ x \circ e = e \circ x = x;\ (x \circ y) \circ z = x \circ (y \circ z);\ x \circ x^{-1} = x^{-1} \circ x = e.$$

This means that $(N, \circ) = \text{ret}(N, *)$ is an Abelian group.

(2) For any $x, y \in \text{der}(\text{ret}(N, *)) = \text{der}(N, \circ) = (N, \bullet)$,

$$x \bullet y = (-x) \circ y = ((-x)*e)*y = ((x*e)*e)*y = ((e*e)*x)*y = (e*x)*y = x*y.$$

That is, $\text{der}(\text{ret}(N, *)) = (N, \bullet) = (N, *)$. □

Definition 6. *Let $(N, *)$ be an AG-group. Then, $(N, *)$ is called a generalized involution AG-group if it satisfies: for any $x \in N$, $(x*x)*(x*x) = e$.*

Obviously, every involution AG-group is a generalized involution AG-group. The inverse is not true, see the following example.

Example 7. *Denote $N = \{e, a, b, c\}$, and define the operations $*$ on N as shown in Table 3. We can verify that $(N, *)$ is a generalized involution AG-group, but it is not an involution AG-group.*

Table 3. Generalized involution AG-group $(N, *)$.

*	e	a	b	c
e	e	a	b	c
a	a	e	c	b
b	c	b	a	e
c	b	c	e	a

Theorem 8. *Let $(N, *)$ be a generalized involution AG-group. Define binary relation $\approx$ on N as follows:*

$$x \approx y \Leftrightarrow x * x = y * y,\ \text{for any}\ x, y \in N.$$

Then,

(1) *$\approx$ is an equvalent relation on N, and we denote the equivalent class contained x by $[x]_{\approx}$.*
(2) *The equivalent class contained e by $[e]_{\approx}$ is an involution sub-AG-group.*
(3) *For any $x, y, z \in N$, $x \approx y$ implies $x*z \approx y*z$ and $z*x \approx z*y$.*
(4) *The quotient $(N/\approx, *)$ is an involution AG-group.*

Proof. (1) For any $a \in N$, we have $a^*a = a^*a$, thus $a \approx a$.

If $a \approx b$, then $a^*a = b^*b$; it is obvious that $b \approx a$.

If $a \approx b$ and $b \approx c$, then $a^*a = b^*b$ and $b^*b = c^*c$; it is obvious that $a^*a = c^*c$, that is, $a \approx c$.

Therefore, $\approx$ is an equivalent relation on N.

(2) $\forall x, y \in [e]_{\approx}$, we have $x^*x = y^*y = e^*e = e$, thus

$$(x^*y)^*(x^*y) = (x^*x)^*(y^*y) = e^*e = e.$$

This means that $[e]_{\approx}$ is a subalgebra of N. Thus, $[e]_{\approx}$ is an involution sub-AG-group of N.

(3) Assume that $x \approx y$, then $x^*x = y^*y$. Thus,

$$(x^*z)^*(x^*z) = (x^*x)^*(z^*z) = (y^*y)^*(z^*z) = (y^*z)^*(y^*z);$$

$$(z^*x)^*(z^*x) = (z^*z)^*(x^*x) = (z^*z)^*(y^*y) = (z^*y)^*(z^*y).$$

It follows that $x^*z \approx y^*z$ and $z^*x \approx z^*y$.

(4) By (3), we know that $(N/\approx, ^*)$ is an AG-group. Moreover, for any

$$x \in [a]_{\approx} * [a]_{\approx} = [a*a]_{\approx},\ x * x = (a * a) * (a * a)$$

By Definition 6,

$$(a^*a)^*(a^*a) = e.$$

Then,

$$x * x = e \text{ for any } x \in [a*a]_{\approx}.$$

From this, we have $x \in [e]_{\approx}$, $[a*a]_{\approx} \subseteq [e]_{\approx}$. Hence, $[a*a]_{\approx} = [e]_{\approx}$. That is, $[a]_{\approx} * [a]_{\approx} = [e]_{\approx}$. Therefore, $(N/\approx, ^*)$ is an involution AG-group. □

Theorem 9. *Let $(N, ^*)$ be an AG-group, denote*

$$I(N) = \{x \in N | x^*x = e\},\ GI(N) = \{x \in N | (x^*x)^*(x^*x) = e\}.$$

Then, $I(N)$ and $GI(N)$ are sub-algebra of N. $I(N)$ is an involution AG-group and $GI(N)$ is a generalized involution AG-group.

Proof. (1) It is obvious that $e \in I(N)$. For any $x, y \in I(X)$, we have $x^*x = e$ and $y^*y = e$. By medial law, $(x^*y)^*(x^*y) = (x^*x)^*(y^*y) = e^*e = e$. Hence, $I(N)$ is a sub-algebra of N and $I(N)$ is an involution AG-group.

(2) Obviously, $e \in GI(N)$. Assume that $x, y \in GI(X)$, then

$$(x^*x)^*(x^*x) = (y^*y)^*(y^*y) = e.$$

Thus,

$$((x^*y)^*(x^*y))^*((x^*y)^*(x^*y)) = ((x^*x)^*(y^*y))^*((x^*x)^*(y^*y)) = ((x^*x)^*(x^*x))^*((y^*y)^*(y^*y)) = e^*e = e.$$

It follows that $x^*y \in GI(N)$, and $GI(N)$ is a subalgebra of N. Moreover, from $((x^*x)^*x)^*x = (x^*x)^*(x^*x) = e$, we get that $a = (x^*x)^*x$ is the left inverse of x, and

$$(a^*a)^*(a^*a) = (((x^*x)^*x)^*((x^*x)^*x))^*(a^*a) = ((x^*x)^*(x^*x))^*(x^*x))^*(a^*a) = (e^*(x^*x))^*(a^*a) = (x^*x)^*(a^*a) = (x^*x)^*(x^*x) = e.$$

That is, $a = (x^*x)^*x \in GI(N)$. It follows that $GI(N)$ is an AG-group. By the definition of $GI(N)$, we get that $GI(N)$ is a generalized involution AG-group. □

5. Filter of AG-Groups and Homomorphism Theorems

Definition 7. *Let (N, *) be an AG-group. A non-empty subset F of N is called a filter of N if, for all x, y∈N, F satisfies the following properties,*

(1) $e \in F$;
(2) $x^*x \in F$; *and*
(3) $x \in F$ and $x^*y \in F$ imply that $y \in F$.

If F is a filter and subalgebra of N, then F will be called a closed filter of N.

Theorem 10. *Let (N, *) be a generalized involution AG-group, $I(N) = \{x \in N \mid e = x^*x\}$ be the involution part of N (see Theorem 9). Then, I(N) is a closed filter of N.*

Proof. It is obvious that $e \in I(N)$. $\forall x \in N$, since

$$(x^*x)^*(x^*x) = e,$$

then $x^*x \in I(N)$. Moreover, assuming that $x \in I(N)$ and $x^*y \in I(N)$, then

$$e = x^*x, (x^*y)^*(x^*y) = e.$$

Thus,

$$y^*y = e^*(y^*y) = (x^*x)^*(y^*y) = (x^*y)^*(x^*y) = e.$$

Hence, $y \in I(N)$, and $I(N)$ is a filter of N. By Theorem 9, $I(N)$ is a subalgebra of N. Therefore, $I(N)$ is a closed filter of N. □

Theorem 11. *Let (N, *) be an AG-group and F be a closed filter of N. Define binary relation $\approx_F$ on N as follows:*

$$x \approx_F y \Leftrightarrow (x * y \in F, y * x \in F),\ for\ any\ x, y\ \text{in}\ N.$$

Then,

(1) $\approx_F$ *is an equivalent relation on N.*
(2) $x \approx_F y$ *and* $a \approx_F b$ *imply* $x^*a \approx_F y^*b$.
(3) *f: N→N/F is a homomorphism mapping, where* $N/F = \{[x]_F: x \in N\}$, $[x]_F$ *denote the equivalent class contained x.*

Proof. (1) $\forall x \in N$, by Definition 7(2), $x^*x \in F$. Thus, $x \approx_F x$.

Assume $x \approx_F y$, then $x^*y \in F$, $y^*x \in F$. It follows that $y \approx_F x$.

Suppose that $x \approx_F y$ and $y \approx_F z$. We have $x^*y \in F$, $y^*x \in F$, $y^*z \in F$ and $z^*y \in F$. By medial law and Definition 7,

$$(y * y) * (z * x) = (y * z) * (y * x) \in F, \text{ then } (z * x) \in F;$$

$$(y * y) * (x * z) = (y * x) * (y * z) \in F, \text{ then } (x * z) \in F.$$

It follows that $x \approx_F z$.

Therefore, $\approx_F$ is an equivalent relation on N.

(2) Suppose that $x \approx_F y$ and $a \approx_F b$. We have $x^*y \in F$, $y^*x \in F$, $a^*b \in F$ and $b^*a \in F$. By medial law and Definition 7,

$$(x * a) * (y * b) = (x * y) * (a * b) \in F;\ (y * b) * (x * a) = (y * x) * (b * a) \in F.$$

It follows that $x^*a \approx_F y^*b$.

(3) Combining (1) and (2), we can obtain (3).

The proof complete. □

Theorem 12. *Let (N, *) be a generalized involution AG-group, I(N) the involution part of N (defined as Theorem 9). Then, f: N→N/I(N) is a homomorphism mapping, and N/I(N) is involutive, where N/I(N) = {[x]| x∈N}, [x] is the equivalent class contained x by closed filter I(N).*

Proof. It follows from Theorem 10 and Theorem 11. □

Theorem 13. *Let (N, *) be an AG-group, P(N) = {x∈N|∃a∈N, s.t x =a*a} be the power part of N (see Theorem 5). Then, P(N) is a closed filter of N.*

Proof. It is obvious that $e = e*e \in P(N)$. For any $x \in N$, $x*x \in P(N)$.

Moreover, assume that $x \in P(N)$ and $x*y \in P(N)$, then there exists $a, b \in N$ such that

$$x = a*a,\ x*y = b*b.$$

Denote $c = a^{-1}*b$, where a^{-1} is the left inverse of a in N. Then,

$$c*c = (a^{-1}*b)*(a^{-1}*b) = (a^{-1}*a^{-1})*(b*b) = (a^{-1}*a^{-1})*(x*y) = (a^{-1}*a^{-1})*((a*a)*y) = (a^{-1}*a^{-1})*((y*a)*a) = (a^{-1}*(y*a))*(a^{-1}*a) = (a^{-1}*(y*a))*e = (e*(y*a))*a^{-1} = (y*a)*a^{-1} = (a^{-1}*a)*y = e*y = y.$$

Thus, $y \in P(N)$. It follows that $P(N)$ is a filter of N. By Theorem 5, $P(N)$ is a subalgebra of N, therefore, $P(N)$ is a closed filter of N. □

Theorem 14. *Let (N, *) be an AG-group, P(N) the power part of N (defined as Theorem 13). Then, f: N→N/P(N) is a homomorphism mapping, where N/P(N) = {[x]| x∈N}, [x] is the equivalent class contained x by closed filter P(N).*

Proof. It follows from Theorems 11 and 13. □

6. Conclusions

In the paper, we give some examples of AG-groups, and obtain some new properties of AG-groups: an AG-group is weak commutative (or alternative) if and only if it is an Abelian group; every AG-group is a quasi-cancellative AG-groupoid. We introduce two new concepts of involution AG-group and generalized involution AG-group, establish a one-to-one correspondence between involution AG-groups and Abelian groups, and construct a homomorphism mapping from generalized involution AG-groups to involution AG-groups. Moreover, we introduce the notion of filter in AG-groups, establish quotient algebra by every filter, and obtain some homomorphism theorems. Some results in this paper are expressed in Figure 2. In the future, we can investigate the combination of some uncertainty set theories (fuzzy set, neutrosophic set, etc.) and algebra systems (see [18–22]).

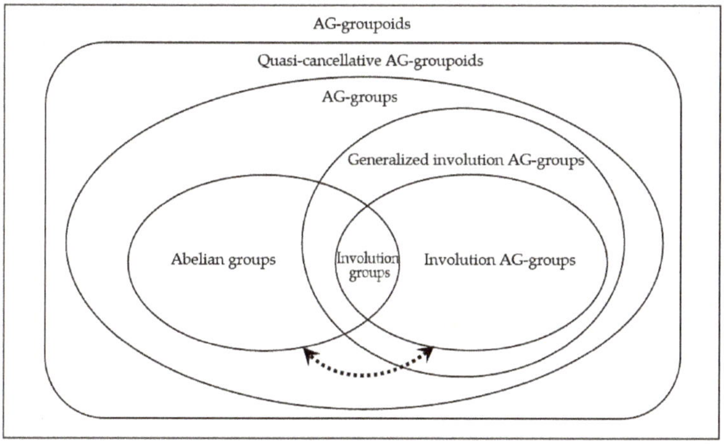

Figure 2. Some results in this paper.

Author Contributions: X.Z. and X.W. initiated the research; and X.Z. wrote final version of the paper.

Funding: This research was funded by National Natural Science Foundation of China grant number 61573240.

Conflicts of Interest: The authors declare no conflict of interest.

References

1. Hall, G.G. *Applied Group Theory*; merican Elsevier Publishing Co., Inc.: New York, NY, USA, 1967.
2. Howie, J.M. *Fundamentals of Semigroup Theory*; Clarendon Press: Oxford, UK, 1995.
3. Lawson, M.V. *Inverse Semigroups: The Theory of Partial Symmetries*; World Scientific: Singapore, 1998.
4. Akinmoyewa, T.J. A study of some properties of generalized groups. *Octogon Math. Mag.* **2009**, *7*, 599–626.
5. Smarandache, F.; Ali, M. Neutrosophic triplet group. *Neural Comput. Appl.* **2018**, *29*, 595–601. [CrossRef]
6. Kazim, M.A.; Naseeruddin, M. On almost semigroups. *Port. Math.* **1977**, *36*, 41–47.
7. Kleinfeld, M.H. Rings with *x(yz) = z(yx)*. *Commun. Algebra* **1978**, *6*, 1369–1373. [CrossRef]
8. Mushtaq, Q.; Iqbal, Q. Decomposition of a locally associative LA-semigroup. *Semigroup Forum* **1990**, *41*, 155–164. [CrossRef]
9. Ahmad, I.; Rashad, M. Constructions of some algebraic structures from each other. *Int. Math. Forum* **2012**, *7*, 2759–2766.
10. Ali, A.; Shah, M.; Ahmad, I. On quasi-cancellativity of AG-groupoids. *Int. J. Contemp. Math. Sci.* **2012**, *7*, 2065–2070.
11. Dudek, W.A.; Gigon, R.S. Completely inverse AG**-groupoids. *Semigroup Forum* **2013**, *87*, 201–229. [CrossRef]
12. Zhang, X.H.; Wu, X.Y.; Mao, X.Y.; Smarandache, F.; Park, C. On neutrosophic extended triplet groups (loops) and Abel-Grassmann's groupoids (AG-groupoids). *J. Intell. Fuzzy Syst.* **2019**, in press.
13. Wu, X.Y.; Zhang, X.H. The decomposition theorems of AG-neutrosophic extended triplet loops and strong AG-(*l,l*)-loops. *Mathematics* **2019**, *7*, 268. [CrossRef]
14. Kamran, M.S. Conditions for LA-Semigroups to Resemble Associative Structures. Ph.D. Thesis, Quaid-i-Azam University, Islamabad, Pakistan, 1993.
15. Shah, M.; Ali, A. Some structure properties of AG-groups. *Int. Math. Forum* **2011**, *6*, 1661–1667.
16. Protić, P.V. Some remarks on Abel-Grassmann's groups. *Quasigroups Relat. Syst.* **2012**, *20*, 267–274.
17. Mushtaq, Q. Abelian groups defined by LA-semigroups. *Stud. Sci. Math. Hung.* **1983**, *18*, 427–428.
18. Ma, Y.C.; Zhang, X.H.; Yang, X.F.; Zhou, X. Generalized neutrosophic extended triplet group. *Symmetry* **2019**, *11*, 327. [CrossRef]
19. Zhang, X.H.; Bo, C.X.; Smarandache, F.; Park, C. New operations of totally dependent-neutrosophic sets and totally dependent-neutrosophic soft sets. *Symmetry* **2018**, *10*, 187. [CrossRef]

20. Zhang, X.H.; Mao, X.Y.; Wu, Y.T.; Zhai, X.H. Neutrosophic filters in pseudo-BCI algebras. *Int. J. Uncertain. Quan.* **2018**, *8*, 511–526. [CrossRef]
21. Zhang, X.H.; Borzooei, R.A.; Jun, Y.B. Q-filters of quantum B-algebras and basic implication algebras. *Symmetry* **2018**, *10*, 573. [CrossRef]
22. Zhan, J.M.; Sun, B.Z.; Alcantud, J.C.R. Covering based multigranulation (I, T)-fuzzy rough set models and applications in multi-attribute group decision-making. *Inf. Sci.* **2019**, *476*, 290–318. [CrossRef]

symmetry

MDPI

Article

Isoperimetric Numbers of Randomly Perturbed Intersection Graphs

Yilun Shang

Department of Computer and Information Sciences, Faculty of Engineering and Environment, Northumbria University, Newcastle NE1 8ST, UK; shylmath@hotmail.com; Tel.: +44-019-1227-3562

Received: 2 February 2019; Accepted: 27 March 2019; Published: 1 April 2019

Abstract: Social networks describe social interactions between people, which are often modeled by intersection graphs. In this paper, we propose an intersection graph model that is induced by adding a sparse random bipartite graph to a given bipartite graph. Under some mild conditions, we show that the vertex–isoperimetric number and the edge–isoperimetric number of the randomly perturbed intersection graph on n vertices are $\Omega(1/\ln n)$ asymptomatically almost surely. Numerical simulations for small graphs extracted from two real-world social networks, namely, the board interlocking network and the scientific collaboration network, were performed. It was revealed that the effect of increasing isoperimetric numbers (i.e., expansion properties) on randomly perturbed intersection graphs is presumably independent of the order of the network.

Keywords: isoperimetric number; random graph; intersection graph; social network

1. Introduction

Complex large-scale network structures arise in a variety of natural and technological settings [1,2], and they pose numerous challenges to computer scientists and applied mathematicians. Many interesting ideas in this area come from the analysis of social networks [3], where each vertex (*actor*) is associated with a set of properties (*attributes*), and pairs of sets with nonempty intersections correspond to edges in the network. Complex and social networks represented by such *intersection graphs* are copious in the real world. Well-known examples include the film actor network [4], where actors are linked by an edge if they performed in the same movie, the academic co-authorship network [5], where two researchers are linked by an edge if they have a joint publication, the circle of friends in online social networks (e.g., Google+), where two users are declared adjacent if they share a common interest, and the Eschenauer–Gligor key predistribution scheme [6] in secure wireless sensor networks, where two sensors establish secure communication over a link if they have at least one common key. Remarkably, it was shown in Reference [7] that all graphs are indeed intersection graphs.

To understand statistical properties of intersection graphs, a probability model was introduced in References [8,9] as a generalization of the classical model $G(n,p)$ of Erdős and Rényi [10]. Formally, let n,m be positive integers and let $p \in [0,1]$. We start with a random bipartite graph $B(n,m,p)$ with independent vertex sets $V = \{v_1, \cdots, v_n\}$ and $W = \{w_1, \cdots, w_m\}$ and edges between V and W existing independently with probability p. In terms of social networks, V is interpreted as a set of actors and W a set of attributes. We then define the *random intersection graph* $G(n,m,p)$ with vertex set V and vertices $v_i, v_j \in V$ adjacent if and only if there exists some $w \in W$ such that both v_i and v_j are adjacent to w in $B(n,m,p)$. Several variant models of random intersection graphs have been proposed, and many graph-theoretic properties of $G(n,m,p)$, such as degree distribution, connected components, fixed subgraphs, independence number, clique number, diameter, Hamiltonicity and clustering, have been extensively studied [8,9,11–14]. We refer the reader to References [15,16] for an updated review of recent results in this prolific field.

In light of the above list of properties studied, it is, perhaps, surprising that there has been little work regarding *isoperimetric numbers* of random intersection graphs. The isoperimetric numbers, which measure the expansion properties of a graph (see Section 2 below for precise definitions), have a long history in random graph theory [17–19] and are strongly related to the graph spectrum and expanders [20]. They have found a wide range of applications in theoretical computer science, including algorithm design, data compression, rapid mixing, error correcting codes, and robust computer networks [21]. Social networks such as co-authorship networks are commonly believed to have poor expansion properties (i.e., small isoperimetric numbers), which indicate the existence of bottlenecks (e.g., cuts with small size) inside the networks, because of their modular and community organization [22,23]. In this paper, we hope to show that it is possible to increase the isoperimetric numbers by a gentle perturbation of the original bipartite graph structure underlying the intersection graphs.

In recent times, there has been an effort to study the effect of random perturbation on graphs. The most mathematically famous example is perhaps the Newman–Watts small-world network [1,24], which is a random instance obtained by adding random edges to a cycle, exhibiting short average distance and high clustering coefficient, namely, the so-called small-world phenomenon. A random graph model $G \cup R$ [25] with general connected base graph G on n vertices and R being a sparse Erdős-Rényi random graph $G(n, \varepsilon/n)$ where $\varepsilon > 0$ is some small constant has been introduced in [26], and its further properties, such as connectivity, fixed subgraphs, Hamiltonicity, diameter, mixing time, vertex and edge expansion, have been intensively examined; see, e.g., [27–34] and references therein. For instance, in Reference [29], a necessary condition for the base graph is given under which the perturbed graph $G \cup R$ is an expander a.a.s. (asymptotically almost surely); for a connected base graph G, it is shown in Reference [30] that, a.a.s. the perturbed graph has an edge–isoperimetric number $\Omega(1/\ln n)$, diameter $O(\ln n)$, and vertex–isoperimetric number $\Omega(1/\ln n)$, where for the last property G is assumed to have bounded maximum degree. Here, we say that $G \cup R$ possesses a graph property $\mathcal{P}$ asymptotically almost surely, or a.a.s. for brevity, if the probability that $G \cup R$ possesses $\mathcal{P}$ tends to 1 as n goes to infinity. In this paper, to go a step further in this line of research, we investigate the bipartite graph type perturbation, where random edges are only added to the base (bipartite) graph between the two independent sets. We provide lower bounds for the isoperimetric numbers of random intersection graphs induced by such perturbations.

The rest of the paper is organized as follows. In Section 2, we state and discuss the main results, with proofs relegated to Section 4. In Section 3, we give numerical examples based upon real network data, complementing our theoretical results in small network sizes. Section 5 contains some concluding remarks.

2. Results

Let $G = (V, E)$ be a graph with vertex set V and edge set E. If $S \subseteq V$ is a set of vertices, then $\partial_G S$ denotes the set of edges of G having one end in S and the other end in $V \backslash S$. Given $S \subseteq V$, write $G[S]$ for the subgraph of G induced by S. We use $N_G(S)$ to denote the collection of vertices of $V \backslash S$ which are adjacent to some vertex of S. For a vertex $v \in V$, $N_G(v)$ is the neighborhood of v, and we denote by $N_G^2(v) = N_G(N_G(v))$ the second neighborhood of v. The above subscript G will be omitted when no ambiguity may arise. For a graph G, its edge–isoperimetric number, $c(G)$ (also called its Cheeger constant), is given by:

$$c(G) = \min_{\substack{S \subseteq V \\ 0 < |S| \le |V|/2}} \frac{|\partial_G S|}{|S|}.$$

The vertex–isoperimetric number of G, $\iota(G)$, can be defined similarly as:

$$\iota(G) = \min_{\substack{S \subseteq V \\ 0 < |S| \le |V|/2}} \frac{|N_G(S)|}{|S|}.$$

It is well-known that $c(G)/\Delta(G) \leq \iota(G) \leq c(G)$ [35], where $\Delta(G)$ is the maximum degree of G.

We will consider the following model of randomly perturbed intersection graphs. Given a fixed bipartite graph $B = B(V, W, E)$ with two independent vertex sets V ($|V| = n$) and W ($|W| = m$), the intersection graph derived from B is denoted by $G(B)$. That is, $G(B)$ is a graph on the vertex set V with two vertices adjacent if they have a common neighbor in B. For each pair of vertices $v \in V$ and $w \in W$, we add the edge $\{v, w\}$ to B independently with probability p. The resulting bipartite graph, denoted $B \cup R$, can be viewed as the union of B and a bipartite graph $R \sim B(n, m, p)$, meaning that R is a random graph distributed according to $B(n, m, p)$. We write $G(B \cup R)$, the intersection graph derived from $B \cup R$. Clearly, if the base graph $B(V, W, E)$ is taken to be the empty bipartite graph, our model $G(B \cup R)$ reduces to the random intersection graph $G(n, m, p)$.

Throughout the paper, the standard Landau asymptotic notations will be utilized (see, e.g., [10]). Let $\lfloor \cdot \rfloor$ be the round-down operator. As customary in the theory of random intersection graphs, we take $m = \lfloor n^\alpha \rfloor$ for a fixed real $\alpha \in (0, \infty)$, which allows for a natural progression from sparse to dense graphs. Recall that we say that $G(B \cup R)$ possesses a graph property $\mathcal{P}$ a.a.s. if the probability that $G(B \cup R)$ possesses $\mathcal{P}$ tends to 1 as n goes to infinity.

We are now ready to formulate the main results of this paper.

Theorem 1. *Let $B = B(V, W, E)$ be a bipartite graph with $|V| = n$ and $|W| = m = \lfloor n^\alpha \rfloor$ such that any two vertices in V are connected by a path and $\Delta := \max_{v \in V} N_B^2(v)$ is a constant (i.e., independent of n). For any $\varepsilon > 0$, let $R \sim B(n, m, p)$ with $p = \varepsilon/n$ if $\alpha \leq 1$ and $p = \varepsilon/\sqrt{nm}$ if $\alpha > 1$. Then there exists some constant $\delta > 0$ satisfying $\iota(G(B \cup R)) \geq \delta/\ln n$ a.a.s.*

A couple of remarks are in order.

Remark 1. *The local effects of the perturbation are quite mild, as a small ε is of interest. Nonetheless, the global influence on the vertex–isoperimetric number can be prominent. To see this, note that any connected (intersection) graph G has $\iota(G) = \Omega(1/n)$. In particular, if G is a tree, we have $\iota(G) = \Theta(1/n)$ (see e.g., [36]).*

Remark 2. *It is easy to check that the maximum degree of $G(B)$ is Δ. In fact, $v \in V$ and $v_1 \in V$ are adjacent in $G(B)$ if and only if they have a common neighbor $w \in W$, namely, $w \in N_B(v)$ and $v_1 \in N_B(w)$. Hence, the degree of v is $N_B(N_B(v))$. The assumption that Δ is a constant cannot be removed in general. Indeed, when $\alpha \geq 1$, consider the bipartite graph $B(V, W, E)$ with $V = \{v_1, \cdots, v_n\}$, $W = \{w_1, \cdots, w_m\}$, and the edge set $E = \{\{v_1, w_i\}, \{v_j, w_{j-1}\} | i = 1, \cdots, n-1, j = 2, \cdots, n\}$. It is clear that $G(B)$ is a star with center v_1 over the vertex set V. There are no more than $n^2 p$ edges over $V \backslash \{v_1\}$ in the graph $G(B \cup R)$, which covers at most $2n^2 p$ vertices. In $G(B \cup R)$, there will be an independent set S (meaning that $G(B \cup R)[S]$ is empty) of order at least:*

$$n - 2n^2 p = n\left(1 - 2\varepsilon\sqrt{\frac{n}{m}}\right)$$

and $N_{G(B \cup R)}(S) = 1$. Therefore, $\iota(G(B \cup R)) \leq 1/(n(1 - 2\varepsilon\sqrt{n/m})) = O(1/n)$. When $\alpha < 1$, consider the bipartite graph $B(V, W, E)$ with the edge set $E = \{\{v_1, w_i\}, \{v_j, w_{j-1}\}, \{v_l, w_m\} | i = 1, \cdots, m$, $j = 2, \cdots, m+1, l = m+2, \cdots, n\}$. Then $G(B)$ can be thought of as the joining of a star $K_{1,m}$ having center v_1 and a complete graph K_{n-m+1} by identifying v_1 with any vertex of K_{n-m+1}. After adding $nm\varepsilon/n = m\varepsilon$ edges to B, in $G(B \cup R)$, there will be an independent set S of order at least $m - 1 - 2m\varepsilon$ and $N_{G(B \cup R)}(S) = 1$. Therefore, $\iota(G(B \cup R)) \leq 1/(m - 1 - 2m\varepsilon) = O(1/m)$.

Recall that the inequality $c(G) \geq \iota(G)$ holds for any graph G. Therefore, a direct corollary of Theorem 1 reads $c(G(B \cup R)) \geq \delta/\ln n$ a.a.s. for some $\delta > 0$. The following theorem shows that this lower bound for edge–isoperimetric number actually holds without any assumption on Δ.

Theorem 2. *Let $B = B(V, W, E)$ be a bipartite graph with $|V| = n$ and $|W| = m = \lfloor n^{\alpha} \rfloor$ such that any two vertices in V are connected by a path. For any $\varepsilon > 0$, let $R \sim B(n, m, p)$ with $p = \varepsilon/\sqrt{nm}$. Then there exists some constant $\delta > 0$ satisfying $c(G(B \cup R)) \geq \delta/(1 + \ln n)$ a.a.s.*

Theorems 1 and 2 hold in the sense of large n limit. In the next section, we shall demonstrate that the isoperimetric numbers can be improved as well for small randomly perturbed intersection graphs based upon real network data.

3. Illustration on Small Networks

To find the exact isoperimetric numbers, one needs to calculate the minimum fraction of neighboring vertices or edges over the nodes inside the subset for all possible subsets of vertices with order at most $|V|/2$. Since this is an NP-hard problem, it is intractable to compute the exact values for general graphs [21,35]. It is well known that Cheeger's inequality, also known as the Alon–Milman inequality, provides bounds for the isoperimetric numbers using graph Laplacian eigenvalues. On the other hand, standard algorithms in linear algebra can be used to efficiently compute the spectrum of a given large graph. Here, instead of evaluating "approximate" values involving other parameters such as eigenvalues, we are interested in obtaining exact values of $\iota(G(B \cup R))$ and $c(G(B \cup R))$ for small networks.

Two intersection-based social networks are considered here: (i) The Norwegian interlocking directorate network Nor-Boards [37], where two directors are adjacent if they are sitting on the board of the same company based on the Norwegian Business Register on 5 August 2009. The underlying bipartite graph $\bar{B}(\bar{V}, \bar{W}, \bar{E})$ contains $|\bar{V}| = 1495$ directors, $|\bar{W}| = 367$ companies, and $|\bar{E}| = 1834$ edges indicating the affiliation relations; (ii) the co-authorship network ca-CondMat [5] based on preprints posted to the Condensed Matter Section of arXiv E-Print Archive between 1995 and 1999. The underlying bipartite graph $\bar{B}(\bar{V}, \bar{W}, \bar{E})$ contains $|\bar{V}|$ =16,726 authors, $|\bar{W}|$ = 22,016 papers, and $|\bar{E}|$ = 58,596 edges indicating authorship.

Figures 1 and 2 report the vertex–isoperimetric numbers and edge–isoperimetric numbers for subsets of Nor-Boards and ca-CondMat, respectively. For a given $n \in [20, 30]$, we first take a subgraph $B = B(V, W, E)$ from $\bar{B}(\bar{V}, \bar{W}, \bar{E})$ with $|V| = n$ so that $G(B)$ is connected and calculate its vertex–isoperimetric and edge–isoperimetric numbers. Each data point (blue square) in Figures 1 and 2 is obtained by means of an ensemble averaging of 30 independently taken graphs. For each chosen bipartite graph B, we then perturb it following the rules specified in Theorems 1 and 2 with $\varepsilon = 1$ to get the perturbed intersection graph $G(B \cup R)$. Each data point (red circle) in Figures 1 and 2 is obtained by means of a mixed ensemble averaging of 50 independently-implemented perturbations for 30 graphs. From a statistics viewpoint, it is clear that our random perturbation scheme increases both the vertex–isoperimetric and the edge–isoperimetric number for both cases. This, together with the theoretical results, suggests that the quantitative effect of random perturbations is independent of the order of the network.

Remark 3. *It is worth stressing that the theoretical results (Theorems 1 and 2) are in the large limit of the network size n. In other words, the form $\frac{1}{\ln n}$ only makes sense as $n \to \infty$. The simulation results presented in Figures 1 and 2 are for very small networks. Therefore, these results have no bearing on the $\frac{1}{\ln n}$ dependence (although a slight decline tendency for $\iota(G(B \cup R))$ can be seen in Figure 1a). The main phenomenon we observe from Figures 1 and 2 is that the random perturbation increases both vertex– and edge–isoperimetric numbers for all the cases considered. The numerical results (for small finite graphs) are a nice complement to the theoretical results (for infinite graphs). However, our numerical observations neither prove the $\frac{1}{\ln n}$ dependence would hold for small graphs nor show that such an increase of isoperimetric numbers would be universal in any sense. (A practical issue stems from graph sampling. To establish a proper model fit to the data, Akaike information criteria and Bayesian information criteria need to be applied.) The establishment of correlation between isoperimetric numbers and graph size n for finite intersection graphs is an interesting future work.*

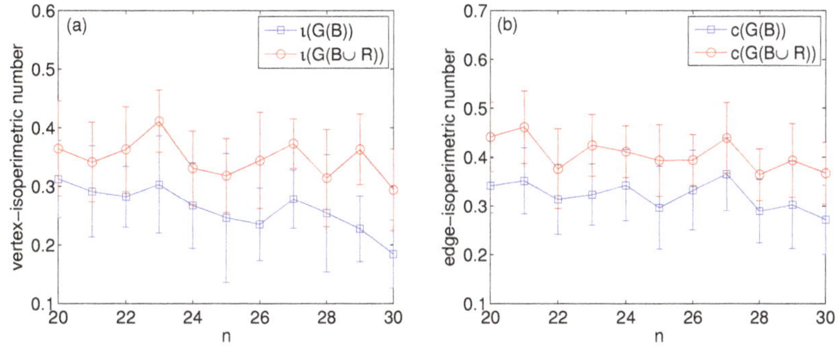

Figure 1. Vertex–isoperimetric number (panel (**a**)) and edge–isoperimetric number (panel (**b**)) versus $n = |G(B)|$ for subgraphs $G(B)$ (and its randomly perturbed version $G(B \cup R)$) taken from Nor-Boards.

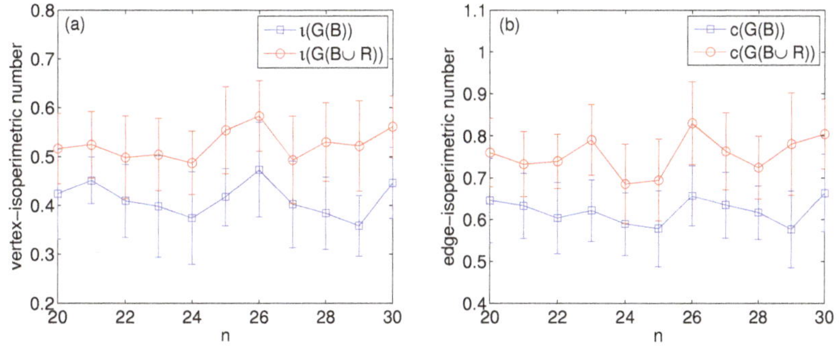

Figure 2. Vertex–isoperimetric number (panel (**a**)) and edge–isoperimetric number (panel (**b**)) versus $n = |G(B)|$ for subgraphs $G(B)$ (and its randomly perturbed version $G(B \cup R)$) taken from ca-CondMat.

4. Proofs

In this section, we prove Theorems 1 and 2. Our idea behind this is somewhat simple: If the network can be carefully decomposed into some subnetworks so that the resulting super-network (with these subnetworks being super-vertices) is sparse and highly connected, then its isoperimetric numbers are expected to be high. Similar approaches have been applied in, e.g., References [29–31].

Proof of Theorem 1. Set $s = C\Delta(\ln n)/\varepsilon$ for some constant $C = C(\varepsilon) > 0$ to be determined. By assumption, $G(B)$ is connected. Following Reference [38] (Proposition 4.5), we can divide the vertex set V into disjoint sets $V_1, V_2, \cdots, V_\theta$ satisfying $s \le |V_i| \le \Delta s$ and $G(B)[V_i]$ connected for each i. Clearly, $n/(\Delta s) \le \theta \le n/s$. Let $[\theta] = \{1, 2, \cdots, \theta\}$. For a graph $G = (V, E)$, we say two sets $S_1, S_2 \subseteq V$ have common neighbors in G if there exist $v_1 \in S_1$, $v_2 \in S_2$, and $v \in V$ such that $\{v_1, v\} \in E$ and $\{v_2, v\} \in E$ hold.

We will first show the following property for the random bipartite graph R holds a.a.s.: For every $\Theta \subseteq [\theta]$ with $0 < |\Theta| \le \theta/2$, there exist at least $|\Theta|/2$ many of V_i ($i \in [\theta]\backslash\Theta$) which have common neighbors with $\cup_{i\in\Theta} V_i$ in R.

Indeed, the probability that two sets V_i and V_j have no common neighbors in R can be computed as $\left\{1 - [1 - (1-p)^{|V_i|}][1 - (1-p)^{|V_j|}]\right\}^m$. Hence, the probability that there exists a set $\Theta \subseteq [\theta]$ with $0 < |\Theta| \le \theta/2$ such that no more than $|\Theta|/2$ many of V_i ($i \in [\theta]\backslash\Theta$) have common neighbors with $\cup_{i\in\Theta} V_i$ in R is upper bounded by:

$$\sum_{1\le j\le \theta/2}\binom{\theta}{j}\binom{\theta-j}{\lfloor\frac{j}{2}\rfloor}\left\{1-[1-(1-p)^{js}][1-(1-p)^{(\theta-\lfloor 3j/2\rfloor)s}]\right\}^m,$$
$$\le \sum_{1\le j\le \theta/2}\left(\frac{e\theta}{j}\right)^j\left(\frac{2e(\theta-j)}{j}\right)^{j/2}\left(1-\frac{s^2j(\theta-3j/2)p^2}{4}\right)^m,$$

where $\binom{\theta}{j}\binom{\theta-j}{\lfloor\frac{j}{2}\rfloor}$ counts the choice of Θ (with $|\Theta|=j$) and the corresponding sets $\{V_i\}$ described above, the estimate $|V_i|\ge s$ for all $i\in[\theta]$ is utilized in the multiplicative probabilities (i.e., there are at least $(\theta-\lfloor 3j/2\rfloor)$ sets in the union $\cup_{i\in\Theta}V_i$), and the upper bound comes from a direct application of inequalities ([10], p. 386). The above probability is further upper bounded by $(C(\ln n)/n)^m\sum_{1\le j\le\theta/2}(2\theta/j)^{3j/2}=o(1)$ when $\alpha\le 1$, and is upper bounded by $\sum_{1\le j\le\theta/2}\theta^{3j/2+2}\exp(-C\varepsilon\Delta j\ln n)=o(1)$ when $\alpha>1$ for a sufficiently large C. Therefore, the above property for the random bipartite graph R holds a.a.s. In the following, we will condition on such an R.

Fix a set $S\subseteq V$ with $|S|\le n/2$. Define three sets of indices: $\Theta_0=\{i\in[\theta]\,|\,V_i\subseteq S\}$, $\Theta_1=\{i\in[\theta]\,|\,0<|V_i\cap S|<|V_i|\}$, and $\Theta_2=\{i\in[\theta]\backslash\Theta_0\,|\,N_{G(B\cup R)}(V_i)\cap S\neq\emptyset\}$. Note that Θ_0 and Θ_1 are deterministic, but Θ_2 is a random set. If $|\Theta_0|\le\theta/2$, $|\Theta_2|\ge|\Theta_0|/2$ a.a.s. by the above assumed property of R. Similarly, if $|\Theta_0|>\theta/2$, we have $|\Theta_2|\ge|\Theta|/2=(\theta-|\Theta_0|-|\Theta_2|)/2$ a.a.s., where $\Theta=[\theta]\backslash(\Theta_0\cup\Theta_2)$. Hence, $|\Theta_2|\ge\min\{|\Theta_0|/2,(\theta-|\Theta_0|)/3\}$ a.a.s. Recall that $|S|\le n/2$. We derive that $n/2\le|V\backslash S|\le|\cup_{i\notin\Theta_0}V_i|\le(\theta-|\Theta_0|)\Delta s\le(\theta-|\Theta_0|)\Delta n/\theta$, and thus, $\theta-|\Theta_0|\ge\theta/(2\Delta)$. Therefore, we have a.a.s.:

$$|\Theta_2|\ge\min\left\{\frac{|\Theta_0|}{2},\frac{\theta}{6\Delta}\right\}\ge\frac{|\Theta_0|}{6\Delta}.$$

By definition, we have $S\subseteq\cup_{i\in\Theta_0\cup\Theta_1}V_i$. Thus, $|S|\le(|\Theta_0|+|\Theta_1|)\Delta s$. Since $G(B)[V_i]$ for $i\in\Theta_1$ is connected, $|N_{G(B\cup R)}(S)|\ge|\Theta_1\cup\Theta_2|$. Now we consider two cases. If $|\Theta_1|\ge|\Theta_0|$, then $|N_{G(B\cup R)}(S)|\ge|\Theta_1|\ge|S|/(2\Delta s)$. If $|\Theta_1|\le|\Theta_0|$, then $|N_{G(B\cup R)}(S)|\ge|\Theta_2|\ge|\Theta_0|/(6\Delta)\ge|S|/(12\Delta^2 s)$ a.a.s. Therefore:

$$\frac{|N_{G(B\cup R)}(S)|}{|S|}\ge\min\left\{\frac{1}{2\Delta s},\frac{1}{12\Delta^2 s}\right\}\qquad a.a.s.$$

Recall the definition of s at the beginning of the proof, and we complete the proof by taking $\delta=\varepsilon/(12\Delta^3C)$. □

We have made no attempt to optimize the constants in the proof. It is easy to check that the condition that $G(B)$ is connected in Theorem 1 can be weakened. For example, the above proof holds if each connected component of $G(B)$ is of order at least $C\Delta(\ln n)/\varepsilon$.

Let $G=(V,E)$ be a graph of order n. For integers a, b, and c, define $\mathcal{S}(a,b,c)$ as a collection of all sets $S\subseteq V$ such that $|S|=a$ and there exists a partition $S=S_1\cup\cdots\cup S_b$, where each $G[S_i]$ is connected, there are no edges in E connecting different S_i, and $|N_G(S_1)|+\cdots+|N_G(S_b)|=c$. The next lemma gives an upper bound of the size of $\mathcal{S}(a,b,c)$.

Lemma 1. *([30])*

$$|\mathcal{S}(a,b,c)|\le\left(\frac{en}{b}\right)^b\left(\frac{ea}{b}\right)^b\left(\frac{ec}{b}\right)^b\left(\frac{e(a+c)}{c}\right)^c.$$

Proof of Theorem 2. Consider the family $\mathcal{S}(a,b,c)$ of sets defined in graph $G(B)$. Since $G(B)$ is connected, we have for each $S\in\mathcal{S}(a,b,c)$, $|\partial_{G(B)}S|\ge c\ge b$. Note that $|\partial_{G(B\cup R)}S|\ge|\partial_{G(R)}S|$

holds. It suffices to show that the following property for the random bipartite graph R holds a.a.s.: There are constants $K, \delta > 0$ such that for any $K \ln n \le a \le n/2$, we have:

$$|\partial_{G(R)} S| \ge \frac{\delta a}{1 + \ln n},$$

for each $S \in \mathcal{S}(a,b,c)$ with $b \le c \le \delta a/(1+\ln n)$. Indeed, when $|S| = a \le K \ln n$, we can choose a small δ such that $2K\delta \le 1$. Thus, $|\partial_{G(B\cup R)} S| \ge |\partial_{G(B)}(S)| \ge 1 \ge \delta a/(1+\ln n)$.

It follows from Lemma 1 and $b \le c \le \delta a/(1+\ln n) \le a$ that:

$$|\mathcal{S}(a,b,c)| \le \left(\frac{2e^4 na^2}{c^3}\right)^c \le \left(\frac{2e^4 n(1+\ln n)^3}{\delta^3 a}\right)^{\delta a/(1+\ln n)} \le e^{C\delta a \ln(1/\delta)},$$

for some constant $C > 0$, where the first inequality holds since $f(x) = (e\rho/x)^x$ is increasing on $(0, \rho]$ and the second inequality holds since $g(x) = (\rho/x^3)^x$ is increasing on $(0, \rho^{1/3}]$.

Note that $mp^2 \to 0$ and $1-(1-p^2)^m \sim mp^2$. For a fixed S with $|S| = a \le n/2$, we obtain:

$$\begin{aligned} \mathbb{P}(|\partial_{G(R)} S| < \delta a) &\lesssim \mathbb{P}(\mathrm{Bin}(a(n-a), mp^2) < \delta a) \le \mathbb{P}\left(\mathrm{Bin}\left(\frac{na}{2}, mp^2\right) < \delta a\right) \\ &\le \exp\left(-\frac{a\varepsilon^2}{16}\right), \end{aligned}$$

provided $\delta < \varepsilon^2/4$, where the first inequality relies on Reference [9] (Theorem 2.2) and the last line uses a standard Chernoff's bound (e.g., [10]). Hence:

$$\begin{aligned} &\mathbb{P}\left(|\partial_{G(R)} S| < \frac{\delta a}{1+\ln n}, \exists S \in \mathcal{S}(a,b,c), b \le c \le \frac{\delta a}{1+\ln n}, K\ln n \le a \le \frac{n}{2}\right) \\ \le\; &\mathbb{P}\left(|\partial_{G(R)} S| < \delta a, \exists S \in \mathcal{S}(a,b,c), b \le c \le n, K \ln n \le a \le n\right) \\ \le\; &n^3 \exp\left(C\delta a \ln\left(\frac{1}{\delta}\right) - \frac{a\varepsilon^2}{16}\right). \end{aligned}$$

By taking $C\delta \ln(1/\delta) \le \varepsilon^2/32$ and $K \ge 100/\varepsilon^2$, the last line above is upper bounded by $n^3 \exp(-\varepsilon^2 a/32) \le n^3 \exp(-\varepsilon^2 K(\ln n)/32) \le n^3 \exp(-25(\ln n)/8) = o(1)$ as $n \to \infty$. The proof is complete. □

5. Concluding Remarks

In this paper, we presented a model of randomly perturbed intersection graphs. The intersection graph is induced by a given bipartite graph (base graph) plus a binomial random bipartite graph. We proved that a.a.s., the vertex–isoperimetric number and the edge–isoperimetric number of the randomly perturbed intersection graphs are of order $\Omega(1/\ln n)$ under some mild conditions. It would be interesting to investigate path length, diameter, and clustering coefficient of this model, which are important characteristics of real-life complex and social networks.

Another intriguing direction is to examine more general intersection graph models, such as active and passive intersection graphs [39]. In particular, if two vertices in one independent set V are declared adjacent when they have at least $k \ge 1$ common neighbors in the other independent set W, what role will k play in estimating the isoperimetric numbers, clustering, and path length of the resulting perturbed intersection graphs? Other perturbation mechanisms are also of research interest.

Acknowledgments: I am very grateful to the editor and two anonymous referees for their valuable comments that have greatly improved the presentation of the paper. The author was supported by a Starting Grant of Northumbria University.

Conflicts of Interest: The author declares no conflict of interest.

References

1. Newman, M.E.J. *Networks: An Introduction*; Oxford University Press: Oxford, UK, 2010.
2. Guillaume, J.L.; Latapy, M. Bipartite structure of all complex networks. *Inf. Process. Lett.* **2004**, *90*, 215–221. [CrossRef]
3. Kadushin, C. *Understanding Social Networks: Theories, Concepts, and Findings*; Oxford University Press: Oxford, UK, 2012
4. Watts, D.J.; Strogatz, S.H. Collective dynamics of 'small-world' networks. *Nature* **1998**, *393*, 440–442. [CrossRef]
5. Newman, M.E.J. The structure of scientific collaboration networks. *Proc. Natl. Acad. Sci. USA* **2001**, *98*, 404–409. [CrossRef] [PubMed]
6. Eschenauer, L.; Gligor, V. A key-management scheme for distributed sensor networks. In Proceedings of the 9th ACM Conference on Computer and Communications Security, Washington, DC, USA, 18–22 November 2002; pp. 41–47.
7. Marczewski, E. Sur deux propriétés des classes d'ensembles. *Fund. Math.* **1945**, *33*, 303–307. [CrossRef]
8. Karoński, M.; Scheinerman, E.R.; Singer-Cohen, K.B. On random intersection graphs: The subgraph problem. *Comb. Probab. Comput.* **1999**, *8*, 131–159. [CrossRef]
9. Singer-Cohen, K.B. Random Intersection Graphs. Ph.D. Thesis, Johns Hopkins University, Baltimore, MD, USA, 1995.
10. Frieze, A.; Karoński, M. *Introduction to Random Graphs*; Cambridge University Press: New York, NY, USA, 2016.
11. Shang, Y. Degree distributions in general random intersection graphs. *Electron. J. Combin.* **2010**, *17*, R23.
12. Kendall, M.; Martin, K.M. Graph-theoretic design and analysis of key predistribution schemes. *Des. Codes Cryptogr.* **2016**, *81*, 11–34. [CrossRef]
13. Zhao, J.; Yagan, O.; Gligor, V. On connectivity and robustness in random intersection graphs. *IEEE Trans. Autom. Contr.* **2017**, *62*, 2121–2136. [CrossRef]
14. Rybarczyk, K. The chromatic number of random intersection graphs. *Discuss. Math. Graph Theory* **2017**, *37*, 465–476. [CrossRef]
15. Bloznelis, M.; Godehardt, E.; Jaworski, J.; Kurauskas, V.; Rybarczyk, K. Recent progress in complex network analysis: Models of random intersection graphs. In *European Conference on Data Analysis*; Lausen, B., Krolak-Schwerdt, S., Boehmer, M., Eds.; Springer: Berlin, Germany, 2015; pp. 69–78.
16. Bloznelis, M.; Godehardt, E.; Jaworski, J.; Kurauskas, V.; Rybarczyk, K. Recent progress in complex network analysis: Properties of random intersection graphs. In *European Conference on Data Analysis*; Lausen, B., Krolak-Schwerdt, S., Boehmer, M., Eds.; Springer: Berlin, Germany, 2015; pp. 79–88.
17. Bollobás, B. The isoperimetric number of random regular graphs. *Eur. J. Comb.* **1988**, *9*, 241–244. [CrossRef]
18. Alon, N. On the edge-expansion of graphs. *Comb. Probab. Comput.* **1997**, *6*, 145–152. [CrossRef]
19. Benjamini, I.; Haber, S.; Krivelevich, M.; Lubetzky, E. The isoperimetric constant of the random graph process. *Random Struct. Algorithms* **2008**, *32*, 101–114. [CrossRef]
20. Puder, D. Expansion of random graphs: New proofs, new results. *Invent. Math.* **2015**, *201*, 845–908. [CrossRef]
21. Hoory, S.; Linial, N.; Wigderson, A. Expander graphs and their applications. *Bull. Am. Math. Soc.* **2006**, *43*, 439–561. [CrossRef]
22. Estrada, E. Spectral scaling and good expansion properties in complex networks. *Europhys. Lett.* **2006**, *73*, 649–655. [CrossRef]
23. Newman, M.E.J. Modularity and community structure in networks. *Proc. Natl. Acad. Sci. USA* **2006**, *103*, 8577–8582. [CrossRef] [PubMed]
24. Newman, M.E.J.; Watts, D.J. Renormalization group analysis of the small-world network model. *Phys. Lett. A* **1999**, *263*, 341–346. [CrossRef]
25. Shang, Y. A sharp threshold for rainbow connection in small-world networks. *Miskolc Math. Notes* **2012**, *13*, 493–497. [CrossRef]
26. Bohman, T.; Frieze, A.; Martin, R. How many random edges make a dense graph Hamiltonian? *Random Struct. Algorithms* **2003**, *22*, 33–42. [CrossRef]
27. Bohman, T.; Frieze, A.; Krivelevich, M.; Martin, R.; Adding random edges to dense graphs. *Random Struct. Algorithms* **2004**, *24*, 105–117. [CrossRef]

28. Krivelevich, M.; Sudakov, B.; Tetali, P. On smoothed analysis in dense graphs and formulas. *Random Struct. Algorithms* **2006**, *29*, 180–193. [CrossRef]
29. Flaxman, A.D. Expansion and lack thereof in randomly perturbed graphs. *Internet Math.* **2007**, *4*, 131–147. [CrossRef]
30. Krivelevich, M.; Reichman, D.; Samotij, W. Smoothed analysis on connected graphs. *SIAM J. Discrete Math.* **2015**, *29*, 1654–1669. [CrossRef]
31. Addario-Berry, L.; Lei, T. The mixing time of the Newman-Watts small-world model. *Adv. Appl. Probab.* **2015**, *47*, 37–56. [CrossRef]
32. Balogh, J.; Treglown, A.; Wagner, A.Z. Tilings in randomly perturbed dense graphs. *Combin. Probab. Comput.* **2019**, *28*, 159–176. [CrossRef]
33. Krivelevich, M.; Kwan, M.; Sudakov, B. Bounded-degree spanning trees in randomly perturbed graphs. *SIAM J. Discrete Math.* **2017**, *31*, 155–171. [CrossRef]
34. Böttcher, J.; Montgomery, R.; Parczyk, O.; Person, Y. Embedding spanning bounded degree subgraphs in randomly perturbed graphs. *Electron. Notes Discrete Math.* **2017**, *61*, 155–161. [CrossRef]
35. Mohar, B. Isoperimetric numbers of graphs. *J. Comb. Theory Ser. B* **1989**, *47*, 274–291. [CrossRef]
36. Grohe, M.; Marx, D. On tree width, bramble size, and expansion. *J. Comb. Theory Ser. B* **2009**, *99*, 218–228. [CrossRef]
37. Seierstad, C.; Opsahl, T. For the few not the many? The effects of affirmative action on presence, prominence, and social capital of women directors in Norway. *Scand. J. Manag.* **2011**, *27*, 44–54. [CrossRef]
38. Krivelevich, M.; Nachmias, A. Coloring complete bipartite graphs from random lists. *Random Struct. Algorithms* **2006**, *29*, 436–449. [CrossRef]
39. Shang, Y. Joint probability generating function for degrees of active/passive random intersection graphs. *Front. Math. China* **2012**, *7*, 117–124. [CrossRef]

symmetry

MDPI

Article

Generalized Permanental Polynomials of Graphs

Shunyi Liu

School of Science, Chang'an University, Xi'an 710064, China; liu@chd.edu.cn

Received: 28 January 2019; Accepted: 14 February 2019; Published: 16 February 2019

Abstract: The search for complete graph invariants is an important problem in graph theory and computer science. Two networks with a different structure can be distinguished from each other by complete graph invariants. In order to find a complete graph invariant, we introduce the generalized permanental polynomials of graphs. Let G be a graph with adjacency matrix $A(G)$ and degree matrix $D(G)$. The generalized permanental polynomial of G is defined by $P_G(x,\mu) = \text{per}(xI - (A(G) - \mu D(G)))$. In this paper, we compute the generalized permanental polynomials for all graphs on at most 10 vertices, and we count the numbers of such graphs for which there is another graph with the same generalized permanental polynomial. The present data show that the generalized permanental polynomial is quite efficient for distinguishing graphs. Furthermore, we can write $P_G(x,\mu)$ in the coefficient form $\sum_{i=0}^{n} c_{\mu i}(G)x^{n-i}$ and obtain the combinatorial expressions for the first five coefficients $c_{\mu i}(G)$ $(i = 0,1,\ldots,4)$ of $P_G(x,\mu)$.

Keywords: generalized permanental polynomial; coefficient; co-permanental

1. Introduction

A graph invariant f is a function from the set of all graphs into any commutative ring, such that f has the same value for any two isomorphic graphs. Graph invariants can be used to check whether two graphs are not isomorphic. If a graph invariant f satisfies the condition that $f(G) = f(H)$ implies G and H are isomorphic, then f is called a complete graph invariant. The problem of finding complete graph invariants is closely related to the graph isomorphism problem. Up to now, no complete graph invariant for general graphs has been found. However, some complete graph invariants have been identified for special cases and graph classes (see, for example, [1]).

Graph polynomials are graph invariants whose values are polynomials, which have been developed for measuring the structural information of networks and for characterizing graphs [2]. Noy [3] surveyed results for determining graphs that can be characterized by graph polynomials. In a series of papers [1,4–6], Dehmer et al. studied highly discriminating descriptors to distinguish graphs (networks) based on graph polynomials. In [5], it was found that the graph invariants based on the zeros of permanental polynomials are quite efficient in distinguishing graphs. Balasubramanian and Parthasarathy [7,8] introduced the bivariate permanent polynomial of a graph and conjectured that this graph polynomial is a complete graph invariant. In [9], Liu gave counterexamples to the conjecture by a computer search.

In order to find almost complete graph invariants, we introduce a graph polynomial by employing graph matrices and the permanent of a square matrix. We will see that this graph polynomial turns out to be quite efficient when we use it to distinguish graphs (networks).

The permanent of an $n \times n$ matrix M with entries m_{ij} $(i,j = 1,2,\ldots,n)$ is defined by

$$\text{per}(M) = \sum_{\sigma} \prod_{i=1}^{n} m_{i\sigma(i)},$$

where the sum is over all permutations σ of $\{1,2,\ldots,n\}$. Valiant [10] proved that computing the permanent is #P-complete, even when restricted to (0,1)-matrices. The permanental polynomial of M, denoted by $\pi(M,x)$, is defined to be the permanent of the characteristic matrix of M; that is,

$$\pi(M,x) = \mathrm{per}(xI_n - M),$$

where I_n is the identity matrix of size n.

Let $G = (V(G), E(G))$ be a graph with adjacency matrix $A(G)$ and degree matrix $D(G)$. The Laplacian matrix and signless Laplacian matrix of G are defined by $L(G) = D(G) - A(G)$ and $Q(G) = D(G) + A(G)$, respectively. The ordinary permanental polynomial of a graph G is defined as the permanental polynomial of the adjacency matrix $A(G)$ of G (i.e., $\pi(A(G),x)$). We call $\pi(L(G),x)$ (respectively, $\pi(Q(G),x)$) the Laplacian (respectively, the signless Laplacian) permanental polynomial of G.

The permanental polynomial $\pi(A(G),x)$ of a graph G was first studied in mathematics by Merris et al. [11], and it was first studied in the chemical literature by Kasum et al. [12]. It was found that the coefficients and roots of $\pi(A(G),x)$ encode the structural information of a (chemical) graph G (see, e.g., [13,14]). Characterization of graphs by the permanental polynomial has been investigated, see [15–19]. The Laplacian permanental polynomial of a graph was first considered by Merris et al. [11], and the signless Laplacian permanental polynomial was first studied by Faria [20]. For more on permanental polynomials of graphs, we refer the reader to the survey [21].

We consider a bivariate graph polynomial of a graph G on n vertices, defined by

$$P_G(x,\mu) = \mathrm{per}(xI_n - (A(G) - \mu D(G))).$$

It is easy to see that $P_G(x,\mu)$ generalizes some well-known permanental polynomials of a graph G. For example, the ordinary permanental polynomial of G is $P_G(x,0)$, the Laplacian permanental polynomial of G is $(-1)^{|V(G)|}P_G(-x,1)$, and the signless Laplacian permanental polynomial of G is $P_G(x,-1)$. We call $P_G(x,\mu)$ the generalized permanental polynomial of G.

We can write the generalized permanental polynomial $P_G(x,\mu)$ in the coefficient form

$$P_G(x,\mu) = \sum_{i=0}^{n} c_{\mu i}(G)x^{n-i}.$$

The general problem is to achieve a better understanding of the coefficients of $P_G(x,\mu)$. For any graph polynomial, it is interesting to determine its ability to characterize or distinguish graphs. A natural question is how well the generalized permanental polynomial distinguishes graphs.

The rest of the paper is organized as follows. In Section 2, we obtain the combinatorial expressions for the first five coefficients $c_{\mu 0}$, $c_{\mu 1}$, $c_{\mu 2}$, $c_{\mu 3}$, and $c_{\mu 4}$ of $P_G(x,\mu)$, and we compute the first five coefficients of $P_G(x,\mu)$ for some specific graphs. In Section 3, we compute the generalized permanental polynomials for all graphs on at most 10 vertices, and we count the numbers of such graphs for which there is another graph with the same generalized permanental polynomial. The presented data shows that the generalized permanental polynomial is quite efficient in distinguishing graphs. It may serve as a powerful tool for dealing with graph isomorphisms.

2. Coefficients

In Section 2.1, we obtain a general relation between the generalized and the ordinary permanental polynomials of graphs. Explicit expressions for the first five coefficients of the generalized permanental polynomial are given in Section 2.2. As an application, we obtain the explicit expressions for the first five coefficients of the generalized permanental polynomials of some specific graphs in Section 2.3.

2.1. Relation between the Generalized and the Ordinary Permanental Polynomials

First, we present two properties of the permanent.

Lemma 1. *Let A, B, and C be three $n \times n$ matrices. If A, B, and C differ only in the rth row (or column), and the rth row (or column) of C is the sum of the rth rows (or columns) of A and B, then* $\mathrm{per}(C) = \mathrm{per}(A) + \mathrm{per}(B)$.

Lemma 2. *Let $M = (m_{ij})$ be an $n \times n$ matrix. Then, for any $i \in \{1,2,\dots,n\}$,*

$$\mathrm{per}(M) = \sum_{j=1}^{n} m_{ij}\,\mathrm{per}(M(i,j)),$$

where $M(i,j)$ denotes the matrix obtained by deleting the ith row and jth column from M.

Since Lemmas 1 and 2 can be easily verified using the definition of the permanent, the proofs are omitted.

We need the following notations. Let $G = (V(G), E(G))$ be a graph with vertex set $V(G) = \{v_1, v_2, \dots, v_n\}$ and edge set $E(G)$. Let $d_i = d_G(v_i)$ be the degree of v_i in G. The degree matrix $D(G)$ of G is the diagonal matrix whose (i,i)th entry is $d_G(v_i)$. Let $v_{r_1}, v_{r_2}, \dots, v_{r_k}$ be k distinct vertices of G. Then $G_{r_1,r_2,\dots,r_k}$ denotes the subgraph obtained by deleting vertices $v_{r_1}, v_{r_2}, \dots, v_{r_k}$ from G. We use $G[h_r]$ to denote the graph obtained from G by attaching to the vertex v_r a loop of weight h_r. Similarly, $G[h_r, h_s]$ stands for the graph obtained by attaching to both v_r and v_s loops of weight h_r and h_s, respectively. Finally, $G[h_1, h_2, \dots, h_n]$ is the graph obtained by attaching a loop of weight h_r to vertex v_r for each $r = 1, 2, \dots, n$. The adjacency matrix $A(G[h_{r_1}, h_{r_2}, \dots, h_{r_s}])$ of $G[h_{r_1}, h_{r_2}, \dots, h_{r_s}]$ is defined as the $n \times n$ matrix (a_{ij}) with

$$a_{ij} = \begin{cases} h_r, & \text{if } i = j = r \text{ and } r \in \{r_1, r_2, \dots, r_s\}, \\ 1, & \text{if } i \neq j \text{ and } v_i v_j \in E(G), \\ 0, & \text{otherwise.} \end{cases}$$

By Lemmas 1 and 2, expanding along the rth column, we can obtain the recursion relation

$$\pi(A(G[h_r]), x) = \pi(A(G), x) - h_r \pi(A(G_r), x). \tag{1}$$

For example, expanding along the first column of $\pi(A(G[h_1]), x)$, we have

$$\begin{aligned} \pi(A(G[h_1]), x) &= \mathrm{per}(xI_n - A(G[h_1])) \\ &= \mathrm{per}\begin{bmatrix} x - h_1 & u \\ v & xI_{n-1} - A(G_1) \end{bmatrix} \\ &= \mathrm{per}\begin{bmatrix} x & u \\ v & xI_{n-1} - A(G_1) \end{bmatrix} + \mathrm{per}\begin{bmatrix} -h_1 & u \\ \mathbf{0} & xI_{n-1} - A(G_1) \end{bmatrix} \\ &= \pi(A(G), x) - h_1 \mathrm{per}(xI_{n-1} - A(G_1)) \\ &= \pi(A(G), x) - h_1 \pi(A(G_1), x). \end{aligned}$$

By repeated application of (1) for $G[h_r, h_s]$, we have

$$\begin{aligned} &\pi(A(G[h_r, h_s]), x) \\ &= \pi(A(G[h_r]), x) - h_s \pi(A(G_s[h_r]), x) \\ &= \pi(A(G), x) - h_r \pi(A(G_r), x) - h_s(\pi(A(G_s), x) - h_r \pi(A(G_{r,s}), x)) \\ &= \pi(A(G), x) - h_r \pi(A(G_r), x) - h_s \pi(A(G_s), x) + h_r h_s \pi(A(G_{r,s}), x). \end{aligned}$$

Additional iterations can be made to take into account loops on additional vertices. For loops on all n vertices, the expression becomes

$$\pi(A(G[h_1,h_2,\ldots,h_n]),x) = \pi(A(G),x) + \sum_{k=1}^{n}(-1)^k \sum_{1\le r_1<\cdots<r_k\le n} h_{r_1}\cdots h_{r_k}\pi(A(G_{r_1,\ldots,r_k}),x). \quad (2)$$

Let $A_\mu(G) := A(G) - \mu D(G)$. We see that the generalized permanental polynomial $P_G(x,\mu)$ of G is the permanental polynomial of $A_\mu(G)$; that is, $\pi(A_\mu(G),x)$. If the degree sequence of G is $(d_1,d_2,\ldots,d_n)$, then $A_\mu(G)$ is precisely the adjacency matrix of $G[-\mu d_1,-\mu d_2,\ldots,-\mu d_n]$. Hence, we obtain a relation between the generalized and ordinary permanental polynomials as an immediate consequence of (2).

Theorem 1. *Let G be a graph on n vertices. Then,*

$$P_G(x,\mu) = \pi(A_\mu(G),x) = \pi(A(G),x) + \sum_{k=1}^{n}\mu^k \sum_{1\le r_1<\cdots<r_k\le n} d_{r_1}\cdots d_{r_k}\pi(A(G_{r_1,\ldots,r_k}),x).$$

Theorem 1 was inspired by Gutman's method [22] for obtaining a general relation between the Laplacian and the ordinary characteristic polynomials of graphs. From Theorem 1, one can easily give a coefficient formula between the generalized and the ordinary permanental polynomials.

Theorem 2. *Suppose that $\pi(A(G),x) = \sum_{i=0}^{n} a_i(G)x^{n-i}$ and $P_G(x,\mu) = \sum_{i=0}^{n} c_{\mu i}(G)x^{n-i}$. Then,*

$$c_{\mu i}(G) = a_i(G) + \sum_{k=1}^{n}\mu^k \sum_{1\le r_1<\cdots<r_k\le n} d_{r_1}\cdots d_{r_k}a_{i-k}(G_{r_1,\ldots,r_k}), \quad 1\le i\le n.$$

2.2. The First Five Coefficients of $P_G(x,\mu)$

In what follows, we use t_G and q_G to denote respectively the number of triangles (i.e., cycles of length 3) and quadrangles (i.e., cycles of length 4) of G, and $t_G(v)$ denotes the number of triangles containing the vertex v of G.

Liu and Zhang [15] obtained combinatorial expressions for the first five coefficients of the permanental polynomial of a graph.

Lemma 3 ([15])**.** *Let G be a graph with n vertices and m edges, and let $(d_1,d_2,\ldots,d_n)$ be the degree sequence of G. Suppose that $\pi(A(G),x) = \sum_{i=0}^{n} a_i(G)x^{n-i}$. Then,*

$$a_0(G)=1,\ a_1(G)=0,\ a_2(G)=m,\ a_3(G)=-2t_G,\ a_4(G)=\binom{m}{2}-\sum_{i=1}^{n}\binom{d_i}{2}+2q_G.$$

Theorem 3. *Let G be a graph with n vertices and m edges, and let $(d_1,d_2,\ldots,d_n)$ be the degree sequence of G. Suppose that $P_G(x,\mu) = \sum_{i=0}^{n} c_{\mu i}(G)x^{n-i}$. Then*

$$c_{\mu 0}(G)=1,\quad c_{\mu 1}(G)=2\mu m,\quad c_{\mu 2}(G)=2\mu^2m^2+m-\frac{1}{2}\mu^2\sum_{i=1}^{n}d_i^2,$$

$$c_{\mu 3}(G)=\frac{1}{3}\mu^3\sum_{i=1}^{n}d_i^3-(\mu^3m+\mu)\sum_{i=1}^{n}d_i^2+\frac{4}{3}\mu^3m^3+2\mu m^2-2t_G,$$

$$c_{\mu 4}(G) = -\frac{1}{4}\mu^4 \sum_{i=1}^{n} d_i^4 + \left(\frac{2}{3}\mu^4 m + \mu^2\right) \sum_{i=1}^{n} d_i^3 - \frac{1}{2}(2\mu^4 m^2 + 5\mu^2 m + 1) \sum_{i=1}^{n} d_i^2$$
$$+ \frac{1}{8}\mu^4 \left(\sum_{i=1}^{n} d_i^2\right)^2 + \mu^2 \sum_{v_i v_j \in E(G)} d_i d_j + 2\mu \sum_{i=1}^{n} d_i t_G(v_i) + 2q_G - 4\mu m\, t_G$$
$$+ \frac{2}{3}\mu^4 m^4 + 2\mu^2 m^3 + \frac{1}{2}m^2 + \frac{1}{2}m.$$

Proof. It is obvious that $c_{\mu 0}(G) = 1$. By Theorem 2 and Lemma 3, we have

$$\begin{aligned}
c_{\mu 1}(G) &= a_1(G) + \mu \textstyle\sum_i d_i a_0(G_i) = 0 + \mu \sum_i d_i = 2\mu m, \\
c_{\mu 2}(G) &= a_2(G) + \mu \textstyle\sum_i d_i a_1(G_i) + \mu^2 \sum_{i<j} d_i d_j a_0(G_{i,j}) = m + 0 + \mu^2 \sum_{i<j} d_i d_j \\
&= m + \tfrac{1}{2}\mu^2 \left((\textstyle\sum_i d_i)^2 - \sum_i d_i^2\right) = 2\mu^2 m^2 + m - \tfrac{1}{2}\mu^2 \sum_i d_i^2, \\
c_{\mu 3}(G) &= a_3(G) + \mu \textstyle\sum_i d_i a_2(G_i) + \mu^2 \sum_{i<j} d_i d_j a_1(G_{i,j}) + \mu^3 \sum_{i<j<k} d_i d_j d_k a_0(G_{i,j,k}) \\
&= -2t_G + \mu \textstyle\sum_i d_i (m - d_i) + 0 + \mu^3 \sum_{i<j<k} d_i d_j d_k \\
&= -2t_G + \mu m \textstyle\sum_i d_i - \mu \sum_i d_i^2 + \frac{1}{6}\mu^3 \left((\sum_i d_i)^3 - 3\sum_i \sum_{j \neq i} d_i^2 d_j - \sum_i d_i^3\right) \\
&= -2t_G + 2\mu m^2 - \mu \textstyle\sum_i d_i^2 + \frac{4}{3}\mu^3 m^3 - \frac{1}{2}\mu^3 \left((\sum_i d_i^2)(\sum_j d_j) - \sum_i d_i^3\right) - \frac{1}{6}\mu^3 \sum_i d_i^3 \\
&= \tfrac{1}{3}\mu^3 \textstyle\sum_i d_i^3 - (\mu^3 m + \mu) \sum_i d_i^2 + \frac{4}{3}\mu^3 m^3 + 2\mu m^2 - 2t_G, \\
c_{\mu 4}(G) &= a_4(G) + \mu \textstyle\sum_i d_i a_3(G_i) + \mu^2 \sum_{i<j} d_i d_j a_2(G_{i,j}) + \mu^3 \sum_{i<j<k} d_i d_j d_k a_1(G_{i,j,k}) \\
&\quad + \mu^4 \textstyle\sum_{i<j<k<l} d_i d_j d_k d_l a_0(G_{i,j,k,l}) \\
&= \tbinom{m}{2} - \textstyle\sum_i \binom{d_i}{2} + 2q_G - 2\mu \sum_i d_i (t_G - t_G(v_i)) + \mu^2 \sum_{i<j} d_i d_j |E(G_{i,j})| + 0 \\
&\quad + \mu^4 \textstyle\sum_{i<j<k<l} d_i d_j d_k d_l.
\end{aligned} \tag{3}$$

By a straightforward calculation, we have

$$\begin{aligned}
\textstyle\sum_{i<j} d_i d_j |E(G_{i,j})| &= \sum_{\substack{i<j \\ v_i v_j \in E(G)}} d_i d_j |E(G_{i,j})| + \sum_{\substack{i<j \\ v_i v_j \notin E(G)}} d_i d_j |E(G_{i,j})| \\
&= \textstyle\sum_{\substack{i<j \\ v_i v_j \in E(G)}} d_i d_j (m - d_i - d_j + 1) + \sum_{\substack{i<j \\ v_i v_j \notin E(G)}} d_i d_j (m - d_i - d_j) \\
&= \textstyle\sum_{i<j} d_i d_j (m - d_i - d_j) + \sum_{v_i v_j \in E(G)} d_i d_j \\
&= m \textstyle\sum_{i<j} d_i d_j - \sum_i \sum_{j \neq i} d_i^2 d_j + \sum_{v_i v_j \in E(G)} d_i d_j \\
&= \tfrac{m}{2} \left(4m^2 - \textstyle\sum_i d_i^2\right) - \left(2m \sum_i d_i^2 - \sum_i d_i^3\right) + \sum_{v_i v_j \in E(G)} d_i d_j \\
&= \textstyle\sum_i d_i^3 - \frac{5}{2} m \sum_i d_i^2 + \sum_{v_i v_j \in E(G)} d_i d_j + 2m^3,
\end{aligned} \tag{4}$$

and

$$\begin{aligned}
&\textstyle\sum_{i<j<k<l} d_i d_j d_k d_l \\
&= \tfrac{1}{24}\left((\sum_i d_i)^4 - 12 \sum_i \sum_{\substack{j \\ j\neq i}} \sum_{\substack{k \\ k\neq i, k\neq j}} d_i^2 d_j d_k - 4 \sum_i \sum_{\substack{j \\ j\neq i}} d_i^3 d_j - 6 \sum_{i<j} d_i^2 d_j^2 - \sum_i d_i^4 \right) \\
&= \tfrac{2}{3}m^4 - \tfrac{1}{2} \times \tfrac{1}{2}\left((\sum_i d_i^2)(\sum_i d_i)^2 - \sum_i d_i^4 - 2\sum_{i<j} d_i^2 d_j^2 - 2 \sum_i \sum_{\substack{j \\ j\neq i}} d_i^3 d_j \right) \\
&\quad -\tfrac{1}{6} \sum_i \sum_{\substack{j \\ j\neq i}} d_i^3 d_j - \tfrac{1}{4} \sum_{i<j} d_i^2 d_j^2 - \tfrac{1}{24} \sum_i d_i^4 \\
&= \tfrac{2}{3}m^4 - m^2 \sum_i d_i^2 + \tfrac{5}{24} \sum_i d_i^4 + \tfrac{1}{4} \sum_{i<j} d_i^2 d_j^2 + \tfrac{1}{3} \sum_i \sum_{\substack{j \\ j\neq i}} d_i^3 d_j \\
&= \tfrac{2}{3}m^4 - m^2 \sum_i d_i^2 + \tfrac{5}{24} \sum_i d_i^4 + \tfrac{1}{4} \times \tfrac{1}{2}\left((\sum_i d_i^2)^2 - \sum_i d_i^4 \right) + \tfrac{1}{3}\left((\sum_i d_i^3)(\sum_i d_i) - \sum_i d_i^4 \right) \\
&= -\tfrac{1}{4} \sum_i d_i^4 + \tfrac{2}{3} m \sum_i d_i^3 - m^2 \sum_i d_i^2 + \tfrac{1}{8} (\sum_i d_i^2)^2 + \tfrac{2}{3} m^4.
\end{aligned} \tag{5}$$

Substituting (4) and (5) into (3), we obtain

$$\begin{aligned}
c_{\mu 4}(G) = &-\frac{1}{4}\mu^4 \sum_{i=1}^{n} d_i^4 + \left(\frac{2}{3}\mu^4 m + \mu^2\right) \sum_{i=1}^{n} d_i^3 - \frac{1}{2}(2\mu^4 m^2 + 5\mu^2 m + 1) \sum_{i=1}^{n} d_i^2 \\
&+ \frac{1}{8}\mu^4 \left(\sum_{i=1}^{n} d_i^2 \right)^2 + \mu^2 \sum_{v_i v_j \in E(G)} d_i d_j + 2\mu \sum_{i=1}^{n} d_i t_G(v_i) + 2q_G - 4\mu m\, t_G \\
&+ \frac{2}{3}\mu^4 m^4 + 2\mu^2 m^3 + \frac{1}{2} m^2 + \frac{1}{2} m.
\end{aligned}$$

This completes the proof. □

Since $\pi(L(G), x) = (-1)^{|V(G)|} P_G(-x, 1)$ and $\pi(Q(G), x) = P_G(x, -1)$, we immediately obtain the combinatorial expressions for the first five coefficients of $\pi(L(G), x)$ and $\pi(Q(G), x)$ by Theorem 3.

Corollary 1. *Let G be a graph with n vertices and m edges, and let $(d_1, d_2, \ldots, d_n)$ be the degree sequence of G. Suppose that $\pi(L(G), x) = \sum_{i=0}^{n} p_i(G) x^{n-i}$, then*

$$p_0(G) = 1, \quad p_1(G) = -2m, \quad p_2(G) = 2m^2 + m - \frac{1}{2}\sum_{i=1}^{n} d_i^2,$$

$$p_3(G) = -\frac{1}{3}\sum_{i=1}^{n} d_i^3 + (m+1)\sum_{i=1}^{n} d_i^2 - \frac{4}{3}m^3 - 2m^2 + 2t_G,$$

$$\begin{aligned}
p_4(G) = &-\frac{1}{4}\sum_{i=1}^{n} d_i^4 + \left(\frac{2}{3}m + 1\right)\sum_{i=1}^{n} d_i^3 - \frac{1}{2}(2m^2 + 5m + 1)\sum_{i=1}^{n} d_i^2 + \frac{1}{8}\left(\sum_{i=1}^{n} d_i^2\right)^2 \\
&+ \sum_{v_i v_j \in E(G)} d_i d_j + 2\sum_{i=1}^{n} d_i t_G(v_i) + 2q_G - 4m\, t_G + \frac{2}{3}m^4 + 2m^3 + \frac{1}{2}m^2 + \frac{1}{2}m.
\end{aligned}$$

Corollary 2. *Let G be a graph with n vertices and m edges, and let $(d_1, d_2, \ldots, d_n)$ be the degree sequence of G. Suppose that $\pi(Q(G), x) = \sum_{i=0}^{n} q_i(G) x^{n-i}$. Then,*

$$\begin{aligned}
q_0(G) &= 1, \qquad q_1(G) = -2m, \qquad q_2(G) = 2m^2 + m - \frac{1}{2}\sum_{i=1}^{n} d_i^2, \\
q_3(G) &= -\frac{1}{3}\sum_{i=1}^{n} d_i^3 + (m+1)\sum_{i=1}^{n} d_i^2 - \frac{4}{3}m^3 - 2m^2 - 2t_G, \\
q_4(G) &= -\frac{1}{4}\sum_{i=1}^{n} d_i^4 + \left(\frac{2}{3}m+1\right)\sum_{i=1}^{n} d_i^3 - \frac{1}{2}(2m^2+5m+1)\sum_{i=1}^{n} d_i^2 + \frac{1}{8}\left(\sum_{i=1}^{n} d_i^2\right)^2 \\
&\quad + \sum_{v_iv_j \in E(G)} d_i d_j - 2\sum_{i=1}^{n} d_i t_G(v_i) + 2q_G + 4m\, t_G + \frac{2}{3}m^4 + 2m^3 + \frac{1}{2}m^2 + \frac{1}{2}m.
\end{aligned}$$

2.3. Examples

In this subsection, by applying Theorem 3, we obtain the first five coefficients of the generalized permanental polynomials of some specific graphs: Paths, cycles, complete graphs, complete bipartite graphs, star graphs, and wheel graphs.

Example 1. *Let P_n ($n \geq 3$) be the path on n vertices. We see at once that $t_{P_n} = q_{P_n} = 0$, and $t_{P_n}(v) = 0$ for each vertex v of P_n. By Theorem 3, we have*

$$\begin{aligned}
c_{\mu 0}(P_n) &= 1, \quad c_{\mu 1}(P_n) = 2(n-1)\mu, \quad c_{\mu 2}(P_n) = (2n^2-6n+5)\mu^2 + n - 1, \\
c_{\mu 3}(P_n) &= \frac{2}{3}(2n^2-8n+9)(n-2)\mu^3 + 2(n-2)^2\mu, \\
c_{\mu 4}(P_n) &= \frac{2}{3}(n^2-5n+7)(n-3)(n-2)\mu^4 + (2n^2-10n+13)(n-3)\mu^2 + \frac{1}{2}(n-3)(n-2).
\end{aligned}$$

Example 2. *Let C_n ($n \geq 5$) be the cycle on n vertices. We see at once that $t_{C_n} = q_{C_n} = 0$, and $t_{C_n}(v) = 0$ for each vertex v of C_n. By Theorem 3, we have*

$$\begin{aligned}
c_{\mu 0}(C_n) &= 1, \quad c_{\mu 1}(C_n) = 2n\mu, \quad c_{\mu 2}(C_n) = 2n(n-1)\mu^2 + n, \\
c_{\mu 3}(C_n) &= \frac{4}{3}n(n-1)(n-2)\mu^3 + 2n(n-2)\mu, \\
c_{\mu 4}(C_n) &= \frac{2}{3}n(n-1)(n-2)(n-3)\mu^4 + 2n(n-2)(n-3)\mu^2 + \frac{1}{2}n(n-3).
\end{aligned}$$

Example 3. *Let K_n ($n \geq 4$) be the complete graph on n vertices. It is easy to check that $t_{K_n} = \binom{n}{3} = n(n-1)(n-2)/6$, $q_{K_n} = 3\binom{n}{4} = n(n-1)(n-2)(n-3)/8$, and $t_{K_n}(v) = \binom{n-1}{2} = (n-1)(n-2)/2$ for each vertex v of K_n. By Theorem 3, we have*

$$\begin{aligned}
c_{\mu 0}(K_n) &= 1, \quad c_{\mu 1}(K_n) = n(n-1)\mu, \quad c_{\mu 2}(K_n) = \frac{1}{2}n(n-1)^3\mu^2 + \frac{1}{2}n(n-1), \\
c_{\mu 3}(K_n) &= \frac{1}{6}n(n-2)(n-1)^4\mu^3 + \frac{1}{2}n(n-2)(n-1)^2\mu - \frac{1}{3}n(n-1)(n-2), \\
c_{\mu 4}(K_n) &= \frac{1}{24}n(n-2)(n-3)(n-1)^5\mu^4 + \frac{1}{4}n(n-2)(n-3)(n-1)^3\mu^2 - \\
&\quad \frac{1}{3}n(n-2)(n-3)(n-1)^2\mu + \frac{3}{8}n(n-1)(n-2)(n-3).
\end{aligned}$$

Example 4. *Let $K_{a,b}$ ($a \geq b \geq 2$) be the complete bipartite graph with partition sets of sizes a and b. We see at once that $t_{K_{a,b}} = 0$, $q_{K_{a,b}} = \binom{a}{2}\binom{b}{2} = ab(a-1)(b-1)/4$, and $t_{K_{a,b}}(v) = 0$ for each vertex v of $K_{a,b}$. By Theorem 3, we have*

$$c_{\mu 0}(K_{a,b}) = 1, \quad c_{\mu 1}(K_{a,b}) = 2ab\mu, \quad c_{\mu 2}(K_{a,b}) = \frac{1}{2}ab(4ab - a - b)\mu^2 + ab,$$
$$c_{\mu 3}(K_{a,b}) = \frac{1}{3}ab(4a^2b^2 - 3a^2b - 3ab^2 + a^2 + b^2)\mu^3 + ab(2ab - a - b)\mu,$$
$$c_{\mu 4}(K_{a,b}) = \frac{1}{24}ab(16a^3b^3 - 24a^3b^2 - 24a^2b^3 + 19a^3b + 6a^2b^2 + 19ab^3 - 6a^3 - 6b^3)\mu^4 + \frac{1}{2}ab(4a^2b^2 - 5a^2b - 5ab^2 + 2a^2 + 2ab + 2b^2)\mu^2 + ab(a-1)(b-1).$$

Example 5. *Let S_n ($n \geq 3$) be the star graph with $n+1$ vertices and n edges. We see at once that $t_{S_n} = q_{S_n} = 0$, and $t_{S_n}(v) = 0$ for each vertex v of S_n. By Theorem 3, we have*

$$c_{\mu 0}(S_n) = 1, \quad c_{\mu 1}(S_n) = 2n\mu, \quad c_{\mu 2}(S_n) = \frac{1}{2}n(3n-1)\mu^2 + n,$$
$$c_{\mu 3}(S_n) = \frac{1}{3}n(2n-1)(n-1)\mu^3 + n(n-1)\mu,$$
$$c_{\mu 4}(S_n) = \frac{1}{24}n(n-1)(n-2)(5n-3)\mu^4 + \frac{1}{2}n(n-1)(n-2)\mu^2.$$

Example 6. *Let W_n ($n \geq 5$) be the wheel graph with $n+1$ vertices and $2n$ edges. It is obvious that $t_{W_n} = q_{W_n} = n$. Let v_0 be the hub (i.e., the vertex of degree n) of W_n. We see that $t_{W_n}(v_0) = n$ and $t_{W_n}(v) = 2$ for other vertices v of W_n. By Theorem 3, we have*

$$c_{\mu 0}(W_n) = 1, \quad c_{\mu 1}(W_n) = 4n\mu, \quad c_{\mu 2}(W_n) = \frac{3}{2}n(5n-3)\mu^2 + 2n,$$
$$c_{\mu 3}(W_n) = 9n(n-1)^2\mu^3 + n(7n-9)\mu - 2n,$$
$$c_{\mu 4}(W_n) = \frac{9}{8}n(n-1)(n-2)(7n-9)\mu^4 + 6n(2n-3)(n-2)\mu^2 - 6n(n-2)\mu + \frac{3}{2}n(n-1).$$

3. Numerical Results

In this section, by computer we enumerate the generalized permanental polynomials for all graphs on at most 10 vertices, and we count the numbers of such graphs for which there is another graph with the same generalized permanental polynomial.

Two graphs G and H are said to be generalized co-permanental if they have the same generalized permanental polynomial. If a graph H is generalized co-permanental but non-isomorphic to G, then H is called a generalized co-permanental mate of G.

In order to compute the generalized permanental polynomials of graphs, we, first of all, have to generate the graphs by computer. We use nauty and Traces [23] to generate all graphs on at most 10 vertices. Next, the generalized permanental polynomials of these graphs are calculated by a Maple procedure. Finally, we count the numbers of generalized co-permanental graphs.

The results are summarized in Table 1. Table 1 lists, for $n \leq 10$, the total number of graphs on n vertices, the total number of distinct generalized permanental polynomials of such graphs, the number of such graphs with a generalized co-permanental mate, the fraction of such graphs with a generalized co-permanental mate, and the size of the largest family of generalized co-permanental graphs.

In Table 1, we see that the smallest generalized co-permanental graphs, with respect to the order, contain 10 vertices. Even more striking is that out of 12,005,168 graphs with 10 vertices, only 106 graphs could not be discriminated by the generalized permanental polynomial.

From Table 1 in [9], we see that the smallest graphs that cannot be distinguished by the bivariate permanent polynomial, introduced by Balasubramanian and Parthasarathy, contain 8 vertices. By comparing the present data of Table 1 with that of Table 1 in [9], we find that the generalized permanental polynomial is more efficient than the bivariate permanent polynomial when we use them to distinguish graphs. From Tables 2 and 3 in [5], it is seen that the generalized permanental polynomial is more efficient than the graph invariants based on the zeros of permanental polynomials of graphs. Comparing the present data of Table 1 with that of Table 1 in [24], we see that the generalized permanental polynomial is also superior to the the generalized characteristic polynomial when distinguishing graphs. So, the generalized permanental polynomial is quite efficient in distinguishing graphs.

Table 1. Graphs on at most 10 vertices.

n	# Graphs	# Generalized Perm. Pols	# with Mate	Frac. with Mate	Max. Family
1	1	1	0	0	1
2	2	2	0	0	1
3	4	4	0	0	1
4	11	11	0	0	1
5	34	34	0	0	1
6	156	156	0	0	1
7	1044	1044	0	0	1
8	12,346	12,346	0	0	1
9	274,668	274,668	0	0	1
10	12,005,168	12,005,115	106	8.83×10^{-6}	2

We enumerate all graphs on 10 vertices with a generalized co-permanental mate for each possible number of edges in Appendix A. We see that the generalized co-permanental graphs G_1 and H_1 with 10 edges are disconnected (see Figure 1), the generalized co-permanental graphs G_2 and H_2 with 11 edges, and G_3 and H_3 with 12 edges are all bipartite (see Figures 2 and 3), and two pairs (G_4, H_4) and (G_5, H_5) of generalized co-permanental graphs with 14 edges are all non-bipartite (see Figure 4). The common generalized permanental polynomial of the smallest generalized co-permanental graphs G_1 and H_1 is

$$\begin{aligned}
P_{G_1}(x,\mu) &= P_{H_1}(x,\mu) \\
&= x^{10} + 20\mu x^9 + (178\mu^2 + 10)x^8 + (928\mu^3 + 156\mu)x^7 + (3137\mu^4 + 1050\mu^2 + 37)x^6 \\
&+ (7180\mu^5 + 3980\mu^3 + 416\mu)x^5 + (11260\mu^6 + 9284\mu^4 + 1912\mu^2 + 60)x^4 \\
&+ (11936\mu^7 + 13632\mu^5 + 4592\mu^3 + 416\mu)x^3 + (8176\mu^8 + 12288\mu^6 + 6068\mu^4 + 1048\mu^2 + 36)x^2 \\
&+ (3264\mu^9 + 6208\mu^7 + 4176\mu^5 + 1136\mu^3 + 96\mu)x + 576\mu^{10} + 1344\mu^8 + 1168\mu^6 + 448\mu^4 + 64\mu^2.
\end{aligned}$$

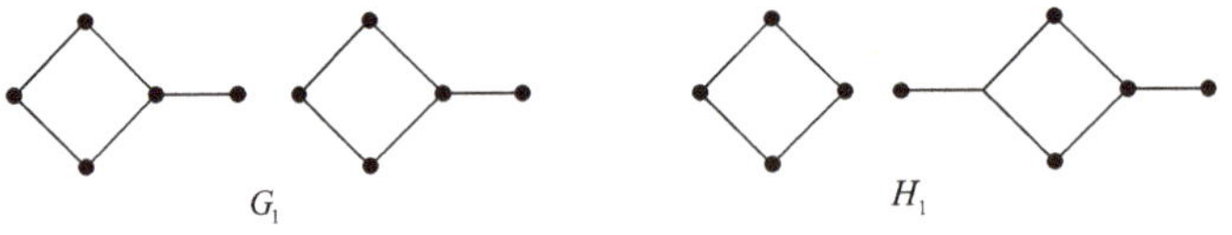

Figure 1. Two generalized co-permanental graphs with 10 vertices and 10 edges.

Figure 2. Two generalized co-permanental graphs with 10 vertices and 11 edges.

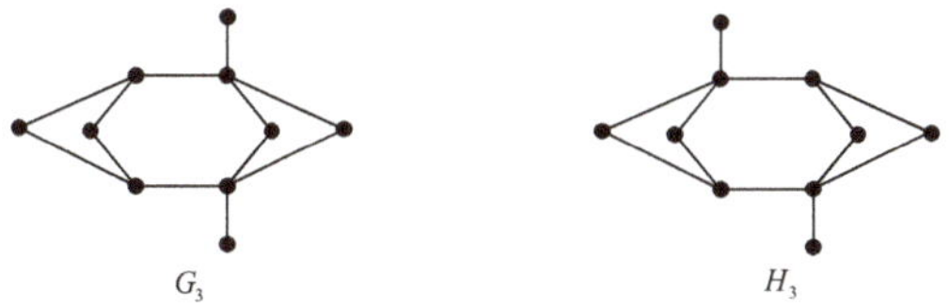

Figure 3. Two generalized co-permanental graphs with 10 vertices and 12 edges.

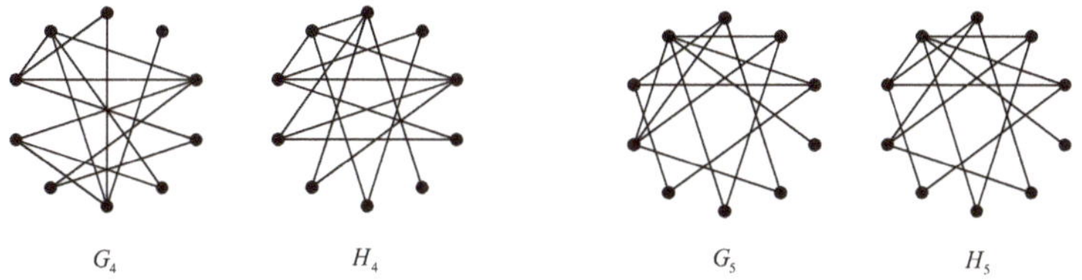

Figure 4. Two pairs of generalized co-permanental graphs with 10 vertices and 14 edges.

4. Conclusions

This paper is a continuance of the research relating to the search of almost-complete graph invariants. In order to find an almost-complete graph invariant, we introduce the generalized permanental polynomials of graphs. As can be seen, the generalized permanental polynomial is quite efficient in distinguishing graphs (networks). It may serve as a powerful tool for dealing with graph isomorphisms. We also obtain the combinatorial expressions for the first five coefficients of the generalized permanental polynomials of graphs.

Funding: This work was supported by the National Natural Science Foundation of China (Grant No. 11501050) and the Fundamental Research Funds for the Central Universities (Grant Nos. 300102128201, 300102128104).

Conflicts of Interest: The author declares no conflict of interest.

Appendix A

In the Appendix, we enumerate all graphs on 10 vertices with a generalized co-permanental mate for each possible number m of edges. Since the coefficient of x^{n-1} in $P_G(x,\mu)$ is $2\mu m$, two graphs with a distinct number of edges must have distinct generalized permanental polynomials. So, the enumeration can be implemented for each possible number of edges. We list the numbers of graphs with 10 vertices for all numbers m of edges, the numbers of distinct generalized permanental polynomials of such graphs, the numbers of such graphs with a generalized co-permanental mate, and the maximum size of a family of generalized co-permanental graphs (see Table A1).

Table A1. Graphs on 10 vertices.

m	# Graphs	# Generalized Perm. Pols	# with Mate	Max. Family
0	1	1	0	1
1	1	1	0	1
2	2	2	0	1
3	5	5	0	1
4	11	11	0	1
5	26	26	0	1
6	66	66	0	1
7	165	165	0	1
8	428	428	0	1

Table A1. *Cont.*

m	# Graphs	# Generalized Perm. Pols	# with Mate	Max. Family
9	1103	1103	0	1
10	2769	2768	2	2
11	6759	6758	2	2
12	15,772	15,771	2	2
13	34,663	34,663	0	1
14	71,318	71,316	4	2
15	136,433	136,429	8	2
16	241,577	241,575	4	2
17	395,166	395,162	8	2
18	596,191	596,183	16	2
19	828,728	828,723	10	2
20	1,061,159	1,061,154	10	2
21	1,251,389	1,251,381	16	2
22	1,358,852	1,358,848	8	2
23	1,358,852	1,358,850	4	2
24	1,251,389	1,251,385	8	2
25	1,061,159	1,061,157	4	2
26	828,728	828,728	0	1
27	596,191	596,191	0	1
28	395,166	395,166	0	1
29	241,577	241,577	0	1
30	136,433	136,433	0	1
31	71,318	71,318	0	1
32	34,663	34,663	0	1
33	15,772	15,772	0	1
34	6759	6759	0	1
35	2769	2769	0	1
36	1103	1103	0	1
37	428	428	0	1
38	165	165	0	1
39	66	66	0	1
40	26	26	0	1
41	11	11	0	1
42	5	5	0	1
43	2	2	0	1
44	1	1	0	1
45	1	1	0	1

References

1. Dehmer, M.; Emmert-Streib, F.; Grabner, M. A computational approach to construct a multivariate complete graph invariant. *Inf. Sci.* **2014**, *260*, 200–208. [CrossRef]
2. Shi, Y.; Dehmer, M.; Li, X.; Gutman, I. *Graph Polynomials*; CRC Press: Boca Raton, FL, USA, 2017.
3. Noy, M. Graphs determined by polynomial invariants. *Theor. Comput. Sci.* **2003**, *307*, 365–384. [CrossRef]
4. Dehmer, M.; Moosbrugger, M.; Shi, Y. Encoding structural information uniquely with polynomial-based descriptors by employing the Randić matrix. *Appl. Math. Comput.* **2015**, *268*, 164–168. [CrossRef]
5. Dehmer, M.; Emmert-Streib, F.; Hu, B.; Shi, Y.; Stefu, M.; Tripathi, S. Highly unique network descriptors based on the roots of the permanental polynomial. *Inf. Sci.* **2017**, *408*, 176–181. [CrossRef]
6. Dehmer, M.; Chen, Z.; Emmert-Streib, F.; Shi, Y.; Tripathi, S. Graph measures with high discrimination power revisited: A random polynomial approach. *Inf. Sci.* **2018**, *467*, 407–414. [CrossRef]
7. Balasubramanian, K.; Parthasarathy, K.R. In search of a complete invariant for graphs. In *Lecture Notes in Mathematics*; Springer: Berlin/Heidelberg, Germany, 1981; Volume 885, pp. 42–59.
8. Parthasarathy, K.R. Graph characterising polynomials. *Discret. Math.* **1999**, *206*, 171–178. [CrossRef]
9. Liu, S. On the bivariate permanent polynomials of graphs. *Linear Algebra Appl.* **2017**, *529*, 148–163. [CrossRef]
10. Valiant, L.G. The complexity of computing the permanent. *Theor. Comput. Sci.* **1979**, *8*, 189–201. [CrossRef]

11. Merris, R.; Rebman, K.R.; Watkins, W. Permanental polynomials of graphs. *Linear Algebra Appl.* **1981**, *38*, 273–288. [CrossRef]
12. Kasum, D.; Trinajstić, N.; Gutman, I. Chemical graph theory. III. On permanental polynomial. *Croat. Chem. Acta* **1981**, *54*, 321–328.
13. Cash, G.G. Permanental polynomials of smaller fullerenes. *J. Chem. Inf. Comput. Sci.* **2000**, *40*, 1207–1209. [CrossRef]
14. Tong, H.; Liang, H.; Bai, F. Permanental polynomials of the larger fullerenes. *MATCH Commun. Math. Comput. Chem.* **2006**, *56*, 141–152.
15. Liu, S.; Zhang, H. On the characterizing properties of the permanental polynomials of graphs. *Linear Algebra Appl.* **2013**, *438*, 157–172. [CrossRef]
16. Liu, S.; Zhang, H. Characterizing properties of permanental polynomials of lollipop graphs. *Linear Multilinear Algebra* **2014**, *62*, 419–444. [CrossRef]
17. Wu, T.; Zhang, H. Per-spectral characterization of graphs with extremal per-nullity. *Linear Algebra Appl.* **2015**, *484*, 13–26. [CrossRef]
18. Wu, T.; Zhang, H. Per-spectral and adjacency spectral characterizations of a complete graph removing six edges. *Discret. Appl. Math.* **2016**, *203*, 158–170. [CrossRef]
19. Zhang, H.; Wu, T.; Lai, H. Per-spectral characterizations of some edge-deleted subgraphs of a complete graph. *Linear Multilinear Algebra* **2015**, *63*, 397–410. [CrossRef]
20. Faria, I. Permanental roots and the star degree of a graph. *Linear Algebra Appl.* **1985**, *64*, 255–265. [CrossRef]
21. Li, W.; Liu, S.; Wu, T.; Zhang, H. On the permanental polynomials of graphs. In *Graph Polynomials*; Shi, Y., Dehmer, M., Li, X., Gutman, I., Eds.; CRC Press: Boca Raton, FL, USA, 2017; pp. 101–122.
22. Gutman, I. Relation between the Laplacian and the ordinary characteristic polynomial. *MATCH Commun. Math. Comput. Chem.* **2003**, *47*, 133–140.
23. McKay, B.D.; Piperno, A. Practical graph isomorphism, II. *J. Symb. Comput.* **2014**, *60*, 94–112. [CrossRef]
24. Van Dam, E.R.; Haemers, W.H.; Koolen, J.H. Cospectral graphs and the generalized adjacency matrix. *Linear Algebra Appl.* **2007**, *423*, 33–41. [CrossRef]

symmetry

MDPI

Article

Dynamics Models of Synchronized Piecewise Linear Discrete Chaotic Systems of High Order

Sergei Sokolov [1], Anton Zhilenkov [2], Sergei Chernyi [1,*], Anatoliy Nyrkov [1] and David Mamunts [1]

1 Department of Integrated Information Security, Admiral Makarov State University of Maritime and Inland Shipping, Saint-Petersburg 198035, Russia; sokolovss@gumrf.ru (S.S.); NyrkowAP@gumrf.ru (A.N.); MamuntsDG@gumrf.ru (D.M.)

2 Faculty of management systems and robotics, ITMO University, Saint-Petersburg 197101, Russia; zhilenkovanton@gmail.com

* Correspondence: sergiiblack@gmail.com; Tel.: +7-921-383-5322

Received: 17 January 2019; Accepted: 11 February 2019; Published: 15 February 2019

Abstract: This paper deals with the methods for investigating the nonlinear dynamics of discrete chaotic systems (DCS) applied to piecewise linear systems of the third order. The paper proposes an approach to the analysis of the systems under research and their improvement. Thus, effective and mathematically sound methods for the analysis of nonlinear motions in the models under consideration are proposed. It makes it possible to obtain simple calculated relations for determining the basic dynamic characteristics of systems. Based on these methods, the authors developed algorithms for calculating the dynamic characteristics of discrete systems, i.e. areas of the existence of steady-state motion, areas of stability, capture band, and parameters of transients. By virtue of the developed methods and algorithms, the dynamic modes of several models of discrete phase synchronization systems can be analyzed. They are as follows: Pulsed and digital different orders, dual-ring systems of various types, including combined ones, and systems with cyclic interruption of auto-tuning. The efficiency of various devices for information processing, generation and stabilization could be increased by using the mentioned discrete synchronization systems on the grounds of the results of the analysis. We are now developing original software for analyzing the dynamic characteristics of various classes of discrete phase synchronization systems, based on the developed methods and algorithms.

Keywords: nonlinear; synchronized; linear discrete; chaotic system; algorithm

1. Introduction

The nonlinear dynamics of discrete chaotic systems are not new for research, but they have not lost their relevance, due to a number of unresolved issues. As it is known [1], the implementation of chaotic systems on digital computers with finite-precision arithmetic (i.e., on real computers) has significant difficulties. It results in the fact that we get pseudochaotic systems [2]. This problem has led to the need for further development of analytical methods in the theory of nonlinear dynamics of discrete chaotic systems. New effective approaches to the synthesis and analysis of chaotic systems have appeared. Thus, Reference [3] shows the increasing importance that the fractional calculus of meromorphic functions has in chaotic systems. References [4,5] show the prospect of solving a number of problems using wavelet analysis.

There are a limited number of papers devoted to the study of nonlinear dynamics of discrete discrete chaotic systems (DCS) of the third order, in which fairly complete and accurate results are represented. This mainly concerns studies in which periodic motions and the acquisition band of synchronization systems [1,2] and numerical studies [3,4] are examined numerically. The purpose of this paper is to summarize theoretically the results of investigating the nonlinear dynamics of a third order phase synchronization systems (SPS), both in terms of the development of qualitatively-numerical methods of analysis, and in part of the study of specific systems described by the generalized model, Equation (1):

$$\begin{cases} \varphi_{n+1} = \varphi_n - \alpha F(\varphi_n) + x_n + g_n \\ x_{n+1} = dx_n - \beta F(\varphi_n) + y_n + g \\ y_{n+1} = hx_n - \eta F(\varphi_n) \end{cases} \tag{1}$$

where φ_n, x_n, y_n are the generalized coordinates of the system, $\alpha, \beta, \eta, d, h, g$ are generalized parameters, g_n is a variable component of the input frequency.

The expression in Equation (1) reduces to a general expression, as written:

$$\vec{q}_{n+1} = A\left(\vec{q}_n\right) + B \cdot \vec{u}_n, \tag{2}$$

where $\vec{q}_n = (\varphi_n, x_n^i)$ is the state vector of the system at the n-th time moment, the dimension of the vector is determined by the order of the system; φ_n is the phase difference of the impulse or code sequences at the inputs of the detector; $A\left(\vec{q}_n\right)$ is a nonlinear transition matrix whose properties depend on the kind of characteristics of the phase detector $F(\varphi_n)$; $\vec{u}_n$ is the exposure vector; and B is the exposure matrix.

2. Phase Portraits of the Onset of Instability of Fixed Points of Piecewise Linear Expressions of the Third Order

The study of steady motions of piecewise linear 3D DCS of the third order is based on the study of typical bifurcations of phase portraits of the mapping (Equation (1)). These include [5–10]:

(1) The loss of stability by k-fold fixed points associated with the transition of local stability boundaries G_1, G_{-1}, G_φ;

(2) The loss of stability by k-fold fixed points associated with the transition of limiting points of nonlinearity ($\varphi_i = \pm c$ for $F_c(\varphi)$ and $\varphi_i = \pm 1$ for $F_1(\varphi)$);

(3) The bifurcations of phase portraits caused by the intersection of separatrix invariant manifolds of k-fold saddle points.

At the qualitative level, the basic regularities of the appearance of fixed k-fold points for mappings of the second and third orders are repeated [5]. The transition of the boundaries of the areas of local stability G_1, G_{-1}, G_φ leads to the loss of stability of the fixed points and to qualitatively similar motions. The boundaries of the existence of fixed points of piecewise linear mappings of both orders in the general case for non-zero frequency detunings do not coincide with the boundaries of local stability. In the phase space, the boundaries of existence correspond to the boundaries of linear sections. This allows the condition for the k-fold fixed point to hit the linearity boundary as one of the necessary conditions for the appearance of periodic motions of the period k [6].

The cross sections of the local stability body of the mapping (Equation (1)) for various values of the generalized parameter n have a shape close to a triangular one. They are bounded by the curves G_1, G_{-1}, G_φ corresponding to the transition of one of the eigenvalues of the linearized matrix of the map (Equation (1)) through the values $\pm\, 1$ or $e^{\pm j\varphi}$. The line R on the sections bounds the region of existence of a simple fixed point. Its equation is obtained from Equation (1) and it is written as follows:

$$\beta = g - (1 - d - h)\alpha - \eta. \tag{3}$$

The singularity of the transition of stability boundaries, in the case of a third order mapping, consists of a large variety of possible combinations of the eigenvalues of the linearized matrix A corresponding to the boundary.

In accordance with Table 1, when the boundary G_{-1} is crossed, there are also three types of nodes and with the transition of the vibrational boundary G_{φ}, there are four types of foci. The transition through the boundaries G_{φ}, G_{-1} occurs in the linear sections of the functions $F_1(\varphi)$ and $F_c(\varphi)$. It is accompanied respectively by such bifurcations as a stable focus-complex saddle and a stable node, i.e. a real saddle. As in the case of second order mappings, the bifurcation data leads to the appearance of invariant closed curves, which are quasiperiodic motions. By virtue of the existence of a boundary R (*for* $g \neq 0$) that does not coincide with G_1 for piecewise linear mappings, in the general case the bifurcations of the appearance of both simple and k-fold fixed points occur on this boundary. In this case, a fixed point (one of the types of a stable node or focus) is generated simultaneously with one of the types of a saddle fixed point. The disappearance of a fixed stable point also occurs at the boundary R because of the fusion of a stable node or focus with a saddle point, followed by the formation of a stream of densified trajectories. The condition for the appearance of a pair of fixed points on the boundaries of piecewise linear mappings will be laid down below as the basis for the method of calculating bifurcation parameters.

Table 1. Parameters for solving.

The Eigenvalues of the Matrix A	Type of a Stable Point
1) $0 < p < 1, 0 < p_2 < 1, 0 < p_3 < 1, p_1, p_2, p_3$ are real-valued	stable node of the 1st type
2) $-1 < p_3 < 0, p_1, p_2, p_3$ are real-valued.	stable node of the 2nd type
3) $0 < p < 1$, are real-valued	stable node of the 3rd type
4) $0 < p < 1, 0 < p_2 < 1, -1 < p_b < 0, p_l, p_2, p_3$ are real-valued	stable node of the 4th type
5) 0 < Re < l are real-valued, p_2, p_3	stable focus of the 1st type
6) −1 < Re < 0	stable focus of the 2nd type
7) 0 < R < l; p_x are real-valued, p_2, p_3	stable focus of the 3rd type
8) $-1 < p < 0$, are real-valued, p_2, p_3	stable focus of the 4th type

In Figure 1, sections of the local stability body of the mapping (Equation (1)) for various values of the generalized parameter n are given on the plane of generalized parameters a, b.

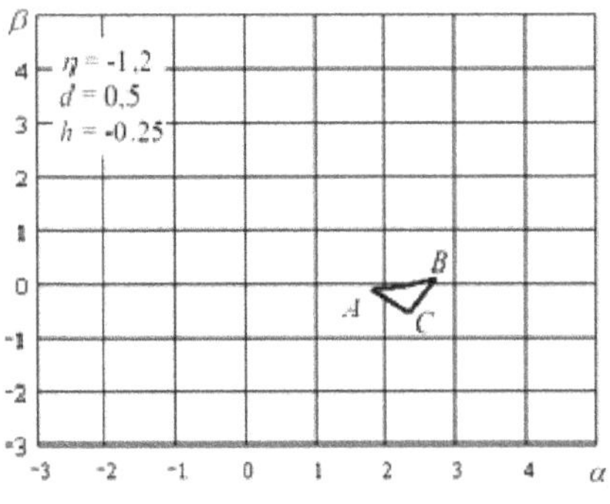

(a)

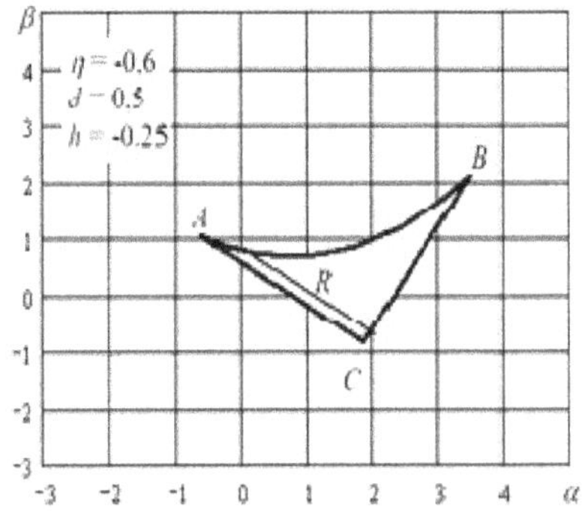

(b)

Figure 1. *Cont.*

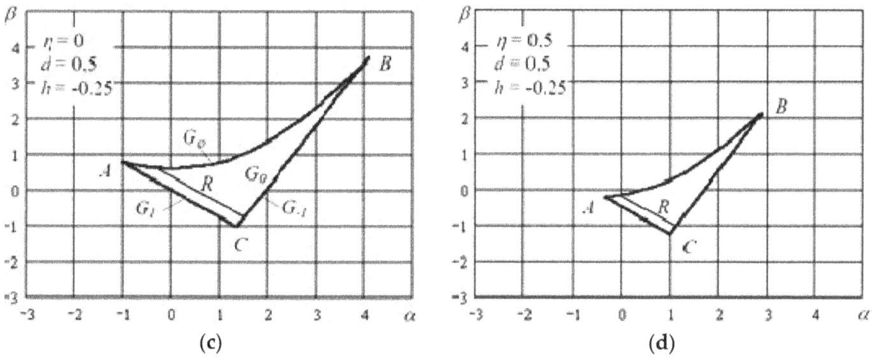

Figure 1. Cross-sections of the body with the local stability synchronization systems of the third order ((**a**) $\eta = -1.2$; (**b**) $\eta = -0.6$; (**c**) $\eta = 0$; (**d**) $\eta = 0.5$).

The formation of quasiperiodic motions under the conditions of the existence of fixed points is determined by the mutual arrangement of invariant separatrix manifolds of a simple or k-fold saddle fixed point. The difference from the second order mapping consists of a greater number of typical phase portraits near the saddle point, determined by the variety of the point itself.

3. SPS Model with Saw-Tooth Nonlinearity

The proposed method for calculating the bifurcation parameters of piecewise linear mappings of the third order is based on the assertions that fixed points on the boundaries of the linear sections of the functions $F_c(\varphi)$ and $F_1(\varphi)$ can arise. For the occurrence of simple fixed points of data, the assertions are sufficient. For the appearance of k-fold fixed points, the formulated assertions appear as necessary ones.

Let $F(\varphi) = F_1(\varphi)$. Since $F_1(\varphi)$ is periodic, the phase space of the mapping (Equation (1)) is a three-dimensional cylinder, whose scan cross-sections are shown in Figure 2.

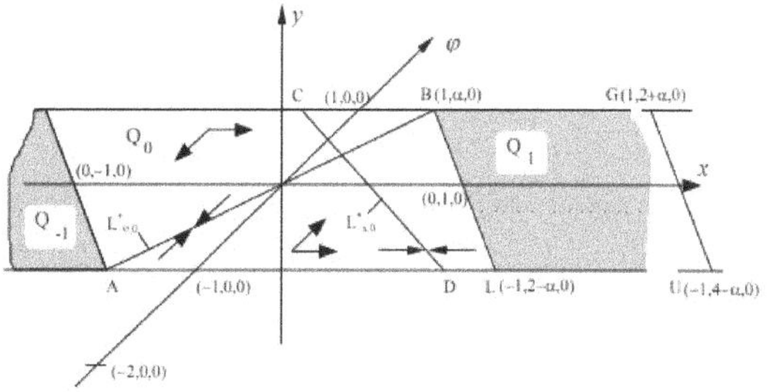

Figure 2. Phase cylinder cross section.

Figure 2 shows the section of the phase space by the plane $y_n = 0$. The lines $L_{\varphi,0}(AB)$, $L_{x,0}(CD)$ and $L_{y,0}$ are sections of the surfaces of the map preserving the coordinates φ, x and y, respectively. The equations of these surfaces can be obtained from (Equation (1)) respectively with $\varphi_{n+1} = \varphi_n$, $x_{n+1} = x_n$, $y_{n+1} = y_n$:

$$
\begin{aligned}
&L_{\varphi,0} : x = \alpha\varphi, \\
&L_{\varphi,0} : x = (y - \gamma\varphi + g)/(1 - d), \\
&L_{\varphi,0} : y = hx - \sigma\varphi.
\end{aligned} \tag{4}
$$

The mapping (display) surface with the preserved coordinate y is defined under the condition $x_n = x_{01}$. It should be noted that the coordinate y is not included in the equation for $L_{\varphi,0}$, so the surface under consideration is perpendicular to the plane $y = 0$. Moreover, the surface b passes through the origin of coordinates, and like in the second order system, it does not depend on the normalized initial detuning g. The point of intersection of these surfaces is the equilibrium state of the system (at the same time, the conditions $\varphi_{n+1} = \varphi_n$, $x_{n+1} = x_n$ and $y_{n+1} = y_n$ are satisfied, and has the coordinates $O(\varphi_{01}, x_{01}, y_{01})$.

By analogy with a system of the second order, domains of space can be found starting from which the solution (Equation (1)) falls on the boundary of the nonlinearity period $F_1(\varphi)$ $\varphi_n = 1$ and $\varphi_n = -1$. The required domains are a set of planes $G_{Q.m}$ (the index m is the number of the period $F_1(\varphi)$) on whose boundary the solution falls) whose equations are written as follows [11–17]:

$$G_{Q.m} : x = (\alpha - 1)\varphi + 2m - 1, \; m = 1, 2, 3 \ldots \tag{5}$$

And

$$G_{Q.m} : x = (\alpha - 1)\varphi + 2m + 1, \; m = -1, -2, -3 \ldots \tag{6}$$

In Equations (5) and (6), like in the expression for $L_{\varphi,0}$, the coordinate y is not included, and consequently these planes are perpendicular to the plane $y_n = 0$.

The arrows show the directions of the motion of the state vector $\vec{q}_n(\varphi_n, x_n, y_n)$ along the directions φ_n and x_n under the mapping, in each of the four zones formed by the segments AB and CD. For some saddle points shown in Figure 2, quasiperiodic motions do not take place.

In Figure 2, the domains of the nonlinear mapping are shaded with output correspondingly to the boundaries $\varphi_n = +1$ and $\varphi_n = -1$ of the phase cylinder scan $-Q_1$ and Q_{-1}. On both sides, the domains Q_1 and Q_{-1} are bounded by the planes $\varphi_n = \pm 1$, with the third one, by the plane $x = 1 - (1 - \alpha)\varphi$ for Q_1 (mapping in the direction of increasing x_n) and the plane $x = -1 - (1 - \alpha)\varphi$ for Q_{-1} (the mapping in the direction of decreasing φ_n). In the directions y_n and one of the directions x_n, the domains Q_1 and Q_{-1} are unbounded. Between the domains Q_1 and Q_{-1} there is a domain Q_0, the map from which occurs linearly.

For a nonlinear mapping, the domain Q_1 passes to the domain Q'_1. Moreover, the point $B(1, \alpha, 0)$ is mapped to the point $B'(-1,\; \alpha d - \beta + g,\; \alpha h - \sigma)$; $L(-1, 2 - \alpha, 0)$ to the point $L'(-1,\; d(2 - \alpha) + \beta + g,\; h(2 - \alpha) - \sigma)$, and so on.

Changing the coordinate $[\vec{q}_n]_y$ of the state vector in the Q_1, a domain leads to a change in the coordinate of the vector $[\vec{q}_{n+1}]_x$ in Q'_1: With increasing (decreasing) $[\vec{q}_n]_y$ increases (decreases) $[\vec{q}_{n+1}]_x$. Thus, the entire domain Q_1 is mapped into an infinite strip along the x_n -axis bounded along the φ_n axis by the planes $\varphi_n = \pm 1$ and, in addition, by two parallel planes that are mappings of the planes $\varphi_n = \pm 1$. Analogous arguments lead to the construction of the domain Q'_{-1}, which is a mapping of Q_{-1}. It should be noted that there is an intersection of the domains Q'_1 and Q_{-1}, as well as Q'_{-1} and Q_1, which fundamentally distinguishes the considered system from the second order system.

Let us consider iterations with initial conditions from an arbitrary state vector $\vec{q}_0 = (\varphi_0, x_0, y_0)$. According to (Equation (1)) vector $\vec{q}_n$ may be expressed by means of $\vec{q}_0$ as follows:

$$\vec{q}_n = A^n \cdot \vec{q}_0 + \sum_{j=0}^{n-1} A^j \cdot \left(\vec{r} + \vec{p}_{n-j-1}\right) \tag{7}$$

where A is linearized matrix corresponding to (Equation (1)) under the linear mapping $\vec{p}_j = (0, 0, 0)^T$, in the case of a nonlinear mapping $\vec{p}_j = (\pm 2, 0, 0)^T$, with this, the sign "+" corresponds to going abroad $\varphi = -1$, the sign "-" corresponds to going abroad $x = +1$. The vector $\vec{p}_j$ returns the state vector of the system to the $(j+1)$-step in the interval $[-1; 1]$ for the coordinate x. We rewrite (Equation (7)) as:

$$\vec{q}_n = A_n \vec{q}_0 + (E - A^n)(E - A)^{-1} \vec{r} + \sum_{j=0}^{n-1} A^j \vec{p}_{n-j-1} \tag{8}$$

For a cycle of period k existing, it is necessary that the closure condition $-\vec{q}_k = \vec{q}_0$ be satisfied. Taking this condition into account, the expression for the initial point of the cycle follows from (Equation (8)):

$$\vec{q}_n = \left(E - A^k\right)^{-1}\left(\sum_{j=0}^{k-1} A^j \vec{p}_{k-j-1}\right) + (E - A)^{-1}\vec{r} \tag{9}$$

The expression (Equation (9)) may be considered as the first necessary condition for the existence of a cycle, or the closure condition. The second condition is to find all the state vectors of a cycle of a required structure within the interval $|\varphi| \leq 1$ (i.e., the structural condition). Implementation of this condition means that all state vectors of the cycle are in the corresponding domains Q_1, Q_0, Q_{-1}. Otherwise, Equation (9) can formally lead to some state that is not a point of the cycle. The formulated conditions are necessary and sufficient for the existence of a cycle with a certain structure.

Similar to a discrete SPS of the second order, it can be shown that an arbitrary cycle existing in a system with nonlinearity $F_1(\varphi)$ is stable under the conditions of local stability of the mapping (Equation (1)).

Consider the structure cycle (u/k), where u is the number of nonlinear mappings on the cycle period, k is the cycle period. For the limit cycle of the first kind $u = 0$, for the limit cycle of the second kind in the case of rotation along the coordinate p in the direction of increasing $u > 0$, in the case of rotation in the direction of decreasing the coordinate $\varphi - u < 0$. In accordance with (9), the vector of an arbitrary point of the cycle can be represented as follows:

$$\vec{q}_j = \vec{l}_j + g\vec{b},\ j = 1,\ldots,k \tag{10}$$

where $\vec{l}_j = \left(E - A^k\right)^{-1}\left(\sum_{j=0}^{k-1} A^j \vec{p}_{k-j-1}\right)$, $\vec{b} = (E - A)^{-1}(0,1,0)^T$; $\vec{l}_j$ is a vector, depending on the structure of the cycle and the choice of the starting point, $\vec{b}$ is a vector depending neither on the structure of the cycle nor its initial state.

When g is changed, all points of the cycle in the phase domain are displaced along the vector $\vec{b}$. This can lead to both the occurrence and destruction of the cycle due to the transition of cycle points between the domains Q_1, Q_0, Q_{-1}, and also when the points of the plane cycle $\varphi_n = \pm 1$ intersect the vectors.

Let us find conditions for the generalized detuning g for which there exists a cycle of a certain structure (u/κ). To do this, we use the above conditions for the existence of a cycle. From (7–9) we assess the values of the generalized detuning $g^-{}_j$ and $g^+{}_j$ for which the state vector $\vec{q}_j$ intersects the boundaries $\varphi_n = -1$ and $\varphi_n = +1$, respectively:

$$g^-{}_j = \frac{-1 - \left[\vec{l_j}\right]_1}{\left[\vec{b_j}\right]_1},\ g^+{}_j = \frac{1 - \left[\vec{l_j}\right]_1}{\left[\vec{b_j}\right]_1} \tag{11}$$

All points of the cycle intersect the plane $\varphi_n = -1$ if the condition $g > \max\limits_{j=1\ldots k}\left(g_j^-\right)$ is satisfied, at least one cycle point intersects the plane $\varphi_n = 1$ for $g < \min\limits_{j=1\ldots k}\left(g_j^+\right)$. A cycle can exist when:

$$\max_{j=1\ldots k}\left(g_j^-\right) < \min_{j=1\ldots k}\left(g_j^+\right), \tag{12}$$

in the detuning range $\max\limits_{j=1\ldots k}\left(g_j^-\right) < g < \min\limits_{j=1\ldots k}\left(g_j^+\right)$.

4. Algorithm for Determining the Acquisition Bandwidth

We constructed an algorithm for determining the acquisition band. It is based on the condition of the occurrence of the simplest limit cycles of the first and second kind. In the general case, it is necessary to determine two values of the initial detuning $\gamma_{\min}, \gamma_{\max}$. With $\gamma < \gamma_{\max}$ all PC2 disappear, with $\gamma > \gamma_{\min}$ all PC1 disappear [16–22].

Let us find $\gamma_{\max}$, for this we define the value of γ^k, at which the PC2 structures $(1/k)$ appear. To be exact, we consider the initial state on the cycle to be the state into which the system comes after the nonlinear mapping through the boundary $\varphi_n = 1$. For this case $\vec{p}_j = (0,0,0)^T, 0 \leq j < k-1$; $\vec{p}_{k-1} = (-2,0,0)^T$. According to Equations (10)–(12), the initial state vector $\vec{q}_0$ will be as follows [11–17]:

$$\vec{q}_0 = \frac{\vec{P}_{k-1}}{E - A^k} + \frac{\vec{r}}{E - A} \tag{13}$$

The cycle of the second kind of period k will exist when conditions (Equation (13)) are fulfilled and will occur, taking into account Equations (1) and (5) with frequency detuning

$$\gamma_2^k = \frac{-1 + 2\left[\left(E - A^k\right)^{-1}\right]_{11}}{\xi\left[\left(E - A\right)^{-1}\right]_{12}} \tag{14}$$

The boundary of the cycle generation may be expressed as follows:

$$\gamma = \gamma_{\max} = \min_{k=1\ldots k\max}\left(\gamma_2^k\right). \tag{15}$$

It remains to find k, for which the founded value of the initial detuning will be the smallest, which determines the boundary condition for the occurrence of PC2. The algorithm proposes the assignment of some $k_{\max}$, which obviously exceeds the desired value. Recommendations for choosing $k_{\max}$ are similar to the second order system and are as follows. In the case of complex eigenvalues of the matrix A, the behavior of the vector $\vec{l}$ is oscillatory in parameter k (the end of the vector with an increase in the cycle period k describes a twisting spiral around a point $(-2,0,0)$ and it is enough to take half the oscillation period as $k_{\max}$.

The analysis of the above dependencies from the standpoint of global stability of the FAS leads to the following conclusions:

1. With increasing α_1, α_2, the stability domains with respect to the amplification D expand. The most significant increase is observed for large m_1. For example, for $m_1 = 0.8$ with increasing α_1, α_2 from values 0.5–0.8 (Figure 3) to values 2–4 (Figure 3), the stability domains in parameter D increase 2–4 times.
2. The boundary of the areas with global stability on the initial mismatch β also expands significantly with increasing α_1, α_2. However, dependence on m_1 is more complex. A decrease in the upper bound β with increasing m_1 is observed near the limits of the local stability. On the contrary, in the farther zone from the boundary of local stability (medium D), there is a significant increase in the upper boundary β with increasing m_1.
3. Limiting the stability of the bottom of the frequency detuning (limiting with the cycles of the first kind) is most expressed with small m_1 and, as stated above, is non-monotonous. The most significant restriction is observed for large D (Figure 3) and can reach values of 0.3–0.4.

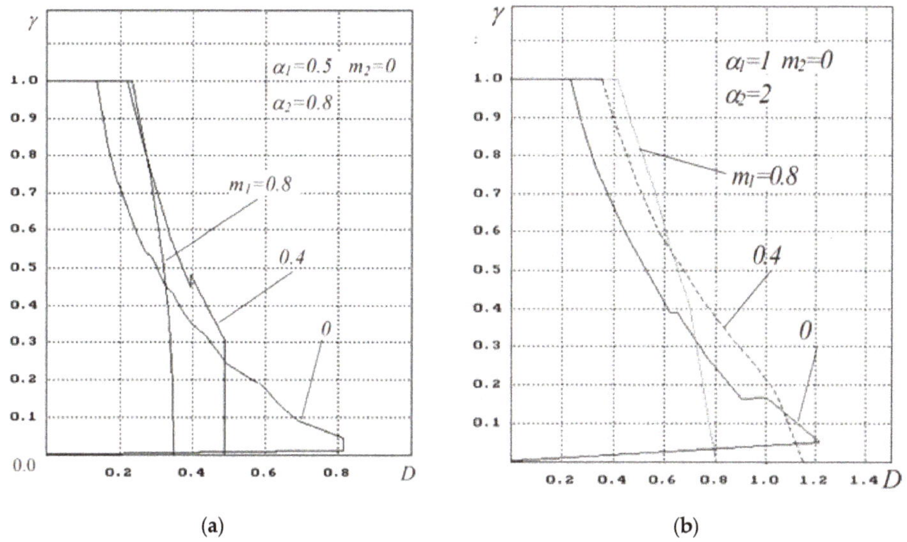

Figure 3. Acquisition band SFS of the third ((**a**) $\alpha_1 = 0.5$; (**b**) $\alpha_1 = 1$) order with $F_1(\varphi)$.

5. SPS Model with a Triangular Nonlinearity

Let $F(\varphi) = F_c(\varphi)$. The basic laws of the appearance of periodic motions of the second kind and quasiperiodic motions in a third order system with a triangular nonlinearity repeat at a qualitative level in the results obtained for a second order system [12,19,21–25]. In this case, both the final dependencies and the mechanisms explaining them are qualitatively repeated. In this regard, we will not dwell on them in detail below. The quantitative differences will be demonstrated on a number of graphs devoted to the analysis of the acquisition band.

The situation with cycles of the first kind, whose existence has been established in a system with a saw-tooth nonlinearity, is completely different. Their analysis is important because they have occurred with small initial detuning and limit the acquisition domain in frequency from below.

A quantitative estimate of the boundary of first kind cycles can be obtained by considering the change in the area of their existence in the parameter space with a change in the shape of the characteristic. Figure 5 shows the region of existence of PC1 on the plane D, γ for different values of c. The boundaries of the areas are almost straight lines, the slope of which depends only on the filter parameters (α_1, m_1, α_2, m_2). Changing c does not change the shape of these curves, but shifts them along the abscissa. PC1 cycles disappear in two cases: Firstly, at a certain maximum value of the parameter c_{max}; and secondly, with a saw-tooth characteristic of the detector and certain filter parameters (Figure 4).

The authors can note the strong influence of parameters on these dependencies.

From a practical point of view, the filter parameters are of a certain interest. Limit cycles of the first kind (quasi-synchronism mode) are impossible for them. Figure 5 shows the regions of existence of PC1 on the plane α_1, α_2 with equal forcing coefficients m_1, m_2. For $m_1 = m_2 = 0$, there is a boundary close to a straight line, above which there are no cycles. With increasing m_1, m_2, the area of existence of cycles is symmetrically limited by α_1, α_2, and disappears when $m_1 = m_2 \geq 0.165$.

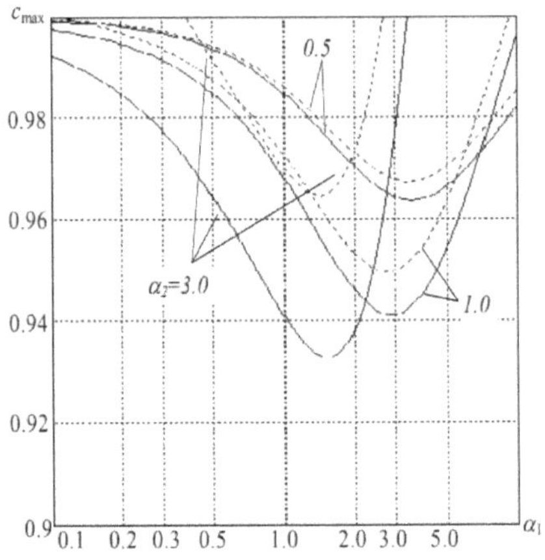

Figure 4. Shows the dependences of c_{max} on the time constant of one of the links of the filter.

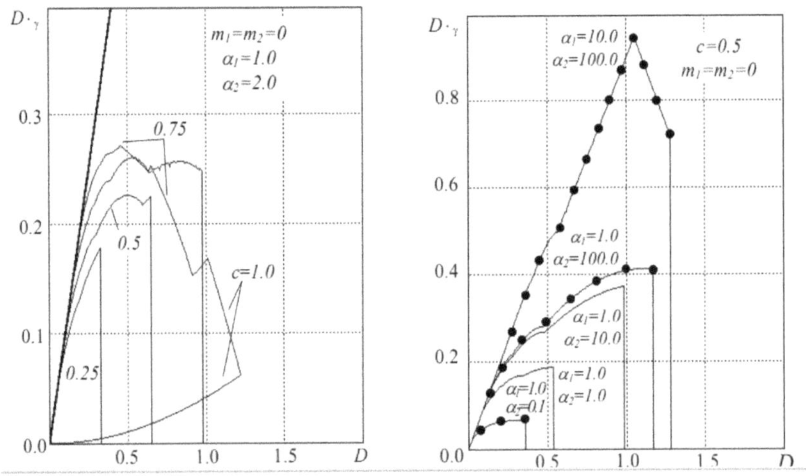

Figure 5. The acquisition band of the pulsed SPS ((**a**) $\alpha_1 = 1$; (**b**) $\alpha_1 = 10$) with $F_c(\varphi)$.

Analysis of the above results shows that the existence of cycles of the first kind with a saw-tooth-like, or close to it, detector characteristic is determined only by the filter parameters and is not related to the gain of the system.

The dependencies in Figure 5 allow us to analyze the acquisition band when changing the duration of the stable branch of the detector characteristics and to answer the question about its optimal value. As in the case of pulsed DSC of the second order, for small D, a weak dependence of the acquisition band on the shape of the characteristics is observed. There is some loss for $F_c(\varphi)$, increasing with the steepness of the stable branch.

With increasing D (to the boundary of local stability) due to a shift in the boundary of the onset of quasiperiodic motions towards large β, the maximum of the acquisition band is provided in the case of $F_c(\varphi)$. With different ratios of filter parameters, the gain in the acquisition band can reach up to 50% as compared with $F_c(\varphi)$. Figure 5 also shows the limitations of the acquisition area from below due to PC1. It is possible to get rid of such restrictions by increasing the steepness (decreasing the duration) of a stable part of the characteristics.

6. Results

The dual-ring synthesizer was simulated. The computer model made it possible to take into account a number of factors additionally, which were not considered in the mathematical model. They included the inconstancy of the discretization periods and the difference between the detector model and the zero order extrapolator.

Allowance for variations of the sampling period (epoch) resulted in corrections of the dynamic characteristics, primarily the stability domain. However, this applied mainly to the range of large gains ($\alpha > 1$, $\beta > 1$). For operating gains, the results of mathematical and computer simulation coincided with high accuracy [25–28].

Figure 6 shows the dependences of the capture band and the transient time of the dual-ring SPS, taking into account the variable nature of the sampling periods, for k1 / k2 = 8. To compare, Figure 6a shows the results for a constant sampling period (upper curves). With positive detuning, allowance for the variations led to a certain decrease in the capture band, repeating the known result for single-ring systems. A change in the capture band as a function of μ repeated similar changes for a model with a constant sampling period. A decrease in the capture band with increasing μ was observed. Changing the sign of μ to the opposite, resulted in an increase in the capture band, partially offsetting the loss from the variations in the sampling period.

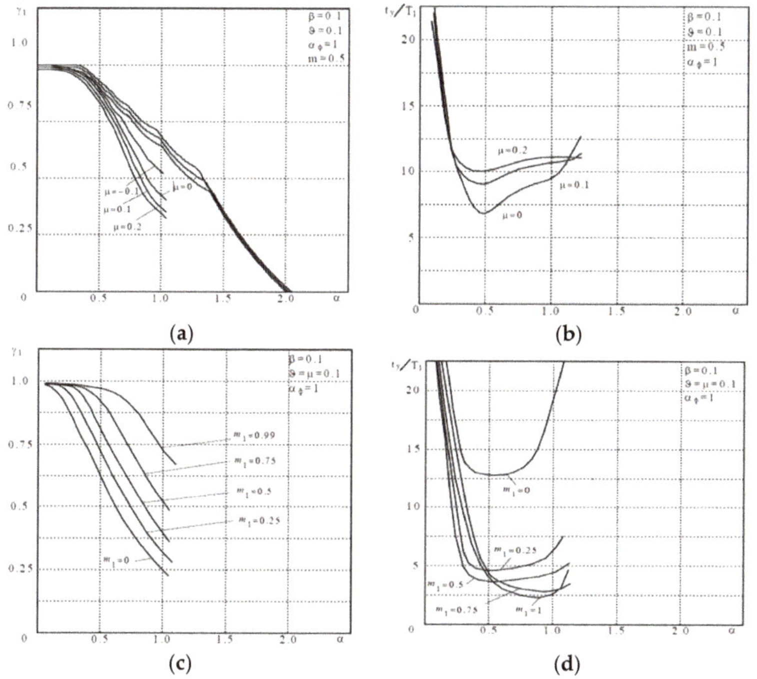

Figure 6. Dependencies of the capture band and the settling time in the dual-ring SPS ((**a**) $\mu = -0.1$; (**b**) $\mu = 0.1$; (**c**) $m_1 = 0$; (**d**) $m_1 = 0.5$).

An analysis of the average time for the frequency settling in a dual-ring SPS (Figure 6b–d) suggested a qualitative coincidence with the results of the model analysis [6,8].

In particular, there is was a fairly wide range of parameters (shown in the Figure for gains), where the frequency setting time was rather small and almost invariable. It confirmed the stabilizing effect of mutual bonds (the results are given to establish the frequency with an accuracy of 0.01 F).

A certain range shift to the left was explained by an increase in the equivalent gain due to the variable sampling period. According to the above results, for filter parameters that provided suppression at a sampling frequency close to 10 dB (m = 0.5, $\alpha\Phi$ = 0.5–1.0), the time for setting the frequency did not exceed 10 samples of the output ring in a wide range of gains.

7. Conclusions

Based on the general provisions of the theory of bifurcations, the directions for analyzing the conditions for the occurrence of periodic and quasiperiodic motions in a third order DCS with a piecewise linear characteristic of the detector are defined. As in the case of the second order DCS, the basis for the occurrence and loss of stability of k-multiple fixed points is the condition that the linear sections fall on the boundaries of linear sections. The mandatory requirement for the occurrence of quasiperiodic motions is the contact of the incoming and outgoing separatrix manifolds by a saddle point. The difference from the second order systems is in a large number of different types of saddle points and, accordingly, the number of possible scenarios of motions in the neighborhood of separatrix manifolds.

The method of estimating the bifurcation parameters of piecewise linear mappings of the third order has been developed. This makes it possible to find the boundaries of areas of the existence of various types of periodic and quasiperiodic motions. The method is based on the mandatory and sufficient conditions for the occurrence of a k-multiple fixed point through the formation of an intermediate complex point node-saddle or focus-saddle and the conditions for tangency of the incoming and outgoing separatrix manifolds at the boundaries of the linear sections of the characteristics.

Author Contributions: S.S. investigated the mathematical structure of the model; A.Z. described mathematical models and their formalization; S.C. carried out the analysis of mathematical structures and components and analysis of references; A.N. described and analyzed the analysis of mathematical components of algorithmization; D.M. investigated graphic structures and components.

Funding: This research was funded by Admiral Makarov State University of Maritime and Inland Shipping.

Conflicts of Interest: The authors declare no conflict of interest.

References

1. Li, S.; Chen, G.; Mou, X. On the Dynamical Degradation of Digital Piecewise Linear Chaotic Maps. *Int. J. Bifurc. Chaos* **2005**, *15*, 3119–3151. [CrossRef]
2. Nie, X.; Coca, D. A matrix-based approach to solving the inverse Frobenius–Perron problem using sequences of density functions of stochastically perturb e d dynamical systems. *Commun. Nonlinear Sci. Numer. Simulat.* **2018**, *54*, 248–266. [CrossRef] [PubMed]
3. Guariglia, E. Harmonic symmetry of the Riemann zeta fractional derivative. *AIP Conf. Proc.* **2018**, *2046*, 020035.
4. Guido, R.C.; Addison, P.; Walker, J. Introducing wavelets and time-frequency analysis. *IEEE Eng. Biol. Med. Mag.* **2009**, *28*, 13. [CrossRef] [PubMed]
5. Guido, R.C. Effectively interpretating discrete wavelet transforms signals. *IEEE Signal Process. Mag.* **2017**, *34*, 89–100. [CrossRef]
6. Gutkin, L.S. Designing Radio Systems and Radio Devices: Proc. Manual for Universities. Radio and Communication. 1986. Available online: https://www.rsl.ru/Gutkin/14579_ls (accessed on 10 April 2018).
7. Shahgildyan, V.V. Radio Transmitting Devices. Radio and Communication. 1990. Available online: https://publ.lib.ru/ARCHIVES/SH/SHAHGIL\T1\textquoterightDYAN_Vagan (accessed on 15 April 2018).
8. Akimov, V.N. Phase-synchronization Systems. Radio and Communication. 1982. Available online: https://www.rsl.ru/127 (accessed on 11 May 2018).
9. Best, R.E. *Phase-Locked Loops: Design, Simulation, and Application*, 3rd ed.; McGrow-Hill: New York, NY, USA, 1997; 360p.

10. Kazakov, L.N.; Paley, D.E.; Ponomarev, N.Y. Comparative Analysis of the Nonlinear Dynamics of Discrete Autonomous SPS of the 2nd and 3rd Orders. 1999. Available online: https://www.elar.uniyar.ac.ru/jspui/bitstream/19/kln_5 (accessed on 20 January 2017).
11. Kazakov, L.N.; Paley, D.E. Analysis of the Capture Band of a Third-order Pulse Phase-locking System with a Sawtooth Characteristic of the Detector. Radiotekhnika. 1998. Available online: https://www.elar.uniyar.ac.ru/jspui/bitstream/129/1/kln_1_a (accessed on 10 March 2017).
12. Fomin, A.F.; Khoroshavin, A.I.; Shelukhin, O.I. Analog and Digital Synchronous-phase Meters and Demodulators. Radio and communication. 1987. Available online: https://www.bsuir.by/m/12_119786_1_86932 (accessed on 30 April 2018).
13. Manchur, G.; Erven, C. Development of a Mode1 for Predicting Flicker from Electric Are Fumaces. *IEEE Trans. Power Deliv.* **1992**, *7*, 416–426. [CrossRef]
14. Zhilenkov, A.; Chernyi, S. Investigation performance of marine equipment with specialized information technology. *Procedia Eng.* **2015**, *100*, 1247–1252. [CrossRef]
15. Zhilenkov, A.; Chernyi, S. Models and Algorithms of the Positioning and Trajectory Stabilization System with Elements of Structural Analysis for Robotic Applications. *Int. J. Embed. Syst.* **2019**, in press.
16. Chernyi, S.; Zhilenkov, A. Modeling of complex structures for the ship's power complex using XILINX system. *Transp. Telecommun.* **2015**, *16*, 73–82. [CrossRef]
17. IEEE Working Group on Power System Harmonic. Bibliography of Power System Harmonics, Part 1 and Ii. Papers 84WM 214-3. In Proceedings of the IEEE PES Winter Meeting, Dallas, TX, USA, 29 January–3 February 1984.
18. IEEE Working Group on Power System Harmonic. Power System Harmonics: An Overview. *IEEE Trans. Power Appar. Syst.* **1983**, *PAS-102*, 2455–2460. [CrossRef]
19. Singh, B.; Al-Haddad, K.; Chandra, A. A Review of Active Filters for Power Quality Improvements. *IEEE Trans. Ind. Electron.* **1999**, *46*, 960–971. [CrossRef]
20. Govindarajan, S.N. Survey of Harmonic Levels on the Southwestern Electric power Company System. *IEEE Trans. Power Deliv.* **1991**, *6*, 1869–1873. [CrossRef]
21. Ding, Q.; Yang, Z.H. *Secure Communication System Based on Hardware Logic Encryption*; Posts & Telecom Press: Beijing, China, 2015.
22. Sokolov, S.; Zhilenkov, A.; Chernyi, S.; Nyrkov, A. Assessment of the Impact of Destabilizing Factors in the Main Engine Shaft of the Adaptive Speed Controller. *Procedia Comput. Sci.* **2018**, *125*, 420–426. [CrossRef]
23. Malioutov, D.; Corum, A.; Cetin, M. Covariance Matrix Estimation for Interest-Rate Risk Modeling via Smooth and Monotone Regularization. *IEEE J. Sel. Top. Signal Process.* **2016**, *10*, 1006–1014. [CrossRef]
24. Chaudhuri, A.; Stenger, H. Survey Sampling Theory and Methods. 2005. Available online: https://www.rsl.ru/ (accessed on 2 March 2018).
25. Nyrkov, A.; Zhilenkov, A.; Sokolov, S.; Chernyi, S. Hard- and Software Implementation of Emergency Prevention System for Maritime Transport. *Autom. Remote Control* **2018**, *79*, 195–202. [CrossRef]
26. Alekseev, G.D.; Karpovich, V.A. Power installations of fishing vessels. Shipbuilding. 1972. Available online: https://www.taylorfrancis.com/books/9781420028638 (accessed on 5 May 2017).
27. Sokolov, S.; Zhilenkov, A.; Nyrkov, A.; Chernyi, S. The Use Robotics for Underwater Research Complex Objects. *Adv. Intell. Syst. Comput.* **2017**, *556*, 421–427.
28. Chernyi, S.; Zhilenkov, A.; Sokolov, S.; Nyrkov, A. Algorithmic approach of destabilizing factors of improving the technical systems efficiency. *Vibroengineering PROCEDIA* **2017**, *13*, 261–265. [CrossRef]

symmetry MDPI

Article

Determining Crossing Number of Join of the Discrete Graph with Two Symmetric Graphs of Order Five

Michal Staš

Faculty of Electrical Engineering and Informatics, Technical University of Košice, 040 01 Košice, Slovakia; michal.stas@tuke.sk

Received: 19 December 2018; Accepted: 18 January 2019; Published: 22 January 2019

Abstract: The main aim of the paper is to give the crossing number of the join product $G + D_n$ for the disconnected graph G of order five consisting of one isolated vertex and of one vertex incident with some vertex of the three-cycle, and D_n consists of n isolated vertices. In the proofs, the idea of the new representation of the minimum numbers of crossings between two different subgraphs that do not cross the edges of the graph G by the graph of configurations $\mathcal{G}_D$ in the considered drawing D of $G + D_n$ will be used. Finally, by adding some edges to the graph G, we are able to obtain the crossing numbers of the join product with the discrete graph D_n and with the path P_n on n vertices for three other graphs.

Keywords: graph; good drawing; crossing number; join product; cyclic permutation

---

1. Introduction

The investigation of the crossing number of graphs is a classical and very difficult problem provided that computing of the crossing number of a given graph in general is an NP-complete problem. It is well known that the problem of reducing the number of crossings in the graph has been studied in many areas, and the most prominent area is very large-scale integration technology.

In the paper, we will use notations and definitions of the crossing numbers of graphs like in [1]. We will often use Kleitman's result [2] on crossing numbers of the complete bipartite graphs. More precisely, he proved that:

$$\mathrm{cr}(K_{m,n}) = \left\lfloor \frac{m}{2} \right\rfloor \left\lfloor \frac{m-1}{2} \right\rfloor \left\lfloor \frac{n}{2} \right\rfloor \left\lfloor \frac{n-1}{2} \right\rfloor, \quad \text{if} \quad m \leq 6.$$

Using Kleitman's result [2], the crossing numbers for join of two paths, join of two cycles, and for join of path and cycle were studied in [1]. Moreover, the exact values for crossing numbers of $G + D_n$ and $G + P_n$ for all graphs G of order at most four are given in [3]. Furthermore, the crossing numbers of the graphs $G + D_n$ are known for a few graphs G of order five and six in [4–10]. In all of these cases, the graph G is connected and contains at least one cycle. Further, the exact values for the crossing numbers $G + P_n$ and $G + C_n$ have been also investigated for some graphs G of order five and six in [5,7,11,12].

The methods presented in the paper are new, and they are based on multiple combinatorial properties of the cyclic permutations. It turns out that if the graph of configurations is used like a graphical representation of the minimum numbers of crossings between two different subgraphs, then the proof of the main theorem will be simpler to understand. Similar methods were partially used for the first time in the papers [8,13]. In [4,9,10,14], the properties of cyclic permutations were also verified with the help of software in [15]. In our opinion, the methods used in [3,5,7] do not allow establishing the crossing number of the join product $G + D_n$.

2. Cyclic Permutations and Configurations

Let G be the disconnected graph of order five consisting of one isolated vertex and of one vertex incident with some vertex of the three-cycle. We will consider the join product of the graph G with the discrete graph on n vertices denoted by D_n. The graph $G + D_n$ consists of one copy of the graph G and of n vertices $t_1, \dots, t_n$, where any vertex t_i, $i = 1, \dots, n$, is adjacent to every vertex of G. Let T^i, $1 \leq i \leq n$, denote the subgraph induced by the five edges incident with the vertex t_i. Thus, the graph $T^1 \cup \dots \cup T^n$ is isomorphic with the complete bipartite graph $K_{5,n}$ and:

$$G + D_n = G \cup K_{5,n} = G \cup \left(\bigcup_{i=1}^{n} T^i \right). \tag{1}$$

In the paper, we will use the same notation and definitions for cyclic permutations and the corresponding configurations for a good drawing D of the graph $G + D_n$ like in [9,14]. Let D be a drawing of the graph $G + D_n$. The rotation $\text{rot}_D(t_i)$ of a vertex t_i in the drawing D like the cyclic permutation that records the (cyclic) counter-clockwise order in which the edges leave t_i has been defined by Hernández-Vélez, Medina, and Salazar [13]. We use the notation (12345) if the counter-clockwise order the edges incident with the vertex t_i is t_iv_1, t_iv_2, t_iv_3, t_iv_4, and t_iv_5. We have to emphasize that a rotation is a cyclic permutation. In the paper, each cyclic permutation will be represented by the permutation with one in the first position. Let $\overline{\text{rot}}_D(t_i)$ denote the inverse permutation of $\text{rot}_D(t_i)$. We will deal with the minimal necessary number of crossings between the edges of T^i and the edges of T^j in a subgraph $T^i \cup T^j$ depending on the rotations $\text{rot}_D(t_i)$ and $\overline{\text{rot}}_D(t_j)$.

We will separate all subgraphs T^i, $i = 1, \dots, n$, of the graph $G + D_n$ into three mutually-disjoint subsets depending on how many of the considered T^i cross the edges of G in D. For $i = 1, \dots, n$, let $R_D = \{T^i : \text{cr}_D(G, T^i) = 0\}$ and $S_D = \{T^i : \text{cr}_D(G, T^i) = 1\}$. Every other subgraph T^i crosses the edges of G at least twice in D. Moreover, let F^i denote the subgraph $G \cup T^i$ for $T^i \in R_D$, where $i \in \{1, \dots, n\}$. Thus, for a given subdrawing of G, any subgraph F^i is exactly represented by $\text{rot}_D(t_i)$.

Let us suppose first a good drawing D of the graph $G + D_n$ in which the edges of G do not cross each other. In this case, without loss of generality, we can choose the vertex notation of the graph in such a way as shown in Figure 1a. Our aim is to list all possible rotations $\text{rot}_D(t_i)$ that can appear in D if the edges of T^i do not cross the edges of G. Since there is only one subdrawing of $F^i \setminus \{v_2, v_5\}$ represented by the rotation (143), there are two possibilities for how to obtain the subdrawing of $F^i \setminus v_5$ depending on in which region the edge t_iv_2 is placed. Of course, the vertex v_5 can be placed in one of four regions of the subdrawing $F^i \setminus v_5$ with the vertex t_i on their boundaries. These $2 \times 4 = 8$ possibilities under our consideration will be denoted by A_k and B_l, for $k = 1, 2$ and $l = 1, \dots, 6$. The configuration is of type A or B in the considered drawing D, if the vertex v_5 is placed in the quadrangular or in the triangular region in the subdrawing $D(F^i \setminus v_5)$, respectively. As for our considerations, it does not play a role in which of the regions is unbounded; assume the drawings shown in Figure 2. Thus, the configurations A_1, A_2, B_1, B_2, B_3, B_4, B_5, and B_6 are represented by the cyclic permutations (15432), (12435), (14532), (12453), (14325), (15243), (12543), and (14352), respectively. In a fixed drawing of the graph $G + D_n$, some configurations from $\mathcal{M}$ need not appear. We denote by $\mathcal{M}_D$ the subset of $\mathcal{M} = \{A_1, A_2, B_1, B_2, B_3, B_4, B_5, B_6\}$ consisting of all configurations that exist in the drawing D.

We remark that if two different subgraphs F^i and F^j with their configurations from $\mathcal{M}_D$ cross in a considered drawing D of the graph $G + D_n$, then the edges of T^i are crossed only by the edges of T^j. Let X, Y be the configurations from $\mathcal{M}_D$. We briefly denote by $\text{cr}_D(X, Y)$ the number of crossings in D between T^i and T^j for two different $T^i, T^j \in R_D$ such that F^i, F^j have configurations X, Y, respectively. Finally, let $\text{cr}(X, Y) = \min\{\text{cr}_D(X, Y)\}$ over all good drawings of the graph $G + D_n$ with $X, Y \in \mathcal{M}_D$. Our aim shall be to establish $\text{cr}(X, Y)$ for all pairs $X, Y \in \mathcal{M}$.

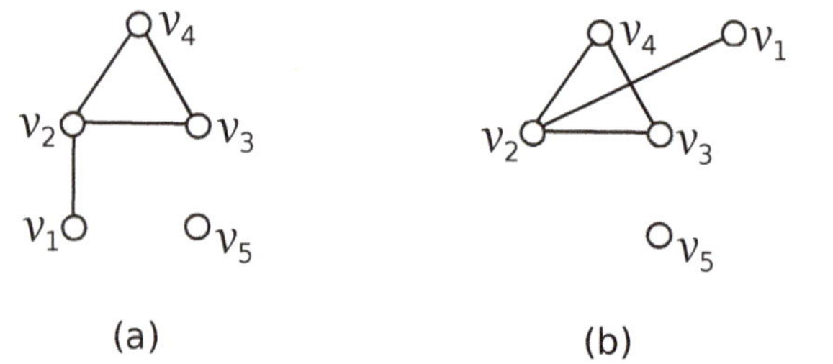

Figure 1. Two good drawings of the graph G. (**a**): the planar drawing of G; (**b**): the drawing of G with $\mathrm{cr}_D(G) = 1$.

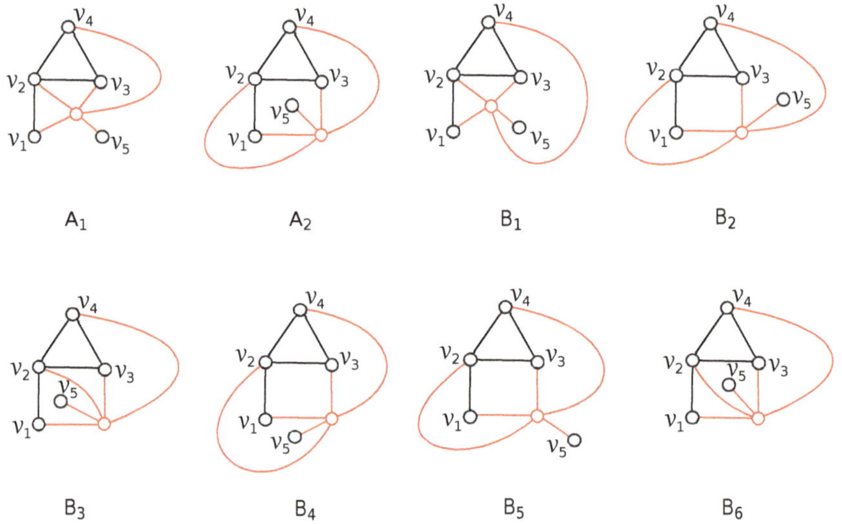

Figure 2. Drawings of eight possible configurations from $\mathcal{M}$ of the subgraph F^i.

The configurations A_1 and A_2 are represented by the cyclic permutations (15432) and (12435), respectively. Since the minimum number of interchanges of adjacent elements of (15432) required to produce cyclic permutation $\overline{(12435)} = (15342)$ is one, any subgraph T^j with the configuration A_2 of F^j crosses the edges of T^i at least once, i.e., $\mathrm{cr}(A_1, A_2) \geq 1$. Details have been worked out by Woodall [16]. The same reason gives $\mathrm{cr}(A_1, B_2) \geq 2$, $\mathrm{cr}(A_1, B_4) \geq 2$, $\mathrm{cr}(A_1, B_6) \geq 2$, $\mathrm{cr}(A_2, B_1) \geq 2$, $\mathrm{cr}(A_2, B_3) \geq 2$, $\mathrm{cr}(A_2, B_5) \geq 2$, $\mathrm{cr}(B_i, B_j) \geq 2$, and $\mathrm{cr}(A_i, B_j) \geq 3$ for $i \equiv j \pmod 2$. Moreover, by a discussion of possible subdrawings, we can verify that $\mathrm{cr}(B_1, B_5) \geq 4$, $\mathrm{cr}(B_3, B_5) \geq 4$, $\mathrm{cr}(B_2, B_6) \geq 4$, and $\mathrm{cr}(B_4, B_6) \geq 4$. Let F^i be the subgraph having the configuration B_5, and let T^j be a subgraph from R_D with $j \neq i$. Using Woodall's result $\mathrm{cr}_D(T^i, T^j) = Q(\mathrm{rot}_D(t_i), \mathrm{rot}_D(t_j)) + 2k$ for some nonnegative integer k, let us also suppose that $Q(\mathrm{rot}_D(t_i), \mathrm{rot}_D(t_j)) = 2$. Of course, any subgraph F^j having the configuration B_1 or B_3 satisfies the mentioned condition. One can easily see that if $t_j \in \omega_{1,2} \cup \omega_{3,4} \cup \omega_{1,2,3}$, then $\mathrm{cr}(T^i, T^j) > 2$. If $t_j \in \omega_{2,4,5}$ and $\mathrm{cr}(T^i, T^j) = 2$, then the subdrawing $D(F^j)$ induced by the edges incident with the vertices v_1 and v_3 crosses the edges of T^i exactly once, and once, respectively. Thus, $\mathrm{rot}_D(t_j) = (12435)$, i.e., the subgraph F^j has the configuration A_2. This forces $\mathrm{cr}(B_5, B_1) \geq 4$ and $\mathrm{cr}(B_5, B_3) \geq 4$. Similar arguments are applied for $\mathrm{cr}(B_6, B_2) \geq 4$ and $\mathrm{cr}(B_6, B_4) \geq 4$. Clearly, also $\mathrm{cr}(A_k, A_k) \geq 4$ and $\mathrm{cr}(B_l, B_l) \geq 4$ for any $k = 1, 2$ and $l = 1, \ldots, 6$. Thus, all lower bounds of the number of crossing of configurations from $\mathcal{M}$ are summarized in the symmetric Table 1 (here, X_k and Y_l are configurations of the subgraphs F^i and F^j, where k, l are integers from $\{1, 2\}$ or $\{1, \ldots, 6\}$, and $X, Y \in \{A, B\}$).

Table 1. The necessary number of crossings between T^i and T^j for the configurations X_k, Y_l.

−	A_1	A_2	B_1	B_2	B_3	B_4	B_5	B_6
A_1	4	1	3	2	3	2	3	2
A_2	1	4	2	3	2	3	2	3
B_1	3	2	4	3	2	3	4	3
B_2	2	3	3	4	3	2	3	4
B_3	3	2	2	3	4	3	4	3
B_4	2	3	3	2	3	4	3	4
B_5	3	2	4	3	4	3	4	3
B_6	2	3	3	4	3	4	3	4

Assume a good drawing D of the graph $G + D_n$ with one crossing among edges of the graph G (in which there is a subgraph $T^i \in R_D$). In this case, without loss of generality, we can choose also the vertex notations of the graph in such a way as shown in Figure 1b. Since there is only one subdrawing of $F^i \setminus \{v_5\}$ represented by the rotation (1324), we have four possibilities for how to obtain the subdrawing of F^i depending on in which region the vertex v_5 is placed. Thus, there are four different possible configurations of the subgraph F^i denoted as A_1, A_2, A_3, and A_4, with the corresponding rotations (13245), (13524), (13254), and (15324), respectively. We denote by $\mathcal{N}_D$ the subset of $\mathcal{N} = \{A_1, A_2, A_3, A_4\}$ consisting of all configurations that exist in the drawing D. The same way as above can be applied for the verification of the lower bounds of the number of crossings of two different configurations from $\mathcal{N}$. Thus, all lower bounds of the numbers of crossings of two configurations from $\mathcal{N}$ are summarized in the symmetric Table 2 (here, A_k and A_l are configurations of the subgraphs F^i and F^j, where $k, l \in \{1, 2, 3, 4\}$).

Table 2. The necessary number of crossings between T^i and T^j for the configurations A_k, A_l.

−	A_1	A_2	A_3	A_4
A_1	4	2	3	3
A_2	2	4	3	3
A_3	3	3	4	2
A_4	3	3	2	4

3. The Graph of Configurations $\mathcal{G}_D$

In general, the low possible number of crossings between two different subgraphs in a good subdrawing of $G + D_n$ is one of the main problems in the proofs on the crossing number of the join of the graph G with the discrete graphs D_n. The lower bounds of the numbers of crossings between two subgraphs, which do not cross the edges of G, were summarized in the symmetric Table 1. Since some configurations from the set $\mathcal{M}$ need not appear in the fixed drawing of $G + D_n$, we will first deal with the smallest possible values in Table 1 as with the worst possible case in the mentioned proofs. Thus, a new graphical representation of Table 1 by the graph of configurations will be useful.

Let us suppose that D is a good drawing of the graph $G + D_n$ with $\mathrm{cr}_D(G) = 0$, and let $\mathcal{M}_D$ be the nonempty set of all configurations that exist in the drawing D belonging to the set $\mathcal{M} = \{A_1, A_2, B_1, B_2, B_3, B_4, B_5, B_6\}$. A graph of configurations $\mathcal{G}_D$ is an ordered triple (V_D, E_D, w_D), where V_D is the set of vertices, E_D is the set of edges, which is formed by all unordered pairs of distinct vertices, and a weight function $w : E_D \to \mathbb{N}$ that associates with each edge of E_D an unordered pair of two vertices of V_D. The vertex $x_k \in V_D$ for some $x \in \{a, b\}$ if the corresponding configuration $X_k \in \mathcal{M}_D$ for some $X \in \{A, B\}$, where $k \in \{1, 2\}$ or $k \in \{1, \ldots, 6\}$. The edge $e = x_k y_l \in E_D$ if x_k and y_l are two different vertices of the graph $\mathcal{G}_D$. Finally, $w_D(e) = m \in \mathbb{N}$ for the edge $e = x_k y_l$, if m is the associated lower bound between two different configurations X_k, and Y_l in Table 1. Of course, $\mathcal{G}_D$ is the simple undirected edge-weighted graph uniquely determined by the drawing D. Moreover, if we define the graph $\mathcal{G} = (V, E, w)$ in the same way over the set $\mathcal{M}$, then $\mathcal{G}_D$ is the subgraph of $\mathcal{G}$

induced by V_D for the considered drawing D. Since the graph $\mathcal{G} = (V, E, w)$ can be represented like the edge-weighted complete graph K_8, it will be more transparent to follow the subcases in the proof of the main theorem; see Figure 3.

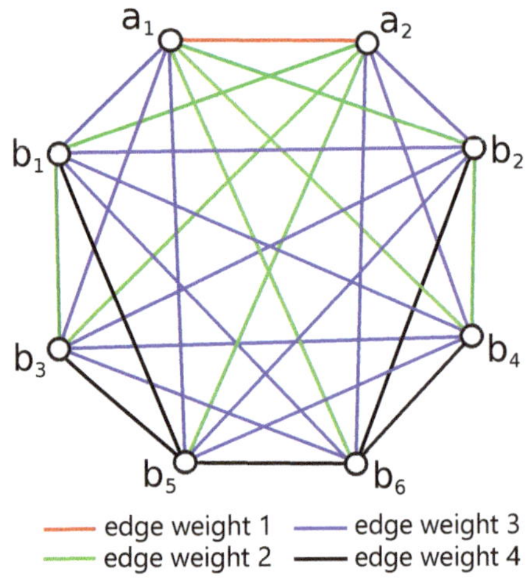

Figure 3. Representation of the lower bounds of Table 1 by the graph $\mathcal{G} = (V, E, w)$.

4. The Crossing Number of $G + D_n$

Two vertices t_i and t_j of $G + D_n$ are antipodal in a drawing of $G + D_n$ if the subgraphs T^i and T^j do not cross. A drawing is antipodal-free if it has no antipodal vertices. In the rest of the paper, each considered drawing of the graph $G + D_n$ will be assumed antipodal-free. In the proof of the main theorem, the following lemma related to some restricted subdrawings of the graph $G + D_n$ is helpful.

Lemma 1. *Let D be a good and antipodal-free drawing of $G + D_n$, $n > 2$. If T^i, $T^j \in R_D$ are different subgraphs such that F^i, F^j have different configurations from any of the sets $\{A_1, B_2\}$, $\{A_1, B_6\}$, $\{A_2, B_1\}$, and $\{A_2, B_5\}$, then:*

$$\mathrm{cr}_D(G \cup T^i \cup T^j, T^k) \geq 4 \qquad \text{for any } T^k \in S_D.$$

Proof of Lemma 1. Let us suppose the configuration A_1 of the subgraph F^i, and note that it is exactly represented by $\mathrm{rot}_D(t_i) = (15432)$. The unique drawing of the subgraph F^i contains four regions with the vertex t_i on their boundaries (Figure 2). If there is a $T^k \in S_D$ with $\mathrm{cr}_D(T^i, T^k) = 1$, then one can easily see that $t_k \in \omega_{1,2,4,5}$. Of course, the edge $t_k v_3$ must cross one edge of the graph G. If $t_k v_3$ crosses the edge $v_1 v_2$, then the subgraph F^k is represented by $\mathrm{rot}_D(t_k) = (13245)$. If the edge $t_k v_3$ crosses the edge $v_2 v_4$, then there are only three possibilities for the considered subdrawing of F^k, i.e., the subgraph F^k can be represented by three possible cyclic permutations (13452), (15234), or (12354).

For the remaining configurations A_2, B_1, B_2, B_5, and B_6 of F^i, using the same arguments, one can easily verify that the rotations of the vertex t_k are from the sets $\{(15324), (12534), (13425), (13542)\}$, $\{(12345), (14235)\}$, $\{(15342), (15423)\}$, $\{(12345)\}$, and $\{(15342)\}$, respectively. This forces that there is no subgraph $T^k \in S_D$ with $\mathrm{cr}_D(T^i \cup T^j, T^k) = 2$, where the subgraph F^j has the configuration B_2 or B_6. The same reason is given for the case of A_2 with the configurations B_1 and B_5. Finally, $\mathrm{cr}_D(G \cup T^i \cup T^j, T^k) \geq 1 + 3 = 4$ for any $T^k \in S_D$. This completes the proof. □

We have to emphasize that we cannot generalize Lemma 1 for all pairs of different configurations from $\mathcal{M}$. Let us assume the configurations A_1 of F^i and B_4 of F^j. For $T^k \in S_D$, the reader can easily find a subdrawing of $G \cup T^i \cup T^j \cup T^k$ in which $\mathrm{cr}_D(T^i, T^k) = \mathrm{cr}_D(T^j, T^k) = 1$. The same remark holds for pairs A_2 with B_3, B_1 with B_3, and B_2 with B_4.

Theorem 1. $\mathrm{cr}(G + D_n) = 4\left\lfloor \frac{n}{2} \right\rfloor \left\lfloor \frac{n-1}{2} \right\rfloor + \left\lfloor \frac{n}{2} \right\rfloor$ *for* $n \geq 1$.

Proof of Theorem 1. The drawing in Figure 4b shows that $\mathrm{cr}(G + D_n) \leq 4\left\lfloor \frac{n}{2} \right\rfloor \left\lfloor \frac{n-1}{2} \right\rfloor + \left\lfloor \frac{n}{2} \right\rfloor$. We prove the reverse inequality by contradiction. The graph $G + D_1$ is planar; hence, $\mathrm{cr}(G + D_1) = 0$. Since the graph $G + D_2$ contains a subdivision of the complete bipartite graph $K_{3,3}$, we have $\mathrm{cr}(G + D_2) \geq 1$. Thus, $\mathrm{cr}(G + D_2) = 1$ by the good drawing of $G + D_2$ in Figure 4a. Suppose now that for $n \geq 3$, there is a drawing D with:

$$\mathrm{cr}_D(G + D_n) < 4\left\lfloor \frac{n}{2} \right\rfloor \left\lfloor \frac{n-1}{2} \right\rfloor + \left\lfloor \frac{n}{2} \right\rfloor, \tag{2}$$

and let

$$\mathrm{cr}(G + D_m) \geq 4\left\lfloor \frac{m}{2} \right\rfloor \left\lfloor \frac{m-1}{2} \right\rfloor + \left\lfloor \frac{m}{2} \right\rfloor \quad \text{for any integer } m < n. \tag{3}$$

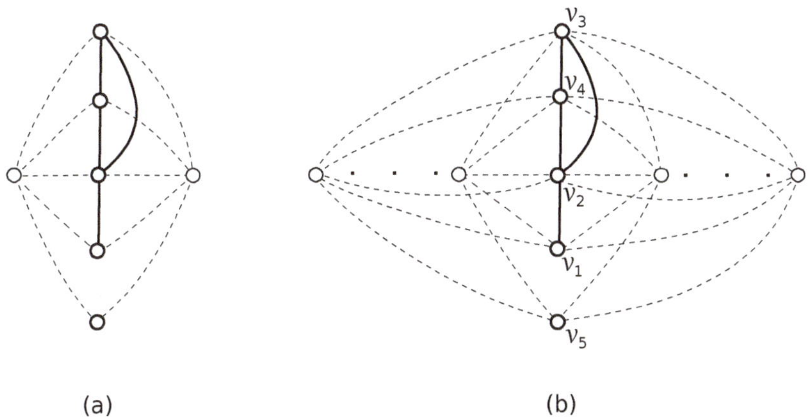

Figure 4. The good drawings of $G + D_2$ and of $G + D_n$. (**a**): the drawing of $G + D_2$ with one crossing; (**b**): the drawing of $G + D_n$ with $4\left\lfloor \frac{n}{2} \right\rfloor \left\lfloor \frac{n-1}{2} \right\rfloor + \left\lfloor \frac{n}{2} \right\rfloor$ crossings.

Let us first show that the considered drawing D must be antipodal-free. As a contradiction, suppose that, without loss of generality, $\mathrm{cr}_D(T^n, T^{n-1}) = 0$. Using positive values in Tables 1 and 2, one can easily verify that both subgraphs T^n and T^{n-1} are not from the set R_D, i.e., $\mathrm{cr}_D(G, T^n \cup T^{n-1}) \geq 1$. The known fact that $\mathrm{cr}(K_{5,3}) = 4$ implies that any T^k, $k = 1, \ldots, n-2$, crosses the edges of the subgraph $T^n \cup T^{n-1}$ at least four times. Therefore, for the number of crossings in the considered drawing D, we have:

$$\mathrm{cr}_D(G + D_n) = \mathrm{cr}_D\,(G + D_{n-2}) + \mathrm{cr}_D(G, T^n \cup T^{n-1}) + \mathrm{cr}_D(T^n \cup T^{n-1}) + \mathrm{cr}_D(K_{5,n-2}, T^n \cup T^{n-1})$$

$$\geq 4\left\lfloor \frac{n-2}{2} \right\rfloor \left\lfloor \frac{n-3}{2} \right\rfloor + \left\lfloor \frac{n-2}{2} \right\rfloor + 1 + 0 + 4(n-2) = 4\left\lfloor \frac{n}{2} \right\rfloor \left\lfloor \frac{n-1}{2} \right\rfloor + \left\lfloor \frac{n}{2} \right\rfloor.$$

This contradiction with the assumption (2) confirms that D must be an antipodal-free drawing. Moreover, if $r = |R_D|$ and $s = |S_D|$, the assumption (3) together with the well-known fact $\mathrm{cr}(K_{5,n}) = 4\left\lfloor \frac{n}{2} \right\rfloor \left\lfloor \frac{n-1}{2} \right\rfloor$ imply that in D, there are at least $\left\lceil \frac{n}{2} \right\rceil + 1$ subgraphs T^i, which do not cross the edges of G. More precisely:

$$\mathrm{cr}_D(G) + \mathrm{cr}_D(G, K_{5,n}) \leq \mathrm{cr}_D(G) + 0r + 1s + 2(n - r - s) < \left\lfloor \frac{n}{2} \right\rfloor,$$

i.e.,

$$s + 2(n - r - s) < \left\lfloor \frac{n}{2} \right\rfloor. \tag{4}$$

This forces that $r \geq 2$, and $r \geq \left\lceil \frac{n}{2} \right\rceil + 1$. Now, for $T^i \in R_D$, we will discuss the existence of possible configurations of subgraphs $F^i = G \cup T^i$ in the drawing D.

Case 1. $\mathrm{cr}_D(G) = 0$. Without loss of generality, we can choose the vertex notation of the graph G in such a way as shown in Figure 1a. Thus, we will deal with the configurations belonging to the nonempty set $\mathcal{M}_D$. According to the minimum value of the weights of edges in the graph $\mathcal{G}_D = (V_D, E_D, w_D)$, we will fix one, or two, or three subgraphs with a contradiction with the condition (2) in the following subcases:

i. $\{A_1, A_2\} \subseteq \mathcal{M}_D$, i.e., $w_D(a_1 a_2) = 1$. Without loss of generality, let us consider two different subgraphs T^n, $T^{n-1} \in R_D$ such that F^n and F^{n-1} have configurations A_1 and A_2, respectively. Then, $\mathrm{cr}_D(G \cup T^n \cup T^{n-1}, T^i) \geq 5$ for any $T^i \in R_D$ with $i \neq n-1, n$ by summing the values in all columns in the considered two rows of Table 1. Moreover, $\mathrm{cr}_D(T^n \cup T^{n-1}, T^i) \geq 3$ for any subgraph T^i with $i \neq n-1, n$ due to the properties of the cyclic permutations. Hence, by fixing the graph $G \cup T^n \cup T^{n-1}$,

$$\mathrm{cr}_D(G + D_n) \geq 4\left\lfloor \frac{n-2}{2} \right\rfloor \left\lfloor \frac{n-3}{2} \right\rfloor + 5(r-2) + 4(n-r) + 1 = 4\left\lfloor \frac{n-2}{2} \right\rfloor \left\lfloor \frac{n-3}{2} \right\rfloor + 4n + r - 9$$

$$\geq 4\left\lfloor \frac{n-2}{2} \right\rfloor \left\lfloor \frac{n-3}{2} \right\rfloor + 4n + \left(\left\lceil \frac{n}{2} \right\rceil + 1 \right) - 9 \geq 4\left\lfloor \frac{n}{2} \right\rfloor \left\lfloor \frac{n-1}{2} \right\rfloor + \left\lfloor \frac{n}{2} \right\rfloor.$$

ii. $\{A_1, A_2\} \not\subseteq \mathcal{M}_D$, i.e., $w_D(e) \geq 2$ for any $e \in E_D$.

Let us assume that $\{A_1, B_2, B_4\} \subseteq \mathcal{M}_D$ or $\{A_2, B_1, B_3\} \subseteq \mathcal{M}_D$, i.e., there is a three-cycle in the graph $\mathcal{G}_D$ with weights of two of all its edges. Without loss of generality, let us consider three different subgraphs T^n, T^{n-1} $T^{n-2} \in R_D$ such that F^n, F^{n-1}m and F^{n-2} have different configurations from $\{A_1, B_2, B_4\}$. Then, $\mathrm{cr}_D(G \cup T^n \cup T^{n-1} \cup T^{n-2}, T^i) \geq 8$ for any $T^i \in R_D$ with $i \neq n-1, n$ by Table 1, and $\mathrm{cr}_D(G \cup T^n \cup T^{n-1} \cup T^{n-2}, T^i) \geq 5$ for any subgraph $T^i \in S_D$ by Lemma 1. Thus, by fixing the graph $G \cup T^n \cup T^{n-1} \cup T^{n-2}$,

$$\mathrm{cr}_D(G + D_n) \geq 4\left\lfloor \frac{n-3}{2} \right\rfloor \left\lfloor \frac{n-4}{2} \right\rfloor + 8(r-3) + 5(n-r) + 6 \geq 4\left\lfloor \frac{n-3}{2} \right\rfloor \left\lfloor \frac{n-4}{2} \right\rfloor + 5n + 3r - 18$$

$$\geq 4\left\lfloor \frac{n-3}{2} \right\rfloor \left\lfloor \frac{n-4}{2} \right\rfloor + 5n + 3\left(\left\lceil \frac{n}{2} \right\rceil + 1 \right) - 18 \geq 4\left\lfloor \frac{n}{2} \right\rfloor \left\lfloor \frac{n-1}{2} \right\rfloor + \left\lfloor \frac{n}{2} \right\rfloor.$$

In the next part, let us suppose that $\{A_1, B_2, B_4\} \not\subseteq \mathcal{M}_D$ and $\{A_2, B_1, B_3\} \not\subseteq \mathcal{M}_D$,

(1) $\{A_j, B_k\} \subseteq \mathcal{M}_D$ for some $k \equiv j+1 \pmod 2$ or $\{B_j, B_{j+2}\} \subseteq \mathcal{M}_D$, where $j \in \{1,2\}$. Without loss of generality, let us consider two different subgraphs T^n, $T^{n-1} \in R_D$ such that F^n and F^{n-1} have configurations A_1 and B_2, respectively. Then, $\mathrm{cr}_D(G \cup T^n \cup T^{n-1}, T^i) \geq 6$ for any $T^i \in R_D$ with $i \neq n-1, n$ by Table 1. Moreover, $\mathrm{cr}_D(T^n \cup T^{n-1}, T^i) \geq 2$ for any subgraph T^i with $i \neq n-1, n$ due to properties of the cyclic permutations. Hence, if we fix the graph $G \cup T^n \cup T^{n-1}$,

$$\mathrm{cr}_D(G + D_n) \geq 4\left\lfloor \frac{n-2}{2} \right\rfloor \left\lfloor \frac{n-3}{2} \right\rfloor + 6(r-2) + 3s + 4(n-r-s) + 2 = 4\left\lfloor \frac{n-2}{2} \right\rfloor \left\lfloor \frac{n-3}{2} \right\rfloor$$

$$+4n + r + r - s - 10 \geq 4\left\lfloor \frac{n-2}{2} \right\rfloor \left\lfloor \frac{n-3}{2} \right\rfloor + 4n + \left\lceil \frac{n}{2} \right\rceil + 1 + 1 - 10 \geq 4\left\lfloor \frac{n}{2} \right\rfloor \left\lfloor \frac{n-1}{2} \right\rfloor + \left\lfloor \frac{n}{2} \right\rfloor.$$

(2) $\{A_j, B_k\} \not\subseteq \mathcal{M}_D$ for any $k \equiv j+1 \pmod 2$ and $\{B_j, B_{j+2}\} \not\subseteq \mathcal{M}_D$, where $j = 1, 2$, i.e., $w_D(e) \geq 3$ for any $e \in E_D$. Without loss of generality, we can assume that $T^n \in R_D$. Then, $\mathrm{cr}_D(T^n, T^i) \geq 3$ for any $T^i \in R_D$ with $i \neq n$. Thus, by fixing the graph $G \cup T^n$,

$$\mathrm{cr}_D(G+D_n) \geq 4\left\lfloor\frac{n-1}{2}\right\rfloor\left\lfloor\frac{n-2}{2}\right\rfloor + 3(r-1) + 2(n-r) + 0 = 4\left\lfloor\frac{n-1}{2}\right\rfloor\left\lfloor\frac{n-2}{2}\right\rfloor + 2n + r - 3$$
$$\geq 4\left\lfloor\frac{n-1}{2}\right\rfloor\left\lfloor\frac{n-2}{2}\right\rfloor + 2n + \left(\left\lceil\frac{n}{2}\right\rceil + 1\right) - 3 \geq 4\left\lfloor\frac{n}{2}\right\rfloor\left\lfloor\frac{n-1}{2}\right\rfloor + \left\lfloor\frac{n}{2}\right\rfloor.$$

Case 2. $\mathrm{cr}_D(G) = 1$. Without loss of generality, we can choose the vertex notation of the graph G in such a way as shown in Figure 1b. Thus, we will deal with the configurations belonging to the nonempty set $\mathcal{N}_D$ in the following two cases:

i. $\{A_i, A_{i+1}\} \subseteq \mathcal{N}_D$ for some $i \in \{1, 2\}$. Without loss of generality, let us consider two different subgraphs T^n, $T^{n-1} \in R_D$ such that F^n and F^{n-1} have different configurations from the set $\{A_1, A_2\}$. Then, $\mathrm{cr}_D(G \cup T^n \cup T^{n-1}, T^i) \geq 6$ for any $T^i \in R_D$ with $i \neq n-1, n$ by Table 2. Moreover, $\mathrm{cr}_D(T^n \cup T^{n-1}, T^i) \geq 2$ for any subgraph T^i with $i \neq n-1, n$ due to the properties of the cyclic permutations. Hence, by fixing the graph $G \cup T^n \cup T^{n-1}$,

$$\mathrm{cr}_D(G+D_n) \geq 4\left\lfloor\frac{n-2}{2}\right\rfloor\left\lfloor\frac{n-3}{2}\right\rfloor + 6(r-2) + 3s + 4(n-r-s) + 2 + 1 = 4\left\lfloor\frac{n-2}{2}\right\rfloor\left\lfloor\frac{n-3}{2}\right\rfloor$$
$$+4n + r + r - s - 9 \geq 4\left\lfloor\frac{n-2}{2}\right\rfloor\left\lfloor\frac{n-3}{2}\right\rfloor + 4n + \left\lceil\frac{n}{2}\right\rceil + 1 + 1 - 9 \geq 4\left\lfloor\frac{n}{2}\right\rfloor\left\lfloor\frac{n-1}{2}\right\rfloor + \left\lfloor\frac{n}{2}\right\rfloor.$$

If F^n and F^{n-1} have different configurations from the set $\{A_3, A_4\}$, then the same argument can be applied.

ii. $\{A_i, A_{i+1}\} \not\subseteq \mathcal{N}_D$ for any $i = 1, 2$. Without loss of generality, we can assume that $T^n \in R_D$. Then, $\mathrm{cr}_D(T^n, T^i) \geq 3$ for any $T^i \in R_D$ with $i \neq n$. Thus, by fixing the graph $G \cup T^n$,

$$\mathrm{cr}_D(G+D_n) \geq 4\left\lfloor\frac{n-1}{2}\right\rfloor\left\lfloor\frac{n-2}{2}\right\rfloor + 3(r-1) + 2(n-r) + 1 = 4\left\lfloor\frac{n-1}{2}\right\rfloor\left\lfloor\frac{n-2}{2}\right\rfloor + 2n + r - 2$$
$$\geq 4\left\lfloor\frac{n-1}{2}\right\rfloor\left\lfloor\frac{n-2}{2}\right\rfloor + 2n + \left(\left\lceil\frac{n}{2}\right\rceil + 1\right) - 2 \geq 4\left\lfloor\frac{n}{2}\right\rfloor\left\lfloor\frac{n-1}{2}\right\rfloor + \left\lfloor\frac{n}{2}\right\rfloor.$$

Thus, it was shown that there is no good drawing D of the graph $G + D_n$ with less than $4\lfloor\frac{n}{2}\rfloor\lfloor\frac{n-1}{2}\rfloor + \lfloor\frac{n}{2}\rfloor$ crossings. This completes the proof of Theorem 1. □

5. Three Other Graphs

Finally, in Figure 4b, we are able to add the edges v_3v_5 and v_1v_5 to the graph G without additional crossings, and we obtain three new graphs H_i for $i = 1, 2, 3$ in Figure 5. Therefore, the drawing of the graphs $H_1 + D_n$, $H_2 + D_n$, and $H_3 + D_n$ with $4\lfloor\frac{n}{2}\rfloor\lfloor\frac{n-1}{2}\rfloor + \lfloor\frac{n}{2}\rfloor$ crossings is obtained. On the other hand, $G + D_n$ is a subgraph of each $H_i + D_n$, and therefore, $\mathrm{cr}(H_i + D_n) \geq \mathrm{cr}(G + D_n)$ for any $i = 1, 2, 3$. Thus, the next results are obvious.

Corollary 1. $\mathrm{cr}(H_i + D_n) = 4\left\lfloor\frac{n}{2}\right\rfloor\left\lfloor\frac{n-1}{2}\right\rfloor + \left\lfloor\frac{n}{2}\right\rfloor$ *for* $n \geq 1$, *where* $i = 1, 2, 3$.

We remark that the crossing numbers of the graphs $H_1 + D_n$ and $H_3 + D_n$ were already obtained by Berežný and Staš [4], and Klešč and Schrötter [7], respectively. Moreover, into the drawing in Figure 4b, it is possible to add n edges, which form the path P_n, $n \geq 2$ on the vertices of D_n without another crossing. Thus, the next results are also obvious.

Theorem 2. $\mathrm{cr}(G + P_n) = \mathrm{cr}(H_2 + P_n) = 4\left\lfloor\frac{n}{2}\right\rfloor\left\lfloor\frac{n-1}{2}\right\rfloor + \left\lfloor\frac{n}{2}\right\rfloor$ *for* $n \geq 2$.

The crossing number of the graph $H_1 + P_n$ has been investigated in [12].

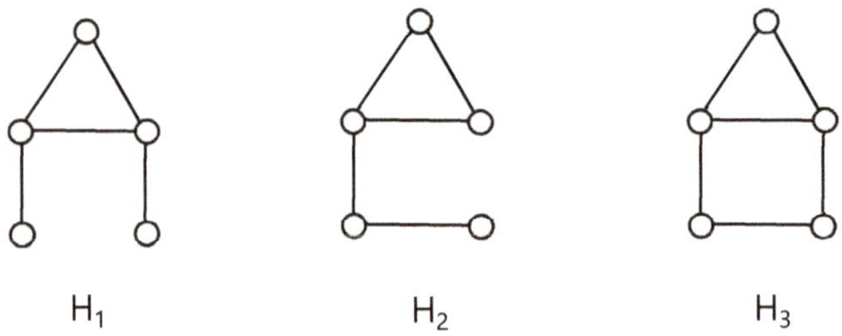

Figure 5. Three graphs H_1, H_2, and H_3 by adding new edges to the graph G.

Funding: This research received no external funding.

Acknowledgments: This work was supported by the internal faculty research Project No. FEI-2017-39.

Conflicts of Interest: The author declares no conflict of interest.

References

1. Klešč, M. The join of graphs and crossing numbers. *Electron. Notes Discret. Math.* **2007**, *28*, 349–355. [CrossRef]
2. Kleitman, D.J. The crossing number of $K_{5,n}$. *J. Comb. Theory* **1970**, *9*, 315–323. [CrossRef]
3. Klešč, M.; Schrötter, Š. The crossing numbers of join products of paths with graphs of order four. *Discuss. Math. Graph Theory* **2011**, *31*, 312–331. [CrossRef]
4. Berežný, Š.; Staš, M. On the crossing number of the join of five vertex graph G with the discrete graph D_n. *Acta Electrotech. Inform.* **2017**, *17*, 27–32. [CrossRef]
5. Klešč, M. The crossing numbers of join of the special graph on six vertices with path and cycle. *Discret. Math.* **2010**, *310*, 1475–1481. [CrossRef]
6. Klešč, M.; Petrillová, J.; Valo, M. On the crossing numbers of Cartesian products of wheels and trees. *Discuss. Math. Graph Theory* **2017**, *37*, 339–413. [CrossRef]
7. Klešč, M.; Schrötter, Š. The crossing numbers of join of paths and cycles with two graphs of order five. In *Lecture Notes in Computer Science: Mathematical Modeling and Computational Science;* Springer: Berlin/Heidelberg, Germany, 2012; Volume 7125, pp. 160–167.
8. Staš, M. On the crossing number of the join of the discrete graph with one graph of order five. *Math. Model. Geom.* **2017**, *5*, 12–19. [CrossRef]
9. Staš, M. Cyclic permutations: Crossing numbers of the join products of graphs. In Proceedings of the Aplimat 2018: 17th Conference on Applied Mathematics, Bratislava, Slovak, 6–8 February 2018; pp. 979–987.
10. Staš, M. Determining crossing numbers of graphs of order six using cyclic permutations. *Bull. Aust. Math. Soc.* **2018**, *98*, 353–362. [CrossRef]
11. Klešč, M.; Valo, M. Minimum crossings in join of graphs with paths and cycles. *Acta Electrotech. Inform.* **2012**, *12*, 32–37. [CrossRef]
12. Staš, M.; Petrillová, J. On the join products of two special graphs on five vertices with the path and the cycle. *Math. Model. Geom.* **2018**, *6*, 1–11.
13. Hernández-Vélez, C.; Medina, C.; Salazar, G. The optimal drawing of $K_{5,n}$. *Electron. J. Comb.* **2014**, *21*, 29.
14. Berežný, Š.; Staš, M. Cyclic permutations and crossing numbers of join products of symmetric graph of order six. *Carpathian J. Math.* **2018**, *34*, 143–155.
15. Berežný, Š.; Buša, J., Jr.; Staš, M. Software solution of the algorithm of the cyclic-order graph. *Acta Electrotech. Inform.* **2018**, *18*, 3–10. [CrossRef]
16. Woodall, D.R. Cyclic-order graphs and Zarankiewicz's crossing number conjecture. *J. Graph Theory* **1993**, *17*, 657–671. [CrossRef]

symmetry

MDPI

Article

Edge Even Graceful Labeling of Polar Grid Graphs

Salama Nagy Daoud [1,2]

[1] Department of Mathematics, Faculty of Science, Taibah University, Al-Madinah 41411, Saudi Arabia; salamadaoud@gmail.com

[2] Department of Mathematics and Computer Science, Faculty of Science, Menoufia University, Shebin El Kom 32511, Egypt

Received: 6 December 2018; Accepted: 23 December 2018; Published: 2 January 2019

Abstract: Edge Even Graceful Labelingwas first defined byElsonbaty and Daoud in 2017. An edge even graceful labeling of a simple graph G with p vertices and q edges is a bijection f from the edges of the graph to the set $\{2, 4, \ldots, 2q\}$ such that, when each vertex is assigned the sum of all edges incident to it mod$2r$ where $r = \max\{p, q\}$, the resulting vertex labels are distinct. In this paper we proved necessary and sufficient conditions for the polar grid graph to be edge even graceful graph.

Keywords: graceful labeling; edge graceful labeling; edge even graceful labeling; polar grid graph

1. Introduction

The field of Graph Theory plays an important role in various areas of pure and applied sciences. One of the important areas in graph theory is Graph Labeling of a graph G which is an assignment of integers either to the vertices or edges or both subject to certain conditions. Graph labeling is a very powerful tool that eventually makes things in different fields very ease to be handled in mathematical way. Nowadays graph labeling has much attention from different brilliant researches ingraph theory which has rigorous applications in many disciplines, e.g., communication networks, coding theory, x-raycrystallography, radar, astronomy, circuit design, communication network addressing, data base management and graph decomposition problems. More interesting applications of graph labeling can be found in [1–10].

Let $G = (V(G), E(G))$ with $p = |V(G)|$ and $q = |E(G)|$ be a simple, connected, finite, undirected graph.

A function f is called a graceful labeling of a graph G if $f : V(G) \to \{0, 1, 2, \ldots, q\}$ is injective and the induced function $f^* : E(G) \to \{1, 2, \ldots, q\}$ defined as $f^*(e = uv) = |f(u) - f(v)|$ is bijective. This type of graph labeling first introducedby Rosa in 1967 [11] as a $\beta-$ valuation, later on Solomon W. Golomb [12] called as graceful labeling.

A function f is called an odd graceful labeling of a graph G if $f : V(G) \to \{0, 1, 2, \ldots, 2q - 1\}$ is injective and the induced function $f^* : E(G) \to \{1, 3, \ldots, 2q - 1\}$ defined as $f^*(e = uv) = |f(u) - f(v)|$ is bijective. This type of graph labeling first introducedby Gnanajothi in 1991 [13]. For more results on this type of labeling see [14,15].

A function f is called an edge graceful labeling of a graph G if $f : E(G) \to \{1, 2, \ldots, q\}$ is bijective and the induced function $f^* : V(G) \to \{0, 1, 2, \ldots, p - 1\}$ defined as $f^*(u) = \sum_{e=uv \in E(G)} f(e) \bmod p$ is bijective. This type of graph labeling first introducedby Lo in 1985 [16]. For more results on this labeling see [17,18].

A function f is called an edge odd graceful labeling of a graph G if $f : E(G) \to \{1, 3, \ldots, 2q - 1\}$ is bijective and the induced function $f^* : V(G) \to \{0, 1, 2, \ldots, 2q - 1\}$ defined as $f^*(u) = \sum_{e=uv \in E(G)} f(e) \bmod 2q$ is injective. This type of graph labeling first introducedby Solairaju and Chithra in 2009 [19]. See also Daoud [20].

A function f is called an edge even graceful labeling of a graph G if $f : E(G) \to \{2, 4, \ldots, 2q\}$ is bijective and the induced function $f^* : V(G) \to \{0, 2, 4, \ldots, 2q - 2\}$ defined as $f^*(u) = \sum_{e=uv \in E(G)} f(e) \bmod 2r$, where $r = \max\{p, q\}$ is injective. This type of graph labeling first introduced by Elsonbaty and Daoud in 2017 [21].

For a summary of the results on these five types of graceful labels as well as all known labels so far, see [22].

2. Polar Grid Graph $P_{m,n}$

The polargrid graph $P_{m,n}$ is the graph consists of n copies of circles C_m which will be numbered from the inner most circle to the outer circle as $C_m^{(1)}, C_m^{(2)}, \ldots, C_m^{(n-1)}, C_m^{(n)}$ and m copies of paths P_{n+1} intersected at the center vertex v_0 which will be numbered as $P_{n+1}^{(1)}, P_{n+1}^{(2)}, \ldots, P_{n+1}^{(m-1)}, P_{n+1}^{(m)}$. See Figure 1.

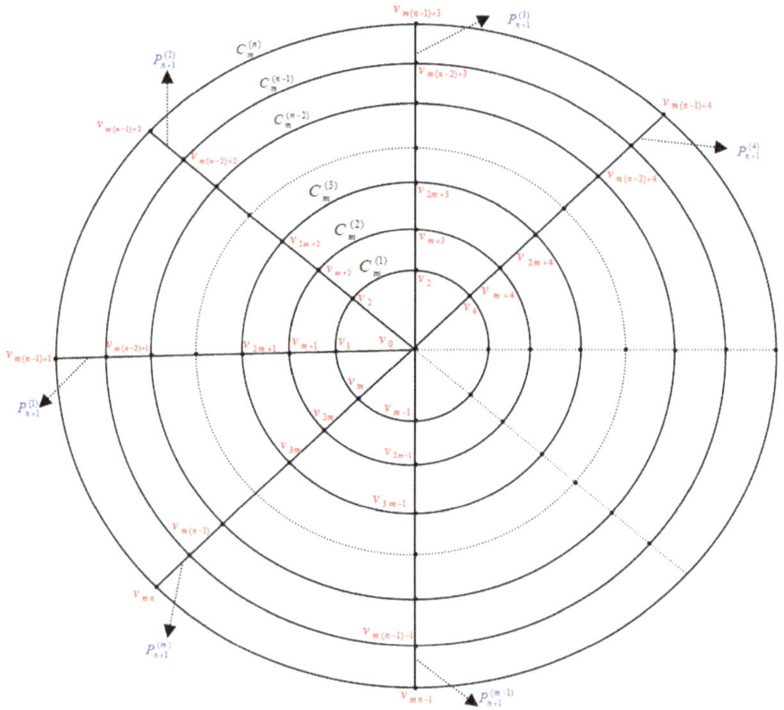

Figure 1. Polar grid graph $P_{m,n}$.

Theorem 1. *If m and n are even positive integes such that $m \geq 4$ and $n \geq 2$, then the polar grid graph $P_{m,n}$ is an edge even graceful graph.*

Proof. Using standard notation $p = |V(P_{m,n})| = mn + 1$, $q = |E(P_{m,n})| = 2mn$ and $r = \max\{p, q\} = 2mn$ Let the polar grid graph $P_{m,n}$ be labeled as in Figure 2. Let $f : E(G) \to \{2, 4, \ldots, 2q\}$. □

First we label the edges of paths $P_{n+1}^{(k)}, \leq k \leq m$ begin with the edges of the path $P_{n+1}^{(1)}$ to the edges of the path $P_{n+1}^{(m)}$ as follows: Move clockwise to label the edges $v_0v_1, v_0v_2, \ldots, v_0v_{m-1}, v_0v_m$ by $2, 4, \ldots, 2m - 2, 2m$, then move anticlockwise to label the edges

$v_1v_{m+1}, v_mv_{2m}, v_{m-1}v_{2m-1}, \ldots, v_3v_{m+3}, v_2v_{m+2}$ by $2m+2, 2m+4, 2m+6, \ldots, 4m-2, 4m,$ then move clockwise to label the edges $v_{m+1}v_{2m+1}, v_{m+2}v_{2m+2}, v_{m+3}v_{2m+3}, \ldots, v_{2m-1}v_{3m-1}, v_mv_{3m}$ by $4m+2, 4m+4, 4m+6, \ldots, 6m-2, 6m$ and so on. Finally move anticlockwise to label the edges $v_{m(n-2)+1}v_{m(n-1)+1}, v_{m(n-1)-1}v_{mn}, v_{m(n-1)-1}v_{mn-1}, \ldots, v_{m(n-2)+2}v_{m(n-1)+2}$ by $2m(n-1)+2, 2m(n-1)+4, 2m(n-1)+6, \ldots, 2mn-2, 2mn$. Second we label the edges of the circles $C_m^{(k)}, 1 \le k \le n$ begin with the edges of the inner most circle $C_m^{(1)}$ to the edges of the circle $C_m^{(\frac{n}{2})}$, then the edges of the outer circle $C_m^{(n)}$. Finally the edges of circles $C_m^{(\frac{n}{2}+1)}, C_m^{(\frac{n}{2}+2)}, \ldots, C_m^{(n-1)}$ respectively as follows: $f(v_{m(k-1)+i}v_{m(k-1)+i+1}) = 2mn+2m(k-1)+2i$, $f(v_{km}v_{m(k-1)+1}) = 2mn+2mk$, $1 \le i \le m-1, 1 \le k \le \frac{n}{2}$; $f(v_{m(n-1)+i}v_{m(n-1)+i+1}) = 3mn+2i$, $1 \le i \le m-1$; $f(v_{km}v_{m(k-1)+1}) = 2mn+(2k+1)m$, $\frac{n}{2}+1 \le k \le n-1$.

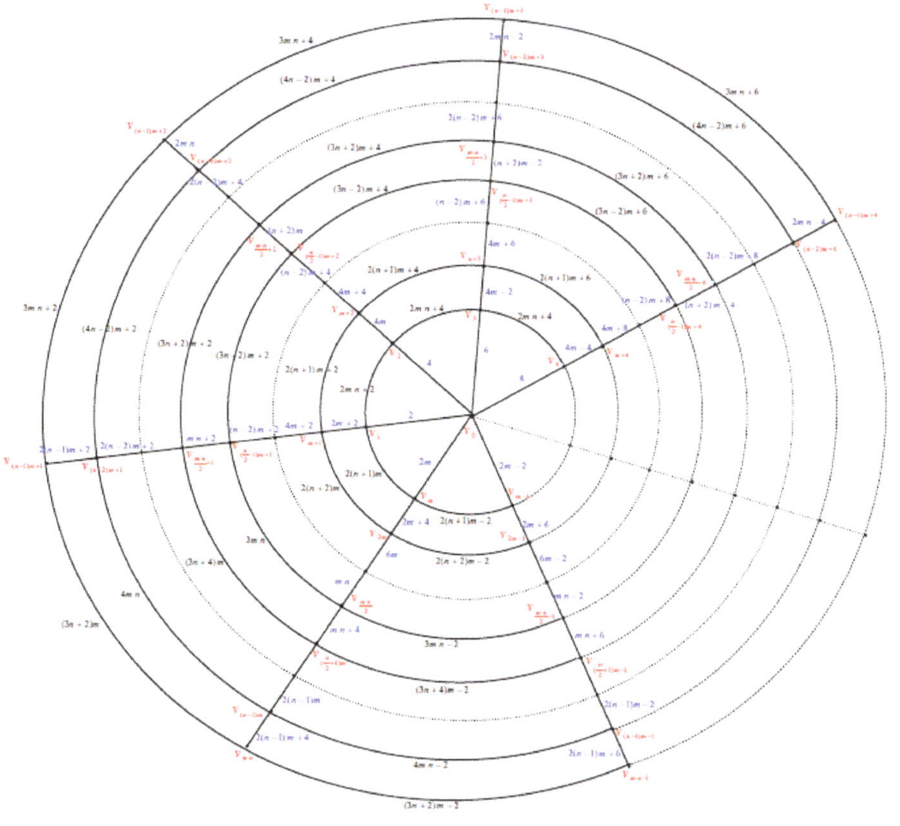

Figure 2. Labeling ofthe polar grid graph $P_{m,n}$ when n is even, $n \ge 2$.

Now the corresponding labels of vertices mod$4mn$ are assigned as follows:

Case (1) $m \equiv 4k\text{mod}4n$, $1 \le k \le n-1$ and $m = 2\text{mod}4n$.

The labels of the vertices of the inner most circle $C_m^{(1)}$ to the circle $C_m^{(\frac{n}{2})}$ are given by $f^*(v_{(k-1)m+i}) \equiv 4m(2k-1)+4i$, $1 \le k \le \frac{n}{2}, 1 \le i \le m$, the labels of the vertices of the outre circle $C_m^{(n)}$ are given by $f^*(v_{(n-1)m+i}) \equiv 2i+2$, $1 \le i \le m$ and the labels of the vertices of the circles $C_m^{(\frac{n}{2}+1)}, \ldots, C_m^{(n-1)}$ are given by $f^*(v_{(k-1)m+i}) \equiv 8m(k-\frac{n}{2})+4i+2$, $1 \le i \le m$, $\frac{n}{2}+1 \le k \le n-1$.

The label of the center vertex v_0 is assigned as follows: when $m = 4k\text{mod}4n$, $1 \le k \le n-1$, $f^*(v_0) \equiv \frac{m}{2}(2m+2) = m^2+m$, since $m = 4k\text{mod}4n$ then $m = 4nh+4k$, thus $f^*(v_0) = m(4k+1)$ and when $m = 2\text{mod}4n$, we have $f^*(v_0) = 3m$.

Case (2) $m \equiv (8k-2) \bmod 4n$, $1 \leq k \leq \frac{n}{2}$. In this case the vertex $v_{km-(\frac{m+2}{4})}$ in the circle $C_m^{(k)}$ will repeat with the center vertex v_0. To avoid this problem we replace the labels of the two edges $v_{km-(\frac{m+2}{4})}v_{km-(\frac{m-2}{4})}$ and $v_{km-(\frac{m-2}{4})}v_{km-(\frac{m-6}{4})}$. That is $f(v_{km-(\frac{m+2}{4})}v_{km-(\frac{m-2}{4})}) = 2mn + m(2k-1) + \frac{m+2}{2}$ and $f(v_{km-(\frac{m-2}{4})}v_{km-(\frac{m-6}{4})}) = 2mn + m(2k-1) + \frac{m-2}{2}$ and we obtain the labels of the corresponding vertices as follows $f^*(v_{km-(\frac{m+2}{4})}) \equiv m(8k-1)+2$, $f^*(v_{km-(\frac{m-2}{4})}) \equiv m(8k-1)+4$, $f^*(v_{km-(\frac{m-6}{4})}) \equiv m(8k-1)+6$ and the label of the center vertex v_0 is assigened as $f^*(v_0) \equiv m(8k-1)$.The rest vertices are labeled as in case(1).

Case (3) $m \equiv (8k+2) \bmod 4n$, $1 \leq k \leq \frac{n}{2} - 1$. In this case the vertex $v_{km-(\frac{m+2}{2})}$ in the circle $C_m^{(\frac{n}{2}+k)}$ will repeat with the center vertex v_0. To avoid this problem we replace the labels of the two edges $v_{km-(\frac{m+2}{2})}v_{km-(\frac{m}{2})}$ and $v_{km-(\frac{m}{2})}v_{km-(\frac{m-2}{2})}$. That is $f(v_{km-(\frac{m+2}{2})}v_{km-(\frac{m}{2})}) = 2mn + m(2k+1) + \frac{m+2}{2}$ and $f(v_{km-(\frac{m}{2})}v_{km-(\frac{m-2}{2})}) = 2mn + m(2k+1) + \frac{m-2}{2}$ and we obtain the labels of the corresponding vertices as follows $f^*(v_{km-(\frac{m+2}{2})}) \equiv m(8k+3)+2$, $f^*(v_{km-(\frac{m}{2})}) \equiv m(8k+3)+4$, $f^*(v_{km-(\frac{m-2}{2})}) \equiv m(8k+3)+6$ and the label of the center vertex v_0 is assigened $f^*(v_0) \equiv m(8k-29)$ as. The rest vertices are labeled as in case(1).

Case (4) $m \equiv 0 \bmod 4n$. In this case the vertex $v_{mn-(\frac{m+2}{2})}$ in the outer circle will repeat with the center vertex v_0. To avoid this problem we replace the labels of the two edges $v_{mn-(\frac{m+4}{2})}v_{mn-(\frac{m+2}{2})}$ and $v_{mn}v_{m(n-1)+1}$. That is $f(v_{mn-(\frac{m+4}{2})}v_{mn-(\frac{m+2}{2})}) = m(3n+2)$ and $f(v_{mn}v_{m(n-1)+1}) = 3mn + m - 4$ and we obtain the labels of the corresponding vertices as follows $f^*(v_{mn-(\frac{m+4}{2})}) \equiv 2m+2$, $f^*(v_{mn-(\frac{m+2}{2})}) \equiv 2m+4$, $f^*(v_{mn}) \equiv m-2$ and $f^*(v_{m(n-1)+1}) \equiv 4mn - m$ and the label of the center vertex v_0 is assigened as $f^*(v_0) \equiv m$. The rest vertices are labeled as in case (1).

Illustration. *The edge even graceful labeling of the polar grid graphs* $P_{14,6}$, $P_{16,6}$, $P_{18,6}$, $P_{24,6}$ *and* $P_{26,6}$ *respectively are shown in Figure 3.*

Remark 1. *In case* $m = 2$ *and* n *is even,* $n > 2$.

Let the label of edges of the polar grid graph be as in Figure 4. Thus we have the label of the corresponding vertices are as follows:

$f^*(v_1) \equiv 4n+12$; $f^*(v_i) \equiv 16i-2$, $2 \leq i \leq \frac{n}{2}$; $f^*(v_{\frac{n}{2}+i}) \equiv 16i+6$, $1 \leq i \leq \frac{n}{2}-1$;
$f^*(v_n) \equiv 4$; $f^*(v'_i) \equiv 16i+2$, $1 \leq i \leq \frac{n}{2}-1$; $f^*(v'_{\frac{n}{2}+i}) \equiv 2$; $f^*(v'_{\frac{n}{2}+i}) \equiv 16i+10$, $1 \leq i \leq \frac{n}{2}-2$;
$f^*(v'_{n-1}) \equiv 4n-4$; $f^*(v'_n) \equiv 4n+8$ and $f^*(v_0) \equiv 4n+4$.

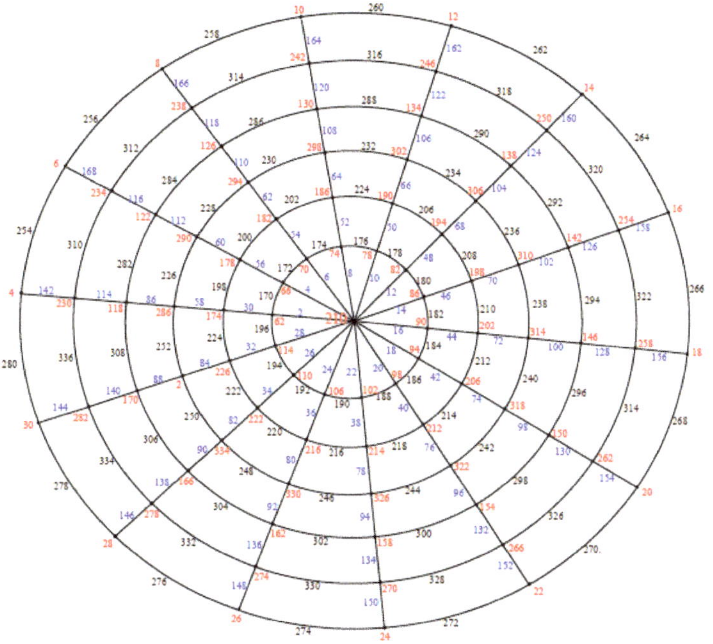

(**a**) $P_{14,6}$

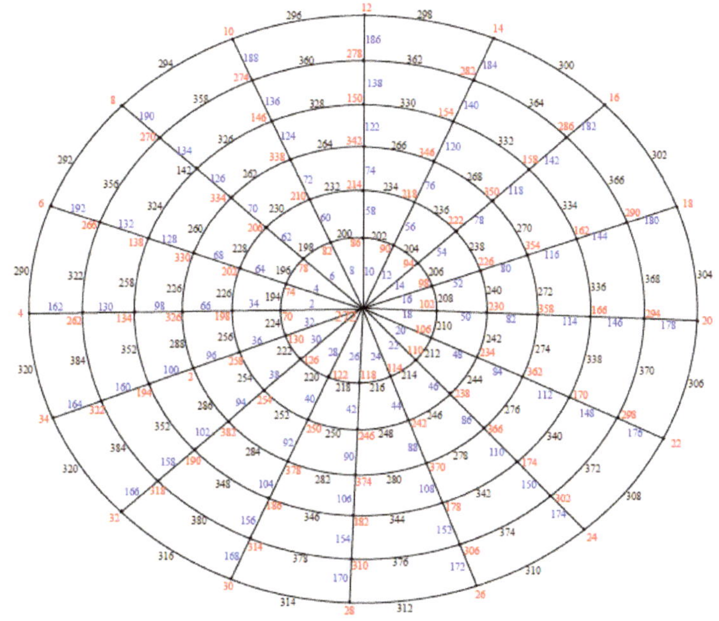

(**b**) $P_{16,6}$

Figure 3. *Cont.*

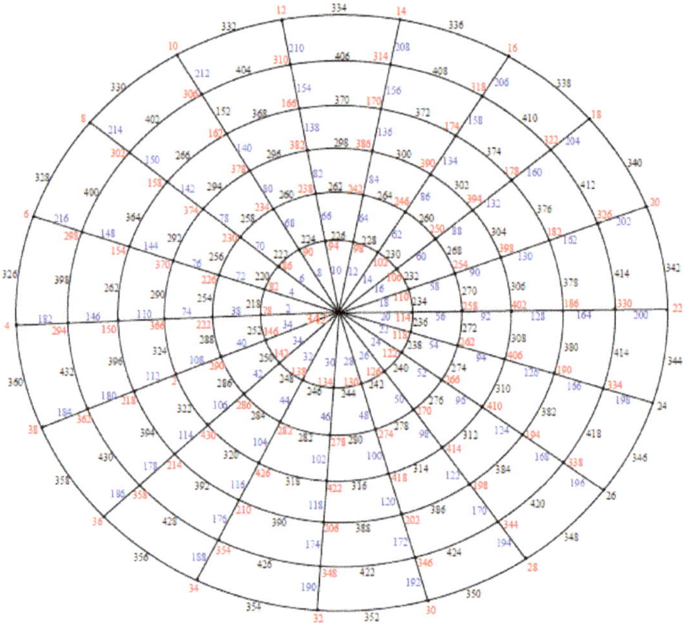

(c) $P_{18,6}$

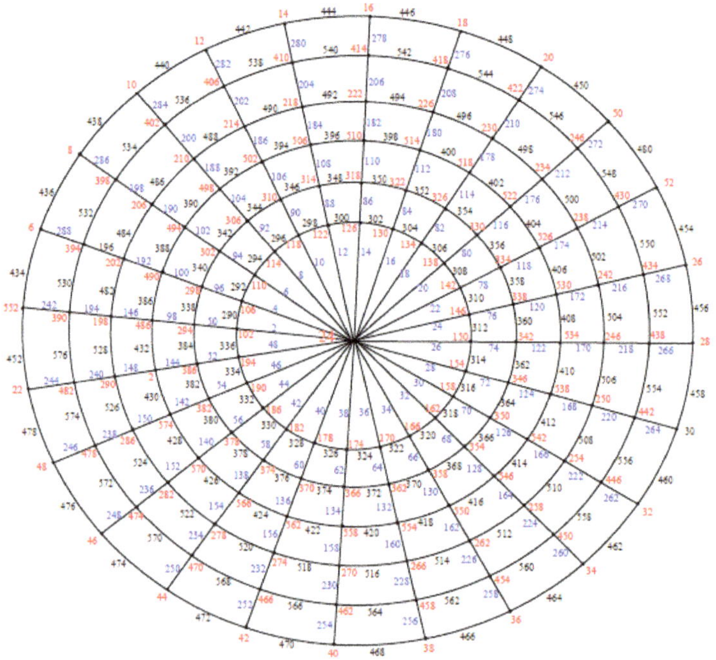

(d) $P_{24,6}$

Figure 3. *Cont.*

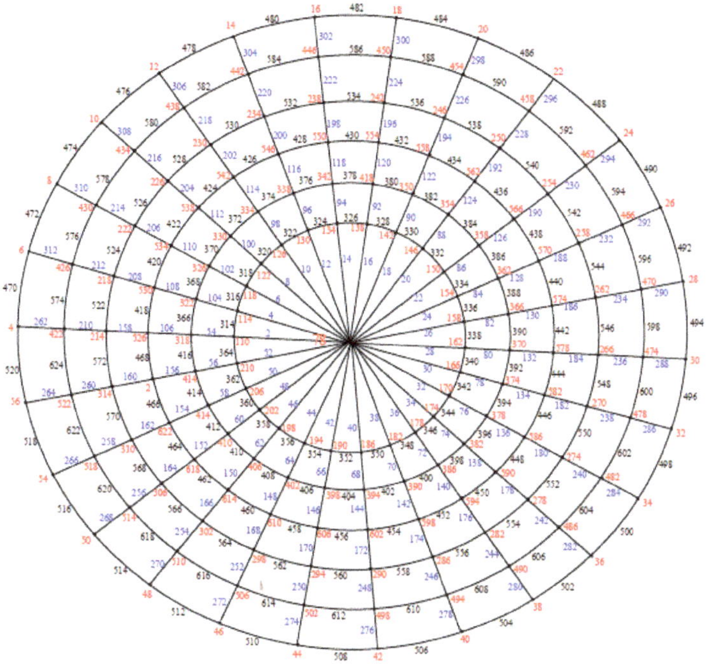

(**e**) $P_{26,6}$

Figure 3. The edge even graceful labeling of the polar grid graphs $P_{14,6}$, $P_{16,6}$, $P_{18,6}$, $P_{24,6}$ and $P_{26,6}$.

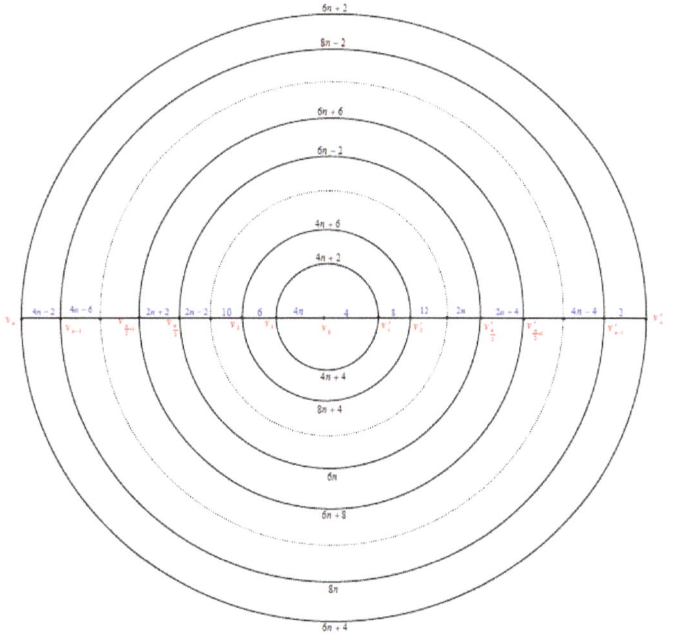

Figure 4. Labeling of the polar grid graph $P_{2,n}$, n is even integer greater than 2.

Note that $P_{2,2}$ is an edge even graceful graph but not follow this rule. See Figure 5.

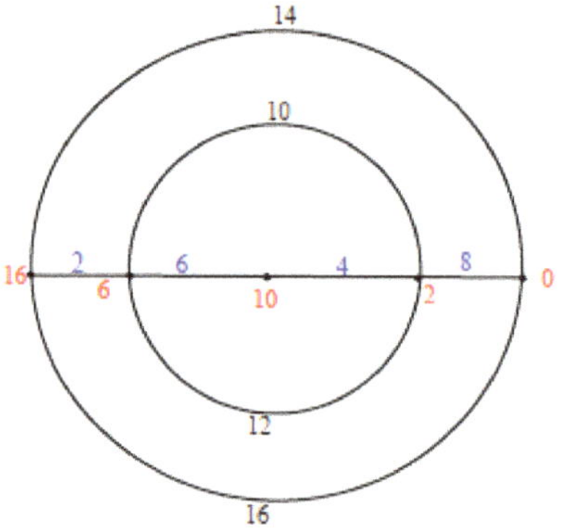

Figure 5. The polar grid graph $P_{2,2}$.

Theorem 2. *If m is an odd positive integer greater than* 1 *and n an is even positive integer greater than or equal* 2*, then the polar grid graph $P_{m,n}$ is an edge even graceful graph.*

Proof. Let the edges of the polar grid graph $P_{m,n}$ be labeled as in Figure 2. □

Now the corresponding labels of vertices mod$4mn$ are assigned as follows: There are four cases
Case (1): $m \equiv (4k-1)\text{mod}4n,\ 1 \le k \le n$

The labels of the vertices of the inner most circle $C_m^{(1)}$ to the circle $C_m^{(\frac{n}{2})}$ are given by $f^*(v_{(k-1)m+i}) \equiv 4m(2k-1)+4i+2,\ 1 \le k \le \frac{n}{2},\ 1 \le i \le m$, the labels of the vertices of the outer circle $C_m^{(n)}$ are given by $f^*(v_{(n-1)m+i}) \equiv 2i+2,\ 1 \le i \le m$ and the labels of the vertices of the circles $C_m^{(\frac{n}{2}+1)}, \ldots, C_m^{(n-1)}$ are given by $f^*(v_{(k-1)m+i}) \equiv 8m(k-\frac{n}{2})+4i+2,\ 1 \le i \le m,\ \frac{n}{2}+1 \le k \le n-1$.

The center vertex v_0 is labeled as $f^*(v_0) \equiv 4mk$, and if $k=n$, we have $f^*(v_0) \equiv 0$.

Case (2): $m \equiv (8k-3)\text{mod}4n,\ 1 \le k \le \frac{n}{2}$.

In this case the vertex $v_{km(\frac{m+1}{2})}$ in the circle $C_m^{(k)}$ will repeat with the center vertex v_0. To avoid this problem we replace the labels of the two edges $v_{km-(\frac{m+1}{2})}v_{km-(\frac{m-1}{2})}$ and $v_{km-(\frac{m-1}{2})}v_{km-(\frac{m-3}{2})}$. That is $f(v_{km-(\frac{m+1}{2})}v_{km-(\frac{m-1}{2})}) = 2mn+m(2k-1)+1$ and $f(v_{km-(\frac{m-1}{2})}v_{km-(\frac{m-3}{2})}) = 2mn+m(2k-1)-1$ and we obtain the labels of the corresponding vertices as follows $f^*(v_{km-(\frac{m+1}{2})}) \equiv 2m(4k-1)+2,\ f^*(v_{km-(\frac{m-1}{2})}) \equiv 2m(4k-1)+4$ and $f^*(v_{km-(\frac{m-3}{2})}) \equiv 2m(4k-1)+6$. The center vertex v_0 is labeled as $f^*(v_0) = 2m(4k-1)$. The rest vertices are labeled as in case (1).

Case (3): $m \equiv (8k+1)\text{mod}4n,\ 1 \le k \le \frac{n}{2}-1$.

In this case the vertex $v_{m(\frac{n}{2}+k)(\frac{m+1}{2})}$ in the circle $C_m^{(\frac{n}{2}+k)}$ will repeat with the center vertex v_0. To avoid this problem we replace the labels of the two edges $v_{m(\frac{n}{2}+k)-(\frac{m+1}{2})}v_{m(\frac{n}{2}+k)(\frac{m-1}{2})}$ and $v_{m(\frac{n}{2}+k)-(\frac{m-1}{2})}v_{m(\frac{n}{2}+k)-(\frac{m-3}{2})}$. That is $f(v_{m(\frac{n}{2}+k)-(\frac{m+1}{2})}v_{m(\frac{n}{2}+k)-(\frac{m-1}{2})}) = 3mn+m(2k-1)+1$ and $f(v_{m(\frac{n}{2}+k)-(\frac{m-1}{2})}v_{m(\frac{n}{2}+k)-(\frac{m-3}{2})}) = 3mn+m(2k+1)-1$ and we obtain the labels of the corresponding vertices as follows $f^*(v_{m(\frac{n}{2}+k)-(\frac{m+1}{2})}) \equiv 2m(4k+1)+2,\ f^*(v_{m(\frac{n}{2}+k)-(\frac{m-1}{2})}) \equiv 2m(4k+1)+4$, and $f^*(v_{m(\frac{n}{2}+k)-(\frac{m-3}{2})}) \equiv 2m(4k+1)+6$ and in this case the center vertex v_0 is labeled as $f^*(v_0) = 2m(4k+1)$. The rest vertices are labeled as in case (1).

Case (4): $m \equiv 1\text{mod}4n$

In this case the vertex v_{mn-1} in the outer circle $C_m^{(n)}$ will repeat with the center vertex v_0. To avoid this problem we replace the labels of the two edges $v_{mn-2}v_{mn-1}$ and $v_{mn}v_{m(n-1)+1}$. That is $f(v_{mn-2}v_{mn-1}) = m(3n+2)$ and $f(v_{mn}v_{m(n-1)+1}) = m(3n+2)-4$ and we obtain the labels of the corresponding vertices as follows $f^*(v_{mn-2}) = 2m+2,\ f^*(v_{mn-1}) \equiv 2m+4,\ f^*(v_{mn}) \equiv 2m-2$ and $f^*(v_{m(n-1)+1}) \equiv 0$, the center vertex v_0 is labeled as $f^*(v_0) \equiv 2m$. The rest vertices are labeled as in case (1).

Illustration. *The edge even graceful labeling of the polar grid graphs* $P_{13,6}$, $P_{15,6}$, $P_{17,6}$ *and* $P_{25,6}$ *respectively are shown in Figure 6.*

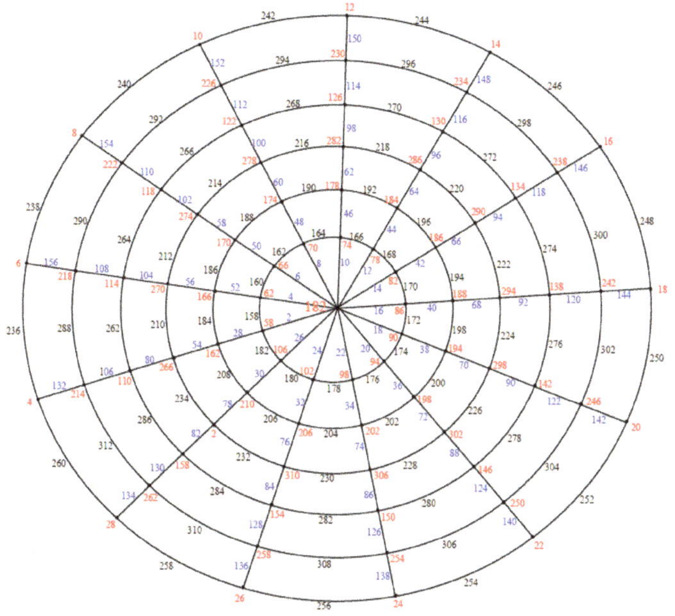

(**a**) $P_{13,6}$

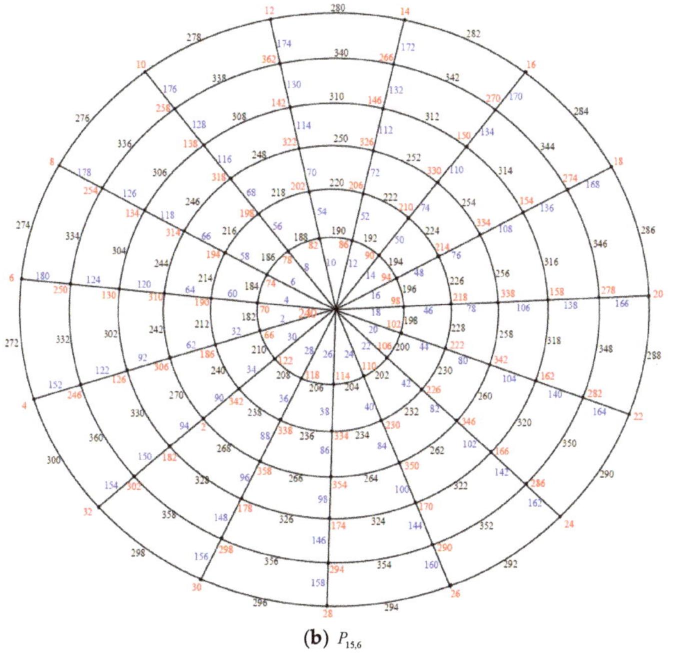

(**b**) $P_{15,6}$

Figure 6. *Cont.*

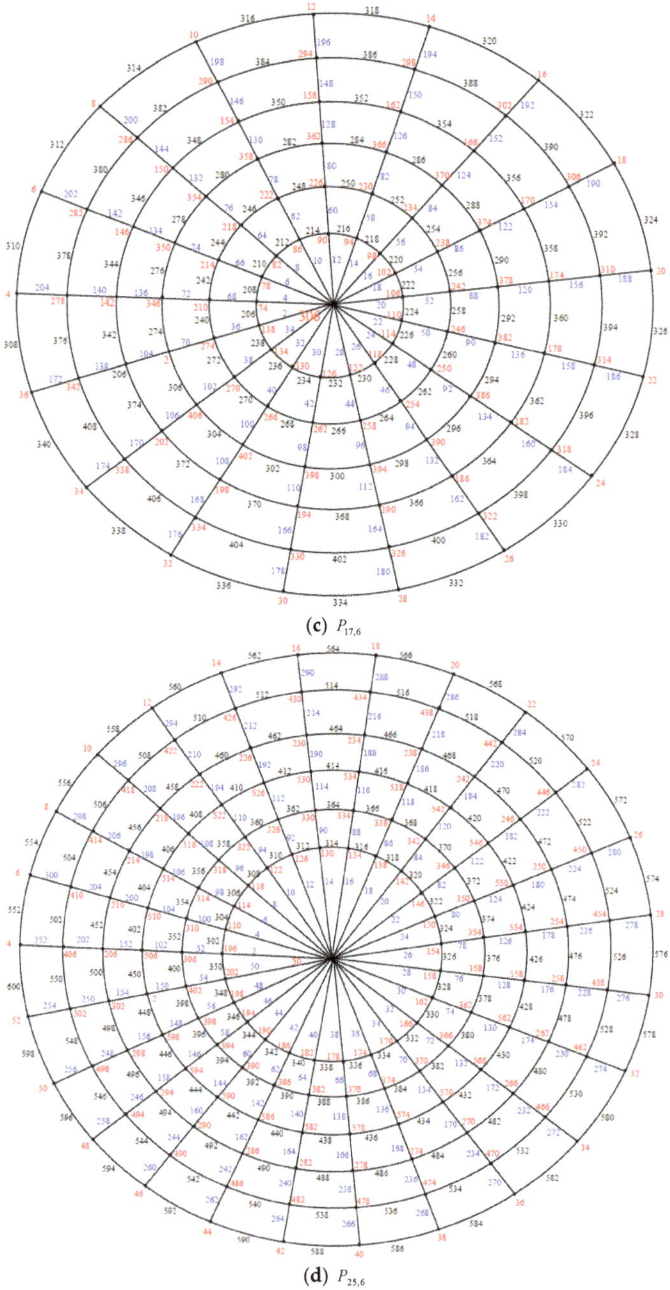

(c) $P_{17,6}$

(d) $P_{25,6}$

Figure 6. The edge even graceful labeling of the polar grid graphs $P_{13,6}$, $P_{15,6}$, $P_{17,6}$ and $P_{25,6}$.

Theorem 3. *If m is an even positive integer greater than or equal 4 and n is an odd positive integer greater than or equal 3. Then the polar grid graph $P_{m,n}$ is an edge even graceful graph.*

Proof. Let the polar grid graph $P_{m,n}$ be labeled as in Figure 7. Let $f : E(G) \to \{2, 4, \ldots, 2q\}$. □

First we label the edges of the circles $C_m^{(k)}$, $1 \leq k \leq n$ begin with the edges of the inner most circle $C_m^{(1)}$ to edges of the outer circle $C_m^{(n)}$ as follows:

$$f(v_{m(k-1)+i}v_{m(k-1)+i+1}) = 2m(k-1) + 2i,\ f(v_{km}v_{(k-1)m+1}) = 2km,\ 1 \leq i \leq m-1,\ 1 \leq k \leq n.$$

Second we label the edges of paths $P_{n+1}^{(k)}$, $1 \leq k \leq m$ begin with the edges of the path $P_{n+1}^{(1)}$ as follows: Move anticlockwise to label the edges $v_0v_1, v_0v_m, v_0v_{m-1}, \ldots, v_0v_3, v_0v_2$ by $2mn+2, 2mn+4, 2mn+6, \ldots, 2m(n+1)-2, 2m(n+1)$, then move clockwise to label the edges $v_1v_{m+1}, v_2v_{m+2}, v_3v_{m+3}, \ldots, v_{m-1}v_{2m-1}, v_mv_{2m}$ by $2m(n+1)+2, 2m(n+1)+4, 2m(n+1)+6, \ldots, 2m(n+2)-2, 2m(n+2)$, then move anticlockwise to label the edges $v_{m+1}v_{2m+1}, v_{2m}v_{3m}, v_{2m-1}v_{3m-1}, \ldots, v_{m+3}v_{2m+3}, v_{m+2}v_{2m+2}$ by $2m(n+2)+2, 2m(n+2)+4, 2m(n+2)+6, \ldots, 2m(n+3)-2, 2m(n+3)$ and so on. Finally move anticlockwise to label the edges $v_{m(n-2)+1}v_{m(n-1)+1}, v_{m(n-1)+1}v_{mn}, v_{m(n-1)-1}v_{mn-1}, \ldots, v_{m(n-2)+3}v_{m(m-1)+3}, v_{m(n-2)+2}v_{m(m-1)+2}$ by $2m(2n-1)+2, 2m(2n-1)+4, 2m(2n-1)+6, \ldots, 4mn-2, 4mn$.

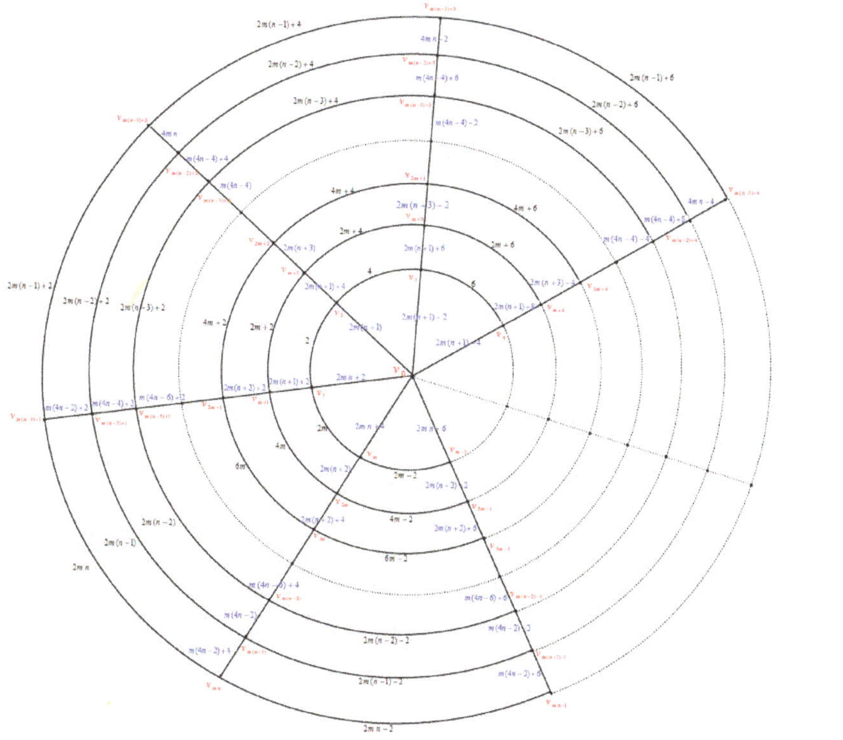

Figure 7. Labeling of the polar grid graph $P_{m,n}$ when n is odd and $n \geq 3$.

The corresponding labels of vertices mod$4mn$ are assigned as follows: There are four cases

Case (1) $m \equiv 4k \bmod 4n$, $1 \leq k \leq n-2$; $m \equiv (4n-2) \bmod 4n$ and $m \equiv 0 \bmod 4n$

$f^*(v_{(k-1)m+i}) \equiv 4m(2k-1) + 4i + 2$, $1 \leq i \leq m$, $1 \leq k \leq n-1$. That is the labels of the vertices in the most inner circle $C_m^{(1)}$ are assigned by $f^*(v_i) \equiv 4m + 4i + 2$, $1 \leq i \leq m$, the labels of the vertices in the circle $C_m^{(2)}$ are assigned by $f^*(v_{m+i}) \equiv 12m + 4i + 2$, the labels of vertices of the circle $C_m^{(\frac{n-1}{2})}$ are assigned by $f^*(\frac{v_{m(n-3)}}{2} + i) \equiv 4mn - 8m + 4i + 2$, the labels of vertices of the circle $C_m^{(\frac{n+1}{2})}$ are assigned by $f^*(\frac{v_{m(n-1)}}{2} + i) \equiv 4i + 2$, the labels of vertices of the circle $C_m^{(\frac{n+3}{2})}$ are assigned by $f^*(\frac{v_{m(n+1)}}{2}+i) \equiv 8m+4i+2, \ldots,$ the labels of the vertices in the circle $C_m^{(n-1)}$ are assigned by $f^*(v_{m(n-2)+i}) \equiv 4mn - 12m + 4i + 2, 1 \le i \le m$ and the labels of the vertices of the outer circle $C_m^{(n)}$ are assigned by $f^*(v_{m(n-1)+i}) \equiv 4mn - 4m + 2i + 2, 1 \le i \le m$. The labels of the center vertex v_0 is assigned by $f^*(v_0) \equiv m^2(2n+1)$ when $m \equiv 4k \bmod 4n$, we have $f^*(v_0) \equiv m(4k+1)$, when $m \equiv (4n-2) \bmod 4n$, $f^*(v_0) \equiv 4mn - m$ and when $m \equiv 0 \bmod 4n$, $f^*(v_0) \equiv m$.

Case (2) $m \equiv (8k-2) \bmod 4n$, $1 \le k \le \frac{n-1}{2}$.

In this case the vertex $v_{km-(\frac{m+2}{4})}$ in the circle $C_m^{(k)}$ will repeat with the center vertex v_0. To avoid this problem we replace the label of two edges $v_{km-(\frac{m+2}{4})}v_{km-(\frac{m-2}{4})}$ and $v_{km-(\frac{m-2}{4})}v_{km-(\frac{m-6}{4})}$. That is $f(v_{km-(\frac{m+2}{4})}v_{km-(\frac{m-2}{4})}) = m(2k-1) + \frac{m+2}{2}$ and $f(v_{km-(\frac{m-2}{4})}v_{km-(\frac{m-6}{2})}) = m(2k-1) + \frac{m-2}{2}$ and we obtain the labels of the corresponding vertices as follows $f^*(v_{km-(\frac{m+2}{4})}) \equiv m(8k-1)+2$, $f^*(v_{km-(\frac{m-2}{4})}) \equiv m(8k-1)+4$ and $f^*(v_{km-(\frac{m-6}{2})}) \equiv m(8k-1)+6$. In this case the center vertex v_0 is labeled as $f^*(v_0) \equiv m(2mn+m+1) \equiv m(8k-1)$. The rest vertices are labeled as in case (1).

Case (3) $m \equiv (8k-6) \bmod 4n$, $1 \le k \le \frac{n-1}{2}$ and $m \neq 2$. In this case the vertex $v_{m(\frac{n+2k-1}{2})-(\frac{m+2}{4})}$ in the circle $C_m^{(\frac{n+2k-1}{2})}$ will repeat with the center vertex v_0. To avoid this problem we replace the labels of the two edges $v_{m(\frac{n+2k-1}{2})-(\frac{m+2}{4})}v_{m(\frac{n+2k-1}{2})-(\frac{m-2}{4})}$ and $v_{m(\frac{n+2k-1}{2})-(\frac{m-2}{4})}v_{m(\frac{n+2k-1}{2})-(\frac{m-6}{4})}$. That is $f(v_{m(\frac{n+2k-1}{2})-(\frac{m+2}{4})}v_{m(\frac{n+2k-1}{2})-(\frac{m-2}{4})}) = mn + 2m(k-1) + \frac{m+2}{2}$ and $f(v_{m(\frac{n+2k-1}{2})-(\frac{m-2}{4})}v_{m(\frac{n+2k-1}{2})-(\frac{m-6}{4})}) = mn + 2m(k-1) + \frac{m-2}{2}$ and we obtain the labels of the corresponding vertices as follows $f^*(v_{m(\frac{n+2k-1}{2})-(\frac{m+2}{4})}) \equiv m(8k-5)+2$, $f^*(v_{m(\frac{n+2k-1}{2})-(\frac{m-2}{4})}) \equiv m(8k-5)+4$ and $f^*(v_{m(\frac{n+2k-1}{2})-(\frac{m-6}{4})}) \equiv m(8k-5)+6$ and in this case the center vertex v_0 is labeled as $f^*(v_0) \equiv m(8k-5)$. The rest vertices are labeled as in case (1).

Case (4) $m \equiv (4n-4) \bmod 4n$. In this case the vertex $v_{mn-(\frac{m+2}{2})}$ in the outer circle $C_m^{(n)}$ will repeat with the center vertex v_0. To avoid this problem we replace the labels of the two edges $v_{mn-(\frac{m+6}{4})}v_{mn-(\frac{m+2}{4})}$ and $v_{mn}v_{m(n-1)+1}$. That is $f(v_{mn-(\frac{m+6}{4})}v_{mn-(\frac{m+2}{4})}) = 2mn$ and $f(v_{mn}v_{m(n-1)+1}) = m(2n-1) - 4$ and we obtain the labels of the corresponding vertices as follows $f^*(v_{mn-(\frac{m+6}{4})}) \equiv 4mn - 2m + 2$, $f^*(v_{mn-(\frac{m+2}{4})}) \equiv 4mn - 2m + 8$, $f^*(v_{mn}) \equiv 4mn - 3m - 2$ and $f^*(v_{m(n-1)+1}) \equiv 4mn - 5m$ and in this case the center vertex v_0 is labeled as $f^*(v_0) \equiv m(4n-3)$. The rest vertices are labeled as in case (1).

Illustration. *The edge even graceful labeling of the polar grid graphs $P_{10,5}$, $P_{12,5}$, $P_{14,5}$, $P_{16,5}$, $P_{18,5}$ and $P_{20,5}$ respectively are shown in Figure 8.*

Remark 2. *In case $m = 2$, n is odd, $n \ge 3$.*

Let the label of edges of the polar grid graph $P_{2,n}$ be as in Figure 9. Thus we have the labels of the corresponding vertices as follows: $f^*(v_1) \equiv 12$; $f^*(v_i) \equiv 16i - 2, 2 \le i \le \frac{n-1}{2}$; $f^*(v_{\frac{n+2i-1}{2}}) \equiv 16i - 10, 1 \le i \le \frac{n-1}{2}$; $f^*(v_n) \equiv 8n - 2$; $f^*(v'_i) \equiv 16i + 2, 1 \le i \le \frac{n-1}{2}$; $f^*(v'_{\frac{n+1}{2}}) \equiv 2$; $f^*(v'_{\frac{n+2i-1}{2}}) \equiv 16i - 6, 2 \le i \le \frac{n-1}{2}$; $f^*(v'_n) \equiv 0$ and $f^*(v_0) \equiv 4$.

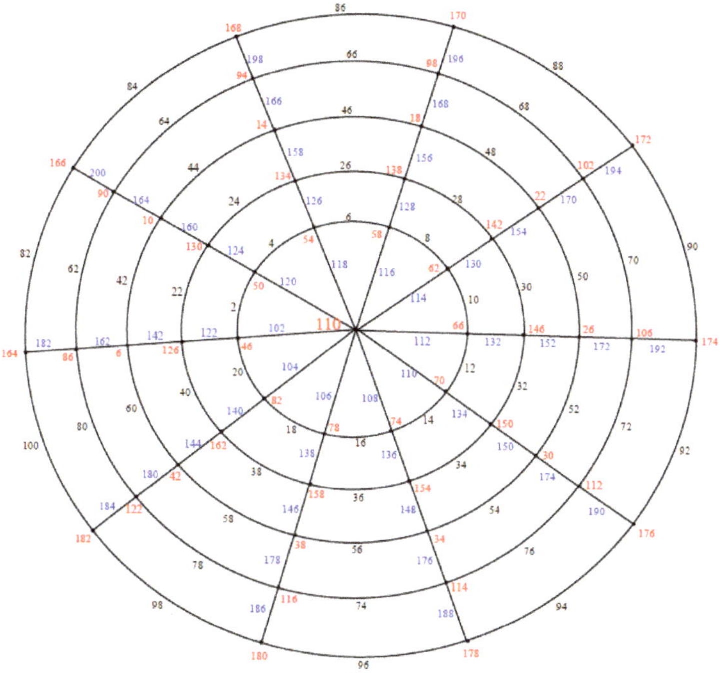

(**a**) $P_{10,5}$

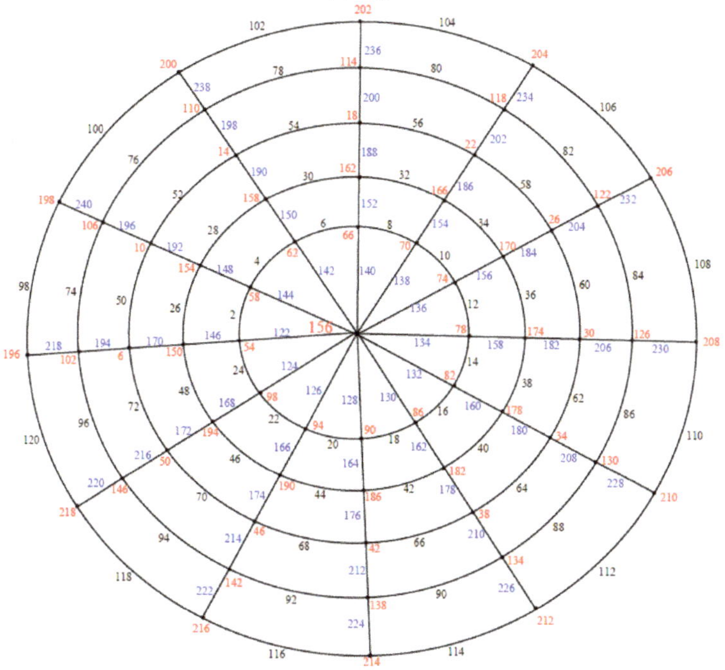

(**b**) $P_{12,5}$

Figure 8. *Cont.*

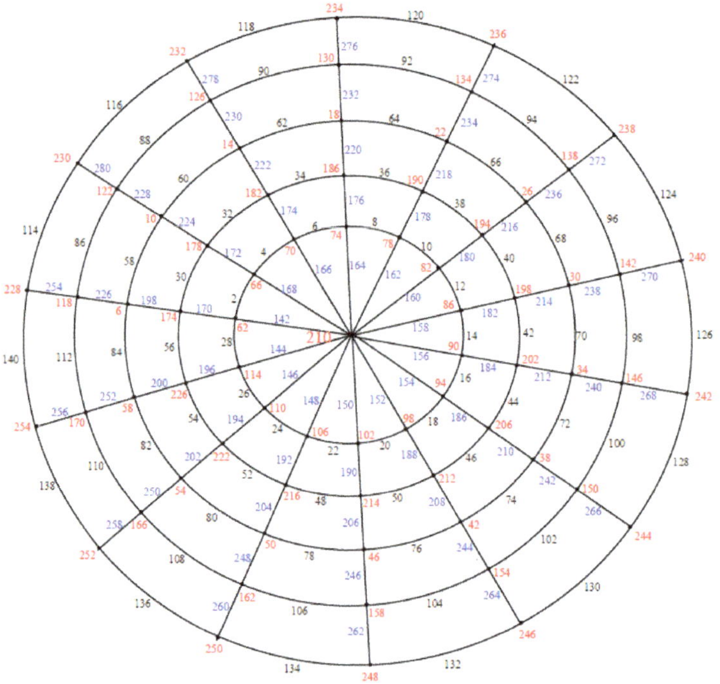

(**c**) $P_{14,5}$

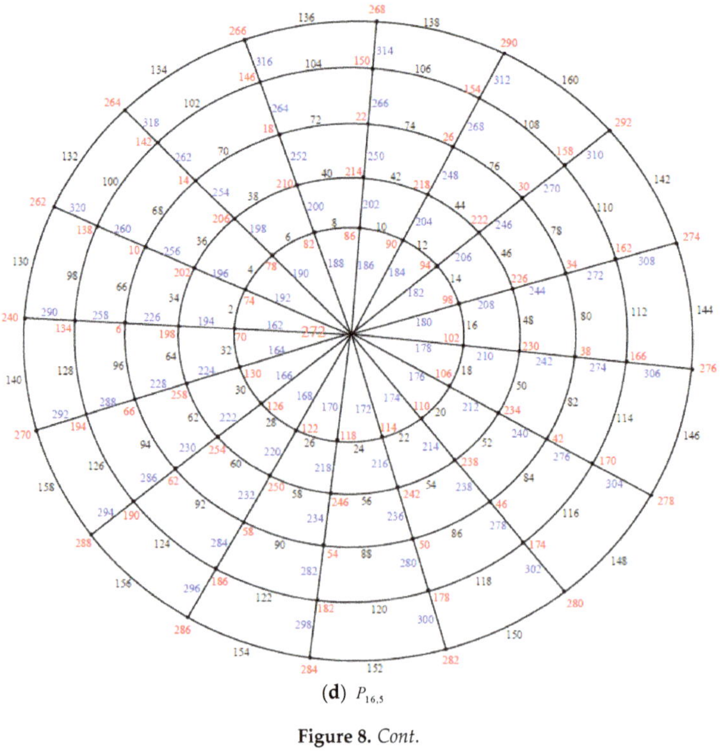

(**d**) $P_{16,5}$

Figure 8. *Cont.*

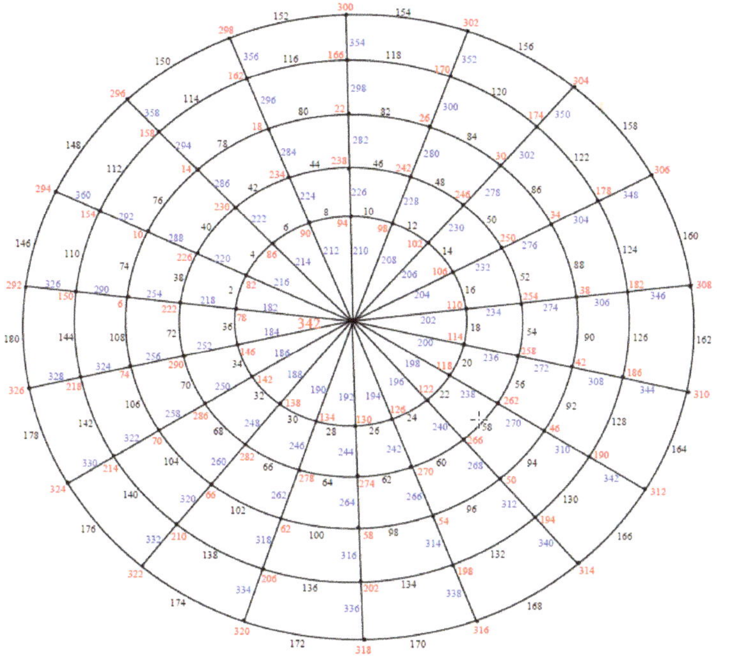

(e) $P_{18,5}$

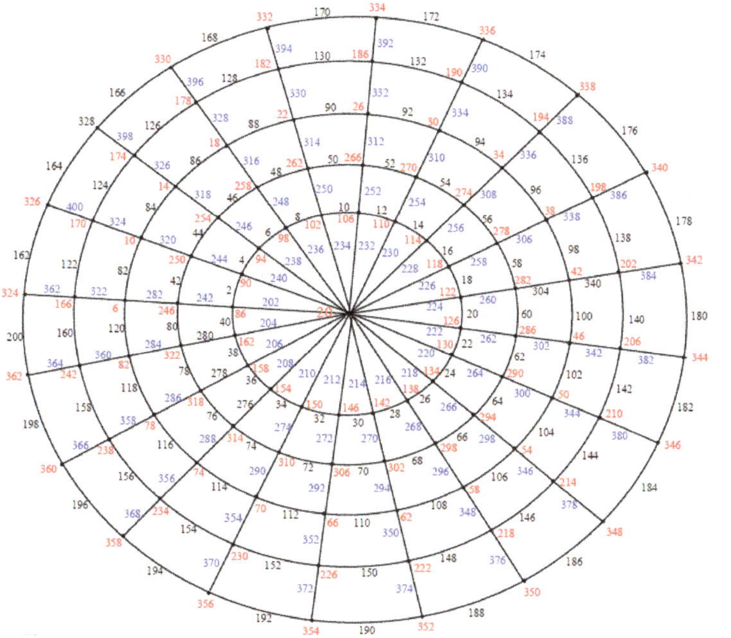

(f) $P_{20,5}$

Figure 8. The edge even graceful labeling of the polar grid graphs $P_{10,5}$, $P_{12,5}$, $P_{14,5}$, $P_{16,5}$, $P_{18,5}$ and $P_{20,5}$.

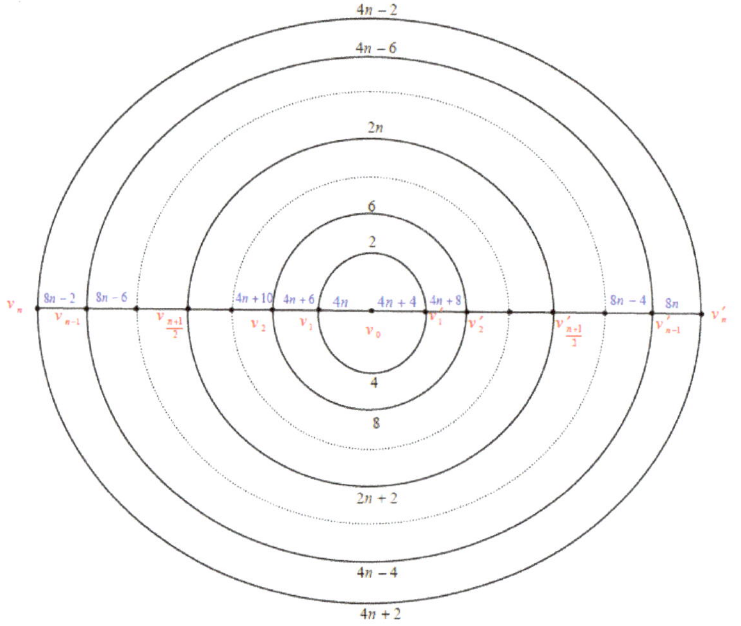

Figure 9. The labeling of the polar grid graph $P_{2,n}$, $n \geq 3$.

Illustration. *The edge even graceful labeling of the polar grid graphs* $P_{2,5}$ *is shown in Figure 10.*

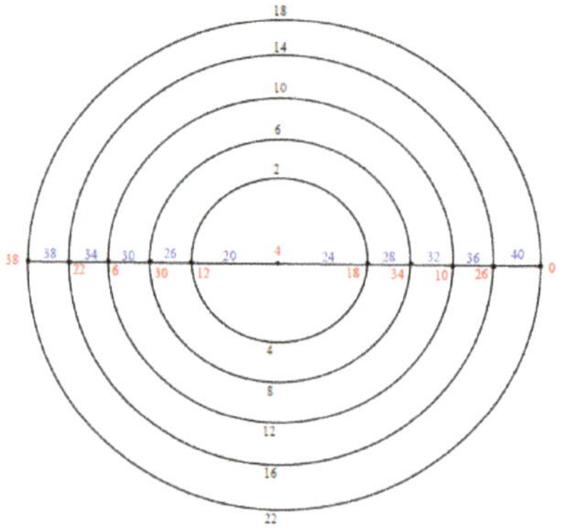

Figure 10. The labeling of the polar grid graph $P_{2,5}$.

Theorem 4. *If m and n are odd positive integers greater than* 1. *Then the polar grid graph* $P_{m,n}$ *is an edge even graceful graph.*

Proof. Let the polar grid graph $P_{m,n}$ be labeled as in Figure 7. Let $f : E(G) \rightarrow \{2, 4, \ldots, 2q\}$. □

The corresponding labels of vertices mod$4mn$ are assigned as follows: There are two cases:
Case (1) $n \equiv 1 \bmod 4$, this case contains five subcases as follows:
SubCase (i) $m \equiv (4k-3) \bmod 4n$, $2 \leq k \leq n$
$f^*(v_{(k-1)m+i}) \equiv 4m(2k-1) + 4i + 2, 1 \leq k \leq n-1, 1 \leq i \leq m$. That is the labels of vertices of the most inner circle $C_m^{(1)}$ are assigned by $f^*(v_i) = 4m + 4i + 2$, the labels of vertices of the

circle $C_m^{(2)}$ are assigned by $f^*(v_{m+i}) \equiv 12m + 4i + 2$, the labels of the vertices of the circle $C_m^{(\frac{n-1}{2})}$ are assigned by $f^*(v_{\frac{m(n-3)}{2}+i}) \equiv 4mn - 8m + 4i + 2$, the labels of the vertices of the circle $C_m^{(\frac{n+1}{2})}$ are assigned by $f^*(v_{\frac{m(n-1)}{2}+i}) \equiv 4i + 2$, the labels of the vertices of the circle $C_m^{(\frac{n+3}{2})}$ are assigned by $f^*(v_{\frac{m(n+1)}{2}+i}) \equiv 8m + 4i + 2, \ldots$, the labels of the vertices of the circle $C_m^{(n-1)}$ are assigned by $f^*(v_{m(n-2)+i}) \equiv 4mn - 12m + 4i + 2$, $1 \leq i \leq m$ and the labels of the vertices of the outer circle $C_m^{(n)}$ are assigned by $f^*(v_{m(n-1)+i}) \equiv 4mn - 12m + 2i + 2$, $1 \leq i \leq m$. The label of the center vertex v_0 is assigned by $f^*(v_0) \equiv 2mn + 2m(2k-1)$, when $k = \frac{n+1}{2}$, we have $f^*(v_0) \equiv 0$.

SubCase (ii) $m \equiv (8k-5) \bmod 4n$, $1 \leq k \leq \frac{n+1}{2}$, $m \neq 3$. In this subcase the vertex $v_{m(\frac{n+4k-1}{4})-(\frac{m+1}{2})}$ in the circle $C_m^{(\frac{n+4k-1}{4})}$ will repeat with the center vertex v_0. To avoid this problem we replace the labels of the two edges $v_{m(\frac{n+4k-1}{4})-(\frac{m+1}{2})}v_{m(\frac{n+4k-1}{4})-(\frac{m-1}{2})}$ and $v_{m(\frac{n+4k-1}{4})-(\frac{m-1}{2})}v_{m(\frac{n+4k-1}{4})-(\frac{m-3}{2})}$. That is $f(v_{m(\frac{n+4k-1}{4})-(\frac{m+1}{2})}v_{m(\frac{n+4k-1}{4})-(\frac{m-1}{2})}) = m[2k + \frac{n-3}{2}] + 1$ and $f(v_{m(\frac{n+4k-1}{4})-(\frac{m-1}{2})}v_{m(\frac{n+4k-1}{4})-(\frac{m-3}{2})}) = m[2k + \frac{n-3}{2}] - 1$ and we obtain the labels of the corresponding vertices as follows $f^*(v_{m(\frac{n+4k-1}{4})-(\frac{m+1}{2})}) \equiv 2mn + 4m(2k-1) + 2$, $f^*(v_{m(\frac{n+4k-1}{4})-(\frac{m-1}{2})}) \equiv 2mn + 4m(2k-1) + 4$, $f^*(v_{m(\frac{n+4k-1}{4})-(\frac{m-3}{2})}) \equiv 2mn + 4m(2k-1) + 6$, and in this case the center vertex v_0 is labeled as $f^*(v_0) \equiv 2mn + 4m(2k-1)$. The rest vertices will be labeled as in subCase (i).

Remark 3. *When $n \equiv 1 \bmod 4$ and $m = 3$, in this case the vertex $v_{3(\frac{n-1}{4})+1}$ in the circle $C_3^{(\frac{n-1}{4}+1)}$ will repeat with the center vertex v_0. To avoid this problem we replace the labels of the two edges $v_{3(\frac{n-1}{4})+2}v_{3(\frac{n-1}{4})+3}$ and $v_{3(\frac{n-1}{4})+1}v_{3(\frac{n-1}{4})+3}$. That is $f(v_{3(\frac{n-1}{4})+2}v_{3(\frac{n-1}{4})+3}) = 3(\frac{n-1}{2}) + 6$ and $f(v_{3(\frac{n-1}{4})+1}v_{3(\frac{n-1}{4})+3}) = 3(\frac{n-1}{2}) + 4$ and we obtain the labels of the corresponding vertices as follows $f^*(v_{3(\frac{n-1}{4})+1}) \equiv 6n + 10$, $f^*(v_{3(\frac{n-1}{4})+2}) \equiv 6n + 18$, $f^*(v_{3(\frac{n-1}{4})+3}) \equiv 6n + 16$ and the center vertex v_0 is labeld as $f^*(v_0) \equiv 6n + 12$.*

SubCase (iii) $m \equiv (8k-1) \bmod 4n$, $1 \leq k \leq \frac{n-5}{4}$. In this subcase the vertex $v_{m(\frac{3n+4k+1}{4})-(\frac{m+1}{2})}$ in the circle $C_m^{(\frac{3n+4k-1}{4})}$ will repeat with the center vertex v_0. To avoid this problem we replace the labels of the two edges $v_{m(\frac{3n+4k+1}{4})-(\frac{m+1}{2})}v_{m(\frac{3n+4k+1}{4})-(\frac{m-1}{2})}$ and $v_{m(\frac{3n+4k+1}{4})-(\frac{m-1}{2})}v_{m(\frac{3n+4k+1}{4})-(\frac{m-3}{2})}$. That is $f(v_{m(\frac{3n+4k+1}{4})-(\frac{m+1}{2})}v_{m(\frac{3n+4k+1}{4})-(\frac{m-1}{2})}) = 2mn + m[2k - \frac{n+1}{2}] + 1$ and $f(v_{m(\frac{3n+4k-1}{4})-(\frac{m-1}{2})}v_{m(\frac{3n+4k-1}{4})-(\frac{m-3}{2})}) = 2mn + m[2k - \frac{n+1}{2}] - 1$ and we obtain the labels of the corresponding vertices as follows $f^*(v_{m(\frac{3n+4k+1}{4})-(\frac{m+1}{2})}) \equiv 2mn + 8km + 2$, $f^*(v_{m(\frac{3n+4k-1}{4})-(\frac{m-1}{2})}) \equiv 2mn + 8km + 4$, $f^*(v_{m(\frac{3n+4k+1}{4})-(\frac{m-3}{2})}) \equiv 2mn + 8km + 6$, and in this case the center vertex v_0 is labeled as $f^*(v_0) \equiv 2mn + 8km$. The rest vertices will be labeled as in subCase (i).

SubCase (iv) $m \equiv (8k-1) \bmod 4n$, $\frac{n+3}{4} \leq k \leq \frac{n-1}{2}$. In this case the vertex $v_{m(\frac{4k-n+1}{4})-(\frac{m+1}{2})}$ in the circle $C_m^{(\frac{4n-k+1}{4})}$ will repeat with the center vertex v_0. To avoid this problem we replace the labels of the two edges $v_{m(\frac{4k-n+1}{4})-(\frac{m+1}{2})}v_{m(\frac{4k-n+1}{4})-(\frac{m-1}{2})}$ and $v_{m(\frac{4k-n+1}{4})-(\frac{m-1}{2})}v_{m(\frac{4k-n+1}{4})-(\frac{m-3}{2})}$. That is $f(v_{m(\frac{4k-n+1}{4})-(\frac{m+1}{2})}v_{m(\frac{4k-n+1}{4})-(\frac{m-1}{2})}) = m[2k - \frac{n+1}{2}] + 1$ and $f(v_{m(\frac{4k-n+1}{4})-(\frac{m-1}{2})}v_{m(\frac{4k-n+1}{4})-(\frac{m-3}{2})}) = m[2k - \frac{n+1}{2}] - 1$ and we obtain the labels of the corresponding vertices as follows $f^*(v_{m(\frac{4k-n+1}{4})-(\frac{m+1}{2})}) \equiv 2mn + 8km + 2$, $f^*(v_{m(\frac{4k-n+1}{4})-(\frac{m-1}{2})}) \equiv 8km - 2mn + 4$, $f^*(v_{m(\frac{4k-n+1}{4})-(\frac{m-3}{2})}) \equiv 8km - 2mn + 6$, and in this case the center vertex v_0 is labeled as $f^*(v_0) \equiv 8km - 2mn$. The rest vertices will be labeled as in subCase (i).

SubCase (v) $m \equiv (2n-3) \bmod 4n$. In this case the vertex v_{mn-1} in the outer circle $C_m^{(n)}$ will repeat with the center vertex v_0. To avoid this problem we replace the labels of the two edges $v_{mn-2}v_{mn-1}$ and $v_{mn}v_{m(n-1)+1}$. That is $f(v_{mn-2}\,v_{mn-1}) = 2mn - 4$, $f(v_{mn}v_{m(n-1)+1}) = 2mn$ and we obtain the labels of the corresponding vertices are as follows $f^*(v_{mn-2}) \equiv 4mn - 2m + 2$, $f^*(v_{mn-1}) \equiv 4mn - 2m + 4$,

$f^*(v_{mn}) \equiv 4mn - 2m - 2$ and $f^*(v_{m(n-1)+1}) \equiv 4mn - 4m$, and in this case the center vertex v_0 is labeled as $f^*(v_0) \equiv 4mn - 2m$. The rest vertices will be labeled as in subCase (i).

Illustration. *The edge odd graceful labeling of the polar grid graphs $P_{3,5}, P_{13,5}, P_{11,5}, P_{7,9}, P_{15,5}$ and $P_{7,5}$ respectively are shown in Figure 11.*

Case (2) $n \equiv 3 \bmod 4$. This case contains also five subcases as follows:

SubCase (i) $m \equiv (4k-3) \bmod 4n$, $2 \le k \le n$

$f^*(v_{(k-1)m+i}) \equiv 4m(2k-1) + 4i + 2, 1 \le k \le n-1, 1 \le i \le m$. That is the labels of vertices of the most inner circle $C_m^{(1)}$ are assigned by $f^*(v_i) \equiv 4m + 4i + 2$, the label of vertices of the circle $C_m^{(2)}$ are assigned by $f^*(v_{m+i}) \equiv 12m + 4i + 2$, the labels of vertices of the circle $C_m^{(\frac{n-1}{2})}$ are assigned by $f^*(v_{\frac{m(n-1)}{2}+i}) \equiv 4mn - 8m + 4i + 2$, the labels of vertices of the circle $C_m^{(\frac{n+1}{2})}$ are assigned by $f^*(v_{\frac{m(n-3)}{2}+i}) \equiv 4i + 2$, the labels of vertices of the circle $C_m^{(\frac{n+3}{2})}$ are assigned by $f^*(v_{\frac{m(n+1)}{2}+i}) \equiv 8m + 4i + 2, \ldots$, the labels of vertices of the circle $C_m^{(n-1)}$ are assigned by $f^*(v_{m(n-2)+i}) \equiv 4mn - 12m + 4i + 2$, $1 \le i \le m$ and the labels of the vertices of the outer circle $C_m^{(n)}$ are assigned by $f^*(v_{m(n-1)+i}) \equiv 4mn - 12m + 2i + 2$, $1 \le i \le m$. The label of the center vertex v_0 is assigned by $f^*(v_0) \equiv 2mn + 2m(2k-1)$, when $k = \frac{n+1}{2}$, we have $f^*(v_0) \equiv 0$.

SubCase (ii) $m \equiv (8k-5) \bmod 4n$, $1 \le k \le \frac{n-3}{4}$, $m \ne 3$

In this subcase the vertex $v_{m(\frac{3n+4k-1}{4})-(\frac{m+1}{2})}$ in the circle $C_m^{(\frac{3n+4k-1}{4})}$ will repeat with the center vertex v_0. To avoid this problem we replace the labels of the two edges $v_{m(\frac{3n+4k-1}{4})-(\frac{m+1}{2})}v_{m(\frac{3n+4k-1}{4})-(\frac{m-1}{2})}$ and $v_{m(\frac{3n+4k-1}{4})-(\frac{m-1}{2})}v_{m(\frac{3n+4k-1}{4})-(\frac{m-3}{2})}$. That is $f(v_{m(\frac{3n+4k-1}{4})-(\frac{m+1}{2})}v_{m(\frac{3n+4k-1}{4})-(\frac{m-1}{2})}) = 2mn + m[2k - \frac{n+3}{2}] + 1$ and $f(v_{m(\frac{3n+4k-1}{4})-(\frac{m-1}{2})}v_{m(\frac{3n+4k-1}{4})-(\frac{m-3}{2})}) = 2mn + m[2k - \frac{n+3}{2}] - 1$ and we obtain the labels of the corresponding vertices as follows $f^*(v_{m(\frac{3n+4k-1}{4})-(\frac{m+1}{2})}) \equiv 2mn + 4m(2k-1) + 2$, $f^*(v_{m(\frac{3n+4k-1}{4})-(\frac{m-1}{2})}) \equiv 2mn + 4m(2k-1) + 4$, $f^*(v_{m(\frac{3n+4k-1}{4})-(\frac{m-3}{2})}) \equiv 2mn + 4m(2k-1) + 6$, and the label of the center vertex v_0 is assigned by $f^*(v_0) \equiv 2mn + 4m(2k-1)$. That rest vertices will be labeled as in subcase (i).

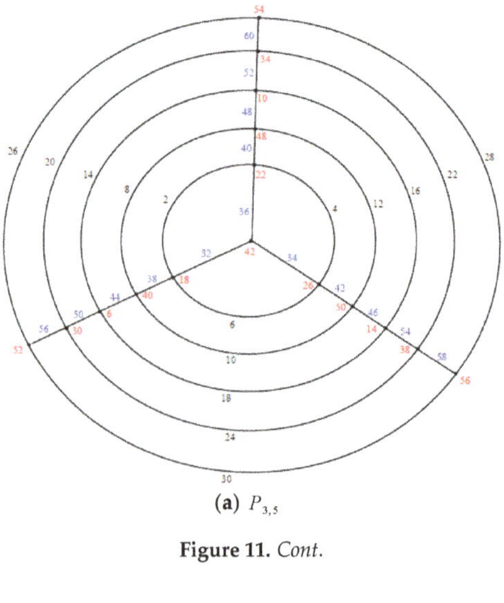

(a) $P_{3,5}$

Figure 11. *Cont.*

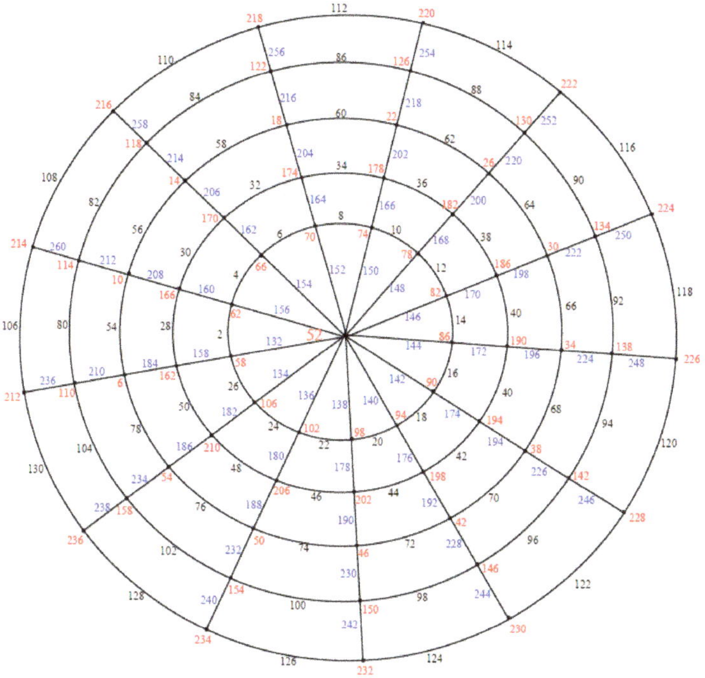

(**b**) $P_{13,5}$

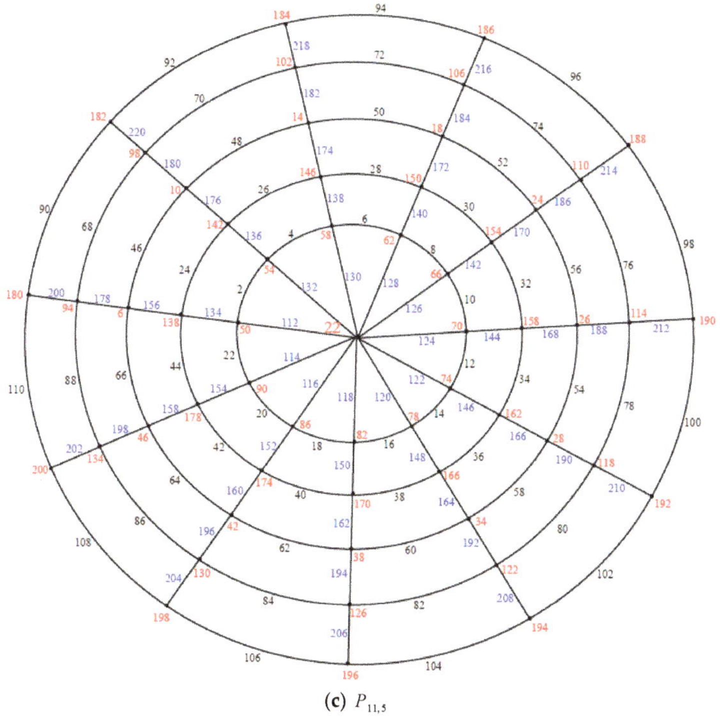

(**c**) $P_{11,5}$

Figure 11. *Cont.*

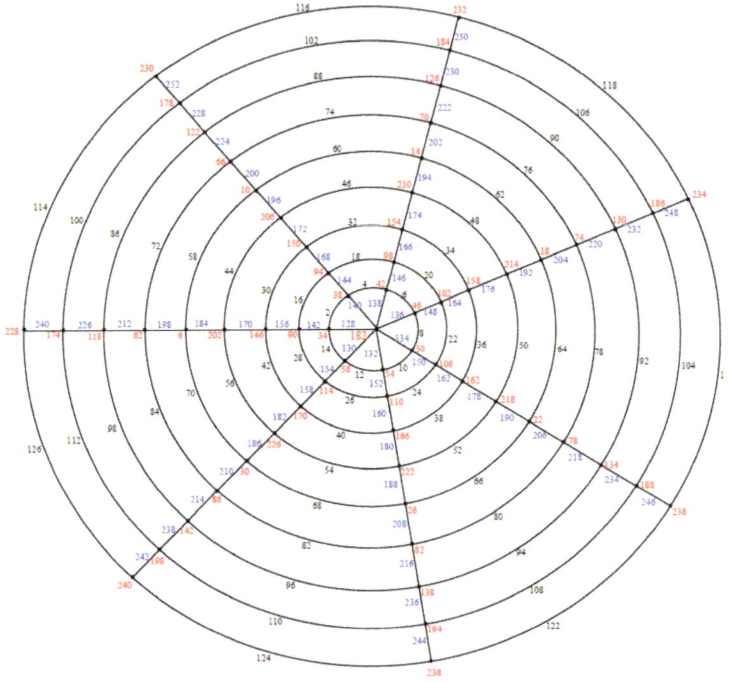

(**d**) $P_{7,9}$

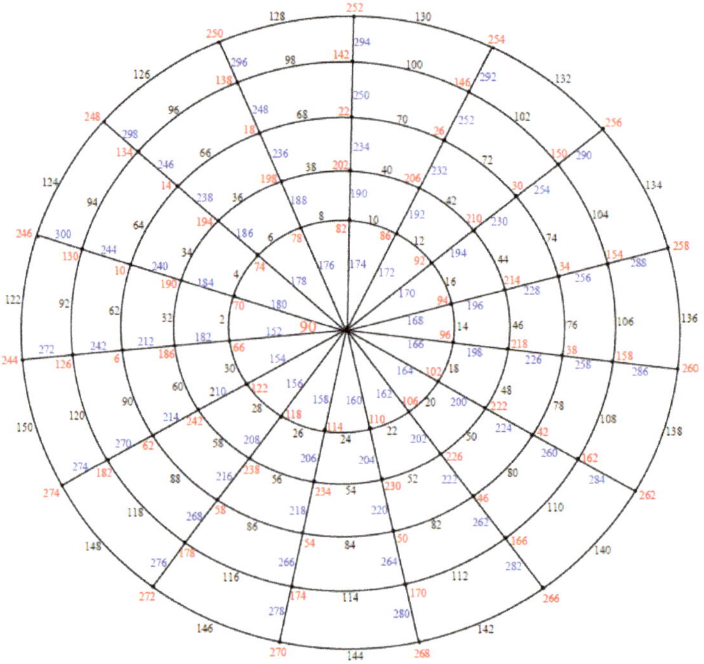

(**e**) $P_{15,5}$

Figure 11. *Cont.*

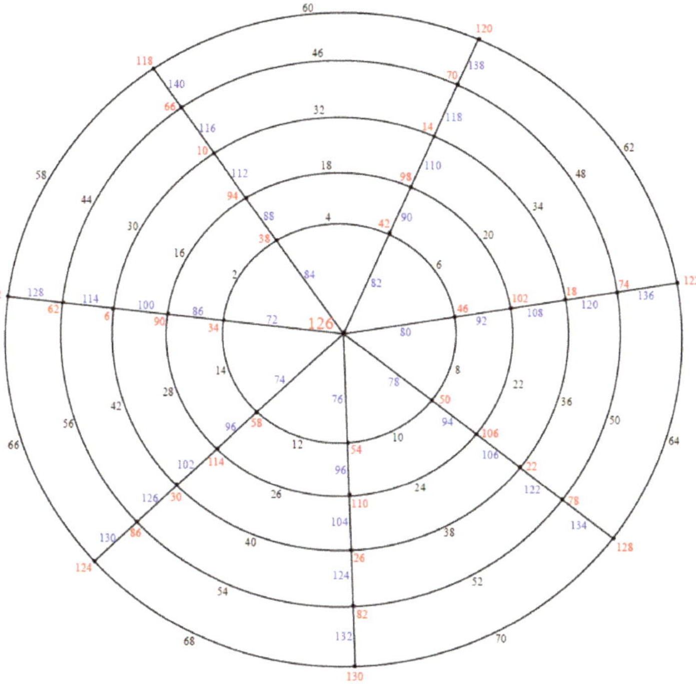

(f) $P_{7,5}$

Figure 11. The polar grid graphs $P_{3,5}$, $P_{13,5}$, $P_{11,5}$, $P_{7,9}$, $P_{15,5}$ and $P_{7,5}$.

Remark 4. *When* $n \equiv 3 \bmod 4$ *and* $m = 3$*, we have the vertex* $v_{3(\frac{3n-1}{4})+1}$ *in the circle* $C_m^{(\frac{3n-1}{4}+1)}$ *will repeat with the center vertex* v_0*. To avoid this problem we replace the labels of the two edges* $v_{3(\frac{3n-1}{4})+2}v_{3(\frac{3n-1}{4})+3}$ *and* $v_{3(\frac{3n-1}{4})+1}v_{3(\frac{3n-1}{4})+3}$*. That is* $f(v_{3(\frac{3n-1}{4})+2}v_{3(\frac{3n-1}{4})+3}) = 3(\frac{3n-1}{2}) + 6$ *and* $f(v_{3(\frac{3n-1}{4})+1}v_{3(\frac{3n-1}{4})+3}) = 3(\frac{3n-1}{2}) + 4$ *and we obtain the labes of the corresponding vertices mod4mn are as follows:* $f^*(v_{3(\frac{3n-1}{4})+1}) \equiv 6n + 10$, $f^*(v_{3(\frac{3n-1}{4})+2}) \equiv 6n + 18$, $f^*(v_{3(\frac{3n-1}{4})+3}) \equiv 6n + 20$ *and the label of the center vertex* v_0 *is assigned by* $f^*(v_0) \equiv 6n + 12$*.*

Note that $P_{3,3}$ is an edge even graceful grapg but not follow this rule. See Figure 12.

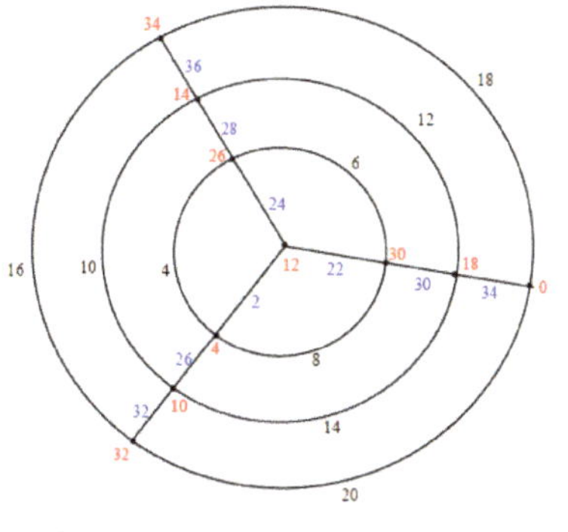

Figure 12. The polar grid graphs $P_{3,3}$.

SubCase (iii) $m \equiv (8k-1)\text{mod}4n$, $\frac{n+5}{4} \le k \le \frac{n+1}{2}$,

In this subcase the vertex $v_{m(\frac{4k-n-1}{4})-(\frac{m+1}{2})}$ in the circle $C_m^{(\frac{4k-n-1}{4})}$ will repeat with the center vertex v_0. To avoid this problem we replace the labels of the two edges $v_{m(\frac{4k-n-1}{4})-(\frac{m+1}{2})}v_{m(\frac{4k-n-1}{4})-(\frac{m-1}{2})}$ and $v_{m(\frac{4k-n-1}{4})-(\frac{m-1}{2})}v_{m(\frac{4k-n-1}{4})-(\frac{m-3}{2})}$ That is $f(v_{m(\frac{4k-n-1}{4})-(\frac{m+1}{2})}v_{m(\frac{4k-n-1}{4})-(\frac{m-1}{2})}) = (2k-1)m+1$ and $f(v_{m(\frac{4k-n-1}{4})-(\frac{m-1}{2})}v_{m(\frac{4k-n-1}{4})-(\frac{m-3}{2})}) = (2k-1)m-1$ and we obtain the labels of the corresponding vertices as follows: $f^*(v_{m(\frac{4k-n-1}{4})-(\frac{m+1}{2})}) \equiv 2m(4k-1)+2$, $f^*(v_{m(\frac{4k-n-1}{4})-(\frac{m-1}{2})}) \equiv 2m(4k-1)+4$ and $f^*(v_{m(\frac{4k-n-1}{4})-(\frac{m-3}{2})}) = 2m(4k-1)+6$. The label of the center vertex v_0 is assigned by $f^*(v_0) \equiv 2m(4k-1)$. The rest vertices will be labeled as in subCase (i).

SubCase (iv) $m \equiv (8k-1)\text{mod}4n$, $1 \le k \le \frac{n-1}{2}$

In this subcase the vertex $v_{m(\frac{n+4k+1}{4})-(\frac{m+1}{2})}$ in the circle $C_m^{(\frac{n+4k+1}{4})}$ will repeat with the center vertex v_0. To avoid this problem we replace the labels of the two edges $v_{m(\frac{n+4k+1}{4})-(\frac{m+1}{2})}v_{m(\frac{n+4k+1}{4})-(\frac{m-1}{2})}$ and $v_{m(\frac{n+4k+1}{4})-(\frac{m-1}{2})}v_{m(\frac{n+4k+1}{4})-(\frac{m-3}{2})}$ That is $f(v_{m(\frac{n+4k+1}{4})-(\frac{m+1}{2})}v_{m(\frac{n+4k+1}{4})-(\frac{m-1}{2})}) = (2k+\frac{n-1}{2})m+1$ and $f(v_{m(\frac{n+4k+1}{4})-(\frac{m-1}{2})}v_{m(\frac{n+4k+1}{4})-(\frac{m-3}{2})}) = (2k+\frac{n-1}{2})m-1$ and we obtain the labels of the corresponding vertices as follows $f^*(v_{m(\frac{n+4k+1}{4})-(\frac{m+1}{2})}) = 2mn+8km+2$, $f^*(v_{m(\frac{n+4k+1}{4})-(\frac{m-1}{2})}) = 2mn+8km+4$, $f^*(v_{m(\frac{n+4k+1}{4})-(\frac{m-3}{2})}) = 2mn+8km+6$ and the label of the center vertex v_0 is labeled as $f^*(v_0) \equiv 2mn+8km$. The rest vertices will be labeled as in subCase (i).

Remark 5. *If* $k = \frac{n+1}{4}$ *we have* $f^*(v_{m(\frac{n+4k+1}{4})-(\frac{m+1}{2})}) = 8km-2mn+2$, $f^*(v_{m(\frac{n+4k+1}{4})-(\frac{m-1}{2})}) = 8km-2mn+4$, $f^*(v_{m(\frac{n+4k+1}{4})-(\frac{m-3}{2})}) = 8km-2mn+6$ *and the center vertex* v_0 *is labeled as* $f^*(v_0) \equiv 8km-2mn$.

SubCase (v) $m \equiv (2n-3)\text{mod}4n$

In this subcase the vertex v_{mn-1} in the outer circle $C_m^{(n)}$ will repeat with the center vertex v_0. To avoid this problem we replace the labels of the two edges $v_{mn-2}v_{mn-1}$ and $v_{mn}v_{m(n-1)+1}$. That is $f(v_{mn-2}v_{mn-1}) = 2mn-4$, $f(v_{mn}v_{m(n-1)+1}) = 2mn$ and we obtain the labes of the corresponding vertices as follows:

$f^*(v_{mn-2}) = 4mn-2m+2$, $f^*(v_{mn-1}) = 4mn-2m+4$, $f^*(v_{mn}) = 4mn-2m-2$ and $f^*(v_{m(n-1)+1}) = 4m(n-1)$ and the label of the center vertex v_0 is assigned by $f^*(v_0) \equiv 2m(2n-1)$.

The rest vertices will be labeled as in subCase (i).

Illustration. *The edge odd graceful labeling of the polar grid graphs* $P_{3,7}$, $P_{13,7}$, $P_{19,7}$, $P_{11,7}$ *and* $P_{15,7}$ *respectively are shown in Figure 13.*

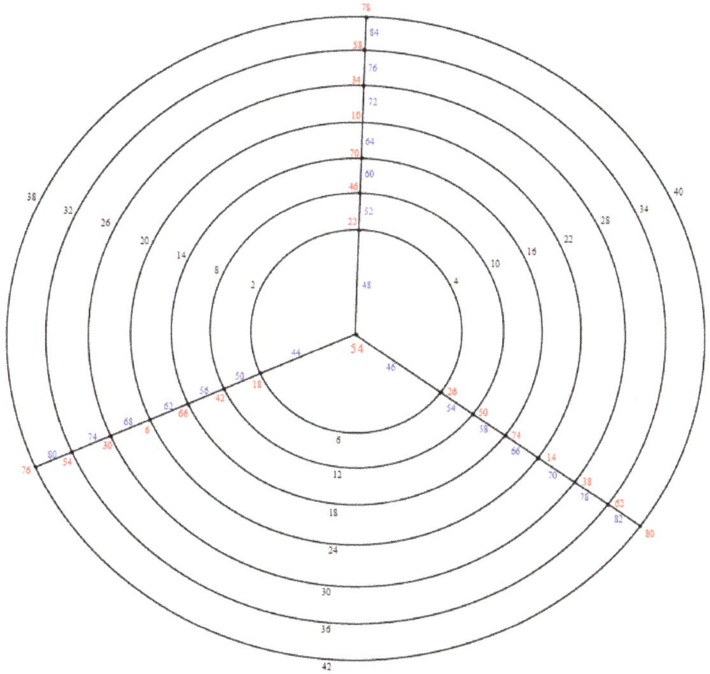

(**a**) $P_{3,7}$

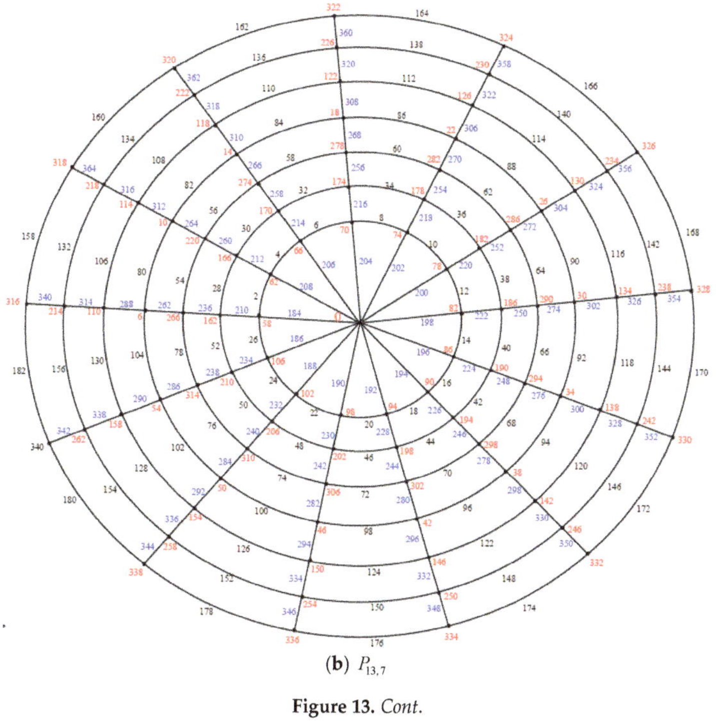

(**b**) $P_{13,7}$

Figure 13. *Cont.*

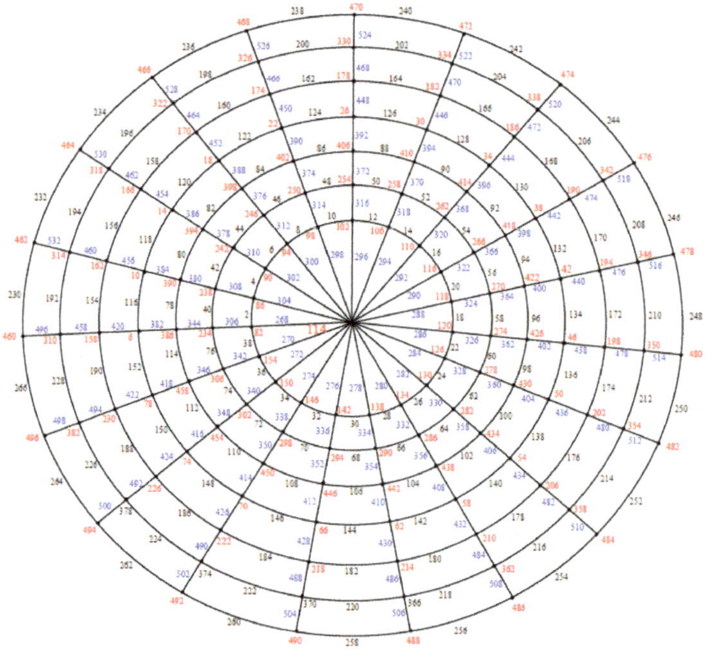

(**c**) $P_{19,7}$

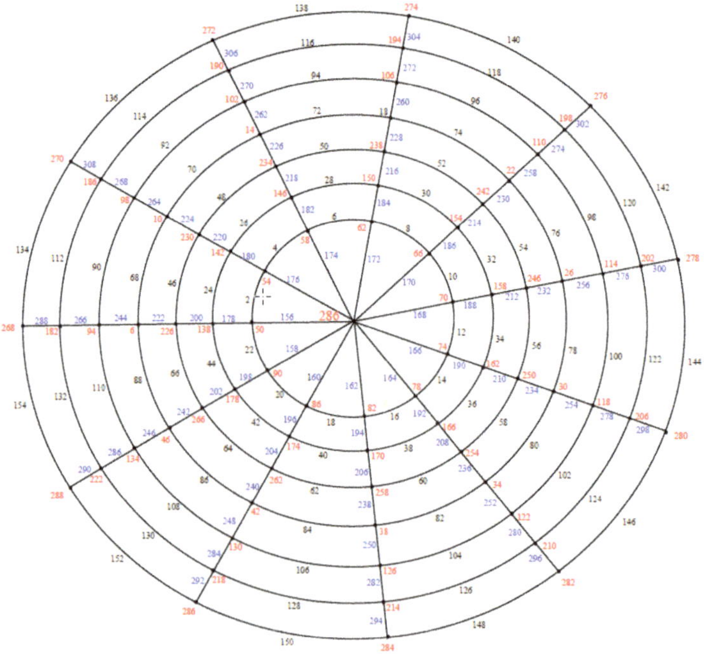

(**d**) $P_{11,7}$

Figure 13. *Cont.*

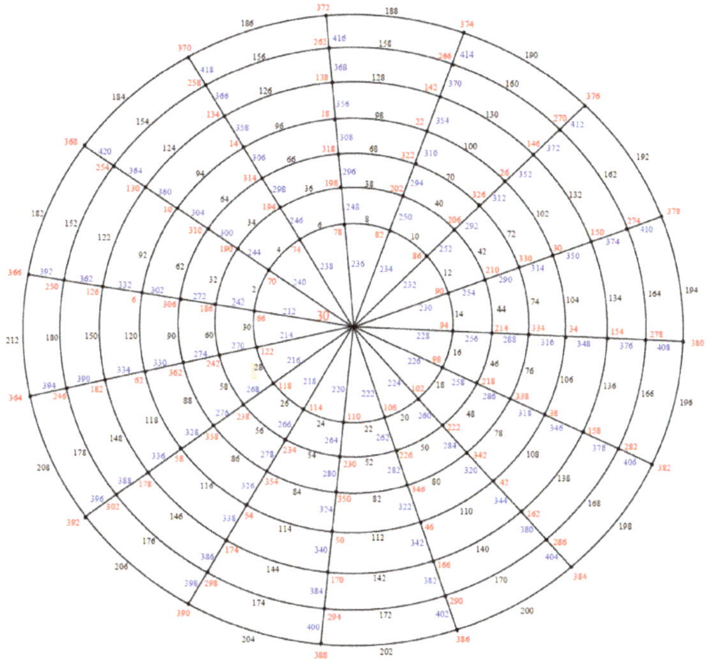

(e) $P_{15,7}$

Figure 13. The polar grid graphs $P_{3,7}$, $P_{13,7}$, $P_{19,7}$, $P_{11,7}$ $P_{15,7}$.

3. Conclusions

This paper gives some basic knowledge about the application of Graph labeling and Graph Theory in real life which is the one branch of mathematics. It is designed for the researcher who research in graph labeling and graph Theory. In this paper, we give necessary and sufficient conditions for a polar grid graph to admit edge even labeling. In future work we will study the necessary and sufficient conditions for the cylinder $P_m \times C_n$, torus $C_m \times C_n$ and rectangular $P_m \times P_n$ grid graphs to be edge even graceful.

Funding: This work was supported by the deanship of Scientific Research, Taibah University, Al-Madinah Al-Munawwarah, Saudi Arabia.

Conflicts of Interest: The authors declare that there are no conflicts of interest regarding the publication of this paper.

References

1. Acharya, B.D.; Arumugam, S.; Rosa, A. *Labeling of Discrete Structures and Applications*; Narosa Publishing House: New Delhi, India, 2008; pp. 1–14.
2. Bloom, G.S. Numbered Undirected Graphs and Their Uses, a Survey of a Unifying Scientific and Engineering Concept and Its Use in Developing a Theory of Non-Redundant Homometric Sets Relating to Some Ambiguities in X-ray Diffraction Analysis. Ph.D. Thesis, University Southern California, Los Angeles, CA, USA, 1975.
3. Bloom, G.S.; Golomb, S.W. Numbered complete graphs, unusual rulers, and assorted applications. In *Theory and Applications of Graphs*; Lecture Notes in Math; Springer: New York, NY, USA, 1978; Volume 642, pp. 53–65.
4. Bloom, G.S.; Golomb, S.W. Applications of numbered undirected graphs. *Proc. IEEE* **1977**, *65*, 562–570. [CrossRef]

5. Bloom, G.S.; Hsu, D.F. On graceful digraphs and a problem in network addressing. *Congr. Numer.* **1982**, *35*, 91–103.
6. Graham, R.L.; Pollak, H.O. On the addressing problem for loop switching. *Bell Syst. Tech. J.* **1971**, *50*, 2495–2519. [CrossRef]
7. Sutton, M. Summable Graphs Labellings and their Applications. Ph.D. Thesis, Department Computer Science, The University of Newcastle, Callaghan, Australia, 2001.
8. Shang, Y. More on the normalized Laplacian Estrada index. *Appl. Anal. Discret. Math.* **2014**, *8*, 346–357. [CrossRef]
9. Shang, Y. Geometric Assortative Growth Model for Small-World Networks. *Sci. World J.* **2014**, *2014*, 759391. [CrossRef] [PubMed]
10. Gross, J.; Yellen, J. *Graph Theory and Its Applications*; CRC Press: London, UK, 1999.
11. Rosa, A. *Certain Valuations of the Vertices of a Graph*; Theory of Graphs (Internat. Symp, Rome, July 1966); Gordan and Breach: New York, NY, USA; Paris, France, 1967; pp. 349–355.
12. Golomb, S.W. How to Number a Graph. In *Graph Theory and Computing*; Read, R.C., Ed.; Academic Press: NewYork, NY, USA, 1972; pp. 23–37.
13. Gnanajothi, R.B. Topics in Graph Theory. Ph.D. Thesis, Madurai Kamaraj University, Tamil Nadu, India, 1991.
14. Seoud, M.A.; Abdel-Aal, M.E. On odd graceful graphs. *Ars Comb.* **2013**, *108*, 161–185.
15. Gao, Z. Odd graceful labelings of some union graphs. *J. Nat. Sci. Heilongjiang Univ.* **2007**, *24*, 35–39.
16. Lo, S.P. On edge-gracefullabelings of graphs. *Congr. Numerantium* **1985**, *50*, 231–241.
17. Kuan, Q.; Lee, S.; Mitchem, J.; Wang, A. On Edge-Graceful Unicyclic Graphs. *Congr. Numerantium* **1988**, *61*, 65–74.
18. Lee, L.; Lee, S.; Murty, G. On Edge-Graceful Labelings of Complete Graphs: Solutions of Lo's Conjecture. *Congr. Numerantium* **1988**, *62*, 225–233.
19. Solairaju, A.; Chithra, K. Edge-Odd Graceful Graphs. *Electron. Notes Discret. Math.* **2009**, *33*, 15–20. [CrossRef]
20. Daoud, S.N. Edge odd graceful labeling of some path and cycle related graphs. *AKCE Int. J. Graphs Comb.* **2017**, *14*, 178–203. [CrossRef]
21. Elsonbaty, A.; Daoud, S.N. Edge even graceful labeling of some path and cycle related graphs. *Ars Comb.* **2017**, *130*, 79–96.
22. Gallian, J.A. A Dynamic Survey of Graph Labeling. *Electron. J. Comb.* **2017**, *22*, #DS6.

symmetry

MDPI

Article

Edge-Version Atom-Bond Connectivity and Geometric Arithmetic Indices of Generalized Bridge Molecular Graphs

Xiujun Zhang [1], Xinling Wu [2], Shehnaz Akhter [3], Muhammad Kamran Jamil [4], Jia-Bao Liu [5,*] and Mohammad Reza Farahani [6]

1 Key Laboratory of Pattern Recognition and Intelligent Information Processing, Institutions of Higher Education of Sichuan Province, Chengdu University, Chengdu 610106, China; woodszhang@cdu.edu.cn
2 South China Business College, Guang Dong University of Foreign Studies, Guangzhou 510545, China; xinlingwu.guangzhou@gmail.com
3 Department of Mathematics, School of Natural Sciences (SNS), National University of Sciences and Technology (NUST), Sector H-12, Islamabad 44000, Pakistan; shehnazakhter36@yahoo.com
4 Department of Mathematics, Riphah Institute of Computing and Applied Sciences, Riphah International University Lahore, Lahore 54660, Pakistan; m.kamran.sms@gmail.com
5 School of Mathematics and Physics, Anhui Jianzhu University, Hefei 230601, China
6 Department of Applied Mathematics, Iran University of Science and Technology, Narmak, Tehran 16844, Iran; mrfarahani88@gmail.com
* Correspondence: liujiabao@ahjzu.edu.cn

Received: 29 November 2018; Accepted: 13 December 2018; Published: 14 December 2018

Abstract: Topological indices are graph invariants computed by the distance or degree of vertices of the molecular graph. In chemical graph theory, topological indices have been successfully used in describing the structures and predicting certain physicochemical properties of chemical compounds. In this paper, we propose a definition of generalized bridge molecular graphs that can model more kinds of long chain polymerization products than the bridge molecular graphs, and provide some results of the edge versions of atom-bond connectivity (ABC_e) and geometric arithmetic (GA_e) indices for some generalized bridge molecular graphs, which have regular, periodic and symmetrical structures. The results of this paper offer promising prospects in the applications for chemical and material engineering, especially in chemical industry research.

Keywords: atom-bond connectivity index; geometric arithmetic index; line graph; generalized bridge molecular graph

1. Introduction

Let G be an undirected simple graph without loops or multiple edges. We denote by $V(G)$ the vertex set of G and we denote by $E(G)$ the edge set of G. We denote by $e = uv$ the edge connect vertices u and v or vertices u and v adjacent. We denote by P_n, C_n, and S_n the path, cycle, and star of n vertices, respectively. We denote by $N(v)$ the open neighborhood of vertex v, i.e., $N(v) = \{u|uv \in E(G)\}$. We denote by $d(v)$ or $d_G(v)$ the degree of a vertex v of a graph G, i.e., $d(v) = |\{u \in N(v)\}|$. Let $L(G)$ or G^L be a line graph of G, so each vertex of $L(G)$ corresponds an edge of G. Two vertices of $L(G)$ are adjacent if and only if a common endpoint is shared by their corresponding edges in G [1]. The degree of edge e in G is denoted by $d_{L(G)}(e)$, which is the number of edges that share common endpoint with edge e in G; it is also the degree of vertex e in $L(G)$. We give simple a illustration to explain the relationship of original graph and corresponding line graph in Figure 1. We can see u, v, w denote corresponding vertexes, and e, f, g, h, i, j denote corresponding edges in original graph G and denote corresponding vertices in line graph $L(G)$. We get $d(u) = d(v) = 3, d(w) = 2$. $d_{L(G)}(e)$, which is the degree of vertex e in $L(G)$, is also the degree of edge e in G, thus $d_{L(G)}(e) = 4$ in Figure 1.

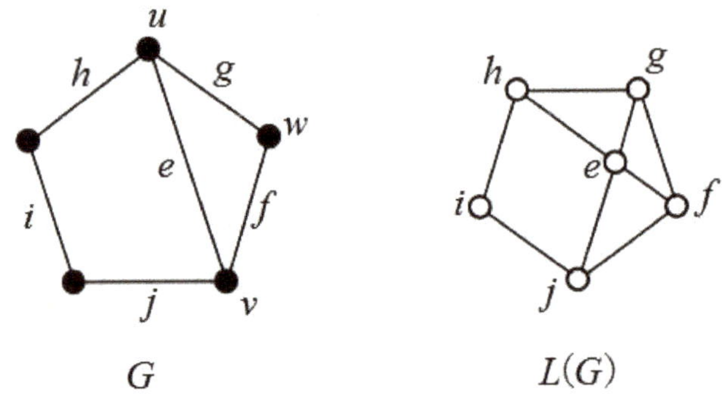

Figure 1. The original graph G and corresponding line graph $L(G)$.

Topological indices are graph invariants, which are obtained by performing some numerical operations on the distance or degree of vertices of the molecular graph. In chemical graph theory, topological indices are the molecular descriptors. They have been successfully used in describing the structures and predicting certain physicochemical properties of chemical compounds. To study the relationship between molecular structure and physical properties of saturated hydrocarbons, Wiener index was first published in 1947 [2], and the edge version of Wiener index, which can be considered as the Wiener index of line graph of G, was proposed by Iranmanesh et al. in 2009 [3]. As the important role of topological indices in chemical research has been confirmed, more topological indices appeared, which include atom-bond connectivity index and geometric arithmetic index.

In chemical graph theory, hydrogen atoms are usually ignored when the topological indices are calculated, which is very similar to how organic chemists usually simply write a benzene ring as a hexagon [4]. Now, three types of graphs of $C_{24}H_{28}$ are illustrated in Figure 2.

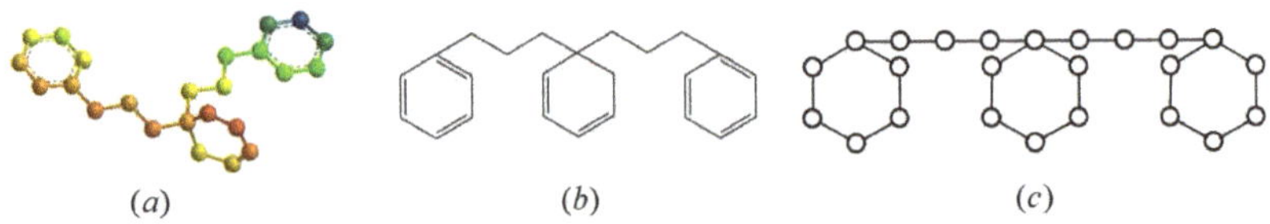

Figure 2. (**a**) $C_{24}H_{28}$ ball and stick model graph in 3D; (**b**) $C_{24}H_{28}$ chemical structure graph; and (**c**) $C_{24}H_{28}$ model graph in chemical graph theory.

To explore the properties of simple short chain compound products, Gao et al. [5] defined some join graphs such as $P_n + C_m$, $P_n + S_m$, $C_m + P_n + C_m$, $S_m + P_n + S_m$, and $C_m + P_n + S_r$, created by P_n, C_n and S_n and obtained the ABC_e and GA_e indices of these graphs. In another paper, Gao et al. [6] defined the bridge molecular structures, which can be used to research some long chain polymerization products, and the forgotten indices ($F(G)$) formulae of some simple bridge molecular structures constructed by P_2, C_6 or K_3 are presented. The forgotten index is defined as $F(G) = \sum_{v \in V(G)} (d(v)^3)$ [7]. In this paper, we define generalized bridge molecular graphs that could cover more kinds of long chain polymerization products, and the edge-version atom-bond connectivity and geometric arithmetic indices of generalized bridge molecular graphs are calculated.

To facilitate the reader, the topological indices discussed in this thesis are all given in Table 1.

Table 1. The definition of topological indices.

Index Name	Definition	Proposed	Recent Studied
atom-bond connection index	$ABC(G) = \sum_{uv\in E(G)} \sqrt{\frac{d(u)+d(v)-2}{d(u)d(v)}}$	[8]	[9–11]
edge version of ABC index	$ABC_e(G) = \sum_{e_1e_2\in E(L(G))} \sqrt{\frac{d_{L(G)}(e_1)+d_{L(G)}(e_2)-2}{d_{L(G)}(e_1)\times d_{L(G)}(e_2)}}$	[12]	[5,13,14]
geometric arithmetic index	$GA(G) = \sum_{uv\in E(G)} \frac{2\sqrt{d_G(u)d_G(v)}}{d_G(u)+d_G(v)}$	[15]	[16–18]
edge version of GA index	$GA_e(G) = \sum_{e_1e_2\in E(L(G))} \frac{2\sqrt{d_{L(G)}(e_1)d_{L(G)}(e_2)}}{d_{L(G)}(e_1)+d_{L(G)}(e_2)}$	[19]	[5,12,19–21]

In Table 1, $d_G(u)$ and $d_G(v)$ are the degrees of the vertices u and v in G, and $d_{L(G)}(e_1)$ and $d_{L(G)}(e_2)$ are the degrees of the edges e_1 and e_2 in G.

2. Main Results and Proofs

2.1. Definition of the Generalized Bridge Molecular Graph

Before we start a discussion, we give the definition of the generalized bridge molecular graph as follows. For a positive integer d, d pairwise disjoint molecular graphs $\{G^{(1)}, G^{(2)}, \cdots, G^{(d)}\}$ with $v^{(i)} \in V(G^{(i)})$ for each $i = 1, 2, \cdots, d$, and $d-1$ pairwise disjoint path molecular graphs $P^{(1)}, P^{(2)}, \cdots, P^{(d-1)}$ (called bridges), the generalized bridge molecular graph $GBG(G^{(1)}, v^{(1)}, G^{(2)}, v^{(2)}, \cdots, G^{(d)}, v^{(d)}; P^{(1)}, P^{(2)}, \cdots, P^{(d-1)})$ is the graph obtained by connecting the vertices $v^{(i)}$ and $v^{(i+1)}$ by a path $P^{(i)}$ for which two end vertices are identified with $v^{(i)}$ and $v^{(i+1)}$ for $i = 1, 2, ..., d-1$ (See Figure 3). When $G := G^{(i)}$, $P := P^{(i)}$, $v := v^{(i)}$ for each i, we simplify $GBG(G^{(1)}, v^{(1)}, G^{(2)}, v^{(2)}, \cdots, G^{(d)}, v^{(d)}; P^{(1)}, P^{(2)}, \cdots, P^{(d-1)})$ to be $GBG(G, v; P; d)$. In this paper, if G is a star, then v is the central vertex and if G is a cycle, v is considered as any vertex. In such cases, we further simplify $GBG(G, v; P; d)$ to be $GBG(G, P; d)$. The bridge molecular graph's bridge is strictly P_2 in [6], which limits the scope of modeling objects. The generalized bridge molecular graphs can model more kinds of long chain polymerization products than the bridge molecular graphs, because the bridge can be either P_2 or P_n and $n \geq 3$.

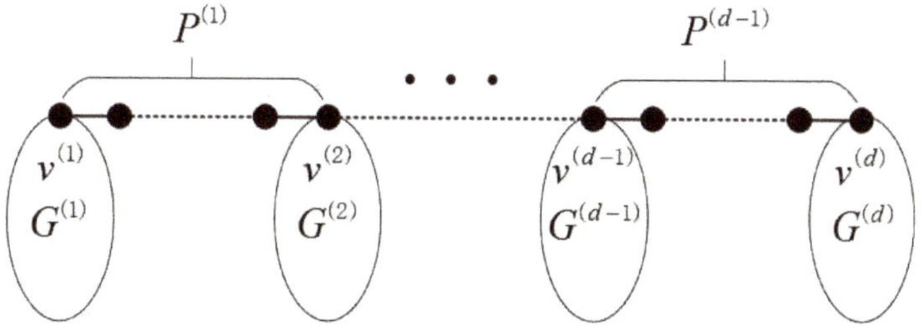

Figure 3. The generalized bridge molecular graph $GBG(G^{(1)}, v^{(1)}, G^{(2)}, v^{(2)}, ..., G^{(d)}, v^{(d)}; P^{(1)}, P^{(2)}, ..., P^{(d-1)})$.

2.2. Results and Discussion

In the following, we discuss the edge-version atom-bond connectivity and geometric arithmetic indices of some generalized bridge molecular graph. The line graph $GBG^L(S_m, P_n; d)$ of $GBG(S_m, P_n; d)$ is illustrated in Figure 4.

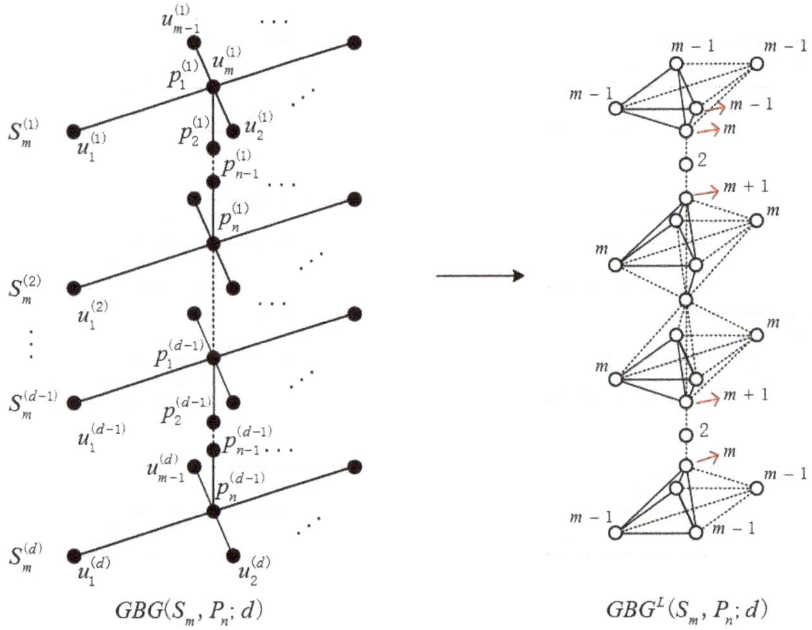

Figure 4. The generalized bridge molecular graph of $GBG(S_m, P_n; d)$ and $GBG^L(S_m, P_n; d)$.

Theorem 1. *Let $GBG(S_m, P_n; d)$ be the generalized bridge molecular graph for $n \geq 4$, $d \geq 2$ and $m \geq 2$ (see Figure 4), then the ABC_e and GA_e of $GBG(S_m, P_n; d)$ are*

$$\begin{aligned} ABC_e(GBG(S_m, P_n; d)) &= \frac{\sqrt{2}}{2}(d-1)(n-2) + \frac{d-2}{m+1}\sqrt{2m} \\ &\quad +2(m-1)\sqrt{\frac{2m-3}{m^2-m}} + 2(d-2)(m-1)\sqrt{\frac{2m-1}{m^2+m}} \\ &\quad +(m-2)\sqrt{2m-4} + \frac{(d-2)(m-1)(m-2)\sqrt{2m-2}}{2m}, \end{aligned}$$

$$\begin{aligned} GA_e(GBG(S_m, P_n; d)) &= \frac{4\sqrt{2m}}{m+2} + \frac{4(d-2)\sqrt{2m+2}}{m+3} + (d-1)(n-4) + (d-2) \\ &\quad +4(m-1)\frac{\sqrt{m^2-m}}{2m-1} + 4(d-2)(m-1)\frac{\sqrt{m^2+m}}{2m+1} \\ &\quad +\frac{d}{2}(m-1)(m-2). \end{aligned}$$

Proof. This line graph has $2 - 2m - n + \frac{d}{2}(m^2 + m + 2n - 4)$ edges. If $d_{L(G)}(e_1)$ and $d_{L(G)}(e_2)$ are the degree of edge of e_1 and e_2, then there are 2 edges of type $d_{L(G)}(e_1) = m, d_{L(G)}(e_2) = 2$, $2(d-2)$ edges of type $d_{L(G)}(e_1) = m+1, d_{L(G)}(e_2) = 2$, $(d-1)(n-4)$ edges of type $d_{L(G)}(e_1) = d_{L(G)}(e_2) = 2$, $d-2$ edges of type $d_{L(G)}(e_1) = d_{L(G)}(e_2) = m+1$, $2(m-1)$ edges of type $d_{L(G)}(e_1) = m, d_{L(G)}(e_2) = m-1$, $2(d-2)(m-1)$ edges of type $d_{L(G)}(e_1) = m$, $d_{L(G)}(e_2) = m+1$, $(m-1)(m-2)$ edges of type $d_{L(G)}(e_1) = d_{L(G)}(e_2) = m-1$, and $\frac{d-2}{2}(m-1)(m-2)$ edges of type $d_{L(G)}(e_1) = d_{L(G)}(e_2) = m$. Hence, we get

$$
\begin{aligned}
ABC_e(GBG(S_m, P_n; d)) &= 2\left(\sqrt{\frac{m+2-2}{m\times 2}}\right) + 2(d-2)\left(\sqrt{\frac{m+1+2-2}{(m+1)\times 2}}\right) \\
&\quad +(d-1)(n-4)\left(\sqrt{\frac{2+2-2}{2\times 2}}\right) \\
&\quad +(d-2)\left(\sqrt{\frac{m+1+m+1-2}{(m+1)\times(m+1)}}\right) \\
&\quad +2(m-1)\left(\sqrt{\frac{m+m-1-2}{m\times(m-1)}}\right) \\
&\quad +2(d-2)(m-1)\left(\sqrt{\frac{m+m+1-2}{m\times(m+1)}}\right) \\
&\quad +(m-1)(m-2)\left(\sqrt{\frac{m-1+m-1-2}{(m-1)\times(m-1)}}\right) \\
&\quad +\frac{d-2}{2}(m-1)(m-2)\left(\sqrt{\frac{m+m-2}{m\times m}}\right) \\
&= \frac{\sqrt{2}}{2}(d-1)(n-2) + \frac{d-2}{m+1}\sqrt{2m} \\
&\quad +2(m-1)\sqrt{\frac{2m-3}{m^2-m}} + 2(d-2)(m-1)\sqrt{\frac{2m-1}{m^2+m}} \\
&\quad +(m-2)\sqrt{2m-4} + \frac{(d-2)(m-1)(m-2)\sqrt{2m-2}}{2m},
\end{aligned}
$$

$$
\begin{aligned}
GA_e(GBG(S_m, P_n; d)) &= 2\left(\frac{2\sqrt{m\times 2}}{m+2}\right) + 2(d-2)\left(\frac{2\sqrt{(m+1)\times 2}}{m+1+2}\right) \\
&\quad +(d-1)(n-4)\left(\frac{2\sqrt{2\times 2}}{2+2}\right) \\
&\quad +(d-2)\left(\frac{2\sqrt{(m+1)\times(m+1)}}{m+1+m+1}\right) \\
&\quad +2(m-1)\left(\frac{2\sqrt{m\times(m-1)}}{m+m-1}\right) \\
&\quad +2(d-2)(m-1)\left(\frac{2\sqrt{m\times(m+1)}}{m+m+1}\right) \\
&\quad +(m-1)(m-2)\left(\frac{2\sqrt{(m-1)\times(m-1)}}{m-1+m-1}\right) \\
&\quad +\frac{d-2}{2}(m-1)(m-2)\left(\frac{2\sqrt{m\times m}}{m+m}\right) \\
&= \frac{4\sqrt{2m}}{m+2} + \frac{4(d-2)\sqrt{2m+2}}{m+3} + (d-1)(n-4) + (d-2) \\
&\quad +4(m-1)\frac{\sqrt{m^2-m}}{2m-1} + 4(d-2)(m-1)\frac{\sqrt{m^2+m}}{2m+1} \\
&\quad +\frac{d}{2}(m-1)(m-2).
\end{aligned}
$$

The proof is complete. □

For Example 1, in Figure 5, $2,7,7,12-tetramethyltridecane$ can be modeled by $GBG(S_3, P_6; 3)$, so $ABC_e(GBG(S_3, P_6; 3)) \approx 13.76052$ and $GA_e(GBG(S_3, P_6; 3)) \approx 19.72337$.

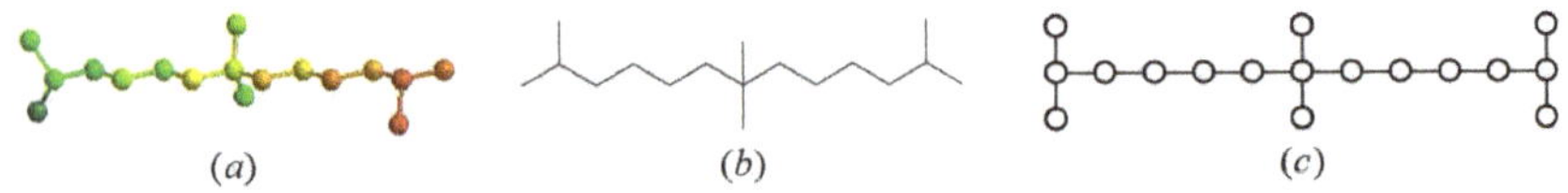

Figure 5. (**a**) 2,7,7,12-*tetramethyltridecane* ball and stick model graph in 3*D*; (**b**) 2,7,7,12-*tetramethyltridecane* chemical structure graph; and (**c**) 2,7,7,12-*tetramethyltridecane* model graph in chemical graph theory.

Theorem 2. *Let $GBG(S_m, P_3; d)$ be the generalized bridge molecular graph for $n = 3$, $d \geq 3$ and $m \geq 2$ (see Figure 6), then the ABC_e and GA_e of $GBG(S_m, P_3; d)$ are*

$$\begin{aligned} ABC_e(GBG(S_m, P_3; d)) &= 2\sqrt{\frac{2m-1}{m^2+m}} + \frac{2d-5}{m+1}\sqrt{2m} + 2(m-1)\sqrt{\frac{2m-3}{m^2-m}} \\ &+ 2(d-2)(m-1)\sqrt{\frac{2m-1}{m^2+m}} + (m-2)\sqrt{2m-4} \\ &+ \frac{(d-2)(m-1)(m-2)}{2m}\sqrt{2m-2}, \end{aligned}$$

$$\begin{aligned} GA_e(GBG(S_m, P_3; d)) &= \frac{4}{2m+1}\sqrt{m^2+m} + (2d-5) \\ &+ \frac{4}{2m-1}(m-1)\sqrt{m^2-m} \\ &+ \frac{4}{2m+1}(d-2)(m-1)\sqrt{m^2+m} \\ &+ \frac{d}{2}(m-1)(m-2). \end{aligned}$$

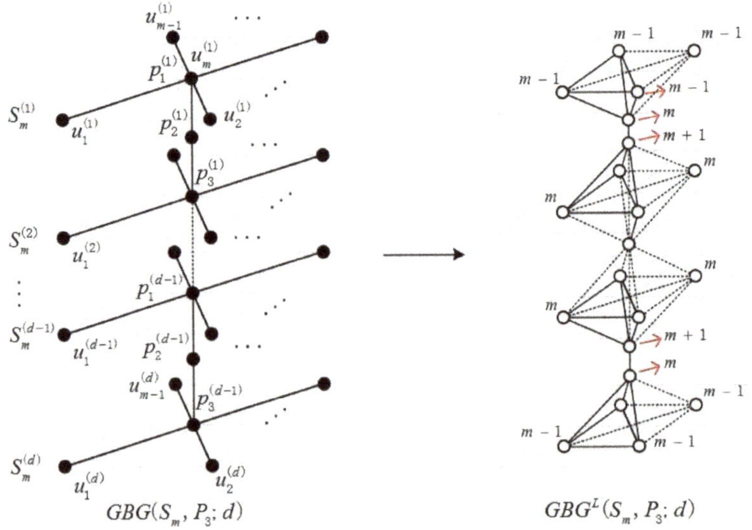

Figure 6. The generalized bridge molecular graph of $GBG(S_m, P_3; d)$ and $GBG^L(S_m, P_3; d)$.

Proof. This line graph has $\frac{d}{2}(m^2+m+2)-2m-1$ edges. If $d_{L(G)}(e_1)$ and $d_{L(G)}(e_2)$ are the degree of edge of e_1 and e_2, then there are 2 edges of type $d_{L(G)}(e_1)=m$, $d_{L(G)}(e_2)=m+1$, $2d-5$ edges of type $d_{L(G)}(e_1)=m+1$, $d_{L(G)}(e_2)=m+1$, $2(m-1)$ edges of type $d_{L(G)}(e_1)=m$, $d_{L(G)}(e_2)=m-1$, $2(d-2)(m-1)$ edges of type $d_{L(G)}(e_1)=m$,$d_{L(G)}(e_2)=m+1$, $(m-1)(m-2)$ edges of type $d_{L(G)}(e_1)=d_{L(G)}(e_2)=m-1$, and $\frac{d-2}{2}(m-1)(m-2)$ edges of type $d_{L(G)}(e_1)=d_{L(G)}(e_2)=m$. Hence, we get

$$\begin{aligned}
ABC_e(GBG(S_m,P_3;d)) &= 2\left(\sqrt{\frac{m+m+1-2}{m\times(m+1)}}\right)+(2d-5)\left(\sqrt{\frac{m+1+m+1-2}{(m+1)\times(m+1)}}\right)\\
&\quad +2(m-1)\left(\sqrt{\frac{m+m-1-2}{m\times(m-1)}}\right)\\
&\quad +2(d-2)(m-1)\left(\sqrt{\frac{m+m+1-2}{m\times(m+1)}}\right)\\
&\quad +(m-1)(m-2)\left(\sqrt{\frac{m-1+m-1-2}{(m-1)\times(m-1)}}\right)\\
&\quad +\frac{d-2}{2}(m-1)(m-2)\left(\sqrt{\frac{m+m-2}{m\times m}}\right)\\
&= 2\sqrt{\frac{2m-1}{m^2+m}}+\frac{2d-5}{m+1}\sqrt{2m}+2(m-1)\sqrt{\frac{2m-3}{m^2-m}}\\
&\quad +2(d-2)(m-1)\sqrt{\frac{2m-1}{m^2+m}}+(m-2)\sqrt{2m-4}\\
&\quad +\frac{(d-2)(m-1)(m-2)}{2m}\sqrt{2m-2},
\end{aligned}$$

$$\begin{aligned}
GA_e(GBG(S_m,P_3;d)) &= 2\left(\frac{2\sqrt{m\times(m+1)}}{m+m+1}\right)+(2d-5)\left(\frac{2\sqrt{(m+1)\times(m+1)}}{m+1+m+1}\right)\\
&\quad +2(m-1)\left(\frac{2\sqrt{m\times(m-1)}}{m+m-1}\right)\\
&\quad +2(d-2)(m-1)\left(\frac{2\sqrt{m\times(m+1)}}{m+m+1}\right)\\
&\quad +(m-1)(m-2)\left(\frac{2\sqrt{(m-1)\times(m-1)}}{m-1+m-1}\right)\\
&\quad +\frac{d-2}{2}(m-1)(m-2)\left(\frac{2\sqrt{m\times m}}{m+m}\right)\\
&= \frac{4}{2m+1}\sqrt{m^2+m}+(2d-5)\\
&\quad +\frac{4}{2m-1}(m-1)\sqrt{m^2-m}\\
&\quad +\frac{4}{2m+1}(d-2)(m-1)\sqrt{m^2+m}\\
&\quad +\frac{d}{2}(m-1)(m-2).
\end{aligned}$$

The proof is complete. □

For Example 2, in Figure 7, $2,4,4,6-$ *tetramethylheptane* can be modeled by $GBG(S_3,P_3;3)$, so $ABC_e(GBG(S_3,P_3;3))\approx 9.394663$ and $GA_e(GBG(S_3,P_3;3))\approx 13.85764$.

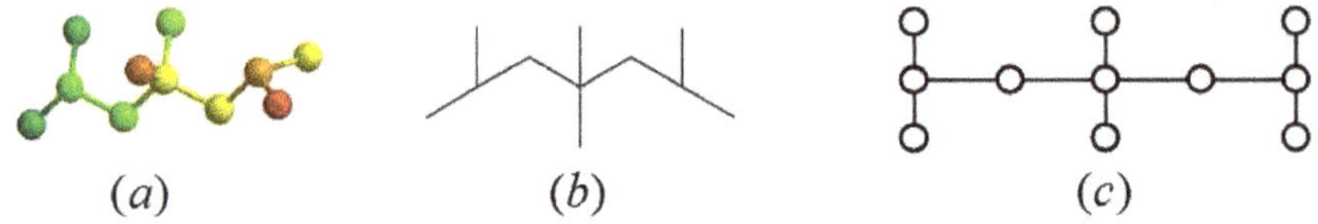

Figure 7. (**a**) 2, 4, 4, 6-*tetramethylheptane* ball and stick model graph in 3D; (**b**) 2, 4, 4, 6-*tetramethylheptane* chemical structure graph; and (**c**) 2, 4, 4, 6-*tetramethylheptane* model graph in chemical graph theory.

Theorem 3. *Let $GBG(S_m, P_2; d)$ be the generalized bridge molecular graph for $n = 2$, $d \geq 4$ and $m \geq 2$ (see Figure 8), then the ABC_e and GA_e of $GBG(S_m, P_2; d)$ are*

$$\begin{aligned} ABC_e(GBG(S_m, P_2; d)) &= 2(m-1)\sqrt{\frac{3m-4}{2m^2-3m+1}} + 2(m-1)\sqrt{\frac{3m-3}{2m^2-m}} \\ &+ 2\frac{(d-3)(m-1)}{m}\sqrt{\frac{3m-2}{2}} + (m-2)\sqrt{2m-4} \\ &+ \frac{(d-2)(m-1)(m-2)}{2m}\sqrt{2m-2} \\ &+ 2\sqrt{\frac{4m-3}{4m^2-2m}} + \frac{d-4}{2m}\sqrt{4m-2}, \end{aligned}$$

$$\begin{aligned} GA_e(GBG(S_m, P_2; d)) &= \frac{4(m-1)}{3m-2}\sqrt{2m^2-3m+1} + \frac{4(m-1)}{3m-1}\sqrt{2m^2-m} \\ &+ \frac{4\sqrt{2}}{3}(d-3)(m-1) + \frac{d}{2}(m-1)(m-2) \\ &+ \frac{4(\sqrt{4m^2-2m})}{4m-1} + (d-4). \end{aligned}$$

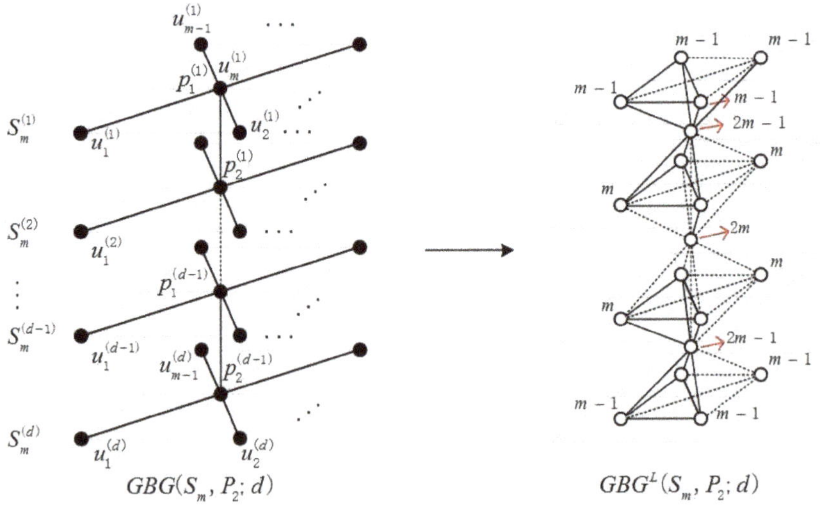

Figure 8. The generalized bridge molecular graph of $GBG(S_m, P_2; d)$ and $GBG^L(S_m, P_2; d)$.

Proof. This line graph has $\frac{1}{2}m(dm + d - 4)$ edges. If $d_{L(G)}(e_1)$ and $d_{L(G)}(e_2)$ are the degree of edge of e_1 and e_2, then there are $2(m-1)$ edges of type $d_{L(G)}(e_1) = 2m - 1, d_{L(G)}(e_2) = m - 1$, $2(m-1)$ edges of type $d_{L(G)}(e_1) = 2m-1, d_{L(G)}(e_2) = m$, $2(d-3)(m-1)$ edges of type $d_{L(G)}(e_1) = 2m, d_{L(G)}(e_2) = m$, $(m-1)(m-2)$ edges of type $d_{L(G)}(e_1) = d_{L(G)}(e_2) = m-1$, $\frac{d-2}{2}(m-1)(m-2)$ edges of type $d_{L(G)}(e_1) = d_{L(G)}(e_2) = m$, 2 edges of type $d_{L(G)}(e_1) = 2m-1, d_{L(G)}(e_2) = 2m$, and $d-4$ edges of type $d_{L(G)}(e_1) = d_{L(G)}(e_2) = 2m$. Hence, we get

$$\begin{aligned}
ABC_e(GBG(S_m, P_2; d)) &= 2(m-1)\left(\sqrt{\frac{2m-1+m-1-2}{(2m-1)\times(m-1)}}\right) \\
&+2(m-1)\left(\sqrt{\frac{2m-1+m-2}{(2m-1)\times m}}\right) \\
&+2(d-3)(m-1)\left(\sqrt{\frac{2m+m-2}{2m\times m}}\right) \\
&+(m-1)(m-2)\left(\sqrt{\frac{m-1+m-1-2}{(m-1)\times(m-1)}}\right) \\
&+\frac{d-2}{2}(m-1)(m-2)\left(\sqrt{\frac{m+m-2}{m\times m}}\right) \\
&+2\left(\sqrt{\frac{2m-1+2m-2}{(2m-1)\times 2m}}\right) \\
&+(d-4)\left(\sqrt{\frac{2m+2m-2}{2m\times 2m}}\right) \\
&= 2(m-1)\sqrt{\frac{3m-4}{2m^2-3m+1}} + 2(m-1)\sqrt{\frac{3m-3}{2m^2-m}} \\
&+2\frac{(d-3)(m-1)}{m}\sqrt{\frac{3m-2}{2}} + (m-2)\sqrt{2m-4} \\
&+\frac{(d-2)(m-1)(m-2)}{2m}\sqrt{2m-2} \\
&+2\sqrt{\frac{4m-3}{4m^2-2m}} + \frac{d-4}{2m}\sqrt{4m-2},
\end{aligned}$$

$$\begin{aligned}
GA_e(GBG(S_m, P_2; d)) &= 2(m-1)\left(\frac{2\sqrt{(2m-1)\times(m-1)}}{2m-1+m-1}\right) \\
&+2(m-1)\left(\frac{2\sqrt{(2m-1)\times m}}{2m-1+m}\right) \\
&+2(d-3)(m-1)\left(\frac{2\sqrt{2m\times m}}{2m+m}\right) \\
&+(m-1)(m-2)\left(\frac{2\sqrt{(m-1)\times(m-1)}}{m-1+m-1}\right) \\
&+\frac{d-2}{2}(m-1)(m-2)\left(\frac{2\sqrt{m\times m}}{m+m}\right) \\
&+2\left(\frac{2\sqrt{(2m-1)\times 2m}}{2m-1+2m}\right) \\
&+(d-4)\left(\frac{2\sqrt{2m\times 2m}}{2m+2m}\right) \\
&= \frac{4(m-1)}{3m-2}\sqrt{2m^2-3m+1} + \frac{4(m-1)}{3m-1}\sqrt{2m^2-m} \\
&+\frac{4\sqrt{2}}{3}(d-3)(m-1) + \frac{d}{2}(m-1)(m-2) \\
&+\frac{4(\sqrt{4m^2-2m})}{4m-1} + (d-4).
\end{aligned}$$

The proof is complete. □

For Example 3, in Figure 9, 2,3,3,4-*tetramethylpentane* can be modeled by $GBG(S_3, P_2; 4)$, so $ABC_e(GBG(S_3, P_2; 4)) \approx 11.69568$ and $GA_e(GBG(S_3, P_2; 4)) \approx 17.24996952$.

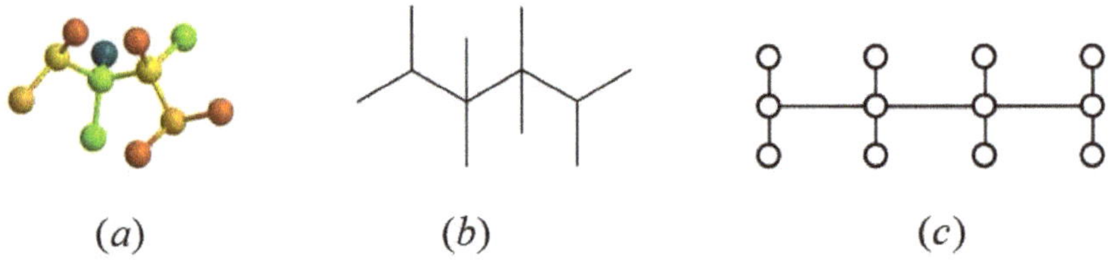

Figure 9. (**a**) 2,3,3,4-*tetramethylpentane* ball and stick model graph in 3D; (**b**) 2,3,3,4-*tetramethylpentane* chemical structure graph; and (**c**) 2,3,3,4-*tetramethylpentane* model graph in chemical graph theory.

Theorem 4. *Let $GBG(C_m, P_n; d)$ be the generalized bridge molecular graph for $n \geq 4$, $d \geq 2$ and $m \geq 3$ (see Figure 10), then the ABC_e and GA_e of $GBG(C_m, P_n; d)$ are*

$$\begin{aligned} ABC_e(GBG(C_m, P_n; d)) &= \frac{\sqrt{2}}{2}(d(m-3) + (d-1)(n-4)) + (2\sqrt{2} + \frac{3\sqrt{6}}{2})d \\ &\quad -\sqrt{2} - 3\sqrt{6} + 4, \end{aligned}$$

$$\begin{aligned} GA_e(GBG(C_m, P_n; d)) &= d(m-3) + (d-1)(n-4) \\ &\quad +(\frac{8\sqrt{2}}{3} + 6)(d-2) + \frac{12\sqrt{6}}{5} + 5. \end{aligned}$$

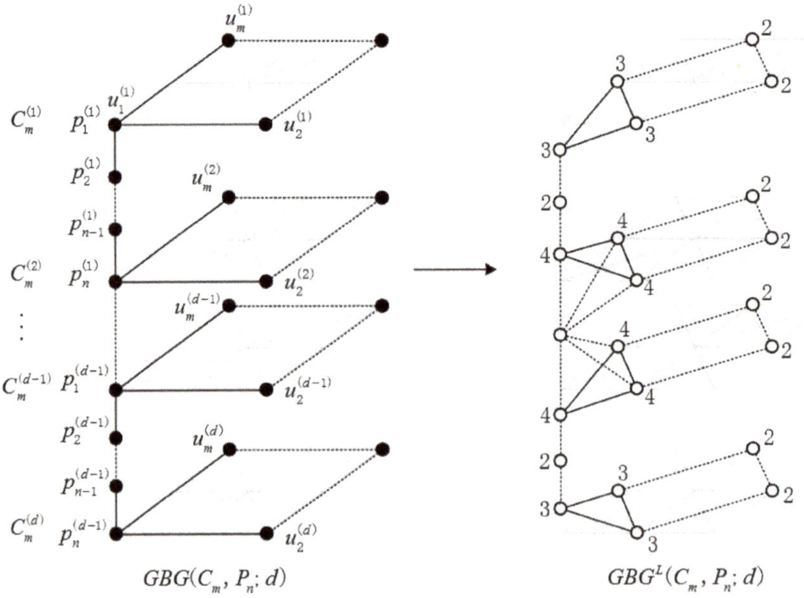

Figure 10. The generalized bridge molecular graph of $GBG(C_m, P_n; d)$ and $GBG^L(C_m, P_n; d)$.

Proof. In Figure 10, the degrees of vertices in line graph $G^L(G_d(C_m+P_n))$ are displayed near by the corresponding vertices. This line graph has $d(m+n+3)-n-4$ edges. In addition, there are $d(m-3)+(d-1)(n-4)$ edges of type $d_{L(G)}(e_1)=d_{L(G)}(e_2)=2$, 6 edges of type $d_{L(G)}(e_1)=2$ and $d_{L(G)}(e_2)=3$, 6 edges of type $d_{L(G)}(e_1)=d_{L(G)}(e_2)=3$, $4(d-2)$ edges of type $d_{L(G)}(e_1)=2$ and $d_{L(G)}(e_2)=4$, and $6(d-2)$ edges of type $d_{L(G)}(e_1)=d_{L(G)}(e_2)=4$. Hence, we have

$$\begin{aligned}
ABC_e(GBG(C_m,P_n;d)) &= \Big(d(m-3)+(d-1)(n-4)\Big)\Big(\sqrt{\frac{2+2-2}{2\times 2}}\Big) \\
&\quad +6\Big(\sqrt{\frac{2+3-2}{2\times 3}}\Big)+6\Big(\sqrt{\frac{3+3-2}{3\times 3}}\Big) \\
&\quad +4(d-2)\Big(\sqrt{\frac{2+4-2}{2\times 4}}\Big)+6(d-2)\Big(\sqrt{\frac{4+4-2}{4\times 4}}\Big) \\
&= \frac{\sqrt{2}}{2}(d(m-3)+(d-1)(n-4))+(2\sqrt{2}+\frac{3\sqrt{6}}{2})d \\
&\quad -\sqrt{2}-3\sqrt{6}+4,
\end{aligned}$$

$$\begin{aligned}
GA_e(GBG(C_m,P_n;d)) &= \Big(d(m-3)+(d-1)(n-4)\Big)\Big(\frac{2\sqrt{2\times 2}}{2+2}\Big) \\
&\quad +6\Big(\frac{2\sqrt{2\times 3}}{2+3}\Big)+6\Big(\frac{2\sqrt{3\times 3}}{3+3}\Big) \\
&\quad +4(d-2)\Big(\frac{2\sqrt{2\times 4}}{2+4}\Big)+6(d-2)\Big(\frac{2\sqrt{4\times 4}}{4+4}\Big) \\
&= d(m-3)+(d-1)(n-4) \\
&\quad +(\frac{8\sqrt{2}}{3}+6)(d-2)+\frac{12\sqrt{6}}{5}+5.
\end{aligned}$$

The proof is complete. □

For Example 4, in Figure 2, $C_{24}H_{28}$ is *(cyclohexa-2,4-diene-1,1-diylbis(propane-3,1-diyl))dibenzene*, which can be modeled by $GBG(C_6,P_5;3)$, so $ABC_e(GBG(C_6,P_5;3)) \approx 22.52347702$ and $GA_e(GBG(C_6,P_5;3)) \approx 31.65001155$.

Theorem 5. *Let $GBG(C_m,P_3;d)$ be the generalized bridge molecular graph for $n=3$, $d\geq 3$, and $m\geq 3$ (see Figure 11), then the ABC_e and GA_e of $GBG(C_m,P_3;d)$ are*

$ABC_e(GBG(C_m,P_3;d)) = \frac{\sqrt{2}}{2}d(m-3)+(\sqrt{2}+7\frac{\sqrt{6}}{4})d+4+\frac{\sqrt{15}}{3}-\frac{15\sqrt{6}}{4}$,

$GA_e(GBG(C_m,P_3;d)) = d(m-3)+(\frac{4\sqrt{2}}{3}+7)d+\frac{8\sqrt{6}}{5}+6+\frac{8\sqrt{3}}{7}-\frac{8\sqrt{2}}{3}-15$.

Proof. In Figure 11, the degrees of vertices in line graph $G^L(GBG(C_m,P_3;d))$ are displayed near by the corresponding vertices. This line graph has $d(m+6)-7$ edges. In addition, there are $d(m-3)$ edges of type $d_{L(G)}(e_1)=d_{L(G)}(e_2)=2$, 4 edges of type $d_{L(G)}(e_1)=2$ and $d_{L(G)}(e_2)=3$, $2(d-2)$ edges of type $d_{L(G)}(e_1)=2$ and $d_{L(G)}(e_2)=4$, 6 edges of type $d_{L(G)}(e_1)=d_{L(G)}(e_2)=3$, 2 edges of type $d_{L(G)}(e_1)=3$ and $d_{L(G)}(e_2)=4$, and $7d-15$ edges of type $d_{L(G)}(e_1)=d_{L(G)}(e_2)=4$. Hence, we have

$$
\begin{aligned}
ABC_e(GBG(C_m, P_3; d)) &= d(m-3)\Big(\sqrt{\frac{2+2-2}{2\times 2}}\Big) + 4\Big(\sqrt{\frac{2+3-2}{2\times 3}}\Big) \\
&\quad +2(d-2)\Big(\sqrt{\frac{2+4-2}{2\times 4}}\Big) + 6\Big(\sqrt{\frac{3+3-2}{3\times 3}}\Big) \\
&\quad +2\Big(\sqrt{\frac{3+4-2}{3\times 4}}\Big) + (7d-15)\Big(\sqrt{\frac{4+4-2}{4\times 4}}\Big) \\
&= \frac{\sqrt{2}}{2}d(m-3) + (\sqrt{2}+7\frac{\sqrt{6}}{4})d + 4 \\
&\quad +\frac{\sqrt{15}}{3} - \frac{15\sqrt{6}}{4},
\end{aligned}
$$

$$
\begin{aligned}
GA_e(GBG(C_m, P_3; d)) &= d(m-3)\Big(\frac{2\sqrt{2\times 2}}{2+2}\Big) + 4\Big(\frac{2\sqrt{2\times 3}}{2+3}\Big) \\
&\quad +2(d-2)\Big(\frac{2\sqrt{2\times 4}}{2+4}\Big) + 6\Big(\frac{2\sqrt{3\times 3}}{3+3}\Big) \\
&\quad +2\Big(\frac{2\sqrt{3\times 4}}{3+4}\Big) + (7d-15)\Big(\frac{2\sqrt{4\times 4}}{4+4}\Big) \\
&= d(m-3) + (\frac{4\sqrt{2}}{3}+7)d + \frac{8\sqrt{6}}{5} + 6 \\
&\quad +\frac{8\sqrt{3}}{7} - \frac{8\sqrt{2}}{3} - 15.
\end{aligned}
$$

The proof is complete. □

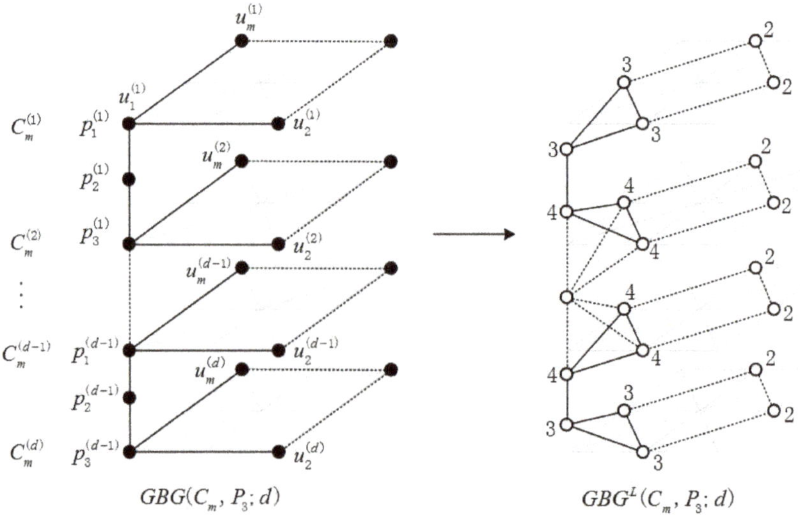

Figure 11. The generalized bridge molecular graph of $GBG(C_m, P_3; d)$ and $GBG^L(C_m, P_3; d)$.

For Example 5, in Figure 12, (*cyclohexane*-1, 1-*diylbis*(*methylene*))*dicyclohexane* can be modeled by $GBG(C_6, P_3; 3)$, so $ABC_e(GBG(C_6, P_3; 3)) \approx 19.57183078$ and $GA_e(GBG(C_6, P_3; 3)) \approx 28.78428831$.

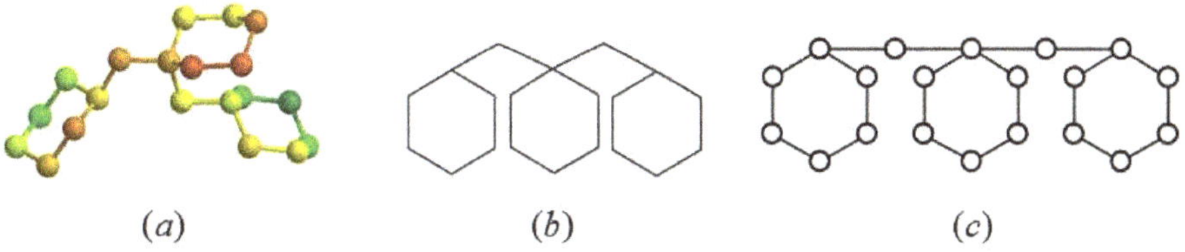

Figure 12. (**a**) *(cyclohexane-1,1-diylbis(methylene))dicyclohexane* ball and stick model graph in *3D*; (**b**) *(cyclohexane-1,1-diylbis(methylene))dicyclohexane* chemical structure graph; and (**c**) *(cyclohexane-1,1-diylbis(methylene))dicyclohexane* model graph in chemical graph theory.

Theorem 6. *Let $GBG(C_m, P_2; d)$ be the generalized bridge molecular graph for $n = 2$, $d \geq 4$, and $m \geq 3$ (see Figure 13), then the ABC_e and GA_e of $GBG(C_m, P_2; d)$ are*

$$\begin{aligned} ABC_e(GBG(C_m, P_2; d)) &= \frac{\sqrt{2}}{2}dm + (\sqrt{2} + \frac{\sqrt{6}}{4} + \frac{4\sqrt{3}}{3} + \frac{\sqrt{10}}{6} - \frac{3\sqrt{2}}{2})d \\ &\quad + \frac{4}{3} + 4\sqrt{\frac{2}{5}} + 2\sqrt{\frac{7}{5}} + 2\sqrt{\frac{3}{10}} - \frac{\sqrt{6}}{2} - 4\sqrt{3} - \frac{2\sqrt{10}}{3}, \end{aligned}$$

$$\begin{aligned} GA_e(GBG(C_m, P_2; d)) &= dm + (\frac{4\sqrt{2}}{3} + \frac{8\sqrt{6}}{5} - 1)d + \sqrt{15} + \frac{16\sqrt{5}}{9} + \frac{4\sqrt{30}}{11} \\ &\quad - \frac{16\sqrt{6}}{5} - \frac{8\sqrt{2}}{3} - 4. \end{aligned}$$

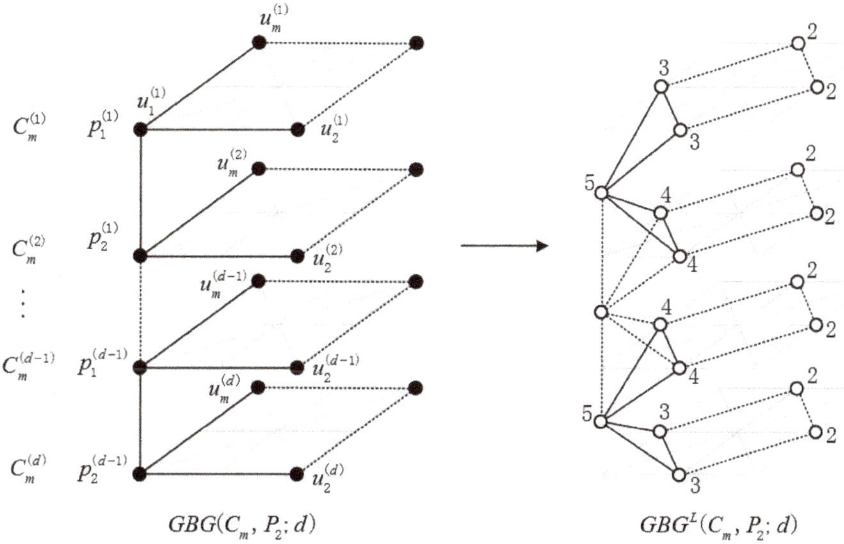

Figure 13. The generalized bridge molecular graph of $GBG(C_m, P_2; d)$ and $GBG^L(C_m, P_2; d)$.

Proof. In Figure 13, the degrees of vertices in line graph $G^L(GBG(C_m, P_2; d))$ are displayed near by the corresponding vertices. This line graph has $d(m-5) - 6$ edges. In addition, there are $d(m-3)$ edges of type $d_{L(G)}(e_1) = d_{L(G)}(e_2) = 2$, 4 edges of type $d_{L(G)}(e_1) = 2$, $d_{L(G)}(e_2) = 3$, $2(d-2)$ edges of type $d_{L(G)}(e_1) = 2$,$d_{L(G)}(e_2) = 4$, 2 edges of type $d_{L(G)}(e_1) = d_{L(G)}(e_2) = 3$, 4 edges of type $d_{L(G)}(e_1) = 3$, $d_{L(G)}(e_2) = 5$, $d-2$ edges of type $d_{L(G)}(e_1) = d_{L(G)}(e_2) = 4$, 4 edges of type

$d_{L(G)}(e_1) = 4$, $d_{L(G)}(e_2) = 5$, $4(d-3)$ edges of type $d_{L(G)}(e_1) = 4$, $d_{L(G)}(e_2) = 6$, 2 edges of type $d_{L(G)}(e_1) = 5$, $d_{L(G)}(e_2) = 6$, and $d-4$ edges of type $d_{L(G)}(e_1) = d_{L(G)}(e_2) = 6$. Hence, we have

$$\begin{aligned}
ABC_e(GBG(C_m, P_2; d)) &= d(m-3)\left(\sqrt{\frac{2+2-2}{2\times 2}}\right) + 4\left(\sqrt{\frac{2+3-2}{2\times 3}}\right) \\
&\quad +2(d-2)\left(\sqrt{\frac{2+4-2}{2\times 4}}\right) + 2\left(\sqrt{\frac{3+3-2}{3\times 3}}\right) \\
&\quad +4\left(\sqrt{\frac{3+5-2}{3\times 5}}\right) + (d-2)\left(\sqrt{\frac{4+4-2}{4\times 4}}\right) \\
&\quad +4\left(\sqrt{\frac{4+5-2}{4\times 5}}\right) + 4(d-3)\left(\sqrt{\frac{4+6-2}{4\times 6}}\right) \\
&\quad +2\left(\sqrt{\frac{5+6-2}{5\times 6}}\right) + (d-4)\left(\sqrt{\frac{6+6-2}{6\times 6}}\right) \\
&= \frac{\sqrt{2}}{2}dm + (\sqrt{2} + \frac{\sqrt{6}}{4} + \frac{4\sqrt{3}}{3} + \frac{\sqrt{10}}{6} - \frac{3\sqrt{2}}{2})d \\
&\quad +\frac{4}{3} + 4\sqrt{\frac{2}{5}} + 2\sqrt{\frac{7}{5}} + 2\sqrt{\frac{3}{10}} - \frac{\sqrt{6}}{2} - 4\sqrt{3} - \frac{2\sqrt{10}}{3},
\end{aligned}$$

$$\begin{aligned}
GA_e(GBG(C_m, P_2; d)) &= d(m-3)\left(\frac{2\sqrt{2\times 2}}{2+2}\right) + 4\left(\frac{2\sqrt{2\times 3}}{2+3}\right) \\
&\quad +2(d-2)\left(\frac{2\sqrt{2\times 4}}{2+4}\right) + 2\left(\frac{2\sqrt{3\times 3}}{3+3}\right) \\
&\quad +4\left(\frac{2\sqrt{3\times 5}}{3+5}\right) + (d-2)\left(\frac{2\sqrt{4\times 4}}{4+4}\right) \\
&\quad +4\left(\frac{2\sqrt{4\times 5}}{4+5}\right) + 4(d-3)\left(\frac{2\sqrt{4\times 6}}{4+6}\right) \\
&\quad +2\left(\frac{2\sqrt{5\times 6}}{5+6}\right) + (d-4)\left(\frac{2\sqrt{6\times 6}}{6+6}\right) \\
&= dm + (\frac{4\sqrt{2}}{3} + \frac{8\sqrt{6}}{5} - 1)d + \sqrt{15} + \frac{16\sqrt{5}}{9} + \frac{4\sqrt{30}}{11} \\
&\quad -\frac{16\sqrt{6}}{5} - \frac{8\sqrt{2}}{3} - 4.
\end{aligned}$$

The proof is complete.$\gamma_i(C_6 \square C_n)$ □

For Example 6, in Figure 14, $2'H,2''H$-$1,1' : 1',1'' : 1'',1'''$-*quaterphenyl* can be modeled by $GBG(C_6, P_2; 4)$, so $ABC_e(GBG(C_6, P_2; 4)) \approx 25.00131406$ and $GA_e(GBG(C_6, P_2; 4)) \approx 37.44953704$.

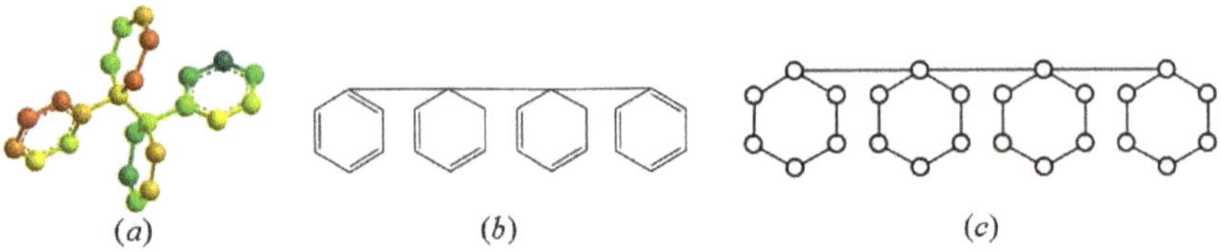

Figure 14. (**a**) $2'H,2''H$-$1,1' : 1',1'' : 1'',1'''$-*quaterphenyl* ball and stick model graph in 3*D*; (**b**) $2'H,2''H$-$1,1' : 1',1'' : 1'',1'''$-*quaterphenyl* chemical structure graph; and (**c**) $2'H,2''H$-$1,1' : 1',1'' : 1'',1'''$-*quaterphenyl* model graph in chemical graph theory.

3. Conclusions

Topological indices are proven to be very helpful to test the chemical properties of new chemical or physical materials. To describe more kinds of long chain polymerization products than the bridge molecular graphs, we propose the generalized bridge molecular graph structures. In this paper, we focus on some generalized bridge molecular graphs such as $GBG(S_m, P_n; d)$ and $GBG(C_m, P_n; d)$ and give the formulas of the edge version ABC and GA indices of these generalized bridge molecular graphs. By demonstrating the calculation of real molecules, we find that some long chain molecular graphs can be quickly modeled and their topological indices can be calculated using generalized bridge molecular graphs. The results of this paper also offer promising prospects in the applications for chemical and material engineering, especially in chemical industry research.

Author Contributions: X.Z. contribute for conceptualization, designing the experiments, wrote original draft preparation and funding; J.-B.L. contribute for supervision, methodology, project administration and formal analysing; X.W. contribute for performed experiments, validation. M.K.J. for resources, software, some computations; S.A. reviewed and edited initial draft and wrote the final draft; M.R.F. analyzed the data curation; All authors read and approved the final version of the paper.

Funding: This work was supported by the key project of the Sichuan Provincial Department of Education under Grant No. 17ZA0079 and 18ZA0118, partially supported by China Postdoctoral Science Foundation under Grant No. 2017M621579 and Postdoctoral Science Foundation of Jiangsu Province under Grant No. 1701081B, Project of Anhui Jianzhu University under Grant No. 2016QD116 and 2017dc03, and the Soft Scientific Research Foundation of Sichuan Provincial Science and Technology Department under Grant No. 2018ZR0265.

Acknowledgments: The authors are grateful to the anonymous reviewers and the editor for the valuable comments and suggestions.

Conflicts of Interest: The authors declare no conflict of interest.

References

1. Harary, F.; Norman, R.Z. Some properties of line digraphs. *Rendiconti del Circolo Matematico di Palermo* **1960**, *9*, 161–169. [CrossRef]
2. Wiener, H. Structural determination of paraffin boiling points. *J. Am. Chem. Soc.* **1947**, *69*, 7–20. [CrossRef]
3. Iranmanesh, A.; Gutman, I.; Khormali, O.; Mahmiani, A. The edge versions of the wiener index. *MATCH Commun. Math. Comput. Chem.* **2009**, *61*, 663–672.
4. Devillers, J.; Balaban, A.T. *Topological Indices and Related Descriptors in QSAR and QSPR*; CRC Press: Boca Raton, FL, USA, 1999.
5. Gao, W.; Farahani, M.R.; Wang, S.H.; Husin, M.N. On the edge-version atom-bond connectivity and geometric arithmetic indices of certain graph operations. *Appl. Math. Comput.* **2017**, *308*, 11–17. [CrossRef]
6. Gao, W.; Siddiqui, M.K.; Imran, M.; Jamil, M.K.; Farahani, M. R. Forgotten topological index of chemical structure in drugs. *Saudi Pharm. J.* **2016**, *24*, 258. [CrossRef] [PubMed]
7. Nikolić, S.; Kovačević, G.; Miličević, A.; Trinajstić, N. The Zagreb indices 30 years after. *Croat. Chem. Acta* **2003**, *76*, 113–124.
8. Estrada, E.; Torres, L.; Rodriguez, L.; Gutman, I. An atom-bond connectivity index: Modelling the enthalpy of formation of Alkanes. *Indian J. Chem.* **1998**, *37A*, 849–855.
9. Dimitrov, D. Efficient computation of trees with minimal atom–bond connectivity index. *Appl. Math. Comput.* **2013**, *224*, 663–670. [CrossRef]
10. Shao, Z.; Wu, P.; Gao, Y.; Gutman, I.; Zhang, X.J. On the maximum ABC index of graphs without pendent vertices. *Appl. Math. Comput.* **2017**, *315*, 298–312. [CrossRef]
11. Shao, Z.; Wu, P.; Zhang, X.; Dimitrov, D.; Liu, J.B. On the maximum ABC index of graphs with prescribed size and without pendent vertices. *IEEE Access* **2018**, *6*, 27604–27616. [CrossRef]
12. Farahani, M.R. The edge version of atom bond connectivity index of connected graph. *Acta Univ. Apul.* **2012**, *36*, 277–284.
13. Gao, W.; Husin, M.N.; Farahani, M.R.; Imran, M. On the Edges Version of Atom-Bond Connectivity Index of Nanotubes. *J. Comput. Theor. Nanosci.* **2017**, *13*, 6733–6740. [CrossRef]
14. Farahani, M.R. The Edge Version of Geometric-Arithmetic Index of Benzenoid Graph. *Proc. Romanian Acad. Ser. B* **2013**, 15, 95–98.

15. Vukicevic, D.; Furtula, B. Topological index based on the ratios of geometric and arithmetical means of end-vertex degrees of edges. *J. Math. Chem.* **2009**, *46*, 1369–1376. [CrossRef]
16. Martińez–Peŕez, A.; Rodriǵuez, J.M,; Sigarreta, J.M. CMMSE: A new approximation to the geometric–arithmetic index. *J. Math. Chem.* **2017**, *4*, 1–19.
17. Baig, A.Q.; Imran, M.; Khalid, W.; Naeem, M. Molecular description of carbon graphite and crystal cubic carbon structures. *Can. J. Chem.* **2017**, *95*, 674–686. [CrossRef]
18. Gutman, I.; Furtula, B.; Das, K.C. Extended energy and its dependence on molecular structure. *Can. J. Chem.* **2017**, *95*, 526–529. [CrossRef]
19. Mahmiani, A.; Khormali, O.; Iranmanesh, A. On the edge version of geometric-arithmetic index. *Dig. J. Nanomater. Biostruct.* **2012**, *7*, 411–414.
20. Zafar, S.; Nadeem, M.F.; Zahid, Z. On the edge version of geometric-arithmetic index of nanocones. *Stud. Univ. Babes-Bolyai Chem.* **2016**, *61*, 273–282.
21. Gao, W.; Husin, M.N.; Farahani,M.R.; Imran, M. On the Edges Version of Atom-Bond Connectivity and Geometric Arithmetic Indices of Nanocones CNCk. *J. Comput. Theor. Nanosci.* **2016**, *13*, 6741–6746. [CrossRef]

symmetry

MDPI

Article

Sufficient Conditions for Triangular Norms Preserving $\otimes-$Convexity

Lifeng Li [1,2,*] and Qinjun Luo [3]

1 School of Science, Xi'an University of Posts and Telecommunications, Xi'an 710121, China
2 Shaanxi Key Laboratory of Network Data Analysis and Intelligent Processing, Xi'an University of Posts and Telecommunications, Xi'an 710121, China
3 School of Statistics, Xi'an University of Finance and Economics, Xi'an 710100, China; tjxy@mail.xaufe.edu.cn
* Correspondence: lilifeng@xupt.edu.cn; Tel.: +86-029-8816-6087

Received: 5 November 2018; Accepted: 30 November 2018; Published: 7 December 2018

Abstract: The convexity in triangular norm (for short, $\otimes-$convexity) is a generalization of Zadeh's quasiconvexity. The aggregation of two $\otimes-$convex sets is under the aggregation operator $\otimes$ is also $\otimes-$convex, but the aggregation operator $\otimes$ is not unique. To solve it in complexity, in the present paper, we give some sufficient conditions for aggregation operators preserve $\otimes-$convexity. In particular, when aggregation operators are triangular norms, we have that several results such as arbitrary triangular norm preserve $\otimes_D-$convexity and $\otimes_a-$convexity on bounded lattices, $\otimes_M$ preserves $\otimes_H-$convexity in the real unite interval $[0,1]$.

Keywords: aggregation operator; triangular norm; $\otimes-$convex set

1. Introduction

Fuzzy set theory introduced by Zadeh in 1965, as an mathematical tool to deal with uncertainty in information system and knowledge base, has been widely used in various fields of science and technology. By applying fuzzy set theory, Zadeh in [1] proposed the concept of quasiconvex fuzzy set, and has attracted wide attention of researchers and practitioners from many different areas such as fuzzy mathematics, optimization and engineering. Subsequently, Zadeh's quasiconvex fuzzy set was generalized with a lattice L instead of the interval $[0,1]$. A fuzzy set $\mu : \mathbb{R}^n \to L$ is quasiconvex if for any $x, y \in \mathbb{R}^n$ and all $\lambda \in [0,1]$ the inequality

$$\mu(\lambda x + (1-\lambda)y) \geq \mu(x) \wedge \mu(y) \tag{1}$$

holds.

A quasiconvex fuzzy set has an important property: intersection of quasiconvex fuzzy sets is a quasiconvex fuzzy set, i.e., let $X \subseteq \mathbb{R}^n$, for any fuzzy sets μ and ν,

$$\mu \textit{ and } \nu \textit{ are quasiconvex} \Rightarrow \min\{\mu, \nu\} \textit{ is quasiconvex}. \tag{2}$$

The above condition is called intersection preserving quasiconvexity. This property is also true for lattice valued fuzzy sets.

The theory of aggregation operators [2], has been successfully used in mathematics, complex networks and decision making etc (e.g., see [3–6]). The arithmetic mean, the ordered weighted averaging operator and the probabilistic aggregation are widely used examples. In reference [7] Janiš, Král and Renčová pointed that the intersection of fuzzy sets is not the only operator preserving quasiconvexity in general, and they gave someconditions in order that an aggregation operator preserves quasiconvexity.

Triangular norms are kinds of binary aggregation operations that become an essential tool in fuzzy logic, information science and computer sciences. By using triangular norms, properties of fuzzy convexity and various generalizations of fuzzy convexity were considered by many authors (for example, see [8–11]). Suppose $\otimes : [0,1]^2 \to [0,1]$ is a triangular norm, Nourouzi [10] given the concept of $\otimes-$convex set which generalized Zadeh's quasiconvex fuzzy set. A $\otimes-$convex set as defined in [10] can also be generalized as being lattice-valued in the following sense. Let L be a lattice and let $\otimes : L^2 \to L$ be a triangular norm. A fuzzy set $\mu : \mathbb{R}^n \to L$ is called $\otimes-$convex if for any $x, y \in \mathbb{R}^n$ and all $\lambda \in [0,1]$ the inequality

$$\mu(\lambda x + (1-\lambda)y) \geq \mu(x) \otimes \mu(y) \tag{3}$$

holds.

Following [7,10], in the present paper, we continue to study sufficient conditions for aggregation operators and triangular norms that preserve $\otimes-$convexity on a bounded lattice. In Section 3, we give some sufficient conditions for aggregation operator preserving $\otimes-$convexity, those results are generalizations of Propositions 2 and 3 (in [7]). Triangular norm is a kind of important aggregation operator, we give some sufficient conditions for triangular norm preserving $\otimes-$convexity in Section 4. And Section 5 is conclusion.

2. Preliminaries

We first give the basic definitions and results from the existing literature. In following, we use L denote a bounded lattice $(L \leq, 0_L, 1_L)$.

Definition 1. *[2] An aggregation operation is a function $A : L^n \to L$ which satisfies*

(i) $A(a_1, a_2, \ldots, a_n) \leq A(a'_1, a'_2, \ldots, a'_n)$ *whenever* $a_i \leq a'_i$ *for* $1 \leq i \leq n$.
(ii) $A(0_L, 0_L, \ldots, 0_L) = 0_L$ *and* $A(1_L, 1_L, \ldots, 1_L) = 1_L$.

A binary aggregation operation is said to be symmetric if for any $a_1, a_2 \in L$, $A(a_1, a_2) = A(a_2, a_1)$. A special aggregation function is a triangular norm defined as following.

Definition 2. *[12] A map $\otimes : L^2 \to L$ is called a triangular norm if*

(T1) $a \otimes b = b \otimes a$.
(T2) $a_1 \otimes b \leq a_2 \otimes b$ *if* $a_1 \leq a_2$.
(T3) $a \otimes (b \otimes c) = (a \otimes b) \otimes c$.
(T4) $a \otimes 1_L = a$.

Example 1. *The two basic triangular norms $\otimes_M$ and $\otimes_D$ defined as the following are the strongest and the weakest triangular norms on L, respectively.*

$$a \otimes_M b = a \wedge b,$$

$$a \otimes_D b = \begin{cases} a \wedge b, & a, b \in \{1_L\}, \\ 0, & otherwise. \end{cases}$$

Example 2. *Suppose $H = (0, \lambda) \subseteq [0,1)$ and let $* : H^2 \to H$ be an operation on H which satisfies (T1)–(T3) and*

$$a * b \leq \min\{a, b\},$$

$$a \otimes_H b = \begin{cases} a * b, & (a,b) \in H^2; \\ \min\{a,b\} & otherwise. \end{cases}$$

Then $\otimes_H$ is a kind of triangular norms on $[0,1]$ follows from Proposition 3.60 in [13].

3. Sufficient Conditions for an Aggregation Operator Preserving $\otimes-$Convexity

In this Section, we generalize Propositions 2 and 3 (in [7]), and give some sufficient conditions for an aggregation operator which preserves $\otimes-$convexity.

Theorem 1. *Let $A : L^2 \to L$ be an aggregation operator on L, let $\mu, \nu : \mathbb{R}^n \to L$ be arbitrarily $\otimes-$convex fuzzy sets. If $A(a \otimes b, c \otimes d) = A(a,c) \otimes A(b,d)$ for each $a,b,c,d \in L$, then $A(\mu,\nu)$ is $\otimes-$convex.*

Proof. Let $\mu, \nu : \mathbb{R}^n \to L$ be arbitrarily $\otimes-$convex fuzzy sets, and $x, y \in \mathbb{R}^n$. Then we see

$$\begin{aligned} & A(\mu,\nu)(\lambda x + (1-\lambda)y) \\ = \ & A(\mu(\lambda x + (1-\lambda)y), \nu(\lambda x + (1-\lambda)y)) \\ \geq \ & A(\mu(x) \otimes \mu(y), \nu(x) \otimes \nu(y)) \\ = \ & A(\mu(x), \nu(x)) \otimes A(\mu(y), \nu(y)) \\ = \ & A(\mu,\nu)(x) \otimes A(\mu,\nu)(y). \end{aligned}$$

Thus, $A(\mu,\nu)$ is $\otimes-$convex. □

The converse of Theorem 1, however, is in general not true. For example,

Example 3. *Consider a lattice $L = (0_L, a, b, 1_L)$, where $0_L \leq a \leq 1_L$, $0_L \leq b \leq 1_L$, and a, b are incomparable elements and the aggregation operator defined in Table 1. Let $\mu, \nu : \mathbb{R}^n \to L$ be arbitrarily $\otimes_D-$convex fuzzy sets. For any $x, y \in \mathbb{R}^n$ and all $\lambda \in [0,1]$*

$$\begin{aligned} & A(\mu,\nu)(\lambda x + (1-\lambda)y) \\ = \ & A(\mu(\lambda x + (1-\lambda)y), \nu(\lambda x + (1-\lambda)y)) \\ \geq \ & A(\mu(x) \otimes_D \mu(y), \nu(x) \otimes_D \nu(y)) \\ = \ & \begin{cases} A(\mu(y), \nu(y)), & \mu(x) = \nu(x) = 1_L, \\ A(\mu(y), \nu(x)), & \mu(x) = \nu(y) = 1_L, \\ A(\mu(x), \nu(x)), & \mu(y) = \nu(y) = 1_L, \\ A(\mu(x), \nu(y)), & \mu(y) = \nu(x) = 1_L, \\ 0_L, & otherwise, \end{cases} \end{aligned}$$

we have

$$\begin{aligned} & A(\mu,\nu)(x) \otimes_D A(\mu,\nu)(y) \\ = \ & \begin{cases} A(\mu(y), \nu(y)), & A(\mu,\nu)(x) = 1_L, \\ A(\mu(x), \nu(x)), & A(\mu,\nu)(y) = 1_L, \\ 0_L, & otherwise, \end{cases} \\ = \ & \begin{cases} A(\mu(y), \nu(y)), & \mu(x) = \nu(x) = 1_L, \\ A(\mu(x), \nu(x)), & \mu(y) = \nu(y) = 1_L, \\ 0_L, & otherwise. \end{cases} . \end{aligned}$$

Hence, $A(\mu,\nu)$ is $\otimes_D-$convex. And $A(1_L \otimes_D b, a \otimes_D 1_L) = A(b,a) = a$, $A(1_L, a) \otimes_D A(b, 1_L) = a \otimes_D b = 0_L$.

Table 1. Aggregation operator A.

A	0_L	a	b	1_L
0_L	0_L	0_L	0_L	0_L
a	0_L	0	b	b
b	0_L	a	b	b
1_L	0_L	a	b	1_L

Theorem 2. *Let $A : L^2 \to L$ be an aggregation operator on L, let $\mu, \nu : \mathbb{R}^n \to L$ be arbitrary $\otimes-$convex fuzzy sets. If $A(\mu, \nu)$ is $\otimes-$convex, then $A(a \otimes b, c \otimes d) \geq A(a, c) \otimes A(b, d)$ for each $a, b, c, d \in L$. Moreover if the triangular norm $\otimes$ is idempotent, then $A(a \otimes b, c \otimes d) = A(a, c) \otimes A(b, d)$ for each $a, b, c, d \in L$.*

Proof. Suppose that $A(\mu, \nu)$ is $\otimes-$convex. Let a, b, c, d be arbitrary elements of L. For $x, y \in \mathbb{R}^n$ and $z = \lambda x + (1 - \lambda)y$, define

$$\mu(t) = \begin{cases} a, & t = z + \theta(y - z), \theta < 0; \\ a \otimes b, & t = z; \\ b, & t = z + \theta(y - z), \theta > 0; \\ 0_L, & otherwise, \end{cases} \qquad \nu(t) = \begin{cases} c, & t = z + \theta(y - z), \theta < 0; \\ c \otimes d, & t = z; \\ d, & t = z + \theta(y - z), \theta > 0; \\ 0_L, & otherwise. \end{cases}$$

Clearly μ, ν are $\otimes-$convex. And

$$A(\mu, \nu)(t) = \begin{cases} A(a, c), & t = z + \theta(y - z), \theta < 0; \\ A(a \otimes b, c \otimes d), & t = z; \\ A(b, d), & t = z + \theta(y - z), \theta > 0; \\ 0_L, & otherwise. \end{cases}$$

As $A(\mu, \nu)$ has to be a $\otimes-$convex fuzzy set, we have

$$A(a \otimes b, c \otimes d) \geq A(a, c) \otimes A(b, d).$$

From the monotonicity of A it follows that $A(a \otimes b, c \otimes d) \leq A(a, c)$ and $A(a \otimes b, c \otimes d) \leq A(b, d)$. Hence

$$A(a \otimes b, c \otimes d) \otimes A(a \otimes b, c \otimes d) \leq A(a, c) \otimes A(b, d).$$

Therefore, since the operator $\otimes$ is idempotent it follows that

$$A(a \otimes b, c \otimes d) \leq A(a, c) \otimes A(b, d).$$

□

Since the triangular norm $a \otimes_M b = a \wedge b$ is idempotent, Proposition 2 (in [7]) follows from Theorems 1 and 2.

Theorem 3. *Let $A : L^2 \to L$ be an aggregation operator on L, and let $\mu, \nu : \mathbb{R}^n \to L$ be arbitrary $\otimes-$convex fuzzy sets. If $A(a, b) = A(a, a) \otimes A(b, b) = A(a \otimes b, a \otimes b)$ for each $a, b \in L$, then $A(\mu, \nu)$ is $\otimes-$convex.*

Proof. Let $\mu, \nu : R^n \to L$ be arbitrary $\otimes-$convex fuzzy sets. For any $x, y \in \mathbb{R}^n$ and all $\lambda \in [0,1]$

$$\begin{aligned}
& A(\mu,\nu)(\lambda x + (1-\lambda)y) \\
= \; & A(\mu(\lambda x + (1-\lambda)y), \nu(\lambda x + (1-\lambda)y)) \\
= \; & A(\mu(\lambda x + (1-\lambda)y), \mu(\lambda x + (1-\lambda)y)) \otimes A(\nu(\lambda x + (1-\lambda)y), \nu(\lambda x + (1-\lambda)y)) \\
\geq \; & A(\mu(x) \otimes \mu(y), \mu(x) \otimes \mu(y)) \otimes A(\nu(x) \otimes \nu(y), \nu(x) \otimes \nu(y)) \\
= \; & A(\mu(x), \mu(y)) \otimes A(\nu(x), \nu(y)) \\
= \; & (A(\mu(x), \mu(x)) \otimes A(\mu(y), \mu(y))) \otimes (A(\nu(x), \nu(x)) \otimes A(\nu(y), \nu(y))) \\
= \; & (A(\mu(x), \mu(x)) \otimes A(\nu(x), \nu(x))) \otimes (A(\mu(y), \mu(y)) \otimes A(\nu(y), \nu(y))) \\
= \; & A(\mu,\nu)(x) \otimes A(\mu,\nu)(y).
\end{aligned}$$

Thus, $A(\mu,\nu)$ is $\otimes-$convex. □

The following shows that the converse of Theorem 3 is in general not true.

Example 4. *Consider a lattice $L = (0_L, a, b, 1_L)$, where $0_L \leq a \leq 1_L, 0_L \leq b \leq 1_L$, and a, b are incomparable elements and the binary symmetric aggregation operator A defined in Table 2. Let $\mu, \nu : \mathbb{R}^n \to L$ be arbitrary $\otimes_D-$convex fuzzy sets. For any $x, y \in \mathbb{R}^n$ and all $\lambda \in [0,1]$, can prove that $A(\mu,\nu)$ is $\otimes_D-$convex. And $A(b,a) = a$, $A(b,b) \otimes_D A(a,a) = a \otimes_D a = 0_L$, and $A(b \otimes_D b, a \otimes_D a) = A(0_L, 0_L) = 0_L$.*

Table 2. Aggregation operator A.

A	0_L	a	b	1_L
0_L	0_L	0_L	0_L	0_L
a	0_L	a	a	a
b	0_L	a	a	b
1_L	0_L	a	b	1_L

Theorem 4. *Let $A : L^2 \to L$ be an symmetric aggregation operator on L, let $\mu, \nu : \mathbb{R}^n \to L$ be arbitrary $\otimes-$convex fuzzy sets. If $A(\mu,\nu)$ is $\otimes-$convex, then $A(a,b) \geq A(a,a) \otimes A(b,b)$ for each $a, b \in L$. Moreover if the triangular norm $\otimes$ is idempotent, then $A(a,b) = A(a,a) \otimes A(b,b) = A(a \otimes b, a \otimes b)$ for each $a, b \in L$.*

Proof. Suppose that $A(\mu,\nu)$ is $\otimes-$convex. Let a, b be arbitrary elements of L, and put, for $x, y \in \mathbb{R}^n$ and $0 < \lambda < 1$, $z = \lambda x + (1-\lambda)y$. We define

$$\mu(t) = \begin{cases} a, & t = z + \theta(y-z), \theta \leq 0; \\ b, & t = z + \theta(y-z), \theta > 0; \\ 0_L, & otherwise, \end{cases} \qquad \nu(t) = \begin{cases} a, & t = z + \theta(y-z), \theta < 0; \\ b, & t = z + \theta(y-z), \theta \geq 0; \\ 0_L, & otherwise. \end{cases}$$

Clearly μ, ν are $\otimes-$convex and as A preserves $\otimes-$convexity, then we have

$$A(a,b) \geq A(a,a) \otimes A(b,b).$$

Suppose that the triangular norm $\otimes$ is idempotent. Let $x, y \in \mathbb{R}^n$ and $z = \lambda x + (1-\lambda)y$, define

$$\mu(t) = \begin{cases} a, & t = z + \theta(y-z), \theta \leq 0; \\ 1_L, & t = z + \theta(y-z), \theta > 0; \\ 0_L, & otherwise, \end{cases} \qquad \nu(t) = \begin{cases} a, & t = z + \theta(y-z), \theta < 0; \\ 1_L, & t = z + \theta(y-z), \theta \geq 0; \\ 0_L, & otherwise. \end{cases}$$

Clearly μ, ν are $\otimes-$convex. Since, in addition, A preserves $\otimes-$convexity this can be combined with the fact that the triangular norm $\otimes$ is idempotent, we deduce

$$A(a,a) \geq A(a, 1_L) \otimes A(1_L, a) = A(1_L, a) \otimes A(1_L, a) = A(1_L, a).$$

From the monotony of A it follows that $A(a,a) \leq A(1_L,a)$. Hence

$$A(a,a) = A(1_L,a).$$

Therefore

$$A(a,b) \leq A(1_L,b) = A(b,b), A(a,b) \leq A(1_L,a) = A(a,a).$$

Hence

$$A(a,b) = A(a,b) \otimes A(a,b) \leq A(a,a) \otimes A(b,b).$$

Thus

$$A(a,b) = A(a,a) \otimes A(b,b).$$

Let $c = a, d = b$, from Theorem 2 we have

$$A(a,b) = A(a \otimes b, a \otimes b).$$

□

Then Proposition 3 (in [7]) follows from Theorems 3 and 4 due to $a \otimes_M b = a \wedge b$ is idempotent.

Since the triangular norm $a \otimes_M b = \min\{a,b\}$ is the strongest triangular norm on $[0,1]$, from the definition of $\otimes-$convexity we can prove the following theorem.

Theorem 5. *If $f_1, f_2 : [0,1] \to [0,1]$ are both nondecreasing, $\min\{f_1(0), f_2(0)\} = 0, f_1(1) = f_2(1) = 1$. Let $A : [0,1]^2 \to [0,1]$ defined by $A(a,b) = \min\{f_1(a), f_2(b)\}$, then $A(\mu,\nu)$ preserves $\otimes-$convexity for any triangular norm on $[0,1]$. But the converse statement is in general not true.*

Example 5. *Suppose $L = [0,1]$, $A(a,b) = \frac{1}{2}(a+b)$. Then $A(\mu,\nu)(\lambda x + (1-\lambda)y) \geq A(\mu,\nu)(x) \otimes_D A(\mu,\nu)(y)$. i.e., $A(\mu,\nu)$ is $\otimes_D-$convex. And $A(a,b) = \frac{1}{2}(a+b) \neq \min\{f_1(a), f_2(b)\}$.*

4. Sufficient Conditions for Triangular Norm Preserving $\otimes-$Convexity

In this section we give some sufficient conditions which guarantee that a triangular norm preserves $\otimes-$convexity. The following theorem is obvious.

Theorem 6. *Let $\otimes : L^2 \to L$ be a triangular norm on L. If $\mu, \nu : \mathbb{R}^n \to L$ are arbitrary $\otimes-$convex fuzzy sets, then $\mu \otimes \nu$ is $\otimes-$convex.*

Theorem 7. *Let $\otimes : L^2 \to L$ be a triangular norm on L. If $\mu, \nu : \mathbb{R}^n \to L$ are arbitrary $\otimes_D-$convex fuzzy sets, then $\mu \otimes \nu$ is $\otimes_D-$convex.*

Proof. Let $\mu, \nu : \mathbb{R}^n \to L$ be arbitrary $\otimes_D-$convex fuzzy sets. For any $x, y \in \mathbb{R}^n$ and all $\lambda \in [0,1]$

$$\begin{aligned}
& (\mu \otimes \nu)(\lambda x + (1-\lambda)y) \\
= \ & \mu(\lambda x + (1-\lambda)y) \otimes \nu(\lambda x + (1-\lambda)y) \\
\geq \ & (\mu(x) \otimes_D \mu(y)) \otimes (\nu(x) \otimes_D \nu(y)) \\
= \ & \begin{cases} \mu(x) \otimes \nu(x), & \mu(y) = \nu(y) = 1_L, \\ \mu(y) \otimes \nu(y), & \mu(x) = \nu(x) = 1_L, \\ \mu(x) \otimes \nu(y), & \mu(y) = \nu(x) = 1_L, \\ \mu(y) \otimes \nu(x), & \mu(x) = \nu(y) = 1_L, \\ 0_L, & otherwise. \end{cases}
\end{aligned}$$

Then we see

$$\begin{aligned}
& (\mu \otimes \nu)(x) \otimes_D (\mu \otimes \nu)(y) \\
= \ & (\mu(x) \otimes \nu(x)) \otimes_D (\mu(y) \otimes \nu(y)) \\
= \ & \begin{cases} \mu(x) \otimes \nu(x), & \mu(y) \otimes \nu(y) = 1_L, \\ \mu(y) \otimes \nu(y), & \mu(x) \otimes \nu(x) = 1_L, \\ 0_L, & otherwise, \end{cases} \\
= \ & \begin{cases} \mu(x) \otimes \nu(x), & \mu(y) = \nu(y) = 1_L, \\ \mu(y) \otimes \nu(y), & \mu(x) = \nu(x) = 1_L, \\ 0_L, & otherwise. \end{cases}
\end{aligned}$$

Hence

$$(\mu \otimes \nu)(\lambda x + (1-\lambda)y) \geq (\mu \otimes \nu)(x) \otimes_D (\mu \otimes \nu)(y).$$

Thus, $\mu \otimes \nu$ is $\otimes_D-$convex.
□

Let $\otimes$ be a triangular norm on L. Li in [14] given a family triangular norms $(\otimes_a)_{a \in L}$ as follows

$$x \otimes_a y = \begin{cases} 0_L, & x \otimes y \leq a \ and \ x, y \neq 1_L; \\ x \otimes y, & otherwise. \end{cases}$$

Theorem 8. *Let $\otimes : L^2 \to L$ be a triangular norm on L, and $a \in L$. If $\mu, \nu : \mathbb{R}^n \to L$ are arbitrary $\otimes_a-$convex fuzzy sets, then $\mu \otimes \nu$ is $\otimes_a-$convex.*

Proof. Let $\mu, \nu : \mathbb{R}^n \to L$ be arbitrary $\otimes_a-$convex fuzzy sets. For any $x, y \in \mathbb{R}^n$ and all $\lambda \in [0,1]$

$$\begin{aligned}
& (\mu \otimes \nu)(\lambda x + (1-\lambda)y) \\
= \ & \mu(\lambda x + (1-\lambda)y) \otimes \nu(\lambda x + (1-\lambda)y) \\
\geq \ & (\mu(x) \otimes_a \mu(y)) \otimes (\nu(x) \otimes_a \nu(y)) \\
= \ & \begin{cases} 0_L, & \mu(x) \otimes \mu(y) \leq a \ or \ \nu(x) \otimes \nu(y) \leq a, \\ \mu(x) \otimes \mu(y) \otimes \nu(x) \otimes \nu(y), & otherwise. \end{cases}
\end{aligned}$$

Then we have

$$\begin{aligned} & (\mu \otimes \nu)(x) \otimes_a (\mu \otimes \nu)(y) \\ = & (\mu(x) \otimes \nu(x)) \otimes_a (\mu(y) \otimes \nu(y)) \\ = & \begin{cases} 0_L, & \mu(x) \otimes \nu(x) \otimes \mu(y) \otimes \nu(y) \leq a, \\ \mu(x) \otimes \mu(y) \otimes \nu(x) \otimes \nu(y), & otherwise. \end{cases} \end{aligned}$$

Since $\mu(x) \otimes \mu(y) \leq a$ or $\nu(x) \otimes \nu(y) \leq a$ implies $\mu(x) \otimes \nu(x) \otimes \mu(y) \otimes \nu(y) \leq a$, we have

$$(\mu \otimes \nu)(\lambda x + (1-\lambda)y) \geq (\mu \otimes \nu)(x) \otimes_a (\mu \otimes \nu)(y).$$

Thus, $\mu \otimes \nu$ is $\otimes_a-$convex.
□

Example 6. *Consider the lattice $(L = \{0_L, a, b, c, d, 1_L\}, \leq, 0, 1)$ given in Figure 1. Consider the function $\otimes_b$ on L defined by*

$$\alpha \otimes_b \beta = \begin{cases} 0_L, & \alpha \wedge \beta \leq b \text{ and } \alpha, \beta \neq 1_L; \\ \alpha \wedge \beta, & otherwise, \end{cases}$$

then $\otimes_b$ is a triangular norm and $\otimes_b$ is described in Table 3.
Hence, for any $\otimes_b$-convex sets $\mu, \nu : \mathbb{R}^n \to L$, $\mu \otimes_M \nu = \mu \wedge \nu$ is also a $\otimes_b$-convex set.

Table 3. Triangular norm $\otimes_b$.

T_b	0_L	a	b	c	d	1_L
0_L	0_L	0_L	0_L	0_L	0_L	0_L
a	0_L	0_L	0_L	0_L	0_L	a
b	0_L	0_L	0_L	0_L	0_L	b
c	0_L	0_L	0_L	c	0_L	c
d	0_L	0_L	0_L	0_L	d	d
1_L	0_L	a	b	c	d	1_L

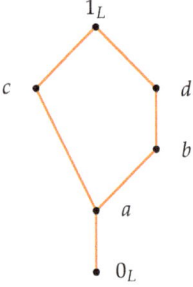

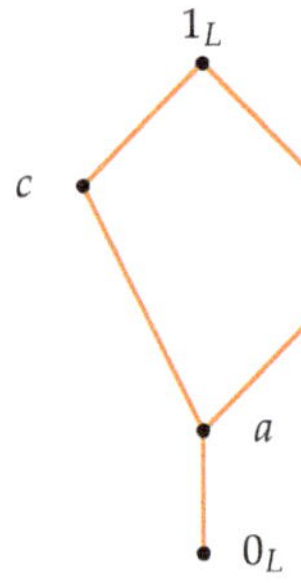

Figure 1. The order $\leq$ on L.

Theorem 9. *Let $\mu, \nu : \mathbb{R}^n \to [0,1]$ be arbitrary $\otimes_H-$convex fuzzy sets. Then $\min\{\mu, \nu\}$ is a $\otimes_H-$convex fuzzy set.*

Proof. Let $\mu, \nu : \mathbb{R}^n \to L$ be arbitrary $\otimes_H-$convex fuzzy sets. For any $x, y \in \mathbb{R}^n$ and all $\lambda \in [0,1]$

$$\begin{aligned} & \min\{\mu,\nu\}(\lambda x + (1-\lambda)y) \\ = & \min\{\mu(\lambda x + (1-\lambda)y), \nu(\lambda x + (1-\lambda)y)\} \\ \geq & \min\{\mu(x) \otimes_H \mu(y), \nu(x) \otimes_H \nu(y)\} \\ = & \begin{cases} \min\{\mu(x) * \mu(y), \nu(x) * \nu(y)\}, & (\mu(x),\mu(y)) \in H^2 \text{ and } (\nu(x),\nu(y)) \in H^2, \\ \min\{\mu(x) * \mu(y), \nu(x), \nu(y)\}, & (\mu(x),\mu(y)) \in H^2 \text{ and } (\nu(x),\nu(y)) \notin H^2, \\ \min\{\mu(x), \mu(y), \nu(x) * \nu(y)\}, & (\mu(x),\mu(y)) \notin H^2 \text{ and } (\nu(x),\nu(y)) \in H^2, \\ \min\{\mu(x), \mu(y), \nu(x), \nu(y)\}, & \text{otherwise.} \end{cases} \end{aligned}$$

Then we deduce

$$\begin{aligned} & \min\{\mu,\nu\}(x) \otimes_H \min\{\mu,\nu\}(y) \\ = & \min\{\mu(x),\nu(x)\} \otimes_H \min\{\mu(y),\nu(y)\} \\ = & \begin{cases} \min\{\mu(x),\nu(x)\} * \min\{\mu(y),\nu(y)\}, & (\min\{\mu(x),\nu(x)\}, \min\{\mu(y),\nu(y)\}) \in H^2, \\ \min\{\mu(x),\mu(y),\nu(x),\nu(y)\}, & \text{otherwise.} \end{cases} \end{aligned}$$

Since $\min\{\mu(x),\mu(y)\} \geq \mu(x) * \mu(y) \geq \min\{\mu(x),\nu(x)\} * \min\{\mu(y),\nu(y)\}, \min\{\nu(x),\nu(y)\} \geq \nu(x) * \nu(y) \geq \min\{\mu(x),\nu(x)\} * \min\{\mu(y)$ we have

$$\min\{\mu,\nu\}(\lambda x + (1-\lambda)y) \geq \min\{\mu,\nu\}(x) \otimes_H \min\{\mu,\nu\}(y).$$

Thus, $\min\{\mu,\nu\}$ is a $\otimes_H-$convex fuzzy set.
□

Example 7. *Suppose $H = (0, \frac{1}{2})$ and the triangular norm $\otimes_H$ is*

$$a \otimes_H b = \begin{cases} \frac{ab}{2}, & (a,b) \in (0,\frac{1}{2})^2; \\ \min\{a,b\} & \text{otherwise,} \end{cases}$$

then, $\min\{\mu,\nu\}$ *is a* $\otimes_H-$*convex fuzzy set.*

5. Conclusions

The authors of the paper [7] discuss properties which are preserved under aggregation for arbitrary lattices and arbitrary pairs of mappings Results in this paper are also discussed under aggregation for an arbitrary lattice and an arbitrary pair of mappings. However, this does not mean that even without these conditions the aggregation of SOME quasiconvex ($\otimes-$convex) mappings to SOME lattices need not be quasiconvex ($\otimes-$convex). Which are the properties of a lattice L and an aggregation A (weaker than those from the paper by Janis, Kral and Rencova in [7]), such that A preserves quasiconvexity ($\otimes-$convex) for mappings into L? We hope to solve this problem in future work.

Author Contributions: Conceptualization, L.L.; methodology, L.L. and Q.L.

Acknowledgments: This work was supported by the National Natural Science Foundation of China (Grant Nos. 11701446, 61702389) and Shaanxi key disciplines of special funds projects.

Conflicts of Interest: The authors declare no conflict of interest.

References

1. Zadeh, L.A. Fuzzy sets. *Inf. Control* **1965**, *8*, 338–353. [CrossRef]
2. Lizasoain, I.; Moreno, C. OWA operators defined on complete lattices. *Fuzzy Sets Syst.* **2013**, *224*, 36–52. [CrossRef]

3. Aggarwal, M. Discriminative aggregation operators for multi criteria decision making. *Appl. Soft Comput.* **2017**, *52*, 1058–1069. [CrossRef]
4. Jiang, W.; Wei, B.; Zhan, J.; Xie, C.; Zhou, D. A Visibility Graph Power Averaging Aggregation Operator: A Methodology Based on Network Analysis. *Comput. Ind. Eng.* **2016**, *101*, 260–268. [CrossRef]
5. Liu, P.; Chen, S.M. Group Decision Making Based on Heronian Aggregation Operators of Intuitionistic Fuzzy Numbers. *IEEE Trans. Cybern.* **2017**, *99*, 2514–2530. [CrossRef] [PubMed]
6. Scellato, S.; Fortuna, L.; Frasca, M.; Gómez-Gardenes, J.; Latora, V. Traffic optimization in transport networks based on local routing. *Eur. Phys. J. B* **2010**, *73*, 303–308. [CrossRef]
7. Janiš, V.; Král, P.; Renčová, M. Aggregation operators preserving quasiconvexity. *Inf. Sci.* **2013**, *228*, 37–44. [CrossRef]
8. Hua, X.J.; Xin, X.L.; Zhu, X. Generalized (convex) fuzzy sublattices. *Comput. Math. Appl.* **2011**, *62*, 699–708. [CrossRef]
9. Pan, X.D.; Meng, D. Triangular norm based graded convex fuzzy sets. *Fuzzy Sets Syst.* **2012**, *209*, 1–13. [CrossRef]
10. Nourouzi, K.; Aghajani, A. Convexity in triangular norm of fuzzy sets. *Chaos Solitons Fractals* **2008**, *36*, 883–889. [CrossRef]
11. Tahayori, H.; Tettamanzi, G.B.; Antoni, G.D.; Visconti, A. On the calculation of extended max and min operations between convex fuzzy sets of the real line. *Fuzzy Sets Syst.* **2009**, *160*, 3103–3114. [CrossRef]
12. De Baets, B.; Mesiar, R. Triangular norms on product lattices. *Fuzzy Sets Syst.* **1999**, *104*, 61–75. [CrossRef]
13. Klement, E.P.; Mesiar, R.; Pap, E. *Triangular Norms*; Kluwer Academic Publishers: Dordrecht, The Netherlands, 2000.
14. Li, L.; Zhang, J.; Zhou, C. Sufficient conditions for a T-partial order obtained from triangular norms to be a lattice. *Kybernetika* **2018**, submitted.

symmetry

MDPI

Article

A Novel Edge Detection Method Based on the Regularized Laplacian Operation

Huilin Xu * and Yuhui Xiao

College of Mathematics and Computer Science, Gannan Normal University, Ganzhou 341000, China; 15707978628@163.com

* Correspondence: xuhuilin@163.com; Tel.: +86-797-839-3663

Received: 12 November 2018; Accepted: 22 November 2018; Published: 3 December 2018

Abstract: In this paper, an edge detection method based on the regularized Laplacian operation is given. The Laplacian operation has been used extensively as a second-order edge detector due to its variable separability and rotation symmetry. Since the image data might contain some noises inevitably, regularization methods should be introduced to overcome the instability of Laplacian operation. By rewriting the Laplacian operation as an integral equation of the first kind, a regularization based on partial differential equation (PDE) can be used to compute the Laplacian operation approximately. We first propose a novel edge detection algorithm based on the regularized Laplacian operation. Considering the importance of the regularization parameter, an unsupervised choice strategy of the regularization parameter is introduced subsequently. Finally, the validity of the proposed edge detection algorithm is shown by some comparison experiments.

Keywords: edge detection; Laplacian operation; regularization; parameter selection; performance evaluation

1. Introduction

In a digital image, edges can be defined as abrupt changes of the image intensity. Edge is one of the most essential features contained in an image. The result of edge detection not only retains the main information of an image, but also reduces the amount of data to be processed drastically. Therefore, edge detection has been used as a front-end step in many image processing and computer vision applications [1].

Since the abrupt changes in an image can be reflected by their derivatives, differentiation-based methods are widely used in edge detection. Generally, edges can be detected by finding the maximum of first-order derivatives or the zero-crossing of second-order derivatives of the image intensity. From the original contribution of Roberts in 1965, there have been a large number of works concerning this topic. Some researchers have paid attention to constructing optimal filters according to some reasonable hypotheses and criteria (see [2–5]), while some others are interested in designing discrete masks, such as the well-known Prewitt, Sobel and Laplacian of Gaussian (LoG) operators. Some recently developed methods can be found in [6–8].

The differentiation-based edge detection methods need to calculate derivatives numerically. As we know, numerical differentiations are unstable since a small perturbation of the data may cause huge errors in its derivatives [9]. In real applications, the image is often corrupted by noise during the processes of collection, acquisition and transmission. In order to calculate derivatives of the noisy data stably, some regularization methods should be introduced. There have been much work into this over the past years, such as the Tikhonov regularization [10], the Lavrentiev regularization [11], the Lanczos method [12], the mollification method [9] and the total variation method [13]. Some of the regularization methods for computing the first-order numerical differentiation have been applied to detecting image edges (see [10,13]).

Compared with the first-order numerical differentiation, the computation of second-order derivatives is more unstable and more likely to be influenced by noises. However, the edge detection based on second-order derivatives has higher localization accuracy and a stronger response to final details [14]. The most common second-order derivative used in edge detection is the Laplacian operation due to its variable separability and rotation invariance. In order to overcome the instability of Laplacian operation, one of the existing works is the LoG [2]. Since the image data is discrete, the sampled representation of the LoG and some related issues have been discussed in [15]. The performance of a LoG detector depends mainly on the choice of the scale parameter. For larger scales, the zero-crossings deviate from the true edges, which may cause poor localization. For small scales, there would be many false zero-crossings produced by noises. Besides the LoG detector, a model for designing a discrete mask of the Laplacian operator is introduced in [7].

In view of the above-mentioned facts, a natural idea is to compute the Laplacian operation by the regularization method and construct a novel edge detection algorithm based on this. By rewriting the Laplacian operation as an integral equation of the first kind, a PDE-based regularization for computing the Laplacian operation has been proposed in [16]. In this paper, the PDE-based regularization method will be generalized to edge detection. Based on the objective parameter selection for edge detection given in [17], we will introduce a new choice strategy of the regularization parameter. Comparative experiments with the LoG detector and the Laplacian-based mask given in [7] are considered.

The paper is organized as follows. In Section 2, the PDE-based regularization method for computing the Laplacian operation of image data is given. The novel edge detection algorithm based on the regularized Laplacian operation is given in Section 3. Comparative experiments are shown in Section 4. Finally, the main conclusions are summarized in Section 5.

2. Regularized Laplacian Operation

Considering the image intensity as a function $f(r)$, $r = (x, y)$ of two variables, the Laplacian operation can be defined as

$$u = \Delta f = \frac{\partial^2 f}{\partial x^2} + \frac{\partial^2 f}{\partial y^2}, \ (x, y) \in \Omega := [0, a] \times [0, b].$$

Without loss of generality, we assume the value of $f(x, y)$ on the boundary of Ω is zero, i.e., $f|_{\partial\Omega} \equiv 0$. Otherwise, denote f_0 as the solution of

$$\begin{cases} \Delta f_0 = 0, & \text{in } \Omega \\ f_0 = f, & \text{on } \partial\Omega \end{cases},$$

and replace f by $f - f_0$. Since the latter satisfies

$$\Delta(f - f_0) = \Delta f = u, \ (f - f_0)|_{\partial\Omega} \equiv 0,$$

it has

$$\begin{cases} \Delta f = u, & \text{in } \Omega \\ f = 0, & \text{on } \partial\Omega \end{cases}. \tag{1}$$

Problem (1) is the Dirichlet problem of the Poisson equation. According to the classic theory of the Poisson equation, the relationship between f and u can be expressed as

$$A[u] := \int_\Omega G(r, r')\, u(r') \mathrm{d}r' = -f, \tag{2}$$

where $G(r, r')$ is the Green function of the Dirichlet problem (see [18]). Since Ω is a rectangular domain, the Green function has the explicit expression

$$G(r, r') = \sum_{k_1, k_2 = 1}^{\infty} p(k_1, k_2)\, u(r; k_1, k_2)\, u(r'; k_1, k_2),$$

where

$$u(r; k_1, k_2) = \sin\frac{k_1 \pi x}{a} \sin\frac{k_2 \pi y}{b},\ p(k_1, k_2) = \frac{4ab}{\pi^2 (k_1^2 b^2 + k_2^2 a^2)}.$$

The calculation of Laplacian operation $u = \Delta f$ is equivalent to solving the integral Equation (2), which can be simplified in the following.

Denote f^δ as the noise data of f; the calculation of the Laplacian operation Δf^δ is unstable, which means the noise may be amplified. A stabilized strategy is to solve the equivalent Equation (2) by the regularization method. Solving the integral Equation (2) by the Lavrentiev regularization method, an efficient method is given in [16]. The Laplacian operation can be computed approximately by solving the regularization equation

$$\alpha u^{\alpha,\delta} + A[u^{\alpha,\delta}] = -f^\delta, \tag{3}$$

where $\alpha > 0$ is the regularization parameter, and $u^{\alpha,\delta}$ is the regularized Laplacian operation. Assuming that $h^{\alpha,\delta}$ is a function satisfying

$$\begin{cases} \Delta h^{\alpha,\delta} = u^{\alpha,\delta}, & \text{in } \Omega \\ h^{\alpha,\delta} = 0, & \text{on } \partial\Omega \end{cases},$$

then it has $A[u^{\alpha,\delta}] = -h^{\alpha,\delta}$. Equation (3) can be rewritten as

$$\begin{cases} \alpha \Delta h^{\alpha,\delta} - h^{\alpha,\delta} = -f^\delta, & \text{in } \Omega \\ h^{\alpha,\delta} = 0, & \text{on } \partial\Omega \end{cases} \tag{4}$$

This boundary value problem of PDE can be solved by classic numerical methods, and then the regularized Laplacian operation $u^{\alpha,\delta}$ can be expressed as

$$u^{\alpha,\delta}(r) = \Delta h^{\alpha,\delta}(r) = \frac{1}{\alpha}[h^{\alpha,\delta}(r) - f^\delta(r)],\ r \in \Omega. \tag{5}$$

From the above rewriting, we can see that (4) and (5) are equivalent to the integral Equation (3). Compared with solving the regularization Equation (3) directly, the computational burden of solving (4) and (5) is reduced drastically.

The work of [16] mainly focuses on the choice of the regularization parameter α and the error estimate of the regularized Laplacian operation $u^{\alpha,\delta}$. Unfortunately, the choice strategy given in [16] depends on the noise level of the noise data, which is unknown in practice. Since the choice strategy of the regularization parameter plays an important role in the regularization method, as the authors stated in [16], the selection of parameter α in the edge detection algorithm should be considered carefully.

3. The Edge Detection Algorithm

In this section, we will construct the novel edge detection algorithm based on the regularized Laplacian operation given in Section 2.

The first thing we are concerned with is the weakness of the Lavrentiev regularization. Notice that $h^{\alpha,\delta}(r) = 0,\ r \in \partial\Omega$, it has $u^{\alpha,\delta}(r) = -\frac{1}{\alpha} f^\delta(r),\ r \in \partial\Omega$. The parameter $\alpha > 0$ is usually a small number, which means the error of the regularized Laplacian operation on the boundary can be amplified $\frac{1}{\alpha}$ times. Thus, the computation is meaningless on $\partial\Omega$. In fact, the validity of the regularized Laplacian operation $u^{\alpha,\delta}(r)$ has been weakened when r is close to the boundary. Experiments in [16] have shown that the weakness only affects the points very close to the boundary. Hence, except a few pixels which are as close as possible to the boundary of the image domain, the edge detection results will be acceptable.

The second thing we are concerned with is the choice strategy of the regularization parameter α. Since the noise level of an image data is unknown, the choice strategy given in [16] cannot be carried out. Considering only the edge detection problem, the objective parameter selection given in [17] can be adopted to choose the regularization parameter.

Once the regularization parameter α is chosen, the regularized Laplacian operation $u^{\alpha,\delta}$ can be obtained by solving Equations (4) and (5), where Equation (4) can be solved by the standard finite difference method or finite element method.

Combined with the objective parameter selection given in [17], the main framework of the choice strategy is summarized as follows:

Step 1: Regularization parameters α_j, $j \in \{1, 2, \ldots, n\}$ are used to generate N different edge maps D_j, $j \in \{1, 2, \ldots, n\}$ by the proposed edge detection algorithm. Then, N potential ground truths (PGTs) are constructed, and each PGT_i includes pixels which have been identified as edges by at least i different edge maps.

Step 2: Each PGT_i is compared with each edge map D_j, and it generates four different probabilities:$\mathrm{TP}_{\mathrm{PGT}_i,\,\mathrm{D}_j}$, $\mathrm{FP}_{\mathrm{PGT}_i,\,\mathrm{D}_j}$, $\mathrm{TN}_{\mathrm{PGT}_i,\,\mathrm{D}_j}$, $\mathrm{FN}_{\mathrm{PGT}_i,\,\mathrm{D}_j}$. Among them, $\mathrm{TP}_{A,B}$ (true positive) means the probability of pixels which have been determined as edges in both edge maps A and B; $\mathrm{FP}_{A,B}$ (false positive) means the probability of pixels determined as edges in A, but non-edges in B; $\mathrm{TN}_{A,B}$ (true negative) means the probability of pixels determined as non-edges in both A and B; and $\mathrm{FN}_{A,B}$ (false negative) means the probability of pixels determined as edges in B, but non-edges in A.

Step 3: For each PGT_i, we average the four probabilities over all edge maps D_j, and get $\overline{\mathrm{TP}}_{\mathrm{PGT}_i}$, $\overline{\mathrm{FP}}_{\mathrm{PGT}_i}$, $\overline{\mathrm{TN}}_{\mathrm{PGT}_i}$, $\overline{\mathrm{FN}}_{\mathrm{PGT}_i}$, where $\overline{\mathrm{TP}}_{\mathrm{PGT}_i} = \frac{1}{N}\sum_{j=1}^{N} \mathrm{TP}_{\mathrm{PGT}_i,\mathrm{D}_j}$, and the expressions of other probabilities are similar. Then, a statistical measurement of each PGT_i is given by the Chi-square test:

$$\chi^2_{\mathrm{PGT}_i} = \frac{\mathrm{TPR} - \mathrm{Q}}{1 - \mathrm{Q}} \cdot \frac{(1 - \mathrm{FPR}) - (1 - \mathrm{Q})}{\mathrm{Q}}, \tag{6}$$

where

$$\mathrm{Q} = \overline{\mathrm{TP}}_{\mathrm{PGT}_i} + \overline{\mathrm{FP}}_{\mathrm{PGT}_i},\ \mathrm{TPR} = \frac{\overline{\mathrm{TP}}_{\mathrm{PGT}_i}}{\overline{\mathrm{TP}}_{\mathrm{PGT}_i} + \overline{\mathrm{FN}}_{\mathrm{PGT}_i}},\ \mathrm{FPR} = \frac{\overline{\mathrm{FP}}_{\mathrm{PGT}_i}}{\overline{\mathrm{FP}}_{\mathrm{PGT}_i} + \overline{\mathrm{TN}}_{\mathrm{PGT}_i}}.$$

The PGT_i with the highest $\chi^2_{\mathrm{PGT}_i}$ is considered as the estimated ground truth (EGT).

Step 4: Each edge map's D_j is then matched to the EGT by four new probabilities: $\mathrm{TP}_{\mathrm{D}_j,\mathrm{EGT}}$, $\mathrm{FP}_{\mathrm{D}_j,\mathrm{EGT}}$, $\mathrm{TN}_{\mathrm{D}_j,\mathrm{EGT}}$, $\mathrm{FN}_{\mathrm{D}_j,\mathrm{EGT}}$. The Chi-square measurements $\chi^2_{\mathrm{D}_j}$ are obtained by the same way as in Step 3. Then, the best edge map is the one which gives the highest $\chi^2_{\mathrm{D}_j}$, and the corresponding regularization parameter α_j is the one we want.

The Chi-square measure (6) can reflect the similarity of two edge maps, and the bigger the value of the Chi-square measurement, the better. As Lopez-Molina et al. stated in [19], the Chi-square measurement can evaluate the errors caused by spurious responses (false positives, FPs) and missing edges (false negatives, FNs), but it cannot work on the localization error when the detected edges deviate from their true position. For example, a reference edge image and three polluted edge maps are given in Figure 1. Compared with the reference edge (Figure 1a), the Chi-square measurements of the three polluted edge maps are the same, yet their localization accuracies are different. In order to reflect the localization error in these edge maps, distance-based error measures should be introduced.

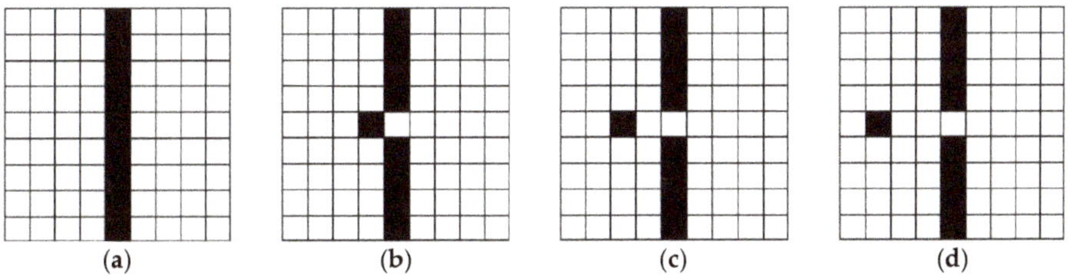

Figure 1. The reference edge image and three polluted edge maps: (**a**) reference edge E_R; (**b**) polluted edge map E_1; (**c**) polluted edge map E_2; (**d**) polluted edge map E_3.

The Baddeley's delta metric (BDM) is one of the most common distance-based measures [20]. It has been proven to be an ideal measure for the comparison of edge detection algorithms [19,21]. Let A and B be two edge maps with the same resolution $M \times N$, and $P = \{1, \ldots, M\} \times \{1, \ldots, N\}$ be the set of pixels in the image. The value of BDM between A and B is defined as

$$\Delta^k(A,B) = \left[\frac{1}{MN} \sum_{p \in P} |w(d(p,A)) - w(d(p,B))|^k \right]^{1/k}, \tag{7}$$

where $d(p,A)$ is the Euclidean distance from $p \in P$ to the closest edge points in A, the parameter k is a given positive integer and $w(d(p,A)) = \min(d(p,A), c)$ for a given constant $c > 0$. Compared with the reference edge E_R in Figure 1, the BDMs of the three polluted edge maps E_i $(i = 1,2,3)$ are given in Table 1 with different parameters c and k. The smaller the value of BDM, the better. As we can see from Table 1, localization errors of the three edge maps are apparently distinguished. Therefore, the Chi-square measure (6) will be replaced by the BDM (7) in the choice strategies of the regularization parameter.

Table 1. The Baddeley's delta metrics (BDMs) between the reference edge image E_R and the polluted edge maps E_i $(i = 1,2,3)$ with the different choices of parameters c and k.

Parameter Sets	$\Delta^k(E_R,E_1)$	$\Delta^k(E_R,E_2)$	$\Delta^k(E_R,E_3)$
$k = 1, c = 2$	0.0566	0.0937	0.1256
$k = 1, c = 3$	0.0950	0.1879	0.2461
$k = 1, c = 4$	0.1397	0.2614	0.3305
$k = 2, c = 2$	0.2182	0.3307	0.3637
$k = 2, c = 3$	0.2753	0.4925	0.6313
$k = 2, c = 4$	0.3317	0.6159	0.8021

4. Experiments and Results

In order to show the validity of the proposed edge detection algorithm, some comparative experiments are given in this section. In the experiments, our regularized edge detector (RED) will be compared with the LoG detector and the Laplacian-based edge detector (LED) proposed in [7].

As Yitzhaky and Peli said in [17], the parameter selection for edge detection depends mainly on the set of parameters used to generate the initial detection results. In order to reduce this influence properly, the range of the parameter is set to be large enough that instead of forming a very sparse edge map it forms a very dense one. The scale parameter of the LoG detector is set from 1.5 to 4 in steps of 0.25. The regularization parameter of the regularized edge detector is set from 0.01 ($\approx$0) to 0.1 in steps of 0.01. The images we used are taken from [22], and some of them are shown in Figure 2. The optimal edge maps given in [22] will be seen as the ground truth in our quantitative comparisons.

Let us first consider the choice strategy of the regularization parameter α, where the parameters in BDM are set as $k = 1$, $c = 2$. Taking the airplane image as an example, the BDM of each PGT_i, $i \in \{1, 2, \ldots 11\}$ is shown in Figure 3a, from which we can see the EGT is PGT_6. Compared with the EGT, the BDM of each edge map D_j is shown in Figure 3b, from which we can see the best edge map is D_6. Hence, the regularization parameter is chosen as $\alpha = 0.05$. The choice of the scale parameter in the LoG detector is carried out similarly. It does not need any parameters in the LED.

For the airplane image, the ground truth and edges detected by the three edge detectors are shown in Figure 4. From Figure 4b, we can see that the influence of the Lavrentiev regularization's weakness on the RED is negligible. From Figure 4b,c, we can see that the RED is better than the LoG detector for noise suppression and maintaining continuous edges. Comparing Figure 4d with Figure 4b,c, we can see the superiority of the parameter-dependent edge detector. Similar results for the elephant image are shown in Figure 5. For some images taken from [22], quantitative comparisons of the edges detected by the LoG detector, the RED and the LED against the ground truth are given in Table 2. Since the smaller the value of BDM, the better, this shows that the RED has better performance than the LoG detector and the LED in most cases.

(**a**) (**b**)

Figure 2. Some images taken from [22]: (**a**) airplane; (**b**) elephant.

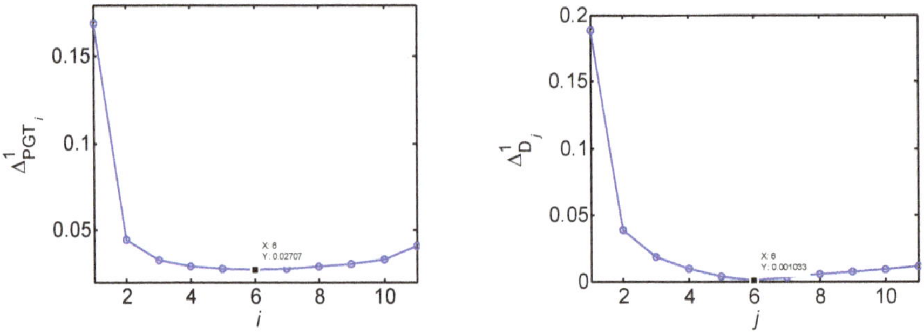

(**a**) (**b**)

Figure 3. The figure of BDMs: (**a**) the BDM of $\Delta^1_{\mathrm{PGT}_i}$, $i \in \{1, 2, \ldots, 11\}$; (**b**) the BDM of $\Delta^1_{\mathrm{D}_j}$, $i \in \{1, 2, \ldots, 11\}$.

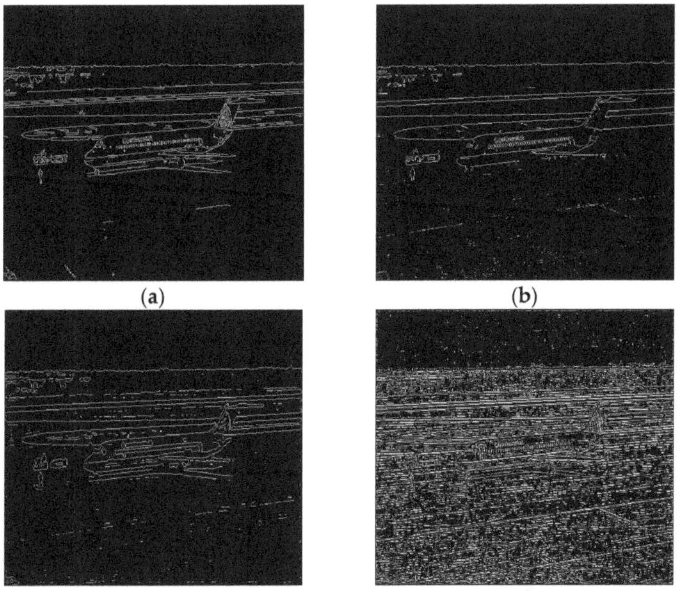

(a) (b) (c) (d)

Figure 4. Edge detection results of the airplane image: (**a**) the ground truth; (**b**) the edge detected by the regularized edge detector (RED); (**c**) the Laplacian of Gaussian (LoG); (**d**) the Laplacian-based edge detector (LED).

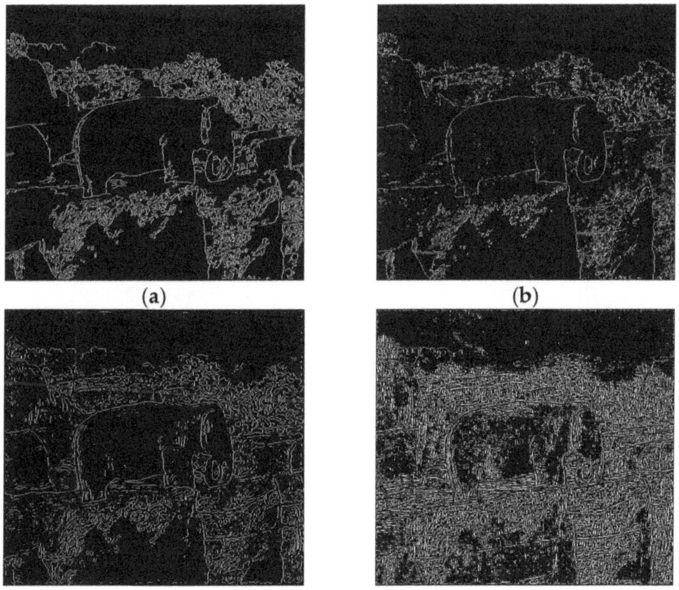

(a) (b) (c) (d)

Figure 5. Edge detection results of the elephant image: (**a**) the ground truth; (**b**) the edge detected by the RED; (**c**) the LoG; (**d**) the LED.

Table 2. Quantitative comparison of the edges detected by the LoG, the RED and the LED.

Images	LED	LoG	RED
Airplane	0.7515	0.1270	0.1232
Elephant	0.6619	0.3041	0.2593
Turtle	0.4430	0.1226	0.1323
Brush	0.5790	0.1883	0.1673
Tiger	0.9239	0.2854	0.2748
Grater	0.5537	0.2353	0.2143
Pitcher	0.5032	0.2584	0.2296

5. Conclusions

In this paper, a novel edge detection algorithm is proposed based on the regularized Laplacian operation. The PDE-based regularization enables us to compute the regularized Laplacian operation in a direct way. Considering the importance of the regularization parameter, an objective choice strategy of the regularization parameter is proposed. Numerical implementations of the regularization parameter and the edge detection algorithm are also given. Based on the image database and ground truth edges taken from [22], the superiority of the RED against the LED and the LoG detector has been shown by the edge images and quantitative comparison.

Author Contributions: All the authors inferred the main conclusions and approved the current version of this manuscript.

Funding: This research was funded by National Natural Science Foundation of China (Grant No. 11661008), Natural Science Foundation of Jianxi Province (Grant No. 20161BAB211025), Science & Technology Project of Jiangxi Educational Committee (Grant No. GJJ150982) and Tendering Subject of Gannan Normal University (Grant No. 15zb03).

Conflicts of Interest: The authors declare no conflict of interest.

References

1. Basu, M. Gaussian-based edge-detection methods—A survey. *IEEE Trans. Syst. Man Cybern. Part C Appl. Rev.* **2002**, *32*, 252–260. [CrossRef]
2. Marr, D.; Hildreth, E. Theory of edge detection. *Proc. R. Soc. Lond. B* **1980**, *207*, 187–217. [CrossRef] [PubMed]
3. Canny, J. A computational approach to edge detection. *IEEE Trans. Pattern Anal. Mach. Intell.* **1986**, *8*, 679–698. [CrossRef] [PubMed]
4. Sarkar, S.; Boyer, K. Optimal infinite impulse response zero crossing based edge detectors. *CVGIP Image Underst.* **1991**, *54*, 224–243. [CrossRef]
5. Demigny, D. On optimal linear filtering for edge detection. *IEEE Trans. Image Process.* **2002**, *11*, 728–737. [CrossRef] [PubMed]
6. Kang, C.C.; Wang, W.J. A novel edge detection method based on the maximizing objective function. *Pattern Recognit.* **2007**, *40*, 609–618. [CrossRef]
7. Wang, X. Laplacian operator-based edge detectors. *IEEE Trans. Pattern Anal. Mach. Intell.* **2007**, *29*, 886–890. [CrossRef] [PubMed]
8. Lopez-Molina, C.; Bustince, H.; Fernandez, J.; Couto, P.; De Baets, B. A gravitational approach to edge detection based on triangular norms. *Pattern Recognit.* **2010**, *43*, 3730–3741. [CrossRef]
9. Murio, D.A. *The Mollification Method and the Numerical Solution of Ill-Posed Problems*; Wiley-Interscience: New York, NY, USA, 1993; pp. 1–5, ISBN 0-471-59408-3.
10. Wan, X.Q.; Wang, Y.B.; Yamamoto, M. Detection of irregular points by regularization in numerical differentiation and application to edge detection. *Inverse Probl.* **2006**, *22*, 1089–1103. [CrossRef]
11. Xu, H.L.; Liu, J.J. Stable numerical differentiation for the second order derivatives. *Adv. Comput. Math.* **2010**, *33*, 431–447. [CrossRef]
12. Huang, X.; Wu, C.; Zhou, J. Numerical differentiation by integration. *Math. Comput.* **2013**, *83*, 789–807. [CrossRef]

13. Wang, Y.C.; Liu, J.J. On the edge detection of an image by numerical differentiations for gray function. *Math. Methods Appl. Sci.* **2018**, *41*, 2466–2479. [CrossRef]
14. Gonzalez, R.C.; Woods, R.E. *Digital Image Processing*, 3rd ed.; Pearson: London, UK, 2007; pp. 158–162, ISBN 978-0-13-168728-8.
15. Gunn, S.R. On the discrete representation of the Laplacian of Gaussian. *Pattern Recognit.* **1999**, *32*, 1463–1472. [CrossRef]
16. Xu, H.L.; Liu, J.J. On the Laplacian operation with applications in magnetic resonance electrical impedance imaging. *Inverse Probl. Sci. Eng.* **2013**, *21*, 251–268. [CrossRef]
17. Yitzhaky, Y.; Peli, E. A method for objective edge detection evaluation and detector parameter selection. *IEEE Trans. Pattern Anal. Mach. Intell.* **2003**, *25*, 1027–1033. [CrossRef]
18. Gu, C.H.; Li, D.Q.; Chen, S.X.; Zheng, S.M.; Tan, Y.J. *Equations of Mathematical Physics*, 2nd ed.; Higher Education Press: Beijing, China, 2002; pp. 80–86, ISBN 7-04-010701-5. (In Chinese)
19. Lopez-Molina, C.; Baets De, B.; Bustince, H. Quantitative error measures for edge detection. *Pattern Recognit.* **2013**, *46*, 1125–1139. [CrossRef]
20. Baddeley, A.J. An error metric for binary images. In Proceedings of the IEEE Workshop on Robust Computer Vision, Bonn, Germany, 9–11 March 1992; Wichmann Verlag: Karlsruhe, Germany, 1992; pp. 59–78.
21. Fernández-García, N.L.; Medina-Carnicer, R.; Carmona-Poyato, A.; Madrid-Cuevas, F.J.; Prieto-Villegas, M. Characterization of empirical discrepancy evaluation measures. *Pattern Recognit. Lett.* **2004**, *25*, 35–47. [CrossRef]
22. Heath, M.D.; Sarkar, S.; Sanocki, T.A.; Bowyer, K.W. A robust visual method for assessing the relative performance of edge detection algorithms. *IEEE Trans. Pattern Anal. Mach. Intell.* **1997**, *19*, 1338–1359. [CrossRef]

symmetry

MDPI

Article

The Complexity of Some Classes of Pyramid Graphs Created from a Gear Graph

Jia-Bao Liu [1] and Salama Nagy Daoud [2,3,*]

1 School of Mathematics and Physics, Anhui Jianzhu University, Hefei 230601, China; liujiabaoad@163.com
2 Department of Mathematics, Faculty of Science, Taibah University, Al-Madinah 41411, Saudi Arabia
3 Department of Mathematics and Computer Science, Faculty of Science, Menoufia University, Shebin El Kom 32511, Egypt
* Correspondence: sdaoud@taibahu.edu.sa; Tel.: +966-598914649

Received: 8 November 2018; Accepted: 23 November 2018; Published: 2 December 2018

Abstract: The methods of measuring the complexity (spanning trees) in a finite graph, a problem related to various areas of mathematics and physics, have been inspected by many mathematicians and physicists. In this work, we defined some classes of pyramid graphs created by a gear graph then we developed the Kirchhoff's matrix tree theorem method to produce explicit formulas for the complexity of these graphs, using linear algebra, matrix analysis techniques, and employing knowledge of Chebyshev polynomials. Finally, we gave some numerical results for the number of spanning trees of the studied graphs.

Keywords: complexity; Chebyshev polynomials; gear graph; pyramid graphs

MSC: 05C05, 05C50

1. Introduction

The graph theory is a theory that combines computer science and mathematics, which can solve considerable problems in several fields (telecom, social network, molecules, computer network, genetics, etc.) by designing graphs and facilitating them through idealistic cases such as the spanning trees, see [1–10].

A spanning tree of a finite connected graph G is a maximal subset of the edges that contains no cycle, or equivalently a minimal subset of the edges that connects all the vertices. The history of enumerating the number of spanning trees $\tau(G)$ of a graph G dates back to 1842 when the physicist Kirchhoff [11] offered the matrix tree theorem established on the determinants of a certain matrix gained from the Laplacian matrix L defined by the difference between the degree matrix D and adjacency matrix A, where D is a diagonal matrix, $D = dig(d_1, d_2, \dots, d_n)$ corresponding to a graph G with n vertices that has the vertex degree of d_i in the ith position of a graph G and A is a matrix with rows and columns labeled by graph vertices, with a 1 or 0 in position (u_i, u_j) according to whether u_i and u_j are adjacent or not. That is

$$L_{ij} = \begin{cases} a_i & \text{if } i = j \\ -1 & \text{if } i \neq j \text{ and } i \text{ is adjacent to } j \\ 0 & \text{otherwise} \end{cases},$$

where a_i denotes the degree of the vertex i.

This method allows beneficial results for a graph comprising a small number of vertices, but is not feasible for large graphs. There is one more method for calculating $\tau(G)$. Let $\lambda_1 \geq \lambda_2 \geq \dots \geq \lambda_k = 0$

denote the eigenvalues of the matrix L of a graph G with n vertices. "Kelmans" and "Chelnokov" [12] have derived that

$$\tau(G) = \frac{1}{k}\prod_{i=1}^{k-1} \lambda_k.$$

One of the favorite methods of calculating the complexity is the contraction–deletion theorem. For any graph G, the complexity $\tau(G)$ of G is equal to $\tau(G) = \tau(G-e) + \tau(G/e)$, where e is any edge of G, and where $G-e$ is the deletion of e from G, and G/e is the contraction of e in G. This gives a recursive method to calculate the complexity of a graph [13,14].

Another important method is using electrically equivalent transformations of networks. Yilun Shang [15] derived a closed-form formula for the enumeration of spanning trees the subdivided-line graph of a simple connected graph using the theory of electrical networks.

Many works have conceived techniques to derive the number of spanning trees of a graph, some of which can be found at [16–18].

Now, we give the following Lemma:

Lemma 1 [19]. $\tau(G) = \frac{1}{k^2}\det(kI - D^c + A^c)$ *where* A^c *and* D^c *are the adjacency and degree matrices of* G^c*, the complement of* G*, respectively, and* I *is the* $k \times k$ *identity matrix.*

The characteristic of this formula is to express $\tau(G)$ straightway as a determinant rather than in terms of cofactors as in Kirchhoff theorem or eigenvalues as in Kelmans and Chelnokov formula.

2. Chebyshev Polynomial

In this part we insert some relations regarding Chebyshev polynomials of the first and second types which we use in our calculations.

We start from their definitions, see Yuanping, et al. [20].

Let $A_n(x)$ be $n \times n$ matrix such that

$$A_n(x) = \begin{pmatrix} 2x & -1 & 0 & \cdots & 0 \\ -1 & 2x & -1 & \ddots & \vdots \\ 0 & \ddots & \ddots & \ddots & 0 \\ \vdots & \ddots & \ddots & \ddots & -1 \\ 0 & \cdots & 0 & -1 & 2x \end{pmatrix}.$$

Furthermore, we render that the Chebyshev polynomials of the first type are defined by

$$T_n(x) = \cos(n\ \cos^{-1} x) \tag{1}$$

The Chebyshev polynomials of the second type are defined by

$$U_{n-1}(x) = \frac{1}{n}\frac{d}{dx}T_n(x) = \frac{\sin(n\ \cos^{-1} x)}{\sin(\cos^{-1} x)} \tag{2}$$

It is easily confirmed that

$$U_n(x) - 2xU_{n-1}(x) + U_{n-2}(x) = 0 \tag{3}$$

It can then be shown from this recursion that by expanding $\det A_n(x)$ one obtains

$$U_n(x) = \det(A_n(x)),\ n \geq 1 \tag{4}$$

Moreover, by solving the recursion (3), one gets the straightforward formula

$$U_n(x) = \frac{(x+\sqrt{x^2-1})^{n+1} - (x-\sqrt{x^2-1})^{n+1}}{2\sqrt{x^2-1}}, \ n \geq 1, \tag{5}$$

where the conformity is valid for all complex x (except at $x = \pm 1$, where the function can be taken as the limit).

The definition of $U_n(x)$ easily yields its zeros and it can therefore be confirmed that

$$U_{n-1}(x) = 2^{n-1} \prod_{j=1}^{n-1} (x - \cos\frac{j\pi}{n}) \tag{6}$$

One further notes that

$$U_{n-1}(-x) = (-1)^{n-1} U_{n-1}(x) \tag{7}$$

From Equations (6) and (7), we have:

$$U_{n-1}{}^2(x) = 4^{n-1} \prod_{j=1}^{n-1} (x^2 - \cos^2\frac{j\pi}{n}) \tag{8}$$

Finally, straightforward manipulation of the above formula produces the following formula (9), which is highly beneficial to us later:

$$U_{n-1}{}^2(\sqrt{\frac{x+2}{4}}) = \prod_{j=1}^{n-1} (x - 2\cos\frac{2j\pi}{n}) \tag{9}$$

Moreover, one can see that

$$U_{n-1}{}^2(x) = \frac{1 - T_{2n}(x)}{2(1-x^2)} = \frac{1 - T_n(2x^2-1)}{2(1-x^2)} \tag{10}$$

$$T_n(x) = \frac{1}{2}[(x+\sqrt{x^2-1})^n + ((x-\sqrt{x^2-1})^n] \tag{11}$$

Now we introduce the following important two Lemmas.

Lemma 2 [21]. *Let $B_n(x)$ be $n \times n$ Circulant matrix such that*

$$B_n(x) = \begin{pmatrix} x & 0 & 1 & \cdots & 1 & 0 \\ 0 & \ddots & \ddots & \ddots & \ddots & 1 \\ 1 & \ddots & \ddots & \ddots & \ddots & \vdots \\ \vdots & \ddots & \ddots & \ddots & \ddots & 1 \\ 1 & \ddots & \ddots & \ddots & \ddots & 0 \\ 0 & 1 & \cdots & 1 & 0 & x \end{pmatrix}.$$

Then for $n \geq 3, \ x \geq 4$, one has

$$\det(B_n(x)) = \frac{2(x+n-3)}{x-3}[T_n(\frac{x-1}{2}) - 1].$$

Lemma 3 [22]. *If $A \in F^{n\times n}, B \in F^{n\times m}, C \in F^{m\times n}$ and $D \in F^{m\times m}$. Suppose that A and D are nonsingular matrices, then:*

$$\det\begin{pmatrix} A & B \\ C & D \end{pmatrix} = \det(A - BD^{-1}C)\det D = \det A \det(D - CA^{-1}B).$$

This Lemma gives a type of symmetry for some matrices which simplify our calculations of the complexity of graphs studied in this paper.

3. Main Results

Definition 1. *The pyramid graph* $A_n^{(m)}$ is the graph created from the gear graph G_{m+1} with vertices $\{u_0;\ u_1, u_2, \ldots, u_m;\ w_1, w_2, \ldots, w_m\}$ and m sets of vertices, say, $\{v_1^1, v_2^1, \ldots, v_n^1\}$, $\{v_1^2, v_2^2, \ldots, v_n^2\}, \ldots, \{v_1^m, v_2^m, \ldots, v_n^m\}$, such that for all $i = 1, 2, \ldots, n$ the vertex v_i^j is adjacent to u_j and u_{j+1},where $j = 1, 2, \ldots, m-1$, and v_i^m is adjacent to u_1 and u_m. See Figure 1.

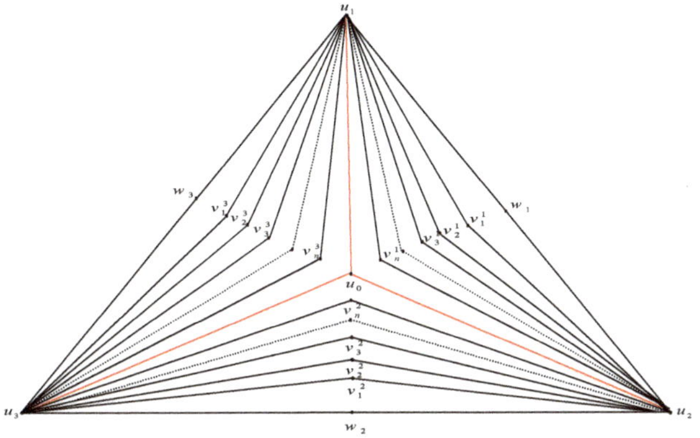

Figure 1. The pyramid graph $A_n^{(3)}$.

Theorem 1. *For* $n \geq 0$, $m \geq 3$, $\tau(A_n^{(m)}) = 2^{mn}[(n+2+\sqrt{2n+3}\,)^m + (n+2-\sqrt{2n+3}\,)^m - 2(n+1)^m]$.

Proof. Using Lemma 1, we have

$$\tau(A_n^{(m)}) = \frac{1}{(mn+2m+1)^2} \times \det((mn+2m+1)\,I - D^c + A^c) = \frac{1}{(mn+2m+1)^2} \times$$

$$\det \begin{pmatrix}
(m+1) & 0 & 0 & \cdots & \cdots & \cdots & 0 & 1 & \cdots & \cdots & \cdots & \cdots & 1 & 1 & \cdots & \cdots & \cdots & \cdots & \cdots & \cdots & \cdots & \cdots & \cdots & \cdots & \cdots & \cdots & \cdots & \cdots & \cdots & 1 \\
0 & 2(n+2) & 1 & \cdots & \cdots & \cdots & 1 & 0 & 1 & \cdots & \cdots & 1 & 0 & 0 & \cdots & 0 & 1 & \cdots & 1 & \cdots & \cdots & 1 & \cdots & 1 & 1 & \cdots & 1 & 0 & \cdots & 0 \\
\vdots & 1 & \ddots & \ddots & \ddots & \ddots & \vdots & 0 & \ddots & \ddots & \ddots & \ddots & 1 & 0 & \cdots & 0 & 0 & \cdots & 0 & 1 & \cdots & 1 & \cdots & \cdots & 1 & \cdots & 1 & 1 & \cdots & 1 \\
\vdots & \vdots & \ddots & \ddots & \ddots & \ddots & \vdots & 1 & \ddots & \ddots & \ddots & \vdots & \vdots & 1 & \cdots & 1 & 0 & \cdots & 0 & 0 & \cdots & 0 & 1 & \cdots & 1 & \cdots & \cdots & 1 & \cdots & 1 \\
\vdots & \vdots & \ddots & \ddots & \ddots & \ddots & \vdots & \vdots & \ddots & \ddots & \ddots & \vdots & \vdots & 1 & \cdots & 1 & 1 & \cdots & 1 & 0 & \cdots & 0 & 0 & \cdots & 0 & 1 & \cdots & 1 & \cdots & \cdots \\
\vdots & \vdots & \ddots & \ddots & \ddots & \ddots & 1 & \vdots & \ddots & \ddots & \ddots & \ddots & 1 & \vdots & \cdots & \cdots & \cdots & \cdots & \cdots & \cdots & \cdots & \cdots & \cdots & \cdots & \cdots & \cdots & \cdots & \cdots & \cdots & \vdots \\
0 & 1 & \cdots & \cdots & \cdots & 1 & 2(n+2) & 1 & \cdots & \cdots & 1 & 0 & 0 & 1 & \cdots & 1 & 1 & \cdots & 1 & \cdots & \cdots & 1 & \cdots & 1 & 0 & \cdots & 0 & 0 & \cdots & 0 \\
1 & 0 & 0 & 1 & \cdots & \cdots & 1 & 3 & 1 & \cdots & \cdots & \cdots & 1 & 1 & \cdots & 1 & 1 & \cdots & 1 & \cdots & \cdots & \cdots & \cdots & \cdots & 1 & \cdots & 1 & 1 & \cdots & 1 \\
\vdots & 1 & \ddots & \ddots & \ddots & \ddots & \vdots & 1 & \ddots & \ddots & \ddots & \ddots & \vdots & 1 & \cdots & 1 & 1 & \cdots & 1 & \cdots & \cdots & \cdots & \cdots & \cdots & 1 & \cdots & 1 & 1 & \cdots & 1 \\
\vdots & \vdots & \ddots & \ddots & \ddots & \ddots & \vdots & \vdots & \ddots & \ddots & \ddots & \ddots & \vdots & \vdots & \cdots & \vdots & \vdots & \cdots & \vdots & \cdots & \cdots & \cdots & \cdots & \cdots & \vdots & \cdots & \vdots & \vdots & \cdots & 1 \\
\vdots & \vdots & \ddots & \ddots & \ddots & \ddots & 1 & \vdots & \ddots & \ddots & \ddots & \ddots & \vdots & \vdots & \cdots & \vdots & \vdots & \cdots & \vdots & \cdots & \cdots & \cdots & \cdots & \cdots & \vdots & \cdots & \vdots & \vdots & \cdots & \vdots \\
\vdots & 1 & \ddots & \ddots & \ddots & \ddots & 0 & \vdots & \ddots & \ddots & \ddots & \ddots & \vdots & 1 & \cdots & 1 & 1 & \cdots & 1 & \cdots & \cdots & \cdots & \cdots & \cdots & 1 & \cdots & 1 & 1 & \cdots & 1 \\
1 & 0 & 1 & \cdots & \cdots & 1 & 0 & 1 & \cdots & \cdots & \cdots & 1 & 3 & 1 & \cdots & 1 & 1 & \cdots & 1 & \cdots & \cdots & \cdots & \cdots & \cdots & 1 & \cdots & 1 & 1 & \cdots & 1 \\
1 & 0 & 0 & 1 & \cdots & 1 & 0 & 1 & 1 & \cdots & \cdots & 1 & 1 & 3 & 1 & \cdots & \cdots & \cdots & \cdots & \cdots & \cdots & \cdots & \cdots & \cdots & \cdots & \cdots & \cdots & \cdots & \cdots & 1 \\
\vdots & \vdots & \vdots & \vdots & \vdots & \vdots & \vdots & \vdots & \vdots & \vdots & \vdots & \vdots & \vdots & 1 & \ddots & \ddots & \ddots & \ddots & \ddots & \ddots & \ddots & \ddots & \ddots & \ddots & \ddots & \ddots & \ddots & \ddots & \ddots & \vdots \\
\vdots & 0 & 0 & 1 & \vdots & 1 & \vdots & 1 & 1 & \cdots & \cdots & 1 & 1 & \vdots & \ddots & \ddots & \ddots & \ddots & \ddots & \ddots & \ddots & \ddots & \ddots & \ddots & \ddots & \ddots & \ddots & \ddots & \ddots & \vdots \\
\vdots & 1 & 0 & 0 & \vdots & 1 & \vdots & 1 & 1 & \cdots & \cdots & 1 & 1 & \vdots & \ddots & \ddots & \ddots & \ddots & \ddots & \ddots & \ddots & \ddots & \ddots & \ddots & \ddots & \ddots & \ddots & \ddots & \ddots & \vdots \\
\vdots & \vdots & \vdots & \vdots & \vdots & \vdots & \vdots & \vdots & \vdots & \vdots & \vdots & \vdots & \vdots & \vdots & \ddots & \ddots & \ddots & \ddots & \ddots & \ddots & \ddots & \ddots & \ddots & \ddots & \ddots & \ddots & \ddots & \ddots & \ddots & \vdots \\
\vdots & 1 & 0 & 0 & \vdots & 1 & \vdots & 1 & 1 & \cdots & \cdots & 1 & 1 & \vdots & \ddots & \ddots & \ddots & \ddots & \ddots & \ddots & \ddots & \ddots & \ddots & \ddots & \ddots & \ddots & \ddots & \ddots & \ddots & \vdots \\
\vdots & 1 & 1 & 0 & \vdots & \vdots & \vdots & \vdots & \vdots & \vdots & \vdots & \vdots & \vdots & \vdots & \ddots & \ddots & \ddots & \ddots & \ddots & \ddots & \ddots & \ddots & \ddots & \ddots & \ddots & \ddots & \ddots & \ddots & \ddots & \vdots \\
\vdots & \vdots & \vdots & \vdots & \vdots & \vdots & \vdots & \vdots & \vdots & \vdots & \vdots & \vdots & \vdots & \vdots & \ddots & \ddots & \ddots & \ddots & \ddots & \ddots & \ddots & \ddots & \ddots & \ddots & \ddots & \ddots & \ddots & \ddots & \ddots & \vdots \\
\vdots & 1 & 1 & 0 & \vdots & 1 & \vdots & \vdots & \vdots & \vdots & \vdots & \vdots & \vdots & \vdots & \ddots & \ddots & \ddots & \ddots & \ddots & \ddots & \ddots & \ddots & \ddots & \ddots & \ddots & \ddots & \ddots & \ddots & \ddots & \vdots \\
\vdots & \vdots & \vdots & 1 & \vdots & \vdots & \vdots & \vdots & \vdots & \vdots & \vdots & \vdots & \vdots & \vdots & \ddots & \ddots & \ddots & \ddots & \ddots & \ddots & \ddots & \ddots & \ddots & \ddots & \ddots & \ddots & \ddots & \ddots & \ddots & \vdots \\
\vdots & \vdots & \vdots & \vdots & \vdots & 1 & \vdots & \vdots & \vdots & \vdots & \vdots & \vdots & \vdots & \vdots & \ddots & \ddots & \ddots & \ddots & \ddots & \ddots & \ddots & \ddots & \ddots & \ddots & \ddots & \ddots & \ddots & \ddots & \ddots & \vdots \\
\vdots & 1 & 1 & 1 & \vdots & 0 & \vdots & \vdots & \vdots & \vdots & \vdots & \vdots & \vdots & \vdots & \ddots & \ddots & \ddots & \ddots & \ddots & \ddots & \ddots & \ddots & \ddots & \ddots & \ddots & \ddots & \ddots & \ddots & \ddots & \vdots \\
\vdots & \vdots & \vdots & \vdots & \vdots & \vdots & \vdots & \vdots & \vdots & \vdots & \vdots & \vdots & \vdots & \vdots & \ddots & \ddots & \ddots & \ddots & \ddots & \ddots & \ddots & \ddots & \ddots & \ddots & \ddots & \ddots & \ddots & \ddots & \ddots & \vdots \\
\vdots & 1 & 1 & \vdots & \vdots & 0 & \vdots & 1 & 1 & \cdots & \cdots & 1 & 1 & \vdots & \ddots & \ddots & \ddots & \ddots & \ddots & \ddots & \ddots & \ddots & \ddots & \ddots & \ddots & \ddots & \ddots & \ddots & \ddots & \vdots \\
\vdots & 0 & 1 & 1 & \vdots & 0 & \vdots & 1 & 1 & \cdots & \cdots & 1 & 1 & \vdots & \ddots & \ddots & \ddots & \ddots & \ddots & \ddots & \ddots & \ddots & \ddots & \ddots & \ddots & \ddots & \ddots & \ddots & \ddots & \vdots \\
\vdots & \vdots & \vdots & \vdots & \vdots & \vdots & \vdots & \vdots & \vdots & \vdots & \vdots & \vdots & \vdots & \vdots & \ddots & \ddots & \ddots & \ddots & \ddots & \ddots & \ddots & \ddots & \ddots & \ddots & \ddots & \ddots & \ddots & \ddots & \ddots & 1 \\
1 & 0 & 1 & 1 & \cdots & 0 & \cdots & 1 & 1 & \cdots & \cdots & 1 & 1 & 1 & \cdots & \cdots & \cdots & \cdots & \cdots & \cdots & \cdots & \cdots & \cdots & \cdots & \cdots & \cdots & \cdots & \cdots & 1 & 3
\end{pmatrix}$$

Let $j = (1 \cdots 1)$ be the $1 \times n$ matrix with all one, and J_n be the $n \times n$ matrix with all one. Set $a = 2n + 4$ and $b = mn + 2m + 1$. Then we obtain:

$$\tau\left(A_n^{(m)}\right) = \frac{1}{b^2} \times \det \begin{pmatrix}
m+1 & 0 & \cdots & \cdots & \cdots & \cdots & 0 & 1 & \cdots & \cdots & \cdots & \cdots & 1 & j & \cdots & \cdots & \cdots & \cdots & j \\
0 & a & 1 & \cdots & \cdots & \cdots & 1 & 0 & 1 & \cdots & \cdots & 1 & 0 & 0 & j & \cdots & \cdots & j & 0 \\
\vdots & 1 & \ddots & \ddots & \ddots & \ddots & \vdots & 0 & 0 & \ddots & \ddots & \ddots & 1 & 0 & \ddots & \ddots & \ddots & \ddots & j \\
\vdots & \vdots & \ddots & \ddots & \ddots & \ddots & \vdots & 1 & 0 & \ddots & \ddots & \ddots & \vdots & j & \ddots & \ddots & \ddots & \ddots & \vdots \\
\vdots & \vdots & \ddots & \ddots & \ddots & \ddots & \vdots & \vdots & \ddots & \ddots & \ddots & \ddots & \vdots & \vdots & \ddots & \ddots & \ddots & \ddots & \vdots \\
\vdots & \vdots & \ddots & \ddots & \ddots & \ddots & 1 & \vdots & \ddots & \ddots & 0 & 0 & 1 & \vdots & \ddots & \ddots & \ddots & \ddots & j \\
0 & 1 & \cdots & \cdots & \cdots & 1 & a & 1 & \cdots & \cdots & 1 & 0 & 0 & j & \cdots & \cdots & j & 0 & 0 \\
1 & 0 & 0 & 1 & \cdots & \cdots & 1 & 3 & 1 & \cdots & \cdots & \cdots & 1 & j & \cdots & \cdots & \cdots & \cdots & j \\
\vdots & 1 & 0 & \ddots & \ddots & \ddots & 1 & 1 & \ddots & \ddots & \ddots & \ddots & \vdots & \vdots & \ddots & \ddots & \ddots & \ddots & \vdots \\
\vdots & \vdots & \ddots & \ddots & \ddots & \ddots & \vdots & \vdots & \ddots & \ddots & \ddots & \ddots & \vdots & \vdots & \ddots & \ddots & \ddots & \ddots & \vdots \\
\vdots & \vdots & \ddots & \ddots & \ddots & \ddots & 1 & \vdots & \ddots & \ddots & \ddots & \ddots & \vdots & \vdots & \ddots & \ddots & \ddots & \ddots & \vdots \\
\vdots & 1 & \ddots & \ddots & \ddots & \ddots & 0 & \vdots & \ddots & \ddots & \ddots & \ddots & 1 & \vdots & \ddots & \ddots & \ddots & \ddots & \vdots \\
1 & 0 & 1 & \cdots & \cdots & 1 & 0 & 1 & \cdots & \cdots & \cdots & 1 & 3 & j & \cdots & \cdots & \cdots & \cdots & j \\
j^t & 0 & 0 & j^t & \cdots & \cdots & j^t & j^t & \cdots & \cdots & \cdots & \cdots & j^t & & & & & & \\
\vdots & j^t & 0 & \ddots & \ddots & \ddots & \vdots & \vdots & \ddots & \ddots & \ddots & \ddots & \vdots & & & & & & \\
\vdots & \vdots & \ddots & \ddots & \ddots & \ddots & \vdots & \vdots & \ddots & \ddots & \ddots & \ddots & \vdots & & & 2I_{mn} + J_{mn} & & & \\
\vdots & \vdots & \ddots & \ddots & \ddots & \ddots & j^t & \vdots & \ddots & \ddots & \ddots & \ddots & \vdots & & & & & & \\
\vdots & j^t & \ddots & \ddots & \ddots & \ddots & 0 & \vdots & \ddots & \ddots & \ddots & \ddots & \vdots & & & & & & \\
j^t & 0 & j^t & \cdots & \cdots & j^t & 0 & j^t & \cdots & \cdots & \cdots & \cdots & j^t & & & & & &
\end{pmatrix}$$

$$= \frac{1}{b^2} \times \det \begin{pmatrix}
b & 0 & \cdots & \cdots & \cdots & \cdots & 0 & 1 & \cdots & \cdots & \cdots & \cdots & 1 & j & \cdots & \cdots & \cdots & \cdots & j \\
b & a & 1 & \cdots & \cdots & \cdots & 1 & 0 & 1 & \cdots & \cdots & 1 & 0 & 0 & j & \cdots & \cdots & j & 0 \\
\vdots & 1 & \ddots & \ddots & \ddots & \ddots & \vdots & 0 & 0 & \ddots & \ddots & \ddots & 1 & 0 & \ddots & \ddots & \ddots & \ddots & j \\
\vdots & \vdots & \ddots & \ddots & \ddots & \ddots & \vdots & 1 & 0 & \ddots & \ddots & \ddots & \vdots & j & \ddots & \ddots & \ddots & \ddots & \vdots \\
\vdots & \vdots & \ddots & \ddots & \ddots & \ddots & \vdots & \vdots & \ddots & \ddots & \ddots & \ddots & \vdots & \vdots & \ddots & \ddots & \ddots & \ddots & \vdots \\
\vdots & \vdots & \ddots & \ddots & \ddots & \ddots & 1 & \vdots & \ddots & \ddots & 0 & 0 & 1 & \vdots & \ddots & \ddots & \ddots & \ddots & j \\
b & 1 & \cdots & \cdots & \cdots & 1 & a & 1 & \cdots & \cdots & 1 & 0 & 0 & j & \cdots & \cdots & j & 0 & 0 \\
b & 0 & 0 & 1 & \cdots & \cdots & 1 & 3 & 1 & \cdots & \cdots & \cdots & 1 & j & \cdots & \cdots & \cdots & \cdots & j \\
\vdots & 1 & 0 & \ddots & \ddots & \ddots & 1 & 1 & \ddots & \ddots & \ddots & \ddots & \vdots & \vdots & \ddots & \ddots & \ddots & \ddots & \vdots \\
\vdots & \vdots & \ddots & \ddots & \ddots & \ddots & \vdots & \vdots & \ddots & \ddots & \ddots & \ddots & \vdots & \vdots & \ddots & \ddots & \ddots & \ddots & \vdots \\
\vdots & \vdots & \ddots & \ddots & \ddots & \ddots & 1 & \vdots & \ddots & \ddots & \ddots & \ddots & \vdots & \vdots & \ddots & \ddots & \ddots & \ddots & \vdots \\
\vdots & 1 & \ddots & \ddots & \ddots & \ddots & 0 & \vdots & \ddots & \ddots & \ddots & \ddots & 1 & \vdots & \ddots & \ddots & \ddots & \ddots & \vdots \\
b & 0 & 1 & \cdots & \cdots & 1 & 0 & 1 & \cdots & \cdots & \cdots & 1 & 3 & j & \cdots & \cdots & \cdots & \cdots & j \\
bj^t & 0 & 0 & j^t & \cdots & \cdots & j^t & j^t & \cdots & \cdots & \cdots & \cdots & j^t & & & & & & \\
\vdots & j^t & 0 & \ddots & \ddots & \ddots & \vdots & \vdots & \ddots & \ddots & \ddots & \ddots & \vdots & & & & & & \\
\vdots & \vdots & \ddots & \ddots & \ddots & \ddots & \vdots & \vdots & \ddots & \ddots & \ddots & \ddots & \vdots & & & 2I_{mn} + J_{mn} & & & \\
\vdots & \vdots & \ddots & \ddots & \ddots & \ddots & j^t & \vdots & \ddots & \ddots & \ddots & \ddots & \vdots & & & & & & \\
\vdots & j^t & \ddots & \ddots & \ddots & \ddots & 0 & \vdots & \ddots & \ddots & \ddots & \ddots & \vdots & & & & & & \\
bj^t & 0 & j^t & \cdots & \cdots & j^t & 0 & j^t & \cdots & \cdots & \cdots & \cdots & j^t & & & & & &
\end{pmatrix}$$

$$= \frac{1}{b} \times \det \begin{pmatrix}
1 & 0 & \cdots & \cdots & \cdots & \cdots & 0 & 1 & \cdots & \cdots & \cdots & \cdots & 1 & j & \cdots & \cdots & \cdots & \cdots & j \\
1 & a & 1 & \cdots & \cdots & \cdots & 1 & 0 & 1 & \cdots & \cdots & 1 & 0 & 0 & j & \cdots & \cdots & j & 0 \\
\vdots & 1 & \ddots & \ddots & \ddots & \ddots & \vdots & 0 & 0 & \ddots & \ddots & \ddots & 1 & 0 & \ddots & \ddots & \ddots & \ddots & j \\
\vdots & \vdots & \ddots & \ddots & \ddots & \ddots & \vdots & 1 & 0 & \ddots & \ddots & \ddots & \vdots & j & \ddots & \ddots & \ddots & \ddots & \vdots \\
\vdots & \vdots & \ddots & \ddots & \ddots & \ddots & \vdots & \vdots & \ddots & \ddots & \ddots & \ddots & \vdots & \vdots & \ddots & \ddots & \ddots & \ddots & \vdots \\
\vdots & \vdots & \ddots & \ddots & \ddots & \ddots & 1 & \vdots & \ddots & \ddots & 0 & 0 & 1 & \vdots & \ddots & \ddots & \ddots & \ddots & j \\
1 & 1 & \cdots & \cdots & \cdots & 1 & a & 1 & \cdots & \cdots & 1 & 0 & 0 & j & \cdots & \cdots & j & 0 & 0 \\
1 & 0 & 0 & 1 & \cdots & \cdots & 1 & 3 & 1 & \cdots & \cdots & \cdots & 1 & j & \cdots & \cdots & \cdots & \cdots & j \\
\vdots & 1 & 0 & \ddots & \ddots & \ddots & 1 & 1 & \ddots & \ddots & \ddots & \ddots & \vdots & \vdots & \ddots & \ddots & \ddots & \ddots & \vdots \\
\vdots & \vdots & \ddots & \ddots & \ddots & \ddots & \vdots & \vdots & \ddots & \ddots & \ddots & \ddots & \vdots & \vdots & \ddots & \ddots & \ddots & \ddots & \vdots \\
\vdots & \vdots & \ddots & \ddots & \ddots & \ddots & 1 & \vdots & \ddots & \ddots & \ddots & \ddots & \vdots & \vdots & \ddots & \ddots & \ddots & \ddots & \vdots \\
\vdots & 1 & \ddots & \ddots & \ddots & \ddots & 0 & \vdots & \ddots & \ddots & \ddots & \ddots & 1 & \vdots & \ddots & \ddots & \ddots & \ddots & \vdots \\
1 & 0 & 1 & \cdots & \cdots & 1 & 0 & 1 & \cdots & \cdots & \cdots & 1 & 3 & j & \cdots & \cdots & \cdots & \cdots & j \\
1j^t & 0 & 0 & j^t & \cdots & \cdots & j^t & j^t & \cdots & \cdots & \cdots & \cdots & j^t & & & & & & \\
\vdots & j^t & 0 & \ddots & \ddots & \ddots & \vdots & \vdots & \ddots & \ddots & \ddots & \ddots & \vdots & & & & & & \\
\vdots & \vdots & \ddots & \ddots & \ddots & \ddots & \vdots & \vdots & \ddots & \ddots & \ddots & \ddots & \vdots & & & 2I_{mn} + J_{mn} & & & \\
\vdots & \vdots & \ddots & \ddots & \ddots & \ddots & j^t & \vdots & \ddots & \ddots & \ddots & \ddots & \vdots & & & & & & \\
\vdots & j^t & \ddots & \ddots & \ddots & \ddots & 0 & \vdots & \ddots & \ddots & \ddots & \ddots & \vdots & & & & & & \\
1j^t & 0 & j^t & \cdots & \cdots & j^t & 0 & j^t & \cdots & \cdots & \cdots & \cdots & j^t & & & & & &
\end{pmatrix}$$

$$
= \frac{1}{b} \times \det \left(\begin{array}{cccccccccccccccccccc}
1 & 0 & \cdots & \cdots & \cdots & \cdots & 0 & 1 & \cdots & \cdots & \cdots & \cdots & 1 & j & \cdots & \cdots & \cdots & \cdots & j \\
0 & a & 1 & \cdots & \cdots & \cdots & 1 & -1 & 0 & \cdots & \cdots & 0 & -1 & -j & 0 & \cdots & \cdots & 0 & -j \\
\vdots & 1 & \ddots & \ddots & \ddots & \ddots & \vdots & -1 & -1 & \ddots & \ddots & \ddots & 0 & -j & \ddots & \ddots & \ddots & \ddots & 0 \\
\vdots & \vdots & \ddots & \ddots & \ddots & \ddots & \vdots & 0 & \ddots & \ddots & \ddots & \ddots & \vdots & 0 & \ddots & \ddots & \ddots & \ddots & \vdots \\
\vdots & \vdots & \ddots & \ddots & \ddots & \ddots & \vdots & \vdots & \ddots & \ddots & \ddots & \ddots & \vdots & \vdots & \ddots & \ddots & \ddots & \ddots & \vdots \\
\vdots & \vdots & \ddots & \ddots & \ddots & \ddots & 1 & \vdots & \ddots & \ddots & \ddots & -1 & 0 & \vdots & \ddots & \ddots & \ddots & \ddots & 0 \\
0 & 1 & \cdots & \cdots & \cdots & 1 & a & 0 & \cdots & \cdots & 0 & -1 & -1 & 0 & \cdots & \cdots & 0 & -j & -j \\
0 & 0 & 0 & 1 & \cdots & \cdots & 1 & 2 & 0 & \cdots & \cdots & \cdots & 0 & 0 & \cdots & \cdots & \cdots & \cdots & 0 \\
\vdots & 1 & 0 & \ddots & \ddots & \ddots & 1 & 0 & \ddots & \ddots & \ddots & \ddots & \vdots & \vdots & \ddots & \ddots & \ddots & \ddots & \vdots \\
\vdots & \vdots & \ddots & \ddots & \ddots & \ddots & \vdots & \vdots & \ddots & \ddots & \ddots & \ddots & \vdots & \vdots & \ddots & \ddots & \ddots & \ddots & \vdots \\
\vdots & \vdots & \ddots & \ddots & \ddots & \ddots & 1 & \vdots & \ddots & \ddots & \ddots & \ddots & \vdots & \vdots & \ddots & \ddots & \ddots & \ddots & \vdots \\
\vdots & 1 & \ddots & \ddots & \ddots & \ddots & 0 & \vdots & \ddots & \ddots & \ddots & \ddots & 0 & \vdots & \ddots & \ddots & \ddots & \ddots & \vdots \\
0 & 0 & 1 & \cdots & \cdots & 1 & 0 & 0 & \cdots & \cdots & \cdots & 0 & 2 & 0 & \cdots & \cdots & \cdots & \cdots & 0 \\
0 & 0 & 0 & j^t & \cdots & \cdots & j^t & 0 & \cdots & \cdots & \cdots & \cdots & 0 & & & & & & \\
\vdots & j^t & 0 & \ddots & \ddots & \ddots & \vdots & \vdots & \ddots & \ddots & \ddots & \ddots & \vdots & & & & & & \\
\vdots & \vdots & \ddots & \ddots & \ddots & \ddots & \vdots & \vdots & \ddots & \ddots & \ddots & \ddots & \vdots & & & & 2I_{mn} & & \\
\vdots & \vdots & \ddots & \ddots & \ddots & \ddots & j^t & \vdots & \ddots & \ddots & \ddots & \ddots & \vdots & & & & & & \\
\vdots & j^t & \ddots & \ddots & \ddots & \ddots & 0 & \vdots & \ddots & \ddots & \ddots & \ddots & \vdots & & & & & & \\
0 & 0 & j^t & \cdots & \cdots & j^t & 0 & 0 & \cdots & \cdots & \cdots & \cdots & 0 & & & & & &
\end{array} \right)
$$

$$
= \frac{1}{b} \times \det \left(\begin{array}{cccccccccccccccccc}
a & 1 & \cdots & \cdots & \cdots & 1 & -1 & 0 & \cdots & \cdots & 0 & -1 & -j & 0 & \cdots & \cdots & 0 & -j \\
1 & \ddots & \ddots & \ddots & \ddots & \vdots & -1 & -1 & \ddots & \ddots & \ddots & 0 & -j & \ddots & \ddots & \ddots & \ddots & 0 \\
\vdots & \ddots & \ddots & \ddots & \ddots & \vdots & 0 & \ddots & \ddots & \ddots & \ddots & \vdots & 0 & \ddots & \ddots & \ddots & \ddots & \vdots \\
\vdots & \ddots & \ddots & \ddots & \ddots & \vdots & \vdots & \ddots & \ddots & \ddots & \ddots & \vdots & \vdots & \ddots & \ddots & \ddots & \ddots & \vdots \\
\vdots & \ddots & \ddots & \ddots & \ddots & 1 & \vdots & \ddots & \ddots & \ddots & -1 & 0 & \vdots & \ddots & \ddots & \ddots & \ddots & 0 \\
1 & \cdots & \cdots & \cdots & 1 & a & 0 & \cdots & \cdots & 0 & -1 & -1 & 0 & \cdots & \cdots & 0 & -j & -j \\
0 & 0 & 1 & \cdots & \cdots & 1 & 2 & 0 & \cdots & \cdots & \cdots & 0 & 0 & \cdots & \cdots & \cdots & \cdots & 0 \\
1 & 0 & \ddots & \ddots & \ddots & 1 & 0 & \ddots & \ddots & \ddots & \ddots & \vdots & \vdots & \ddots & \ddots & \ddots & \ddots & \vdots \\
\vdots & \ddots & \ddots & \ddots & \ddots & \vdots & \vdots & \ddots & \ddots & \ddots & \ddots & \vdots & \vdots & \ddots & \ddots & \ddots & \ddots & \vdots \\
\vdots & \ddots & \ddots & \ddots & \ddots & 1 & \vdots & \ddots & \ddots & \ddots & \ddots & \vdots & \vdots & \ddots & \ddots & \ddots & \ddots & \vdots \\
1 & \ddots & \ddots & \ddots & \ddots & 0 & \vdots & \ddots & \ddots & \ddots & \ddots & 0 & \vdots & \ddots & \ddots & \ddots & \ddots & \vdots \\
0 & 1 & \cdots & \cdots & 1 & 0 & 0 & \cdots & \cdots & \cdots & 0 & 2 & 0 & \cdots & \cdots & \cdots & \cdots & 0 \\
0 & 0 & j^t & \cdots & \cdots & j^t & 0 & \cdots & \cdots & \cdots & \cdots & 0 & & & & & & \\
j^t & 0 & \ddots & \ddots & \ddots & \vdots & \vdots & \ddots & \ddots & \ddots & \ddots & \vdots & & & & & & \\
\vdots & \ddots & \ddots & \ddots & \ddots & \vdots & \vdots & \ddots & \ddots & \ddots & \ddots & \vdots & & & 2I_{mn} & & & \\
\vdots & \ddots & \ddots & \ddots & \ddots & j^t & \vdots & \ddots & \ddots & \ddots & \ddots & \vdots & & & & & & \\
j^t & \ddots & \ddots & \ddots & \ddots & 0 & \vdots & \ddots & \ddots & \ddots & \ddots & \vdots & & & & & & \\
0 & j^t & \cdots & \cdots & j^t & 0 & 0 & \cdots & \cdots & \cdots & \cdots & 0 & & & & & &
\end{array} \right)
$$

Using Lemma 3, yields

$$
\tau(A_n^{(m)}) = \frac{1}{b} \times \det \begin{pmatrix} A & B \\ C & 2I_{mn} \end{pmatrix} = \frac{1}{b} \times \det\left(A - B\frac{1}{2I_{mn}}C\right) \times 2^{mn}
$$

$$
= \frac{1}{b} 2^{mn} \times 2^{-2m} \times \det \left(\begin{array}{cccccccccccc}
2a & n+2 & 2(n+1) & \cdots & 2(n+1) & n+2 & -2 & 0 & \cdots & \cdots & 0 & -2 \\
n+2 & 2a & n+2 & 2(n+1) & \cdots & 2(n+1) & -2 & \ddots & \ddots & \ddots & \ddots & 0 \\
2(n+1) & n+2 & \ddots & \ddots & \ddots & \vdots & 0 & \ddots & \ddots & \ddots & \ddots & \vdots \\
\vdots & \ddots & \ddots & \ddots & \ddots & 2(n+1) & \vdots & \ddots & \ddots & \ddots & \ddots & \vdots \\
2(n+1) & \ddots & \ddots & \ddots & \ddots & n+2 & \vdots & \ddots & \ddots & \ddots & \ddots & 0 \\
n+2 & 2(n+1) & \cdots & 2(n+1) & n+2 & 2a & 0 & \cdots & \cdots & 0 & -2 & -2 \\
0 & 0 & 2 & \cdots & \cdots & 2 & 4 & 0 & \cdots & \cdots & \cdots & 0 \\
2 & \ddots & \ddots & \ddots & \ddots & \vdots & 0 & \ddots & \ddots & \ddots & \ddots & \vdots \\
\vdots & \ddots & \ddots & \ddots & \ddots & \vdots & \vdots & \ddots & \ddots & \ddots & \ddots & \vdots \\
\vdots & \ddots & \ddots & \ddots & \ddots & 2 & \vdots & \ddots & \ddots & \ddots & \ddots & \vdots \\
2 & \ddots & \ddots & \ddots & \ddots & 0 & \vdots & \ddots & \ddots & \ddots & \ddots & 0 \\
0 & 2 & \cdots & \cdots & 2 & 0 & 0 & \cdots & \cdots & \cdots & 0 & 4
\end{array} \right)
$$

Using Lemma 3 again, yields

$$
\tau(A_n^{(m)}) = \frac{2^{mn-2m}}{b} \times \det \begin{pmatrix} D & E \\ F & 4I_m \end{pmatrix} = \frac{2^{mn}}{b} \times \det\left(D - E\frac{1}{4I_m}F\right)
$$

$$\tau\,(A_n^{(m)}) = \frac{2^{mn}}{b} \times \det \begin{pmatrix} 2a & (n+3) & 2(n+2) & \cdots & 2(n+2) & (n+3) \\ (n+3) & 2a & (n+3) & \ddots & \cdots & 2(n+2) \\ 2(n+2) & \ddots & \ddots & \ddots & \ddots & \vdots \\ \vdots & \ddots & \ddots & \ddots & \ddots & 2(n+2) \\ 2(n+2) & \ddots & \ddots & \ddots & \ddots & (n+3) \\ (n+3) & 2(n+2) & \cdots & 2(n+2) & (n+3) & 2a \end{pmatrix}$$

Straightforward inducement using the properties of determinants, one can obtain

$$\tau\,(A_n^{(m)}) = \frac{2^{mn}}{b} \times \frac{2b}{m\,n+m+2} \times \det \begin{pmatrix} (2a-n-3) & 0 & (n+1) & \cdots & (n+1) & 0 \\ 0 & (2a-n-3) & 0 & \ddots & \cdots & (n+1) \\ (n+1) & \ddots & \ddots & \ddots & \ddots & \vdots \\ \vdots & \ddots & \ddots & \ddots & \ddots & (n+1) \\ (n+1) & \ddots & \ddots & \ddots & \ddots & 0 \\ 0 & (n+1) & \cdots & (n+1) & 0 & (2a-n-3) \end{pmatrix}$$

$$= \frac{2^{m\,n+1}\,(n+1)^m}{m\,n+m+2} \times \det \begin{pmatrix} \frac{(2a-n-3)}{(n+1)} & 0 & 1 & \cdots & 1 & 0 \\ 0 & \frac{(2a-n-3)}{(n+1)} & 0 & \ddots & \ddots & 1 \\ 1 & \ddots & \ddots & \ddots & \ddots & \vdots \\ \vdots & \ddots & \ddots & \ddots & \ddots & 1 \\ 1 & \ddots & \ddots & \ddots & \ddots & 0 \\ 0 & 1 & \cdots & 1 & 0 & \frac{(2a-n-3)}{(n+1)} \end{pmatrix}$$

Using Lemma 2, yields

$$\begin{aligned} \tau\,(A_n^{(m)}) &= 2^{m\,n+1} \times \frac{(n+1)^m}{mn+m+2} \times \frac{2(\frac{2a-n-3}{n+1}+m-3)}{\frac{2a-n-3}{n+1}-3} \times [T_m(\frac{\frac{2a-n-3}{n+1}-1}{2}) - 1] \\ &= 2^{mn+1} \times (n+1)^m \times [T_m(\tfrac{n+2}{n+1}) - 1]. \end{aligned}$$

Using Equation (11), yields the result. □

Definition 2. *The pyramid graph $B_n^{(m)}$ is the graph created from the gear graph G_{m+1} with vertices $\{u_0;\, u_1, u_2, \ldots, u_m;\, w_1, w_2, \ldots, w_m\}$ with double internal edges and m sets of vertices, say, $\{v_1^1, v_2^1, \ldots, v_n^1\}$, $\{v_1^2, v_2^2, \ldots, v_n^2\}, \ldots, \{v_1^m, v_2^m, \ldots, v_n^m\}$, such that for all $i = 1, 2, \ldots, n$ the vertex v_i^j is adjacent to u_j and u_{j+1}, where $j = 1, 2, \ldots, m-1$, and v_i^m is adjacent to u_1 and u_m. See Figure 2.*

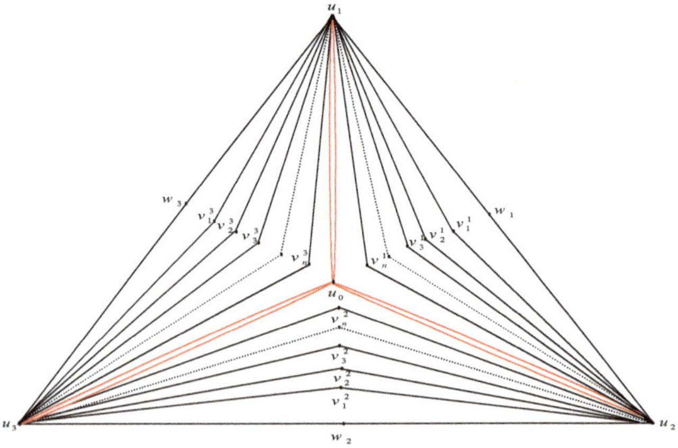

Figure 2. The pyramid graph $B_n^{(3)}$.

Theorem 2. *For $n \geq 0$, $m \geq 3$, $\tau(B_n^{(m)}) = 2^{mn}[(n+3+2\sqrt{n+2}\,)^m + (n+3-2\sqrt{n+2}\,)^m - 2(n+1)^m]$.*

Proof. Using Lemma 1, we get:

$$\tau(B_n^{(m)}) = \frac{1}{(mn+2m+1)^2} \times \det((mn+2m+1)\,I - D^c + A^c) = \frac{1}{(mn+2m+1)^2} \times$$

$$\det\begin{pmatrix}
(2m+1) & -1 & -1 & \cdots & \cdots & \cdots & -1 & 1 & \cdots & \cdots & \cdots & \cdots & 1 & 1 & \cdots & \cdots & \cdots & \cdots & \cdots & \cdots & \cdots & \cdots & \cdots & \cdots & \cdots & \cdots & \cdots & \cdots & \cdots & 1 \\
-1 & (2n+5) & 1 & \cdots & \cdots & \cdots & 1 & 0 & 1 & \cdots & \cdots & 1 & 0 & 0 & \cdots & 0 & 1 & \cdots & 1 & \cdots & \cdots & 1 & \cdots & 1 & 1 & \cdots & 1 & 0 & \cdots & 0 \\
\vdots & 1 & \ddots & \ddots & \ddots & \ddots & \vdots & 0 & \ddots & \ddots & \ddots & \ddots & 1 & 0 & \cdots & 0 & 0 & \cdots & 0 & 1 & \cdots & 1 & \cdots & \cdots & 1 & \cdots & 1 & 1 & \cdots & 1 \\
\vdots & \vdots & \ddots & \ddots & \ddots & \ddots & \vdots & 1 & \ddots & \ddots & \ddots & \vdots & \vdots & 1 & \cdots & 1 & 0 & \cdots & 0 & 0 & \cdots & 0 & 1 & \cdots & 1 & \cdots & \cdots & 1 & \cdots & 1 \\
\vdots & \vdots & \ddots & \ddots & \ddots & \ddots & \vdots & \vdots & \ddots & \ddots & \ddots & \vdots & \vdots & 1 & \cdots & 1 & 1 & \cdots & 1 & 0 & \cdots & 0 & 0 & \cdots & 0 & 1 & \cdots & 1 & \cdots & \cdots \\
\vdots & \vdots & \ddots & \ddots & \ddots & \ddots & 1 & \vdots & \ddots & \ddots & \ddots & \ddots & 1 & \vdots & \cdots & \cdots & \cdots & \cdots & \cdots & \cdots & \cdots & \cdots & \cdots & \cdots & \cdots & \cdots & \cdots & \cdots & \cdots & \vdots \\
-1 & 1 & \cdots & \cdots & \cdots & 1 & (2n+5) & 1 & \cdots & \cdots & 1 & 0 & 0 & 1 & \cdots & 1 & 1 & \cdots & 1 & \cdots & \cdots & 1 & \cdots & 1 & 0 & \cdots & 0 & 0 & \cdots & 0 \\
1 & 0 & 0 & 1 & \cdots & \cdots & 1 & 3 & 1 & \cdots & \cdots & \cdots & 1 & 1 & \cdots & 1 & 1 & \cdots & 1 & \cdots & \cdots & \cdots & \cdots & \cdots & 1 & \cdots & 1 & 1 & \cdots & 1 \\
\vdots & 1 & \ddots & \ddots & \ddots & \ddots & \vdots & 1 & \ddots & \ddots & \ddots & \ddots & \vdots & 1 & \cdots & 1 & 1 & \cdots & 1 & \cdots & \cdots & \cdots & \cdots & \cdots & 1 & \cdots & 1 & 1 & \cdots & 1 \\
\vdots & \vdots & \ddots & \ddots & \ddots & \ddots & \vdots & \vdots & \ddots & \ddots & \ddots & \ddots & \vdots & \vdots & \cdots & \vdots & \vdots & \cdots & \vdots & \cdots & \cdots & \cdots & \cdots & \cdots & \vdots & \cdots & \vdots & \vdots & \cdots & 1 \\
\vdots & \vdots & \ddots & \ddots & \ddots & \ddots & 1 & \vdots & \ddots & \ddots & \ddots & \ddots & \vdots & \vdots & \cdots & \vdots & \vdots & \cdots & \vdots & \cdots & \cdots & \cdots & \cdots & \cdots & \vdots & \cdots & \vdots & \vdots & \cdots & \vdots \\
\vdots & 1 & \ddots & \ddots & \ddots & \ddots & 0 & \vdots & \ddots & \ddots & \ddots & \ddots & \vdots & 1 & \cdots & 1 & 1 & \cdots & 1 & \cdots & \cdots & \cdots & \cdots & \cdots & 1 & \cdots & 1 & 1 & \cdots & 1 \\
1 & 0 & 1 & \cdots & \cdots & 1 & 0 & 1 & \cdots & \cdots & \cdots & 1 & 3 & 1 & \cdots & 1 & 1 & \cdots & 1 & \cdots & \cdots & \cdots & \cdots & \cdots & 1 & \cdots & 1 & 1 & \cdots & 1 \\
1 & 0 & 0 & 1 & \cdots & 1 & 0 & 1 & 1 & \cdots & \cdots & 1 & 1 & 3 & 1 & \cdots & \cdots & \cdots & \cdots & \cdots & \cdots & \cdots & \cdots & \cdots & \cdots & \cdots & \cdots & \cdots & \cdots & 1 \\
\vdots & \vdots & \vdots & \vdots & \vdots & \vdots & \vdots & \vdots & \vdots & \vdots & \vdots & \vdots & \vdots & 1 & \ddots & \ddots & \ddots & \ddots & \ddots & \ddots & \ddots & \ddots & \ddots & \ddots & \ddots & \ddots & \ddots & \ddots & \ddots & \vdots \\
\vdots & 0 & 0 & 1 & \vdots & 1 & \vdots & 1 & 1 & \cdots & \cdots & 1 & 1 & \vdots & \ddots & \ddots & \ddots & \ddots & \ddots & \ddots & \ddots & \ddots & \ddots & \ddots & \ddots & \ddots & \ddots & \ddots & \ddots & \vdots \\
\vdots & 1 & 0 & 0 & \vdots & 1 & \vdots & 1 & 1 & \cdots & \cdots & 1 & 1 & \vdots & \ddots & \ddots & \ddots & \ddots & \ddots & \ddots & \ddots & \ddots & \ddots & \ddots & \ddots & \ddots & \ddots & \ddots & \ddots & \vdots \\
\vdots & \vdots & \vdots & \vdots & \vdots & \vdots & \vdots & \vdots & \vdots & \vdots & \vdots & \vdots & \vdots & \vdots & \ddots & \ddots & \ddots & \ddots & \ddots & \ddots & \ddots & \ddots & \ddots & \ddots & \ddots & \ddots & \ddots & \ddots & \ddots & \vdots \\
\vdots & 1 & 0 & 0 & \vdots & 1 & \vdots & 1 & 1 & \cdots & \cdots & 1 & 1 & \vdots & \ddots & \ddots & \ddots & \ddots & \ddots & \ddots & \ddots & \ddots & \ddots & \ddots & \ddots & \ddots & \ddots & \ddots & \ddots & \vdots \\
\vdots & 1 & 1 & 0 & \vdots & \vdots & \vdots & \vdots & \vdots & \vdots & \vdots & \vdots & \vdots & \vdots & \ddots & \ddots & \ddots & \ddots & \ddots & \ddots & \ddots & \ddots & \ddots & \ddots & \ddots & \ddots & \ddots & \ddots & \ddots & \vdots \\
\vdots & \vdots & \vdots & \vdots & \vdots & \vdots & \vdots & \vdots & \vdots & \vdots & \vdots & \vdots & \vdots & \vdots & \ddots & \ddots & \ddots & \ddots & \ddots & \ddots & \ddots & \ddots & \ddots & \ddots & \ddots & \ddots & \ddots & \ddots & \ddots & \vdots \\
\vdots & 1 & 1 & 0 & \vdots & 1 & \vdots & \vdots & \vdots & \vdots & \vdots & \vdots & \vdots & \vdots & \ddots & \ddots & \ddots & \ddots & \ddots & \ddots & \ddots & \ddots & \ddots & \ddots & \ddots & \ddots & \ddots & \ddots & \ddots & \vdots \\
\vdots & \vdots & \vdots & 1 & \vdots & \vdots & \vdots & \vdots & \vdots & \vdots & \vdots & \vdots & \vdots & \vdots & \ddots & \ddots & \ddots & \ddots & \ddots & \ddots & \ddots & \ddots & \ddots & \ddots & \ddots & \ddots & \ddots & \ddots & \ddots & \vdots \\
\vdots & \vdots & \vdots & \vdots & \vdots & 1 & \vdots & \vdots & \vdots & \vdots & \vdots & \vdots & \vdots & \vdots & \ddots & \ddots & \ddots & \ddots & \ddots & \ddots & \ddots & \ddots & \ddots & \ddots & \ddots & \ddots & \ddots & \ddots & \ddots & \vdots \\
\vdots & 1 & 1 & 1 & \vdots & 0 & \vdots & \vdots & \vdots & \vdots & \vdots & \vdots & \vdots & \vdots & \ddots & \ddots & \ddots & \ddots & \ddots & \ddots & \ddots & \ddots & \ddots & \ddots & \ddots & \ddots & \ddots & \ddots & \ddots & \vdots \\
\vdots & \vdots & \vdots & \vdots & \vdots & \vdots & \vdots & \vdots & \vdots & \vdots & \vdots & \vdots & \vdots & \vdots & \ddots & \ddots & \ddots & \ddots & \ddots & \ddots & \ddots & \ddots & \ddots & \ddots & \ddots & \ddots & \ddots & \ddots & \ddots & \vdots \\
\vdots & 1 & 1 & \vdots & \vdots & 0 & \vdots & 1 & 1 & \cdots & \cdots & 1 & 1 & \vdots & \ddots & \ddots & \ddots & \ddots & \ddots & \ddots & \ddots & \ddots & \ddots & \ddots & \ddots & \ddots & \ddots & \ddots & \ddots & \vdots \\
\vdots & 0 & 1 & 1 & \vdots & 0 & \vdots & 1 & 1 & \cdots & \cdots & 1 & 1 & \vdots & \ddots & \ddots & \ddots & \ddots & \ddots & \ddots & \ddots & \ddots & \ddots & \ddots & \ddots & \ddots & \ddots & \ddots & \ddots & \vdots \\
\vdots & \vdots & \vdots & \vdots & \vdots & \vdots & \vdots & \vdots & \vdots & \vdots & \vdots & \vdots & \vdots & \vdots & \ddots & \ddots & \ddots & \ddots & \ddots & \ddots & \ddots & \ddots & \ddots & \ddots & \ddots & \ddots & \ddots & \ddots & \ddots & 1 \\
1 & 0 & 1 & 1 & \cdots & 0 & \cdots & 1 & 1 & \cdots & \cdots & 1 & 1 & 1 & \cdots & \cdots & \cdots & \cdots & \cdots & \cdots & \cdots & \cdots & \cdots & \cdots & \cdots & \cdots & \cdots & \cdots & 1 & 3
\end{pmatrix}$$

Let $j = (1 \cdots 1)$ be the $1 \times n$ matrix with all one, and J_n be the $n \times n$ matrix with all one. Set $a = 2n+5$ and $b = mn+2m+1$. Then we get:

$$\tau\left(B_n^{(m)}\right) = \frac{1}{b^2} \times \det\begin{pmatrix}
2m+1 & -1 & \cdots & \cdots & \cdots & \cdots & -1 & 1 & \cdots & \cdots & \cdots & \cdots & 1 & j & \cdots & \cdots & \cdots & \cdots & j \\
-1 & a & 1 & \cdots & \cdots & \cdots & 1 & 0 & 1 & \cdots & \cdots & 1 & 0 & 0 & j & \cdots & \cdots & j & 0 \\
\vdots & 1 & \ddots & \ddots & \ddots & \ddots & \vdots & 0 & 0 & \ddots & \ddots & \ddots & 1 & 0 & \ddots & \ddots & \ddots & \ddots & j \\
\vdots & \vdots & \ddots & \ddots & \ddots & \ddots & \vdots & 1 & 0 & \ddots & \ddots & \ddots & \vdots & j & \ddots & \ddots & \ddots & \ddots & \vdots \\
\vdots & \vdots & \ddots & \ddots & \ddots & \ddots & \vdots & \vdots & \ddots & \ddots & \ddots & \ddots & \vdots & \vdots & \ddots & \ddots & \ddots & \ddots & \vdots \\
\vdots & \vdots & \ddots & \ddots & \ddots & \ddots & 1 & \vdots & \ddots & \ddots & 0 & 0 & 1 & \vdots & \ddots & \ddots & \ddots & \ddots & j \\
-1 & 1 & \cdots & \cdots & \cdots & 1 & a & 1 & \cdots & \cdots & 1 & 0 & 0 & j & \cdots & \cdots & j & 0 & 0 \\
1 & 0 & 0 & 1 & \cdots & \cdots & 1 & 3 & 1 & \cdots & \cdots & \cdots & 1 & j & \cdots & \cdots & \cdots & \cdots & j \\
\vdots & 1 & 0 & \ddots & \ddots & \ddots & 1 & 1 & \ddots & \ddots & \ddots & \ddots & \vdots & \vdots & \ddots & \ddots & \ddots & \ddots & \vdots \\
\vdots & \vdots & \ddots & \ddots & \ddots & \ddots & \vdots & \vdots & \ddots & \ddots & \ddots & \ddots & \vdots & \vdots & \ddots & \ddots & \ddots & \ddots & \vdots \\
\vdots & \vdots & \ddots & \ddots & \ddots & \ddots & 1 & \vdots & \ddots & \ddots & \ddots & \ddots & \vdots & \vdots & \ddots & \ddots & \ddots & \ddots & \vdots \\
\vdots & 1 & \ddots & \ddots & \ddots & \ddots & 0 & \vdots & \ddots & \ddots & \ddots & \ddots & 1 & \vdots & \ddots & \ddots & \ddots & \ddots & \vdots \\
1 & 0 & 1 & \cdots & \cdots & 1 & 0 & 1 & \cdots & \cdots & \cdots & 1 & 3 & j & \cdots & \cdots & \cdots & \cdots & j \\
j^t & 0 & 0 & j^t & \cdots & \cdots & j^t & j^t & \cdots & \cdots & \cdots & \cdots & j^t & & & & & & \\
\vdots & j^t & 0 & \ddots & \ddots & \ddots & \vdots & \vdots & \ddots & \ddots & \ddots & \ddots & \vdots & & & & & & \\
\vdots & \vdots & \ddots & \ddots & \ddots & \ddots & \vdots & \vdots & \ddots & \ddots & \ddots & \ddots & \vdots & & & 2I_{mn} + J_{mn} & & & \\
\vdots & \vdots & \ddots & \ddots & \ddots & \ddots & j^t & \vdots & \ddots & \ddots & \ddots & \ddots & \vdots & & & & & & \\
\vdots & j^t & \ddots & \ddots & \ddots & \ddots & 0 & \vdots & \ddots & \ddots & \ddots & \ddots & \vdots & & & & & & \\
j^t & 0 & j^t & \cdots & \cdots & j^t & 0 & j^t & \cdots & \cdots & \cdots & \cdots & j^t & & & & & &
\end{pmatrix}$$

$$
= \frac{1}{b^2} \times \det \begin{pmatrix}
b & -1 & \cdots & \cdots & \cdots & \cdots & -1 & 1 & \cdots & \cdots & \cdots & \cdots & 1 & j & \cdots & \cdots & \cdots & \cdots & j \\
b & a & 1 & \cdots & \cdots & \cdots & 1 & 0 & 1 & \cdots & \cdots & 1 & 0 & 0 & j & \cdots & \cdots & j & 0 \\
\vdots & 1 & \ddots & \ddots & \ddots & \ddots & \vdots & 0 & 0 & \ddots & \ddots & \ddots & 1 & 0 & \ddots & \ddots & \ddots & \ddots & j \\
\vdots & \vdots & \ddots & \ddots & \ddots & \ddots & \vdots & 1 & 0 & \ddots & \ddots & \ddots & \vdots & j & \ddots & \ddots & \ddots & \ddots & \vdots \\
\vdots & \vdots & \ddots & \ddots & \ddots & \ddots & \vdots & \vdots & \ddots & \ddots & \ddots & \ddots & \vdots & \vdots & \ddots & \ddots & \ddots & \ddots & \vdots \\
\vdots & \vdots & \ddots & \ddots & \ddots & \ddots & 1 & \vdots & \ddots & \ddots & 0 & 0 & 1 & \vdots & \ddots & \ddots & \ddots & \ddots & j \\
b & 1 & \cdots & \cdots & \cdots & 1 & a & 1 & \cdots & \cdots & 1 & 0 & 0 & j & \cdots & \cdots & j & 0 & 0 \\
b & 0 & 0 & 1 & \cdots & \cdots & 1 & 3 & 1 & \cdots & \cdots & \cdots & 1 & j & \cdots & \cdots & \cdots & \cdots & j \\
\vdots & 1 & 0 & \ddots & \ddots & \ddots & 1 & 1 & \ddots & \ddots & \ddots & \ddots & \vdots & \vdots & \ddots & \ddots & \ddots & \ddots & \vdots \\
\vdots & \vdots & \ddots & \ddots & \ddots & \ddots & \vdots & \vdots & \ddots & \ddots & \ddots & \ddots & \vdots & \vdots & \ddots & \ddots & \ddots & \ddots & \vdots \\
\vdots & \vdots & \ddots & \ddots & \ddots & \ddots & 1 & \vdots & \ddots & \ddots & \ddots & \ddots & \vdots & \vdots & \ddots & \ddots & \ddots & \ddots & \vdots \\
\vdots & 1 & \ddots & \ddots & \ddots & \ddots & 0 & \vdots & \ddots & \ddots & \ddots & \ddots & 1 & \vdots & \ddots & \ddots & \ddots & \ddots & \vdots \\
b & 0 & 1 & \cdots & \cdots & 1 & 0 & 1 & \cdots & \cdots & \cdots & 1 & 3 & j & \cdots & \cdots & \cdots & \cdots & j \\
bj^t & 0 & 0 & j^t & \cdots & \cdots & j^t & j^t & \cdots & \cdots & \cdots & \cdots & j^t & & & & & & \\
\vdots & j^t & 0 & \ddots & \ddots & \ddots & \vdots & \vdots & \ddots & \ddots & \ddots & \ddots & \vdots & & & & & & \\
\vdots & \vdots & \ddots & \ddots & \ddots & \ddots & \vdots & \vdots & \ddots & \ddots & \ddots & \ddots & \vdots & & & 2I_{m n} + J_{m n} & & & \\
\vdots & \vdots & \ddots & \ddots & \ddots & \ddots & j^t & \vdots & \ddots & \ddots & \ddots & \ddots & \vdots & & & & & & \\
\vdots & j^t & \ddots & \ddots & \ddots & \ddots & 0 & \vdots & \ddots & \ddots & \ddots & \ddots & \vdots & & & & & & \\
bj^t & 0 & j^t & \cdots & \cdots & j^t & 0 & j^t & \cdots & \cdots & \cdots & \cdots & j^t & & & & & &
\end{pmatrix}
$$

$$
= \frac{1}{b} \times \det \begin{pmatrix}
1 & -1 & \cdots & \cdots & \cdots & \cdots & -1 & 1 & \cdots & \cdots & \cdots & \cdots & 1 & j & \cdots & \cdots & \cdots & \cdots & j \\
1 & a & 1 & \cdots & \cdots & \cdots & 1 & 0 & 1 & \cdots & \cdots & 1 & 0 & 0 & j & \cdots & \cdots & j & 0 \\
\vdots & 1 & \ddots & \ddots & \ddots & \ddots & \vdots & 0 & 0 & \ddots & \ddots & \ddots & 1 & 0 & \ddots & \ddots & \ddots & \ddots & j \\
\vdots & \vdots & \ddots & \ddots & \ddots & \ddots & \vdots & 1 & 0 & \ddots & \ddots & \ddots & \vdots & j & \ddots & \ddots & \ddots & \ddots & \vdots \\
\vdots & \vdots & \ddots & \ddots & \ddots & \ddots & \vdots & \vdots & \ddots & \ddots & \ddots & \ddots & \vdots & \vdots & \ddots & \ddots & \ddots & \ddots & \vdots \\
\vdots & \vdots & \ddots & \ddots & \ddots & \ddots & 1 & \vdots & \ddots & \ddots & 0 & 0 & 1 & \vdots & \ddots & \ddots & \ddots & \ddots & j \\
1 & 1 & \cdots & \cdots & \cdots & 1 & a & 1 & \cdots & \cdots & 1 & 0 & 0 & j & \cdots & \cdots & j & 0 & 0 \\
1 & 0 & 0 & 1 & \cdots & \cdots & 1 & 3 & 1 & \cdots & \cdots & \cdots & 1 & j & \cdots & \cdots & \cdots & \cdots & j \\
\vdots & 1 & 0 & \ddots & \ddots & \ddots & 1 & 1 & \ddots & \ddots & \ddots & \ddots & \vdots & \vdots & \ddots & \ddots & \ddots & \ddots & \vdots \\
\vdots & \vdots & \ddots & \ddots & \ddots & \ddots & \vdots & \vdots & \ddots & \ddots & \ddots & \ddots & \vdots & \vdots & \ddots & \ddots & \ddots & \ddots & \vdots \\
\vdots & \vdots & \ddots & \ddots & \ddots & \ddots & 1 & \vdots & \ddots & \ddots & \ddots & \ddots & \vdots & \vdots & \ddots & \ddots & \ddots & \ddots & \vdots \\
\vdots & 1 & \ddots & \ddots & \ddots & \ddots & 0 & \vdots & \ddots & \ddots & \ddots & \ddots & 1 & \vdots & \ddots & \ddots & \ddots & \ddots & \vdots \\
1 & 0 & 1 & \cdots & \cdots & 1 & 0 & 1 & \cdots & \cdots & \cdots & 1 & 3 & j & \cdots & \cdots & \cdots & \cdots & j \\
1j^t & 0 & 0 & j^t & \cdots & \cdots & j^t & j^t & \cdots & \cdots & \cdots & \cdots & j^t & & & & & & \\
\vdots & j^t & 0 & \ddots & \ddots & \ddots & \vdots & \vdots & \ddots & \ddots & \ddots & \ddots & \vdots & & & & & & \\
\vdots & \vdots & \ddots & \ddots & \ddots & \ddots & \vdots & \vdots & \ddots & \ddots & \ddots & \ddots & \vdots & & & 2I_{m n} + J_{m n} & & & \\
\vdots & \vdots & \ddots & \ddots & \ddots & \ddots & j^t & \vdots & \ddots & \ddots & \ddots & \ddots & \vdots & & & & & & \\
\vdots & j^t & \ddots & \ddots & \ddots & \ddots & 0 & \vdots & \ddots & \ddots & \ddots & \ddots & \vdots & & & & & & \\
1j^t & 0 & j^t & \cdots & \cdots & j^t & 0 & j^t & \cdots & \cdots & \cdots & \cdots & j^t & & & & & &
\end{pmatrix}
$$

$$
= \frac{1}{b} \times \det \begin{pmatrix}
1 & -1 & \cdots & \cdots & \cdots & \cdots & -1 & 1 & \cdots & \cdots & \cdots & \cdots & 1 & j & \cdots & \cdots & \cdots & \cdots & j \\
0 & (a+1) & 2 & \cdots & \cdots & \cdots & 2 & -1 & 0 & \cdots & \cdots & 0 & -1 & -j & 0 & \cdots & \cdots & 0 & -j \\
\vdots & 2 & \ddots & \ddots & \ddots & \ddots & \vdots & -1 & -1 & \ddots & \ddots & \ddots & 0 & -j & \ddots & \ddots & \ddots & \ddots & 0 \\
\vdots & \vdots & \ddots & \ddots & \ddots & \ddots & \vdots & 0 & \ddots & \ddots & \ddots & \ddots & \vdots & 0 & \ddots & \ddots & \ddots & \ddots & \vdots \\
\vdots & \vdots & \ddots & \ddots & \ddots & \ddots & \vdots & \vdots & \ddots & \ddots & \ddots & \ddots & \vdots & \vdots & \ddots & \ddots & \ddots & \ddots & \vdots \\
\vdots & \vdots & \ddots & \ddots & \ddots & \ddots & 2 & \vdots & \ddots & \ddots & \ddots & -1 & 0 & \vdots & \ddots & \ddots & \ddots & \ddots & 0 \\
0 & 2 & \cdots & \cdots & \cdots & 2 & (a+1) & 0 & \cdots & \cdots & 0 & -1 & -1 & 0 & \cdots & \cdots & 0 & -j & -j \\
0 & 1 & 1 & 2 & \cdots & \cdots & 2 & 2 & 0 & \cdots & \cdots & \cdots & 0 & 0 & \cdots & \cdots & \cdots & \cdots & 0 \\
\vdots & 2 & 1 & \ddots & \ddots & \ddots & 2 & 0 & \ddots & \ddots & \ddots & \ddots & \vdots & \vdots & \ddots & \ddots & \ddots & \ddots & \vdots \\
\vdots & \vdots & \ddots & \ddots & \ddots & \ddots & \vdots & \vdots & \ddots & \ddots & \ddots & \ddots & \vdots & \vdots & \ddots & \ddots & \ddots & \ddots & \vdots \\
\vdots & \vdots & \ddots & \ddots & \ddots & \ddots & 2 & \vdots & \ddots & \ddots & \ddots & \ddots & \vdots & \vdots & \ddots & \ddots & \ddots & \ddots & \vdots \\
\vdots & 2 & \ddots & \ddots & \ddots & \ddots & 1 & \vdots & \ddots & \ddots & \ddots & \ddots & 0 & \vdots & \ddots & \ddots & \ddots & \ddots & \vdots \\
0 & 1 & 2 & \cdots & \cdots & 2 & 1 & 0 & \cdots & \cdots & \cdots & 0 & 2 & 0 & \cdots & \cdots & \cdots & \cdots & 0 \\
0 & j^t & j^t & 2j^t & \cdots & \cdots & 2j^t & 0 & \cdots & \cdots & \cdots & \cdots & 0 & & & & & & \\
\vdots & 2j^t & j^t & \ddots & \ddots & \ddots & \vdots & \vdots & \ddots & \ddots & \ddots & \ddots & \vdots & & & & & & \\
\vdots & \vdots & \ddots & \ddots & \ddots & \ddots & \vdots & \vdots & \ddots & \ddots & \ddots & \ddots & \vdots & & & 2I_{m n} & & & \\
\vdots & \vdots & \ddots & \ddots & \ddots & \ddots & 2j^t & \vdots & \ddots & \ddots & \ddots & \ddots & \vdots & & & & & & \\
\vdots & 2j^t & \ddots & \ddots & \ddots & \ddots & j^t & \vdots & \ddots & \ddots & \ddots & \ddots & \vdots & & & & & & \\
0 & j^t & 2j^t & \cdots & \cdots & 2j^t & j^t & 0 & \cdots & \cdots & \cdots & \cdots & 0 & & & & & &
\end{pmatrix}
$$

$$= \frac{1}{b} \times \det \left(\begin{array}{cccccccccccccccccc}
(a+1) & 2 & \cdots & \cdots & \cdots & 2 & -1 & 0 & \cdots & \cdots & 0 & -1 & -j & 0 & \cdots & \cdots & 0 & -j \\
2 & \ddots & \ddots & \ddots & \ddots & \vdots & -1 & -1 & \ddots & \ddots & \ddots & 0 & -j & \ddots & \ddots & \ddots & \ddots & 0 \\
\vdots & \ddots & \ddots & \ddots & \ddots & \vdots & 0 & \ddots & \ddots & \ddots & \ddots & \vdots & 0 & \ddots & \ddots & \ddots & \ddots & \vdots \\
\vdots & \ddots & \ddots & \ddots & \ddots & \vdots & \vdots & \ddots & \ddots & \ddots & \ddots & \vdots & \vdots & \ddots & \ddots & \ddots & \ddots & \vdots \\
\vdots & \ddots & \ddots & \ddots & \ddots & 2 & \vdots & \ddots & \ddots & \ddots & -1 & 0 & \vdots & \ddots & \ddots & \ddots & \ddots & 0 \\
2 & \cdots & \cdots & \cdots & 2 & (a+1) & 0 & \cdots & \cdots & 0 & -1 & -1 & 0 & \cdots & \cdots & 0 & -j & -j \\
1 & 1 & 2 & \cdots & \cdots & 2 & 2 & 0 & \cdots & \cdots & \cdots & 0 & 0 & \cdots & \cdots & \cdots & \cdots & 0 \\
2 & 1 & \ddots & \ddots & \ddots & 2 & 0 & \ddots & \ddots & \ddots & \ddots & \vdots & \vdots & \ddots & \ddots & \ddots & \ddots & \vdots \\
\vdots & \ddots & \ddots & \ddots & \ddots & \vdots & \vdots & \ddots & \ddots & \ddots & \ddots & \vdots & \vdots & \ddots & \ddots & \ddots & \ddots & \vdots \\
\vdots & \ddots & \ddots & \ddots & \ddots & 2 & \vdots & \ddots & \ddots & \ddots & \ddots & \vdots & \vdots & \ddots & \ddots & \ddots & \ddots & \vdots \\
2 & \ddots & \ddots & \ddots & \ddots & 1 & \vdots & \ddots & \ddots & \ddots & \ddots & 0 & \vdots & \ddots & \ddots & \ddots & \ddots & \vdots \\
1 & 2 & \cdots & \cdots & 2 & 1 & 0 & \cdots & \cdots & \cdots & 0 & 2 & 0 & \cdots & \cdots & \cdots & \cdots & 0 \\
j^t & j^t & 2j^t & \cdots & \cdots & 2j^t & 0 & \cdots & \cdots & \cdots & \cdots & 0 & & & & & & \\
2j^t & j^t & \ddots & \ddots & \ddots & \vdots & \vdots & \ddots & \ddots & \ddots & \ddots & \vdots & & & & & & \\
\vdots & \ddots & \ddots & \ddots & \ddots & \vdots & \vdots & \ddots & \ddots & \ddots & \ddots & \vdots & & & & 2I_{mn} & & \\
\vdots & \ddots & \ddots & \ddots & \ddots & 2j^t & \vdots & \ddots & \ddots & \ddots & \ddots & \vdots & & & & & & \\
2j^t & \ddots & \ddots & \ddots & \ddots & j^t & \vdots & \ddots & \ddots & \ddots & \ddots & \vdots & & & & & & \\
j^t & 2j^t & \cdots & \cdots & 2j^t & j^t & 0 & \cdots & \cdots & \cdots & \cdots & 0 & & & & & &
\end{array} \right)$$

Using Lemma 3, yields

$$\tau\,(B_n^{(m)}) = \frac{1}{b} \times \det \left(\begin{array}{cc} A & B \\ C & 2I_{mn} \end{array} \right) = \frac{1}{b} \times \det(A - B\frac{1}{2I_{mn}}\ C) \times 2^{mn}$$

$$= \frac{1}{b} 2^{mn} \times 2^{-2m} \times \det \left(\begin{array}{cccccccccccc}
(2a+2n+2) & 3n+4 & 4(n+1) & \cdots & 4(n+1) & 3n+4 & -2 & 0 & \cdots & \cdots & 0 & -2 \\
3n+4 & (2a+2n+2) & 3n+4 & 4(n+1) & \cdots & 4(n+1) & -2 & \ddots & \ddots & \ddots & \ddots & 0 \\
4(n+1) & 3n+4 & \ddots & \ddots & \ddots & \vdots & 0 & \ddots & \ddots & \ddots & \ddots & \vdots \\
\vdots & \ddots & \ddots & \ddots & \ddots & 4(n+1) & \vdots & \ddots & \ddots & \ddots & \ddots & \vdots \\
4(n+1) & \ddots & \ddots & \ddots & \ddots & 3n+4 & \vdots & \ddots & \ddots & \ddots & \ddots & 0 \\
3n+4 & 4(n+1) & \cdots & 4(n+1) & 3n+4 & (2a+2n+2) & 0 & \cdots & \cdots & 0 & -2 & -2 \\
2 & 2 & 4 & \cdots & \cdots & 4 & 4 & 0 & \cdots & \cdots & \cdots & 0 \\
4 & \ddots & \ddots & \ddots & \ddots & \vdots & 0 & \ddots & \ddots & \ddots & \ddots & \vdots \\
\vdots & \ddots & \ddots & \ddots & \ddots & \vdots & \vdots & \ddots & \ddots & \ddots & \ddots & \vdots \\
\vdots & \ddots & \ddots & \ddots & \ddots & 4 & \vdots & \ddots & \ddots & \ddots & \ddots & \vdots \\
4 & \ddots & \ddots & \ddots & \ddots & 2 & \vdots & \ddots & \ddots & \ddots & \ddots & 0 \\
2 & 4 & \cdots & \cdots & 4 & 2 & 0 & \cdots & \cdots & \cdots & 0 & 4
\end{array} \right)$$

Using Lemma 3 again, yields

$$\tau\,(B_n^{(m)}) = \frac{2^{mn-2m}}{b} \times \det \left(\begin{array}{cc} D & E \\ F & 4I_m \end{array} \right) = \frac{2^{mn}}{b} \times \det(D - E\frac{1}{4I_m}\ F)$$

$$\tau\,(B_n^{(m)}) = \frac{2^{mn}}{b} \times \det \left(\begin{array}{cccccc}
(2a+2n+4) & (3n+7) & 4(n+2) & \cdots & 4(n+2) & (3n+7) \\
(3n+7) & (2a+2n+4) & (3n+7) & \ddots & \cdots & 4(n+2) \\
4(n+2) & \ddots & \ddots & \ddots & \ddots & \vdots \\
\vdots & \ddots & \ddots & \ddots & \ddots & 4(n+2) \\
4(n+2) & \ddots & \ddots & \ddots & \ddots & (3n+7) \\
(3n+7) & 4(n+2) & \cdots & 4(n+2) & (3n+7) & (2a+2n+4)
\end{array} \right)$$

With a straightforward inducement using properties of determinants, we obtain

$$\tau\,(B_n^{(m)}) = \frac{2^{mn}}{b} \times \frac{4b}{mn+m+4} \times \det \left(\begin{array}{cccccc}
(2a-n-3) & 0 & (n+1) & \cdots & (n+1) & 0 \\
0 & (2a-n-3) & 0 & \ddots & \cdots & (n+1) \\
(n+1) & \ddots & \ddots & \ddots & \ddots & \vdots \\
\vdots & \ddots & \ddots & \ddots & \ddots & (n+1) \\
(n+1) & \ddots & \ddots & \ddots & \ddots & 0 \\
0 & (n+1) & \cdots & (n+1) & 0 & (2a-n-3)
\end{array} \right)$$

$$= \frac{2^{mn+2} \times (n+1)^m}{mn+m+4} \times \det \begin{pmatrix} \frac{(2a-n-3)}{(n+1)} & 0 & 1 & \cdots & 1 & 0 \\ 0 & \frac{(2a-n-3)}{(n+1)} & 0 & \ddots & \ddots & 1 \\ 1 & \ddots & \ddots & \ddots & \ddots & \vdots \\ \vdots & \ddots & \ddots & \ddots & \ddots & 1 \\ 1 & \ddots & \ddots & \ddots & \ddots & 0 \\ 0 & 1 & \cdots & 1 & 0 & \frac{(2a-n-3)}{(n+1)} \end{pmatrix}$$

Using Lemma 2, yields

$$\tau(B_n^{(m)}) = 2^{mn+2} \times \frac{(n+1)^m}{mn+m+4} \times \frac{2(\frac{2a-n-3}{n+1}+m-3)}{\frac{2a-n-3}{n+1}-3} \times [T_m(\frac{\frac{2a-n-3}{n+1}-1}{2}) - 1]$$
$$= 2^{mn+1} \times (n+1)^m \times [T_m(\tfrac{n+3}{n+1}) - 1].$$

Using Equation (11), yields the result. □

Definition 3. *The pyramid graph $C_n^{(m)}$ is the graph created from the gear graph G_{m+1} with vertices $\{u_0;\ u_1, u_2, \ldots, u_m;\ w_1, w_2, \ldots, w_m\}$ with double external edges and m sets of vertices, say, $\{v_1^1, v_2^1, \ldots, v_n^1\}$, $\{v_1^2, u_2^2, \ldots, v_n^2\}, \ldots, \{v_1^m, v_2^m, \ldots, v_n^m\}$, such that for all $i = 1, 2, \ldots, n$ the vertex v_i^j is adjacent to u_j and u_{j+1}, where $j = 1, 2, \ldots, m-1$, and v_i^m is adjacent to u_1 and u_m. See Figure 3.*

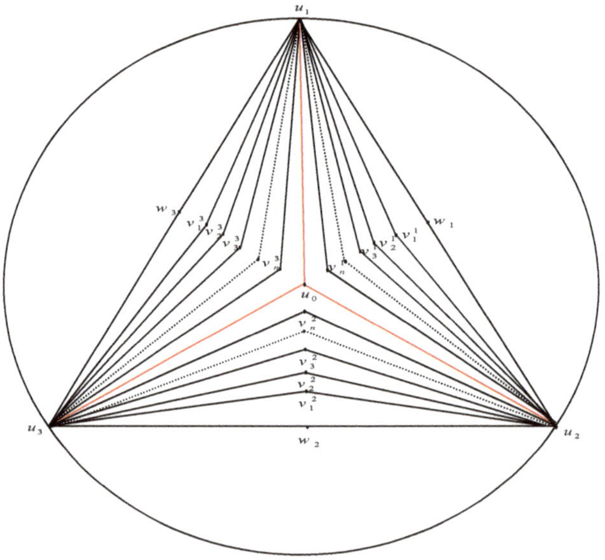

Figure 3. The pyramid graph $C_n^{(3)}$.

Theorem 3. *For $n \geq 0$, $m \geq 3$, $\tau(C_n^{(m)}) = 2^{mn}[(n+4+\sqrt{2n+7})^m + (n+4-\sqrt{2n+7})^m - 2(n+3)^m]$.*

Proof. Using Lemma 1, we have:

$$\tau(C_n^{(m)}) = \frac{1}{(mn+2m+1)^2} \times \det((mn+2m+1)\, I - D^c + A^c) = \frac{1}{(mn+2m+1)^2} \times$$

$$\det \begin{pmatrix}
(m+1) & 0 & 0 & \cdots & \cdots & \cdots & 0 & 1 & \cdots & \cdots & \cdots & \cdots & 1 & 1 & \cdots & \cdots & \cdots & \cdots & \cdots & \cdots & \cdots & \cdots & \cdots & \cdots & \cdots & \cdots & \cdots & \cdots & \cdots & 1 \\
0 & 2(n+3) & 0 & 1 & \cdots & 1 & 0 & 0 & 1 & \cdots & \cdots & 1 & 0 & 0 & \cdots & 0 & 1 & \cdots & 1 & \cdots & \cdots & 1 & \cdots & 1 & 1 & \cdots & 1 & 0 & \cdots & 0 \\
\vdots & 0 & \ddots & \ddots & \ddots & \ddots & 1 & 0 & \ddots & \ddots & \ddots & \ddots & 1 & 0 & \cdots & 0 & 0 & \cdots & 0 & 1 & \cdots & 1 & \cdots & \cdots & 1 & \cdots & 1 & 1 & \cdots & 1 \\
\vdots & 1 & \ddots & \ddots & \ddots & \ddots & \vdots & 1 & \ddots & \ddots & \ddots & \vdots & \vdots & 1 & \cdots & 1 & 0 & \cdots & 0 & 0 & \cdots & 0 & 1 & \cdots & 1 & \cdots & \cdots & 1 & \cdots & 1 \\
\vdots & \vdots & \ddots & \ddots & \ddots & \ddots & 1 & \vdots & \ddots & \ddots & \ddots & \vdots & \vdots & 1 & \cdots & 1 & 1 & \cdots & 1 & 0 & \cdots & 0 & 0 & \cdots & 0 & 1 & \cdots & 1 & \cdots & \cdots \\
\vdots & 1 & \ddots & \ddots & \ddots & \ddots & 0 & \vdots & \ddots & \ddots & \ddots & \ddots & 1 & \vdots & \cdots & \cdots & \cdots & \cdots & \cdots & \cdots & \cdots & \cdots & \cdots & \cdots & \cdots & \cdots & \cdots & \cdots & \cdots & \vdots \\
0 & 0 & 1 & \cdots & 1 & 0 & 2(n+3) & 1 & \cdots & \cdots & 1 & 0 & 0 & 1 & \cdots & 1 & 1 & \cdots & 1 & \cdots & \cdots & 1 & \cdots & 1 & 0 & \cdots & 0 & 0 & \cdots & 0 \\
1 & 0 & 0 & 1 & \cdots & \cdots & 1 & 3 & 1 & \cdots & \cdots & \cdots & 1 & 1 & \cdots & 1 & 1 & \cdots & 1 & \cdots & \cdots & \cdots & \cdots & \cdots & 1 & \cdots & 1 & 1 & \cdots & 1 \\
\vdots & 1 & \ddots & \ddots & \ddots & \ddots & \vdots & 1 & \ddots & \ddots & \ddots & \ddots & \vdots & 1 & \cdots & 1 & 1 & \cdots & 1 & \cdots & \cdots & \cdots & \cdots & \cdots & 1 & \cdots & 1 & 1 & \cdots & 1 \\
\vdots & \vdots & \ddots & \ddots & \ddots & \ddots & \vdots & \vdots & \ddots & \ddots & \ddots & \ddots & \vdots & \vdots & \cdots & \vdots & \vdots & \cdots & \vdots & \cdots & \cdots & \cdots & \cdots & \cdots & \vdots & \cdots & \vdots & \vdots & \cdots & 1 \\
\vdots & \vdots & \ddots & \ddots & \ddots & \ddots & 1 & \vdots & \ddots & \ddots & \ddots & \ddots & \vdots & \vdots & \cdots & \vdots & \vdots & \cdots & \vdots & \cdots & \cdots & \cdots & \cdots & \cdots & \vdots & \cdots & \vdots & \vdots & \cdots & \vdots \\
\vdots & 1 & \ddots & \ddots & \ddots & \ddots & 0 & \vdots & \ddots & \ddots & \ddots & \ddots & \vdots & 1 & \cdots & 1 & 1 & \cdots & 1 & \cdots & \cdots & \cdots & \cdots & \cdots & 1 & \cdots & 1 & 1 & \cdots & 1 \\
1 & 0 & 1 & \cdots & \cdots & 1 & 0 & 1 & \cdots & \cdots & \cdots & 1 & 3 & 1 & \cdots & 1 & 1 & \cdots & 1 & \cdots & \cdots & \cdots & \cdots & \cdots & 1 & \cdots & 1 & 1 & \cdots & 1 \\
1 & 0 & 0 & 1 & \cdots & 1 & 0 & 1 & 1 & \cdots & \cdots & 1 & 1 & 3 & 1 & \cdots & \cdots & \cdots & \cdots & \cdots & \cdots & \cdots & \cdots & \cdots & \cdots & \cdots & \cdots & \cdots & \cdots & 1 \\
\vdots & \vdots & \vdots & \vdots & \vdots & \vdots & \vdots & \vdots & \vdots & \vdots & \vdots & \vdots & \vdots & 1 & \ddots & \ddots & \ddots & \ddots & \ddots & \ddots & \ddots & \ddots & \ddots & \ddots & \ddots & \ddots & \ddots & \ddots & \ddots & \vdots \\
\vdots & 0 & 0 & 1 & \vdots & 1 & \vdots & 1 & 1 & \cdots & \cdots & 1 & 1 & \vdots & \ddots & \ddots & \ddots & \ddots & \ddots & \ddots & \ddots & \ddots & \ddots & \ddots & \ddots & \ddots & \ddots & \ddots & \ddots & \vdots \\
\vdots & 1 & 0 & 0 & \vdots & 1 & \vdots & 1 & 1 & \cdots & \cdots & 1 & 1 & \vdots & \ddots & \ddots & \ddots & \ddots & \ddots & \ddots & \ddots & \ddots & \ddots & \ddots & \ddots & \ddots & \ddots & \ddots & \ddots & \vdots \\
\vdots & \vdots & \vdots & \vdots & \vdots & \vdots & \vdots & \vdots & \vdots & \vdots & \vdots & \vdots & \vdots & \vdots & \ddots & \ddots & \ddots & \ddots & \ddots & \ddots & \ddots & \ddots & \ddots & \ddots & \ddots & \ddots & \ddots & \ddots & \ddots & \vdots \\
\vdots & 1 & 0 & 0 & \vdots & 1 & \vdots & 1 & 1 & \cdots & \cdots & 1 & 1 & \vdots & \ddots & \ddots & \ddots & \ddots & \ddots & \ddots & \ddots & \ddots & \ddots & \ddots & \ddots & \ddots & \ddots & \ddots & \ddots & \vdots \\
\vdots & 1 & 1 & 0 & \vdots & \vdots & \vdots & \vdots & \vdots & \vdots & \vdots & \vdots & \vdots & \vdots & \ddots & \ddots & \ddots & \ddots & \ddots & \ddots & \ddots & \ddots & \ddots & \ddots & \ddots & \ddots & \ddots & \ddots & \ddots & \vdots \\
\vdots & \vdots & \vdots & \vdots & \vdots & \vdots & \vdots & \vdots & \vdots & \vdots & \vdots & \vdots & \vdots & \vdots & \ddots & \ddots & \ddots & \ddots & \ddots & \ddots & \ddots & \ddots & \ddots & \ddots & \ddots & \ddots & \ddots & \ddots & \ddots & \vdots \\
\vdots & 1 & 1 & 0 & \vdots & 1 & \vdots & \vdots & \vdots & \vdots & \vdots & \vdots & \vdots & \vdots & \ddots & \ddots & \ddots & \ddots & \ddots & \ddots & \ddots & \ddots & \ddots & \ddots & \ddots & \ddots & \ddots & \ddots & \ddots & \vdots \\
\vdots & \vdots & \vdots & 1 & \vdots & \vdots & \vdots & \vdots & \vdots & \vdots & \vdots & \vdots & \vdots & \vdots & \ddots & \ddots & \ddots & \ddots & \ddots & \ddots & \ddots & \ddots & \ddots & \ddots & \ddots & \ddots & \ddots & \ddots & \ddots & \vdots \\
\vdots & \vdots & \vdots & \vdots & \vdots & 1 & \vdots & \vdots & \vdots & \vdots & \vdots & \vdots & \vdots & \vdots & \ddots & \ddots & \ddots & \ddots & \ddots & \ddots & \ddots & \ddots & \ddots & \ddots & \ddots & \ddots & \ddots & \ddots & \ddots & \vdots \\
\vdots & 1 & 1 & 1 & \vdots & 0 & \vdots & \vdots & \vdots & \vdots & \vdots & \vdots & \vdots & \vdots & \ddots & \ddots & \ddots & \ddots & \ddots & \ddots & \ddots & \ddots & \ddots & \ddots & \ddots & \ddots & \ddots & \ddots & \ddots & \vdots \\
\vdots & \vdots & \vdots & \vdots & \vdots & \vdots & \vdots & \vdots & \vdots & \vdots & \vdots & \vdots & \vdots & \vdots & \ddots & \ddots & \ddots & \ddots & \ddots & \ddots & \ddots & \ddots & \ddots & \ddots & \ddots & \ddots & \ddots & \ddots & \ddots & \vdots \\
\vdots & 1 & 1 & \vdots & \vdots & 0 & \vdots & 1 & 1 & \cdots & \cdots & 1 & 1 & \vdots & \ddots & \ddots & \ddots & \ddots & \ddots & \ddots & \ddots & \ddots & \ddots & \ddots & \ddots & \ddots & \ddots & \ddots & \ddots & \vdots \\
\vdots & 0 & 1 & 1 & \vdots & 0 & \vdots & 1 & 1 & \cdots & \cdots & 1 & 1 & \vdots & \ddots & \ddots & \ddots & \ddots & \ddots & \ddots & \ddots & \ddots & \ddots & \ddots & \ddots & \ddots & \ddots & \ddots & \ddots & \vdots \\
\vdots & \vdots & \vdots & \vdots & \vdots & \vdots & \vdots & \vdots & \vdots & \vdots & \vdots & \vdots & \vdots & \vdots & \ddots & \ddots & \ddots & \ddots & \ddots & \ddots & \ddots & \ddots & \ddots & \ddots & \ddots & \ddots & \ddots & \ddots & \ddots & 1 \\
1 & 0 & 1 & 1 & \cdots & 0 & \cdots & 1 & 1 & \cdots & \cdots & 1 & 1 & 1 & \cdots & \cdots & \cdots & \cdots & \cdots & \cdots & \cdots & \cdots & \cdots & \cdots & \cdots & \cdots & \cdots & \cdots & 1 & 3
\end{pmatrix}$$

Let $j = (1 \cdots 1)$ be the $1 \times n$ matrix with all one, and J_n be the $n \times n$ matrix with all one. Set $a = 2n + 6$ and $b = mn + 2m + 1$. Then we have:

$$\tau\left(C_n^{(m)}\right) = \frac{1}{b^2} \times \det \begin{pmatrix}
m+1 & 0 & \cdots & \cdots & \cdots & \cdots & 0 & 1 & \cdots & \cdots & \cdots & \cdots & 1 & j & \cdots & \cdots & \cdots & \cdots & j \\
0 & a & 0 & 1 & \cdots & 1 & 0 & 0 & 1 & \cdots & \cdots & 1 & 0 & 0 & j & \cdots & \cdots & j & 0 \\
\vdots & 0 & \ddots & \ddots & \ddots & \ddots & 1 & 0 & 0 & \ddots & \ddots & \ddots & 1 & 0 & \ddots & \ddots & \ddots & \ddots & j \\
\vdots & 1 & \ddots & \ddots & \ddots & \ddots & \vdots & 1 & 0 & \ddots & \ddots & \ddots & \vdots & j & \ddots & \ddots & \ddots & \ddots & \vdots \\
\vdots & \vdots & \ddots & \ddots & \ddots & \ddots & 1 & \vdots & \ddots & \ddots & \ddots & \ddots & \vdots & \vdots & \ddots & \ddots & \ddots & \ddots & \vdots \\
\vdots & 1 & \ddots & \ddots & \ddots & \ddots & 0 & \vdots & \ddots & \ddots & 0 & 0 & 1 & \vdots & \ddots & \ddots & \ddots & \ddots & j \\
0 & 0 & 1 & \cdots & 1 & 0 & a & 1 & \cdots & \cdots & 1 & 0 & 0 & j & \cdots & \cdots & j & 0 & 0 \\
1 & 0 & 0 & 1 & \cdots & \cdots & 1 & 3 & 1 & \cdots & \cdots & \cdots & 1 & j & \cdots & \cdots & \cdots & \cdots & j \\
\vdots & 1 & 0 & \ddots & \ddots & \ddots & 1 & 1 & \ddots & \ddots & \ddots & \ddots & \vdots & \vdots & \ddots & \ddots & \ddots & \ddots & \vdots \\
\vdots & \vdots & \ddots & \ddots & \ddots & \ddots & \vdots & \vdots & \ddots & \ddots & \ddots & \ddots & \vdots & \vdots & \ddots & \ddots & \ddots & \ddots & \vdots \\
\vdots & \vdots & \ddots & \ddots & \ddots & \ddots & 1 & \vdots & \ddots & \ddots & \ddots & \ddots & \vdots & \vdots & \ddots & \ddots & \ddots & \ddots & \vdots \\
\vdots & 1 & \ddots & \ddots & \ddots & \ddots & 0 & \vdots & \ddots & \ddots & \ddots & \ddots & 1 & \vdots & \ddots & \ddots & \ddots & \ddots & \vdots \\
1 & 0 & 1 & \cdots & \cdots & 1 & 0 & 1 & \cdots & \cdots & \cdots & 1 & 3 & j & \cdots & \cdots & \cdots & \cdots & j \\
j^t & 0 & 0 & j^t & \cdots & \cdots & j^t & j^t & \cdots & \cdots & \cdots & \cdots & j^t & & & & & & \\
\vdots & j^t & 0 & \ddots & \ddots & \ddots & \vdots & \vdots & \ddots & \ddots & \ddots & \ddots & \vdots & & & & & & \\
\vdots & \vdots & \ddots & \ddots & \ddots & \ddots & \vdots & \vdots & \ddots & \ddots & \ddots & \ddots & \vdots & & & 2I_{mn} + J_{mn} & & & \\
\vdots & \vdots & \ddots & \ddots & \ddots & \ddots & j^t & \vdots & \ddots & \ddots & \ddots & \ddots & \vdots & & & & & & \\
\vdots & j^t & \ddots & \ddots & \ddots & \ddots & 0 & \vdots & \ddots & \ddots & \ddots & \ddots & \vdots & & & & & & \\
j^t & 0 & j^t & \cdots & \cdots & j^t & 0 & j^t & \cdots & \cdots & \cdots & \cdots & j^t & & & & & &
\end{pmatrix}$$

$$
= \frac{1}{b^2} \times \det \begin{pmatrix}
b & 0 & \cdots & \cdots & \cdots & \cdots & 0 & 1 & \cdots & \cdots & \cdots & \cdots & 1 & j & \cdots & \cdots & \cdots & \cdots & j \\
b & a & 0 & 1 & \cdots & 1 & 0 & 0 & 1 & \cdots & \cdots & 1 & 0 & 0 & j & \cdots & \cdots & j & 0 \\
\vdots & 1 & \ddots & \ddots & \ddots & \ddots & 1 & 0 & 0 & \ddots & \ddots & \ddots & 1 & 0 & \ddots & \ddots & \ddots & \ddots & j \\
\vdots & \vdots & \ddots & \ddots & \ddots & \ddots & \vdots & 1 & 0 & \ddots & \ddots & \ddots & \vdots & j & \ddots & \ddots & \ddots & \ddots & \vdots \\
\vdots & \vdots & \ddots & \ddots & \ddots & \ddots & 1 & \vdots & \ddots & \ddots & \ddots & \ddots & \vdots & \vdots & \ddots & \ddots & \ddots & \ddots & \vdots \\
\vdots & 1 & \ddots & \ddots & \ddots & \ddots & 0 & \vdots & \ddots & \ddots & 0 & 0 & 1 & \vdots & \ddots & \ddots & \ddots & \ddots & j \\
b & 0 & 1 & \cdots & 1 & 0 & a & 1 & \cdots & \cdots & 1 & 0 & 0 & j & \cdots & \cdots & j & 0 & 0 \\
b & 0 & 0 & 1 & \cdots & \cdots & 1 & 3 & 1 & \cdots & \cdots & \cdots & 1 & j & \cdots & \cdots & \cdots & \cdots & j \\
\vdots & 1 & 0 & \ddots & \ddots & \ddots & 1 & 1 & \ddots & \ddots & \ddots & \ddots & \vdots & \vdots & \ddots & \ddots & \ddots & \ddots & \vdots \\
\vdots & \vdots & \ddots & \ddots & \ddots & \ddots & \vdots & \vdots & \ddots & \ddots & \ddots & \ddots & \vdots & \vdots & \ddots & \ddots & \ddots & \ddots & \vdots \\
\vdots & \vdots & \ddots & \ddots & \ddots & \ddots & 1 & \vdots & \ddots & \ddots & \ddots & \ddots & \vdots & \vdots & \ddots & \ddots & \ddots & \ddots & \vdots \\
\vdots & 1 & \ddots & \ddots & \ddots & \ddots & 0 & \vdots & \ddots & \ddots & \ddots & \ddots & 1 & \vdots & \ddots & \ddots & \ddots & \ddots & \vdots \\
b & 0 & 1 & \cdots & \cdots & 1 & 0 & 1 & \cdots & \cdots & \cdots & 1 & 3 & j & \cdots & \cdots & \cdots & \cdots & j \\
bj^t & 0 & 0 & j^t & \cdots & \cdots & j^t & j^t & \cdots & \cdots & \cdots & \cdots & j^t & & & & & & \\
\vdots & j^t & 0 & \ddots & \ddots & \ddots & \vdots & \vdots & \ddots & \ddots & \ddots & \ddots & \vdots & & & & & & \\
\vdots & \vdots & \ddots & \ddots & \ddots & \ddots & \vdots & \vdots & \ddots & \ddots & \ddots & \ddots & \vdots & & & 2I_{mn}+J_{mn} & & & \\
\vdots & \vdots & \ddots & \ddots & \ddots & \ddots & j^t & \vdots & \ddots & \ddots & \ddots & \ddots & \vdots & & & & & & \\
\vdots & j^t & \ddots & \ddots & \ddots & \ddots & 0 & \vdots & \ddots & \ddots & \ddots & \ddots & \vdots & & & & & & \\
bj^t & 0 & j^t & \cdots & \cdots & j^t & 0 & j^t & \cdots & \cdots & \cdots & \cdots & j^t & & & & & &
\end{pmatrix}
$$

$$
= \frac{1}{b} \times \det \begin{pmatrix}
1 & 0 & \cdots & \cdots & \cdots & \cdots & 0 & 1 & \cdots & \cdots & \cdots & \cdots & 1 & j & \cdots & \cdots & \cdots & \cdots & j \\
1 & a & 0 & 1 & \cdots & 1 & 0 & 0 & 1 & \cdots & \cdots & 1 & 0 & 0 & j & \cdots & \cdots & j & 0 \\
\vdots & 0 & \ddots & \ddots & \ddots & \ddots & 1 & 0 & 0 & \ddots & \ddots & \ddots & 1 & 0 & \ddots & \ddots & \ddots & \ddots & j \\
\vdots & 1 & \ddots & \ddots & \ddots & \ddots & \vdots & 1 & 0 & \ddots & \ddots & \ddots & \vdots & j & \ddots & \ddots & \ddots & \ddots & \vdots \\
\vdots & \vdots & \ddots & \ddots & \ddots & \ddots & 1 & \vdots & \ddots & \ddots & \ddots & \ddots & \vdots & \vdots & \ddots & \ddots & \ddots & \ddots & \vdots \\
\vdots & 1 & \ddots & \vdots & \ddots & \ddots & 0 & \vdots & \ddots & \ddots & 0 & 0 & 1 & \vdots & \ddots & \ddots & \ddots & \ddots & j \\
1 & 0 & 1 & \cdots & 1 & 0 & a & 1 & \cdots & \cdots & 1 & 0 & 0 & j & \cdots & \cdots & j & 0 & 0 \\
1 & 0 & 0 & 1 & \cdots & \cdots & 1 & 3 & 1 & \cdots & \cdots & \cdots & 1 & j & \cdots & \cdots & \cdots & \cdots & j \\
\vdots & 1 & 0 & \ddots & \ddots & \ddots & 1 & 1 & \ddots & \ddots & \ddots & \ddots & \vdots & \vdots & \ddots & \ddots & \ddots & \ddots & \vdots \\
\vdots & \vdots & \ddots & \ddots & \ddots & \ddots & \vdots & \vdots & \ddots & \ddots & \ddots & \ddots & \vdots & \vdots & \ddots & \ddots & \ddots & \ddots & \vdots \\
\vdots & \vdots & \ddots & \ddots & \ddots & \ddots & 1 & \vdots & \ddots & \ddots & \ddots & \ddots & \vdots & \vdots & \ddots & \ddots & \ddots & \ddots & \vdots \\
\vdots & 1 & \ddots & \ddots & \ddots & \ddots & 0 & \vdots & \ddots & \ddots & \ddots & \ddots & 1 & \vdots & \ddots & \ddots & \ddots & \ddots & \vdots \\
1 & 0 & 1 & \cdots & \cdots & 1 & 0 & 1 & \cdots & \cdots & \cdots & 1 & 3 & j & \cdots & \cdots & \cdots & \cdots & j \\
1j^t & 0 & 0 & j^t & \cdots & \cdots & j^t & j^t & \cdots & \cdots & \cdots & \cdots & j^t & & & & & & \\
\vdots & j^t & 0 & \ddots & \ddots & \ddots & \vdots & \vdots & \ddots & \ddots & \ddots & \ddots & \vdots & & & & & & \\
\vdots & \vdots & \ddots & \ddots & \ddots & \ddots & \vdots & \vdots & \ddots & \ddots & \ddots & \ddots & \vdots & & & 2I_{mn}+J_{mn} & & & \\
\vdots & \vdots & \ddots & \ddots & \ddots & \ddots & j^t & \vdots & \ddots & \ddots & \ddots & \ddots & \vdots & & & & & & \\
\vdots & j^t & \ddots & \ddots & \ddots & \ddots & 0 & \vdots & \ddots & \ddots & \ddots & \ddots & \vdots & & & & & & \\
1j^t & 0 & j^t & \cdots & \cdots & j^t & 0 & j^t & \cdots & \cdots & \cdots & \cdots & j^t & & & & & &
\end{pmatrix}
$$

$$
= \frac{1}{b} \times \det \begin{pmatrix}
1 & 0 & \cdots & \cdots & \cdots & \cdots & 0 & 1 & \cdots & \cdots & \cdots & \cdots & 1 & j & \cdots & \cdots & \cdots & \cdots & j \\
0 & a & 0 & 1 & \cdots & 1 & 0 & -1 & 0 & \cdots & \cdots & 0 & -1 & -j & 0 & \cdots & \cdots & 0 & -j \\
\vdots & 1 & \ddots & \ddots & \ddots & \ddots & 1 & -1 & -1 & \ddots & \ddots & \ddots & 0 & -j & \ddots & \ddots & \ddots & \ddots & 0 \\
\vdots & \vdots & \ddots & \ddots & \ddots & \ddots & \vdots & 0 & \ddots & \ddots & \ddots & \ddots & \vdots & 0 & \ddots & \ddots & \ddots & \ddots & \vdots \\
\vdots & \vdots & \ddots & \ddots & \ddots & \ddots & 1 & \vdots & \ddots & \ddots & \ddots & \ddots & \vdots & \vdots & \ddots & \ddots & \ddots & \ddots & \vdots \\
\vdots & 1 & \ddots & \ddots & \ddots & \ddots & 0 & \vdots & \ddots & \ddots & \ddots & -1 & 0 & \vdots & \ddots & \ddots & \ddots & \ddots & 0 \\
0 & 0 & 1 & \cdots & 1 & 0 & a & 0 & \cdots & \cdots & 0 & -1 & -1 & 0 & \cdots & \cdots & 0 & -j & -j \\
0 & 0 & 0 & 1 & \cdots & \cdots & 1 & 2 & 0 & \cdots & \cdots & \cdots & 0 & 0 & \cdots & \cdots & \cdots & \cdots & 0 \\
\vdots & 1 & 0 & \ddots & \ddots & \ddots & 1 & 0 & \ddots & \ddots & \ddots & \ddots & \vdots & \vdots & \ddots & \ddots & \ddots & \ddots & \vdots \\
\vdots & \vdots & \ddots & \ddots & \ddots & \ddots & \vdots & \vdots & \ddots & \ddots & \ddots & \ddots & \vdots & \vdots & \ddots & \ddots & \ddots & \ddots & \vdots \\
\vdots & \vdots & \ddots & \ddots & \ddots & \ddots & 1 & \vdots & \ddots & \ddots & \ddots & \ddots & \vdots & \vdots & \ddots & \ddots & \ddots & \ddots & \vdots \\
\vdots & 1 & \ddots & \ddots & \ddots & \ddots & 0 & \vdots & \ddots & \ddots & \ddots & \ddots & 0 & \vdots & \ddots & \ddots & \ddots & \ddots & \vdots \\
0 & 0 & 1 & \cdots & \cdots & 1 & 0 & 0 & \cdots & \cdots & \cdots & 0 & 2 & 0 & \cdots & \cdots & \cdots & \cdots & 0 \\
0 & 0 & 0 & j^t & \cdots & \cdots & j^t & 0 & \cdots & \cdots & \cdots & \cdots & 0 & & & & & & \\
\vdots & j^t & 0 & \ddots & \ddots & \ddots & \vdots & \vdots & \ddots & \ddots & \ddots & \ddots & \vdots & & & & & & \\
\vdots & \vdots & \ddots & \ddots & \ddots & \ddots & \vdots & \vdots & \ddots & \ddots & \ddots & \ddots & \vdots & & & 2I_{mn} & & & \\
\vdots & \vdots & \ddots & \ddots & \ddots & \ddots & j^t & \vdots & \ddots & \ddots & \ddots & \ddots & \vdots & & & & & & \\
\vdots & j^t & \ddots & \ddots & \ddots & \ddots & 0 & \vdots & \ddots & \ddots & \ddots & \ddots & \vdots & & & & & & \\
0 & 0 & j^t & \cdots & \cdots & j^t & 0 & 0 & \cdots & \cdots & \cdots & \cdots & 0 & & & & & &
\end{pmatrix}
$$

$$
=\frac{1}{b}\times\det\left(\begin{array}{cccccccccccccccccc}
a & 0 & 1 & \cdots & 1 & 0 & -1 & 0 & \cdots & \cdots & 0 & -1 & -j & 0 & \cdots & \cdots & 0 & -j \\
0 & \ddots & \ddots & \ddots & \ddots & 1 & -1 & -1 & \ddots & \ddots & \ddots & 0 & -j & \ddots & \ddots & \ddots & \ddots & 0 \\
1 & \ddots & \ddots & \ddots & \ddots & \vdots & 0 & \ddots & \ddots & \ddots & \ddots & \vdots & 0 & \ddots & \ddots & \ddots & \ddots & \vdots \\
\vdots & \ddots & \ddots & \ddots & \ddots & 1 & \vdots & \ddots & \ddots & \ddots & \ddots & \vdots & \vdots & \ddots & \ddots & \ddots & \ddots & \vdots \\
1 & \ddots & \ddots & \ddots & \ddots & 0 & \vdots & \ddots & \ddots & \ddots & -1 & 0 & \vdots & \ddots & \ddots & \ddots & \ddots & 0 \\
0 & 1 & \cdots & 1 & 0 & a & 0 & \cdots & \cdots & 0 & -1 & -1 & 0 & \cdots & \cdots & 0 & -j & -j \\
0 & 0 & 1 & \cdots & \cdots & 1 & 2 & 0 & \cdots & \cdots & \cdots & 0 & 0 & \cdots & \cdots & \cdots & \cdots & 0 \\
1 & 0 & \ddots & \ddots & \ddots & 1 & 0 & \ddots & \ddots & \ddots & \ddots & \vdots & \vdots & \ddots & \ddots & \ddots & \ddots & \vdots \\
\vdots & \ddots & \ddots & \ddots & \ddots & \vdots & \vdots & \ddots & \ddots & \ddots & \ddots & \vdots & \vdots & \ddots & \ddots & \ddots & \ddots & \vdots \\
\vdots & \ddots & \ddots & \ddots & \ddots & 1 & \vdots & \ddots & \ddots & \ddots & \ddots & \vdots & \vdots & \ddots & \ddots & \ddots & \ddots & \vdots \\
1 & \ddots & \ddots & \ddots & \ddots & 0 & \vdots & \ddots & \ddots & \ddots & \ddots & 0 & \vdots & \ddots & \ddots & \ddots & \ddots & \vdots \\
0 & 1 & \cdots & \cdots & 1 & 0 & 0 & \cdots & \cdots & \cdots & 0 & 2 & 0 & \cdots & \cdots & \cdots & \cdots & 0 \\
0 & 0 & j^t & \cdots & \cdots & j^t & 0 & \cdots & \cdots & \cdots & \cdots & 0 & & & & & & \\
j^t & 0 & \ddots & \ddots & \ddots & \vdots & \vdots & \ddots & \ddots & \ddots & \ddots & \vdots & & & & & & \\
\vdots & \ddots & \ddots & \ddots & \ddots & \vdots & \vdots & \ddots & \ddots & \ddots & \ddots & \vdots & & & 2I_{mn} & & & \\
\vdots & \ddots & \ddots & \ddots & \ddots & j^t & \vdots & \ddots & \ddots & \ddots & \ddots & \vdots & & & & & & \\
j^t & \ddots & \ddots & \ddots & \ddots & 0 & \vdots & \ddots & \ddots & \ddots & \ddots & \vdots & & & & & & \\
0 & j^t & \cdots & \cdots & j^t & 0 & 0 & \cdots & \cdots & \cdots & \cdots & 0 & & & & & &
\end{array}\right)
$$

Using Lemma 3, yields

$$
\tau(C_n^{(m)})=\frac{1}{b}\times\det\begin{pmatrix} A & B \\ C & 2I_{mn}\end{pmatrix}=\frac{1}{b}\times\det(A-B\frac{1}{2I_{mn}}C)\times 2^{mn}
$$

$$
=\frac{1}{b}\times 2^{mn}\times 2^{-2m}\times\det\left(\begin{array}{cccccccccccc}
2a & n & 2(n+1) & \cdots & 2(n+1) & n & -2 & 0 & \cdots & \cdots & 0 & -2 \\
n & 2a & n+2 & 2(n+1) & \cdots & 2(n+1) & -2 & \ddots & \ddots & \ddots & \ddots & 0 \\
2(n+1) & n & \ddots & \ddots & \ddots & \vdots & 0 & \ddots & \ddots & \ddots & \ddots & \vdots \\
\vdots & \ddots & \ddots & \ddots & \ddots & 2(n+1) & \vdots & \ddots & \ddots & \ddots & \ddots & \vdots \\
2(n+1) & \ddots & \ddots & \ddots & \ddots & n & \vdots & \ddots & \ddots & \ddots & \ddots & 0 \\
n & 2(n+1) & \cdots & 2(n+1) & n & 2a & 0 & \cdots & \cdots & 0 & -2 & -2 \\
0 & 0 & 2 & \cdots & \cdots & 2 & 4 & 0 & \cdots & \cdots & \cdots & 0 \\
2 & \ddots & \ddots & \ddots & \ddots & \vdots & 0 & \ddots & \ddots & \ddots & \ddots & \vdots \\
\vdots & \ddots & \ddots & \ddots & \ddots & \vdots & \vdots & \ddots & \ddots & \ddots & \ddots & \vdots \\
\vdots & \ddots & \ddots & \ddots & \ddots & 2 & \vdots & \ddots & \ddots & \ddots & \ddots & \vdots \\
2 & \ddots & \ddots & \ddots & \ddots & 0 & \vdots & \ddots & \ddots & \ddots & \ddots & 0 \\
0 & 2 & \cdots & \cdots & 2 & 0 & 0 & \cdots & \cdots & \cdots & 0 & 4
\end{array}\right)
$$

Using Lemma 3 again, yields

$$
\tau(C_n^{(m)})=\frac{2^{mn-2m}}{b}\times\det\begin{pmatrix} D & E \\ F & 4I_m\end{pmatrix}=\frac{2^{mn}}{b}\times\det(D-E\frac{1}{4I_m}F)
$$

$$
\tau(C_n^{(m)})=\frac{2^{mn}}{b}\times\det\begin{pmatrix}
2a & (n+1) & 2(n+2) & \cdots & 2(n+2) & (n+1) \\
(n+1) & 2a & (n+3) & \ddots & \cdots & 2(n+2) \\
2(n+2) & \ddots & \ddots & \ddots & \ddots & \vdots \\
\vdots & \ddots & \ddots & \ddots & \ddots & 2(n+2) \\
2(n+2) & \ddots & \ddots & \ddots & \ddots & (n+1) \\
(n+1) & 2(n+2) & \cdots & 2(n+2) & (n+1) & 2a
\end{pmatrix}
$$

Using properties of determinants, we have:

$$\tau\,(C_n^{(m)}) = \frac{2^{mn}}{b} \times \frac{2b}{m\,n+3m+2} \times \det\begin{pmatrix} (2a-n-1) & 0 & (n+3) & \cdots & (n+3) & 0 \\ 0 & (2a-n-1) & 0 & \ddots & \cdots & (n+3) \\ (n+3) & \ddots & \ddots & \ddots & \ddots & \vdots \\ \vdots & \ddots & \ddots & \ddots & \ddots & (n+3) \\ (n+3) & \ddots & \ddots & \ddots & \ddots & 0 \\ 0 & (n+3) & \cdots & (n+3) & 0 & (2a-n-1) \end{pmatrix}$$

$$= \frac{2^{m\,n+1}\,(n+3)^m}{m\,n+3m+2} \times \det\begin{pmatrix} \frac{(2a-n-1)}{(n+3)} & 0 & 1 & \cdots & 1 & 0 \\ 0 & \frac{(2a-n-1)}{(n+3)} & 0 & \ddots & \ddots & 1 \\ 1 & \ddots & \ddots & \ddots & \ddots & \vdots \\ \vdots & \ddots & \ddots & \ddots & \ddots & 1 \\ 1 & \ddots & \ddots & \ddots & \ddots & 0 \\ 0 & 1 & \cdots & 1 & 0 & \frac{(2a-n-1)}{(n+3)} \end{pmatrix}$$

Using Lemma 2, yields:

$$\begin{aligned}\tau\,(C_n^{(m)}) &= 2^{m\,n+1} \times \frac{(n+3)^m}{mn+3m+2} \times \frac{2(\frac{2a-n-1}{n+3}+m-3)}{\frac{2a-n-1}{n+3}-3} \times [T_m(\frac{\frac{2a-n-1}{n+3}-1}{2})-1] \\ &= 2^{mn+1} \times (n+3)^m \times [T_m(\frac{n+4}{n+3})-1].\end{aligned}$$

Using Equation (11), yields the result. □

Definition 4. *The pyramid graph $D_n^{(m)}$ is the graph created from the gear graph G_{m+1} with vertices $\{u_0\,;\ u_1, u_2, \ldots, u_m\,;\ w_1, w_2, \ldots, w_m\}$ with double internal and external edges and m sets of vertices, say, $\{v_1^1, v_2^1, \ldots, v_n^1\}$, $\{v_1^2, v_2^2, \ldots, v_n^2\}, \ldots, \{v_1^m, v_2^m, \ldots, v_n^m\}$, such that for all $i = 1, 2, \ldots, n$ the vertex v_i^j is adjacent to u_j and u_{j+1}, where $j = 1, 2, \ldots, m-1$, and v_i^m is adjacent to u_1 and u_m. See Figure 4.*

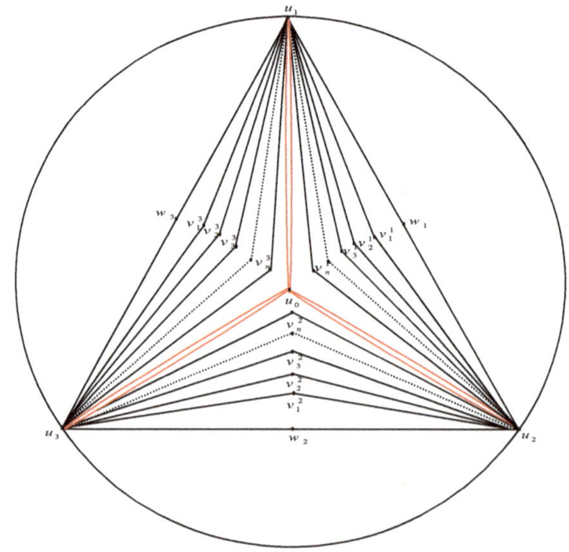

Figure 4. The pyramid graph $D_n^{(3)}$.

Theorem 4. *For $n \geq 0$, $m \geq 3$, $\tau(D_n^{(m)}) = 2^{mn}[(n+5+2\sqrt{n+4})^m + (n+5-2\sqrt{n+4})^m - 2(n+3)^m]$.*

Proof. Applying Lemma 1, we have:

$$\tau(D_n^{(m)}) = \frac{1}{(mn+2m+1)^2} \times \det((mn+2m+1)\,I - D^c + A^c) = \frac{1}{(mn+2m+1)^2} \times$$

$$\det\begin{pmatrix}
(2m+1) & -1 & -1 & \cdots & \cdots & \cdots & -1 & 1 & \cdots & \cdots & \cdots & \cdots & 1 & 1 & \cdots & \cdots & \cdots & \cdots & \cdots & \cdots & \cdots & \cdots & \cdots & \cdots & \cdots & \cdots & \cdots & \cdots & \cdots & \cdots & 1 \\
-1 & (2n+7) & 0 & 1 & \cdots & 1 & 0 & 0 & 1 & \cdots & \cdots & 1 & 0 & 0 & \cdots & 0 & 1 & \cdots & 1 & \cdots & \cdots & 1 & \cdots & 1 & 1 & \cdots & 1 & 0 & \cdots & 0 \\
\vdots & 0 & \ddots & \ddots & \ddots & \ddots & 1 & 0 & \ddots & \ddots & \ddots & \ddots & 1 & 0 & \cdots & 0 & 0 & \cdots & 0 & 1 & \cdots & 1 & \cdots & \cdots & 1 & \cdots & 1 & 1 & \cdots & 1 \\
\vdots & 1 & \ddots & \ddots & \ddots & \ddots & \vdots & 1 & \ddots & \ddots & \ddots & \vdots & \vdots & 1 & \cdots & 1 & 0 & \cdots & 0 & 0 & \cdots & 0 & 1 & \cdots & 1 & \cdots & \cdots & 1 & \cdots & 1 \\
\vdots & \vdots & \ddots & \ddots & \ddots & \ddots & 1 & \vdots & \ddots & \ddots & \ddots & \vdots & \vdots & 1 & \cdots & 1 & 1 & \cdots & 1 & 0 & \cdots & 0 & 0 & \cdots & 0 & 1 & \cdots & 1 & \cdots & \cdots \\
\vdots & 1 & \ddots & \ddots & \ddots & \ddots & 0 & \vdots & \ddots & \ddots & \ddots & \ddots & 1 & \vdots & \cdots & \cdots & \cdots & \cdots & \cdots & \cdots & \cdots & \cdots & \cdots & \cdots & \cdots & \cdots & \cdots & \cdots & \cdots & \vdots \\
-1 & 0 & 1 & \cdots & 1 & 0 & (2n+7) & 1 & \cdots & \cdots & 1 & 0 & 0 & 1 & \cdots & 1 & 1 & \cdots & 1 & \cdots & \cdots & 1 & \cdots & 1 & 0 & \cdots & 0 & 0 & \cdots & 0 \\
1 & 0 & 0 & 1 & \cdots & \cdots & 1 & 3 & 1 & \cdots & \cdots & \cdots & 1 & 1 & \cdots & 1 & 1 & \cdots & 1 & \cdots & \cdots & \cdots & \cdots & \cdots & 1 & \cdots & 1 & 1 & \cdots & 1 \\
\vdots & 1 & \ddots & \ddots & \ddots & \ddots & \vdots & 1 & \ddots & \ddots & \ddots & \ddots & \vdots & 1 & \cdots & 1 & 1 & \cdots & 1 & \cdots & \cdots & \cdots & \cdots & \cdots & 1 & \cdots & 1 & 1 & \cdots & 1 \\
\vdots & \vdots & \ddots & \ddots & \ddots & \ddots & \vdots & \vdots & \ddots & \ddots & \ddots & \ddots & \vdots & \vdots & \cdots & \vdots & \vdots & \cdots & \vdots & \cdots & \cdots & \cdots & \cdots & \cdots & \vdots & \cdots & \vdots & \vdots & \cdots & 1 \\
\vdots & \vdots & \ddots & \ddots & \ddots & \ddots & 1 & \vdots & \ddots & \ddots & \ddots & \ddots & \vdots & \vdots & \cdots & \vdots & \vdots & \cdots & \vdots & \cdots & \cdots & \cdots & \cdots & \cdots & \vdots & \cdots & \vdots & \vdots & \cdots & \vdots \\
1 & 1 & \ddots & \ddots & \ddots & \ddots & 0 & \vdots & \ddots & \ddots & \ddots & \ddots & \vdots & 1 & \cdots & 1 & 1 & \cdots & 1 & \cdots & \cdots & \cdots & \cdots & \cdots & 1 & \cdots & 1 & 1 & \cdots & 1 \\
1 & 0 & 1 & \cdots & \cdots & 1 & 0 & 1 & \cdots & \cdots & \cdots & 1 & 3 & 1 & \cdots & 1 & 1 & \cdots & 1 & \cdots & \cdots & \cdots & \cdots & \cdots & 1 & \cdots & 1 & 1 & \cdots & 1 \\
\vdots & 0 & 0 & 1 & \cdots & 1 & 0 & 1 & 1 & \cdots & \cdots & 1 & 1 & 3 & 1 & \cdots & \cdots & \cdots & \cdots & \cdots & \cdots & \cdots & \cdots & \cdots & \cdots & \cdots & \cdots & \cdots & \cdots & 1 \\
\vdots & \vdots & \vdots & \vdots & \vdots & \vdots & \vdots & \vdots & \vdots & \vdots & \vdots & \vdots & \vdots & 1 & \ddots & \ddots & \ddots & \ddots & \ddots & \ddots & \ddots & \ddots & \ddots & \ddots & \ddots & \ddots & \ddots & \ddots & \ddots & \vdots \\
\vdots & 0 & 0 & 1 & \vdots & 1 & \vdots & 1 & 1 & \cdots & \cdots & 1 & 1 & \vdots & \ddots & \ddots & \ddots & \ddots & \ddots & \ddots & \ddots & \ddots & \ddots & \ddots & \ddots & \ddots & \ddots & \ddots & \ddots & \vdots \\
\vdots & 1 & 0 & 0 & \vdots & 1 & \vdots & 1 & 1 & \cdots & \cdots & 1 & 1 & \vdots & \ddots & \ddots & \ddots & \ddots & \ddots & \ddots & \ddots & \ddots & \ddots & \ddots & \ddots & \ddots & \ddots & \ddots & \ddots & \vdots \\
\vdots & \vdots & \vdots & \vdots & \vdots & \vdots & \vdots & \vdots & \vdots & \vdots & \vdots & \vdots & \vdots & \vdots & \ddots & \ddots & \ddots & \ddots & \ddots & \ddots & \ddots & \ddots & \ddots & \ddots & \ddots & \ddots & \ddots & \ddots & \ddots & \vdots \\
\vdots & 1 & 0 & 0 & \vdots & 1 & \vdots & 1 & 1 & \cdots & \cdots & 1 & 1 & \vdots & \ddots & \ddots & \ddots & \ddots & \ddots & \ddots & \ddots & \ddots & \ddots & \ddots & \ddots & \ddots & \ddots & \ddots & \ddots & \vdots \\
\vdots & 1 & 1 & 0 & \vdots & \vdots & \vdots & \vdots & \vdots & \vdots & \vdots & \vdots & \vdots & \vdots & \ddots & \ddots & \ddots & \ddots & \ddots & \ddots & \ddots & \ddots & \ddots & \ddots & \ddots & \ddots & \ddots & \ddots & \ddots & \vdots \\
\vdots & \vdots & \vdots & \vdots & \vdots & \vdots & \vdots & \vdots & \vdots & \vdots & \vdots & \vdots & \vdots & \vdots & \ddots & \ddots & \ddots & \ddots & \ddots & \ddots & \ddots & \ddots & \ddots & \ddots & \ddots & \ddots & \ddots & \ddots & \ddots & \vdots \\
\vdots & 1 & 1 & 0 & \vdots & 1 & \vdots & \vdots & \vdots & \vdots & \vdots & \vdots & \vdots & \vdots & \ddots & \ddots & \ddots & \ddots & \ddots & \ddots & \ddots & \ddots & \ddots & \ddots & \ddots & \ddots & \ddots & \ddots & \ddots & \vdots \\
\vdots & \vdots & \vdots & 1 & \vdots & \vdots & \vdots & \vdots & \vdots & \vdots & \vdots & \vdots & \vdots & \vdots & \ddots & \ddots & \ddots & \ddots & \ddots & \ddots & \ddots & \ddots & \ddots & \ddots & \ddots & \ddots & \ddots & \ddots & \ddots & \vdots \\
\vdots & \vdots & \vdots & \vdots & \vdots & 1 & \vdots & \vdots & \vdots & \vdots & \vdots & \vdots & \vdots & \vdots & \ddots & \ddots & \ddots & \ddots & \ddots & \ddots & \ddots & \ddots & \ddots & \ddots & \ddots & \ddots & \ddots & \ddots & \ddots & \vdots \\
\vdots & 1 & 1 & 1 & \vdots & 0 & \vdots & \vdots & \vdots & \vdots & \vdots & \vdots & \vdots & \vdots & \ddots & \ddots & \ddots & \ddots & \ddots & \ddots & \ddots & \ddots & \ddots & \ddots & \ddots & \ddots & \ddots & \ddots & \ddots & \vdots \\
\vdots & \vdots & \vdots & \vdots & \vdots & \vdots & \vdots & \vdots & \vdots & \vdots & \vdots & \vdots & \vdots & \vdots & \ddots & \ddots & \ddots & \ddots & \ddots & \ddots & \ddots & \ddots & \ddots & \ddots & \ddots & \ddots & \ddots & \ddots & \ddots & \vdots \\
\vdots & 1 & 1 & \vdots & \vdots & 0 & \vdots & 1 & 1 & \cdots & \cdots & 1 & 1 & \vdots & \ddots & \ddots & \ddots & \ddots & \ddots & \ddots & \ddots & \ddots & \ddots & \ddots & \ddots & \ddots & \ddots & \ddots & \ddots & \vdots \\
\vdots & 0 & 1 & 1 & \vdots & 0 & \vdots & 1 & 1 & \cdots & \cdots & 1 & 1 & \vdots & \ddots & \ddots & \ddots & \ddots & \ddots & \ddots & \ddots & \ddots & \ddots & \ddots & \ddots & \ddots & \ddots & \ddots & \ddots & \vdots \\
\vdots & \vdots & \vdots & \vdots & \vdots & \vdots & \vdots & \vdots & \vdots & \vdots & \vdots & \vdots & \vdots & \vdots & \ddots & \ddots & \ddots & \ddots & \ddots & \ddots & \ddots & \ddots & \ddots & \ddots & \ddots & \ddots & \ddots & \ddots & \ddots & 1 \\
1 & 0 & 1 & 1 & \cdots & 0 & \cdots & 1 & 1 & \cdots & \cdots & 1 & 1 & 1 & \cdots & \cdots & \cdots & \cdots & \cdots & \cdots & \cdots & \cdots & \cdots & \cdots & \cdots & \cdots & \cdots & \cdots & 1 & 3
\end{pmatrix}$$

Let $j = (1 \cdots 1)$ be the $1 \times n$ matrix with all one, and J_n the $n \times n$ matrix with all one. Set $a = 2n + 7$ and $b = mn + 2m + 1$. Then we have:

$$\tau\left(D_n^{(m)}\right) = \frac{1}{b^2}\det\begin{pmatrix}
2m+1 & -1 & \cdots & \cdots & \cdots & \cdots & -1 & 1 & \cdots & \cdots & \cdots & \cdots & 1 & j & \cdots & \cdots & \cdots & \cdots & j \\
-1 & a & 0 & 1 & \cdots & 1 & 0 & 0 & 1 & \cdots & \cdots & 1 & 0 & 0 & j & \cdots & \cdots & j & 0 \\
\vdots & 0 & \ddots & \ddots & \ddots & \ddots & 1 & 0 & 0 & \ddots & \ddots & \ddots & 1 & 0 & \ddots & \ddots & \ddots & \ddots & j \\
\vdots & 1 & \ddots & \ddots & \ddots & \ddots & \vdots & 1 & 0 & \ddots & \ddots & \ddots & \vdots & j & \ddots & \ddots & \ddots & \ddots & \vdots \\
\vdots & \vdots & \ddots & \ddots & \ddots & \ddots & 1 & \vdots & \ddots & \ddots & \ddots & \ddots & \vdots & \vdots & \ddots & \ddots & \ddots & \ddots & \vdots \\
\vdots & 1 & \ddots & \ddots & \ddots & \ddots & 0 & \vdots & \ddots & \ddots & 0 & 0 & 1 & \vdots & \ddots & \ddots & \ddots & \ddots & j \\
-1 & 0 & 1 & \cdots & 1 & 0 & a & 1 & \cdots & \cdots & 1 & 0 & 0 & j & \cdots & \cdots & j & 0 & 0 \\
1 & 0 & 0 & 1 & \cdots & \cdots & 1 & 3 & 1 & \cdots & \cdots & \cdots & 1 & j & \cdots & \cdots & \cdots & \cdots & j \\
\vdots & 1 & 0 & \ddots & \ddots & \ddots & 1 & 1 & \ddots & \ddots & \ddots & \ddots & \vdots & \vdots & \ddots & \ddots & \ddots & \ddots & \vdots \\
\vdots & \vdots & \ddots & \ddots & \ddots & \ddots & \vdots & \vdots & \ddots & \ddots & \ddots & \ddots & \vdots & \vdots & \ddots & \ddots & \ddots & \ddots & \vdots \\
\vdots & \vdots & \ddots & \ddots & \ddots & \ddots & 1 & \vdots & \ddots & \ddots & \ddots & \ddots & \vdots & \vdots & \ddots & \ddots & \ddots & \ddots & \vdots \\
\vdots & 1 & \ddots & \ddots & \ddots & \ddots & 0 & \vdots & \ddots & \ddots & \ddots & \ddots & 1 & \vdots & \ddots & \ddots & \ddots & \ddots & \vdots \\
1 & 0 & 1 & \cdots & \cdots & 1 & 0 & 1 & \cdots & \cdots & \cdots & 1 & 3 & j & \cdots & \cdots & \cdots & \cdots & j \\
j^t & 0 & 0 & j^t & \cdots & \cdots & j^t & j^t & \cdots & \cdots & \cdots & \cdots & j^t & & & & & & \\
\vdots & j^t & 0 & \ddots & \ddots & \ddots & \vdots & \vdots & \ddots & \ddots & \ddots & \ddots & \vdots & & & & & & \\
\vdots & \vdots & \ddots & \ddots & \ddots & \ddots & \vdots & \vdots & \ddots & \ddots & \ddots & \ddots & \vdots & & & 2I_{mn} + J_{mn} & & & \\
\vdots & \vdots & \ddots & \ddots & \ddots & \ddots & j^t & \vdots & \ddots & \ddots & \ddots & \ddots & \vdots & & & & & & \\
\vdots & j^t & \ddots & \ddots & \ddots & \ddots & 0 & \vdots & \ddots & \ddots & \ddots & \ddots & \vdots & & & & & & \\
j^t & 0 & j^t & \cdots & \cdots & j^t & 0 & j^t & \cdots & \cdots & \cdots & \cdots & j^t & & & & & &
\end{pmatrix}$$

$$
= \frac{1}{b^2} \det \begin{pmatrix}
b & -1 & \cdots & \cdots & \cdots & \cdots & -1 & 1 & \cdots & \cdots & \cdots & \cdots & 1 & j & \cdots & \cdots & \cdots & \cdots & j \\
b & a & 0 & 1 & \cdots & 1 & 0 & 0 & 1 & \cdots & \cdots & 1 & 0 & 0 & j & \cdots & \cdots & j & 0 \\
\vdots & 0 & \ddots & \ddots & \ddots & \ddots & 1 & 0 & 0 & \ddots & \ddots & \ddots & 1 & 0 & \ddots & \ddots & \ddots & \ddots & j \\
\vdots & 1 & \ddots & \ddots & \ddots & \ddots & \vdots & 1 & 0 & \ddots & \ddots & \ddots & \vdots & j & \ddots & \ddots & \ddots & \ddots & \vdots \\
\vdots & \vdots & \ddots & \ddots & \ddots & \ddots & 1 & \vdots & \ddots & \ddots & \ddots & \ddots & \vdots & \vdots & \ddots & \ddots & \ddots & \ddots & \vdots \\
\vdots & 1 & \ddots & \ddots & \ddots & \ddots & 0 & \vdots & \ddots & \ddots & 0 & 0 & 1 & \vdots & \ddots & \ddots & \ddots & \ddots & j \\
b & 0 & 1 & \cdots & 1 & 0 & a & 1 & \cdots & \cdots & 1 & 0 & 0 & j & \cdots & \cdots & j & 0 & 0 \\
b & 0 & 0 & 1 & \cdots & \cdots & 1 & 3 & 1 & \cdots & \cdots & \cdots & 1 & j & \cdots & \cdots & \cdots & \cdots & j \\
\vdots & 1 & 0 & \ddots & \ddots & \ddots & 1 & 1 & \ddots & \ddots & \ddots & \ddots & \vdots & \vdots & \ddots & \ddots & \ddots & \ddots & \vdots \\
\vdots & \vdots & \ddots & \ddots & \ddots & \ddots & \vdots & \vdots & \ddots & \ddots & \ddots & \ddots & \vdots & \vdots & \ddots & \ddots & \ddots & \ddots & \vdots \\
\vdots & \vdots & \ddots & \ddots & \ddots & \ddots & 1 & \vdots & \ddots & \ddots & \ddots & \ddots & \vdots & \vdots & \ddots & \ddots & \ddots & \ddots & \vdots \\
\vdots & 1 & \ddots & \ddots & \ddots & \ddots & 0 & \vdots & \ddots & \ddots & \ddots & \ddots & 1 & \vdots & \ddots & \ddots & \ddots & \ddots & \vdots \\
b & 0 & 1 & \cdots & \cdots & 1 & 0 & 1 & \cdots & \cdots & \cdots & 1 & 3 & j & \cdots & \cdots & \cdots & \cdots & j \\
bj^t & 0 & 0 & j^t & \cdots & \cdots & j^t & j^t & \cdots & \cdots & \cdots & \cdots & j^t & & & & & & \\
\vdots & j^t & 0 & \ddots & \ddots & \ddots & \vdots & \vdots & \ddots & \ddots & \ddots & \ddots & \vdots & & & & & & \\
\vdots & \vdots & \ddots & \ddots & \ddots & \ddots & \vdots & \vdots & \ddots & \ddots & \ddots & \ddots & \vdots & & & 2I_{mn} + J_{mn} & & & \\
\vdots & \vdots & \ddots & \ddots & \ddots & \ddots & j^t & \vdots & \ddots & \ddots & \ddots & \ddots & \vdots & & & & & & \\
\vdots & j^t & \ddots & \ddots & \ddots & \ddots & 0 & \vdots & \ddots & \ddots & \ddots & \ddots & \vdots & & & & & & \\
bj^t & 0 & j^t & \cdots & \cdots & j^t & 0 & j^t & \cdots & \cdots & \cdots & \cdots & j^t & & & & & &
\end{pmatrix}
$$

$$
= \frac{1}{b} \det \begin{pmatrix}
1 & -1 & \cdots & \cdots & \cdots & \cdots & -1 & 1 & \cdots & \cdots & \cdots & \cdots & 1 & j & \cdots & \cdots & \cdots & \cdots & j \\
1 & a & 0 & 1 & \cdots & 1 & 0 & 0 & 1 & \cdots & \cdots & 1 & 0 & 0 & j & \cdots & \cdots & j & 0 \\
\vdots & 0 & \ddots & \ddots & \ddots & \ddots & 1 & 0 & 0 & \ddots & \ddots & \ddots & 1 & 0 & \ddots & \ddots & \ddots & \ddots & j \\
\vdots & 1 & \ddots & \ddots & \ddots & \ddots & \vdots & 1 & 0 & \ddots & \ddots & \ddots & \vdots & j & \ddots & \ddots & \ddots & \ddots & \vdots \\
\vdots & \vdots & \ddots & \ddots & \ddots & \ddots & 1 & \vdots & \ddots & \ddots & \ddots & \ddots & \vdots & \vdots & \ddots & \ddots & \ddots & \ddots & \vdots \\
\vdots & 1 & \ddots & \ddots & \ddots & \ddots & 0 & \vdots & \ddots & \ddots & 0 & 0 & 1 & \vdots & \ddots & \ddots & \ddots & \ddots & j \\
1 & 0 & 1 & \cdots & 1 & 0 & a & 1 & \cdots & \cdots & 1 & 0 & 0 & j & \cdots & \cdots & j & 0 & 0 \\
1 & 0 & 0 & 1 & \cdots & \cdots & 1 & 3 & 1 & \cdots & \cdots & \cdots & 1 & j & \cdots & \cdots & \cdots & \cdots & j \\
\vdots & 1 & 0 & \ddots & \ddots & \ddots & 1 & 1 & \ddots & \ddots & \ddots & \ddots & \vdots & \vdots & \ddots & \ddots & \ddots & \ddots & \vdots \\
\ddots \; \vdots & \vdots & \ddots & \ddots & \ddots & \ddots & \vdots & \vdots & \ddots & \ddots & \ddots & \ddots & \vdots & \vdots & \ddots & \ddots & \ddots & \ddots & \vdots \\
\vdots & \vdots & \ddots & \ddots & \ddots & \ddots & 1 & \vdots & \ddots & \ddots & \ddots & \ddots & \vdots & \vdots & \ddots & \ddots & \ddots & \ddots & \vdots \\
\vdots & 1 & \ddots & \ddots & \ddots & \ddots & 0 & \vdots & \ddots & \ddots & \ddots & \ddots & 1 & \vdots & \ddots & \ddots & \ddots & \ddots & \vdots \\
1 & 0 & 1 & \cdots & \cdots & 1 & 0 & 1 & \cdots & \cdots & \cdots & 1 & 3 & j & \cdots & \cdots & \cdots & \cdots & j \\
1j^t & 0 & 0 & j^t & \cdots & \cdots & j^t & j^t & \cdots & \cdots & \cdots & \cdots & j^t & & & & & & \\
\vdots & j^t & 0 & \ddots & \ddots & \ddots & \vdots & \vdots & \ddots & \ddots & \ddots & \ddots & \vdots & & & & & & \\
\vdots & \vdots & \ddots & \ddots & \ddots & \ddots & \vdots & \vdots & \ddots & \ddots & \ddots & \ddots & \vdots & & & 2I_{mn} + J_{mn} & & & \\
\vdots & \vdots & \ddots & \ddots & \ddots & \ddots & j^t & \vdots & \ddots & \ddots & \ddots & \ddots & \vdots & & & & & & \\
\vdots & j^t & \ddots & \ddots & \ddots & \ddots & 0 & \vdots & \ddots & \ddots & \ddots & \ddots & \vdots & & & & & & \\
1j^t & 0 & j^t & \cdots & \cdots & j^t & 0 & j^t & \cdots & \cdots & \cdots & \cdots & j^t & & & & & &
\end{pmatrix}
$$

$$
= \frac{1}{b} \det \begin{pmatrix}
1 & -1 & \cdots & \cdots & \cdots & \cdots & -1 & 1 & \cdots & \cdots & \cdots & \cdots & 1 & j & \cdots & \cdots & \cdots & \cdots & j \\
0 & (a+1) & 1 & 2 & \cdots & 2 & 1 & -1 & 0 & \cdots & \cdots & 0 & -1 & -j & 0 & \cdots & \cdots & 0 & -j \\
\vdots & 1 & \ddots & \ddots & \ddots & \ddots & 2 & -1 & -1 & \ddots & \ddots & \ddots & 0 & -j & \ddots & \ddots & \ddots & \ddots & 0 \\
\vdots & 2 & \ddots & \ddots & \ddots & \ddots & \vdots & 0 & \ddots & \ddots & \ddots & \ddots & \vdots & 0 & \ddots & \ddots & \ddots & \ddots & \vdots \\
\vdots & \vdots & \ddots & \ddots & \ddots & \ddots & 2 & \vdots & \ddots & \ddots & \ddots & \ddots & \vdots & \vdots & \ddots & \ddots & \ddots & \ddots & \vdots \\
\vdots & 2 & \ddots & \ddots & \ddots & \ddots & 1 & \vdots & \ddots & \ddots & \ddots & -1 & 0 & \vdots & \ddots & \ddots & \ddots & \ddots & 0 \\
0 & 1 & 2 & \cdots & 2 & 1 & (a+1) & 0 & \cdots & \cdots & 0 & -1 & -1 & 0 & \cdots & \cdots & 0 & -j & -j \\
0 & 1 & 1 & 2 & \cdots & \cdots & 2 & 2 & 0 & \cdots & \cdots & \cdots & 0 & 0 & \cdots & \cdots & \cdots & \cdots & 0 \\
\vdots & 2 & 1 & \ddots & \ddots & \ddots & 2 & 0 & \ddots & \ddots & \ddots & \ddots & \vdots & \vdots & \ddots & \ddots & \ddots & \ddots & \vdots \\
\vdots & \vdots & \ddots & \ddots & \ddots & \ddots & \vdots & \vdots & \ddots & \ddots & \ddots & \ddots & \vdots & \vdots & \ddots & \ddots & \ddots & \ddots & \vdots \\
\vdots & \vdots & \ddots & \ddots & \ddots & \ddots & 2 & \vdots & \ddots & \ddots & \ddots & \ddots & \vdots & \vdots & \ddots & \ddots & \ddots & \ddots & \vdots \\
\vdots & 2 & \ddots & \ddots & \ddots & \ddots & 1 & \vdots & \ddots & \ddots & \ddots & \ddots & 0 & \vdots & \ddots & \ddots & \ddots & \ddots & \vdots \\
0 & 1 & 2 & \cdots & \cdots & 2 & 1 & 0 & \cdots & \cdots & \cdots & 0 & 2 & 0 & \cdots & \cdots & \cdots & \cdots & 0 \\
0 & j^t & j^t & 2j^t & \cdots & \cdots & 2j^t & 0 & \cdots & \cdots & \cdots & \cdots & 0 & & & & & & \\
\vdots & 2j^t & j^t & \ddots & \ddots & \ddots & \vdots & \vdots & \ddots & \ddots & \ddots & \ddots & \vdots & & & & & & \\
\vdots & \vdots & \ddots & \ddots & \ddots & \ddots & \vdots & \vdots & \ddots & \ddots & \ddots & \ddots & \vdots & & & 2I_{mn} & & & \\
\vdots & \vdots & \ddots & \ddots & \ddots & \ddots & 2j^t & \vdots & \ddots & \ddots & \ddots & \ddots & \vdots & & & & & & \\
\vdots & 2j^t & \ddots & \ddots & \ddots & \ddots & j^t & \vdots & \ddots & \ddots & \ddots & \ddots & \vdots & & & & & & \\
0 & j^t & 2j^t & \cdots & \cdots & 2j^t & j^t & 0 & \cdots & \cdots & \cdots & \cdots & 0 & & & & & &
\end{pmatrix}
$$

$$
= \frac{1}{b}\det\begin{pmatrix}
(a+1) & 1 & 2 & \cdots & 2 & 1 & -1 & 0 & \cdots & \cdots & 0 & -1 & -j & 0 & \cdots & \cdots & 0 & -j \\
1 & \ddots & \ddots & \ddots & \ddots & 2 & -1 & -1 & \ddots & \ddots & \ddots & 0 & -j & \ddots & \ddots & \ddots & \ddots & 0 \\
2 & \ddots & \ddots & \ddots & \ddots & \vdots & 0 & \ddots & \ddots & \ddots & \ddots & \vdots & 0 & \ddots & \ddots & \ddots & \ddots & \vdots \\
\vdots & \ddots & \ddots & \ddots & \ddots & 2 & \vdots & \ddots & \ddots & \ddots & \ddots & \vdots & \vdots & \ddots & \ddots & \ddots & \ddots & \vdots \\
2 & \ddots & \ddots & \ddots & \ddots & 1 & \vdots & \ddots & \ddots & \ddots & -1 & 0 & \vdots & \ddots & \ddots & \ddots & \ddots & 0 \\
1 & 2 & \cdots & 2 & 1 & (a+1) & 0 & \cdots & \cdots & 0 & -1 & -1 & 0 & \cdots & \cdots & 0 & -j & -j \\
1 & 1 & 2 & \cdots & \cdots & 2 & 2 & 0 & \cdots & \cdots & \cdots & 0 & 0 & \cdots & \cdots & \cdots & \cdots & 0 \\
2 & 1 & \ddots & \ddots & \ddots & 2 & 0 & \ddots & \ddots & \ddots & \ddots & \vdots & \vdots & \ddots & \ddots & \ddots & \ddots & \vdots \\
\vdots & \ddots & \ddots & \ddots & \ddots & \vdots & \vdots & \ddots & \ddots & \ddots & \ddots & \vdots & \vdots & \ddots & \ddots & \ddots & \ddots & \vdots \\
\vdots & \ddots & \ddots & \ddots & \ddots & 2 & \vdots & \ddots & \ddots & \ddots & \ddots & \vdots & \vdots & \ddots & \ddots & \ddots & \ddots & \vdots \\
2 & \ddots & \ddots & \ddots & \ddots & 1 & \vdots & \ddots & \ddots & \ddots & \ddots & 0 & \vdots & \ddots & \ddots & \ddots & \ddots & \vdots \\
1 & 2 & \cdots & \cdots & 2 & 1 & 0 & \cdots & \cdots & \cdots & 0 & 2 & 0 & \cdots & \cdots & \cdots & \cdots & 0 \\
j^t & j^t & 2j^t & \cdots & \cdots & 2j^t & 0 & \cdots & \cdots & \cdots & \cdots & 0 & & & & & & \\
2j^t & j^t & \ddots & \ddots & \ddots & \vdots & \vdots & \ddots & \ddots & \ddots & \ddots & \vdots & & & & & & \\
\vdots & \ddots & \ddots & \ddots & \ddots & \vdots & \vdots & \ddots & \ddots & \ddots & \ddots & \vdots & & & & 2I_{mn} & & \\
\vdots & \ddots & \ddots & \ddots & \ddots & 2j^t & \vdots & \ddots & \ddots & \ddots & \ddots & \vdots & & & & & & \\
2j^t & \ddots & \ddots & \ddots & \ddots & j^t & \vdots & \ddots & \ddots & \ddots & \ddots & \vdots & & & & & & \\
j^t & 2j^t & \cdots & \cdots & 2j^t & j^t & 0 & \cdots & \cdots & \cdots & \cdots & 0 & & & & & &
\end{pmatrix}
$$

Using Lemma 3, yields

$$
\tau(D_n^{(m)}) = \frac{1}{b}\times\det\begin{pmatrix} A & B \\ C & 2I_{mn}\end{pmatrix} = \frac{1}{b}\times\det(A - B\frac{1}{2I_{mn}}C)\times 2^{mn}
$$

$$
= \frac{1}{b}2^{mn}\times 2^{-2m}\times\det\begin{pmatrix}
(2a+2n+2) & 3n+2 & 4(n+1) & \cdots & 4(n+1) & 3n+2 & -2 & 0 & \cdots & \cdots & 0 & -2 \\
3n+2 & (2a+2n+2) & 3n+2 & 4(n+1) & \cdots & 4(n+1) & -2 & \ddots & \ddots & \ddots & \ddots & 0 \\
4(n+1) & 3n+4 & \ddots & \ddots & \ddots & \vdots & 0 & \ddots & \ddots & \ddots & \ddots & \vdots \\
\vdots & \ddots & \ddots & \ddots & \ddots & 4(n+1) & \vdots & \ddots & \ddots & \ddots & \ddots & \vdots \\
4(n+1) & \ddots & \ddots & \ddots & \ddots & 3n+2 & \vdots & \ddots & \ddots & \ddots & \ddots & 0 \\
3n+2 & 4(n+1) & \cdots & 4(n+1) & 3n+2 & (2a+2n+2) & 0 & \cdots & \cdots & 0 & -2 & -2 \\
2 & 2 & 4 & \cdots & \cdots & 4 & 4 & 0 & \cdots & \cdots & \cdots & 0 \\
4 & \ddots & \ddots & \ddots & \ddots & \vdots & 0 & \ddots & \ddots & \ddots & \ddots & \vdots \\
\vdots & \ddots & \ddots & \ddots & \ddots & \vdots & \vdots & \ddots & \ddots & \ddots & \ddots & \vdots \\
\vdots & \ddots & \ddots & \ddots & \ddots & 4 & \vdots & \ddots & \ddots & \ddots & \ddots & \vdots \\
4 & \ddots & \ddots & \ddots & \ddots & 2 & \vdots & \ddots & \ddots & \ddots & \ddots & 0 \\
2 & 4 & \cdots & \cdots & 4 & 2 & 0 & \cdots & \cdots & \cdots & 0 & 4
\end{pmatrix}
$$

Using Lemma 3, yields

$$
\tau(D_n^{(m)}) = \frac{2^{mn-2m}}{b}\times\det\begin{pmatrix} A & B \\ C & 4I_m\end{pmatrix} = \frac{2^{mn}}{b}\times\det(A - B\frac{1}{4I_m}C)
$$

$$
\tau(D_n^{(m)}) = \frac{2^{mn}}{b}\times\det\begin{pmatrix}
(2a+2n+4) & (3n+5) & 4(n+2) & \cdots & 4(n+2) & (3n+5) \\
(3n+5) & (2a+2n+4) & (3n+5) & \ddots & \cdots & 4(n+2) \\
4(n+2) & \ddots & \ddots & \ddots & \ddots & \vdots \\
\vdots & \ddots & \ddots & \ddots & \ddots & 4(n+2) \\
4(n+2) & \ddots & \ddots & \ddots & \ddots & (3n+5) \\
(3n+5) & 4(n+2) & \cdots & 4(n+2) & (3n+5) & (2a+2n+4)
\end{pmatrix}
$$

Straightforward inducement using properties of determinants, we get:

$$
\tau(D_n^{(m)}) = \frac{2^{mn}}{b}\times\frac{4b}{mn+3m+4}\times\det\begin{pmatrix}
(2a-n-1) & 0 & (n+3) & \cdots & (n+3) & 0 \\
0 & (2a-n-1) & 0 & \ddots & \cdots & (n+3) \\
(n+3) & \ddots & \ddots & \ddots & \ddots & \vdots \\
\vdots & \ddots & \ddots & \ddots & \ddots & (n+3) \\
(n+3) & \ddots & \ddots & \ddots & \ddots & 0 \\
0 & (n+3) & \cdots & (n+3) & 0 & (2a-n-1)
\end{pmatrix}
$$

$$= \frac{2^{mn+2}(n+3)^m}{mn+3m+4} \times \det \begin{pmatrix} \frac{(2a-n-1)}{(n+3)} & 0 & 1 & \cdots & 1 & 0 \\ 0 & \frac{(2a-n-1)}{(n+3)} & 0 & \ddots & \ddots & 1 \\ 1 & \ddots & \ddots & \ddots & \ddots & \vdots \\ \vdots & \ddots & \ddots & \ddots & \ddots & 1 \\ 1 & \ddots & \ddots & \ddots & \ddots & 0 \\ 0 & 1 & \cdots & 1 & 0 & \frac{(2a-n-1)}{(n+3)} \end{pmatrix}$$

Using Lemma 2, yields:

$$\tau(D_n^{(m)}) = 2^{mn+2} \times \frac{(n+3)^m}{mn+3m+4} \times \frac{2(\frac{2a-n-1}{n+3}+m-3)}{\frac{2a-n-1}{n+3}-3} \times [T_m(\frac{\frac{2a-n-1}{n+3}-1}{2}) - 1] = 2^{mn+1} \times (n+3)^m \times [T_m(\tfrac{n+5}{n+3}) - 1].$$

Using Equation (11), yields the result. □

4. Numerical Results

The following Table 1 illustrates some values of the number of spanning trees of studied pyramid graphs.

Table 1. Some values of the number of spanning trees of studied pyramid graphs.

m	n	$\tau(P_n^{(m)})$	$\tau(A_n^{(m)})$	$\tau(B_n^{(m)})$	$\tau(C_n^{(m)})$
3	0	50	196	242	676
3	1	1024	3200	3136	8192
3	2	15,488	43,264	36,992	92,416
3	3	200,704	524,288	409,600	991,232
3	4	2,367,488	5,914,624	4,333,568	10,240,000
3	5	26,214,400	63,438,848	44,302,336	102,760,448
4	0	192	1152	1792	6400
4	1	11,520	49,152	57,600	184,320
4	2	458,752	1,638,400	1,622,016	4,816,896
4	3	14,745,600	47,185,920	41,746,432	117,440,512
4	4	415,236,096	1,233,125,376	1,006,632,960	2,717,908,992
4	5	10,687,086,592	30,064,771,072	23,102,226,432	60,397,977,600
5	0	722	6724	12,482	58,564
5	1	123,904	739,328	984,064	3,964,928
5	2	12,781,568	59,969,536	65,619,968	237,899,776
5	3	1,007,681,536	4,060,086,272	3,901,751,296	13,088,325,632
5	4	67,194,847,232	243,609,370,624	213,408,284,672	674,448,277,504
5	5	3,995,393,327,104	243,609,370,624	10,953,240,346,624	33,019,708,571,648

5. Conclusions

The number of spanning trees $\tau(G)$ in graphs (networks) is an important invariant. The computation of this number is not only beneficial from a mathematical (computational) standpoint, but it is also an important measure of the reliability of a network and electrical circuit layout. Some computationally laborious problems such as the traveling salesman problem can be resolved approximately by using spanning trees. In this paper, we define some classes of pyramid graphs created by a gear graph and we have studied the problem of computing the number of spanning trees of these graphs.

Author Contributions: All authors contributed equally to this work. Funding Acquisition, J.-B.L.; Methodology, J.-B.L. and S.N.D. Daoud; Writing—Original Draft, J.-B.L. and S.N.D. Daoud; All authors read and approved the final manuscript.

Funding: The work was partially supported by the China Postdoctoral Science Foundation under grant No. 2017M621579 and the Postdoctoral Science Foundation of Jiangsu Province under grant No. 1701081B, Project of Anhui Jianzhu University under Grant no. 2016QD116 and 2017dc03, Anhui Province Key Laboratory of Intelligent Building & Building Energy Saving.

Acknowledgments: The authors are grateful to the anonymous reviewers for their helpful comments and suggestions for improving the original version of the paper.

Conflicts of Interest: The authors declare that there are no conflicts of interest regarding the publication of this paper.

References

1. Applegate, D.L.; Bixby, R.E.; Chvátal, V.; Cook, W.J. *The Traveling Salesman Problem: A Computational Study*; Princeton University Press: Princeton, NJ, USA, 2006.
2. Cvetkoviĕ, D.; Doob, M.; Sachs, H. *Spectra of Graphs: Theory and Applications*, 3rd ed.; Johann Ambrosius Barth: Heidelberg, Germany, 1995.
3. Kirby, E.C.; Klein, D.J.; Mallion, R.B.; Pollak, P.; Sachs, H. A theorem for counting spanning trees in general chemical graphs and its particular application to toroidal fullerenes. *Croat. Chem. Acta* **2004**, *77*, 263–278.
4. Boesch, F.T.; Satyanarayana, A.; Suffel, C.L. A survey of some network reliability analysis and synthesis results. *Networks* **2009**, *54*, 99–107. [CrossRef]
5. Boesch, F.T. On unreliability polynomials and graph connectivity in reliable network synthesis. *J. Graph Theory* **1986**, *10*, 339–352. [CrossRef]
6. Wu, F.Y. Number of spanning trees on a Lattice. *J. Phys. A* **1977**, *10*, 113–115. [CrossRef]
7. Zhang, F.; Yong, X. Asymptotic enumeration theorems for the number of spanning trees and Eulerian trail in circulant digraphs & graphs. *Sci. China Ser. A* **1999**, *43*, 264–271.
8. Chen, G.; Wu, B.; Zhang, Z. Properties and applications of Laplacian spectra for Koch networks. *J. Phys. A Math. Theor.* **2012**, *45*, 025102.
9. Atajan, T.; Inaba, H. Network reliability analysis by counting the number of spanning trees. In Proceedings of the IEEE International Symposium on Communications and Information Technology, ISCIT 2004, Sapporo, Japan, 26–29 October 2004; pp. 601–604.
10. Brown, T.J.N.; Mallion, R.B.; Pollak, P.; Roth, A. Some methods for counting the spanning trees in labelled molecular graphs, examined in relation to certain fullerenes. *Discret. Appl. Math.* **1996**, *67*, 51–66. [CrossRef]
11. Kirchhoff, G.G. Uber die Auflosung der Gleichungen, auf welche man be ider Untersuchung der Linearen Verteilung galvanischer Storme gefuhrt wird. *Ann. Phys. Chem.* **1847**, *72*, 497–508. [CrossRef]
12. Kelmans, A.K.; Chelnokov, V.M. A certain polynomials of a graph and graphs with an extermal number of trees. *J. Comb. Theory B* **1974**, *16*, 197–214. [CrossRef]
13. Biggs, N.L. *Algebraic Graph Theory*, 2nd ed.; Cambridge University Press: Cambridge, UK, 1993; p. 205.
14. Daoud, S.N. The deletion-contraction method for counting the number of spanning trees of graphs. *Eur. Phys. J. Plus* **2015**, *130*, 217. [CrossRef]
15. Shang, Y. On the number of spanning trees, the Laplacian eigenvalues, and the Laplacian Estrada index of subdivided-line graphs. *Open Math.* **2016**, *14*, 641–648. [CrossRef]
16. Bozkurt, Ş.B.; Bozkurt, D. On the Number of Spanning Trees of Graphs. *Sci. World J.* **2014**, *2014*, 294038. [CrossRef] [PubMed]
17. Daoud, S.N. Number of Spanning Trees in Different Product of Complete and Complete Tripartite Graphs. *ARS Comb.* **2018**, *139*, 85–103.
18. Daoud, S.N. Complexity of Graphs Generated by Wheel Graph and Their Asymptotic Limits. *J. Egypt. Math. Soc.* **2017**, *25*, 424–433. [CrossRef]
19. Daoud, S.N. Chebyshev polynomials and spanning tree formulas. *Int. J. Math. Comb.* **2012**, *4*, 68–79.
20. Zhang, Y.; Yong, X.; Golin, M.J. Chebyshev polynomials and spanning trees formulas for circulant and related graphs. *Discret. Math.* **2005**, *298*, 334–364. [CrossRef]

21. Daoud, S.N. On a class of some pyramid graphs and Chebyshev polynomials. *J. Math. Probl. Eng. Hindawi Publ. Corp.* **2013**, *2013*, 820549.
22. Marcus, M. *A Survey of Matrix Theory and Matrix Inequalities*; University Allyn and Bacon. Inc.: Boston, MA, USA, 1964.

symmetry

MDPI

Article

Novel Three-Way Decisions Models with Multi-Granulation Rough Intuitionistic Fuzzy Sets

Zhan-Ao Xue [1,2,*], Dan-Jie Han [1,2,*], Min-Jie Lv [1,2] and Min Zhang [1,2]

[1] College of Computer and Information Engineering, Henan Normal University, Xinxiang 453007, China; lmj2921419592@163.com (M.-J.L.); zhang_min95@163.com (M.Z.)

[2] Engineering Lab of Henan Province for Intelligence Business & Internet of Things, Henan Normal University, Xinxiang 453007, China

* Correspondence: 121017@htu.edu.cn (Z.-A.X.); handanjie2017@163.com (D.-J.H.)

Received: 27 October 2018; Accepted: 16 November 2018; Published: 21 November 2018

Abstract: The existing construction methods of granularity importance degree only consider the direct influence of single granularity on decision-making; however, they ignore the joint impact from other granularities when carrying out granularity selection. In this regard, we have the following improvements. First of all, we define a more reasonable granularity importance degree calculating method among multiple granularities to deal with the above problem and give a granularity reduction algorithm based on this method. Besides, this paper combines the reduction sets of optimistic and pessimistic multi-granulation rough sets with intuitionistic fuzzy sets, respectively, and their related properties are shown synchronously. Based on this, to further reduce the redundant objects in each granularity of reduction sets, four novel kinds of three-way decisions models with multi-granulation rough intuitionistic fuzzy sets are developed. Moreover, a series of concrete examples can demonstrate that these joint models not only can remove the redundant objects inside each granularity of the reduction sets, but also can generate much suitable granularity selection results using the designed comprehensive score function and comprehensive accuracy function of granularities.

Keywords: three-way decisions; intuitionistic fuzzy sets; multi-granulation rough intuitionistic fuzzy sets; granularity importance degree

---

1. Introduction

Pawlak [1,2] proposed rough sets theory in 1982 as a method of dealing with inaccuracy and uncertainty, and it has been developed into a variety of theories [3–6]. For example, the multi-granulation rough sets (MRS) model is one of the important developments [7,8]. The MRS can also be regarded as a mathematical framework to handle granular computing, which is proposed by Qian et al. [9]. Thereinto, the problem of granularity reduction is a vital research aspect of MRS. Considering the test cost problem of granularity structure selection in data mining and machine learning, Yang et al. constructed two reduction algorithms of cost-sensitive multi-granulation decision-making system based on the definition of approximate quality [10]. Through introducing the concept of distribution reduction [11] and taking the quality of approximate distribution as the measure in the multi-granulation decision rough sets model, Sang et al. proposed an α-lower approximate distribution reduction algorithm based on multi-granulation decision rough sets, however, the interactions among multiple granularities were not considered [12]. In order to overcome the problem of updating reduction, when the large-scale data vary dynamically, Jing et al. developed an incremental attribute reduction approach based on knowledge granularity with a multi-granulation view [13]. Then other multi-granulation reduction methods have been put forward one after another [14–17].

The notion of intuitionistic fuzzy sets (IFS), proposed by Atanassov [18,19], was initially developed in the framework of fuzzy sets [20,21]. Within the previous literature, how to get reasonable membership and non-membership functions is a key issue. In the interest of dealing with fuzzy information better, many experts and scholars have expanded the IFS model. Huang et al. combined IFS with MRS to obtain intuitionistic fuzzy MRS [22]. On the basis of fuzzy rough sets, Liu et al. constructed covering-based multi-granulation fuzzy rough sets [23]. Moreover, multi-granulation rough intuitionistic fuzzy cut sets model was structured by Xue et al. [24]. In order to reduce the classification errors and the limitation of ordering by single theory, they further combined IFS with graded rough sets theory based on dominance relation and extended them to a multi-granulation perspective. [25]. Under the optimistic multi-granulation intuitionistic fuzzy rough sets, Wang et al. proposed a novel method to solve multiple criteria group decision-making problems [26]. However, the above studies rarely deal with the optimal granularity selection problem in intuitionistic fuzzy environments. The measure of similarity between intuitionistic fuzzy sets is also one of the hot areas of research for experts, and some similarity measures about IFS are summarized in references [27–29], whereas these metric formulas cannot measure the importance degree of multiple granularities in the same IFS.

For further explaining the semantics of decision-theoretic rough sets (DTRS), Yao proposed a three-way decisions theory [30,31], which vastly pushed the development of rough sets. As a risk decision-making method, the key strategy of three-way decisions is to divide the domain into acceptance, rejection, and non-commitment. Up to now, researchers have accumulated a vast literature on its theory and application. For instance, in order to narrow the applications limits of three-way decisions model in uncertainty environment, Zhai et al. extended the three-way decisions models to tolerance rough fuzzy sets and rough fuzzy sets, respectively, the target concepts are relatively extended to tolerance rough fuzzy sets and rough fuzzy sets [32,33]. To accommodate the situation where the objects or attributes in a multi-scale decision table are sequentially updated, Hao et al. used sequential three-way decisions to investigate the optimal scale selection problem [34]. Subsequently, Luo et al. applied three-way decisions theory to incomplete multi-scale information systems [35]. With respect to multiple attribute decision-making, Zhang et al. study the inclusion relations of neutrosophic sets in their case in reference [36]. For improving the classification correct rate of three-way decisions, Zhang et al. proposed a novel three-way decisions model with DTRS by considering the new risk measurement functions through the utility theory [37]. Yang et al. combined three-way decisions theory with IFS to obtain novel three-way decision rules [38]. At the same time, Liu et al. explored the intuitionistic fuzzy three-way decision theory based on intuitionistic fuzzy decision systems [39]. Nevertheless, Yang et al. [38] and Liu et al. [39] only considered the case of a single granularity, and did not analyze the decision-making situation of multiple granularities in an intuitionistic fuzzy environment. The DTRS and three-way decisions theory are both used to deal with decision-making problems, so it is also enlightening for us to study three-way decisions theory through DTRS. An extension version that can be used to multi-periods scenarios has been introduced by Liang et al. using intuitionistic fuzzy decision- theoretic rough sets [40]. Furthermore, they introduced the intuitionistic fuzzy point operator into DTRS [41]. The three-way decisions are also applied in multiple attribute group decision making [42], supplier selection problem [43], clustering analysis [44], cognitive computer [45], and so on. However, they have not applied the three-way decisions theory to the optimal granularity selection problem. To solve this problem, we have expanded the three-way decisions models.

The main contributions of this paper include four points:

(1) The new granularity importance degree calculating methods among multiple granularities (i.e., $sig\prime_{in}^{\Delta}(A_i, A\prime, D)$ and $sig\prime_{out}^{\Delta}(A_i, A\prime, D)$) are given respectively, which can generate more discriminative granularities.

(2) Optimistic optimistic multi-granulation rough intuitionistic fuzzy sets (OOMRIFS) model, optimistic pessimistic multi-granulation rough intuitionistic fuzzy sets (OIMRIFS) model,

pessimistic optimistic multi-granulation rough intuitionistic fuzzy sets (IOMRIFS) model and pessimistic pessimistic multi-granulation rough intuitionistic fuzzy sets (IIMRIFS) model are constructed by combining intuitionistic fuzzy sets with the reduction of the optimistic and pessimistic multi-granulation rough sets. These four models can reduce the subjective errors caused by a single intuitionistic fuzzy set.

(3) We put forward four kinds of three-way decisions models based on the proposed four multi-granulation rough intuitionistic fuzzy sets (MRIFS), which can further reduce the redundant objects in each granularity of reduction sets.

(4) Comprehensive score function and comprehensive accuracy function based on MRIFS are constructed. Based on this, we can obtain the optimal granularity selection results.

The rest of this paper is organized as follows. In Section 2, some basic concepts of MRS, IFS, and three-way decisions are briefly reviewed. In Section 3, we propose two new granularity importance degree calculating methods and a granularity reduction Algorithm 1. At the same time, a comparative example is given. Four novel MRIFS models are constructed in Section 4, and the properties of the four models are verified by Example 2. Section 5 proposes some novel three-way decisions models based on above four new MRIFS, and the comprehensive score function and comprehensive accuracy function based on MRIFS are built. At the same time, through Algorithm 2, we make the optimal granularity selection. In Section 6, we use Example 3 to study and illustrate the three-way decisions models based on new MRIFS. Section 7 concludes this paper.

2. Preliminaries

The basic notions of MRS, IFS, and three-way decisions theory are briefly reviewed in this section. Throughout the paper, we denote U as a nonempty object set, i.e., the universe of discourse and $A = \{A_1, A_2, \cdots, A_m\}$ is an attribute set.

Definition 1 ([9]). *Suppose $IS =< U, A, V, f >$ is a consistent information system, $A = \{A_1, A_2, \cdots, A_m\}$ is an attribute set. And R_{A_i} is an equivalence relation generated by A. $[x]_{A_i}$ is the equivalence class of R_{A_i}, $\forall X \subseteq U$, the lower and upper approximations of optimistic multi-granulation rough sets (OMRS) of X are defined by the following two formulas:*

$$\underline{\sum_{i=1}^{m} A_i}^{O}(X) = \{x \in U | [x]_{A_1} \subseteq X \vee [x]_{A_2} \subseteq X \vee [x]_{A_3} \subseteq X \ldots \vee [x]_{A_m} \subseteq X\};$$
$$\overline{\sum_{i=1}^{m} A_i}^{O}(X) = \sim (\underline{\sum_{i=1}^{m} A_i}^{O}(\sim X)).$$

where $\vee$ is a disjunction operation, $\sim X$ is a complement of X, if $\underline{\sum_{i=1}^{m} A_i}^{O}(X) \neq \overline{\sum_{i=1}^{m} A_i}^{O}(X)$, the pair $(\underline{\sum_{i=1}^{m} A_i}^{O}(X), \overline{\sum_{i=1}^{m} A_i}^{O}(X))$ is referred to as an optimistic multi-granulation rough set of X.

Definition 2 ([9]). *Let $IS =< U, A, V, f >$ be an information system, where $A = \{A_1, A_2, \cdots, A_m\}$ is an attribute set, and R_{A_i} is an equivalence relation generated by A. $[x]_{A_i}$ is the equivalence class of R_{A_i}, $\forall X \subseteq U$, the pessimistic multi-granulation rough sets (IMRS) of X with respect to A are defined as follows:*

$$\underline{\sum_{i=1}^{m} A_i}^{I}(X) = \{x \in U | [x]_{A_1} \subseteq X \wedge [x]_{A_2} \subseteq X \wedge [x]_{A_3} \subseteq X \wedge \ldots \wedge [x]_{A_m} \subseteq X\};$$
$$\overline{\sum_{i=1}^{m} A_i}^{I}(X) = \sim (\underline{\sum_{i=1}^{m} A_i}^{I}(\sim X)).$$

where $[x]_{A_i}(1 \leq i \leq m)$ *is equivalence class of x for* A_i, $\wedge$ *is a conjunction operation, if* $\underline{\sum_{i=1}^{m} A_i}^{I}(X) \neq \overline{\sum_{i=1}^{m} A_i}^{I}(X)$, *the pair* $(\underline{\sum_{i=1}^{m} A_i}^{I}(X), \overline{\sum_{i=1}^{m} A_i}^{I}(X))$ *is referred to as a pessimistic multi-granulation rough set of X.*

Definition 3 ([18,19]). *Let U be a finite non-empty universe set, then the IFS E in U are denoted by:*

$$E = \{< x, \mu_E(x), \nu_E(x) > | x \in U\},$$

where $\mu_E(x) : U \to [0,1]$ *and* $\nu_E(x) : U \to [0,1]$. $\mu_E(x)$ *and* $\nu_E(x)$ *are called membership and non-membership functions of the element x in E with* $0 \leq \mu_E(x) + \nu_E(x) \leq 1$. *For* $\forall\, x \in U$, *the hesitancy degree function is defined as* $\pi_E(x) = 1 - \mu_E(x) - \nu_E(x)$, *obviously,* $\pi_E(x) : U \to [0,1]$. *Suppose* $\forall\, E_1, E_2 \in IFS(U)$, *the basic operations of* E_1 *and* E_2 *are given as follows:*

(1) $E_1 \subseteq E_2 \Leftrightarrow \mu_{E_1}(x) \leq \mu_{E_2}(x),\ \nu_{E_1}(x) \geq \nu_{E_2}(x),\ \forall x \in U$;
(2) $A = B \Leftrightarrow \mu_A(x) = \mu_B(x),\ \nu_A(x) = \nu_B(x),\ \forall x \in U$;
(3) $E_1 \cup E_2 = \{< x, \max\{\mu_{E_1}(x), \mu_{E_2}(x)\}, \min\{\nu_{E_1}(x), \nu_{E_2}(x)\} > | x \in U\}$;
(4) (4) $E_1 \cap E_2 = \{< x, \min\{\mu_{E_1}(x), \mu_{E_2}(x)\}, \max\{\nu_{E_1}(x), \nu_{E_2}(x)\} > | x \in U\}$;
(5) (5) $\sim E_1 = \{< x, \nu_{E_1}(x), \mu_{E_1}(x) > | x \in U\}$.

Definition 4 ([30,31]). *Let* $U = \{x_1, x_2, \cdots, x_n\}$ *be a universe of discourse,* $\xi = \{\omega_P, \omega_N, \omega_B\}$ *represents the decisions of dividing an object x into receptive* $POS(X)$, *rejective* $NEG(X)$, *and boundary regions* $BND(X)$, *respectively. The cost functions* λ_{PP}, λ_{NP} *and* λ_{BP} *are used to represent the three decision- making costs of* $\forall x \in U$, *and the cost functions* λ_{PN}, λ_{NN} *and* λ_{BN} *are used to represent the three decision-making costs of* $\forall x \notin U$, *as shown in Table 1.*

Table 1. Cost matrix of decision actions.

Decision Actions	Decision Functions	
	X	**~X**
ω_P	λ_{PP}	λ_{PN}
ω_B	λ_{BP}	λ_{BN}
ω_N	λ_{NP}	λ_{NN}

According to the minimum-risk principle of Bayesian decision procedure, three-way decisions rules can be obtained as follows:

(P): If $P(X|[x]) \geq \alpha$, then $x \in POS(X)$;
(N): If $P(X|[x]) \leq \beta$, then $x \in NEG(X)$;
(B): If $\beta < P(X|[x]) < \alpha$, then $x \in BND(X)$.
Here α, β and γ represent respectively:

$$\alpha = \frac{\lambda_{PN} - \lambda_{BN}}{(\lambda_{PN} - \lambda_{BN}) + (\lambda_{BP} - \lambda_{PP})};$$

$$\beta = \frac{\lambda_{BN} - \lambda_{NN}}{(\lambda_{BN} - \lambda_{NN}) + (\lambda_{NP} - \lambda_{BP})};$$

$$\gamma = \frac{\lambda_{PN} - \lambda_{NN}}{(\lambda_{PN} - \lambda_{NN}) + (\lambda_{NP} - \lambda_{PP})}.$$

3. Granularity Reduction Algorithm Derives from Granularity Importance Degree

Definition 5 ([10,12]). *Let $DIS = (U, C \cup D, V, f)$ be a decision information system, $A = \{A_1, A_2, \cdots, A_m\}$ are m sub-attributes of condition attributes C. $U/D = \{X_1, X_2, \cdots, X_s\}$ is the partition induced by the decision attributes D, then approximation quality of U/D about granularity set A is defined as:*

$$\gamma(A, D) = \frac{\left| \cup \left\{ \underline{\sum_{i=1}^{m} A_i}^{\Delta} (X_t) | 1 \leq t \leq s \right\} \right|}{|U|}.$$

where $|X|$ denotes the cardinal number of set X. $\Delta \in \{O, I\}$ represents two cases of optimistic and pessimistic multi-granulation rough sets, the same as the following.

Definition 6 ([12]). *Let $DIS = (U, C \cup D, V, f)$ be a decision information system, $A = \{A_1, A_2, \cdots, A_m\}$ are m sub-attributes of C, $A\prime \subseteq A$, $X \in U/D$,*

(1) If $\underline{\sum_{i=1, A_i \in A}^{m} A_i}^{\Delta} (X) \neq \underline{\sum_{i=1, A_i \in A - A'}^{m} A_i}^{\Delta} (X)$, then A' is important in A for X;

(2) If $\underline{\sum_{i=1, A_i \in A}^{m} A_i}^{\Delta} (X) = \underline{\sum_{i=1, A_i \in A - A'}^{m} A_i}^{\Delta} (X)$, then A' is not important in A for X.

Definition 7 ([10,12]). *Suppose $DIS = (U, C \cup D, V, f)$ is a decision information system, $A = \{A_1, A_2, \cdots, A_m\}$ are m sub-attributes of C, $A\prime \subseteq A$. $\forall A_i \in A\prime$, on the granularity sets $A\prime$, the internal importance degree of A_i for D can be defined as follows:*

$$sig_{in}^{\Delta}(A_i, A\prime, D) = |\gamma(A\prime, D) - \gamma(A\prime - \{A_i\}, D)|.$$

Definition 8 ([10,12]). *Let $DIS = (U, C \cup D, V, f)$ be a decision information system, $A = \{A_1, A_2, \cdots, A_m\}$ are m sub-attributes of C, $A\prime \subseteq A$. $\forall A_i \in A - A\prime$, on the granularity sets $A\prime$, the external importance degree of A_i for D can be defined as follows:*

$$sig_{out}^{\Delta}(A_i, A\prime, D) = |\gamma(A_i \cup A\prime, D) - \gamma(A\prime, D)|.$$

Theorem 1. *Let $DIS = (U, C \cup D, V, f)$ be a decision information system, $A = \{A_1, A_2, \cdots, A_m\}$ are m sub-attributes of C, $A\prime \subseteq A$.*

(1) For $\forall A_i \in A\prime$, on the basis of attribute subset family A', the granularity importance degree of A_i in $A\prime$ with respect to D is expressed as follows:

$$sig_{in}^{\Delta}(A_i, A\prime, D) = \frac{1}{m-1} \sum |sig_{in}^{\Delta}(\{A_k, A_i\}, A\prime, D) - sig_{in}^{\Delta}(A_k, A\prime - \{A_i\}, D)|.$$

where $1 \leq k \leq m, k \neq i$, the same as the following.

(2) For $\forall A_i \in A - A\prime$, on the basis of attribute subset family A', the granularity importance degree of A_i in $A - A\prime$ with respect to D, we have:

$$sig_{out}^{\Delta}(A_i, A\prime, D) = \frac{1}{m-1} \sum |sig_{out}^{\Delta}(\{A_k, A_i\}, \{A_i\} \cup A\prime, D) - sig_{out}^{\Delta}(A_k, A\prime, D)|.$$

Proof. (1) According to Definition 7, then

$$\begin{aligned} sig_{in}^{\Delta}(A_i, A\prime, D) &= |\gamma(A\prime, D) - \gamma(A\prime - \{A_i\}, D)| \\ &= \tfrac{m-1}{m-1}|\gamma(A\prime, D) - \gamma(A\prime - \{A_i\}, D)| + \sum|\gamma(A\prime - \{A_k, A_i\}, D) - \gamma(A\prime - \{A_k, A_i\}, D)| \\ &= \tfrac{1}{m-1}\sum(|\gamma(A\prime, D) - \gamma(A\prime - \{A_k, A_i\}, D) - (\gamma(A\prime - \{A_i\}, D) - \gamma(A\prime - \{A_k, A_i\}, D)|) \\ &= \tfrac{1}{m-1}\sum|sig_{in}^{\Delta}(\{A_k, A_i\}, A\prime, D) - sig_{in}^{\Delta}(A_k, A\prime - \{A_i\}, D)|. \end{aligned}$$

(2) According to Definition 8, we can get:

$$\begin{aligned} sig_{out}^{\Delta}(A_i, A\prime, D) &= |\gamma(\{A_i\} \cup A\prime, D) - \gamma(A\prime, D)| \\ &= \tfrac{m-1}{m-1}|\gamma(\{A_i\} \cup A\prime, D) - \gamma(A\prime, D)| - \sum|\gamma(A\prime - \{A_k\}, D) - \gamma(A\prime - \{A_k\}, D)| \\ &= \tfrac{1}{m-1}\sum(|\gamma(\{A_i\} \cup A\prime, D) - \gamma(A\prime - \{A_k\}, D)| - |(\gamma(A\prime - \{A_k\}, D) - \gamma(A\prime, D)|) \\ &= \tfrac{1}{m-1}\sum|sig_{out}^{\Delta}(\{A_k, A_i\}, \{A_i\} \cup A\prime, D) - sig_{out}^{\Delta}(A_k, A\prime, D)|. \end{aligned}$$

□

In Definitions 7 and 8, only the direct effect of a single granularity on the whole granularity sets is given, without considering the indirect effect of the remaining granularities on decision-making. The following Definitions 9 and 10 synthetically analyze the interdependence between multiple granularities and present two new methods for calculating granularity importance degree.

Definition 9. *Let $DIS = (U, C \cup D, V, f)$ be a decision information system, $A = \{A_1, A_2, \cdots, A_m\}$ are m sub-attributes of C, $A\prime \subseteq A$. $\forall A_i, A_k \in A\prime$, on the attribute subset family, A, the new internal importance degree of A_i relative to D is defined as follows:*

$$sig\prime_{in}^{\Delta}(A_i, A\prime, D) = sig_{in}^{\Delta}(A_i, A\prime, D) + \frac{1}{m-1}\sum|sig_{in}^{\Delta}(A_k, A\prime - \{A_i\}, D) - sig_{in}^{\Delta}(A_k, A\prime, D)|.$$

$sig_{in}^{\Delta}(A_i, A\prime, D)$ and $\frac{1}{m-1}\sum|sig_{in}^{\Delta}(A_k, A\prime - \{A_i\}, D) - sig_{in}^{\Delta}(A_k, A\prime, D)|$ respectively indicate the direct and indirect effects of granularity A_i on decision-making. When $|sig_{in}^{\Delta}(A_k, A\prime - \{A_i\}, D) - sig_{in}^{\Delta}(A_k, A\prime, D)| > 0$ is satisfied, it is shown that the granularity importance degree of A_k is increased by the addition of A_i in attribute subset $A\prime - \{A_i\}$, so the granularity importance degree of A_k should be added to A_i. Therefore, when there are m sub-attributes, we should add $\frac{1}{m-1}\sum|sig_{in}^{\Delta}(A_k, A\prime - \{A_i\}, D) - sig_{in}^{\Delta}(A_k, A\prime, D)|$ to the granularity importance degree of A_i.

If $|sig_{in}^{\Delta}(A_k, A\prime - \{A_i\}, D) - sig_{in}^{\Delta}(A_k, A\prime, D)| = 0$ and $k \neq i$, then it shows that there is no interaction between granularity A_i and other granularities, which means $sig\prime_{in}^{\Delta}(A_i, A\prime, D) = sig_{in}^{\Delta}(A_i, A\prime, D)$.

Definition 10. *Let $DIS = (U, C \cup D, V, f)$ be a decision information system, $A = \{A_1, A_2, \cdots, A_m\}$ be m sub-attributes of C, $A\prime \subseteq A$. $\forall A_i \in A - A\prime$, the new external importance degree of A_i relative to D is defined as follows:*

$$sig\prime_{out}^{\Delta}(A_i, A\prime, D) = sig_{out}^{\Delta}(A_i, A\prime, D) + \frac{1}{m-1}\sum|sig_{out}^{\Delta}(A_k, A\prime, D) - sig_{out}^{\Delta}(A_k, \{A_i\} \cup A\prime, D)|.$$

Similarly, the new external importance degree calculation formula has a similar effect.

Theorem 2. *Let $DIS = (U, C \cup D, V, f)$ be a decision information system, $A = \{A_1, A_2, \cdots, A_m\}$ be m sub-attributes of C, $A\prime \subseteq A$, $\forall A_i \in A\prime$. The improved internal importance can be rewritten as:*

$$sig\prime_{in}^{\Delta}(A_i, A\prime, D) = \frac{1}{m-1}\sum sig_{in}^{\Delta}(A_i, A\prime - \{A_k\}, D).$$

Proof.

$$\begin{aligned} sig{\prime}_{in}^{\Delta}(A_i, A\prime, D) &= sig_{in}^{\Delta}(A_i, A\prime, D) + \tfrac{1}{m-1}\sum |sig_{in}^{\Delta}(A_k, A\prime - \{A_i\}, D) - sig_{in}^{\Delta}(A_k, A\prime, D)| \\ &= \tfrac{m-1}{m-1}|\gamma(A\prime, D) - \gamma(A\prime - \{A_i\}, D)| + \tfrac{1}{m-1}\sum ||\gamma(A\prime - \{A_i\}, D) - \\ &\gamma(A\prime - \{A_k, A_i\}, D)| - |\gamma(A\prime, D) - \gamma(A\prime - \{A_k\}, D)|| \\ &= \tfrac{1}{m-1}\sum |\gamma(A\prime - \{A_k\}, D) - \gamma(A\prime - \{A_k, A_i\}, D)| \\ &= \tfrac{1}{m-1}\sum sig_{in}^{\Delta}(A_i, A\prime - \{A_k\}, D). \end{aligned}$$

□

Theorem 3. *Let $DIS = (U, C \cup D, V, f)$ be a decision information system, $A = \{A_1, A_2, \cdots, A_m\}$ are m sub-attributes of C, $A\prime \subseteq A$. The improved external importance can be expressed as follows:*

$$sig{\prime}_{out}^{\Delta}(A_i, A\prime, D) = \frac{1}{m-1}\sum sig_{out}^{\Delta}(A_i, \{A_k\} \cup A\prime, D).$$

Proof.

$$\begin{aligned} sig{\prime}_{out}^{\Delta}(A_i, A\prime, D) &= sig_{out}^{\Delta}(A_i, A\prime, D) + \tfrac{1}{m-1}\sum |(sig_{out}^{\Delta}(A_k, A\prime, D) - sig_{out}^{\Delta}(A_k, \{A_i\} \cup A\prime, D))| \\ &= \tfrac{m-1}{m-1}|\gamma(\{A_i\} \cup A\prime, D) - \gamma(A\prime, D)| + \tfrac{1}{m-1}\sum ||\gamma(A\prime, D) - \gamma(\{A_k\} \cup A\prime, D)| - \\ &|\gamma(\{A_i\} \cup A\prime, D)|| \\ &= \tfrac{1}{m-1}\sum |\gamma(\{A_i, A_k\} \cup A\prime, D) - \gamma(\{A_i\} \cup A\prime, D)| \\ &= \tfrac{1}{m-1}\sum sig_{out}^{\Delta}(A_i, \{A_k\} \cup A\prime, D). \end{aligned}$$

□

Theorems 2 and 3 show that when $sig_{in}^{\Delta}(A_i, A\prime - \{A_k\}, D) = 0$ ($sig_{out}^{\Delta}(A_i, \{A_k\} \cup A\prime, D) = 0$) is satisfied, having $sig{\prime}_{in}^{\Delta}(A_i, A\prime, D) = 0$ ($sig{\prime}_{out}^{\Delta}(A_i, A\prime, D) = 0$). And each granularity importance degree is calculated on the basis of removing A_k from $A\prime$, which makes it more convenient for us to choose the required granularity.

According to [10,12], we can get optimistic and pessimistic multi-granulation lower approximations L^O and L^I. The granularity reduction algorithm based on improved granularity importance degree is derived from Theorems 2 and 3, as shown in Algorithm 1.

Algorithm 1. Granularity reduction algorithm derives from granularity importance degree

---

Input: $DIS = (U, C \cup D, V, f)$, $A = \{A_1, A_2, \cdots, A_m\}$ *are m sub-attributes of* C, $A\prime \subseteq A, \forall A_i \in A\prime$, $U/D = \{X_1, X_2, \cdots, X_s\}$;
Output: A granularity reduction set A_i^{Δ} of this information system.
set up $A_i^{\Delta} \leftarrow \phi, 1 \leq h \leq m$;
compute U/D, optimistic and pessimistic multi-granulation lower approximations L^{Δ};
for $\forall A_i \in A$
  compute $sig\prime_{in}^{\Delta}(A_i, A\prime, D)$ via Definition 9;
  if $(sig\prime_{in}^{\Delta}(A_i, A\prime, D) > 0)$ **then** $A_i^{\Delta} = A_i^{\Delta} \cup A_i$;
  end
  for $\forall A_i \in A - A_i^{\Delta}$
    if $\gamma(A_i^{\Delta}, D) = \gamma(A, D)$ **then** compute $sig_{out}^{\prime,\Delta}(A_i, A\prime, D)$ via Definition 10;
    end
    if $sig\prime_{out}^{\Delta}(A_h, A\prime, D) = \max\{sig\prime_{out}^{\Delta}(A_h, A\prime, D)\}$ **then** $A_i^{\Delta} = A_i^{\Delta} \cup A_h$;
    end
  end
  for $\forall A_i \in A_i^{\Delta}$,
    if $\gamma(A_i^{\Delta} - A_i, D) = \gamma(A, D)$ **then** $A_i^{\Delta} = A_i^{\Delta} - A_i$;
    end
  end
return granularity reduction set A_i^{Δ};
end

Therefore, we can obtain two reductions by utilizing Algorithm 1.

Example 1. *This paper calculates the granularity importance of 10 on-line investment schemes given in Reference [12]. After comparing and analyzing the obtained granularity importance degree, we can obtain the reduction results of 5 evaluation sites through Algorithm 1, and the detailed calculation steps are as follows.*

According to [12], we can get $A = \{A_1, A_2, A_3, A_4, A_5\}$, $A\prime \subseteq A$, $U/D = \{\{x_1, x_2, x_4, x_6, x_8\}, \{x_3, x_5, x_7, x_9, x_{10}\}\}$.

(1) Reduction set of OMRS

First of all, we can calculate the internal importance degree of OMRS by Theorem 2 as shown in Table 2.

Table 2. Internal importance degree of optimistic multi-granulation rough sets (OMRS).

	A_1	A_2	A_3	A_4	A_5
$sig_{in}^{O}(A_i, A\prime, D)$	0	0.15	0.05	0	0.05
$sig\prime_{in}^{O}(A_i, A\prime, D)$	0.025	0.375	0.225	0	0

Then, according to Algorithm 1, we can deduce the initial granularity set is $\{A_1, A_2, A_3\}$. Inspired by Definition 5, we obtain $r^O(\{A_2, A_3\}, D) = r^O(A, D) = 1$. So, the reduction set of the OMRS is $A_i^O = \{A_2, A_3\}$.

As shown in Table 2, when using the new method to calculate internal importance degree, more discriminative granularities can be generated, which are more convenient for screening out the required granularities. In literature [12], the approximate quality of granularity A_2 in the reduction set is different from that of the whole granularity set, so it is necessary to calculate the external importance degree again. When calculating the internal and external importance degree, References [10,12] only considered the direct influence of the single granularity on the granularity A_2, so the influence of the granularity A_2 on the overall decision-making can't be fully reflected.

(2) Reduction set of IMRS

Similarly, by using Theorem 2, we can get the internal importance degree of each site under IMRS, as shown in Table 3.

Table 3. Internal importance degree of pessimistic multi-granulation rough sets (IMRS).

	A_1	A_2	A_3	A_4	A_5
$sig^I_{in}(A_i, A\prime, D)$	0	0.05	0	0	0
$sig\prime^I_{in}(A_i, A\prime, D)$	0	0.025	0	0.025	0.025

According to Algorithm 1, the sites 2, 4, and 5 with internal importance degrees greater than 0, which are added to the granularity reduction set as the initial granularity set, and then the approximate quality of it can be calculated as follows:

$$r^I(\{A_2, A_4\}, D) = r^I(\{A_4, A_5\}, D) = r^I(A, D) = 0.2.$$

Namely, the reduction set of IMRS is $A^I_i = \{A_2, A_4\}$ or $A^I_i = \{A_4, A_5\}$ without calculating the external importance degree.

In this paper, when calculating the internal and external importance degree of each granularity, the influence of removing other granularities on decision-making is also considered. According to Theorem 2, after calculating the internal importance degree of OMRS and IMRS, if the approximate quality of each granularity in the reduction sets are the same as the overall granularities, it is not necessary to calculate the external importance degree again, which can reduce the amount of computation.

4. Novel Multi-Granulation Rough Intuitionistic Fuzzy Sets Models

In Example 1, two reduction sets are obtained under IMRS, so we need a novel method to obtain more accurate granularity reduction results by calculating granularity reduction.

In order to obtain the optimal determined site selection result, we combine the optimistic and pessimistic multi-granulation reduction sets based on Algorithm 1 with IFS, respectively, and construct the following four new MRIFS models.

Definition 11 ([22,25]). *Suppose $IS = (U, A, V, f)$ is an information system, $A = \{A_1, A_2, \cdots, A_m\}$. $\forall E \subseteq U$, E are IFS. Then the lower and upper approximations of optimistic MRIFS of A_i are respectively defined by:*

$$\underline{\sum_{i=1}^{m} R_{A_i}}^{O}(E) = \{< x, \mu_{\underline{\sum_{i=1}^{m} R_{A_i}}^{O}(E)}(x), \nu_{\underline{\sum_{i=1}^{m} R_{A_i}}^{O}(E)}(x) > | x \in U\};$$

$$\overline{\sum_{i=1}^{m} R_{A_i}}^{O}(E) = \{< x, \mu_{\overline{\sum_{i=1}^{m} R_{A_i}}^{O}(E)}(x), \nu_{\overline{\sum_{i=1}^{m} R_{A_i}}^{O}(E)}(x) > | x \in U\}.$$

where

$$\mu_{\underline{\sum_{i=1}^{m} R_{A_i}}^{O}(E)}(x) = \bigvee_{i=1}^{m} \inf_{y \in [x]_{A_i}} \mu_E(y),\ \nu_{\underline{\sum_{i=1}^{m} R_{A_i}}^{O}(E)}(x) = \bigwedge_{i=1}^{m} \sup_{y \in [x]_{A_i}} \nu_E(y);$$

$$\mu_{\overline{\sum_{i=1}^{m} R_{A_i}}^{O}(E)}(x) = \bigwedge_{i=1}^{m} \sup_{y \in [x]_{A_i}} \mu_E(y),\ \nu_{\overline{\sum_{i=1}^{m} R_{A_i}}^{O}(E)}(x) = \bigvee_{i=1}^{m} \inf_{y \in [x]_{A_i}} \nu_E(y).$$

where R_{A_i} is an equivalence relation of x in A, $[x]_{A_i}$ is the equivalence class of R_{A_i}, and $\vee$ is a disjunction operation.

Definition 12 ([22,25]). *Suppose $IS =< U, A, V, f >$ is an information system, $A = \{A_1, A_2, \cdots, A_m\}$. $\forall E \subseteq U$, E are IFS. Then the lower and upper approximations of pessimistic MRIFS of A_i can be described as follows:*

$$\underline{\sum_{i=1}^{m} R_{A_i}}^{I}(E) = \{< x, \mu_{\underline{\sum_{i=1}^{m} R_{A_i}}^{I}(E)}(x), \nu_{\underline{\sum_{i=1}^{m} R_{A_i}}^{I}(E))}(x) > | x \in U\};$$

$$\overline{\sum_{i=1}^{m} R_{A_i}}^{I}(E) = \{< x, \mu_{\overline{\sum_{i=1}^{m} R_{A_i}}^{I}(E)}(x), \nu_{\overline{\sum_{i=1}^{m} R_{A_i}}^{I}(E)}(x) > | x \in U\}.$$

where

$$\mu_{\underline{\sum_{i=1}^{m} R_{A_i}}^{I}(E)}(x) = \bigwedge_{i=1}^{m} \inf_{y\in[x]_{A_i}} \mu_E(y), \ \nu_{\underline{\sum_{i=1}^{m} R_{A_i}}^{I}(E)}(x) = \bigvee_{i=1}^{m} \sup_{y\in[x]_{A_i}} \nu_E(y);$$

$$\mu_{\overline{\sum_{i=1}^{m} R_{A_i}}^{I}(E)}(x) = \bigvee_{i=1}^{m} \sup_{y\in[x]_{A_i}} \mu_E(y), \ \nu_{\overline{\sum_{i=1}^{m} R_{A_i}}^{I}(E)}(x) = \bigwedge_{i=1}^{m} \inf_{y\in[x]_{A_i}} \nu_E(y).$$

where $[x]_{A_i}$ is the equivalence class of x about the equivalence relation R_{A_i}, and $\wedge$ is a conjunction operation.

Definition 13. *Suppose $IS =< U, A, V, f >$ is an information system, $A_i^O = \{A_1, A_2, \cdots, A_r\} \subseteq A$, $A = \{A_1, A_2, \cdots, A_m\}$. And $R_{A_i^O}$ is an equivalence relation of x with respect to the attribute reduction set A_i^O under OMRS, $[x]_{A_i^O}$ is the equivalence class of $R_{A_i^O}$. Let E be IFS of U and they can be characterized by a pair of lower and upper approximations:*

$$\underline{\sum_{i=1}^{r} R_{A_i^O}}^{O}(E) = \{< x, \mu_{\underline{\sum_{i=1}^{r} R_{A_i^O}}^{O}(E)}(x), \nu_{\underline{\sum_{i=1}^{r} R_{A_i^O}}^{O}(E)}(x) > | x \in U\};$$

$$\overline{\sum_{i=1}^{r} R_{A_i^O}}^{O}(E) = \{< x, \mu_{\overline{\sum_{i=1}^{r} R_{A_i^O}}^{O}(E)}(x), \nu_{\overline{\sum_{i=1}^{r} R_{A_i^O}}^{O}(E)}(x) > | x \in U\}.$$

where

$$\mu_{\underline{\sum_{i=1}^{r} R_{A_i^O}}^{O}(E)}(x) = \bigvee_{i=1}^{r} \inf_{y\in[x]_{A_i^O}} \mu_E(y), \ \nu_{\underline{\sum_{i=1}^{r} R_{A_i^O}}^{O}(E)}(x) = \bigwedge_{i=1}^{r} \sup_{y\in[x]_{A_i^O}} \nu_E(y);$$

$$\mu_{\overline{\sum_{i=1}^{r} R_{A_i^O}}^{O}(E)}(x) = \bigwedge_{i=1}^{r} \sup_{y\in[x]_{A_i^O}} \mu_E(y), \ \nu_{\overline{\sum_{i=1}^{r} R_{A_i^O}}^{O}(E)}(x) = \bigvee_{i=1}^{r} \inf_{y\in[x]_{A_i^O}} \nu_E(y).$$

If $\underline{\sum_{i=1}^{r} R_{A_i^O}}^{O}(E) \neq \overline{\sum_{i=1}^{r} R_{A_i^O}}^{O}(E)$, then E can be called OOMRIFS.

Definition 14. *Suppose $IS =< U, A, V, f >$ is an information system, $\forall E \subseteq U$, E are IFS. $A_i^O = \{A_1, A_2, \cdots, A_r\} \subseteq A$, $A = \{A_1, A_2, \cdots, A_m\}$. where A_i^O is an optimistic multi-granulation attribute reduction set. Then the lower and upper approximations of pessimistic MRIFS under optimistic multi-granulation environment can be defined as follows:*

$$\underline{\sum_{i=1}^{r} R_{A_i^O}}^{I}(E) = \{< x, \mu_{\underline{\sum_{i=1}^{r} R_{A_i^O}}^{I}(E)}(x), \nu_{\underline{\sum_{i=1}^{r} R_{A_i^O}}^{I}(E)}(x) > | x \in U\};$$

$$\overline{\sum_{i=1}^{r} R_{A_i^O}}^{I}(E) = \{< x, \mu_{\overline{\sum_{i=1}^{r} R_{A_i^O}}^{I}(E)}(x), \nu_{\overline{\sum_{i=1}^{r} R_{A_i^O}}^{I}(E)}(x) > | x \in U\}.$$

where

$$\mu_{\underline{\sum_{i=1}^{r} R_{A_i^O}}^{I}(E)}(x) = \bigwedge_{i=1}^{r} \inf_{y\in[x]_{A_i^O}} \mu_E(y), \nu_{\underline{\sum_{i=1}^{r} R_{A_i^O}}^{I}(E)}(x) = \bigvee_{i=1}^{r} \sup_{y\in[x]_{A_i^O}} \nu_E(y);$$

$$\mu_{\overline{\sum_{i=1}^{r} R_{A_i^O}}^{I}(E)}(x) = \bigvee_{i=1}^{r} \sup_{y\in[x]_{A_i^O}} \mu_E(y), \nu_{\overline{\sum_{i=1}^{r} R_{A_i^O}}^{I}(E)}(x) = \bigwedge_{i=1}^{r} \inf_{y\in[x]_{A_i^O}} \nu_E(y).$$

The pair $(\underline{\sum_{i=1}^{r} R_{A_i^O}}^{I}(E), \overline{\sum_{i=1}^{r} R_{A_i^O}}^{I}(E))$ *are called OIMRIFS, if* $\underline{\sum_{i=1}^{r} R_{A_i^O}}^{I}(E) \neq \overline{\sum_{i=1}^{r} R_{A_i^O}}^{I}(E)$.

According to Definitions 13 and 14, the following theorem can be obtained.

Theorem 4. *Let* $IS = < U, A, V, f >$ *be an information system,* $A_i^O = \{A_1, A_2, \cdots, A_r\} \subseteq A$, $A = \{A_1, A_2, \cdots, A_m\}$, *and* E_1, E_2 *be IFS on U. Comparing with Definitions 13 and 14, the following proposition is obtained.*

(1) $\underline{\sum_{i=1}^{r} R_{A_i^O}}^{O}(E_1) = \bigcup_{i=1}^{r} \underline{R_{A_i^O}}^{O}(E_1)$;

(2) $\overline{\sum_{i=1}^{r} R_{A_i^O}}^{O}(E_1) = \bigcap_{i=1}^{r} \overline{R_{A_i^O}}^{O}(E_1)$;

(3) $\underline{\sum_{i=1}^{r} R_{A_i^O}}^{I}(E_1) = \bigcap_{i=1}^{r} \underline{R_{A_i^O}}^{I}(E_1)$;

(4) $\underline{\sum_{i=1}^{r} R_{A_i^O}}^{I}(E_1) = \bigcup_{i=1}^{r} \underline{R_{A_i^O}}^{I}(E_1)$;

(5) $\underline{\sum_{i=1}^{r} R_{A_i^O}}^{I}(E_1) \subseteq \underline{\sum_{i=1}^{r} R_{A_i^O}}^{O}(E_1)$;

(6) $\overline{\sum_{i=1}^{r} R_{A_i^O}}^{O}(E_1) \subseteq \overline{\sum_{i=1}^{r} R_{A_i^O}}^{I}(E_1)$;

(7) $\underline{\sum_{i=1}^{r} R_{A_i^O}}^{O}(E_1 \cap E_2) = \underline{\sum_{i=1}^{r} R_{A_i^O}}^{O}(E_1) \cap \underline{\sum_{i=1}^{r} R_{A_i^O}}^{O}(E_2), \underline{\sum_{i=1}^{r} R_{A_i^O}}^{I}(E_1 \cap E_2) = \underline{\sum_{i=1}^{r} R_{A_i^O}}^{I}(E_1) \cap \underline{\sum_{i=1}^{r} R_{A_i^O}}^{I}(E_2)$;

(8) $\overline{\sum_{i=1}^{r} R_{A_i^O}}^{O}(E_1 \cup E_2) = \overline{\sum_{i=1}^{r} R_{A_i^O}}^{O}(E_1) \cup \overline{\sum_{i=1}^{r} R_{A_i^O}}^{O}(E_2), \overline{\sum_{i=1}^{r} R_{A_i^O}}^{I}(E_1 \cup E_2) = \overline{\sum_{i=1}^{r} R_{A_i^O}}^{I}(E_1) \cup \overline{\sum_{i=1}^{r} R_{A_i^O}}^{I}(E_2)$;

(9) $\underline{\sum_{i=1}^{r} R_{A_i^O}}^{O}(E_1 \cup E_2) \supseteq \underline{\sum_{i=1}^{r} R_{A_i^O}}^{O}(E_1) \cup \underline{\sum_{i=1}^{r} R_{A_i^O}}^{O}(E_2), \underline{\sum_{i=1}^{r} R_{A_i^O}}^{I}(E_1 \cup E_2) \supseteq \underline{\sum_{i=1}^{r} R_{A_i^O}}^{I}(E_1) \cup \underline{\sum_{i=1}^{r} R_{A_i^O}}^{I}(E_2)$;

(10) $\overline{\sum_{i=1}^{r} R_{A_i^O}}^{O}(E_1 \cap E_2) \subseteq \overline{\sum_{i=1}^{r} R_{A_i^O}}^{O}(E_1) \cap \overline{\sum_{i=1}^{r} R_{A_i^O}}^{O}(E_2), \overline{\sum_{i=1}^{r} R_{A_i^O}}^{I}(E_1 \cap E_2) \subseteq \overline{\sum_{i=1}^{r} R_{A_i^O}}^{I}(E_1) \cap \overline{\sum_{i=1}^{r} R_{A_i^O}}^{I}(E_2)$.

Proof. It is easy to prove by the Definitions 13 and 14. □

Definition 15. *Let* $IS = < U, A, V, f >$ *be an information system, and E be IFS on U.* $A_i^I = \{A_1, A_2, \cdots, A_r\} \subseteq A$, $A = \{A_1, A_2, \cdots, A_m\}$, *where* A_i^I *is a pessimistic multi-granulation attribute reduction set. Then, the pessimistic optimistic lower and upper approximations of E with respect to equivalence relation* $R_{A_i^I}$ *are defined by the following formulas:*

$$\underline{\sum_{i=1}^{r} R_{A_i^I}}^{O}(E) = \{< x, \mu_{\underline{\sum_{i=1}^{r} R_{A_i^I}}^{O}(E)}(x), \nu_{\underline{\sum_{i=1}^{r} R_{A_i^I}}^{O}(E)}(x) > | x \in U\};$$

$$\overline{\sum_{i=1}^{r} R_{A_i^I}}^{O}(E) = \{< x, \mu_{\overline{\sum_{i=1}^{r} R_{A_i^I}}^{O}(E)}(x), \nu_{\overline{\sum_{i=1}^{r} R_{A_i^I}}^{O}(E)}(x) > | x \in U\}.$$

where

$$\mu_{\underline{\sum_{i=1}^{r} R_{A_i^I}}^{O}(E)}(x) = \bigvee_{i=1}^{r} \inf_{y\in[x]_{A_i^I}} \mu_E(y), \nu_{\underline{\sum_{i=1}^{r} R_{A_i^I}}^{O}(E)}(x) = \bigwedge_{i=1}^{r} \sup_{y\in[x]_{A_i^I}} \nu_E(y);$$
$$\mu_{\overline{\sum_{i=1}^{r} R_{A_i^I}}^{O}(E)}(x) = \bigwedge_{i=1}^{r} \sup_{y\in[x]_{A_i^I}} \mu_E(y), \nu_{\overline{\sum_{i=1}^{r} R_{A_i^I}}^{O}(E)}(x) = \bigvee_{i=1}^{r} \inf_{y\in[x]_{A_i^I}} \nu_E(y).$$

If $\underline{\sum_{i=1}^{r} R_{A_i^I}}^{O}(E) \neq \overline{\sum_{i=1}^{r} R_{A_i^I}}^{O}(E)$*, then E can be called IOMRIFS.*

Definition 16. *Let* $IS =< U, A, V, f >$ *be an information system, and E be IFS on U.* $A_i^I = \{A_1, A_2, \cdots, A_r\} \subseteq A$, $A = \{A_1, A_2, \cdots, A_m\}$, *where* A_i^I *is a pessimistic multi-granulation attribute reduction set. Then, the pessimistic lower and upper approximations of E under IMRS are defined by the following formulas:*

$$\underline{\sum_{i=1}^{r} R_{A_i^I}}^{I}(E) = \{< x, \mu_{\underline{\sum_{i=1}^{r} R_{A_i^I}}^{I}(E)}(x), \nu_{\underline{\sum_{i=1}^{r} R_{A_i^I}}^{I}(E)}(x) > | x \in U\};$$
$$\overline{\sum_{i=1}^{r} R_{A_i^I}}^{I}(E) = \{< x, \mu_{\overline{\sum_{i=1}^{r} R_{A_i^I}}^{I}(E)}(x), \nu_{\overline{\sum_{i=1}^{r} R_{A_i^I}}^{I}(E)}(x) > | x \in U\}.$$

where

$$\mu_{\underline{\sum_{i=1}^{r} R_{A_i^I}}^{I}(E)}(x) = \bigwedge_{i=1}^{r} \inf_{y\in[x]_{A_i^I}} \mu_E(y), \nu_{\underline{\sum_{i=1}^{r} R_{A_i^I}}^{I}(E)}(x) = \bigvee_{i=1}^{r} \sup_{y\in[x]_{A_i^I}} \nu_E(y);$$
$$\mu_{\overline{\sum_{i=1}^{r} R_{A_i^I}}^{I}(E)}(x) = \bigvee_{i=1}^{r} \sup_{y\in[x]_{A_i^I}} \mu_E(y), \nu_{\overline{\sum_{i=1}^{r} R_{A_i^I}}^{I}(E)}(x) = \bigwedge_{i=1}^{r} \inf_{y\in[x]_{A_i^I}} \nu_E(y).$$

where $R_{A_i^I}$ *is an equivalence relation of x about the attribute reduction set* A_i^I *under IMRS,* $[x]_{A_i^O}$ *is the equivalence class of* $R_{A_i^I}$.

If $\underline{\sum_{i=1}^{r} R_{A_i^I}}^{I}(E) \neq \overline{\sum_{i=1}^{r} R_{A_i^I}}^{I}(E)$*, then the pair* $(\underline{\sum_{i=1}^{r} R_{A_i^I}}^{I}(E), \overline{\sum_{i=1}^{r} R_{A_i^I}}^{I}(E))$ *is said to be IIMRIFS.*

According to Definitions 15 and 16, the following theorem can be captured.

Theorem 5. *Let* $IS =< U, A, V, f >$ *be an information system,* $A_i^I = \{A_1, A_2, \cdots, A_r\} \subseteq A$, $A = \{A_1, A_2, \cdots, A_m\}$, *and* E_1, E_2 *be IFS on U. Then IOMRIFS and IIOMRIFS models have the following properties:*

(1) $\underline{\sum_{i=1}^{r} R_{A_i^I}}^{O}(E_1) = \bigcup_{i=1}^{r} \underline{R_{A_i^I}}^{O}(E_1)$;

(2) $\overline{\sum_{i=1}^{r} R_{A_i^I}}^{O}(E_1) = \bigcap_{i=1}^{r} \overline{R_{A_i^I}}^{O}(E_1)$;

(3) $\underline{\sum_{i=1}^{r} R_{A_i^I}}^{I}(E_1) = \bigcup_{i=1}^{r} \underline{R_{A_i^I}}^{I}(E_1)$;

(4) $\underline{\sum_{i=1}^{r} R_{A_i^I}}^{I}(E_1) = \bigcup_{i=1}^{r} \underline{R_{A_i^I}}^{I}(E_1)$;

(5) $\underline{\sum_{i=1}^{r} R_{A_i^I}}^{I}(E_1) \subseteq \underline{\sum_{i=1}^{r} R_{A_i^I}}^{O}(E_1)$;

(6) $\overline{\sum_{i=1}^{r} R_{A_i^I}}^{O}(E_1) \subseteq \overline{\sum_{i=1}^{r} R_{A_i^I}}^{I}(E_1)$.

(7) $\underline{\sum_{i=1}^{r} R_{A_i^I}}^{O}(E_1 \cap E_2) = \underline{\sum_{i=1}^{r} R_{A_i^I}}^{O}(E_1) \cap \underline{\sum_{i=1}^{r} R_{A_i^I}}^{O}(E_2), \underline{\sum_{i=1}^{r} R_{A_i^I}}^{I}(E_1 \cap E_2) = \underline{\sum_{i=1}^{r} R_{A_i^I}}^{I}(E_1) \cap \underline{\sum_{i=1}^{r} R_{A_i^I}}^{I}(E_2)$;

(8) $\overline{\sum_{i=1}^{r} R_{A_i^I}}^{O}(E_1 \cup E_2) = \overline{\sum_{i=1}^{r} R_{A_i^I}}^{O}(E_1) \cup \overline{\sum_{i=1}^{r} R_{A_i^I}}^{O}(E_2), \overline{\sum_{i=1}^{r} R_{A_i^I}}^{I}(E_1 \cup E_2) = \overline{\sum_{i=1}^{r} R_{A_i^I}}^{I}(E_1) \cup \overline{\sum_{i=1}^{r} R_{A_i^I}}^{I}(E_2)$;

(9) $\underline{\sum_{i=1}^{r} R_{A_i^I}}^{O}(E_1 \cup E_2) \supseteq \underline{\sum_{i=1}^{r} R_{A_i^I}}^{O}(E_1) \cup \underline{\sum_{i=1}^{r} R_{A_i^I}}^{O}(E_2), \underline{\sum_{i=1}^{r} R_{A_i^I}}^{I}(E_1 \cup E_2) \supseteq \underline{\sum_{i=1}^{r} R_{A_i^I}}^{I}(E_1) \cup \underline{\sum_{i=1}^{r} R_{A_i^I}}^{I}(E_2)$;

(10) $\overline{\sum_{i=1}^{r} R_{A_i^I}}^{O}(E_1 \cap E_2) \subseteq \overline{\sum_{i=1}^{r} R_{A_i^I}}^{O}(E_1) \cap \overline{\sum_{i=1}^{r} R_{A_i^I}}^{O}(E_2), \overline{\sum_{i=1}^{r} R_{A_i^I}}^{I}(E_1 \cap E_2) \subseteq \overline{\sum_{i=1}^{r} R_{A_i^I}}^{I}(E_1) \cap \overline{\sum_{i=1}^{r} R_{A_i^I}}^{I}(E_2)$.

Proof. It can be derived directly from Definitions 15 and 16. □

The characteristics of the proposed four models are further verified by Example 2 below.

Example 2. (Continued with Example 1). *From Example 1, we know that these 5 sites are evaluated by 10 investment schemes respectively. Suppose they have the following IFS with respect to 10 investment schemes*

$$E = \left\{ \frac{[0.25,0.43]}{x_1}, \frac{[0.51,0.28]}{x_2}, \frac{[0.54,0.38]}{x_3}, \frac{[0.37,0.59]}{x_4}, \frac{[0.49,0.35]}{x_5}, \frac{[0.92,0.04]}{x_6}, \frac{[0.09,0.86]}{x_7}, \frac{[0.15,0.46]}{x_8}, \frac{[0.72,0.12]}{x_9}, \frac{[0.67,0.23]}{x_{10}} \right\}.$$

(1) *In OOMRIFS, the lower and upper approximations of OOMRIFS can be calculated as follows:*

$$\underline{\sum_{i=1}^{r} R_{A_i^O}}^{O}(E) = \left\{ \frac{[0.25,0.59]}{x_1}, \frac{[0.49,0.38]}{x_2}, \frac{[0.49,0.38]}{x_3}, \frac{[0.25,0.59]}{x_4}, \frac{[0.49,0.38]}{x_5}, \frac{[0.25,0.46]}{x_6}, \frac{[0.09,0.86]}{x_7}, \frac{[0.15,0.46]}{x_8}, \frac{[0.15,0.46]}{x_9}, \frac{[0.67,0.23]}{x_{10}} \right\},$$

$$\overline{\sum_{i=1}^{r} R_{A_i^O}}^{O}(E) = \left\{ \frac{[0.51,0.28]}{x_1}, \frac{[0.51,0.28]}{x_2}, \frac{[0.54,0.35]}{x_3}, \frac{[0.51,0.28]}{x_4}, \frac{[0.54,0.35]}{x_5}, \frac{[0.92,0.04]}{x_6}, \frac{[0.54,0.35]}{x_7}, \frac{[0.15,0.46]}{x_8}, \frac{[0.72,0.12]}{x_9}, \frac{[0.67,0.23]}{x_{10}} \right\}.$$

(2) *Similarly, in OIMRIFS, we have:*

$$\underline{\sum_{i=1}^{r} R_{A_i^O}}^{I}(E) = \left\{ \frac{[0.25,0.59]}{x_1}, \frac{[0.25,0.59]}{x_2}, \frac{[0.09,0.86]}{x_3}, \frac{[0.25,0.59]}{x_4}, \frac{[0.09,0.86]}{x_5}, \frac{[0.15,0.59]}{x_6}, \frac{[0.09,0.86]}{x_7}, \frac{[0.15,0.46]}{x_8}, \frac{[0.09,0.86]}{x_9}, \frac{[0.09,0.86]}{x_{10}} \right\},$$

$$\overline{\sum_{i=1}^{r} R_{A_i^O}}^{I}(E) = \left\{ \frac{[0.92,0.04]}{x_1}, \frac{[0.54,0.28]}{x_2}, \frac{[0.54,0.28]}{x_3}, \frac{[0.92,0.04]}{x_4}, \frac{[0.54,0.28]}{x_5}, \frac{[0.92,0.04]}{x_6}, \frac{[0.72,0.12]}{x_7}, \frac{[0.92,0.04]}{x_8}, \frac{[0.92,0.04]}{x_9}, \frac{[0.72,0.12]}{x_{10}} \right\}.$$

From the above results, Figure 1 can be drawn as follows:

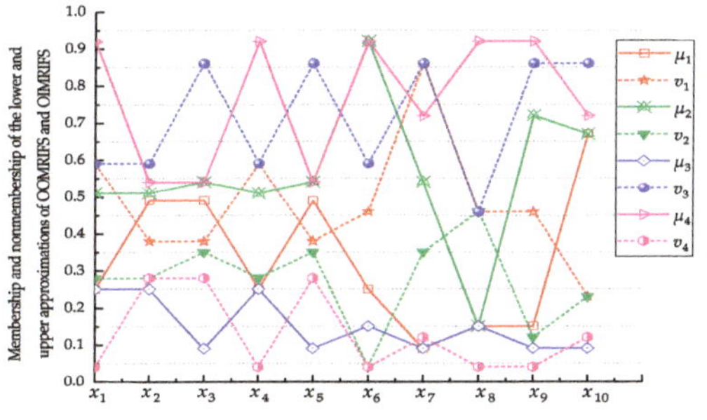

Figure 1. The lower and upper approximations of OOMRIFS and OIMRIFS.

Note that

$\mu_1 = \underline{\mu^{OO}}(x_j)$ *and* $v_1 = \underline{v^{OO}}(x_j)$ *represent the lower approximation of OOMRIFS;*

$\mu_2 = \overline{\mu^{OO}}(x_j)$ *and* $v_2 = \overline{v^{OO}}(x_j)$ *represent the upper approximation of OOMRIFS;*

$\mu_3 = \underline{\mu^{OI}}(x_j)$ *and* $\nu_3 = \underline{\nu^{OI}}(x_j)$ *represent the lower approximation of OIMRIFS;*
$\mu_4 = \overline{\mu^{OI}}(x_j)$ *and* $\nu_4 = \overline{\nu^{OI}}(x_j)$ *represent the upper approximation of OIMRIFS.*
Regarding Figure 1, we can get,

$$\overline{\mu^{OI}}(x_j) \geq \overline{\mu^{OO}}(x_j) \geq \underline{\mu^{OO}}(x_j) \geq \underline{\mu^{OI}}(x_j);\ \underline{\nu^{OI}}(x_j) \geq \underline{\nu^{OO}}(x_j) \geq \overline{\nu^{OO}}(x_j) \geq \overline{\nu^{OI}}(x_j).$$

As shown in Figure 1, the rules of Theorem 4 are satisfied. By constructing the OOMRIFS and OIMRIFS models, we can reduce the subjective scoring errors of experts under intuitionistic fuzzy conditions.

(3) *Similar to* (1), *in IOMRIFS, we have:*

$$\underline{\sum_{i=1}^{r} R_{A_i^I}}^{O}(E) = \left\{\frac{[0.25,0.43]}{x_1}, \frac{[0.25,0.43]}{x_2}, \frac{[0.25,0.43]}{x_3}, \frac{[0.37,0.59]}{x_4}, \frac{[0.25,0.43]}{x_5}, \frac{[0.25,0.46]}{x_6}, \frac{[0.09,0.86]}{x_7}, \frac{[0.15,0.46]}{x_8}, \frac{[0.67,0.23]}{x_9}, \frac{[0.67,0.23]}{x_{10}}\right\},$$

$$\overline{\sum_{i=1}^{r} R_{A_i^I}}^{O}(E) = \left\{\frac{[0.51,0.28]}{x_1}, \frac{[0.51,0.28]}{x_2}, \frac{[0.54,0.35]}{x_3}, \frac{[0.37,0.59]}{x_4}, \frac{[0.49,0.35]}{x_5}, \frac{[0.92,0.04]}{x_6}, \frac{[0.51,0.35]}{x_7}, \frac{[0.49,0.35]}{x_8}, \frac{[0.72,0.12]}{x_9}, \frac{[0.67,0.23]}{x_{10}}\right\}.$$

(4) *The same as* (1), *in IIMRIFS, we can get:*

$$\underline{\sum_{i=1}^{r} R_{A_i^I}}^{I}(E) = \left\{\frac{[0.25,0.59]}{x_1}, \frac{[0.09,0.86]}{x_2}, \frac{[0.09,0.86]}{x_3}, \frac{[0.25,0.59]}{x_4}, \frac{[0.09,0.86]}{x_5}, \frac{[0.09,0.86]}{x_6}, \frac{[0.09,0.86]}{x_7}, \frac{[0.09,0.86]}{x_8}, \frac{[0.15,0.46]}{x_9}, \frac{[0.67,0.23]}{x_{10}}\right\},$$

$$\overline{\sum_{i=1}^{r} R_{A_i^I}}^{I}(E) = \left\{\frac{[0.92,0.04]}{x_1}, \frac{[0.54,0.28]}{x_2}, \frac{[0.92,0.04]}{x_3}, \frac{[0.92,0.04]}{x_4}, \frac{[0.54,0.28]}{x_5}, \frac{[0.92,0.04]}{x_6}, \frac{[0.92,0.04]}{x_7}, \frac{[0.92,0.04]}{x_8}, \frac{[0.92,0.04]}{x_9}, \frac{[0.72,0.12]}{x_{10}}\right\}.$$

From (3) *and* (4), *we can obtain Figure 2 as shown:*

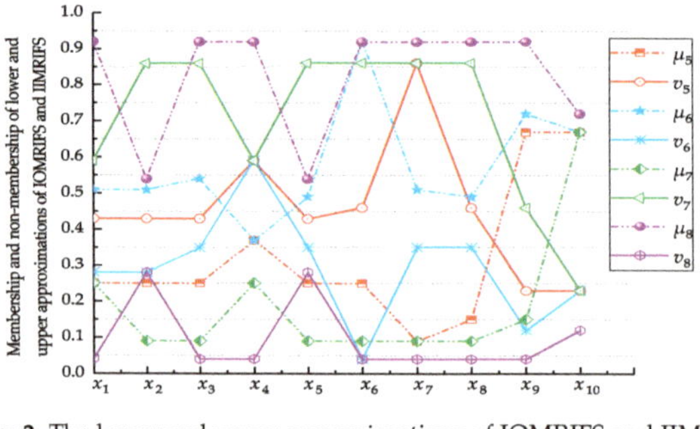

Figure 2. The lower and upper approximations of IOMRIFS and IIMRIFS.

Note that
$\mu_5 = \underline{\mu^{IO}}(x_j)$ *and* $\nu_5 = \underline{\nu^{IO}}(x_j)$ *represent the lower approximation of IOMRIFS;*
$\mu_6 = \overline{\mu^{IO}}(x_j)$ *and* $\nu_6 = \overline{\nu^{IO}}(x_j)$ *represent the upper approximation of IOMRIFS;*
$\mu_7 = \underline{\mu^{II}}(x_j)$ *and* $\nu_7 = \underline{\nu^{II}}(x_j)$ *represent the lower approximation of IIMRIFS;*
$\mu_8 = \overline{\mu^{II}}(x_j)$ *and* $\nu_8 = \overline{\nu^{II}}(x_j)$ *represent the upper approximation of IIMRIFS.*
For Figure 2, we can get,

$$\overline{\mu^{II}}(x_j) \geq \overline{\mu^{IO}}(x_j) \geq \underline{\mu^{IO}}(x_j) \geq \underline{\mu^{II}}(x_j);\ \underline{\nu^{II}}(x_j) \geq \underline{\nu^{IO}}(x_j) \geq \overline{\nu^{IO}}(x_j) \geq \overline{\nu^{II}}(x_j).$$

As shown in Figure 2, the rules of Theorem 5 are satisfied.

Through the Example 2, we can obtain four relatively more objective MRIFS models, which are beneficial to reduce subjective errors.

5. Three-Way Decisions Models Based on MRIFS and Optimal Granularity Selection

In order to obtain the optimal granularity selection results in the case of optimistic and pessimistic multi-granulation sets, it is necessary to further distinguish the importance degree of each granularity in the reduction sets. We respectively combine the four MRIFS models mentioned above with three-way decisions theory to get four new three-way decisions models. By extracting the rules, the redundant objects in the reduction sets are removed, and the decision error is further reduced. Then the optimal granularity selection results in two cases are obtained respectively by constructing the comprehensive score function and comprehensive accuracy function measurement formulas of each granularity of the reduction sets.

5.1. Three-Way Decisions Model Based on OOMRIFS

Suppose A_i^O is the reduction set under OMRS. According to reference [46], the expected loss function $R^{OO}(\omega_*|[x]_{A_i^O})(* = P, B, N)$ of object x can be obtained:

$$R^{OO}(\omega_P|[x]_{A_i^O}) = \lambda_{PP} \cdot \mu^{OO}(x) + \lambda_{PN} \cdot \nu^{OO}(x) + \lambda_{PB} \cdot \pi^{OO}(x);$$
$$R^{OO}(\omega_N|[x]_{A_i^O}) = \lambda_{NP} \cdot \mu^{OO}(x) + \lambda_{NN} \cdot \nu^{OO}(x) + \lambda_{NB} \cdot \pi^{OO}(x);$$
$$R^{OO}(\omega_B|[x]_{A_i^O}) = \lambda_{BP} \cdot \mu^{OO}(x) + \lambda_{BN} \cdot \nu^{OO}(x) + \lambda_{BB} \cdot \pi^{OO}(x).$$

where

$$\mu^{OO}(x) = \mu_{\underline{\sum_{i=1}^{r} R_{A_i^O}}^{O}(E)}(x) = \bigvee_{i=1}^{r} \inf_{y \in [x]_{A_i^O}} \mu_E(y),\ \nu^{OO}(x) = \nu_{\underline{\sum_{i=1}^{r} R_{A_i^O}}^{O}(E)}(x) = \bigwedge_{i=1}^{r} \sup_{y \in [x]_{A_i^O}} \nu_E(y),\ \pi^{OO}(x) = 1 - \mu_{\underline{\sum_{i=1}^{r} R_{A_i^O}}^{O}(E)}(x) - \nu_{\underline{\sum_{i=1}^{r} R_{A_i^O}}^{O}(E)}(x);$$

or

$$\mu^{OO}(x) = \mu_{\overline{\sum_{i=1}^{r} R_{A_i^O}}^{O}(E)}(x) = \bigwedge_{i=1}^{r} \sup_{y \in [x]_{A_i^O}} \mu_E(y),\ \nu^{OO}(x) = \nu_{\overline{\sum_{i=1}^{r} R_{A_i^O}}^{O}(E)}(x) = \bigvee_{i=1}^{r} \inf_{y \in [x]_{A_i^O}} \nu_E(y),\ \pi^{OO}(x) = 1 - \mu_{\overline{\sum_{i=1}^{r} R_{A_i^O}}^{O}(E)}(x) - \nu_{\overline{\sum_{i=1}^{r} R_{A_i^O}}^{O}(E)}(x).$$

The minimum-risk decision rules derived from the Bayesian decision process are as follows:
(P′): If $R'(\omega_P|[x]_{A_i^O}) \leq R'(\omega_B|[x]_{A_i^O})$ and $R'(\omega_P|[x]_{A_i^O}) \leq R'(\omega_N|[x]_{A_i^O})$, then $x \in POS(X)$;
(N′): If $R'(\omega_N|[x]_{A_i^O}) \leq R'(\omega_P|[x]_{A_i^O})$ and $R'(\omega_N|[x]_{A_i^O}) \leq R'(\omega_B|[x]_{A_i^O})$, then $x \in NEG(X)$;
(B′): If $R'(\omega_B|[x]_{A_i^O}) \leq R'(\omega_N|[x]_{A_i^O})$ and $R'(\omega_B|[x]_{A_i^O}) \leq R'(\omega_P|[x]_{A_i^O})$, then $x \in BND(X)$.
Thus, the decision rules (P′)-(B′) can be re-expressed concisely as:
(P′) rule satisfies:

$$(\mu^{OO}(x) \leq (1 - \pi^{OO}(x)) \cdot \frac{\lambda_{NN} - \lambda_{PN}}{(\lambda_{PP} - \lambda_{NP}) + (\lambda_{PN} - \lambda_{NN})}) \wedge (\mu^{OO}(x) \leq (1 - \pi^{OO}(x)) \cdot \frac{\lambda_{BN} - \lambda_{PN}}{(\lambda_{PP} - \lambda_{BP}) + (\lambda_{PN} - \lambda_{BN})});$$

(N′) rule satisfies:

$$(\mu^{OO}(x) < (1 - \pi^{OO}(x)) \cdot \frac{\lambda_{PN} - \lambda_{NN}}{(\lambda_{NP} - \lambda_{PP}) + (\lambda_{PN} - \lambda_{NN})}) \wedge (\mu^{OO}(x) < (1 - \pi^{OO}(x)) \cdot \frac{\lambda_{BN} - \lambda_{NN}}{(\lambda_{NP} - \lambda_{BP}) + (\lambda_{BN} - \lambda_{NN})});$$

(B′) rule satisfies:

$$(\mu^{OO}(x) > (1 - \pi^{OO}(x)) \cdot \frac{\lambda_{BN} - \lambda_{PN}}{(\lambda_{PN} - \lambda_{BN}) + (\lambda_{BP} - \lambda_{PP})}) \wedge (\mu^{OO}(x) \geq (1 - \pi^{OO}(x)) \cdot \frac{\lambda_{BN} - \lambda_{NN}}{(\lambda_{BN} - \lambda_{NN}) + (\lambda_{NP} - \lambda_{BP})}).$$

Therefore, the three-way decisions rules based on OOMRIFS are as follows:
(P1): If $\mu^{OO}(x) \geq (1 - \pi^{OO}(x)) \cdot \alpha$, then $x \in POS(X)$;
(N1): If $\mu^{OO}(x) \leq (1 - \pi^{OO}(x)) \cdot \beta$, then $x \in NEG(X)$;
(B1): If $(1 - \pi^{OO}(x)) \cdot \beta \leq \mu^{OO}(x)$ and $\mu^{OO}(x) \leq (1 - \pi^{OO}(x)) \cdot \alpha$, then $x \in BND(X)$.

5.2. Three-Way Decisions Model Based on OIMRIFS

Suppose A_i^O is the reduction set under OMRS. According to reference [46], the expected loss functions $R^{OO}(\omega_*|[x]_{A_i^O})(* = P, B, N)$ of an object x are presented as follows:

$$R^{OI}(\omega_P|[x]_{A_i^O}) = \lambda_{PP} \cdot \mu^{OI}(x) + \lambda_{PN} \cdot \nu^{OI}(x) + \lambda_{PB} \cdot \pi^{OI}(x);$$
$$R^{OI}(\omega_N|[x]_{A_i^O}) = \lambda_{NP} \cdot \mu^{OI}(x) + \lambda_{NN} \cdot \nu^{OI}(x) + \lambda_{NB} \cdot \pi^{OI}(x);$$
$$R^{OI}(\omega_B|[x]_{A_i^O}) = \lambda_{BP} \cdot \mu^{OI}(x) + \lambda_{BN} \cdot \nu^{OI}(x) + \lambda_{BB} \cdot \pi^{OI}(x).$$

where

$$\mu^{OI}(x) = \mu_{\underline{\sum_{i=1}^{r} R_{A_i^O}}^{I}(E)}(x) = \bigwedge_{i=1}^{r} \inf_{y\in[x]_{A_i^O}} \mu_E(y),\ \nu^{OI}(x) = \nu_{\underline{\sum_{i=1}^{r} R_{A_i^O}}^{I}(E)}(x) = \bigvee_{i=1}^{r} \sup_{y\in[x]_{A_i^O}} \nu_E(y),\ \pi^{OI}(x) = 1 - \mu_{\underline{\sum_{i=1}^{r} R_{A_i^O}}^{I}(E)}(x) - \nu_{\underline{\sum_{i=1}^{r} R_{A_i^O}}^{I}(E)}(x);$$

or

$$\mu^{OI}(x) = \mu_{\overline{\sum_{i=1}^{r} R_{A_i^O}}^{I}(E)}(x) = \bigvee_{i=1}^{r} \sup_{y\in[x]_{A_i^O}} \mu_E(y),\ \nu^{OI}(x) = \nu_{\overline{\sum_{i=1}^{r} R_{A_i^O}}^{I}(E)}(x) = \bigwedge_{i=1}^{r} \inf_{y\in[x]_{A_i^O}} \nu_E(y),\ \pi^{OI}(x) = 1 - \mu_{\overline{\sum_{i=1}^{r} R_{A_i^O}}^{I}(E)}(x) - \nu_{\overline{\sum_{i=1}^{r} R_{A_i^O}}^{I}(E)}(x).$$

Therefore, the three-way decisions rules based on OIMRIFS are as follows:
(P2): If $\mu^{OI}(x) \geq (1 - \pi^{OI}(x)) \cdot \alpha$, then $x \in POS(X)$;
(N2): If $\mu^{OI}(x) \leq (1 - \pi^{OI}(x)) \cdot \beta$, then $x \in NEG(X)$;
(B2): If $(1 - \pi^{OI}(x)) \cdot \beta \leq \mu^{OI}(x)$ and $\mu^{OI}(x) \leq (1 - \pi^{OI}(x)) \cdot \alpha$, then $x \in BND(X)$.

5.3. Three-Way Decisions Model Based on IOMRIFS

Suppose A_i^I is the reduction set under IMRS. According to reference [46], the expected loss functions $R^{IO}(\omega_*|[x]_{A_i^I})(* = P, B, N)$ of an object x are as follows:

$$R^{IO}(\omega_P|[x]_{A_i^I}) = \lambda_{PP} \cdot \mu^{IO}(x) + \lambda_{PN} \cdot \nu^{IO}(x) + \lambda_{PB} \cdot \pi^{IO}(x);$$
$$R^{IO}(\omega_N|[x]_{A_i^I}) = \lambda_{NP} \cdot \mu^{IO}(x) + \lambda_{NN} \cdot \nu^{IO}(x) + \lambda_{NB} \cdot \pi^{IO}(x);$$
$$R^{IO}(\omega_B|[x]_{A_i^I}) = \lambda_{BP} \cdot \mu^{IO}(x) + \lambda_{BN} \cdot \nu^{IO}(x) + \lambda_{BB} \cdot \pi^{IO}(x).$$

where

$$\mu^{IO}(x) = \mu_{\underline{\sum_{i=1}^{r} R_{A_i^I}}^{O}(E)}(x) = \bigvee_{i=1}^{r} \inf_{y\in[x]_{A_i^I}} \mu_E(y),\ \nu^{IO}(x) = \nu_{\underline{\sum_{i=1}^{r} R_{A_i^I}}^{O}(E)}(x) = \bigwedge_{i=1}^{r} \sup_{y\in[x]_{A_i^I}} \nu_E(y),\ \pi^{IO}(x) = 1 - \mu_{\underline{\sum_{i=1}^{r} R_{A_i^I}}^{O}(E)}(x) - \nu_{\underline{\sum_{i=1}^{r} R_{A_i^I}}^{O}(E)}(x);$$

or

$$\mu^{IO}(x) = \mu_{\overline{\sum_{i=1}^{r} R_{A_i^I}}^{O}(E)}(x) = \bigwedge_{i=1}^{r} \sup_{y\in[x]_{A_i^I}} \mu_E(y),\ \nu^{IO}(x) = \nu_{\overline{\sum_{i=1}^{r} R_{A_i^I}}^{O}(E)}(x) = \bigvee_{i=1}^{r} \inf_{y\in[x]_{A_i^I}} \nu_E(y),\ \pi^{IO}(x) = 1 - \mu_{\overline{\sum_{i=1}^{r} R_{A_i^I}}^{O}(E)}(x) - \nu_{\overline{\sum_{i=1}^{r} R_{A_i^I}}^{O}(E)}(x).$$

Therefore, the three-way decisions rules based on IOMRIFS are as follows:
(P3): If $\mu^{IO}(x) \geq (1 - \pi^{IO}(x)) \cdot \alpha$, then $x \in POS(X)$;
(N3): If $\mu^{IO}(x) \leq (1 - \pi^{IO}(x)) \cdot \beta$, then $x \in NEG(X)$;
(B3): If $(1 - \pi^{IO}(x)) \cdot \beta \leq \mu^{IO}(x)$ and $\mu^{IO}(x) \leq (1 - \pi^{IO}(x)) \cdot \alpha$, then $x \in BND(X)$.

5.4. Three-Way Decisions Model Based on IIMRIFS

Suppose A_i^I is the reduction set under IMRS. Like Section 5.1, the expected loss functions $R^{II}(\omega_*|[x]_{A_i^I})(* = P, B, N)$ of an object x are as follows:

$$R^{II}(\omega_P|[x]_{A_i^I}) = \lambda_{PP} \cdot \mu^{II}(x) + \lambda_{PN} \cdot \nu^{II}(x) + \lambda_{PB} \cdot \pi^{II}(x);$$

$$R^{II}(\omega_N|[x]_{A_i^I}) = \lambda_{NP} \cdot \mu^{II}(x) + \lambda_{NN} \cdot \nu^{II}(x) + \lambda_{NB} \cdot \pi^{II}(x);$$

$$R^{II}(\omega_B|[x]_{A_i^I}) = \lambda_{BP} \cdot \mu^{II}(x) + \lambda_{BN} \cdot \nu^{II}(x) + \lambda_{BB} \cdot \pi^{II}(x).$$

where

$$\mu^{II}(x) = \mu_{\underline{\sum_{i=1}^{r} R_{A_i^I}}^{I}(E)}(x) = \bigwedge_{i=1}^{r} \inf_{y \in [x]_{A_i^I}} \mu_E(y),\ \nu^{II}(x) = \nu_{\underline{\sum_{i=1}^{r} R_{A_i^I}}^{I}(E)}(x) = \bigvee_{i=1}^{r} \sup_{y \in [x]_{A_i^I}} \nu_E(y),\ \pi^{II}(x) = 1 - \mu_{\underline{\sum_{i=1}^{r} R_{A_i^I}}^{I}(E)}(x) - \nu_{\underline{\sum_{i=1}^{r} R_{A_i^I}}^{I}(E)}(x);$$

or

$$\mu^{II}(x) = \mu_{\overline{\sum_{i=1}^{r} R_{A_i^I}}^{I}(E)}(x) = \bigvee_{i=1}^{r} \sup_{y \in [x]_{A_i^I}} \mu_E(y),\ \nu^{II}(x) = \nu_{\overline{\sum_{i=1}^{r} R_{A_i^I}}^{I}(E)}(x) = \bigwedge_{i=1}^{r} \inf_{y \in [x]_{A_i^I}} \nu_E(y),\ \pi^{II}(x) = 1 - \mu_{\overline{\sum_{i=1}^{r} R_{A_i^I}}^{I}(E)}(x) - \nu_{\overline{\sum_{i=1}^{r} R_{A_i^I}}^{I}(E)}(x).$$

Therefore, the three-way decisions rules based on IIMRIFS are captured as follows:
(P4): If $\mu^{II}(x) \geq (1 - \pi^{II}(x)) \cdot \alpha$, then $x \in POS(X)$;
(N4): If $\mu^{II}(x) \leq (1 - \pi^{II}(x)) \cdot \beta$, then $x \in NEG(X)$;
(B4): If $(1 - \pi^{II}(x)) \cdot \beta \leq \mu^{II}(x)$ and $\mu^{II}(x) \leq (1 - \pi^{II}(x)) \cdot \alpha$, then $x \in BND(X)$.

By constructing the above three decision models, the redundant objects in the reduction sets can be removed, which is beneficial to the optimal granular selection.

5.5. Comprehensive Measuring Methods of Granularity

Definition 17 ([40]). *Let an intuitionistic fuzzy number $\widetilde{E}(f_1) = (\mu_{\widetilde{E}}(f_1), \nu_{\widetilde{E}}(f_1))$, $f_1 \in U$, then the score function of $\widetilde{E}(f_1)$ is calculated as:*

$$S(\widetilde{E}(f_1)) = \mu_{\widetilde{E}}(f_1) - \nu_{\widetilde{E}}(f_1).$$

The accuracy function of $\widetilde{E}(f_1)$ is defined as:

$$H(\widetilde{E}(f_1)) = \mu_{\widetilde{E}}(f_1) + \nu_{\widetilde{E}}(f_1).$$

where $-1 \leq S(\widetilde{E}(f_1)) \leq 1$ and $0 \leq H(\widetilde{E}(f_1)) \leq 1$.

Definition 18. *Let $DIS = (U, C \cup D)$ be a decision information system, $A = \{A_1, A_2, \cdots, A_m\}$ are m sub-attributes of C. Suppose E are IFS on the universe $U = \{x_1, x_2, \cdots, x_n\}$, defined by $\mu_{A_i}(x_j)$ and $\nu_{A_i}(x_j)$, where $\mu_{A_i}(x_j)$ and $\nu_{A_i}(x_j)$ are their membership and non-membership functions respectively. $|[x_j]_{A_i}|$ is the number of equivalence classes of x_j on granularity A_i, $U/D = \{X_1, X_2, \cdots, X_s\}$ is the partition induced by the decision attributes D. Then, the comprehensive score function of granularity A_i is captured as:*

$$CSF_{A_i}(E) = \frac{1}{s} \times \sum_{j=1, n \in [x_j]_{A_i}}^{n} \frac{|\mu_{A_i}(x_j) - \nu_{A_i}(x_j)|}{|[x_j]_{A_i}|}.$$

The comprehensive accuracy function of granularity A_i is captured as:

$$CAF_{A_i}(E) = \frac{1}{s} \times \sum_{j=1, n \in [x_j]_{A_i}}^{n} \frac{|\mu_{A_i}(x_j) + \nu_{A_i}(x_j)|}{|[x_j]_{A_i}|}.$$

where $-1 \leq CSF_{A_i}(E) \leq 1$ and $0 \leq CAF_{A_i}(E) \leq 1$.

With respect to Definition 19, according to references [27,39], we can deduce the following rules.

Definition 19. *Let two granularities A_1, A_2, then we have:*

(1) If $CSF_{A_1}(E) > CSF_{A_2}(E)$, then A_2 is smaller than A_1, expressed as $A_1 > A_2$;
(2) If $CSF_{A_1}(E) < CSF_{A_2}(E)$, then A_1 is smaller than A_2, expressed as $A_1 < A_2$;
(3) If $CSF_{A_1}(E) = CSF_{A_2}(E)$, then

(i) If $CSF_{A_1}(E) = CSF_{A_2}(E)$, then A_2 is equal to A_1, expressed as $A_1 = A_2$;
(ii) If $CSF_{A_1}(E) > CSF_{A_2}(E)$, then A_2 is smaller than A_1, expressed as $A_1 > A_2$;
(iii) If $CSF_{A_1}(E) < CSF_{A_2}(E)$, then A_1 is smaller than A_2, expressed as $A_1 < A_2$.

5.6. Optimal Granularity Selection Algorithm to Derive Three-Way Decisions from MRIFS

Suppose the reduction sets of optimistic and IMRS are A_i^O and A_i^I respectively. In this section, we take the reduction set under OMRS as an example to make the result $A_i^{O'}$ of optimal granularity selection.

Algorithm 2. Optimal granularity selection algorithm to derive three-way decisions from MRIFS

Input: $DIS = (U, C \cup D, V, f)$, $A = \{A_1, A_2, \cdots, A_m\}$ be m sub-attributes of condition attributes C, $\forall A_i \in A'$, $U/D = \{X_1, X_2, \cdots, X_s\}$, IFS E;
Output: Optimal granularity selection result $A_i^{O'}$.
1: compute via Algorithm 1;
2: **if** $|A_i^O| > 1$
3:    **for** $\forall A_i \in A_i^O$
4:      compute $\mu_{\sum_{i=1}^r R_{A_i^O}^{\Delta}(E)}(x_j)$, $\nu_{\sum_{i=1}^r R_{A_i^O}^{\Delta}(E)}(x_j)$, $\mu_{\overline{\sum_{i=1}^r R_{A_i^O}^{\Delta}(E)}}(x_j)$ and $\nu_{\overline{\sum_{i=1}^r R_{A_i^O}^{\Delta}(E)}}(x_j)$;
5:      according (P1)-(B1) and (P2)-(B2), compute $POS(\underline{X^{O\Delta}})$, $NEG(\underline{X^{O\Delta}})$, $BND(\underline{X^{O\Delta}})$, $POS(\overline{X^{O\Delta}})$, $NEG(\overline{X^{O\Delta}})$, $BND(\overline{X^{O\Delta}})$;
6:      **if** $NEG(\underline{X^{O\Delta}}) \neq U$ or $NEG(\overline{X^{O\Delta}}) \neq U$
7:         compute $U/\underline{A_i^{O\Delta}}$, $CSF_{\underline{A_i^{O\Delta}}}(E)$, $CAF_{\underline{A_i^{O\Delta}}}(E)$ or $(U/\overline{A_i^{O\Delta}})$, $(CSF_{\overline{A_i^{O\Delta}}}(E), CAF_{\overline{A_i^{O\Delta}}}(E)$;
8:         according to Definition 19 to get $A_i^{O'}$;
9:         **return** $A_i^{O'} = A_i$;
10:      **end**
11:      **else**
12:         **return** NULL;
13:      **end**
14:    **end**
15: **end**
16: **else**
17: **return** $A_i^{O'} = A_i^O$;
18: **end**

---

6. Example Analysis 3 (Continued with Example 2)

In Example 1, only site 1 can be ignored under optimistic and pessimistic multi-granulation conditions, so it can be determined that site 1 does not need to be evaluated, while sites 2 and 3 need to be further investigated under the environment of optimistic multi-granulation. At the same time, with respect to the environment of pessimistic multi-granulation, comprehensive considera- tion site 3 can ignore the assessment and sites 2, 4 and 5 need to be further investigated.

According to Example 1, we can get that the reduction set of OMRS is $\{A_2, A_3\}$, but in the case of IMRS, there are two reduction sets, which are contradictory. Therefore, two reduction sets should be reconsidered simultaneously, so the joint reduction set under IMRS is $\{A_2, A_4, A_5\}$.

Where the corresponding granularity structures of sites 2, 3, 4 and 5 are divided as follows:

$U/A_2 = \{\{x_1, x_2, x_4\}, \{x_3, x_5, x_7\}, \{x_6, x_8, x_9\}, \{x_{10}\}\}$,
$U/A_3 = \{\{x_1, x_4, x_6\}, \{x_2, x_3, x_5\}, \{x_8\}, \{x_7, x_9, x_{10}\}\}$,
$U/A_4 = \{\{x_1, x_2, x_3, x_5\}, \{x_4\}, \{x_6, x_7, x_8\}, \{x_9, x_{10}\}\}$,
$U/A_5 = \{\{x_1, x_3, x_4, x_6\}, \{x_2, x_7\}, \{x_5, x_8\}, \{x_9, x_{10}\}\}$.

According to reference [11], we can get:
$\alpha = \frac{8-2}{(8-2)+(2-0)} = 0.75$; $\beta = \frac{2-0}{(2-0)+(6-2)} = 0.33$.

The optimal site selection process under optimistic and IMRS is as follows:

(1) Optimal site selection based on OOMRIFS

According to the Example 2, we can get the values of evaluation functions $\underline{\mu^{OO}}(x_j)$, $(1-\underline{\pi^{OO}}(x_j))\cdot\alpha$, $(1-\underline{\pi^{OO}}(x_j))\cdot\beta$, $\overline{\mu^{OO}}(x_j)$, $(1-\overline{\pi^{OO}}(x_j))\cdot\alpha$ and $(1-\overline{\pi^{OO}}(x_j))\cdot\beta$ of OOMRIFS, as shown in Table 4.

Table 4. The values of evaluation functions for OOMRIFS.

	$\underline{\mu^{OO}}(x_j)$	$(1-\underline{\pi^{OO}}(x_j))\cdot\alpha$	$(1-\underline{\pi^{OO}}(x_j))\cdot\beta$	$\overline{\mu^{OO}}(x_j)$	$(1-\overline{\pi^{OO}}(x_j))\cdot\alpha$	$(1-\overline{\pi^{OO}}(x_j))\cdot\beta$
x_1	0.25	0.63	0.2772	0.51	0.5925	0.2607
x_2	0.49	0.6525	0.2871	0.51	0.5925	0.2607
x_3	0.49	0.6525	0.2871	0.54	0.6675	0.2937
x_4	0.25	0.63	0.2772	0.51	0.5925	0.2607
x_5	0.49	0.6525	0.2871	0.54	0.6675	0.2937
x_6	0.25	0.5325	0.2343	0.92	0.72	0.3168
x_7	0.09	0.7125	0.3135	0.54	0.6675	0.2937
x_8	0.15	0.4575	0.2013	0.15	0.4575	0.2013
x_9	0.15	0.4575	0.2013	0.72	0.63	0.2772
x_{10}	0.67	0.675	0.297	0.67	0.675	0.297

We can get decision results of the lower and upper approximations of OOMRIFS by three-way decisions of the Section 5.1, as follows:

$POS(\underline{X^{OO}}) = \phi$,
$NEG(\underline{X^{OO}}) = \{x_1, x_4, x_7, x_8, x_9\}$,
$BND(\underline{X^{OO}}) = \{x_2, x_3, x_5, x_6, x_{10}\}$;
$POS(\overline{X^{OO}}) = \{x_6, x_9\}$,
$NEG(\overline{X^{OO}}) = \{x_8\}$,
$BND(\overline{X^{OO}}) = \{x_2, x_3, x_5\}$.

In the light of three-way decisions rules based on OOMRIFS, after getting rid of the objects in the rejection domain, we choose to fuse the objects in the delay domain with those in the acceptance domain for the optimal granularity selection. Therefore, the new granularities A_2, A_3 are as follows:

$U/\underline{A_2^{OI}} = \{\{x_2\}, \{x_3, x_5\}, \{x_6\}, \{x_{10}\}\}$,
$U/\underline{A_3^{OI}} = \{\{x_2, x_3, x_5\}, \{x_6\}, \{x_{10}\}\}$;
$U/\overline{A_2^{OI}} = \{\{x_1, x_2, x_4\}, \{x_3, x_5, x_7\}, \{x_6, x_9\}, \{x_{10}\}\}$,
$U/\overline{A_3^{OI}} = \{\{x_1, x_4, x_6\}, \{x_2, x_3, x_5\}, \{x_7, x_9, x_{10}\}\}$.

Then, according to Definition 18, we can get:

$$\begin{aligned}
CSF_{\underline{A_2^{OO}}}(E) &= \tfrac{1}{s} \times \sum_{j=1, n\in[x_j]_{A_i}}^{n} \frac{|\mu_{A_i}(x_j)-\nu_{A_i}(x_j)|}{|[x_j]_{A_i}|} \\
&= \tfrac{1}{4} \times \sum_{j=1, n\in[x_j]_{\underline{A_2^{OO}}}}^{10} \frac{|\mu_{\underline{A_2^{OO}}}(x_j)-\nu_{\underline{A_2^{OO}}}(x_j)|}{|[x_j]_{\underline{A_2^{OO}}}|} \\
&= \tfrac{1}{4} \times ((0.49-0.38) + \tfrac{(0.49-0.38)+(0.49-0.38)}{2} + (0.25-0.46) + (0.67-0.23)) \\
&= 0.1125,
\end{aligned}$$

$$\begin{aligned}
CSF_{\underline{A_3^{OO}}}(E) &= \tfrac{1}{s} \times \sum_{j=1, n\in[x_j]_{A_i}}^{n} \frac{|\mu_{A_i}(x_j)-\nu_{A_i}(x_j)|}{|[x_j]_{A_i}|} \\
&= \tfrac{1}{3} \times \sum_{j=1, n\in[x_j]_{\underline{A_3^{OO}}}}^{10} \frac{|\mu_{\underline{A_3^{OO}}}(x_j)-\nu_{\underline{A_3^{OO}}}(x_j)|}{|[x_j]_{\underline{A_3^{OO}}}|} \\
&= \tfrac{1}{3} \times ((0.25-0.46) + \tfrac{(0.49-0.38)+(0.49-0.38)+(0.49-0.38)}{3} + (0.81-0.14)) \\
&= 0.1133;
\end{aligned}$$

Similarly, we have:
$CSF_{\overline{A_2^{OO}}}(E) = 0.4$, $CSF_{\overline{A_3^{OO}}}(E) = 0.3533$.

From the above results, in OOMRIFS, we can see that we can't get the selection result of sites 2 and 3 only according to the comprehensive score function of granularities A_2 and A_3. Therefore, we need to further calculate the comprehensive accuracies to get the results as follows:

$$\begin{aligned} CAF_{\underline{A_2^{OO}}}(E) &= \frac{1}{s} \times \sum_{j=1, n \in [x_j]_{A_i}}^{n} \frac{|\mu_{A_i}(x_j) + v_{A_i}(x_j)|}{|[x_j]_{A_i}|} \\ &= \frac{1}{4} \times \sum_{j=1, n \in [x_j]_{\underline{A_2^{OO}}}}^{10} \frac{|\mu_{\underline{A_2^{OO}}}(x_j) + v_{\underline{A_2^{OO}}}(x_j)|}{|[x_j]_{\underline{A_2^{OO}}}|} \\ &= \frac{1}{4} \times ((0.49 + 0.38) + \frac{(0.49+0.38)+(0.49+0.38)}{2} + (0.25 + 0.46) + (0.67 + 0.23)) \\ &= 0.8375, \end{aligned}$$

$$\begin{aligned} CAF_{\underline{A_3^{OO}}}(E) &= \frac{1}{s} \times \sum_{j=1, n \in [x_j]_{A_i}}^{n} \frac{|\mu_{A_i}(x_j) + v_{A_i}(x_j)|}{|[x_j]_{A_i}|} \\ &= \frac{1}{3} \times \sum_{j=1, n \in [x_j]_{\underline{A_3^{OO}}}}^{10} \frac{|\mu_{\underline{A_3^{OO}}}(x_j) + v_{\underline{A_3^{OO}}}(x_j)|}{|[x_j]_{\underline{A_3^{OO}}}|} \\ &= \frac{1}{3} \times ((0.25 + 0.46) + \frac{(0.49+0.38)+(0.49+0.38)+(0.49+0.38)}{3} + (0.81 + 0.14)) \\ &= 0.8267; \end{aligned}$$

Analogously, we have:
$CAF_{\overline{A_2^{OO}}}(E) = 0.87$, $CAF_{\overline{A_3^{OO}}}(E) = 0.86$.
Through calculation above, we know that the comprehensive accuracy of the granularity A_3 is higher, so the site 3 is selected as the selection result.

(2) Optimal site selection based on OIMRIFS

The same as (1), we can get the values of evaluation functions $\underline{\mu^{OI}}(x_j)$, $(1 - \underline{\pi^{OI}}(x_j)) \cdot \alpha$, $(1 - \underline{\pi^{OI}}(x_j)) \cdot \beta$, $\overline{\mu^{OI}}(x_j)$, $(1 - \overline{\pi^{OI}}(x_j)) \cdot \alpha$ and $(1 - \overline{\pi^{OI}}(x_j)) \cdot \beta$ of OIMRIFS listed in Table 5.

Table 5. The values of evaluation functions for OIMRIFS.

	$\underline{\mu^{OI}}(x_j)$	$(1-\underline{\pi^{OI}}(x_j))\cdot\alpha$	$(1-\underline{\pi^{OI}}(x_j))\cdot\beta$	$\overline{\mu^{OI}}(x_j)$	$(1-\overline{\pi^{OI}}(x_j))\cdot\alpha$	$(1-\overline{\pi^{OI}}(x_j))\cdot\beta$
x_1	0.25	0.63	0.2772	0.92	0.72	0.3168
x_2	0.25	0.63	0.2772	0.54	0.615	0.2706
x_3	0.09	0.7125	0.3135	0.54	0.615	0.2706
x_4	0.25	0.63	0.2772	0.92	0.72	0.3168
x_5	0.09	0.7125	0.3135	0.54	0.615	0.2706
x_6	0.15	0.555	0.2442	0.92	0.72	0.3168
x_7	0.09	0.7125	0.3135	0.72	0.63	0.2772
x_8	0.15	0.4575	0.2013	0.92	0.72	0.3168
x_9	0.09	0.7125	0.3135	0.92	0.72	0.3168
x_{10}	0.09	0.7125	0.3135	0.72	0.63	0.2772

We can get decision results of the lower and upper approximations of OIMRIFS by three-way decisions in the Section 5.2, as follows:
$POS(\underline{X^{OI}}) = \phi$,
$NEG(\underline{X^{OI}}) = U$,
$BND(\underline{X^{OI}}) = \phi$;
$POS(\overline{X^{OI}}) = \{x_1, x_4, x_6, x_7, x_8, x_9, x_{10}\}$,
$NEG(\overline{X^{OI}}) = \phi$,
$BND(\overline{X^{OI}}) = \{x_2, x_3, x_5\}$.
Hence, in the upper approximations of OIMRIFS, the new granularities A_2, A_3 are as follows:

$U/\underline{A_2^{OI}} = \{\{x_1, x_2, x_4\}, \{x_3, x_5, x_7\}, \{x_6, x_8, x_9\}, \{x_{10}\}\}$,
$U/\overline{A_3^{OI}} = \{\{x_1, x_4, x_6\}, \{x_2, x_3, x_5\}, \{x_8\}, \{x_7, x_9, x_{10}\}\}$.
According to Definition 18, we can calculate that
$CSF_{\underline{A_2^{OI}}}(E) = CSF_{\underline{A_3^{OI}}}(E) = 0$;
$CAF_{\underline{A_2^{OI}}}(E) = CAF_{\underline{A_3^{OI}}}(E) = 0$;
$CSF_{\overline{A_2^{OI}}}(E) = 0.6317, CSF_{\overline{A_3^{OI}}}(E) = 0.6783$;
$CAF_{\overline{A_2^{OI}}}(E) = 0.885, CAF_{\overline{A_3^{OI}}}(E) = 0.905$.

In OIMRIFS, the comprehensive score and comprehensive accuracy of the granularity A_3 are both higher than the granularity A_2. So, we choose site 3 as the evaluation site.

In reality, we are more inclined to select the optimal granularity in the case of more stringent requirements. According to (1) and (2), we can find that the granularity A_3 is a better choice when the requirements are stricter in four cases of OMRS. Therefore, we choose site 3 as the optimal evaluation site.

(3) Optimal site selection based on IOMRIFS

Similar to (1), we can obtain the values of evaluation functions $\underline{\mu^{IO}}(x_j)$, $(1 - \underline{\pi^{IO}}(x_j)) \cdot \alpha$, $(1 - \underline{\pi^{IO}}(x_j)) \cdot \beta$, $\overline{\mu^{IO}}(x_j)$, $(1 - \overline{\pi^{IO}}(x_j)) \cdot \alpha$ and $(1 - \overline{\pi^{IO}}(x_j)) \cdot \beta$ of IOMRIFS, as described in Table 6.

Table 6. The values of evaluation functions for IOMRIFS.

	$\underline{\mu^{IO}}(x_j)$	$(1-\underline{\pi^{IO}}(x_j))\cdot\alpha$	$(1-\underline{\pi^{IO}}(x_j))\cdot\beta$	$\overline{\mu^{IO}}(x_j)$	$(1-\overline{\pi^{IO}}(x_j))\cdot\alpha$	$(1-\overline{\pi^{IO}}(x_j))\cdot\beta$
x_1	0.25	0.51	0.2244	0.51	0.5925	0.2607
x_2	0.25	0.51	0.2244	0.51	0.5925	0.2607
x_3	0.25	0.51	0.2244	0.54	0.6675	0.2937
x_4	0.37	0.72	0.3168	0.37	0.72	0.3168
x_5	0.25	0.51	0.2244	0.49	0.63	0.2772
x_6	0.25	0.5325	0.2343	0.92	0.72	0.3168
x_7	0.09	0.7125	0.3135	0.51	0.645	0.2838
x_8	0.15	0.4575	0.2013	0.49	0.63	0.2772
x_9	0.67	0.675	0.297	0.72	0.63	0.2772
x_{10}	0.67	0.675	0.297	0.67	0.675	0.297

We can get decision results of the lower and upper approximations of IOMRIFS by three-way decisions in the Section 5.3, as follows:
$POS(\underline{X^{IO}}) = \phi$,
$NEG(\underline{X^{IO}}) = \{x_7, x_8\}$,
$BND(\underline{X^{IO}}) = \{x_1, x_2, x_3, x_4, x_5, x_6, x_9, x_{10}\}$;
$POS(\overline{X^{IO}}) = \{x_6, x_9\}$,
$NEG(\overline{X^{IO}}) = \phi$,
$BND(\overline{X^{IO}}) = \{x_1, x_2, x_3, x_4, x_5, x_7, x_8, x_{10}\}$.
Therefore, the granularities A_2, A_4, A_5 can be rewritten as follows:
$U/\underline{A_2^{IO}} = \{\{x_1, x_2, x_4\}, \{x_3, x_5\}, \{x_6, x_9\}, \{x_{10}\}\}$,
$U/\underline{A_4^{IO}} = \{\{x_1, x_2, x_3, x_5\}, \{x_4\}, \{x_6\}, \{x_9, x_{10}\}\}$,
$U/\underline{A_5^{IO}} = \{\{x_1, x_3, x_4, x_6\}, \{x_2\}, \{x_5\}, \{x_9, x_{10}\}\}$;
$U/\overline{A_2^{IO}} = \{\{x_1, x_2, x_4\}, \{x_3, x_5, x_7\}, \{x_6, x_8, x_9\}, \{x_{10}\}\}$,
$U/\overline{A_4^{IO}} = \{\{x_1, x_2, x_3, x_5\}, \{x_4\}, \{x_6, x_7, x_8\}, \{x_9, x_{10}\}\}$,
$U/\overline{A_5^{IO}} = \{\{x_1, x_3, x_4, x_6\}, \{x_2, x_7\}, \{x_5, x_8\}, \{x_9, x_{10}\}\}$.
According to Definition 18, one can see that the results are captured as follows:
$CSF_{\underline{A_2^{IO}}}(E) = 0.0454, CSF_{\underline{A_4^{IO}}}(E) = -0.0567, CSF_{\underline{A_5^{IO}}}(E) = -0.0294$;
$CSF_{\overline{A_2^{IO}}}(E) = 0.3058, CSF_{\overline{A_4^{IO}}}(E) = 0.2227, CSF_{\overline{A_5^{IO}}}(E) = 0.2813$.

In summary, the comprehensive score function of the granularity A_2 is higher than the granularity A_3 in IOMRIFS, so we choose site 2 as the result of granularity selection.

(4) Optimal site selection based on IIMRIFS

In the same way as (1), we can get the values of evaluation functions $\underline{\mu^{II}}(x_j)$, $(1-\underline{\pi^{II}}(x_j))\cdot\alpha$, $(1-\underline{\pi^{II}}(x_j))\cdot\beta$, $\overline{\mu^{II}}(x_j)$, $(1-\overline{\pi^{II}}(x_j))\cdot\alpha$ and $(1-\overline{\pi^{II}}(x_j))\cdot\beta$ of IIMRIFS, as shown in Table 7.

Table 7. The values of evaluation functions for IIMRIFS.

	$\underline{\mu^{II}}(x_j)$	$(1-\underline{\pi^{II}}(x_j))\cdot\alpha$	$(1-\underline{\pi^{II}}(x_j))\cdot\beta$	$\overline{\mu^{II}}(x_j)$	$(1-\overline{\pi^{II}}(x_j))\cdot\alpha$	$(1-\overline{\pi^{II}}(x_j))\cdot\beta$
x_1	0.25	0.63	0.2772	0.92	0.72	0.3168
x_2	0.09	0.7125	0.3135	0.54	0.615	0.2706
x_3	0.09	0.7125	0.3135	0.92	0.72	0.3168
x_4	0.25	0.63	0.2772	0.92	0.72	0.3168
x_5	0.09	0.7125	0.3135	0.54	0.615	0.2706
x_6	0.09	0.7125	0.3135	0.92	0.72	0.3168
x_7	0.09	0.7125	0.3135	0.92	0.72	0.3168
x_8	0.09	0.7125	0.3135	0.92	0.72	0.3168
x_9	0.15	0.4575	0.2013	0.92	0.72	0.3168
x_{10}	0.67	0.675	0.297	0.72	0.63	0.2772

We can get decision results of the lower and upper approximations of IIMRIFS by three-way decisions in the Section 5.4, as follows:

$POS(\underline{X^{II}}) = \phi$,
$NEG(\underline{X^{II}}) = \{x_1, x_2, x_3, x_4, x_5, x_6, x_7, x_8, x_9\}$,
$BND(\underline{X^{II}}) = \{x_{10}\}$;
$POS(\overline{X^{II}}) = \{x_1, x_3, x_4, x_6, x_7, x_8, x_9, x_{10}\}$,
$NEG(\overline{X^{II}}) = \phi$,
$BND(\overline{X^{II}}) = \{x_2, x_5\}$.

Therefore, the granularity structures of A_2, A_4, A_5 can be rewritten as follows:

$U/\underline{A_2^{II}} = U/\underline{A_4^{II}} = U/\underline{A_5^{II}} = \{x_{10}\}$;
$U/\overline{A_2^{II}} = \{\{x_1, x_2, x_4\}, \{x_3, x_5, x_7\}, \{x_6, x_8, x_9\}, \{x_{10}\}\}$,
$U/\overline{A_4^{II}} = \{\{x_1, x_2, x_3, x_5\}, \{x_4\}, \{x_6, x_7, x_8\}, \{x_9, x_{10}\}\}$,
$U/\overline{A_5^{II}} = \{\{x_1, x_3, x_4, x_6\}, \{x_2, x_7\}, \{x_5, x_8\}, \{x_9, x_{10}\}\}$.

According to Definition 18, one can see that the results are captured as follows:

$CSF_{\underline{A_2^{II}}}(E) = CSF_{\underline{A_4^{II}}}(E) = CSF_{\underline{A_5^{II}}}(E) = 0.44$;
$CAF_{\underline{A_2^{II}}}(E) = CAF_{\underline{A_4^{II}}}(E) = CAF_{\underline{A_5^{II}}}(E) = 0.9$;
$CSF_{\overline{A_2^{II}}}(E) = 0.7067, CSF_{\overline{A_4^{II}}}(E) = 0.7675, CSF_{\overline{A_5^{II}}}(E) = 0.69$;
$CAF_{\overline{A_2^{II}}}(E) = 0.9067,\ CAF_{\overline{A_4^{II}}}(E) = 0.9275, CAF_{\overline{A_5^{II}}}(E) = 0.91$.

In IIMRIFS, the values of the comprehensive score and comprehensive accuracy of granularity A_4 are higher than A_2 and A_5, so site 4 is chosen as the evaluation site.

Considering (3) and (4) synthetically, we find that the results of granularity selection in IOMRIFS and IIMRIFS are inconsistent, so we need to further compute the comprehensive accuracies of IIMRIFS.

$CAF_{\underline{A_2^{IO}}}(E) = 0.7896, CAF_{\underline{A_4^{IO}}}(E) = 0.8125, CAF_{\underline{A_5^{IO}}}(E) = 0.7544$;
$CAF_{\overline{A_2^{IO}}}(E) = 0.8725, CAF_{\overline{A_4^{IO}}}(E) = 0.886,\ CAF_{\overline{A_5^{IO}}}(E) = 0.8588$.

Through the above calculation results, we can see that the comprehensive score and comprehensive accuracy of granularity A_4 are higher than A_2 and A_5 in the case of pessimistic multi- granulation when the requirements are stricter. Therefore, the site 4 is eventually chosen as the optimal evaluation site.

7. Conclusions

In this paper, we propose two new granularity importance degree calculating methods among multiple granularities, and a granularity reduction algorithm is further developed. Subsequently, we design four novel MRIFS models based on reduction sets under optimistic and IMRS, i.e., OOMRIFS, OIMRIFS, IOMRIFS, and IIMRIFS, and further demonstrate their relevant properties. In addition, four three-way decisions models with novel MRIFS for the issue of internal redundant objects in reduction sets are constructed. Finally, we designe the comprehensive score function and the comprehensive precision function for the optimal granularity selection results. Meanwhile, the validity of the proposed models is verified by algorithms and examples. The works of this paper expand the application scopes of MRIFS and three-way decisions theory, which can solve issues such as spam e-mail filtering, risk decision, investment decisions, and so on. A question worth considering is how to extend the methods of this article to fit the big data environment. Moreover, how to combine the fuzzy methods based on triangular or trapezoidal fuzzy numbers with the methods proposed in this paper is also a research problem. These issues will be investigated in our future work.

Author Contributions: Z.-A.X. and D.-J.H. initiated the research and wrote the paper, M.-J.L. participated in some of these search work, and M.Z. supervised the research work and provided helpful suggestions.

Funding: This research received no external funding.

Acknowledgments: This work is supported by the National Natural Science Foundation of China under Grant Nos. 61772176, 61402153, and the Scientific And Technological Project of Henan Province of China under Grant Nos. 182102210078, 182102210362, and the Plan for Scientific Innovation of Henan Province of China under Grant No. 18410051003, and the Key Scientific And Technological Project of Xinxiang City of China under Grant No. CXGG17002.

Conflicts of Interest: The authors declare no conflicts of interest.

References

1. Pawlak, Z. Rough sets. *Int. J. Comput. Inf. Sci.* **1982**, *11*, 341–356. [CrossRef]
2. Pawlak, Z.; Skowron, A. Rough sets: some extensions. *Inf. Sci.* **2007**, *177*, 28–40. [CrossRef]
3. Yao, Y.Y. Probabilistic rough set approximations. *Int. J. Approx. Reason.* **2008**, *49*, 255–271. [CrossRef]
4. Slezak, D.; Ziarko, W. The investigation of the Bayesian rough set model. *Int. J. Approx. Reason.* **2005**, *40*, 81–91. [CrossRef]
5. Ziarko, W. Variable precision rough set model. *J. Comput. Syst. Sci.* **1993**, *46*, 39–59. [CrossRef]
6. Zhu, W. Relationship among basic concepts in covering-based rough sets. *Inf. Sci.* **2009**, *179*, 2478–2486. [CrossRef]
7. Ju, H.R.; Li, H.X.; Yang, X.B.; Zhou, X.Z. Cost-sensitive rough set: A multi-granulation approach. *Knowl.-Based Syst.* **2017**, *123*, 137–153. [CrossRef]
8. Qian, Y.H.; Liang, J.Y.; Dang, C.Y. Incomplete multi-granulation rough set. *IEEE Trans. Syet. Man Cybern. A* **2010**, *40*, 420–431. [CrossRef]
9. Qian, Y.H.; Liang, J.Y.; Yao, Y.Y.; Dang, C.Y. MGRS: A multi-granulation rough set. *Inf. Sci.* **2010**, *180*, 949–970. [CrossRef]
10. Yang, X.B.; Qi, Y.S.; Song, X.N.; Yang, J.Y. Test cost sensitive multigranulation rough set: model and mini-mal cost selection. *Inf. Sci.* **2013**, *250*, 184–199. [CrossRef]
11. Zhang, W.X.; Mi, J.S.; Wu, W.Z. Knowledge reductions in inconsistent information systems. *Chinese J. Comput.* **2003**, *26*, 12–18. (In Chinese)
12. Sang, Y.L.; Qian, Y.H. Granular structure reduction approach to multigranulation decision-theoretic rough sets. *Comput. Sci.* **2017**, *44*, 199–205. (In Chinese)
13. Jing, Y.G.; Li, T.R.; Fujita, H.; Yu, Z.; Wang, B. An incremental attribute reduction approach based on knowledge granularity with a multi-granulation view. *Inf. Sci.* **2017**, *411*, 23–38. [CrossRef]
14. Feng, T.; Fan, H.T.; Mi, J.S. Uncertainty and reduction of variable precision multigranulation fuzzy rough sets based on three-way decisions. *Int. J. Approx. Reason.* **2017**, *85*, 36–58. [CrossRef]
15. Tan, A.H.; Wu, W.Z.; Tao, Y.Z. On the belief structures and reductions of multigranulation spaces with decisions. *Int. J. Approx. Reason.* **2017**, *88*, 39–52. [CrossRef]

16. Kang, Y.; Wu, S.X.; Li, Y.W.; Liu, J.H.; Chen, B.H. A variable precision grey-based multi-granulation rough set model and attribute reduction. *Knowl.-Based Syst.* **2018**, *148*, 131–145. [CrossRef]
17. Xu, W.H.; Li, W.T.; Zhang, X.T. Generalized multigranulation rough sets and optimal granularity selection. *Granul. Comput.* **2017**, *2*, 271–288. [CrossRef]
18. Atanassov, K.T. More on intuitionistic fuzzy sets. *Fuzzy Set Syst.* **1989**, *33*, 37–45. [CrossRef]
19. Atanassov, K.T.; Rangasamy, P. Intuitionistic fuzzy sets. *Fuzzy Sets Syst.* **1986**, *20*, 87–96. [CrossRef]
20. Zadeh, L.A. Fuzzy sets. *Inf. Control* **1965**, *8*, 338–353. [CrossRef]
21. Zhang, X.H. Fuzzy anti-grouped filters and fuzzy normal filters in pseudo-BCI algebras. *J. Intell. Fuzzy Syst.* **2017**, *33*, 1767–1774. [CrossRef]
22. Huang, B.; Guo, C.X.; Zhang, Y.L.; Li, H.X.; Zhou, X.Z. Intuitionistic fuzzy multi-granulation rough sets. *Inf. Sci.* **2014**, *277*, 299–320. [CrossRef]
23. Liu, C.H.; Pedrycz, W. Covering-based multi-granulation fuzzy rough sets. *J. Intell. Fuzzy Syst.* **2016**, *30*, 303–318. [CrossRef]
24. Xue, Z.A.; Wang, N.; Si, X.M.; Zhu, T.L. Research on multi-granularity rough intuitionistic fuzzy cut sets. *J. Henan Normal Univ. (Nat. Sci. Ed.)* **2016**, *44*, 131–139. (In Chinese)
25. Xue, Z.A.; Lv, M.J.; Han, D.J.; Xin, X.W. Multi-granulation graded rough intuitionistic fuzzy sets models based on dominance relation. *Symmetry* **2018**, *10*, 446. [CrossRef]
26. Wang, J.Q.; Zhang, X.H. Two types of intuitionistic fuzzy covering rough sets and an application to multiple criteria group decision making. *Symmetry* **2018**, *10*, 462. [CrossRef]
27. Boran, F.E.; Akay, D. A biparametric similarity measure on intuitionistic fuzzy sets with applications to pattern recognition. *Inf. Sci.* **2014**, *255*, 45–57. [CrossRef]
28. Intarapaiboon, P. A hierarchy-based similarity measure for intuitionistic fuzzy sets. *Soft Comput.* **2016**, *20*, 1–11. [CrossRef]
29. Ngan, R.T.; Le, H.S.; Cuong, B.C.; Mumtaz, A. H-max distance measure of intuitionistic fuzzy sets in decision making. *Appl. Soft Comput.* **2018**, *69*, 393–425. [CrossRef]
30. Yao, Y.Y. The Superiority of Three-way decisions in probabilistic rough set models. *Inf. Sci.* **2011**, *181*, 1080–1096. [CrossRef]
31. Yao, Y.Y. Three-way decisions with probabilistic rough sets. *Inf. Sci.* **2010**, *180*, 341–353. [CrossRef]
32. Zhai, J.H.; Zhang, Y.; Zhu, H.Y. Three-way decisions model based on tolerance rough fuzzy set. *Int. J. Mach. Learn. Cybern.* **2016**, *8*, 1–9. [CrossRef]
33. Zhai, J.H.; Zhang, S.F. Three-way decisions model based on rough fuzzy set. *J. Intell. Fuzzy Syst.* **2018**, *34*, 2051–2059. [CrossRef]
34. Hao, C.; Li, J.H.; Fan, M.; Liu, W.Q.; Tsang, E.C.C. Optimal scale selection in dynamic multi-scale decision tables based on sequential three-way decisions. *Inf. Sci.* **2017**, *415*, 213–232. [CrossRef]
35. Luo, C.; Li, T.R.; Huang, Y.Y.; Fujita, H. Updating three-way decisions in incomplete multi-scale information systems. *Inf. Sci.* **2018**. [CrossRef]
36. Zhang, X.H.; Bo, C.X.; Smarandache, F.; Dai, J.H. New inclusion relation of neutrosophic sets with applications and related lattice structure. *Int. J. Mach. Learn. Cybern.* **2018**, *9*, 1753–1763. [CrossRef]
37. Zhang, Q.H.; Xie, Q.; Wang, G.Y. A novel three-way decision model with decision-theoretic rough sets using utility theory. *Knowl.-Based Syst.* **2018**. [CrossRef]
38. Yang, X.P.; Tan, A.H. Three-way decisions based on intuitionistic fuzzy sets. In Proceedings of the International Joint Conference on Rough Sets, Olsztyn, Poland, 3–7 July 2017.
39. Liu, J.B.; Zhou, X.Z.; Huang, B.; Li, H.X. A three-way decision model based on intuitionistic fuzzy decision systems. In Proceedings of the International Joint Conference on Rough Sets, Olsztyn, Poland, 3–7 July 2017.
40. Liang, D.C.; Liu, D. Deriving three-way decisions from intuitionistic fuzzy decision-theoretic rough sets. *Inf. Sci.* **2015**, *300*, 28–48. [CrossRef]
41. Liang, D.C.; Xu, Z.S.; Liu, D. Three-way decisions with intuitionistic fuzzy decision-theoretic rough sets based on point operators. *Inf. Sci.* **2017**, *375*, 18–201. [CrossRef]
42. Sun, B.Z.; Ma, W.M.; Li, B.J.; Li, X.N. Three-way decisions approach to multiple attribute group decision making with linguistic information-based decision-theoretic rough fuzzy set. *Int. J. Approx. Reason.* **2018**, *93*, 424–442. [CrossRef]

43. Abdel-Basset, M.; Gunasekaran, M.; Mai, M.; Chilamkurti, N. Three-way decisions based on neutrosophic sets and AHP-QFD framework for supplier selection problem. *Future Gener. Comput. Syst.* **2018**, *89*. [CrossRef]
44. Yu, H.; Zhang, C.; Wang, G.Y. A tree-based incremental overlapping clustering method using the three- way decision theory. *Knowl.-Based Syst.* **2016**, *91*, 189–203. [CrossRef]
45. Li, J.H.; Huang, C.C.; Qi, J.J.; Qian, Y.H.; Liu, W.Q. Three-way cognitive concept learning via multi-granulation. *Inf. Sci.* **2017**, *378*, 244–263. [CrossRef]
46. Xue, Z.A.; Zhu, T.L.; Xue, T.Y.; Liu, J. Methodology of attribute weights acquisition based on three-way decision theory. *Comput. Sci.* **2015**, *42*, 265–268. (In Chinese)

symmetry

MDPI

Article

Maximum Detour–Harary Index for Some Graph Classes

Wei Fang [1], Wei-Hua Liu [2,*], Jia-Bao Liu [3], Fu-Yuan Chen [4] and Zhen-Mu Hong [5] and Zheng-Jiang Xia [5]

1 College of Information & Network Engineering, Anhui Science and Technology University, Fengyang 233100, China; fangw@ahstu.edu.cn
2 College of Information and Management Science, Henan Agricultural University, Zhengzhou 450002, China
3 School of Mathematics and Physics, Anhui Jianzhu University, Hefei 230601, China; liujiabaoad@163.com
4 Institute of Statistics and Applied Mathematics, Anhui University of Finance and Economics, Bengbu 233030, China; accfy2016@163.com
5 School of Finance, Anhui University of Finance and Economics, Bengbu 233030, China; zmhong@mail.ustc.edu.cn (Z.-M.H.); 120150025@aufe.edu.cn (Z.-J.X.)
* Correspondence: liuwhnuc@sina.com

Received: 12 September 2018; Accepted: 22 October 2018; Published: 7 November 2018

Abstract: The definition of a Detour–Harary index is $\omega H(G) = \frac{1}{2}\sum_{u,v\in V(G)}\frac{1}{l(u,v|G)}$, where G is a simple and connected graph, and $l(u,v|G)$ is equal to the length of the longest path between vertices u and v. In this paper, we obtained the maximum Detour–Harary index about unicyclic graphs, bicyclic graphs, and cacti, respectively.

Keywords: Detour–Harary index; maximum; unicyclic; bicyclic; cacti

---

1. Introduction

In recent years, chemical graph theory (CGT) has been fast-growing. It helps researchers to understand the structural properties of a molecular graph, for example, References [1–3].

A simple graph is an undirected graph without multiple edges and loops. Let G be a simple and connected graph, and $V(G)$ and $E(G)$ be the vertex set and edge set of G, respectively. For vertices u, v of G, $d_G(v_1, v_2)$ (or $d(v_1, v_2)$ for short) is the distance between v_1 and v_2, which equals to the length of the shortest path between v_1 and v_2 in G; $l(v_1, v_2|G)$ (or $l(v_1, v_2)$ for short) is the detour distance between v_1 and v_2, which equals to the longest path of a shortest path between v_1 and v_2 in G.

$G[S]$ is an induced subgraph of G, the vertex set is S, and the edge set is the set of edges of G and both ends in S. $G - S$ is the induced subgraph $G[V(G) \setminus S]$; when $S = \{w\}$, we write $G - w$ for short.

In 1947, Wiener introduced the first molecular topological index–Wiener index. The Wiener index has applications in many fields, such as chemistry, communication, and cryptology [4–7]. Moreover, the Wiener index was studied from a purely graph-theoretical point of view [8–10]. In Reference [11], Wiener gave the definition of the Wiener index:

$$W(G) = \frac{1}{2}\sum_{u,v\in V(G)} d(u,v).$$

The Harary index was independently introduced by Plavšić et al. [12] and by Ivanciuc et al. [13] in 1993. In References [12,13], they gave the definition of the Harary index:

$$H(G) = \frac{1}{2}\sum_{u,v\in V(G)} \frac{1}{d(u,v)}.$$

In Reference [13], Ivanciuc gave the definition of the Detour index:

$$\omega(G) = \frac{1}{2} \sum_{u,v \in V(G)} l(u,v|G).$$

Lukovits [14] investigated the use of the Detour index in quantitative structure–activity relationship (QSAR) studies. Trinajstić and his collaborators [15] analyzed the use of the Detour index, and compared its application with Wiener index. They found that the Detour index in combination with the Wiener index is very efficient in the structure-boiling point modeling of acyclic and cyclic saturated hydrocarbons.

In this paper, we introduce a new graph invariant reciprocal to the Detour index, namely, the Detour–Harary index, as

$$\omega H(G) = \frac{1}{2} \sum_{u,v \in V(G)} \frac{1}{l(u,v|G)}.$$

Let G be a simple and connected graph, $V(G) = n$ and $E(G) = m$. If $m = n - 1$, then G is a tree; if $m = n$, then G is a unicyclic graph; if $m = n + 1$, then G is a bicyclic graph.

Suppose $\mathcal{U}_n$ ($\mathcal{B}_n$, respectively) is the set of unicyclic (bicyclic, respectively) graphs set with n vertices. Any bicyclic graph G can be obtained from $\theta(p,q,l)$-graph or $\theta(p,q,l)$-graph G_0 by attaching trees to the vertices, where $p, q, l \geq 1$, and at most one of them is equal to 1. We denote G_0 be the kernel of G (Figure 1).

If each block of G is either a cycle or an edge, then we called graph G a *cactus* graph. Suppose $\mathcal{C}_n^k$ be the set of all cacti with n-vertices and k cycles. Obviously, $\mathcal{C}_n^0$ are trees, $\mathcal{C}_n^1$ are unicyclic graphs, and $\mathcal{C}_n^2$ are bicyclic graphs with exactly two cycles.

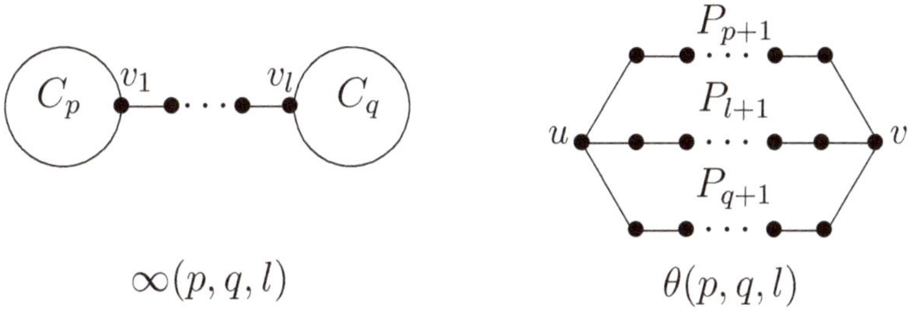

Figure 1. ∞-graph and θ-graph.

There are more results about cacti and bicyclic graphs [16–25]. More results about Harary index can be found in References [26–34], and more results about Detour index can be found in References [14,35–39].

Note that the Detour–Harary index is the same as Harary index for a tree graph; we study the Detour–Harary index of topological structures containing cycles. In this paper, we gave the maximum Detour–Harary index among $\mathcal{U}_n$,$\mathcal{B}_n$ and $\mathcal{C}_n^k$ ($k \geq 3$), respectively.

2. Preliminaries

In this section, we introduce useful lemmas and graph transformations.

Lemma 1. *[40] Let G be a connected graph, x be a cut-vertex of G, and u and v be vertices occurring in different components that arise upon the deletion of vertex x. Then*

$$l(u,v|G) = l(u,x|G) + l(x,v|G).$$

2.1. Edge-Lifting Transformation

The edge-lifting transformation [41]. Let G_1 and G_2 be two graphs with $n_1 \geq 2$ and $n_2 \geq 2$ vertices. $u_0 \in V(G_1)$ and $v_0 \in V(G_2)$, G is the graph obtained from G_1 and G_2 by adding an edge between u_0 and v_0. G' is the graph obtained by identifying u_0 to v_0 and adding a pendent edge to $u_0(v_0)$. We called graph G' the edge-lifting transformation of graph G (see Figure 2).

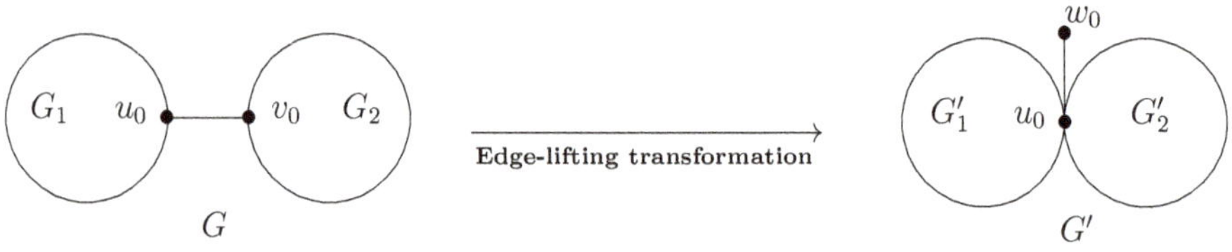

Figure 2. Edge-lifting transformation.

Lemma 2. *Let graph G' be the edge-lifting transformation of graph G. Then $\omega H(G) < \omega H(G')$.*

Proof. By the definition of $\omega H(G)$ and Lemma 1,

$$\begin{aligned}
\omega H(G) = {} & \omega H(G_1) + \omega H(G_2) + \sum_{x \in V(G_1)\setminus\{u_0\}} \frac{1}{l(v_0, x|G)} + \sum_{y \in V(G_2)\setminus\{v_0\}} \frac{1}{l(u_0, y|G)} \\
& + \frac{1}{l(u_0, v_0|G)} + \sum_{\substack{x \in V(G_1)\setminus\{u_0\} \\ y \in V(G_2)\setminus\{v_0\}}} \frac{1}{l(x, y|G)} \\
= {} & \omega H(G_1) + \omega H(G_2) + \sum_{x \in V(G_1)\setminus\{u_0\}} \frac{1}{1 + l(u_0, x|G)} + \sum_{y \in V(G_2)\setminus\{v_0\}} \frac{1}{1 + l(v_0, y|G)} \\
& + 1 + \sum_{\substack{x \in V(G_1)\setminus\{u_0\} \\ y \in V(G_2)\setminus\{v_0\}}} \frac{1}{l(u_0, x|G) + 1 + l(v_0, y|G)}, \\
\omega H(G') = {} & \omega H(G'_1) + \omega H(G'_2) + \sum_{x' \in V(G'_1)\setminus\{u_0\}} \frac{1}{l(w_0, x'|G')} + \sum_{y' \in V(G'_2)\setminus\{u_0\}} \frac{1}{l(w_0, y'|G')} \\
& + \frac{1}{l(u_0, w_0|G')} + \sum_{\substack{x' \in V(G'_1)\setminus\{u_0\} \\ y' \in V(G'_2)\setminus\{u_0\}}} \frac{1}{l(x', y'|G')} \\
= {} & \omega H(G'_1) + \omega H(G'_2) + \sum_{x' \in V(G'_1)\setminus\{u_0\}} \frac{1}{1 + l(u_0, x'|G')} + \sum_{y' \in V(G'_2)\setminus\{u_0\}} \frac{1}{1 + l(u_0, y'|G')} \\
& + 1 + \sum_{\substack{x' \in V(G'_1)\setminus\{u_0\} \\ y' \in V(G'_2)\setminus\{u_0\}}} \frac{1}{l(u_0, x'|G') + l(u_0, y'|G')}.
\end{aligned}$$

Obviously,

$$\begin{aligned}
\omega H(G_1) &= \omega H(G'_1); \\
\omega H(G_2) &= \omega H(G'_2); \\
l(u_0, x|G) &= l(u_0, x'|G'), \text{ where } x \in V(G_1) \setminus \{u_0\} \text{ and } x' \in V(G'_1) \setminus \{u_0\}; \\
l(v_0, y|G) &= l(u_0, y'|G'), \text{ where } y \in V(G_2) \setminus \{v_0\} \text{ and } y' \in V(G'_2) \setminus \{u_0\}.
\end{aligned}$$

Then

$$\omega H(G) - \omega H(G') = \sum_{\substack{x \in V(G_1)\setminus\{u_0\} \\ y \in V(G_2)\setminus\{v_0\}}} \frac{1}{l(x,u_0|G) + 1 + l(v_0,y|G)} - \sum_{\substack{x' \in V(G_1')\setminus\{u_0\} \\ y' \in V(G_2')\setminus\{u_0\}}} \frac{1}{l(x',u_0|G') + l(u_0,y'|G')} < 0.$$

□

2.2. Cycle-Edge Transformation

Suppose $G \in \mathcal{C}_n^l$ is a cactus as shown in Figure 3. $C_p = v_1v_2\cdots v_pv_1$ is a cycle of G; G_i is a cactus, and $v_i \in V(G_i)$, $1 \le i \le p$; $W_{v_i} = N_G(v_i) \cap V(G_i)$, $1 \le i \le p$. G' is the graph obtained from G by deleting the edges from v_i to $W_{v_i}(2 \le i \le p)$, while adding the edges from v_1 to $W_{v_i}(2 \le i \le p)$.

We called graph G' the cycle-edge transformation of graph G (see Figure 3).

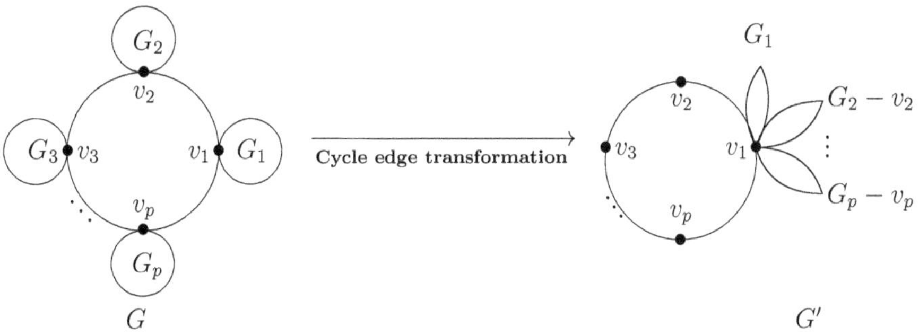

Figure 3. Cycle-edge transformation.

Lemma 3. *Suppose $G \in \mathcal{C}_n^l$ is a cactus, $p \ge 3$, and G' is the cycle-edge transformation of G (see Figure 3). Then, $\omega H(G) \le \omega H(G')$, and the equality holds if and only if $G \cong G'$.*

Proof. Let $V_i = V(G_i - v_i)$, $1 \le i \le p$. By the definition of $\omega H(G)$ and Lemma 1,

$$\begin{aligned}\omega H(G) &= \omega H(C_p) + \frac{1}{2}\sum_{i=1}^{p}\sum_{x,y \in V_i}\frac{1}{l(x,y|G)} + \frac{1}{2}\sum_{i=1}^{p}\sum_{j=1}^{p}\sum_{\substack{x \in V_i \\ y \in V_j \\ i \ne j}}\frac{1}{l(x,y|G)} + \sum_{i=1}^{p}\sum_{\substack{x \in V_i \\ y \in V(C_p)}}\frac{1}{l(x,y|G)} \\ &= \omega H(C_p) + \frac{1}{2}\sum_{i=1}^{p}\sum_{x,y \in V_i}\frac{1}{l(x,y|G)} + \frac{1}{2}\sum_{i=1}^{p}\sum_{j=1}^{p}\sum_{\substack{x \in V_i \\ y \in V_j \\ i \ne j}}\frac{1}{l(x,v_i|G) + l(v_i,v_j|G) + l(v_j,y|G)} \\ &\quad + \sum_{i=1}^{p}\sum_{\substack{x \in V_i \\ y \in V(C_p)}}\frac{1}{l(x,v_i|G) + l(v_i,y|G)},\end{aligned}$$

$$\begin{aligned}\omega H(G') &= \omega H(C_p) + \frac{1}{2}\sum_{i=1}^{p}\sum_{x,y\in V_i}\frac{1}{l(x,y|G')} + \frac{1}{2}\sum_{i=1}^{p}\sum_{j=1}^{p}\sum_{\substack{x\in V_i\\ y\in V_j\\ i\neq j}}\frac{1}{l(x,y|G')} + \sum_{i=1}^{p}\sum_{\substack{x\in V_i\\ y\in V(C_p)}}\frac{1}{l(x,y|G')}\\
&= \omega H(C_p) + \frac{1}{2}\sum_{i=1}^{p}\sum_{x,y\in V_i}\frac{1}{l(x,y|G')} + \frac{1}{2}\sum_{i=1}^{p}\sum_{j=1}^{p}\sum_{\substack{x\in V_i\\ y\in V_j\\ i\neq j}}\frac{1}{l(x,v_1|G')+l(v_1,y|G')}\\
&\quad + \sum_{i=1}^{p}\sum_{\substack{x\in V_i\\ y\in V(C_p)}}\frac{1}{l(x,v_1|G')+l(v_1,y|G')}.\end{aligned}$$

Obviously,

$$\begin{aligned}\sum_{i=1}^{p}\sum_{x,y\in V_i}\frac{1}{l(x,y|G)} &= \sum_{i=1}^{p}\sum_{x,y\in V_i}\frac{1}{l(x,y|G')};\\
l(x,v_i|G) &= l(x,v_1|G'), \text{where } x\in V_i;\\
l(v_j,y|G) &= l(v_1,y|G'), \text{where } y\in V_j;\\
\sum_{i=1}^{p}\sum_{\substack{x\in V_i\\ y\in V(C_p)}}\frac{1}{l(x,v_i|G)+l(v_i,y|G)} &= \sum_{i=1}^{p}\sum_{\substack{x\in V_i\\ y\in V(C_p)}}\frac{1}{l(x,v_1|G')+l(v_1,y|G')}.\end{aligned}$$

Then

$$\begin{aligned}\omega H(G) - \omega H(G') &= \frac{1}{2}\sum_{i=1}^{p}\sum_{j=1}^{p}\sum_{\substack{x\in V_i\\ y\in V_j\\ i\neq j}}\frac{1}{l(x,v_i|G)+l(v_i,v_j|G)+l(v_j,y|G)}\\
&\quad - \frac{1}{2}\sum_{i=1}^{p}\sum_{j=1}^{p}\sum_{\substack{x\in V_i\\ y\in V_j\\ i\neq j}}\frac{1}{l(x,v_1|G')+l(v_1,y|G')} < 0.\end{aligned}$$

The proof is completed. □

2.3. Cycle Transformation

Suppose $G \in \mathcal{C}_n^l$ is a cactus, as shown in Figure 4. $C_p = v_1v_2\cdots v_pv_1$ is a cycle of G, and G_1 is a simple and connected graph, $v_1 \in V(G_1)$. G' is the graph obtained from G by deleting the edges from v_i to $v_{i+1}(2 \le i \le p-1)$, meanwhile, adding the edges from v_1 to $v_i(3 \le i \le p-1)$.

We called graph G' is the *cycle transformation* of G (see Figure 4).

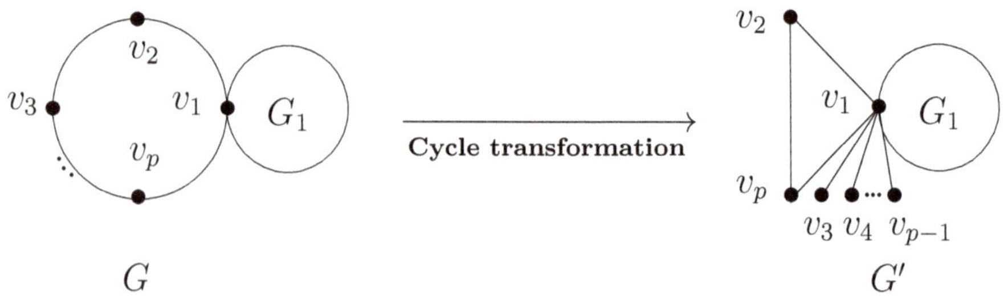

Figure 4. Cycle transformation.

Lemma 4. *Suppose graph G is a simple and connected graph with $p \geq 4$, and G' is the cycle transformation of G(see Figure 4). Then, $\omega H(G) < \omega H(G')$.*

Proof. Let $V(C_p) = \{v_1, v_2, \cdots, v_p\}$, $V_1 = V(C_p - v_1)$, $V_2 = V(G_1 - v_1)$. By the definition of $\omega H(G)$,

$$\begin{aligned}
\omega H(G) &= \omega H(G_1) + \sum_{x,y \in V(C_p)} \frac{1}{l(x,y|G)} + \sum_{\substack{x \in V_1 \\ y \in V_2}} \frac{1}{l(x,y|G)} \\
&= \omega H(G_1) + \sum_{x,y \in V(C_p)} \frac{1}{l(x,y|G)} + \sum_{\substack{x \in V_1, \\ y \in V_2}} \frac{1}{l(x,v_1|G) + l(v_1,y|G)}, \\
\omega H(G') &= \omega H(G_1) + \sum_{x,y \in V(C_p)} \frac{1}{l(x,y|G')} + \sum_{\substack{x \in V_1, \\ y \in V_2}} \frac{1}{l(x,y|G')} \\
&= \omega H(G_1) + \sum_{x,y \in V(C_p)} \frac{1}{l(x,y|G')} + \sum_{\substack{x \in V_1, \\ y \in V_2}} \frac{1}{l(x,v_1|G') + l(v_1,y|G')},
\end{aligned}$$

Obviously,

$$\begin{aligned}
l(x,y|G) &\geq l(x,y|G'), \text{where } x, y \in V_1; \\
l(x,v_1|G) &> 2 \geq l(x,v_1|G'), \text{where } x \in V_1; \\
l(v_1,y|G) &= l(v_1,y|G'), \text{where } y \in V_2.
\end{aligned}$$

Then

$$\begin{aligned}
\omega H(G) - \omega H(G') = &\left(\sum_{x,y \in V(C_p)} \frac{1}{l(x,y|G)} - \sum_{x,y \in V(C_p)} \frac{1}{l(x,y|G')} \right) \\
&+ \left(\sum_{\substack{x \in V_1, \\ y \in V_2}} \frac{1}{l(x,v_1|G) + l(v_1,y|G)} - \sum_{\substack{x \in V_1, \\ y \in V_2}} \frac{1}{l(x,v_1|G') + l(v_1,y|G')} \right) < 0.
\end{aligned}$$

□

3. Maximum Detour–Harary Index of Unicyclic Graphs

For any unicyclic graph $G \in \mathcal{U}_n$, by repeating edge-lifting transformations, cycle-edge transformations, cycle transformations, or any combination of these on G, we get U_1 from G, where graph U_1 is defined in Figure 5.

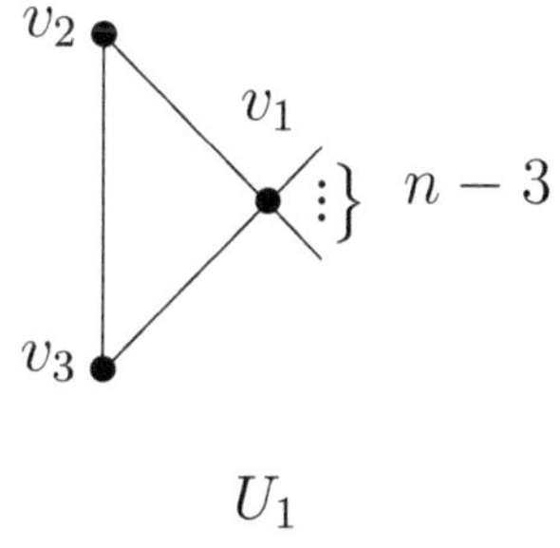

Figure 5. Unicyclic graph U_1.

Theorem 1. *Let U_1 be defined as Figure 5. Then, U_1 is the unique graph that attains the maximum Detour–Harary index among all graphs in $\mathcal{U}_n (n \geq 3)$, and $\omega H(U_1) = \frac{3n^2-n-6}{12}$.*

Proof. By Lemmas 2–4, U_1 is the unique graph which attains the maximum Detour–Harary index of all graphs in $\mathcal{U}_n$. We then calculate the value $\omega H(U_1)$.

Let $V(U_1) = \{v_1, v_2, \cdots, v_n\}$. It can be checked directly that

$$\sum_{i=2}^{n} \frac{1}{l(v_1, v_i|U_1)} = n-2;$$

$$\sum_{1 \leq i \leq n, i \neq 2} \frac{1}{l(v_2, v_i|U_1)} = \sum_{1 \leq j \leq n, j \neq 3} \frac{1}{l(v_3, v_j|U_1)} = \frac{1}{2} + \frac{1}{2} + \frac{n-3}{3} = \frac{n}{3};$$

$$\sum_{1 \leq i \leq n, i \neq 4} \frac{1}{l(v_4, v_i|U_1)} = 1 + \frac{n-4}{2} + \frac{2}{3} = \frac{3n-2}{6}.$$

Then

$$\begin{aligned} \omega H(U_1) &= \frac{1}{2}[\sum_{i=2}^{n} \frac{1}{l(v_1, v_i|U_1)} + 2 \sum_{1 \leq i \leq n, i \neq 2} \frac{1}{l(v_2, v_i|U_1)} + (n-3) \sum_{i=1}^{n} \frac{1}{l(v_4, v_i|U_1)}] \\ &= \frac{3n^2-n-6}{12}. \end{aligned}$$

The proof is completed. □

4. Maximum Detour–Harary Index of Bicyclic Graphs

For any bicyclic graph $G \in \infty(p,q,l)$ with exactly two cycles, by repeating edge-lifting transformations, cycle-edge transformations, cycle transformations, or any combination of these on G, we get B_1 from G, where graph B_1 is defined in Figure 6.

For any bicyclic graph $G \in \theta(p,q,l)$ with n vertices, by repeating edge-lifting transformations on G, we get B_2 from G, where graph B_2 is defined in Figure 7.

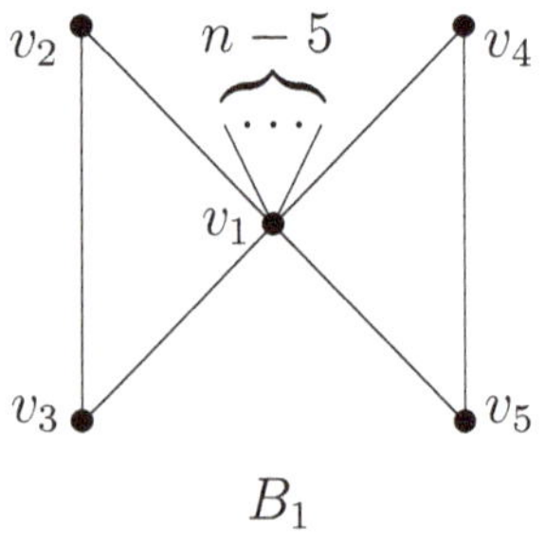

Figure 6. Bicyclic graph B_1.

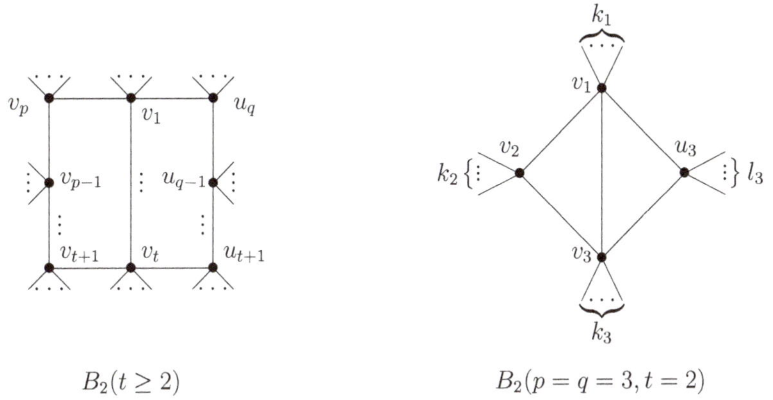

Figure 7. Bicyclic graph $B_2(t \geq 2)$.

Theorem 2. *Let B_2, B_3 be defined as Figures 7 and 8. Then, $\omega H(B_2) \leq \omega H(B_3)$, and the equality holds if and only if $B_2 \cong B_3$.*

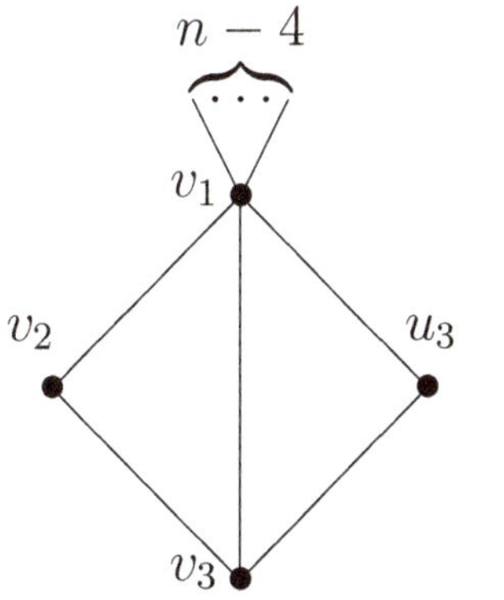

Figure 8. Bicyclic graph $B_2(t \geq 2)$.

Proof. Case 1. $B_2 = B_3$. Obviously, $\omega H(B_2) = \omega H(B_3)$.

Case 2. $B_2 \neq B_3$ and $p = q = 3, t = 2$(see Figures 7 and 8).

Let $V_1 = \{v_1, v_2, v_3, u_3\}$, $W_{v_i} = \{w \mid wv_i \in E(B_2)$ and $d_{B_2}(w) = 1\}$ and $|W_{v_i}| = k_i$, $W_{u_3} = \{w \mid wu_3 \in E(B_2)$ and $d_{B_2}(w) = 1\}$ and $|W_{u_3}| = l_3, k_i + l_3 = n - 4$ for $1 \leq i \leq 3$.

$$\omega H(B_2) = \sum_{x,y \in V_1} \frac{1}{l(x,y|B_2)} + \sum_{\substack{x \in V_1, \\ y \in V(B_2)-V_1}} \frac{1}{l(x,y|B_2)} + \sum_{x,y \in V(B_2)-V_1} \frac{1}{l(x,y|B_2)},$$

$$\omega H(B_3) = \sum_{x,y \in V_1} \frac{1}{l(x,y|B_3)} + \sum_{\substack{x \in V_1, \\ y \in V(B_3)-V_1}} \frac{1}{l(x,y|B_3)} + \sum_{x,y \in V(B_3)-V_1} \frac{1}{l(x,y|B_3)}.$$

Easily,

$$\sum_{x,y \in V_1} \frac{1}{l(x,y|B_2)} = \sum_{x,y \in V_1} \frac{1}{l(x,y|B_3)} \tag{1}$$

$$\begin{aligned}\sum_{\substack{x\in V_1,\\ y\in V(B_2)-V_1}}\frac{1}{l(x,y|B_2)} &= \sum_{w\in V(B_2)-V_1}\frac{1}{l(v_1,w|B_2)}+\sum_{w\in V(B_2)-V_1}\frac{1}{l(v_2,w|B_2)}\\ &\quad+\sum_{w\in V(B_2)-V_1}\frac{1}{l(v_3,w|B_2)}+\sum_{w\in V(B_2)-V_1}\frac{1}{l(u_3,w|B_2)}\\ &= (1\cdot k_1+\frac{1}{4}\cdot k_2+\frac{1}{3}\cdot k_3+\frac{1}{4}\cdot l_3)+(\frac{1}{4}\cdot k_1+1\cdot k_2+\frac{1}{4}\cdot k_3+\frac{1}{4}\cdot l_3)\\ &\quad+(\frac{1}{3}\cdot k_1+\frac{1}{4}\cdot k_2+1\cdot k_3+\frac{1}{4}\cdot l_3)+(\frac{1}{4}\cdot k_1+\frac{1}{4}\cdot k_2+\frac{1}{4}\cdot k_3+1\cdot l_3)\\ &= \frac{11(k_1+k_3)}{6}+\frac{7(k_2+l_3)}{4},\end{aligned}$$

$$\begin{aligned}\sum_{\substack{x\in V_1,\\ y\in V(B_3)-V_1}}\frac{1}{l(x,y|B_3)} &= \sum_{w\in V(B_3)-V_1}\frac{1}{l(v_1,w|B_3)}+\sum_{w\in V(B_3)-V_1}\frac{1}{l(v_2,w|B_3)}\\ &\quad+\sum_{w\in V(B_3)-V_1}\frac{1}{l(v_3,w|B_3)}+\sum_{w\in V(B_3)-V_1}\frac{1}{l(u_3,w|B_3)}\\ &= 1\cdot(n-4)+\frac{1}{4}\cdot(n-4)+\frac{1}{3}\cdot(n-4)+\frac{1}{4}\cdot(n-4)\\ &= \frac{11(n-4)}{6}\\ &= \frac{11(k_1+k_2+k_3+l_3)}{6}, \qquad \text{(since } k_i+l_3=n-4 \text{ for } 1\le i\le 3)\end{aligned}$$

Then,

$$\sum_{\substack{x\in V_1,\\ y\in V(B_2)-V_1}}\frac{1}{l(x,y|B_2)}-\sum_{\substack{x\in V_1,\\ y\in V(B_3)-V_1}}\frac{1}{l(x,y|B_3)}=\frac{1}{12}(k_2+l_3)\ge 0, \tag{2}$$

the equality holds if and only if $k_2=l_3=0$.

On the other hand $\frac{1}{l(x,y|B_2)}\le\frac{1}{l(x,y|B_3)}=\frac{1}{2}$, where $x,y\in V(B_2)-V_1$, then

$$\sum_{x,y\in V(B_3)-V_1}\frac{1}{l(x,y|B_2)}\le\sum_{x,y\in V(B_3)-V_1}\frac{1}{l(x,y|B_3)}, \tag{3}$$

the equality holds if $k_1=n-4$ or $k_2=n-4$ or $k_3=n-4$ or $l_3=n-4$.

By (1)–(3) and $B_2\ne B_3$, we have $\omega H(B_2)<\omega H(B_3)$.

Case 3. $B_2\ne B_3$ and $p+q-t>4$.

It can be checked directly that

$$\begin{aligned}\omega H(B_2) &\le \underbrace{(1+1+\cdots+1)}_{n-p-q+t}+\tfrac{1}{2}\tbinom{n-p-q+t}{2}+\tfrac{1}{4}\left[\tbinom{n}{2}-(n-p-q+t)-\tbinom{n-p-q+t}{2}\right],\\ \omega H(B_3) &= \underbrace{(1+1+\cdots+1)}_{n-4}+\tfrac{1}{2}[1+\tbinom{n-4}{2}]+\tfrac{1}{3}[5+(n-4)]+\tfrac{1}{4}[2(n-4)].\end{aligned}$$

B_2,B_3 are bicyclic graphs and $|V(B_2)|=|V(B_3)|=n$. Since $p+q-t>4$, then $n-p-q+t\le n-5$ and $\binom{n-p-q+t}{2}<\binom{n-4}{2}$, we have $\omega H(B_2)<\omega H(B_3)$.

The proof is completed. □

Theorem 3. *Let B_1, B_3 be defined as Figures 6 and 8. Then,*

$$\max\{\omega H(\mathcal{B}_n)\}=\begin{cases}\omega H(B_3)=\frac{13}{6}, & \text{if } n=4,\\ \omega H(B_1)=\omega H(B_3)=\frac{3n^2-5n-2}{12}, & \text{if } n\ge 5.\end{cases}$$

Proof. Let $G \in \infty(p,q,l)$, by Lemmas 2–4, we have $\omega H(G) \leq \omega H(B_1)$, and the equality holds if and only if $G \cong B_1$.

For any bicyclic graph with $G \in \theta(p,q,l)$, by Lemmas 2–4 and Theorem 2, we have $\omega H(G) \leq \omega H(B_3)$, and the equality holds if and only if $G \cong B_3$. Thus, $\max\{\omega H(\mathcal{B}_n)\} = \max\{\omega H(B_1), \omega H(B_3)\}$.

It can be checked directly that

$$\omega H(B_1) = (n-5) + \frac{1}{2}[\binom{n-5}{2} + 6] + \frac{1}{3}[4(n-5)] + \frac{1}{4}\cdot 4 = \frac{3n^2-5n-2}{12}, n \geq 5;$$

$$\omega H(B_3) = (n-4) + \frac{1}{2}\binom{n-4}{2} + \frac{1}{3}(n-4) + \frac{1}{4}[2(n-4)] = \frac{3n^2-5n-2}{12}, n \geq 4.$$

Therefore

$$\max\{\omega H(\mathcal{B}_n)\} = \begin{cases} \omega H(B_3) = \frac{13}{6}, & \text{if } n = 4, \\ \omega H(B_1) = \omega H(B_3) = \frac{3n^2-5n-2}{12}, & \text{if } n \geq 5. \end{cases}$$

The proof is completed. □

5. Maximum Detour–Harary Index of Cacti

For any cactus graph $G \in \mathcal{C}_n^k (k \geq 3)$, by repeating edge-lifting transformations, cycle-edge transformations, cycle transformations, or any combination of these on G, we get $\mathcal{C}_1$ from G, where graph $\mathcal{C}_1$ is defined in Figure 9.

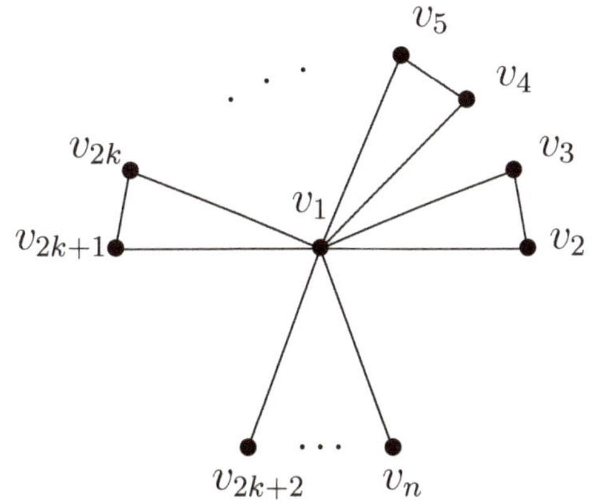

Figure 9. Cactus graph $\mathcal{C}_1 (k \geq 3)$.

Theorem 4. *Let $\mathcal{C}_1$ be defined as Figure 9. Then, $\mathcal{C}_1$ is the unique cactus graph in $\mathcal{C}_n^k (k \geq 3)$ that attains the maximum Detour–Harary index, and* $\omega H(\mathcal{C}_1) = \frac{3n^2+2k^2-4nk+3n-2k-6}{12}$.

Proof. By Lemmas 2–4, $\mathcal{C}_1$ is the unique graph that attains the maximum Detour–Harary index of all graphs in $\mathcal{C}_n^k (k \geq 3)$.

Let $V(\mathcal{C}_1) = \{v_1, v_2, \cdots, v_n\}$, and it can be checked directly that

$$\sum_{i=2}^{n} \frac{1}{l(v_1, v_i|\mathcal{C}_1)} = 1 \cdot (n-2k-1) + \frac{1}{2} \cdot 2k = n-k-1;$$
$$\sum_{1 \le i \le n, i \ne 2} \frac{1}{l(v_2, v_i|\mathcal{C}_1)} = \frac{1}{2} \cdot 2 + \frac{1}{3} \cdot (n-2k-1) + \frac{1}{4} \cdot (2k-2) = \frac{1}{3}n - \frac{1}{6}k + \frac{1}{6};$$
$$\sum_{j=1}^{n-1} \frac{1}{l(v_n, v_j|\mathcal{C}_1)} = 1 + \frac{1}{2} \cdot (n-2k-2) + \frac{1}{3} \cdot 2k = \frac{1}{2}n - \frac{1}{3}k.$$

Then,

$$\begin{aligned} \omega H(\mathcal{C}_1) &= \frac{1}{2}[(n-k-1) + 2k \cdot (\frac{1}{3}n - \frac{1}{6}k + \frac{1}{6}) + (n-2k-1) \cdot (\frac{1}{2}n - \frac{1}{3}k)] \\ &= \frac{3n^2 + 2k^2 - 4nk + 3n - 2k - 6}{12}. \end{aligned}$$

The proof is completed. □

Author Contributions: Conceptualization, W.F. and W.-H.L.; methodology, F.-Y.C.; Z.-J.X. and J.-B.L.; writing—original draft preparation, W.F. and Z.-M.H; writing—review and editing, W.-H.L.

Funding: This research was funded by NSFC Grant (No.11601001, No.11601002, No.11601006).

Conflicts of Interest: The authors declare no conflict of interest.

References

1. Imran, M.; Ali, M.A.; Ahmad, S.; Siddiqui, M.K.; Baig, A.Q. Topological sharacterization of the symmetrical structure of bismuth tri-iodide. *Symmetry* **2018**, *10*, 201. [CrossRef]
2. Liu, J.; Siddiqui, M.K.; Zahid, M.A.; Naeem, M.; Baig, A.Q. Topological Properties of Crystallographic Structure of Molecules. *Symmetry* **2018**, *10*, 265. [CrossRef]
3. Shao, Z.; Siddiqui, M.K.; Muhammad, M.H. Computing zagreb indices and zagreb polynomials for symmetrical nanotubes. *Symmetry* **2018**, *10*, 244. [CrossRef]
4. Dobrynin, A.; Entringer, R.; Gutman, I. Wiener Index of Trees: Theory and Applications. *Acta Appl. Math.* **2001**, *66*, 211–249. [CrossRef]
5. Alizadeh, Y.; Andova, V.; Zar, S.K.; Skrekovski, R.V. Wiener dimension: Fundamental properties and (5,0)-nanotubical fullerenes. *MATCH Commun. Math. Comput. Chem.* **2014**, *72*, 279–294.
6. Needham, D.E.; Wei, I.C.; Seybold, P.G. Molecular modeling of the physical properties of alkanes. *J. Am. Chem. Soc.* **1988**, *110*, 4186–4194. [CrossRef]
7. Vijayabarathi, A.; Anjaneyulu, G.S.G.N. Wiener index of a graph and chemical applications. *Int. J. ChemTech Res.* **2013**, *5*, 1847–1853.
8. Gutman, I.; Cruz, R.; Rada, J. Wiener index of Eulerian graphs. *Discret. Appl. Math.* **2014**, *162*, 247–250. [CrossRef]
9. Lin, H. Extremal Wiener index of trees with given number of vertices of even degree. *MATCH Commun. Math. Comput. Chem.* **2014**, *72*, 311–320.
10. Lin, H. Note on the maximum Wiener index of trees with given number of vertices of maximum degree. *MATCH Commun. Math. Comput. Chem.* **2014**, *72*, 783–790.
11. Wiener, H. Structural determination of paraffin boiling points. *Am. Chem. Soc.* **1947**, *69*, 17–20. [CrossRef]
12. Plavšić, D.; Nikolić, S.; Trinajstić, N.; Mihalić, Z. On the Harary index for the characterization of chemical graphs. *J. Math. Chem.* **1993**, *12*, 235–250. [CrossRef]
13. Ivanciuc, O.; Balaban, T.S.; Balaban, A.T. Reciprocal distance matrix, related local vertex invariants and topological indices. *J. Math. Chem.* **1993**, *12*, 309–318. [CrossRef]
14. Lukovits, I. The Detour index. *Croat. Chem. Acta* **1996**, *69*, 873–882.
15. Trinajstić, N.; Nikolić, S.; Lučić, B.; Amić, D.; Mihalić, Z. The Detour matrix in chemistry. *J. Chem. Inf. Comput. Sci.* **1997**, *37*, 631–638. [CrossRef]

16. Chen, S. Cacti with the smallest, second smallest, and third smallest Gutman index. *J. Comb. Optim.* **2016**, *31*, 327–332. [CrossRef]
17. Chen, Z.; Dehmer, M.; Shi, Y.; Yang, H. Sharp upper bounds for the Balaban index of bicyclic graphs. *MATCH Commun. Math. Comput. Chem.* **2016**, *75*, 105–128.
18. Fang, W.; Gao, Y.; Shao, Y.; Gao, W.; Jing, G.; Li, Z. Maximum Balaban index and sum-Balaban index of bicyclic graphs. *MATCH Commun. Math. Comput. Chem.* **2016**, *75*, 129–156.
19. Fang, W.; Wang, Y.; Liu, J.-B.; Jing, G. Maximum Resistance-Harary index of cacti. *Discret. Appl. Math.* **2018**. [CrossRef]
20. Gutman, I.; Li, S.; Wei, W. Cacti with n-vertices and t-cycles having extremal Wiener index. *Discret. Appl. Math.* **2017**, *232*, 189–200. [CrossRef]
21. Ji, S.; Li, X.; Shi, Y. Extremal matching energy of bicyclic graphs. *MATCH Commun. Math. Comput. Chem.* **2013**, *70*, 697–706.
22. Liu, J.; Pan, X.; Yu, L.; Li, D. Complete characterization of bicyclic graphs with minimal Kirchhoff index. *Discret. Appl. Math.* **2016**, *200*, 95–107. [CrossRef]
23. Wang, H.; Hua, H.; Wang, D. Cacti with minimum, second-minimum, and third-minimum Kirchhoff indices. *Math. Commun.* **2010**, *15*, 347–358.
24. Wang, L.; Fan, Y.; Wang, Y. Maximum Estrada index of bicyclic graphs. *Discret. Appl. Math.* **2015**, *180*, 194–199. [CrossRef]
25. Lu, Y.; Wang, L.; Xiao, P. Complex Unit Gain Bicyclic Graphs with Rank 2,3 or 4. *Linear Algebra Appl.* **2017**, *523*, 169–186. [CrossRef]
26. Furtula, B.; Gutman, I.; Katanić, V. Three-center Harary index and its applications. *Iran. J. Math. Chem.* **2016**, *7*, 61–68.
27. Feng, L.; Lan, Y.; Liu, W.; Wang, X. Minimal Harary index of graphs with small parameters. *MATCH Commun. Math. Comput. Chem.* **2016**, *76*, 23–42.
28. Hua, H.; Ning, B. Wiener index, Harary index and hamiltonicity of graphs. *MATCH Commun. Math. Comput. Chem.* **2017**, *78*, 153–162.
29. Li, X.; Fan, Y. The connectivity and the Harary index of a graph. *Discret. Appl. Math.* **2015**, *181*, 167–173. [CrossRef]
30. Xu, K.; Das, K.C. On Harary index of graphs. *Discret. Appl. Math.* **2011**, *159*, 1631–1640. [CrossRef]
31. Xu, K. Trees with the seven smallest and eight greatest Harary indices. *Discret. Appl. Math.* **2012**, *160*, 321–331. [CrossRef]
32. Xu, K.; Das, K.C. Extremal unicyclic and bicyclic graphs with respect to Harary Index. *Bull. Malaysian Math. Sci. Soc.* **2013**, *36*, 373–383.
33. Xu, K.; Wang, J.; Das, K.C.; Klavžar, S. Weighted Harary indices of apex trees and k-apex trees. *Discret. Appl. Math.* **2015**, *189*, 30–40. [CrossRef]
34. Zhou, B.; Cai, X.; Trinajstić, N. On Harary index. *J. Math. Chem.* **2008**, *44*, 611–618. [CrossRef]
35. Fang, W.; Yu, H.; Gao, Y.; Jing, G.; Li, Z.; Li, X. Minimum Detour index of cactus graphs. *Ars Comb.* **2019**, in press.
36. Qi, X.; Zhou, B. Detour index of a class of unicyclic graphs. *Filomat* **2010**, *24*, 29–40.
37. Qi, X.; Zhou, B. Hyper-Detour index of unicyclic graphs. *MATCH Commun. Math. Comput. Chem.* **2011**, *66*, 329–342.
38. Rücker, G.; Rücker, C. Symmetry-aided computation of the Detour matrix and the Detour index. *J. Chem. Inf. Comput. Sci.* **1998**, *38*, 710–714. [CrossRef]
39. Zhou, B.; Cai, X. On Detour index. *MATCH Commun. Math. Comput. Chem.* **2010**, *44*, 199–210.
40. Qi, X. Detour index of bicyclic graphs. *Util. Math.* **2013**, *90*, 101–113.
41. Deng, H. On the Balaban index of trees. *MATCH Commun. Math. Comput. Chem.* **2011**, *66*, 253–260.

symmetry

MDPI

Article

Q-Filters of Quantum B-Algebras and Basic Implication Algebras

Xiaohong Zhang [1,2,*], Rajab Ali Borzooei [3] and Young Bae Jun [3,4]

1 Department of Mathematics, Shaanxi University of Science and Technology, Xi'an 710021, China
2 Department of Mathematics, Shanghai Maritime University, Shanghai 201306, China
3 Department of Mathematics, Shahid Beheshti University, Tehran 1983963113, Iran; borzooei@sbu.ac.ir (R.A.B.); skywine@gmail.com (Y.B.J.)
4 Department of Mathematics Education, Gyeongsang National University, Jinju 52828, Korea
* Correspondence: zhangxiaohong@sust.edu.cn or zhangxh@shmtu.edu.cn

Received: 26 September 2018; Accepted: 26 October 2018; Published: 1 November 2018

Abstract: The concept of quantum B-algebra was introduced by Rump and Yang, that is, unified algebraic semantics for various noncommutative fuzzy logics, quantum logics, and implication logics. In this paper, a new notion of q-filter in quantum B-algebra is proposed, and quotient structures are constructed by q-filters (in contrast, although the notion of filter in quantum B-algebra has been defined before this paper, but corresponding quotient structures cannot be constructed according to the usual methods). Moreover, a new, more general, implication algebra is proposed, which is called basic implication algebra and can be regarded as a unified frame of general fuzzy logics, including nonassociative fuzzy logics (in contrast, quantum B-algebra is not applied to nonassociative fuzzy logics). The filter theory of basic implication algebras is also established.

Keywords: fuzzy implication; quantum B-algebra; q-filter; quotient algebra; basic implication algebra

1. Introduction

For classical logic and nonclassical logics (multivalued logic, quantum logic, t-norm-based fuzzy logic [1–6]), logical implication operators play an important role. In the study of fuzzy logics, fuzzy implications are also the focus of research, and a large number of literatures involve this topic [7–16]. Moreover, some algebraic systems focusing on implication operators are also hot topics. Especially with the in-depth study of noncommutative fuzzy logics in recent years, some related implication algebraic systems have attracted the attention of scholars, such as pseudo-basic-logic (BL) algebras, pseudo- monoidal t-norm-based logic (MTL) algebras, and pseudo- B, C, K axiom (BCK)/ B, C, I axiom (BCI) algebras [17–23] (see also References [5–7]).

For formalizing the implication fragment of the logic of quantales, Rump and Yang proposed the notion of quantum B-algebras [24,25], which provide a unified semantic for a wide class of nonclassical logics. Specifically, quantum B-algebras encompass many implication algebras, like pseudo-BCK/BCI algebras, (commutative and noncommutative) residuated lattices, pseudo- MV/BL/MTL algebras, and generalized pseudo-effect algebras. New research articles on quantum B-algebras can be found in References [26–28]. Note that all hoops and pseudo-hoops are special quantum B-algebras, and they are closely related to L-algebras [29].

Although the definition of a filter in a quantum B-algebra is given in Reference [30], quotient algebraic structures are not established by using filters. In fact, filters in special subclasses of quantum B-algebras are mainly discussed in Reference [30], and these subclasses require a unital element. In this paper, by introducing the concept of a q-filter in quantum B-algebras, we establish the quotient structures using q-filters in a natural way. At the same time, although quantum B-algebra has generality, it cannot include the implication structure of non-associative fuzzy logics [31,32], so we propose a wider concept, that is, basic implication algebra that can include a wider range of implication operations, establish filter theory, and obtain quotient algebra.

2. Preliminaries

Definition 1. *Let (X, ≤) be partially ordered set endowed with two binary operations → and ⇝ [24,25]. Then, (X, →, ⇝, ≤) is called a quantum B-algebra if it satisfies: ∀x, y, z∈X,*

(1) $y \to z \leq (x \to y) \to (x \to z)$;
(2) $y \rightsquigarrow z \leq (x \rightsquigarrow y) \rightsquigarrow (x \rightsquigarrow z)$;
(3) $y \leq z \Rightarrow x \to y \leq x \to z$;
(4) $x \leq y \to z \Longleftrightarrow y \leq x \rightsquigarrow z$.

If u∈X exists, such that u→x = u⇝x = x for all x in X, then u is called a unit element of X. Obviously, the unit element is unique. When a unit element exists in X, we call X unital.

Proposition 1. *An algebra structure (X, →, ⇝, ≤) endowed with a partially order ≤ and two binary operations → and ⇝ is a quantum B-algebra if and only if it satisfies [4]: ∀x, y, z∈X,*

(1) $x \to (y \rightsquigarrow z) = y \rightsquigarrow (x \to z)$;
(2) $y \leq z \Rightarrow x \to y \leq x \to z$;
(3) $x \leq y \to z \Longleftrightarrow y \leq x \rightsquigarrow z$.

Proposition 2. *Let (X, →, ⇝, ≤) be a quantum B-algebra [24–26]. Then, (∀ x, y, z∈X)*

(1) $y \leq z \Rightarrow x \rightsquigarrow y \leq x \rightsquigarrow z$;
(2) $y \leq z \Rightarrow z \rightsquigarrow x \leq y \rightsquigarrow x$;
(3) $y \leq z \Rightarrow z \to x \leq y \to x$;
(4) $x \leq (x \rightsquigarrow y) \to y$, $x \leq (x \to y) \rightsquigarrow y$;
(5) $x \to y = ((x \to y) \rightsquigarrow y) \to y$, $x \rightsquigarrow y = ((x \rightsquigarrow y) \to y) \rightsquigarrow y$;
(6) $x \to y \leq (y \to z) \rightsquigarrow (x \to z)$;
(7) $x \rightsquigarrow y \leq (y \rightsquigarrow z) \to (x \rightsquigarrow z)$;
(8) *assume that u is the unit of X, then* $u \leq x \rightsquigarrow y \Longleftrightarrow x \leq y \Longleftrightarrow u \leq x \to y$;
(9) *if 0∈X exists, such that 0 ≤ x for all x in X, then* $0 = 0 \rightsquigarrow 0 = 0 \to 0$ *is the greatest element (denote by 1), and* $x \to 1 = x \rightsquigarrow 1 = 1$ *for all x∈X;*
(10) *if X is a lattice, then* $(x \vee y) \to z = (x \to z) \wedge (y \to z)$, $(x \vee y) \rightsquigarrow z = (x \rightsquigarrow z) \vee (y \rightsquigarrow z)$.

Definition 2. *Let (X, ≤) be partially ordered set and Y⊆ X [24]. If x ≥ y∈Y implies x∈Y, then Y is called to be an upper set of X. The smallest upper set containing a given x∈X is denoted by ↑x. For quantum B-algebra X, the set of upper sets is denoted by U(X). For A, B∈U(X), define*

$$A \cdot B = \{x \in X \mid \exists b \in B: b \to x \in A\}.$$

We can verify that $A \cdot B = \{x \in X \mid \exists a \in A: a \rightsquigarrow x \in B\} = \{x \in X \mid \exists a \in A, b \in B: a \leq b \to x\} = \{x \in X \mid \exists a \in A, b \in B: b \leq a \rightsquigarrow x\}$.

Definition 3. *Let A be an empty set, ≤ be a binary relation on A [17,18], → and ⇝ be binary operations on A, and 1 be an element of A. Then, structure (A, →, ⇝, ≤, 1) is called a pseudo-BCI algebra if it satisfies the following axioms: ∀ x, y, z∈A,*

(1) $x \to y \leq (y \to z) \rightsquigarrow (x \to z)$, $x \rightsquigarrow y \leq (y \rightsquigarrow z) \to (x \rightsquigarrow z)$,
(2) $x \leq (x \rightsquigarrow y) \to y$, $x \leq (x \to y) \rightsquigarrow y$;

(3) $x \leq x$;
(4) $x \leq y, y \leq x \Rightarrow x = y$;
(5) $x \leq y \Longleftrightarrow x \rightarrow y = 1 \Longleftrightarrow x \rightsquigarrow y = 1$.

If pseudo-BCI algebra A satisfies: $x \rightarrow 1 = 1$ (or $x \rightsquigarrow 1 = 1$) for all $x \in A$, then A is called a pseudo-BCK algebra.

Proposition 3. *Let $(A, \rightarrow, \rightsquigarrow, \leq, 1)$ be a pseudo-BCI algebra [18–20]. We have $(\forall x, y, z \in A)$*

(1) *if $1 \leq x$, then $x = 1$;*
(2) *if $x \leq y$, then $y \rightarrow z \leq x \rightarrow z$ and $y \rightsquigarrow z \leq x \rightsquigarrow z$;*
(3) *if $x \leq y$ and $y \leq z$, then $x \leq z$;*
(4) $x \rightarrow (y \rightsquigarrow z) = y \rightsquigarrow (x \rightarrow z)$;
(5) $x \leq y \rightarrow z \Longleftrightarrow y \leq x \rightsquigarrow z$;
(6) $x \rightarrow y \leq (z \rightarrow x) \rightarrow (z \rightarrow y)$; $x \rightsquigarrow y \leq (z \rightsquigarrow x) \rightsquigarrow (z \rightsquigarrow y)$;
(7) *if $x \leq y$, then $z \rightarrow x \leq z \rightarrow y$ and $z \rightsquigarrow x \leq z \rightsquigarrow y$;*
(8) $1 \rightarrow x = 1 \rightsquigarrow x = x$;
(9) $y \rightarrow x = ((y \rightarrow x) \rightsquigarrow x) \rightarrow x$, $y \rightsquigarrow x = ((y \rightsquigarrow x) \rightarrow x) \rightsquigarrow x$;
(10) $x \rightarrow y \leq (y \rightarrow x) \rightsquigarrow 1$, $x \rightsquigarrow y \leq (y \rightsquigarrow x) \rightarrow 1$;
(11) $(x \rightarrow y) \rightarrow 1 = (x \rightarrow 1) \rightsquigarrow (y \rightarrow 1)$, $(x \rightsquigarrow y) \rightsquigarrow 1 = (x \rightsquigarrow 1) \rightarrow (y \rightsquigarrow 1)$;
(12) $x \rightarrow 1 = x \rightsquigarrow 1$.

Proposition 4. *Let $(A, \rightarrow, \rightsquigarrow, \leq, 1)$ be a pseudo-BCK algebra [17], then $(\forall x, y \in A)$: $x \leq y \rightarrow x$, $x \leq y \rightsquigarrow x$.*

Definition 4. *Let X be a unital quantum B-algebra [24]. If there exists $x \in X$, such that $x \rightarrow u = x \rightsquigarrow u = u$, then we call that x integral. The subset of integral element in X is denoted by I(X).*

Proposition 5. *Let X be a quantum B-algebra [24]. Then, the following assertions are equivalent:*

(1) *X is a pseudo-BCK algebra;*
(2) *X is unital, and every element of X is integral;*
(3) *X has the greatest element, which is a unit element.*

Proposition 6. *Every pseudo-BCI algebra is a unital quantum B-algebra [25]. And, a quantum B-algebra is a pseudo-BCI algebra if and only if its unit element u is maximal.*

Definition 5. *Let $(A, \rightarrow, \rightsquigarrow, \leq, 1)$ be a pseudo-BCI algebra [20,21]. When the following identities are satisfied, we call X an antigrouped pseudo-BCI algebra:*

$$\forall x \in A, (x \rightarrow 1) \rightarrow 1 = x \text{ or } (x \rightsquigarrow 1) \rightsquigarrow 1 = x.$$

Proposition 7. *Let $(A, \rightarrow, \rightsquigarrow, \leq, 1)$ be a pseudo-BCI algebra [20]. Then, A is antigrouped if and only if the following conditions are satisfied:*

(G1) *for all $x, y, z \in A$, $(x \rightarrow y) \rightarrow (x \rightarrow z) = y \rightarrow z$, and*
(G2) *for all $x, y, z \in A$, $(x \rightsquigarrow y) \rightsquigarrow (x \rightsquigarrow z) = y \rightsquigarrow z$.*

Definition 6. *Let $(A, \rightarrow, \rightsquigarrow, \leq, 1)$ be a pseudo-BCI algebra and $F \subseteq X$ [19,20]. When the following conditions are satisfied, we call F a pseudo-BCI filter (briefly, filter) of X:*

(F1) $1 \in F$;

(F2) $x \in F, x \to y \in F \Longrightarrow y \in F$;
(F3) $x \in F, x \rightsquigarrow y \in F \Longrightarrow y \in F$.

Definition 7. *Let $(A, \to, \rightsquigarrow, \leq, 1)$ be a pseudo-BCI algebra and F be a filter of X [20,21]. When the following condition is satisfied, we call F an antigrouped filter of X:*

$$(GF)\ \forall x \in X,\ (x \to 1) \to 1 \in F \text{ or } (x \rightsquigarrow 1) \rightsquigarrow 1 \in F \Longrightarrow x \in F.$$

Definition 8. *A subset F of pseudo-BCI algebra X is called a p-filter of X if the following conditions are satisfied [20,21]:*

(P1) $1 \in F$,
(P2) $(x \to y) \rightsquigarrow (x \to z) \in F$ *and* $y \in F$ *imply* $z \in F$,
(P3) $(x \rightsquigarrow y) \to (x \rightsquigarrow z) \in F$ *and* $y \in F$ *imply* $z \in F$.

3. Q-Filters in Quantum B-Algebra

In Reference [30], the notion of filter in quantum B-algebra is proposed. If X is a quantum B-algebra and F is a nonempty set of X, then F is called the filter of X if $F \in U(X)$ and $F \cdot F \subseteq F$. That is, F is a filter of X, if and only if: (1) F is a nonempty upper subset of X; (2) $(z \in X, y \in F, y \to z \in F) \Rightarrow z \in F$. We denote the set of all filters of X by $F(X)$.

In this section, we discuss a new concept of q-filter in quantum B-algebra; by using q-filters, we construct the quotient algebras.

Definition 9. *A nonempty subset F of quantum B-algebra X is called a q-filter of X if it satisfies:*

(1) *F is an upper set of X, that is,* $F \in U(X)$;
(2) *for all* $x \in F$, $x \to x \in F$ *and* $x \rightsquigarrow x \in F$;
(3) $x \in F, y \in X, x \to y \in F \Longrightarrow y \in F$.
(4) *A q-filter of X is normal if* $x \to y \in F \Longleftrightarrow x \rightsquigarrow y \in F$.

Proposition 8. *Let F be a q-filter of quantum B-algebra X. Then,*

(1) $x \in F, y \in X, x \rightsquigarrow y \in F \Longrightarrow y \in F$.
(2) $x \in F$ *and* $y \in X \Longrightarrow (x \rightsquigarrow y) \to y \in F$ *and* $(x \to y) \rightsquigarrow y \in F$.
(3) *if X is unital, then Condition (2) in Definition 9 can be replaced by* $u \in F$*, where u is the unit element of X.*

Proof. (1) Assume that $x \in F$, $y \in X$, and $x \rightsquigarrow y \in F$. Then, by Proposition 2 (4), $x \leq (x \rightsquigarrow y) \to y$. Applying Definition 9 (1) and (3), we get that $y \in F$.

(2) Using Proposition 2 (4) and Definition 9 (1), we can get (2).

(3) If X is unital with unit u, then $u \to u = u$. Moreover, applying Proposition 2 (8), $u \leq x \rightsquigarrow x$ and $u \leq x \to x$ from $x \leq x$, for all $x \in X$. Therefore, for unital quantum B-algebra X, Condition (2) in Definition 8 can be replaced by condition "$u \in F$". □

By Definition 6, and Propositions 6 and 8, we get the following result (the proof is omitted).

Proposition 9. *Let $(A, \to, \rightsquigarrow, \leq, 1)$ be a pseudo-BCI algebra. Then, an empty subset of A is a q-filter of A (as a quantum B-algebra) if and only if it is a filter of A (according to Definition 6).*

Example 1. *Let X = {a, b, c, d, e, f}. Define operations $\to$ and $\rightsquigarrow$ on X as per the following Cayley Tables 1 and 2; the order on X is defined as follows: $b \leq a \leq f$; $e \leq d \leq c$. Then, X is a quantum B-algebra (we can verify* *it with the Matlab software (The MathWorks Inc., Natick, MA, USA)), but it is not a pseudo-BCI algebra. Let* $F_1 = \{f\}$, $F_2 = \{a, b, f\}$; *then,* F_1 *is a filter but not a q-filter of X, and* F_2 *is a normal q-filter of X.*

Table 1. Cayley table of operation $\to$.

$\to$	a	b	c	d	e	f
a	f	a	c	c	c	f
b	f	a	c	c	c	f
c	c	c	f	a	b	c
d	c	c	f	f	a	c
e	c	c	f	f	a	c
f	a	b	c	d	e	f

Table 2. Cayley table of operation $\rightsquigarrow$.

$\rightsquigarrow$	a	b	c	d	e	f
a	f	a	c	c	d	f
b	f	f	c	c	c	f
c	c	c	f	a	a	c
d	c	c	f	a	a	c
e	c	c	f	f	a	c
f	a	b	c	d	e	f

Theorem 1. *Let X be a quantum B-algebra and F a normal q-filter of X. Define the binary* $\approx_F$ *on X as follows:*

$$x \approx_F y \Longleftrightarrow x \to y \in F \text{ and } y \to x \in F, \text{ where } x, y \in X.$$

Then,

(1) $\approx_F$ *is an equivalent relation on X;*

(2) $\approx_F$ *is a congruence relation on X, that is,* $x \approx_F y \Longrightarrow (z \to x) \approx_F (z \to y)$, $(x \to z) \approx_F (y \to z)$, $(z \rightsquigarrow x) \approx_F (z \rightsquigarrow y)$, $(x \rightsquigarrow z) \approx_F (y \rightsquigarrow z)$, *for all* $z \in X$.

Proof. (1) For any $x \in X$, by Definition 9 (2), $x \to x \in F$, it follows that $x \approx_F x$.

For all $x, y \in X$, if $x \approx_F y$, we can easily verify that $y \approx_F x$.

Assume that $x \approx_F y$, $y \approx_F z$. Then, $x \to y \in F$, $y \to x \in F$, $y \to z \in F$, and $z \to y \in F$, since

$$y \to z \leq (x \to y) \to (x \to z) \text{ by Definition 1 (1).}$$

From this and Definition 9, we have $x \to z \in F$. Similarly, we can get $z \to x \in F$. Thus, $x \approx_F z$.

Therefore, $\approx_F$ is an equivalent relation on X.

(2) If $x \approx_F y$, then $x \to y \in F$, $y \to x \in F$. Since

$$x \to y \leq (z \to x) \to (z \to y), \text{ by Definition 1 (1).}$$

$$y \to x \leq (z \to y) \to (z \to x), \text{ by Definition 1 (1).}$$

Using Definition 9 (1), $(z \to x) \to (z \to y) \in F$, $(z \to y) \to (z \to x) \in F$. It follows that $(z \to x) \approx_F (z \to y)$.

Moreover, since

$$x \to y \leq (y \to z) \rightsquigarrow (x \to z), \text{ by Proposition 2 (6).}$$

$$y \to x \leq (x \to z) \rightsquigarrow (y \to z), \text{ by Proposition 2 (6).}$$

Then, form $x \to y \in F$ and $y \to x \in F$; using Definition 9 (1), we have $(y \to z) \rightsquigarrow (x \to z) \in F$, $(x \to z) \rightsquigarrow (y \to z) \in F$. Since F is normal, by Definition 9 we get $(y \to z) \to (x \to z) \in F$, $(x \to z) \to (y \to z) \in F$. Thus, $(x \to z) \approx_F (y \to z)$.

Similarly, we can get that $x \approx_F y \Longrightarrow (z \rightsquigarrow x) \approx_F (z \rightsquigarrow y)$ and $(x \rightsquigarrow z) \approx_F (y \rightsquigarrow z)$. □

Definition 10. *A quantum B-algebra X is considered to be perfect, if it satisfies:*

(1) *for any normal q-filter F of X, x, y in X, (there exists an$\in$X, such that $[x \to y]_F = [a \to a]_F$) $\Longleftrightarrow$ (there exists b$\in$X, such that $[x \rightsquigarrow y]_F = [b \rightsquigarrow b]_F$).*

(1) *for any normal q-filter F of X, $(X/\approx_F \to, \rightsquigarrow, \leq)$ is a quantum B-algebra, where quotient operations $\to$ and $\rightsquigarrow$ are defined in a canonical way, and $\leq$ is defined as follows:*

$$[x]_F \leq [y]_F \Longleftrightarrow (\text{there exists } a \in X \text{ such that } [x]_F \to [y]_F = [a \to a]_F)$$
$$\Longleftrightarrow (\text{there exists } b \in X \text{ such that } [x]_F \rightsquigarrow [y]_F = [b \rightsquigarrow b]_F).$$

Theorem 2. *Let $(A, \to, \rightsquigarrow, \leq, 1)$ be a pseudo-BCI algebra, then A is a perfect quantum B-algebra.*

Proof. By Proposition 6, we know that A is a quantum B-algebra.

(1) For any normal q-filter F of A, $x, y \in A$, if there exists $a \in A$, such that $[x \to y]_F = [a \to a]_F$, then

$$[x \to y]_F = [a \to a]_F = [1]_F.$$

It follows that $(x \to y) \to 1 \in F$, $1 \to (x \to y) = x \to y \in F$. Applying Proposition 3 (11) and (12), we have

$$(x \to 1) \rightsquigarrow (y \to 1) = (x \to y) \to 1 \in F.$$

Since F is normal, from $(x \to 1) \rightsquigarrow (y \to 1) \in F$ and $x \to y \in F$ we get that

$$(x \to 1) \to (y \to 1) \in F \text{ and } x \rightsquigarrow y \in F.$$

Applying Proposition 3 (11) and (12) again, $(x \rightsquigarrow y) \to 1 = (x \to 1) \to (y \to 1)$. Thus,

$$(x \rightsquigarrow y) \to 1 = (x \to 1) \to (y \to 1) \in F \text{ and } 1 \to (x \rightsquigarrow y) = x \rightsquigarrow y \in F.$$

This means that $[x \rightsquigarrow y]_F = [1]_F = [1 \rightsquigarrow 1]_F$. Similarly, we can prove that the inverse is true. That is, Definition 10 (1) holds for A.

(2) For any normal q-filter F of pseudo-BCI algebra A, binary $\leq$ on $A/\approx_F$ is defined as the following:

$$[x]_F \leq [y]_F \Longleftrightarrow [x]_F \to [y]_F = [1]_F.$$

We verify that $\leq$ is a partial binary on $A/\approx_F$.

Obviously, $[x]_F \leq [x]_F$ for any $x \in A$.

If $[x]_F \leq [y]_F$ and $[y]_F \leq [x]_F$, then $[x]_F \to [y]_F = [x \to y]_F = [1]_F$, $[y]_F \to [x]_F = [y \to x]_F = [1]_F$. By the definition of equivalent class, $x \to y = 1 \to (x \to y) \in F$, $y \to x = 1 \to (y \to x) \in F$. It follows that $x \approx_F y$; thus, $[x]_F = [y]_F$.

If $[x]_F \leq [y]_F$ and $[y]_F \leq [z]_F$, then $[x]_F \to [y]_F = [x \to y]_F = [1]_F$, $[y]_F \to [z]_F = [y \to z]_F = [1]_F$. Thus,

$$x \to y = 1 \to (x \to y) \in F, (x \to y) \to 1 \in F;$$

$$y \to z = 1 \to (y \to z) \in F, (y \to z) \to 1 \in F.$$

Applying Definition 3 and Proposition 3,

$$y \to z \leq (x \to y) \to (x \to z),$$

$$(x \to y) \to 1 = (x \to 1) \rightsquigarrow (y \to 1) \leq ([(y \to 1) \rightsquigarrow (z \to 1)] \to [(x \to 1) \rightsquigarrow (z \to 1)]) = [(y \to z) \to 1] \to [(x \to z) \to 1].$$

By Definition 9,

$$1 \to (x \to z) = x \to z \in F,\ (x \to z) \to 1 \in F.$$

This means that $(x \to z) \approx_F 1$, $[x \to z]_F = [1]_F$. That is, $[x]_F \to [z]_F = [x \to z]_F = [1]_F$, $[x]_F \leq [z]_F$.

Therefore, applying Theorem 1, we know that $(A/\approx_F \to, \rightsquigarrow, [1]_F)$ is a quantum B-algebra and pseudo-BCI algebra. That is, Definition 10 (2) holds for A.

Hence, we know that A is a perfect quantum B-algebra. □

The following examples show that there are some perfect quantum B-algebras that may not be a pseudo-BCI algebra.

Example 2. *Let $X = \{a, b, c, d, e, 1\}$. Define operations $\to$ and $\rightsquigarrow$ on X as per the following Cayley Tables 3 and 4, the order on X is defined as the following: $b \leq a \leq 1$; $e \leq d \leq c$. Then, X is a pseudo-BCI algebra (we can verify it with Matlab). Denote $F_1 = \{1\}$, $F_2 = \{a, b, 1\}$, $F_3 = X$, then F_i $(i = 1, 2, 3)$ are all normal q-filters of X, and quotient algebras $(X/\approx_{Fi} \to, \rightsquigarrow, [1]_{Fi})$ are pseudo-BCI algebras. Thus, X is a perfect quantum B-algebra.*

Table 3. Cayley table of operation $\to$.

$\to$	a	b	c	d	e	1
a	1	a	c	c	c	1
b	1	1	c	c	c	1
c	c	c	1	a	b	c
d	c	c	1	1	a	c
e	c	c	1	1	1	c
1	a	b	c	d	e	1

Table 4. Cayley table of operation $\rightsquigarrow$.

$\rightsquigarrow$	a	b	c	d	e	1
a	1	a	c	c	d	1
b	1	1	c	c	c	1
c	c	c	1	a	a	c
d	c	c	1	1	a	c
e	c	c	1	1	1	c
1	a	b	c	d	e	1

Example 3. *Let $X = \{a, b, c, d, e, f\}$. Define operations $\to$ and $\rightsquigarrow$ on X as per the following Cayley Tables 5 and 6, the order on X is defined as follows: $b \leq a \leq f$; $e \leq d \leq c$. Then, X is a quantum B-algebra (we can verify it with Matlab), but it is not a pseudo-BCI algebra, since $e \rightsquigarrow e \neq e \to e$. Denote $F = \{a, b, f\}$, then F, X are all normal q-filters of X, quotient algebras $(X/\approx_F \to, \rightsquigarrow, \leq)$, $(X/\approx_X \to, \rightsquigarrow, \leq)$ are quantum B-algebras, and X is a perfect quantum B-algebra.*

Table 5. Cayley table of operation $\to$.

$\to$	a	b	c	d	e	f
a	f	a	c	c	c	f
b	f	f	c	c	c	f
c	c	c	f	a	b	c
d	c	c	f	f	a	c
e	c	c	f	f	f	c
f	a	b	c	d	e	f

Table 6. Cayley table of operation $\rightsquigarrow$.

$\rightsquigarrow$	a	b	c	d	e	f
a	f	a	c	c	d	f
b	f	f	c	c	c	f
c	c	c	f	a	a	c
d	c	c	f	f	a	c
e	c	c	f	f	a	c
f	a	b	c	d	e	f

4. Basic Implication Algebras and Filters

Definition 11. *Let $(A, \vee, \wedge, \otimes, \rightarrow, 0, 1)$ be a type-(2, 2, 2, 2, 0, 0) algebra [32]. A is called a nonassociative residuated lattice, if it satisfies:*

(A1) $(A, \vee, \wedge, 0, 1)$ *is a bounded lattice;*
(A2) $(A, \otimes, 1)$ *is a commutative groupoid with unit element 1;*
(A3) $\forall x, y, z \in A, x \otimes y \leq z \Longleftrightarrow x \leq y \rightarrow z$.

Proposition 10. *Let $(A, \vee, \wedge, \otimes, \rightarrow, 0, 1)$ be a nonassociative residuated lattice [32]. Then, $(\forall x, y, z \in A)$*

(1) $x \leq y \Longleftrightarrow x \rightarrow y = 1$;
(2) $x \leq y \Rightarrow x \otimes z \leq y \otimes z$;
(3) $x \leq y \Rightarrow y \rightarrow z \leq x \rightarrow z$;
(4) $x \leq y \Rightarrow z \rightarrow x \leq z \rightarrow y$;
(5) $x \otimes (y \vee z) = (x \otimes y) \vee (x \otimes z)$;
(6) $x \rightarrow (y \wedge z) = (x \rightarrow y) \wedge (x \rightarrow z)$;
(7) $(y \vee z) \rightarrow x = (y \rightarrow x) \wedge (z \rightarrow x)$;
(8) $(x \rightarrow y) \otimes x \leq x, y$;
(9) $(x \rightarrow y) \rightarrow y \geq x, y$.

Example 4. *Let $A = [0, 1]$, operation $\otimes$ on A is defined as follows:*

$$x \otimes y = 0.5xy + 0.5max\{0, x + y - 1\}, x, y \in A.$$

Then, $\otimes$ is a nonassociative t-norm on A (see Example 1 in Reference [32]). Operation $\rightarrow$ is defined as follows:

$$x \rightarrow y = max\{z \in [0, 1] \mid z \otimes x \leq y\}, x, y \in A.$$

Then, $(A, max, min, \otimes, \rightarrow, 0, 1)$ is a nonassoiative residuated lattice (see Theorem 5 in Reference [32]). Assume that $x = 0.55$, $y = 0.2$, $z = 0.1$, then

$$y \rightarrow z = 0.2 \rightarrow 0.1 = \max\{a \in [0, 1| a \otimes 0.2 \leq 0.1\} = \frac{5}{6}.$$

$$x \rightarrow y = 0.55 \rightarrow 0.2 = \max\{a \in [0, 1| a \otimes 0.55 \leq 0.2\} = \frac{17}{31}.$$

$$x \rightarrow z = 0.55 \rightarrow 0.1 = \max\{a \in [0, 1| a \otimes 0.55 \leq 0.1\} = \frac{4}{11}.$$

$$(x \rightarrow y) \rightarrow (x \rightarrow z) = \frac{17}{31} \rightarrow \frac{4}{11} = \max\{a \in [0, 1| a \otimes \frac{17}{31} \leq \frac{4}{11}\} = \frac{67}{88}.$$

Therefore,

$$y \rightarrow z \nleq (x \rightarrow y) \rightarrow (x \rightarrow z).$$

Example 4 shows that Condition (1) in Definition 1 is not true for general non-associative residuated lattices, that is, quantum B-algebras are not common basic of non-associative fuzzy logics. So, we discuss more general implication algebras in this section.

Definition 12. *A basic implication algebra is a partially ordered set* $(X, \leq)$ *with binary operation* $\rightarrow$*, such that the following are satisfied for x, y, and z in X:*

(1) $x \leq y \Rightarrow z{\rightarrow}x \leq z{\rightarrow}y$;
(2) $x \leq y \Rightarrow y{\rightarrow}z \leq x{\rightarrow}z$.

A basic implication algebra is considered to be normal, if it satisfies:

(3) *for any* $x, y \in X$, $x{\rightarrow}x = y{\rightarrow}y$;
(4) *for any* $x, y \in X$, $x \leq y \Longleftrightarrow x{\rightarrow}y = e$, *where* $e = x{\rightarrow}x = y{\rightarrow}y$.

We can verify that the following results are true (the proofs are omitted).

Proposition 11. *Let* $(X, \rightarrow, \leq)$ *be a basic implication algebra. Then, for all* $x, y, z \in X$,

(1) $x \leq y \Rightarrow y{\rightarrow}x \leq x{\rightarrow}x \leq x{\rightarrow}y$;
(2) $x \leq y \Rightarrow y{\rightarrow}x \leq y{\rightarrow}y \leq x{\rightarrow}y$;
(3) $x \leq y$ *and* $u \leq v \Rightarrow y{\rightarrow}u \leq x{\rightarrow}v$;
(4) $x \leq y$ *and* $u \leq v \Rightarrow v{\rightarrow}x \leq u{\rightarrow}y$.

Proposition 12. *Let* $(X, \rightarrow, \leq, e)$ *be a normal basic implication algebra. Then for all* $x, y, z \in X$,

(1) $x{\rightarrow}x = e$;
(2) $x{\rightarrow}y = y{\rightarrow}x = e \Rightarrow x = y$;
(3) $x \leq y \Rightarrow y{\rightarrow}x \leq e$;
(4) *if e is unit (that is, for all x in X,* $e{\rightarrow}x = x$*), then e is a maximal element (that is,* $e \leq x \Rightarrow e = x$*).*

Proposition 13. *(1) If* $(X, \rightarrow, \rightsquigarrow, \leq)$ *is a a quantum B-algebra, then* $(X, \rightarrow, \leq)$ *and* $(X, \rightsquigarrow, \leq)$ *are basic implication algebras; (2) If* $(A, \rightarrow, \rightsquigarrow, \leq, 1)$ *is a pseudo-BCI algebra, then* $(A, \rightarrow, \leq, 1)$ *and* $(A, \rightsquigarrow, \leq, 1)$ *are normal basic implication algebras with unit 1; (3) If* $(A, \vee, \wedge, \otimes, \rightarrow, 0, 1)$ *is a non-associative residuated lattice, then* $(A, \rightarrow, \leq, 1)$ *is a normal basic implication algebra.*

The following example shows that element *e* may not be a unit.

Example 5. *Let* $X = \{a, b, c, d, 1\}$*. Define* $a \leq b \leq c \leq d \leq 1$ *and operation* $\rightarrow$ *on X as per the following Cayley Table 7. Then, X is a normal basic implication algebra in which element 1 is not a unit.* $(X, \rightarrow, \leq)$ *is not a commutative quantum B-algebra, since*

$$c = 1 \rightarrow c \nleq b = (c \rightarrow d) \rightarrow (1 \rightarrow d).$$

Table 7. Cayley table of operation $\rightarrow$.

$\rightarrow$	*a*	*b*	*c*	*d*	**1**
a	1	1	1	1	1
b	*d*	1	1	1	1
c	*d*	*d*	1	1	1
d	*b*	*c*	*d*	1	1
1	*b*	*b*	*c*	*b*	1

The following example shows that element *e* may be not maximal.

Example 6. *Let X = {a, b, c, d, 1}. Define* $a \leq b \leq c \leq d$, $a \leq b \leq c \leq 1$ *and operation* $\rightarrow$ *on X as per the following Cayley Table 8. Then, X is a normal basic implication algebra, and element 1 is not maximal and is not a unit.*

Table 8. Cayley table of operation $\rightarrow$.

$\rightarrow$	*a*	*b*	*c*	*d*	**1**
a	1	1	1	1	1
b	c	1	1	1	1
c	c	c	1	1	1
d	a	c	a	1	c
1	a	b	b	c	1

Definition 13. *A nonempty subset F of basic implication algebra* $(X, \rightarrow, \leq)$ *is called a filter of X if it satisfies:*

(1) *F is an upper set of X, that is,* $x \in F$ *and* $x \leq y \in X \Longrightarrow y \in F$;
(2) *for all* $x \in F$, $x \rightarrow x \in F$;
(3) $x \in F, y \in X, x \rightarrow y \in F \Longrightarrow y \in F$;
(4) $x \in X, y \rightarrow z \in F \Longrightarrow (x \rightarrow y) \rightarrow (x \rightarrow z) \in F$;
(5) $x \in X, y \rightarrow z \in F \Longrightarrow (z \rightarrow x) \rightarrow (y \rightarrow x) \in F$.

For normal basic implication algebra $(X, \rightarrow, \leq, e)$*, a filter F of X is considered to be regular, if it satisfies:*

(6) $x \in X, (x \rightarrow y) \rightarrow e \in F$ *and* $(y \rightarrow z) \rightarrow e \in F \Longrightarrow (x \rightarrow z) \rightarrow e \in F$.

Proposition 14. *Let* $(X, \rightarrow, \leq, e)$ *be a normal basic implication algebra and* $F \subseteq X$*. Then, F is a filter of X if and only if it satisfies:*

(1) $e \in F$;
(2) $x \in F, y \in X, x \rightarrow y \in F \Longrightarrow y \in F$;
(3) $x \in X, y \rightarrow z \in F \Longrightarrow (x \rightarrow y) \rightarrow (x \rightarrow z) \in F$;
(4) $x \in X, y \rightarrow z \in F \Longrightarrow (z \rightarrow x) \rightarrow (y \rightarrow x) \in F$.

Obviously, if *e* is the maximal element of normal basic implication algebra $(X, \rightarrow, \leq, e)$, then any filter of X is regular.

Theorem 3. *Let X be a basic implication algebra and F a filter of X. Define binary* $\approx_F$ *on X as follows:*

$$x \approx_F y \Longleftrightarrow x \rightarrow y \in F \text{ and } y \rightarrow x \in F, \text{ where } x, y \in X.$$

Then

(1) $\approx_F$ *is a equivalent relation on X;*
(2) $\approx_F$ *is a congruence relation on X, that is,* $x \approx_F y \Longrightarrow (z \rightarrow x) \approx_F (z \rightarrow y)$, $(x \rightarrow z) \approx_F (y \rightarrow z)$, *for all* $z \in X$.

Proof (1) $\forall x \in X$, from Definition 13 (2), $x \rightarrow x \in F$, thus $x \approx_F x$. Moreover, $\forall x, y \in X$, if $x \approx_F y$, then $y \approx_F x$.

If $x \approx_F y$ and $y \approx_F z$. Then $x \rightarrow y \in F$, $y \rightarrow x \in F$, $y \rightarrow z \in F$, and $z \rightarrow y \in F$. Applying Definition 13 (4) and (5), we have

$$(x \rightarrow y) \rightarrow (x \rightarrow z) \in F, \ (z \rightarrow y) \rightarrow (z \rightarrow x) \in F.$$

From this and Definition 13 (3), we have $x \rightarrow z \in F$, $z \rightarrow x \in F$. Thus, $x \approx_F z$.

Hence, $\approx_F$ is a equivalent relation on X.

(2) Assume $x \approx_F y$. By the definition of bianary relation $\approx_F$, we have $x \rightarrow y \in F$, $y \rightarrow x \in F$. Using Definition 13 (4),

$$(z \rightarrow x) \rightarrow (z \rightarrow y) \in F,\ (z \rightarrow y) \rightarrow (z \rightarrow x) \in F.$$

This means that $(z \rightarrow x) \approx_F (z \rightarrow y)$. Moreover, using Definition 13 (5), we have

$$(y \rightarrow z) \rightarrow (x \rightarrow z) \in F,\ (x \rightarrow z) \rightarrow (y \rightarrow z) \in F.$$

Hence, $(x \rightarrow z) \approx_F (y \rightarrow z)$. □

Theorem 4. *Let $(X, \rightarrow, \leq, e)$ be a normal basic implication algebra and F a regular filter of X. Define quotient operation $\rightarrow$ and binary relation $\leq$ on $X/\approx_F$ as follows:*

$$[x]_F \rightarrow [y]_F = [x]_F \rightarrow [y]_F,\ \forall x, y \in X;$$
$$[x]_F \leq [y]_F \Longleftrightarrow [x]_F \rightarrow [y]_F = [e]_F,\ \forall x, y \in X.$$

Then, $(X/\approx_F, \rightarrow, \leq, [e]_F)$ is a normal basic implication algebra, and $(X, \rightarrow, \leq, e) \sim (X/\approx_F, \rightarrow, \leq, [e]_F)$.

Proof. Firstly, we prove that binary relation $\leq$ on $X/\approx_F$ is a partial order.

(1) $\forall x \in X$, obviously, $[x]_F \leq [x]_F$.

(2) Assume that $[x]_F \leq [y]_F$ and $[y]_F \leq [x]_F$, then

$$[x]_F \rightarrow [y]_F = [x \rightarrow y]_F = [e]_F,\ [y]_F \rightarrow [x]_F = [y \rightarrow x]_F = [e]_F.$$

It follows that $e \rightarrow (x \rightarrow y) \in F$, $e \rightarrow (y \rightarrow x) \in F$. Applying Proposition 14 (1) and (2), we get that $(x \rightarrow y) \in F$ and $(y \rightarrow x) \in F$. This means that $[x]_F = [y]_F$.

(3) Assume that $[x]_F \leq [y]_F$ and $[y]_F \leq [z]_F$, then

$$[x]_F \rightarrow [y]_F = [x \rightarrow y]_F = [e]_F,\ [y]_F \rightarrow [z]_F = [y \rightarrow z]_F = [e]_F.$$

Using the definition of equivalent relation $\approx_F$, we have

$$e \rightarrow (x \rightarrow y) \in F,\ (x \rightarrow y) \rightarrow e \in F;\ e \rightarrow (y \rightarrow z) \in F,\ (y \rightarrow z) \rightarrow e \in F.$$

From $e \rightarrow (x \rightarrow y) \in F$ and $e \rightarrow (y \rightarrow z) \in F$, applying Proposition 14 (1) and (2), $(x \rightarrow y) \in F$ and $(y \rightarrow z) \in F$. By Proposition 14 (4), $(x \rightarrow y) \rightarrow (x \rightarrow z) \in F$. It follows that $(x \rightarrow z) \in F$. Hence, $(x \rightarrow x) \rightarrow (x \rightarrow z) \in F$, by Proposition 14 (4). Therefore,

$$e \rightarrow (x \rightarrow z) = (x \rightarrow x) \rightarrow (x \rightarrow z) \in F.$$

Moreover, from $(x \rightarrow y) \rightarrow e \in F$ and $(y \rightarrow z) \rightarrow e \in F$, applying regularity of F and Definition 13 (6), we get that $(x \rightarrow z) \rightarrow e \in F$.

Combining the above $e \rightarrow (x \rightarrow z) \in F$ and $(x \rightarrow z) \rightarrow e \in F$, we have $x \rightarrow z \approx_F e$, that is, $[x \rightarrow z]_F = [e]_F$. This means that $[x]_F \leq [z]_F$. It follows that the binary relation $\leq$ on $X/\approx_F$ is a partially order.

Therefore, applying Theorem 3, we know that $(X/\approx_F \rightarrow, \leq, [e]_F)$ is a normal basic implication algebra, and $(X, \rightarrow, \leq, e) \sim (X/\approx_F \rightarrow, \leq, [e]_F)$ in the homomorphism mapping $f\colon X \rightarrow X/\approx_F; f(x) = [x]_F$. □

Example 7. *Let $X = \{a, b, c, d, 1\}$. Define operations $\rightarrow$ on X as per the following Cayley Table 9, and the order binary on X is defined as follows: $a \leq b \leq c \leq 1$, $b \leq d \leq 1$. Then $(X, \rightarrow, \leq, 1)$ is a normal basic implication algebra (it is not a quantum B-algebra). Denote $F = \{1\}$, then F is regular filters of X, and the quotient algebras $(X, \rightarrow, \leq, 1)$ is isomorphism to $(X/\approx_F, \rightarrow, [1]_F)$.*

Table 9. Cayley table of operation $\to$.

$\to$	a	b	c	d	1
a	1	1	1	1	1
b	d	1	1	1	1
c	b	d	1	d	1
d	a	c	c	1	1
1	a	b	c	d	1

Example 8. *Denote X = {a, b, c, d, 1}. Define operations $\to$ on X as per the following Cayley Table 10, and the order binary on X is defined as follows: $a \le b \le c \le 1$, $b \le d \le 1$. Then (X, $\to$, $\le$, 1) is a normal basic implication algebra (it is not a quantum B-algebra). Let F = {1, d}, then F is a regular filters of X, and the quotient algebras ($X/\approx_F$, $\to$, $[1]_F$) is presented as the following Table 11, where $X/\approx_F$ = {{a}, {b, c}, $[1]_F$ = {1, d}}. Moreover, (X, $\to$, $\le$, 1) $\sim$ ($X/\approx_F$ $\to$, $[1]_F$).*

Table 10. Cayley table of operation $\to$.

$\to$	a	b	c	d	1
a	1	1	1	1	1
b	c	1	1	1	1
c	b	d	1	d	1
d	a	c	c	1	1
1	a	b	c	d	1

Table 11. Quotient algebra ($X/\approx_F$, $\to$, $[1]_F$).

$\to$	{a}	{b,c}	$[1]_F$
{a}	$[1]_F$	$[1]_F$	$[1]_F$
{b,c}	{b,c}	$[1]_F$	$[1]_F$
$[1]_F$	{a}	{b,c}	$[1]_F$

5. Conclusions

In this paper, we introduced the notion of a q-filter in quantum B-algebras and investigated quotient structures; by using q-filters as a corollary, we obtained quotient pseudo-BCI algebras by their filters. Moreover, we pointed out that the concept of quantum B-algebra does not apply to non-associative fuzzy logics. From this fact, we proposed the new concept of basic implication algebra, and established the corresponding filter theory and quotient algebra. In the future, we will study in depth the structural characteristics of basic implication algebras and the relationship between other algebraic structures and uncertainty theories (see References [33–36]). Moreover, we will consider the applications of q-filters for Gentzel's sequel calculus.

Author Contributions: X.Z. initiated the research and wrote the draft. B.R.A., Y.B.J., and X.Z. completed the final version.

Funding: This work was supported by the National Natural Science Foundation of China (Grant No. 61573240).

Conflicts of Interest: The authors declare no conflict of interest.

References

1. Hájek, P. *Metamathematics of Fuzzy Logic*; Kluwer: Dordrecht, The Netherlands, 1998.
2. Klement, E.P.; Mesiar, R.; Pap, E. *Triangular Norm*; Springer: Berlin, Germany, 2000.
3. Esteva, F.; Godo, L. Monoidal t-norm based logic: Towards a logic for left-continuous t-norms. *Fuzzy Sets Syst.* **2001**, *124*, 271–288. [CrossRef]
4. Flaminio, T. Strong non-standard completeness for fuzzy logics. *Soft Comput.* **2008**, *12*, 321–333. [CrossRef]

5. Zhang, X.H. *Fuzzy Logics and Algebraic Analysis*; Science in China Press: Beijing, China, 2008.
6. Zhang, X.H.; Dudek, W.A. BIK+-logic and non-commutative fuzzy logics. *Fuzzy Syst. Math.* **2009**, *23*, 8–20.
7. Iorgulescu, A. *Implicative-Groups vs. Groups and Generalizations*; Matrix Room: Bucuresti, Romania, 2018.
8. Fodor, J.C. Contrapositive symmetry of fuzzy implications. *Fuzzy Sets Syst.* **1995**, *69*, 141–156. [CrossRef]
9. Ruiz-Aguilera, D.; Torrens, J. Distributivity of residual implications over conjunctive and disjunctive uninorms. *Fuzzy Sets Syst.* **2007**, *158*, 23–37. [CrossRef]
10. Yao, O. On fuzzy implications determined by aggregation operators. *Inf. Sci.* **2012**, *193*, 153–162.
11. Baczyński, M.; Beliakov, G.; Bustince Sola, H.; Pradera, A. *Advances in Fuzzy Implication Functions*; Springer: Berlin/Heidelberg, Germany, 2013.
12. Vemuri, N.; Jayaram, B. The ⊗-composition of fuzzy implications: Closures with respect to properties, powers and families. *Fuzzy Sets Syst.* **2015**, *275*, 58–87. [CrossRef]
13. Li, D.C.; Qin, S.J. The quintuple implication principle of fuzzy reasoning based on interval-valued S-implication. *J. Log. Algebraic Meth. Program.* **2018**, *100*, 185–194. [CrossRef]
14. Su, Y.; Liu, H.W.; Pedrycz, W. A method to construct fuzzy implications–rotation construction. *Int. J. Appr. Reason.* **2018**, *92*, 20–31. [CrossRef]
15. Wang, C.Y.; Wan, L.J. Type-2 fuzzy implications and fuzzy-valued approximation reasoning. *Int. J. Appr. Reason.* **2018**, *102*, 108–122. [CrossRef]
16. Luo, M.X.; Zhou, K.Y. Logical foundation of the quintuple implication inference methods. *Int. J. Appr. Reason.* **2018**, *101*, 1–9. [CrossRef]
17. Georgescu, G.; Iorgulescu, A. Pseudo-BCK algebras. In *Combinatorics, Computability and Logic*; Springer: London, UK, 2001; pp. 97–114.
18. Dudek, W.A.; Jun, Y.B. Pseudo-BCI algebras. *East Asian Math. J.* **2008**, *24*, 187–190.
19. Dymek, G. Atoms and ideals of pseudo-BCI-algebras. *Commen. Math.* **2012**, *52*, 73–90.
20. Zhang, X.H.; Jun, Y.B. Anti-grouped pseudo-BCI algebras and anti-grouped filters. *Fuzzy Syst. Math.* **2014**, *28*, 21–33.
21. Zhang, X.H. Fuzzy anti-grouped filters and fuzzy normal filters in pseudo-BCI algebras. *J. Intell. Fuzzy Syst.* **2017**, *33*, 1767–1774. [CrossRef]
22. Emanovský, P.; Kühr, J. Some properties of pseudo-BCK- and pseudo-BCI-algebras. *Fuzzy Sets Syst.* **2018**, *339*, 1–16. [CrossRef]
23. Zhang, X.H.; Park, C.; Wu, S.P. Soft set theoretical approach to pseudo-BCI algebras. *J. Intell. Fuzzy Syst.* **2018**, *34*, 559–568. [CrossRef]
24. Rump, W.; Yang, Y. Non-commutative logical algebras and algebraic quantales. *Ann. Pure Appl. Log.* **2014**, *165*, 759–785. [CrossRef]
25. Rump, W. Quantum B-algebras. *Cent. Eur. J. Math.* **2013**, *11*, 1881–1899. [CrossRef]
26. Han, S.W.; Xu, X.T.; Qin, F. The unitality of quantum B-algebras. *Int. J. Theor. Phys.* **2018**, *57*, 1582–1590. [CrossRef]
27. Han, S.W.; Wang, R.R.; Xu, X.T. On the injective hulls of quantum B-algebras. *Fuzzy Sets Syst.* **2018**. [CrossRef]
28. Rump, W. Quantum B-algebras: Their omnipresence in algebraic logic and beyond. *Soft Comput.* **2017**, *21*, 2521–2529. [CrossRef]
29. Rump, W. L-algebras, self-similarity, and l-groups. *J. Algebra* **2008**, *320*, 2328–2348. [CrossRef]
30. Botur, M.; Paseka, J. Filters on some classes of quantum B-algebras. *Int. J. Theor. Phys.* **2015**, *54*, 4397–4409. [CrossRef]
31. Botur, M.; Halaš, R. Commutative basic algebras and non-associative fuzzy logics. *Arch. Math. Log.* **2009**, *48*, 243–255. [CrossRef]
32. Botur, M. A non-associative generalization of Hájek's BL-algebras. *Fuzzy Sets Syst.* **2011**, *178*, 24–37. [CrossRef]
33. Zhang, X.H.; Smarandache, F.; Liang, X.L. Neutrosophic duplet semi-group and cancellable neutrosophic triplet groups. *Symmetry* **2017**, *9*, 275. [CrossRef]
34. Zhang, X.H.; Bo, C.X.; Smarandache, F.; Park, C. New operations of totally dependent- neutrosophic sets and totally dependent-neutrosophic soft sets. *Symmetry* **2018**, *10*, 187. [CrossRef]

35. Zhang, X.H.; Bo, C.X.; Smarandache, F.; Dai, J.H. New inclusion relation of neutrosophic sets with applications and related lattice structure. *Int. J. Mach. Learn. Cyber.* **2018**, *9*, 1753–1763. [CrossRef]
36. Caponetto, R.; Fortuna, L.; Graziani, S.; Xibilia, M.G. Genetic algorithms and applications in system engineering: A survey. *Trans. Inst. Meas. Control* **1993**, *15*, 143–156. [CrossRef]

symmetry

MDPI

Article

Fuzzy Normed Rings

Aykut Emniyet [1] and Memet Şahin [2,*]

1 Department of Mathematics, Osmaniye Korkut Ata University, 8000 Osmaniye, Turkey; aykutemniyet@osmaniye.edu.tr
2 Department of Mathematics, Gaziantep University, 27310 Gaziantep, Turkey
* Correspondence: mesahin@gantep.edu.tr; Tel.: +90-5432182646

Received: 12 September 2018; Accepted: 8 October 2018; Published: 16 October 2018

Abstract: In this paper, the concept of fuzzy normed ring is introduced and some basic properties related to it are established. Our definition of normed rings on fuzzy sets leads to a new structure, which we call a fuzzy normed ring. We define fuzzy normed ring homomorphism, fuzzy normed subring, fuzzy normed ideal, fuzzy normed prime ideal, and fuzzy normed maximal ideal of a normed ring, respectively. We show some algebraic properties of normed ring theory on fuzzy sets, prove theorems, and give relevant examples.

Keywords: Fuzzy sets; ring; normed space; fuzzy normed ring; fuzzy normed ideal

---

1. Introduction

Normed rings attracted attention of researchers after the studies by Naimark [1], a generalization of normed rings [2] and commutative normed rings [3]. Naimark defined normed rings in an algebraic fashion, while Gel'fand addressed them as complex Banach spaces and introduced the notion of commutative normed rings. In Reference [4], Jarden defined the ultrametric absolute value and studied the properties of normed rings in a more topological perspective. During his invaluable studies, Zadeh [5] presented fuzzy logic theory, changing the scientific history forever by making a modern definition of vagueness and using the sets without strict boundaries. As, in almost every aspect of computational science, fuzzy logic also became a convenient tool in classical algebra. Zimmermann [6] made significant contributions to the fuzzy set theory. Mordeson, Bhutani, and Rosenfeld [7] defined fuzzy subgroups, Liu [8], Mukherjee, and Bhattacharya [9] examined normal fuzzy subgroups. Liu [8] also discussed fuzzy subrings and fuzzy ideals. Wang, Ruan and Kerre [10] studied fuzzy subrings and fuzzy rings. Swamy and Swamy [11] defined and proved major theorems on fuzzy prime ideals of rings. Gupta and Qi [12] are concerned with T-norms, T-conorms and T-operators. In this study, we use the definitions of Kolmogorov, Silverman, and Formin [13] on linear spaces and norms. Uluçay, Şahin, and Olgun [14] worked out on normed Z-Modules and also on soft normed rings [15]. Şahin, Olgun, and Uluçay [16] defined normed quotient rings while Şahin and Kargın [17] presented neutrosophic triplet normed space. In Reference [18], Olgun and Şahin investigated fitting ideals of the universal module and while Olgun [19] found a method to solve a problem on universal modules. Şahin and Kargin proposed neutrosophic triplet inner product [20] and Florentin, Şahin, and Kargin introduced neutrosophic triplet G-module [21]. Şahin and et al defined isomorphism theorems for soft G-module in [22]. Fundamental homomorphism theorems for neutrosophic extended triplet groups [23] were introduced by Mehmet, Moges, and Olgun in 2018. In Reference [24], Bal, Moges, and Olgun introduced neutrosophic triplet cosets and quotient groups, and deal with its application areas in neutrosophic logic.

This paper anticipates a normed ring on R and fuzzy rings are defined in the previous studies. Now, we use that norm on fuzzy sets, hence a fuzzy norm is obtained and by defining our fuzzy norm on fuzzy rings, we get fuzzy normed rings in this study. The organization of this paper is as follows.

In Section 2, we give preliminaries and fuzzy normed rings. In Section 3, consists of further definitions and relevant theorems on fuzzy normed ideals of a normed ring. Fuzzy normed prime and fuzzy normed maximal ideals of a normed ring are introduced in Section 4. The conclusions are summarized in Section 5.

2. Preliminaries

In this section, definition of normed linear space, normed ring, Archimedean strict T-norm and concepts of fuzzy sets are outlined.

Definition 1. *[13] A functional $\|\|$ defined on a linear space L is said to be a norm (in L) if it has the following properties:*
N1: $\|x\| \geq 0$ for all $x \in L$, where $\|x\| = 0$ if and only if $x = 0$;
N2: $\|\alpha \cdot x\| = |\alpha|.\|x\|$; (and hence $\|x\| = \|-x\|$), for all $x \in L$ and for all α;
N3: Triangle inequality: $\|x + y\| \leq \|x\| + \|y\|$ for all $x, y \in L$.
A linear space L, equipped with a norm is called a normed linear space.

Definition 2. *[3] A ring A is said to be a normed ring if A possesses a norm $\|\|$, that is, a non-negative real-valued function $\|\| : A \to R$ such that for any $a, b \in A$,*

1. $\|a\| = 0 \Leftrightarrow a = 0$,
2. $\|a + b\| \leq \|a\| + \|b\|$,
3. $\|-a\| = \|a\|$, *(and hence $\|1_A\| = 1 = \|-1\|$ if identity exists), and*
4. $\|ab\| \leq \|a\|\|b\|$.

Definition 3. *[12] Let $* : [0,1] \times [0,1] \to [0,1]$. $*$ is an Archimedean strict T-norm iff for all $x, y, z \in [0,1]$:*

(1) $*$ *is commutative and associative, that is,* $*\,(x,y) = *\,(y,x)$ *and* $*\,(x, *\,(y,z)) = *\,(*\,(x,y), z)$,
(2) $*$ *is continuous,*
(3) $*\,(x,1) = x$,
(4) $*$ *is monotone, which means* $*\,(x,y) \leq *\,(x,z)$ *if* $y \leq z$,
(5) $*\,(x,x) < x$ *for* $x \in (0,1)$, *and*
(6) *when* $x < z$ *and* $y < t$, $*\,(x,y) < *\,(z,t)$ *for all* $x, y, z, t \in (0,1)$.

For convenience, we use the word t-norm shortly and show it as $x * y$ instead of $*\,(x,y)$. Some examples of t-norms are $x * y = \min\{x,y\}$, $x * y = \max\{x + y - 1, 0\}$ and $x * y = x.y$.

Definition 4. *[12] Let $\diamond : [0,1] \times [0,1] \to [0,1]$. $*$ is an Archimedean strict T-conorm iff for all $x, y, z \in [0,1]$:*

(1) $\diamond$ *is commutative and associative, that is,* $\diamond\,(x,y) = \diamond\,(y,x)$ *and* $\diamond\,(x, \diamond\,(y,z)) = \diamond\,(\diamond\,(x,y), z)$,
(2) $\diamond$ *is continuous,*
(3) $\diamond\,(x,0) = x$,
(4) $\diamond$ *is monotone, which means* $\diamond\,(x,y) \leq \diamond\,(x,z)$ *if* $y \leq z$,
(5) $\diamond\,(x,x) > x$ *for* $x \in (0,1)$, *and*
(6) *when* $x < z$ *and* $y < t$, $\diamond\,(z,t) < \diamond\,(x,y)$ *for all* $x, y, z, t \in (0,1)$.

For convenience, we use the word s-norm shortly and show it as $x \diamond y$ instead of $\diamond\,(x,y)$. Some examples of s-norms are $x \diamond y = \max\{x,y\}$, $x * y = \min\{x + y, 1\}$ and $x \diamond y = x + y - x.y$.

Definition 5. *[6] The fuzzy set B on a universal set X* is a set of ordered pairs

$$B = \{(x, \mu_B(x) : x \in X)\}$$

Here, $\mu_B(x)$ is the membership function or membership grade of x in B. For all $x \in X$, we have $0 \leq \mu_B(x) \leq 1$. If $x \notin B$, $\mu_B(x) = 0$, and if x is entirely contained in B, $\mu_B(x) = 1$. The membership grade of x in B is shown as $B(x)$ in the rest of this paper.

Definition 6. *[6] For the fuzzy sets A and B, the membership functions of the intersection, union and complement are defined pointwise as follows respectively:*

$$(A \cap B)(x) = \min\{A(x), B(x)\},$$

$$(A \cup B)(x) = \max\{A(x), B(x)\},$$

$$\overline{A}(x) = 1 - A(x).$$

Definition 7. *[10] Let $(R, +, .)$ be a ring and $F(R)$ be the set of all fuzzy subsets of R. As $A \in F(R)$, $\wedge$ is the fuzzy intersection and $\vee$ is the fuzzy union functions, for all $x, y \in R$, if A satisfies (1) $A(x-y) \geq A(x) \wedge A(y)$ and (2) $A(x.y) \geq A(x) \wedge A(y)$ then A is called a fuzzy subring of R. If A is a subring of R for all $a \in A$, then A is itself a fuzzy ring.*

Definition 8. *[11] A non-empty fuzzy subset A of R is said to be an ideal (in fact a fuzzy ideal) if and only if, for any $x, y \in R$, $A(x-y) \geq A(x) \wedge A(y)$ and $A(x.y) \geq A(x) \vee A(y)$.*

Note: *The fuzzy operations of the fuzzy subsets $A, B \in F(R)$ on the ring R can be extended to the operations below by t-norms and s-norms:*

For all $z \in R$,

$$(A + B)(z) = \underset{x+y=z}{\diamond}(A(x) * B(y));$$

$$(A - B)(z) = \underset{x-y=z}{\diamond}(A(x) * B(y));$$

$$(A.B)(z) = \underset{x.y=z}{\diamond}(A(x) * B(y)).$$

3. Fuzzy Normed Rings and Fuzzy Normed Ideals

In this section, there has been defined the fuzzy normed ring and some basic properties related to it. Throughout the rest of this paper, R is the set of real numbers, R will denote an associative ring with identity, NR is a normed ring and $F(X)$ is the set of all fuzzy subsets of the set X.

Definition 9. *Let $*$ be a continuous t-norm and $\diamond$ a continuous s-norm, NR a normed ring and let A be a fuzzy set. If the fuzzy set $A = \{(x, \mu_A(x)) : x \in NR\}$ over a fuzzy normed ring $F(NR)$ satisfy the following conditions then A is called a fuzzy normed subring of the normed ring $(NR, +, .)$:*

For all $x, y \in NR$,

(i) $A(x-y) \geq A(x) * A(y)$

(ii) $A(x.y) \geq A(x) * A(y)$.

Let 0 be the zero of the normed ring NR. For any fuzzy normed subring A and for all $x \in NR$, we have $A(x) \leq A(0)$, since $A(x-x) \geq A(x) * A(x) \Rightarrow A(0) \geq A(x)$.

Example 1. *Let A fuzzy set and $R = (Z, +, .)$ be the ring of all integers. Define a mapping $f : A \to F(NR(Z))$ where, for any $a \in A$ and $x \in Z$,*

$$A_f(x) = \begin{cases} 0 \text{ if } x \text{ is odd} \\ \frac{1}{a} \text{ if } x \text{ is even} \end{cases}$$

Corresponding t-norm $(*)$ *and t-conorm* $(\diamond)$ *are defined as* $a * b = min\{a, b\}$, $a \diamond b = max\{a, b\}$; *then,* A *is a fuzzy set as well as a fuzzy normed ring over* $[(Z, +, .), A]$.

Lemma 1. $A \in F(NR)$ *is a fuzzy normed subring of the normed ring* NR *if and only if* $A - A \subseteq A$ *and* $A.A \subseteq A$.

Proof. Let A be a fuzzy normed subring of NR. By [10], it is clear that A is a fuzzy group under addition and so $A - A \subseteq A$. Also for all $z \in NR$,

$$(A.A)(z) = \underset{x.y=z}{\diamond}(A(x) * A(y)) \leq \underset{x.y=z}{\diamond} A(xy) = A(z) \Rightarrow A.A \subseteq A$$

Now we suppose $A - A \subseteq A$ and $A.A \subseteq A$. For all $x, y \in NR$,

$$A(x-y) \geq (A-A)(x-y) = \underset{s-t=x-y}{\diamond}(A(s) * A(t)) \geq A(x) * A(y).$$

Similarly,

$$A(xy) \geq (A.A)(xy) = \underset{st=xy}{\diamond}(A(s) * A(t)) \geq A(x) * A(y).$$

Thus, A is a fuzzy normed subring of NR. □

Lemma 2.

i. *Let* A *be a fuzzy normed subring of the normed ring* NR *and let* $f : NR \to NR'$ *be a ring homomorphism. Then,* $f(A)$ *is a fuzzy normed subring of* NR'.
ii. *Let* $f : NR \to NR'$ *be a normed ring homomorphism. If* B *is a fuzzy normed subring of* NR', *then* $f^{-1}(B)$ *is a fuzzy normed subring of* NR.

Proof. (i) Take $u, v \in NR'$. As f is onto, there exists $x, y \in NR$ such that $f(x) = u$ and $f(y) = v$. So,

$$\begin{aligned}
(f(A))(u) * (f(A))(v) &= \left(\underset{f(x)=u}{\diamond} A(x)\right) * \left(\underset{f(y)=v}{\diamond} A(y)\right) \\
&= \underset{f(x)=u, f(y)=v}{\diamond} (A(x) * A(y)) \\
&\leq \underset{f(x)=u, f(y)=v}{\diamond} (A(x-y)) \text{ (as } A \text{ is a fuzzy normed subring of } NR) \\
&\leq \underset{f(x)-f(y)=u-v}{\diamond} (A(x-y)) \\
&= \underset{f(x-y)=u-v}{\diamond} (A(x-y)) \text{ (since } f \text{ is } a \text{ homomorphism)} \\
&= \underset{f(z)=u-v}{\diamond} A(z) \\
&= (f(A))(u-v).
\end{aligned}$$

Similarly, it is easy to see that

$$(f(A))(u.v) \geq (f(A)(u) * f(A)(v)).$$

Therefore, $f(A)$ is a fuzzy normed subring of NR'.
(ii) Proof is straightforward and similar to the proof of (i). □

Definition 10. *Let* A_1 *and* A_2 *be two fuzzy normed rings over the normed ring* NR. *Then* A_1 *is a fuzzy normed subring of* A_2 if

$$A_1(x) \leq A_2(x)$$

for all $x \in NR$.

Definition 11. *Let NR be a normed ring, $A \in F(NR)$ and let $A \neq \emptyset$. If for all $x, y \in NR$*

(i) $A(x-y) \geq A(x) * A(y)$ *and*
(ii) $A(x.y) \geq A(y)$ $(A(x.y) \geq A(x))$,

then A is called a fuzzy left (right) normed ideal of NR.

Definition 12. *If the fuzzy set A is both a fuzzy normed right and a fuzzy normed left ideal of NR, then A is called a fuzzy normed ideal of NR; i.e., if for all $x, y \in NR$*

(i) $A(x-y) \geq A(x) * A(y)$ *and*
(ii) $A(x.y) \geq A(x) \diamond A(y)$,

then $A \in F(NR)$ is a fuzzy normed ideal of NR.

Remark 1. *Let the multiplicative identity of NR (if exists) be 1_{NR}. As $A(x.y) \geq A(x) \diamond A(y)$ for all $x, y \in NR$, $A(x.1_{NR}) \geq A(x) \diamond A(1_{NR})$ and therefore for all $x \in NR$, $A(x) \geq A(1_{NR})$.*

Example 2. *Let A and B be two (fuzzy normed left, fuzzy normed right) ideals of a normed ring NR. Then, $A \cap B$ is also a (fuzzy normed left, fuzzy normed right) ideal of NR.*

Solution: Let $x, y \in NR$.

$$\begin{aligned}(A \cap B)(x-y) &= \min\{A(x-y), B(x-y)\} \\ &\geq \min\{A(x) * A(y), B(x) * B(y)\} \\ &\geq \min\{(A \cap B)(x), (A \cap B)(y)\}.\end{aligned}$$

On the other hand, as A and B are fuzzy normed left ideals, using $A(x.y) \geq A(y)$ and $B(x.y) \geq B(y)$ we have

$$(A \cap B)(x.y) = \min\{A(x.y), B(x.y)\} \geq \min\{A(y), B(y)\} = (A \cap B)(y).$$

So $A \cap B$ is a fuzzy normed left ideal. Similarly, it is easy to show that $A \cap B$ is a fuzzy normed right ideal. As a result $A \cap B$ is an fuzzy normed ideal of NR.

Example 3. *Let A be a fuzzy ideal of NR. The subring $A^0 = \{x : \mu_A(x) = \mu_A(0_{NR})\}$ is a fuzzy normed ideal of NR, since for all $x \in NR$, $A^0(x) \leq A^0(0)$.*

Theorem 1. *Let A be a fuzzy normed ideal of NR, $X = \{a_1, a_2, \ldots, a_m\} \subseteq NR$, $x, y \in NR$ and let $FN(X)$ be the fuzzy normed ideal generated by the set X in NR.* Then,

(i) $w \in FN(X) \Rightarrow A(w) \geq \underset{1 \leq i \leq m}{*}(A(a_i))$,
(ii) $x \in (y) \Rightarrow A(x) \geq A(y)$,
(iii) $A(0) \geq A(x)$ *and*
(iv) *if 1 is the multiplicative identity of NR, then $A(x) \geq A(1)$.*

Proof. (ii), (iii), and (iv) can be proved using (i). The set $FN(X)$ consists of the finite sums in the form $ra + as + uav + na$ where $a \in X$, $r, s, u, v \in NR$ and n is an integer. Let $w \in FN(X)$. So there exists an integer n and $r, s, u, v \in NR$ such that $w = ra_i + a_is + ua_iv + na_i$ where $1 \leq i \leq m$. As A is a fuzzy normed ideal,

$$A(ra_i + a_is + ua_iv + na_i) \geq A(ra_i) * A(a_is) * A(ua_iv) * A(na_i) \geq A(a_i).$$

Therefore

$$A(w) \geq \underset{1 \leq i \leq m}{*}(A(a_i)).$$

□

4. Fuzzy Normed Prime Ideal and Fuzzy Normed Maximal Ideal

In this section, fuzzy normed prime ideal and fuzzy normed maximal ideal are outlined.

Definition 13. *Let A and B be two fuzzy subsets of the normed ring NR. We define the operation $A \circ B$* as follows:

$$A \circ B(x) = \begin{cases} \underset{x=yz}{\diamond}(A(y) * B(z)) & \text{, if } x \text{ can be defined as } x = yz \\ 0 & \text{, otherwise .} \end{cases}$$

If the normed ring NR has a multiplicative inverse, namely if $NR.NR = NR$, then the second case does not occur.

Lemma 3. *If A and B are a fuzzy normed right and a fuzzy normed left ideal of a normed ring NR, respectively, $A \circ B \subseteq A \cap B$ and hence $(A \circ B)(x) \leq (A \cap B)(x)$ for all $x \in NR$.*

Proof. It is shown in Example 2 that if A and B are fuzzy normed left ideals of NR, then $A \cap B$ is also a fuzzy normed left ideal. Now, let A and B be a fuzzy normed right and a fuzzy normed left ideal of NR, respectively. If $A \circ B(x) = 0$, the proof is trivial.

Let

$$(A \circ B)(x) = \underset{x=yz}{\diamond}(A(y) * B(z)).$$

As A is a fuzzy normed right ideal and B is a fuzzy normed left ideal, we have

$$A(y) \leq A(yz) = A(x)$$

and

$$B(y) \leq B(yz) = B(x)$$

Thus,

$$\begin{aligned} (A \circ B)(x) &= \underset{x=yz}{\diamond}(A(y) * B(z)) \\ &\leq \min(A(x), B(x)) \\ &= (A \cap B)(x) \end{aligned}$$

□

Definition 14. *Let A and B be fuzzy normed ideals of a normed ring NR and let FNP be a non-constant function, which is not an ideal of NR. If*

$$A \circ B \subseteq FNP \Rightarrow A \subseteq FNP \text{ or } B \subseteq FNP,$$

then FNP is called a fuzzy normed prime ideal of NR.

Example 4. *Show that if the fuzzy normed ideal I $(I \neq NR)$ is a fuzzy normed prime ideal of NR, then the characteristic function λ_I* is also a fuzzy normed prime ideal.

Solution: As $I \neq NR$, λ_I is a non-constant function on NR. Let A and B be two fuzzy normed ideals on NR such that $A \circ B \subseteq \lambda_I$, but $A\lambda_I$ and $B\lambda_I$. There exists $x, y \in NR$ such that $A(x) \leq \lambda_I(x)$ and $B(y) \leq \lambda_I(y)$. In this case, $A(x) \neq 0$ and $B(y) \neq 0$, but $\lambda_I(x) = 0$ and $\lambda_I(y) = 0$. Therefore

$x \notin I, y \notin I$. As I is a fuzzy normed prime ideal, there exists an $r \in NR$, such that $xry \notin I$. This is obvious, because if I is fuzzy normed prime, $A \circ B(xry) \subseteq I \Rightarrow A(x) \subseteq I$ or $B(ry) \subseteq I$ and therefore as $(NRxNR)(\underline{NRry}NR) = (NRxNR)(NRyNR) \subseteq I$, we have either $NRxNR \subseteq I$ or $NRyNR \subseteq I$. Assume $NRxNR \subseteq I$. Then $xxx = (x)^3 \in I \Rightarrow x \subseteq I$, but this contradicts with the fact that $\lambda_I(x) = 0$. Now let $a = xry$. $\lambda_I(a) = 0$. Thus, $A \circ B(a) = 0$. On the other hand,

$$\begin{aligned} A \circ B(a) &= \underset{a=cd}{\diamond}(A(c) * B(d)) \\ &\geq A(x) * B(ry) \\ &\geq A(x) * B(y) \\ &\geq 0 \text{ (as } A(x) \neq 0 \text{ and } B(y) \neq 0). \end{aligned}$$

This is a contradiction, since $A \circ B(a) = 0$. Therefore if A and B are fuzzy normed ideals of a normed ring NR, then $A \circ B \subseteq \lambda_I \Rightarrow A \subseteq \lambda_I$ or $B \subseteq \lambda_I$. As a result, the characteristic function λ_I is a fuzzy normed prime ideal.

Theorem 2. *Let FNP be a fuzzy normed prime ideal of a normed ring NR. The ideal defined by* $FNP^0 = \{x : x \in NR, FNP(x) = FNP(0)\}$ *and is also a fuzzy normed prime ideal of NR.*

Proof: Let $x, y \in FNP^0$. As FNP is an fuzzy normed ideal, $FNP(x-y) \geq FNP(x) * FNP(y) = FNP(0)$. On the other hand, by Theorem 1, we have $FNP(0) \geq FNP(x-y)$. So, $FNP(x-y) = FNP(0)$ and $x - y \in FNP^0$. Now, let $x \in FNP^0$ and $r \in NR$. In this case, $FNP(rx) \geq FNP(x) = FNP(0)$ and thus $FNP(rx) = FNP(0)$. Similarly, $FNP(xr) = FNP(0)$. Now, for all $x \in FNP^0$ and $r \in NR$, $rx, xr \in FNP^0$. Therefore, FNP^0 is a fuzzy normed ideal of NR. Let I and J be two ideals of NR, such that $IJ \subseteq FNP^0$. Now, we define fuzzy normed ideals $A = FNP^0\lambda_I$ and $B = FNP^0\lambda_J$. We will show that $(A \circ B)(x) \leq FNP(x)$ for all $x \in NR$. Assume $(A \circ B)(x) \neq 0$. Recall $A \circ B = \underset{x=yz}{\diamond}(A(y) * B(z))$, so we only need to take the cases of $A(y) * B(z) \neq 0$ under consideration. However, in all these cases, $A(y) = FNP(0)$ or $A(y) = 0$ and similarly $B(z) = FNP(0)$ or $B(z) = 0$ and hence $A(y) = B(z) = FNP(0)$. Now, $\lambda_I(y) = 1$ and $\lambda_J(z) = 1$ implies $y \in I$, $z \in J$ and $x \in IJ \subseteq FNP^0$. Thus, $FNP(x) = FNP(0)$ and for all $x \in NR$, we get $(A \circ B)(x) \leq FNP(x)$. As FNP is a fuzzy normed prime ideal and A and B are fuzzy normed ideals, either $A \subseteq FNP$ or $B \subseteq FNP$. Assume $A = FNP^0\lambda_I \subseteq FNP$. We need to show that $I \subseteq FNP^0$. Let $IFNP$. Then, there exists an $a \in I$, such that $a \notin FNP^0$; i.e., $FNP(a) \neq FNP(0)$. It is evident that $FNP(0) \geq FNP(a)$. Thus, $FNP(a)<FNP(0)$. However, $A(a) = FNP^0\lambda_I(a) = FNP(0)>FNP(a)$ and this is a contradiction to the assumption $A \subseteq FNP$. So, $I \subseteq FNP^0$. Similarly, one can show that $B \subseteq FNP$ and $J \subseteq FNP^0$. Thus, FNP^0 is a fuzzy normed prime ideal. □

Definition 15. *Let A be a fuzzy normed ideal of a normed ring NR. If A is non-constant and for all fuzzy normed ideals B of NR,* $A \subseteq B$ *implies* $A^0 = B^0$ *or* $B = \lambda_{NR}$, *A is called a fuzzy normed maximal ideal of the normed ring NR.* Fuzzy normed maximal left(right) ideals are defined similarly.

Example 5. *Let A be a fuzzy normed maximal left (right) ideal of a normed ring NR. Then,* $A^0 = \{x \in NR : A(x) = A(0)\}$ *is a fuzzy normed maximal left (right) ideal of NR.*

Theorem 3. *If A is a fuzzy normed left(right) maximal ideal of a normed ring NR, then* $A(0) = 1$.

Proof. Assume $A(0) \neq 1$. Let $A(0) < t < 1$ and let B be a fuzzy subset of NR such that $B(x) = t$ for all $x \in NR$. B is trivially an ideal of NR. Also it is easy to verify that $A \subset B$, $B \neq \lambda_{NR}$ and $B^0 = \{x \in NR : B(x) = B(0)\} = NR$. But, despite the fact that $A \subset B$, $A^0 \neq B^0$ and $B \neq \lambda_{NR}$ is a contradiction to the fuzzy normed maximality of A. Thus, $A(0) = 1$. □

5. Conclusions

In this paper, we defined a fuzzy normed ring. Here we examine the algebraic properties of fuzzy sets in ring structures. Some related notions, e.g., the fuzzy normed ring homomorphism, fuzzy normed subring, fuzzy normed ideal, fuzzy normed prime ideal and fuzzy normed maximal ideal are proposed. We hope that this new concept will bring a new opportunity in research and development of fuzzy set theory. To extend our work, further research can be done to study the properties of fuzzy normed rings in other algebraic structures such as fuzzy rings and fuzzy fields.

Author Contributions: All authors contributed equally.

Acknowledgments: We thank Vakkas Uluçay for the arrangement.

Conflicts of Interest: The authors declare no conflict of interest.

References

1. Naĭmark, M.A. *Normed Rings*; Noordhoff: Groningen, The Netherlands, 1964.
2. Arens, R. A generalization of normed rings. *Pac. J. Math.* **1952**, *2*, 455–471. [CrossRef]
3. Gel'fand, I.; Raikov, M.D.A.; Shilov, G.E. *Commutative Normed Rings*; Chelsea Publishing Company: New York, NY, USA, 1964.
4. Jarden, M. *Normed Rings-Algebraic Patching*; Springer: Berlin/Heidelberg, Germany, 2011.
5. Zadeh, L.A. Fuzzy Sets. *Inf. Control* **1965**, *8*, 338–353. [CrossRef]
6. Zimmermann, H.J. *Fuzzy Set Theory and Its Application*, 2nd ed.; Kluwer Academic Publishers: Norwell, MA, USA, 1991.
7. Mordeson, J.N.; Bhutani, K.R.; Rosenfeld, A. *Fuzzy Group Theory*; Springer: Heidelberg, Germany, 2005.
8. Liu, W. Fuzzy Invariant Subgroups and Fuzzy Ideals. *Fuzzy Sets Syst.* **1982**, *8*, 133–139. [CrossRef]
9. Mukherjee, N.P.; Bhattacharya, P. Fuzzy Normal Subgroups and Fuzzy Cosets. *Inf. Sci.* **1984**, *34*, 225–239. [CrossRef]
10. Wang, X.; Ruan, D.; Kerre, E.E. *Mathematics of Fuzziness—Basic Issues*; Springer: Berlin/Heidelberg, Germany, 2009.
11. Swamy, U.M.; Swamy, K.L.N. Fuzzy Prime Ideals of Rings. *J. Math. Anal. Appl.* **1988**, *134*, 94–103. [CrossRef]
12. Gupta, M.M.; Qi, J. Theory of T-norms and Fuzzy Inference Methods. *Fuzzy Sets Syst.* **1991**, *40*, 431–450. [CrossRef]
13. Kolmogorov, A.N.; Silverman, R.A.; Fomin, S.V. *Introductory Real Analysis*; Dover Publications, Inc.: New York, NY, USA, 1970.
14. Uluçay, V.; Şahin, M.; Olgun, N. Normed Z-Modules. *Int. J. Pure Appl. Math.* **2017**, *112*, 425–435.
15. Şahin, M.; Olgun, N.; Uluçay, V. Soft Normed Rings. *Springer Plus* **2016**, *5*, 1950–1956.
16. Şahin, M.; Olgun, N.; Uluçay, V. Normed Quotient Rings. *New Trends Math. Sci.* **2018**, *6*, 52–58.
17. Şahin, M.; Kargın, A. Neutrosohic Triplet Normed Space. *Open Phys.* **2017**, *15*, 697–704. [CrossRef]
18. Olgun, N.; Şahin, M. Fitting ideals of universal modules. In Proceedings of the International Conference on Natural Science and Engineering (ICNASE'16), Kilis, Turkey, 19–20 March 2016; pp. 1–8.
19. Olgun, N. A Problem on Universal Modules. *Commun. Algebra* **2015**, *43*, 4350–4358. [CrossRef]
20. Şahin, M.; Kargın, A. Neutrosophic Triplet Inner Product. *Neutrosophic Oper. Res.* **2017**, *2*, 193–205.
21. Smarandache, F.; Şahin, M.; Kargın, A. Neutrosophic Triplet G-Module. *Mathematics* **2018**, *6*, 53. [CrossRef]
22. Şahin, M.; Olgun, N.; Kargın, A.; Uluçay, V. Isomorphism theorems for soft G-modules. *Afr. Mat.* **2018**, 1–8. [CrossRef]
23. Çelik, M.; Shalla, M.; Olgun, N. Fundamental Homomorphism Theorems for Neutrosophic Extended Triplet Groups. *Symmetry* **2018**, *10*, 321–335.
24. Bal, M.; Shalla, M.M.; Olgun, N. Neutrosophic Triplet Cosets and Quotient Groups. *Symmetry* **2018**, *10*, 126. [CrossRef]

symmetry

MDPI

Article

Invariant Graph Partition Comparison Measures

Fabian Ball * and Andreas Geyer-Schulz

Karlsruhe Institute of Technology, Institute of Information Systems and Marketing, Kaiserstr. 12, 76131 Karlsruhe, Germany; andreas.geyer-schulz@kit.edu

* Correspondence: fabian.ball@kit.edu; Tel.: +49-721-6084-8404

Received: 10 September 2018; Accepted: 11 October 2018; Published: 15 October 2018

Abstract: Symmetric graphs have non-trivial automorphism groups. This article starts with the proof that all partition comparison measures we have found in the literature fail on symmetric graphs, because they are not invariant with regard to the graph automorphisms. By the construction of a pseudometric space of equivalence classes of permutations and with Hausdorff's and von Neumann's methods of constructing invariant measures on the space of equivalence classes, we design three different families of invariant measures, and we present two types of invariance proofs. Last, but not least, we provide algorithms for computing invariant partition comparison measures as pseudometrics on the partition space. When combining an invariant partition comparison measure with its classical counterpart, the decomposition of the measure into a structural difference and a difference contributed by the group automorphism is derived.

Keywords: graph partitioning; graph clustering; invariant measures; partition comparison; finite automorphism groups; graph automorphisms

1. Introduction

Partition comparison measures are routinely used in a variety of tasks in cluster analysis: finding the proper number of clusters, assessing the stability and robustness of solutions of cluster algorithms, comparing different solutions of randomized cluster algorithms or comparing optimal solutions of different cluster algorithms in benchmarks [1], or in competitions like the 10th DIMACS graph-clustering challenge [2]. Their development has been for more than a century an active area of research in statistics, data analysis and machine learning. One of the oldest and still very well-known measure is the one of Jaccard [3]; more recent approaches were by Horta and Campello [4] and Romano et al. [5]. For an overview of many of these measures, see Appendix B. Besides the need to compare clustering partitions, there is an ongoing discussion of what actually are the best clusters [6,7]. Another problem often addressed is how to measure cluster validity [8,9].

However, the comparison of graph partitions leads to new challenges because of the need to handle graph automorphisms properly. The following small example shows that standard partition comparison measures have unexpected results when applied to graph partitions: in Figure 1, we show two different ways of partitioning the cycle graph C_4 (Figure 1a,d). Partitioning means grouping the nodes into non-overlapping clusters. The nodes are arbitrarily labeled with 1 to 4 (Figure 1b,e), and then, there are four possibilities of relabeling the nodes so that the edges stay the same. One possibility is relabeling 1 by 2, 2 by 3, 3 by 4 and 4 by 1, and the images resulting from this relabeling are shown in Figure 1c,f. The relabeling corresponds to a counterclockwise rotation of the graph by 90°, and formal details are given in Section 2. The effects of this relabeling on the partitions $\mathcal{P}_1$ and $\mathcal{Q}_1$ are different:

1. Partition $\mathcal{P}_1 = \{\{1,2\},\{3,4\}\}$ is mapped to the structurally equivalent partition $\mathcal{P}_2 = \{\{1,4\},\{2,3\}\}$.
2. Partition $\mathcal{Q}_1 = \{\{1,3\},\{2,4\}\}$ is mapped to the identical partition $\mathcal{Q}_2$.

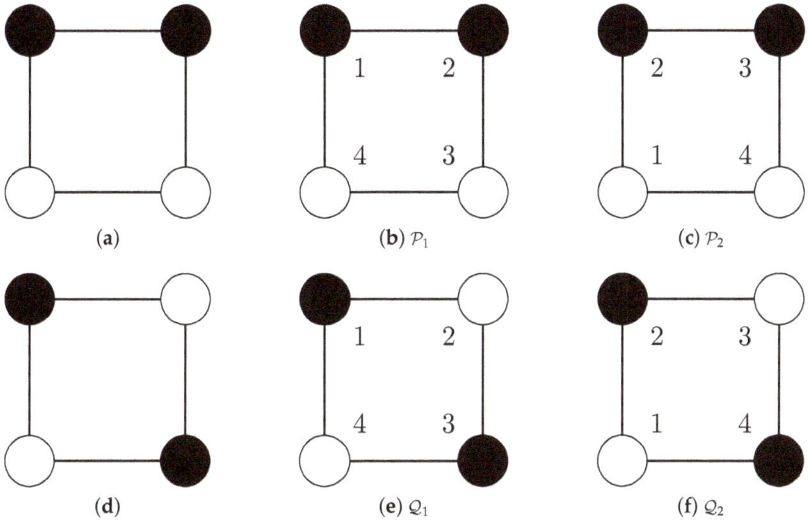

Figure 1. Two structurally different partitions of the cycle graph C_4: grouping pairs of neighbors (**a**) and grouping pairs of diagonals (**d**). Equally-colored nodes represent graph clusters, and the choice of colors is arbitrary. Adding, again arbitrary, but fixed, node labels impacts the node partitions and results in the failure to recognize the structural difference when comparing these partitions with partition comparison measures (see Table 1). The different images (**b**,**c**) ($\mathcal{P}_1 = \{\{1,2\},\{3,4\}\}$, $\mathcal{P}_2 = \{\{1,4\},\{2,3\}\}$) and (**e**,**f**) ($\mathcal{Q}_1 = \mathcal{Q}_2 = \{\{1,3\},\{2,4\}\}$) emerge from the graph's symmetry.

Table 1 illustrates the failure of partition comparison measures (here, the Rand Index (RI)) to recognize structural differences:

1. Because $\mathcal{P}_1$ and $\mathcal{P}_2$ are structurally equivalent, the RI should be one (as for Cases 1, 2 and 3) instead of $1/3$.
2. Comparisons of structurally different different partitions (Cases 4 and 5) and comparisons of structurally equivalent partitions (Case 6) should not result in the same value.

Table 1. The Rand index is $RI = \frac{N_{11}+N_{00}}{N_{11}+N_{10}+N_{01}+N_{00}}$. N_{11} indicates the number of nodes that are in both partitions together in a cluster; N_{10} and N_{01} are the number of nodes that are together in a cluster in one partition, but not in the other; and N_{00} are the number of nodes that are in both partitions in different clusters. See Appendix B for the formal definitions. Partitions $\mathcal{P}_1$ and $\mathcal{P}_2$ are equivalent (yet not equal, denoted "$\sim$"), and partitions $\mathcal{Q}_1$ and $\mathcal{Q}_2$ are identical (thus, also equivalent, denoted "$=$"). However, the comparison of the structurally different partitions (denoted "$\neq$") $\mathcal{P}_i$ and $\mathcal{Q}_j$ yields the same result as the comparison between the equivalent partitions $\mathcal{P}_1$ and $\mathcal{P}_2$. This makes the recognition of structural differences impossible.

Case	Compared Partitions	Relation	N_{11}	N_{10}	N_{01}	N_{00}	RI
1	$\mathcal{P}_1, \mathcal{P}_1$	$=$	2	0	0	4	1
2	$\mathcal{P}_2, \mathcal{P}_2$	$=$	2	0	0	4	1
3	$\mathcal{Q}_1, \mathcal{Q}_1$ or $\mathcal{Q}_1, \mathcal{Q}_2$ or $\mathcal{Q}_2, \mathcal{Q}_2$	$=$	2	0	0	4	1
4	$\mathcal{P}_1, \mathcal{Q}_1$ or $\mathcal{P}_1, \mathcal{Q}_2$	$\neq$	0	2	2	2	$\frac{1}{3}$
5	$\mathcal{P}_2, \mathcal{Q}_1$ or $\mathcal{P}_2, \mathcal{Q}_2$	$\neq$	0	2	2	2	$\frac{1}{3}$
6	$\mathcal{P}_1, \mathcal{P}_2$	$\sim$	0	2	2	2	$\frac{1}{3}$

One may argue that graphs in real applications contain symmetries only rarely. However, recent investigations of graph symmetries in real graph datasets show that a non-negligible proportion of these graphs contain symmetries. MacArthur et al. [10] state that "a certain degree of symmetry is also ubiquitous in complex systems" [10] (p. 3525). Their study includes a small number of biological, technological and social networks. In addition, Darga et al. [11] studied automorphism groups in very large sparse graphs (circuits, road networks and the Internet router network), with up to five million nodes with eight million links with execution times below 10 s. Katebi et al. [12] reported symmetries in 268 of 432 benchmark graphs. A recent large-scale study conducted by the authors of this article for approximately 1700 real-world graphs revealed that about three quarters of these graphs contain symmetries [13].

The rather frequent occurrence of symmetries in graphs and the obvious deficiencies of classic partition comparison measures demonstrated above have motivated our analysis of the effects of graph automorphisms on partition comparison measures.

Our contribution has the following structure: Permutation groups and graph automorphisms are introduced in Section 2. The full automorphism group of the butterfly graph serves as a motivating example for the formal definition of stable partitions, stable with regard to the actions of the automorphism group of a graph. In Section 3, we first provide a definition that captures the property that a measure is invariant with regard to the transformations in an automorphism group. Based on this definition, we first give a simple proof by counterexample for each partition comparison measure in Appendix B, that these measures based on the comparison of two partitions are not invariant to the effects of automorphisms on partitions. The non-existence of partition comparison measures for which the identity and the invariance axioms hold simultaneously is proven subsequently. In Section 4, we construct three families of invariant partition comparison measures by a two-step process: First, we define a pseudometric space by defining equivalence classes of partitions as the orbit of a partition under the automorphism group $Aut(G)$. Second, the definitions of the invariant counterpart of a partition comparison measure are given: we define them as the computation of the maximum, the minimum and the average of the direct product of the two equivalence classes. The section also contains a proof of the equivalence of several variants of the computation of the invariant measures, which—by exploiting the group properties of $Aut(G)$—differ in the complexity of the computation. In Section 5, we introduce the decomposition of the measures into a structurally stable and unstable part, as well as upper bounds for instability. In Section 6, we present an application of the decomposition of measures for analyzing partitions of the Karate graph. The article ends with a short discussion, conclusion and outlook in Section 7.

2. Graphs, Permutation Groups and Graph Automorphisms

We consider connected, undirected, unweighted and loop-free graphs. Let $G = (V, E)$ denote a graph where V is a finite set of nodes and E is a set of edges. An edge is represented as $\{u, v\} \in \{\{x, y\} \mid (x, y) \in V \times V \land x \neq y\}$. Nodes adjacent to $u \in V$ (there exists an edge between u and those nodes) are called neighbors. A partition $\mathcal{P}$ of a graph G is a set of subsets $C_i, i = 1, \ldots, k$ of V with the usual properties: (i) $C_i \cap C_j = \emptyset$ $(i \neq j)$, (ii) $\bigcup_i C_i = V$ and (iii) $C_i \neq \emptyset$. Each subset is called a cluster, and it is identified by its labeled nodes.

As a partition quality criterion, we use the well-known modularity measure Q of Newman and Girvan [14] (see Appendix A). It is a popular optimization criterion for unsupervised graph clustering algorithms, which try to partition the nodes of the graph in a way that the connectivity within the clusters is maximized and the number of edges connecting the clusters is minimized. For a fast and efficient randomized state-of-the-art algorithm, see Ovelgönne and Geyer-Schulz [15].

Partitions are compared by comparison measures, which are functions of the form $m : P(V) \times P(V) \to \mathbb{R}$ where $P(V)$ denotes the set of all possible partitions of the set V. A survey of many of these measures is given in Appendix B.

A permutation on V is a bijection $g : V \to V$. We denote permutations by the symbols f, g and h. Each permutation can be written in cycle form: for a permutation with a single cycle of length r, we write $c = (v_1\, v_2\, \ldots\, v_r)$. c maps v_i to v_{i+1} $(i = 1, \ldots, r-1)$, v_r to v_1 and leave all other nodes fixed. Permutations with more than one cycle are written as a product of disjoint cycles (i.e., no two cycles have a common element). (v_k) means that the element v_k remains fixed, and for brevity, these elements are omitted.

Permutations are applied from the right: The image of u under the permutation g is ug. The composition of g and h is $h \circ g$, with $\circ$ being the permutation composition symbol. For brevity, $h \circ g$ is written as gh, so that $u(gh) = (ug)h$ holds. Computer scientists call this a postfix notation; in prefix notation, we have $h(g(u))$. Often, we also find u^g, which we will use in the following. For k compositions $g \circ g \circ g \circ \ldots$, we write g^k and $g^0 = id$.

A set of permutation functions forms a permutation group H, if the usual group axioms hold [16]:

1. Closure: $\forall g, h \in H : g \circ h \in H$
2. Unit element: The identity function $id \in H$ acts as the neutral element: $\forall g \in H : id \circ g = g \circ id = g$
3. Inverse element: For any g in H, the inverse permutation function $g^{-1} \in H$ is the inverse of g: $\forall g \in H : g \circ g^{-1} = g^{-1} \circ g = id$
4. Associativity: The associative law holds: $\forall f, g, h \in H : f \circ (g \circ h) = (f \circ g) \circ h$

If H_1 is a subset of H and if H_1 is a group, H_1 is a subgroup of H (written $H_1 \leq H$). The set of all permutations of V is denoted by $Sym(V)$. $Sym(V)$ is a group, and it is called the symmetric group (see [17]). $Sym(V) \sim Sym(V')$ iff $|V| = |V'|$ with $\sim$ denoting isomorphism. A generator of a finite permutation group H is a subset of the permutations of H from which all permutations in H can be generated by application of the group axioms [18].

An action of H on V (H acts on V) is called the group action of a set [19] (p. 5):

1. $u^{id} = u$, $\forall u \in V$
2. $(u^g)^h = u^{gh}$, $\forall u \in V$, $\forall g, h \in H$

Groups acting on a set V also act on combinatorial structures defined on V [20] (p. 149), for example the power set 2^V, the set of all partitions $P(V)$ or the set of graphs $G(V)$. We denote combinatorial structures as capital calligraphic letters; in the following, only partitions ($\mathcal{P}$) are of interest because they are the results of graph cluster algorithms. The action of a permutation g on a combinatorial structure is performed by pointwise application of g. For instance, for $\mathcal{P}$, the image of g is $\mathcal{P}^g = \{\{u^g \mid u \in C\} \mid C \in \mathcal{P}\}$.

Let H be a permutation group. When H acts on V, a node u is mapped by the elements of H onto other nodes. The set of these images is called the orbit of u under H:

$$u^H = \left\{u^h \mid h \in H\right\}.$$

The group of permutations H_u that fixes u is called the stabilizer of u under H:

$$H_u = \{h \in H \mid u^h = u\}.$$

The orbit stabilizer theorem is given without proof [16]. It links the order of a permutation group with the cardinality of an orbit and the order of the stabilizer:

Theorem 1. *The relation:*

$$|H| = |u^H| \cdot |H_u|$$

holds.

The action of H on V induces an equivalence relation on the set: for $u_1, u_2 \in V$, let $u_1 \sim u_2$ iff there exists $h \in H$ so that $u_1 = u_2^h$. All elements of an orbit are equivalent, and the orbits of a group partition the set V. An orbit of length one (in terms of set cardinality) is called trivial. Analogously, for a partition $\mathcal{P}$, the definition is:

Definition 1. *The image of the action of H on a partition $\mathcal{P}$ (or the orbit of $\mathcal{P}$ under H) is the set of all equivalent partitions of partition $\mathcal{P}$ under H*

$$\mathcal{P}^H = \left\{\mathcal{P}^h \mid h \in H\right\}.$$

A graph automorphism f is a permutation that preserves edges, i.e., $\{u^f, v^f\} \in E \Leftrightarrow \{u, v\} \in E$, $\forall u, v \in V$.

The automorphism group of a graph contains all permutations of vertices that map edges to edges and non-edges to non-edges. The automorphism group of G is defined as:

$$Aut(G) = \left\{f \in Sym(V) \mid E^f = E\right\}$$

where $E^f = \left\{\{u^f, v^f\} \mid \{u, v\} \in E\right\}$. Of course, $Aut(G) \leq Sym(V)$.

Example 1. *Let G_{bf} be the butterfly graph (Figure 2, e.g., Erdős et al. [21], Burr et al. [22]) whose full automorphism group is given in Table 2 (first column). The permutation $(2\,5)$ is not an automorphism, because it does not preserve the edges from 1 to 2 and from 5 to 4. The butterfly graph has the two orbits $\{1,2,4,5\}$ and $\{3\}$. The group $H = \{id, g_1, g_2, g_3\}$ is a subgroup of $Aut(G_{bf})$.*

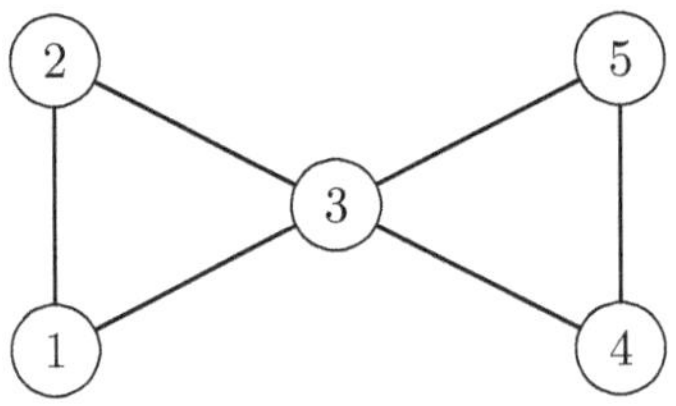

Figure 2. The butterfly graph (five nodes, with two node pairs connected by the bridging node 3).

Table 2. The full automorphism group $Aut(G_{bf}) = \{id, g_1, \ldots, g_7\}$ of the butterfly graph in Figure 2 and its effect on three partitions. Bold partitions are distinct. A possible generator is $\{g_1, g_4\}$.

Permutation	$\mathcal{P}_1$, $Q = 0$	$\mathcal{P}_2$, $Q = \frac{1}{9}$	$\mathcal{P}_3$, $Q = -\frac{1}{18}$
$id = (1)(2)(3)(4)(5)$	$\mathbf{\{1,2\},\{3\},\{4,5\}}$	$\mathbf{\{1,2,3\},\{4,5\}}$	$\mathbf{\{1,2,3,4\},\{5\}}$
$g_1 = (1\,2)$	$\{2,1\},\{3\},\{4,5\}$	$\{2,1,3\},\{4,5\}$	$\{2,1,3,4\},\{5\}$
$g_2 = (4\,5)$	$\{1,2\},\{3\},\{5,4\}$	$\{1,2,3\},\{5,4\}$	$\mathbf{\{1,2,3,5\},\{4\}}$
$g_3 = (1\,2)(4\,5)$	$\{2,1\},\{3\},\{5,4\}$	$\{2,1,3\},\{5,4\}$	$\{2,1,3,5\},\{4\}$
$g_4 = (1\,4)(2\,5)$	$\{4,5\},\{3\},\{1,2\}$	$\mathbf{\{4,5,3\},\{1,2\}}$	$\mathbf{\{4,5,3,1\},\{2\}}$
$g_5 = (1\,5)(2\,4)$	$\{5,4\},\{3\},\{2,1\}$	$\{5,4,3\},\{2,1\}$	$\{5,4,3,2\},\{1\}$
$g_6 = (1\,4\,2\,5)$	$\{4,5\},\{3\},\{2,1\}$	$\{4,5,3\},\{2,1\}$	$\mathbf{\{4,5,3,2\},\{1\}}$
$g_7 = (1\,5\,2\,4)$	$\{5,4\},\{3\},\{1,2\}$	$\{5,4,3\},\{1,2\}$	$\{5,4,3,1\},\{2\}$

Definition 2. *Let $G = (V, E)$ be a graph. A partition $\mathcal{P}$ is called stable, if $|\mathcal{P}^{Aut(G)}| = 1$, otherwise it is called unstable.*

Stability here means that the automorphism group of the graph does not affect the given partition by tearing apart clusters.

Example 2. *Only $\mathcal{P}_1$ in Table 2 is stable because its orbit is trivial. The two modularity optimal partitions (e.g., $\mathcal{P}_2^{id}$ and $\mathcal{P}_2^{g_4}$) are not stable because $|\mathcal{P}_2^{Aut(G_{bf})}| = 2$. Furthermore, $|\mathcal{P}_3^{Aut(G_{bf})}| = 4$.*

For the evaluation of graph clustering solutions, the effects of graph automorphisms on graph partitions are of considerable importance:

1. Automorphisms may lead to multiple equivalent optimal solutions as the butterfly graph shows ($\mathcal{P}_2^{id}$ and $\mathcal{P}_2^{g_4}$ in Table 2).
2. Partition comparison measures are not invariant with regard to automorphisms, as we show in Section 3.

3. Graph Partition Comparison Measures Are Not Invariant

When comparing graph partitions, a natural requirement is that the partition comparison measure is invariant under automorphism.

Definition 3. *A partition comparison measure $m : P(V) \times P(V) \to \mathbb{R}$ is invariant under automorphism, if:*

$$m(\mathcal{P}, \mathcal{Q}) = m(\tilde{\mathcal{P}}, \tilde{\mathcal{Q}})$$

for all $\mathcal{P}, \mathcal{Q} \in P(V)$ and $\tilde{\mathcal{P}} \in \mathcal{P}^{Aut(G)}$, $\tilde{\mathcal{Q}} \in \mathcal{Q}^{Aut(G)}$.

Observe that if $\mathcal{Q} \in \mathcal{P}^{Aut(G)}$, then such a measure m cannot distinguish between $\mathcal{P}$ and $\mathcal{Q}$, since $m(\mathcal{P}, \mathcal{Q}) = m(\mathcal{P}, \mathcal{P})$ by definition.

However, unfortunately, as we show in the rest of this section, such a partition comparison measure does not exist. In the following, we present two proofs of this fact, which differ both in their level of generality and sophistication.

3.1. Variant 1: Construction of a Counterexample

Theorem 2. *The measures for comparing partitions defined in Appendix B do not fulfill Definition 3 in general.*

Proof. We choose the cycle graph C_{36} and compute all modularity maximal partitions with $Q = 2/3$. Each of these six partitions has six clusters, and each of these clusters consists of a chain of six nodes (see Figure 3).

Clearly, since all partitions are equivalent, an invariant partition comparison measure should identify them as equivalent:

$$m(\mathcal{P}_0, \mathcal{P}_0^{g^0}) = \ldots = m(\mathcal{P}_0, \mathcal{P}_0^{g^5}) \tag{1}$$

Computing $m(\mathcal{P}_0, \mathcal{P}_0^{g^k})$ for $k = 0, \ldots, 5$ produces Table 3. Because the values in each row differ (in contrast to the requirements defined by Equation (1)), each row of Table 3 contains the counterexample for the measure used. □

3.2. Variant 2: Inconsistency of the Identity and the Invariance Axiom

Theorem 3. *Let $G = (V, E)$ be a graph with $|V| > 2$ and nontrivial $Aut(G)$. For partition comparison measures $m : P(V) \times P(V) \to \mathbb{R}$, it is impossible to fulfill jointly the identity axiom $m(\mathcal{P}, \mathcal{Q}) = c$, if and only if $\mathcal{P} = \mathcal{Q}$ (e.g., for a distance measure $c = 0$, for a similarity measure $c = 1$, etc.) for all $\mathcal{P}, \mathcal{Q} \in P(V)$ and the axiom of invariance (from Definition 3) $m(\mathcal{P}, \mathcal{Q}) = c, \forall \mathcal{Q} \in \mathcal{P}^{Aut(G)}$.*

Proof.

1. Since $Aut(G)$ is nontrivial, a nontrivial orbit with at least two different partitions, namely $\mathcal{P}$ and $\mathcal{Q}$, exists because $|\mathcal{P}^{Aut(G)}| > 1$. It follows from the invariance axiom that $m(\mathcal{P}, \mathcal{Q}) = c$.
2. The identity axiom implies that it follows from $m(\mathcal{P}, \mathcal{Q}) = c$ that $\mathcal{P} = \mathcal{Q}$.

3. This contradicts the assumption that $\mathcal{P}$ and $\mathcal{Q}$ are different.

□

Table 3. Comparing the modularity maximizing partitions of the cycle graph C_{36} with modularity $Q = \frac{2}{3}$. The six optimal partitions consist of six clusters (see Figure 3). The number of pairs in the same cluster in both partitions is denoted by N_{11}, in different clusters by N_{00} and in the same cluster in one partition, but not in the other, by N_{01} or N_{10}. For the definitions of all partition comparison measures, see Appendix B. To compute this table, the R package `partitionComparison` has been used [23].

Measure	$m(\mathcal{P}_0, \mathcal{P}_0^{g^k})$ with $g = (1\,2\,3\,\ldots\,35\,36)$ for k:					
	0	1	2	3	4	5
	Pair counting measures ($f(N_{11}, N_{00}, N_{01}, N_{10})$; see Tables A1 and A2)					
RI	1.0	0.90476	0.84762	0.82857	0.84762	0.90476
ARI	1.0	0.61111	0.37778	0.3	0.37778	0.61111
H	1.0	0.80952	0.69524	0.65714	0.69524	0.80952
CZ	1.0	0.66667	0.46667	0.4	0.46667	0.66667
K	1.0	0.66667	0.46667	0.4	0.46667	0.66667
MC	1.0	0.33333	−0.06667	−0.2	−0.06667	0.33333
P	1.0	0.61111	0.37778	0.3	0.37778	0.61111
W_I	1.0	0.66667	0.46667	0.4	0.46667	0.66667
W_{II}	1.0	0.66667	0.46667	0.4	0.46667	0.66667
FM	1.0	0.66667	0.46667	0.4	0.46667	0.66667
Γ	1.0	0.61111	0.37778	0.3	0.37778	0.61111
SS1	1.0	0.80556	0.68889	0.65	0.68889	0.80556
B1	1.0	0.91383	0.87084	0.85796	0.87084	0.91383
GL	1.0	0.95	0.91753	0.90625	0.91753	0.95
SS2	1.0	0.33333	0.17949	0.14286	0.17949	0.33333
SS3	1.0	0.62963	0.42519	0.36	0.42519	0.62963
RT	1.0	0.82609	0.73554	0.70732	0.73554	0.82609
GK	1.0	0.94286	0.79937	0.71429	0.79937	0.94286
J	1.0	0.5	0.30435	0.25	0.30435	0.5
RV	1.0	0.61039	0.37662	0.29870	0.37662	0.61039
RR	0.14286	0.09524	0.06667	0.05714	0.06667	0.09524
M	0.0	120.0	192.0	216.0	192.0	120.0
Mi	0.0	0.81650	1.03280	1.09545	1.03280	0.81650
Pe	0.00002	0.00001	0.00001	0.00001	0.00001	0.00001
B2	0.12245	0.07483	0.04626	0.03673	0.04626	0.07483
LI	24.37212	14.89407	9.20724	7.31163	9.20724	14.89407
NLI	1.0	0.61111	0.37778	0.3	0.37778	0.61111
FMG	0.94730	0.61396	0.41396	0.34730	0.41396	0.61396
	Set-based comparison measures (see Table A3)					
LA	1.0	0.83333	0.66667	0.5	0.66667	0.83333
d_{CE}	0.0	0.16667	0.33333	0.5	0.33333	0.16667
D	0.0	12.0	24.0	36.0	24.0	12.0
	Information theory-based measures (see Table A4)					
MI	1.79176	1.34120	1.15525	1.09861	1.15525	1.34120
NMI (max)	1.0	0.74854	0.64475	0.61315	0.64475	0.74854
NMI (min)	1.0	0.74854	0.64475	0.61315	0.64475	0.74854
NMI (Σ)	1.0	0.74854	0.64475	0.61315	0.64475	0.74854
VI	0.0	0.90112	1.27303	1.38629	1.27303	0.90112

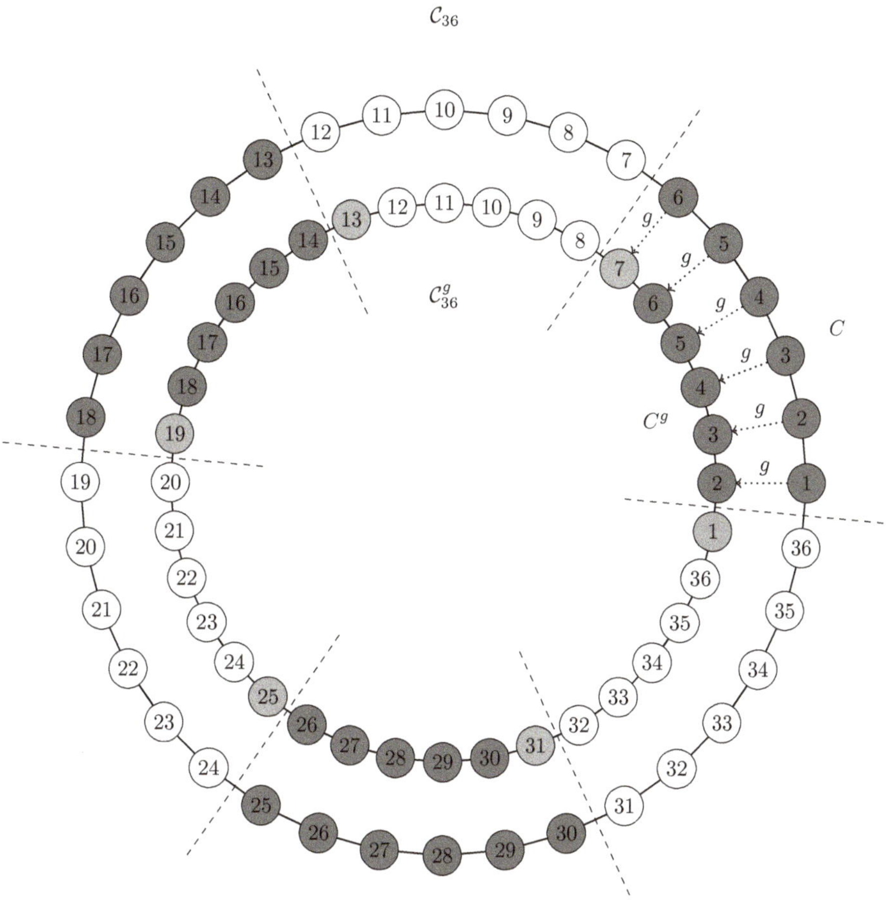

Figure 3. The cycle graph C_{36} (the "outer" cycle) and an initial partition of six clusters (connected nodes of the same color, separated by dashed lines). A single application of $g = (1\,2\,\ldots\,36)$ "rotates" the graph by one node (the "inner" cycle C_{36}^{g}). As a consequence, in each cluster, one node drops out and is added to another cluster: For instance, Node 1 drops out of the "original" cluster $C = \{1, 2, 3, 4, 5, 6\}$, and Node 7 is added, resulting in $C^{g} = \{2, 3, 4, 5, 6, 7\}$. All dropped nodes are shown in light gray.

4. The Construction of Invariant Measures for Finite Permutation Groups

The purpose of this section is to construct invariant counterparts for most of the partition comparison measures in Appendix B. We proceed in two steps:

1. We construct a pseudometric space from the images of the actions of $Aut(G)$ on partitions in $P(V)$ (Definition 1).
2. We extend the metrics for partition comparison by constructing invariant metrics on the pseudo-metric space of partitions.

4.1. The Construction of the Pseudometric Space of Equivalence Classes of Graph Partitions

We use a variant of the idea of Doob's concept of a pseudometric space [24] (p. 5). A metric for a space S (with $s, t, u \in S$) is a function $d : S \times S \to \mathbb{R}^{+}$ for which the following holds:

1. Symmetry: $d(s, t) = d(t, s)$.

2. Identity: $d(s,t) = 0$ if and only if $s = t$.
3. Triangle inequality: $d(s,u) \leq d(s,t) + d(t,u)$.

A pseudometric space (S, d^*) relaxes the identity condition to $d^*(s,s) = 0$. The distance between two elements s_1, s_2 of an equivalence class $[s]$ is defined as $d^*(s_1, s_2) = 0$ by Definition 3.

For graphs, S is the finite set of partitions $P(V)$ and S^* is the partition of $P(V)$ into orbits of $Aut(G)$:

$$S^*(V) = P(V)^{Aut(G)} = \left\{ \mathcal{P}^{Aut(G)} \mid \mathcal{P} \in P(V) \right\}.$$

A partition $\mathcal{P}$ in S corresponds to its orbit $\mathcal{P}^{Aut(G)}$ in S^*. The relations between the spaces used in the following are:

1. (S, d) is a metric space with $S = P(V)$ and with the function $d : P(V) \times P(V) \to \mathbb{R}$.
2. (S^*, d^*) is a metric space with $S^* = P(V)^{Aut(G)} = \{\mathcal{P}^{Aut(G)} \mid \mathcal{P} \in P(V)\}$ and the function d^*: $P(V)^{Aut(G)} \times P(V)^{Aut(G)} \to \mathbb{R}$. We construct three variants of d^* in Section 4.2.
3. (S, d^*) is the pseudometric space with $S = P(V)$ and with the metric d^*. The partitions in S are mapped to arguments of d^* by the transformation $ec : P(V) \to P(V)^{Aut(G)}$, which is defined as $ec(\mathcal{P}) := \mathcal{P}^{Aut(G)}$.

Table 4 illustrates S^* (the space of equivalence classes) of the pseudometric space (S, d^*) of the butterfly graph (shown in Figure 2). S^* is the partition of $P(\{1,2,3,4,5\})$ into 17 equivalence classes. Only the four classes E_1, E_8, E_{12} and E_{17} are stable because they are trivial orbits. The three partitions from Table 2 are contained in the following equivalence classes: $\mathcal{P}_1 \in E_8$, $\mathcal{P}_2 \in E_{14}$, and $\mathcal{P}_3 \in E_{13}$.

Table 4. The equivalence classes of the pseudometric space (S, d^*) of the butterfly graph (see Figure 2). Classes are grouped by their partition type, which is the corresponding integer partition. k is the number of partitions per type; l is the number of clusters the partitions of a type consists of; dia_{1-RI} is the diameter (see Equation (2)) of the equivalence class computed for the distance d_{RI} computed from the Rand Index (RI) by $1 - RI$.

	$\mathcal{P}^{Aut(G)}$	Q	dia_{1-RI}
	Partition type $(1,1,1,1,1)$, $k = 1$, $l = 5$		
E_1	$\{1\},\{2\},\{3\},\{4\},\{5\}$	$-\frac{2}{9}$	0.0
	Partition type $(1,1,1,2)$, $k = 10$, $l = 4$		
E_2	$\{1\},\{2\},\{3\},\{4,5\}$ $\{1,2\},\{3\},\{4\},\{5\}$	$-\frac{1}{9}$	0.2
E_3	$\{1\},\{2\},\{3,4\},\{5\}$ $\{1\},\{2\},\{3,5\},\{4\}$ $\{1\},\{2,3\},\{4\},\{5\}$ $\{1,3\},\{2\},\{4\},\{5\}$	$-\frac{1}{6}$	0.2
E_4	$\{1\},\{2,4\},\{3\},\{5\}$ $\{1\},\{2,5\},\{3\},\{4\}$ $\{1,4\},\{2\},\{3\},\{5\}$ $\{1,5\},\{2\},\{3\},\{4\}$	$-\frac{5}{18}$	0.2
	Partition type $(1,1,3)$ $k = 10$, $l = 3$		
E_5	$\{1\},\{2\},\{3,4,5\}$ $\{4\},\{5\},\{1,2,3\}$	0	0.6
E_6	$\{1\},\{3\},\{2,4,5\}$ $\{3\},\{5\},\{1,2,4\}$ $\{3\},\{4\},\{1,2,5\}$ $\{2\},\{3\},\{1,4,5\}$	$-\frac{2}{9}$	0.4
E_7	$\{1\},\{5\},\{2,3,4\}$ $\{1\},\{4\},\{2,3,5\}$ $\{2\},\{5\},\{1,3,4\}$ $\{2\},\{4\},\{1,3,5\}$	$-\frac{1}{6}$	0.6
	Partition type $(1,2,2)$, $k = 15$, $l = 3$		
E_8	$\{3\},\{1,2\},\{4,5\}$	0	0.0
E_9	$\{3\},\{1,4\},\{2,5\}$ $\{3\},\{1,5\},\{2,4\}$	$-\frac{1}{3}$	0.4
E_{10}	$\{1\},\{2,3\},\{4,5\}$ $\{5\},\{1,2\},\{3,4\}$ $\{4\},\{1,2\},\{3,5\}$ $\{2\},\{1,3\},\{4,5\}$	$-\frac{1}{18}$	0.4
E_{11}	$\{1\},\{2,4\},\{3,5\}$ $\{1\},\{2,5\},\{3,4\}$ $\{5\},\{1,3\},\{2,4\}$ $\{4\},\{1,3\},\{2,5\}$ $\{2\},\{1,4\},\{3,5\}$ $\{5\},\{1,4\},\{2,3\}$ $\{2\},\{1,5\},\{3,4\}$ $\{4\},\{1,5\},\{2,3\}$	$-\frac{2}{9}$	0.4

Table 4. *Cont.*

	$\mathcal{P}^{Aut(G)}$			Q	dia_{1-RI}
	Partition type $(1,4)$, $k=5$, $l=2$				
E_{12}	$\{1,2,4,5\},\{3\}$			$-\frac{2}{9}$	0.0
E_{13}	$\{2,3,4,5\},\{1\}$	$\{1,2,3,4\},\{5\}$	$\{1,2,3,5\},\{4\}$	$-\frac{1}{18}$	0.6
	$\{1,3,4,5\},\{2\}$				
	Partition type $(2,3)$, $k=10$, $l=2$				
E_{14}	$\{1,2\},\{3,4,5\}$	$\{4,5\},\{1,2,3\}$		$\frac{1}{9}$	0.4
E_{15}	$\{3,5\},\{1,2,4\}$	$\{3,4\},\{1,2,5\}$	$\{1,3\},\{2,4,5\}$	$-\frac{1}{6}$	0.6
	$\{2,3\},\{1,4,5\}$				
E_{16}	$\{2,5\},\{1,3,4\}$	$\{2,4\},\{1,3,5\}$	$\{1,4\},\{2,3,5\}$	$-\frac{2}{9}$	0.6
	$\{1,5\},\{2,3,4\}$				
	Partition type (5), $k=1$, $l=1$				
E_{17}	$\{1,2,3,4,5\}$			0	0.0

4.2. *The Construction of Left-Invariant and Additive Measures on the Pseudometric Space of Equivalence Classes of Graph Partitions*

In the following, we consider only partition comparison measures, which are distance functions of a metric space. Note that a normalized similarity measure s can be transformed into a distance by the transformation $d = 1 - s$.

In a pseudometric space (S, d^*), we measure the distance $d^*(\mathcal{P}, \mathcal{Q})$ between equivalence classes (which are sets) of partitions instead of the distance $d(\mathcal{P}, \mathcal{Q})$ between partitions. The partitions $\mathcal{P}$ and $\mathcal{Q}$ are formal arguments of d^*, which are expanded to equivalence classes by $\mathcal{P}^{Aut(G)}$ and $\mathcal{Q}^{Aut(G)}$. The standard construction of a distance measure between sets has been developed for the point set topology and is due to Felix Hausdorff [25] (p. 166) and Kazimierz Kuratowski [26] (p. 209). For finite sets, it requires the computation of the distances for all pairs of the direct product of the two sets. Since for finite permutation groups, we deal with distances between two finite sets of partitions, we use the following definitions for the lower and upper measures, respectively. Both definitions have the form of an optimization problem:

$$d_L^*(\mathcal{P}, \mathcal{Q}) = \min_{\substack{\tilde{\mathcal{P}} \in \mathcal{P}^{Aut(G)}, \\ \tilde{\mathcal{Q}} \in \mathcal{Q}^{Aut(G)}}} d(\tilde{\mathcal{P}}, \tilde{\mathcal{Q}})$$

and:

$$d_U^*(\mathcal{P}, \mathcal{Q}) = \begin{cases} 0 & \text{if } \mathcal{P}^{Aut(G)} = \mathcal{Q}^{Aut(G)} \\ \max_{\substack{\tilde{\mathcal{P}} \in \mathcal{P}^{Aut(G)}, \\ \tilde{\mathcal{Q}} \in \mathcal{Q}^{Aut(G)}}} d(\tilde{\mathcal{P}}, \tilde{\mathcal{Q}}) & \text{else} \end{cases}$$

The diameter of a finite equivalence class of partitions is defined by

$$\text{dia}(\mathcal{P}) = \max_{\substack{\tilde{\mathcal{P}} \in \mathcal{P}^{Aut(G)}, \\ \tilde{\mathcal{Q}} \in \mathcal{P}^{Aut(G)}}} d(\tilde{\mathcal{P}}, \tilde{\mathcal{Q}}). \tag{2}$$

The third option of defining a distance between two finite equivalence classes of partitions of taking the average distance is due to John von Neumann [27]:

$$d_{av}^*(\mathcal{P}, \mathcal{Q}) = \begin{cases} 0 & \text{if } \mathcal{P}^{Aut(G)} = \mathcal{Q}^{Aut(G)} \\ \frac{1}{|\mathcal{P}^{Aut(G)}| \cdot |\mathcal{Q}^{Aut(G)}|} \sum_{\substack{\tilde{\mathcal{P}} \in \mathcal{P}^{Aut(G)}, \\ \tilde{\mathcal{Q}} \in \mathcal{Q}^{Aut(G)}}} d(\tilde{\mathcal{P}}, \tilde{\mathcal{Q}}) & \text{else} \end{cases}$$

Note that the definitions for d_L^*, d_U^* and d_{av}^* require the computation of the minimal, maximal and average distance of all pairs of the direct product $\mathcal{P}^{Aut(G)} \times \mathcal{Q}^{Aut(G)}$. The computational complexity of this is quadratic in the size of the larger equivalence class.

Posed as a measurement problem, we can instead fix one partition in one of the orbits and measure the minimal, maximal and average distance between all pairs of either the direct product of $\{\mathcal{P}\} \times \mathcal{Q}^{Aut(G)}$ or $\{\mathcal{Q}\} \times \mathcal{P}^{Aut(G)}$. The complexity of this is linear in the size of the smaller equivalence class.

Theorems 4 and 5 and their proofs are based on these observations. They are the basis for the development of algorithms for the computation of invariant partition comparison measures of a computational complexity of at most linear order and often of constant order.

Theorem 4. *For all $\mathcal{P}^{Aut(G)} \neq \mathcal{Q}^{Aut(G)}$, the following equations hold:*

$$\begin{aligned} d_L^*(\mathcal{P},\mathcal{Q}) &= \min_{\substack{\tilde{\mathcal{P}}\in\mathcal{P}^{Aut(G)},\\ \tilde{\mathcal{Q}}\in\mathcal{Q}^{Aut(G)}}} d(\tilde{\mathcal{P}},\tilde{\mathcal{Q}}) = \min_{g,h\in Aut(G)} d(\mathcal{P}^h,\mathcal{Q}^g) \\ &= \min_{\tilde{\mathcal{Q}}\in\mathcal{Q}^{Aut(G)}} d(\mathcal{P},\tilde{\mathcal{Q}}) = \min_{g\in Aut(G)} d(\mathcal{P},\mathcal{Q}^g) \\ &= \min_{\tilde{\mathcal{P}}\in\mathcal{P}^{Aut(G)}} d(\tilde{\mathcal{P}},\mathcal{Q}) = \min_{h\in Aut(G)} d(\mathcal{P}^h,\mathcal{Q}) \end{aligned}$$

For $\mathcal{P}^{Aut(G)} \neq \mathcal{Q}^{Aut(G)}$:

$$\begin{aligned} d_U^*(\mathcal{P},\mathcal{Q}) &= \max_{\substack{\tilde{\mathcal{P}}\in\mathcal{P}^{Aut(G)},\\ \tilde{\mathcal{Q}}\in\mathcal{P}^{Aut(G)}}} d(\tilde{\mathcal{P}},\tilde{\mathcal{Q}}) = \max_{g,h\in Aut(G)} d(\mathcal{P}^h,\mathcal{Q}^g) \\ &= \max_{\tilde{\mathcal{Q}}\in\mathcal{Q}^{Aut(G)}} d(\mathcal{P},\tilde{\mathcal{Q}}) = \max_{g\in Aut(G)} d(\mathcal{P},\mathcal{Q}^g) \\ &= \max_{\tilde{\mathcal{P}}\in\mathcal{P}^{Aut(G)}} d(\tilde{\mathcal{P}},\mathcal{Q}) = \max_{h\in Aut(G)} d(\mathcal{P}^h,\mathcal{Q}) \end{aligned}$$

Proof. Let $g,h,f \in Aut(G)$, $\tilde{\mathcal{P}} \in \mathcal{P}^{Aut(G)}$ and $\tilde{\mathcal{Q}} \in \mathcal{Q}^{Aut(G)}$, that is $\tilde{\mathcal{P}} = \mathcal{P}^h$ and $\tilde{\mathcal{Q}} = \mathcal{Q}^g$. Then, since the orbits of both partitions are generated by $Aut(G)$, the following identities between distances hold:

$$\begin{aligned} d(\mathcal{P},\tilde{\mathcal{Q}}) &= d(\mathcal{P},\mathcal{Q}^g) = d(\mathcal{P}^{g^{-1}},\mathcal{Q}), \\ d(\tilde{\mathcal{P}},\mathcal{Q}) &= d(\mathcal{P}^h,\mathcal{Q}) = d(\mathcal{P},\mathcal{Q}^{h^{-1}}) \end{aligned}$$

as well as:

$$d(\tilde{\mathcal{P}},\tilde{\mathcal{Q}}) = d(\mathcal{P}^h,\mathcal{Q}^g) = d(\mathcal{P}^{hg^{-1}},\mathcal{Q}),$$

and:

$$d(\tilde{\mathcal{P}},\tilde{\mathcal{Q}}) = d(\mathcal{P}^h,\mathcal{Q}^g) = d(\mathcal{P},\mathcal{Q}^{gh^{-1}}).$$

Furthermore, let $f = gh^{-1}$.

1. For d_L^*, we have:

$$\begin{aligned} \min_{\tilde{\mathcal{Q}}\in\mathcal{Q}^{Aut(G)}} d(\mathcal{P},\tilde{\mathcal{Q}}) &= \min_{g\in Aut(G)} d(\mathcal{P},\mathcal{Q}^g) \\ &= \min_{g^{-1}\in Aut(G)} d(\mathcal{P}^{g^{-1}},\mathcal{Q}) = \min_{\tilde{\mathcal{P}}\in\mathcal{P}^{Aut(G)}} d(\tilde{\mathcal{P}},\mathcal{Q}) \end{aligned}$$

by switching the reference systems. In the next sequence of equations, we establish that taking the minimum over all reference systems is equivalent to finding the minimum for one arbitrarily fixed reference system.

$$\min_{\substack{\tilde{\mathcal{P}}\in\mathcal{P}^{Aut(G)},\\ \tilde{\mathcal{Q}}\in\mathcal{Q}^{Aut(G)}}} d(\tilde{\mathcal{P}},\tilde{\mathcal{Q}}) = \min_{g,h\in Aut(G)} d(\mathcal{P}^h,\mathcal{Q}^g) = \min_{g,h\in Aut(G)} d(\mathcal{P},\mathcal{Q}^{gh^{-1}})$$

$$= \min_{f\in Aut(G)} d(\mathcal{P},\mathcal{Q}^f) = \min_{\tilde{\mathcal{Q}}\in\mathcal{Q}^{Aut(G)}} d(\mathcal{P},\tilde{\mathcal{Q}})$$

2. For the proof of d_U^* for $\mathcal{P}^{Aut(G)} \neq \mathcal{Q}^{Aut(G)}$ we substitute max for min in the proof of d_L^*.

□

Theorem 5. *For all $\mathcal{P}^{Aut(G)} \neq \mathcal{Q}^{Aut(G)}$, the following equations hold:*

$$d_{av}^*(\mathcal{P},\mathcal{Q}) = \frac{1}{|\mathcal{P}^{Aut(G)}|\cdot|\mathcal{Q}^{Aut(G)}|}\sum_{\substack{\tilde{\mathcal{P}}\in\mathcal{P}^{Aut(G)},\\ \tilde{\mathcal{Q}}\in\mathcal{Q}^{Aut(G)}}} d(\tilde{\mathcal{P}},\tilde{\mathcal{Q}}) \quad (3)$$

$$= \frac{1}{|Aut(G)|^2}\sum_{h,g\in Aut(G)} d(\mathcal{P}^h,\mathcal{Q}^g) \quad (4)$$

$$= \frac{1}{|\mathcal{P}^{Aut(G)}|}\sum_{\tilde{\mathcal{P}}\in\mathcal{P}^{Aut(G)}} d(\tilde{\mathcal{P}},\mathcal{Q}) \quad (5)$$

$$= \frac{1}{|Aut(G)|}\sum_{h\in Aut(G)} d(\mathcal{P}^h,\mathcal{Q}) \quad (6)$$

$$= \frac{1}{|\mathcal{Q}^{Aut(G)}|}\sum_{\tilde{\mathcal{Q}}\in\mathcal{Q}^{Aut(G)}} d(\mathcal{P},\tilde{\mathcal{Q}}) \quad (7)$$

$$= \frac{1}{|Aut(G)|}\sum_{g\in Aut(G)} d(\mathcal{P},\mathcal{Q}^g) \quad (8)$$

Proof. For the proof of the equality of the identities of d_{av}^*, we use the property of an average of n observations $x_{i,j}$ with k identical groups of size m with $i \in 1,\ldots,k$, $j \in 1,\ldots,m$:

$$\frac{1}{n}\sum_{i=1}^{k}\sum_{j=1}^{m} x_{i,j} = \frac{k}{km}\sum_{j=1}^{m} x_{1,j} = \frac{1}{m}\sum_{j=1}^{m} x_{1,j} \quad (9)$$

The computation of an average over the group equals the result of the computation of an average over the orbit, because the orbit stabilizer Theorem 1 implies that each element of the orbit is generated $|Aut(G)_{\mathcal{P}}|$ times, and this means that we average $|Aut(G)_{\mathcal{P}}|$ groups of identical values and that Equation (9) applies. This establishes the equality of Expressions (3) and (4), as well as of Expressions (5) and (6) and of Expressions (7) and (8), respectively.

The two decompositions of the direct product $Aut(G) \times Aut(G)$ establish the equality of Expressions (4) and (6), as well as of Expressions (4) and (8). □

Note that these proofs also show that $d_L^*(\mathcal{P},\mathcal{Q})$, $d_U^*(\mathcal{P},\mathcal{Q})$ and $d_{av}^*(\mathcal{P},\mathcal{Q})$ are invariant. Next, we prove that the three measures $d_L^*(\mathcal{P},\mathcal{Q})$, $d_U^*(\mathcal{P},\mathcal{Q})$ and $d_{av}^*(\mathcal{P},\mathcal{Q})$ are invariant measures.

Theorem 6. *The lower pseudometric space (S, d_L^*) has the following properties:*

1. *Identity: $d_L^*(\mathcal{P},\mathcal{Q}) = 0$, if $\mathcal{P}^{Aut(G)} = \mathcal{Q}^{Aut(G)}$.*
2. *Invariance: $d_L^*(\mathcal{P},\mathcal{Q}) = d_L^*(\tilde{\mathcal{P}},\tilde{\mathcal{Q}})$, for all $\mathcal{P},\mathcal{Q} \in P(V)$ and $\tilde{\mathcal{P}} \in \mathcal{P}^{Aut(G)}$, $\tilde{\mathcal{Q}} \in \mathcal{Q}^{Aut(G)}$.*
3. *Symmetry: $d_L^*(\mathcal{P},\mathcal{Q}) = d_L^*(\mathcal{Q},\mathcal{P})$.*
4. *Triangle inequality: $d_L^*(\mathcal{P},\mathcal{R}) \leq d_L^*(\mathcal{P},\mathcal{Q}) + d_L^*(\mathcal{Q},\mathcal{R})$*

These properties also hold for the upper pseudometric space (S, d_U^*) and the average pseudometric space (S, d_{av}^*).

Proof.

1. Identity holds because of the definition of the distance d^* between two elements in an equivalence class of the pseudometric space (S, d^*).
2. Invariance of $d^*_L(\mathcal{P},\mathcal{Q})$, $d^*_U(\mathcal{P},\mathcal{Q})$ and $d^*_{av}(\mathcal{P},\mathcal{Q})$ is proven by Theorems 4 and 5.
3. Symmetry holds, because d is symmetric, and min, max and the average do not depend on the order of their respective arguments.
4. To proof the triangular inequality, we make use of Theorems 4 and 5 and of the fact that d is a metric for which the triangular inequality holds:

   (a) For d^*_L follows:

$$\begin{aligned}
d^*_L(\mathcal{P},\mathcal{R}) &= \min_{\substack{\tilde{\mathcal{P}}\in\mathcal{P}^{Aut(G)},\\ \tilde{\mathcal{R}}\in\mathcal{R}^{Aut(G)}}} d(\tilde{\mathcal{P}},\tilde{\mathcal{R}}) \\
&\leq \min_{\substack{\tilde{\mathcal{P}}\in\mathcal{P}^{Aut(G)},\\ \tilde{\mathcal{Q}}\in\mathcal{Q}^{Aut(G)},\\ \tilde{\mathcal{R}}\in\mathcal{R}^{Aut(G)}}} \left(d(\tilde{\mathcal{P}},\tilde{\mathcal{Q}}) + d(\tilde{\mathcal{Q}},\tilde{\mathcal{R}})\right) \\
&= \min_{\substack{\tilde{\mathcal{P}}\in\mathcal{P}^{Aut(G)},\\ \tilde{\mathcal{R}}\in\mathcal{R}^{Aut(G)}}} \left(d(\tilde{\mathcal{P}},\mathcal{Q}) + d(\mathcal{Q},\tilde{\mathcal{R}})\right) \\
&= \min_{\tilde{\mathcal{P}}\in\mathcal{P}^{Aut(G)}} d(\tilde{\mathcal{P}},\mathcal{Q}) + \min_{\tilde{\mathcal{R}}\in\mathcal{R}^{Aut(G)}} d(\mathcal{Q},\tilde{\mathcal{R}}) \\
&= d^*_L(\mathcal{P},\mathcal{Q}) + d^*_L(\mathcal{Q},\mathcal{R})
\end{aligned}$$

   (b) For the proof of the triangular inequality for d^*_U, we substitute max for min and d_U for d_L in the proof of the triangular inequality for d^*_L.

   (c) For d^*_{av}, it follows:

$$\begin{aligned}
d^*_{av}(\mathcal{P},\mathcal{R}) &= \frac{1}{|\mathcal{P}^{Aut(G)}|\cdot|\mathcal{R}^{Aut(G)}|}\sum_{\tilde{\mathcal{P}}\in\mathcal{P}^{Aut(G)}}\sum_{\tilde{\mathcal{R}}\in\mathcal{R}^{Aut(G)}} d(\tilde{\mathcal{P}},\tilde{\mathcal{R}}) \\
&\leq \frac{1}{|\mathcal{P}^{Aut(G)}|\cdot|\mathcal{R}^{Aut(G)}|}\sum_{\tilde{\mathcal{P}}}\sum_{\tilde{\mathcal{R}}}[d(\tilde{\mathcal{P}},\mathcal{Q}) + d(\mathcal{Q},\tilde{\mathcal{R}})] \\
&= \frac{1}{|\mathcal{P}^{Aut(G)}|\cdot|\mathcal{R}^{Aut(G)}|}\sum_{\tilde{\mathcal{P}}}\sum_{\tilde{\mathcal{R}}} d(\tilde{\mathcal{P}},\mathcal{Q}) + \frac{1}{|\mathcal{P}^{Aut(G)}|\cdot|\mathcal{R}^{Aut(G)}|}\sum_{\tilde{\mathcal{P}}}\sum_{\tilde{\mathcal{R}}} d(\mathcal{Q},\tilde{\mathcal{R}}) \\
&= \frac{1}{|\mathcal{R}^{Aut(G)}|}\sum_{\tilde{\mathcal{R}}} d^*_{av}(\mathcal{P},\mathcal{Q}) + \frac{1}{|\mathcal{P}^{Aut(G)}|}\sum_{\tilde{\mathcal{P}}} d^*_{av}(\mathcal{Q},\mathcal{R}) \\
&= d^*_{av}(\mathcal{P},\mathcal{Q}) + d(\tilde{\mathcal{Q}},\mathcal{R})
\end{aligned}$$

□

5. Decomposition of Partition Comparison Measures

In this section, we assess the structural (dis)similarity between two partitions and the effect of the group actions by combining a partition comparison measure and its invariant counterpart defined in Section 4. The distances $d(\mathcal{P},\mathcal{Q})$, $d^*_L(\mathcal{P},\mathcal{Q})$, $d^*_U(\mathcal{P},\mathcal{Q})$ and $d^*_{av}(\mathcal{P},\mathcal{Q})$ allow the decomposition of a partition comparison measure (transformed into a distance) into a structural component d_{struc} and the effect $d_{Aut(G)}$ of the automorphism group $Aut(G)$:

$$
\begin{aligned}
d(\mathcal{P},\mathcal{Q}) &= \underbrace{d_L^*(\mathcal{P},\mathcal{Q})}_{d_{struc}} + \underbrace{(d(\mathcal{P},\mathcal{Q}) - d_L^*(\mathcal{P},\mathcal{Q}))}_{d_{Aut(G)}} \\
&= \underbrace{d_U^*(\mathcal{P},\mathcal{Q})}_{d_{struc}} - \underbrace{(d_U^*(\mathcal{P},\mathcal{Q}) - d(\mathcal{P},\mathcal{Q}))}_{d_{Aut(G)}} \\
&= \underbrace{d_{av}^*(\mathcal{P},\mathcal{Q})}_{d_{struc}} - \underbrace{(d_{av}^*(\mathcal{P},\mathcal{Q}) - d(\mathcal{P},\mathcal{Q}))}_{d_{Aut(G)}}
\end{aligned}
$$

$\text{dia}(\mathcal{P})$ measures the effect of the automorphism group $Aut(G)$ on the equivalence class $\mathcal{P}^{Aut(G)}$ (see the last column of Table 4). $e_{\max}^{Aut(G)}$ is an upper bound of the automorphism effect on the distance of two partitions $\mathcal{P}$ and $\mathcal{Q}$:

$$e_{\max}^{Aut(G)} = \min(\text{dia}(\mathcal{P}), \text{dia}(\mathcal{Q})).$$

This follows from Theorem 4. Note that $e_{\max}^{Aut(G)} \geq d_U^* - d_L^*$, as Case 1 in Table 5 shows.

Table 5. Measure decomposition for partitions of the butterfly graph for the Rand distance $d_{RI} = 1 - RI$.

Case	$\mathcal{P}$	$\mathcal{Q}$	d_{RI}	d^*	d_{struc}	$d_{Aut(G)}$
	$\{\{1,2,3,4\}\{5\}\}$	$\{\{4\}\{5\}\{1,2,3\}\}$	0.3	d_L^*	0.3	0.0
1	$\in E_{13}$	$\in E_5$	0.3	d_{av}^*	0.5	-0.2
	$\text{dia}(E_{13}) = 0.6$	$\text{dia}(E_5) = 0.6$	0.3	d_U^*	0.7	-0.4
	$\{\{2,4\}\{1,3,5\}\}$	$\{\{3\}\{1,4\}\{2,5\}\}$	0.6	d_L^*	0.2	0.4
2	$\in E_{16}$	$\in E_9$	0.6	d_{av}^*	0.4	0.2
	$\text{dia}(E_{16}) = 0.6$	$\text{dia}(E_9) = 0.4$	0.6	d_U^*	0.6	0.0
	$\{\{1\}\{2,5\}\{3,4\}\}$	$\{\{1\}\{2,3\}\{4\}\{5\}\}$	0.3	d_L^*	0.1	0.2
3	$\in E_{11}$	$\in E_3$	0.3	d_{av}^*	0.25	0.05
	$\text{dia}(E_{11}) = 0.4$	$\text{dia}(E_3) = 0.2$	0.3	d_U^*	0.3	0.0
	$\{\{3\}\{1,2\}\{4,5\}\}$	$\{\{1\}\{2,3\}\{4,5\}\}$	0.3	d_L^*	0.3	0.0
4	$\in E_8$	$\in E_{10}$	0.3	d_{av}^*	0.3	0.0
	$(\text{dia}(E_8) = 0$, stable)	$\text{dia}(E_{10}) = 0.4$	0.3	d_U^*	0.3	0.0

In Table 5, we show a few examples of measure decomposition for partitions of the butterfly graph for the Rand distance d_{RI}:

1. In Case 1, we compare two partitions from nontrivial equivalence classes: the difference of 0.4 between d_U^* and d_L^* indicates that the potential maximal automorphism effect is larger than the lower measure. In addition, it is also smaller (by 0.2) than the automorphism effect in each of the equivalence classes. That $d_{Aut(G)}$ is zero for the lower measure implies that the pair $(\mathcal{P},\mathcal{Q})$ is a pair with the minimal distance between the equivalence classes. The fact that $d_{av}^* = 0.5$ is the mid-point between the lower and upper measures indicates a symmetric distribution of the distances between the equivalence classes.
2. That $d_{Aut(G)}$ is zero for the upper measure in Case 2 means that we have found a pair with the maximal distance between the equivalence classes.
3. In Case 3, we have also found a pair with maximal distance between the equivalence classes. However, the maximal potential automorphism effect is smaller than for Cases 1 and 2. In addition, the distribution of distances between the equivalence classes is asymmetric.
4. Case 4 shows the comparison of a partition from a trivial with a partition from a non-trivial equivalence class. Note, that in this case, all three invariant measures, as well as d_{RI} coincide and that no automorphism effect exists.

A different approach to measure the potential instability in clustering a graph G is the computation of the Kolmogorov–Sinai entropy of the finite permutation group $Aut(G)$ acting on the graph [28].

Note, that the Kolmogorov–Sinai entropy of a finite permutation group is a measure of the uncertainty of the automorphism group. It cannot be used as a measure to compare two graph partitions.

6. Invariant Measures for the Karate Graph

In this section, we illustrate the use of invariant measures for the three partitions $\mathcal{P}_O$, $\mathcal{P}_1$ and $\mathcal{P}_2$ of the Karate graph K [29], which is shown in Figure 4. $Aut(K)$ is of order 480, and it consists of the three subgroups $G_1 = Sym(\Omega_1)$ with $\Omega_1 = \{15, 16, 19, 21, 23\}$, $G_2 = Sym(\Omega_2)$ with $\Omega_2 = \{18, 22\}$ and $G_3 = \{(), (5\,11), (6\,7)\}$. In addition to the modularity optimal partition $\mathcal{P}_O$ (with its clusters separated by longer and dashed lines in Figure 4), we use the partitions $\mathcal{P}_1$ and $\mathcal{P}_2$:

$$\begin{aligned}\mathcal{P}_1 = \{&\{5,6,7,\mathbf{19},\mathbf{21}\}, \{1,2,3,4,8,12,13,14,18,20,22\},\\ &\{9,10,\mathit{11},\mathbf{15},\mathbf{16},17,\mathbf{23},27,30,31,33,34\}, \{24,25,26,28,29,32\}\}\\ \mathcal{P}_2 = \{&\{5,6,7,8,12,\mathbf{19},\mathbf{21}\}, \{1,2,3,4,13,14,18,20,22\},\\ &\{9,10,\mathit{11},\mathbf{15},\mathbf{16},17,\mathbf{23},27,30,31,33,34\}, \{24,25,26,28,29,32\}\}\end{aligned}$$

Both partitions are affected by the orbits $\{\mathbf{15}, \mathbf{16}, \mathbf{19}, \mathbf{21}, \mathbf{23}\}$ and $\{5, \mathit{11}\}$, each overlapping two clusters. The dissimilarity to $\mathcal{P}_O$ is larger for $\mathcal{P}_2$, which is reflected in Tables 6 and 7.

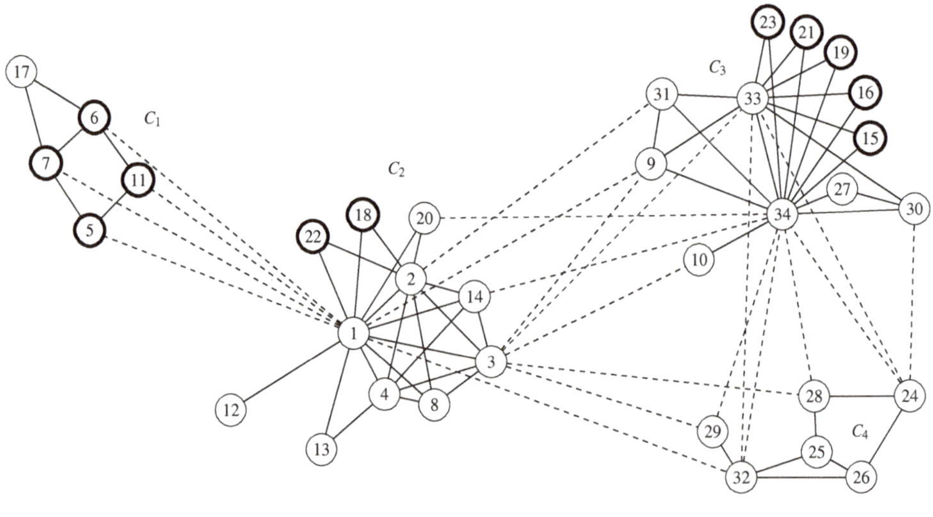

Figure 4. Zachary's Karate graph K with the vertices of the orbits of the three subgroups of $Aut(K)$ in bold and the clusters of P_O separated by dashed edges.

For the optimal partition $\mathcal{P}_O$ of type $(5, 6, 11, 12)$, the upper bound of the size of the equivalence class is 480 [30] (p. 112). The actual size of the equivalence class of $\mathcal{P}_O$ is one, which means the optimal solution is not affected by $Aut(K)$. Partition $\mathcal{P}_1$, which is of the same type as $\mathcal{P}_O$, also has an upper bound of 480 for its equivalence class. The actual size of the equivalence classes of both $\mathcal{P}_1$ and $\mathcal{P}_2$ is 20. Note that the actual size of the equivalence classes that drive the complexity of computing invariant measures is in our example far below the upper bound. Table 6 shows the diameters of the equivalence classes of the partitions.

Table 6. Diameter (computed using d_{RI}), orbit size and stability of partitions $\mathcal{P}_O$, $\mathcal{P}_1$ and $\mathcal{P}_2$.

$\mathcal{X}$	$\mathcal{P}_O$	$\mathcal{P}_1$	$\mathcal{P}_2$
dia($\mathcal{X}$)	0.0000	0.1176	0.1390
$\lvert\mathcal{X}^{Aut(G)}\rvert$	1	20	20
$\mathcal{X}$ stable?	yes	no	no

Table 7 illustrates the decomposition into structural effects and automorphism effects for the three partitions of the Karate graph. We see that for the comparison of a stable partition ($\mathcal{P}_O$) with one of the unstable partitions, the classic partition comparison measures are sufficient. However, when comparing the two unstable partitions $\mathcal{P}_1$ and $\mathcal{P}_2$, the structural effect (0.0499) is dominated by the maximal automorphism effect (0.1176). Furthermore, we note that the distribution of values over the orbit of the automorphism group is asymmetric (by looking at d_L^*, d_U^* and d_{av}^*).

Table 7. Invariant measures and automorphism effects for the Karate graph. The R package `partitionComparison` has been used for the computations [23].

Measure $d = d_{RI}$	$m(\mathcal{P}_O, \mathcal{P}_1)$	$m(\mathcal{P}_O, \mathcal{P}_2)$	$m(\mathcal{P}_1, \mathcal{P}_2)$
d	0.0927	0.1426	0.0499
$d_L^* + d_{Aut(G)}$	0.0927	0.1426	$0.0499 + 0.0000$
$d_U^* - d_{Aut(G)}$	0.0927	0.1426	$0.1676 - 0.1176$
$d_{av}^* - d_{Aut(G)}$	0.0927	0.1426	$0.1280 - 0.0781$
$e_{\max}^{Aut(K)}$	0.0000	0.0000	0.1176

The analysis of the effects of the automorphism group of the Karate network showed that the automorphism group does not affect the stability of the optimal partition. However, the first results show that the situation is different for other networks like the Internet AS graph with 40,164 nodes and 85,123 edges (see Rossi et al. [31], and the data of of the graph `tech-internet-as` are from Rossi and Ahmed [32]): for this graph, several locally optimal solutions with a modularity value above 0.694 exist, all of which are unstable. Further analysis of the structural properties of the solution landscape of this graph is work in progress.

7. Discussion, Conclusions and Outlook

In this contribution, we study the effects of graph automorphisms on partition comparison measures. Our main results are:

1. A formal definition of partition stability, namely $\mathcal{P}$ is stable iff $|\mathcal{P}^{Aut(G)}| = 1$.
2. A proof of the non-invariance of all partition comparison measures if the automorphism group is nontrivial ($|Aut(G)| > 1$).
3. The construction of a pseudometric space of equivalence classes of graph partitions for three classes of invariant measures concerning finite permutation groups of graph automorphisms.
4. The proof that the measures are invariant and that for these measures (after the transformation to a distance), the axioms of a metric space hold.
5. The space of partitions is equipped with a metric (the original partition comparison measure) and a pseudometric (the invariant partition comparison measure).
6. The decomposition of the value of a partition comparison measure into a structural part and a remainder that measures the effect of group actions.

Our definitions of invariant measures have the advantage that any existing partition comparison measure (as long as it is a distance or can be transformed into one) can still be used for the task. Moreover, the decomposition of measures restores the primary purpose of the existing comparison measures, which is to quantify structural difference. However, the construction of these measures leads directly to the classic graph isomorphism problem, whose complexity—despite considerable efforts and hopes to the contrary [33]—is still an open theoretical problem [34,35]. However, from a pragmatic point of view, today, quite efficient and practically usable algorithms exist to tackle the graph isomorphism problem [34]. In addition, for very large and sparse graphs, algorithms for finding generators of the automorphism group exist [11]. Therefore, this dependence on a computationally hard problem in general is not an actual disadvantage and allows one to implement the presented measure decomposition. The efficient implementation of algorithms for the decomposition of graph partition comparison measures is left for further research.

Another constraint is that we have investigated the effects of automorphisms on partition comparison measures in the setting of graph clustering only. The reason for this restriction is that the automorphism group of the graph is already defined by the graph itself and, therefore, is completely contained in the graph data. For arbitrary datasets, the information about the automorphism group is usually not contained in the data, but must be inferred from background theories. However, provided we know the automorphism group, our results on the decomposition of the measures generalize to arbitrary cluster problems.

All in all, this means that this article provides two major assets: first, it provides a theoretic framework that is independent of the preferred measure and the data. Second, we provide insights into a source of possible partition instability that has not yet been discussed in the literature. The downsides (symmetry group must be known and graph clustering only) are in our opinion not too severe, as we discussed above. Therefore, we think that our study indicates that a better understanding of the principle of symmetry is important for future research in data analysis.

Supplementary Materials: The R package `partitionComparison` by the authors of this article that implements the different partition comparison measures is available at https://cran.r-project.org/package=partitionComparison.

Author Contributions: The F.B. implemented the R package mentioned in the Supplementary Materials and conducted the non-invariance proof by counterexample. The more general proof of the non-existence of invariant measures, as well as the idea of creating a pseudometric to repair a measure's deficiency of not being invariant is mainly due to the A.G.-S. Both authors contributed equally to writing the article and revising it multiple times.

Funding: We acknowledge support by Deutsche Forschungsgemeinschaft and the Open Access Publishing Fund of the Karlsruhe Institute of Technology.

Acknowledgments: We thank Andreas Geyer-Schulz (Institute of Analysis, Faculty of Mathematics, KIT, Karlsruhe) for repeated corrections and suggestions for improvement in the proofs.

Conflicts of Interest: The authors declare no conflict of interest.

Appendix A. Modularity

Newman's and Girvan's modularity [14] is defined as:

$$Q = \sum_i \left(e_{ii} - a_i^2\right)$$

with the edge fractions:

$$e_{ij} = \frac{|\{\{u,v\} \in E | u \in C_i \wedge v \in C_j\}|}{2|E|}, \; i \neq j,$$

and the cluster density:

$$e_{ii} = \frac{|\{\{u,v\} \in E | u,v \in C_i\}|}{|E|}.$$

We have to distinguish e_{ij} and e_{ii} because of the set-based definition E. e_{ij} is the fraction of edges from cluster C_i to cluster C_j and e_{ji}, vice versa. Therefore, the edges are counted twice, and thus, the fraction has to be weighted with $\frac{1}{2}$. The second part of Q is the marginal distribution:

$$a_i^2 = \left(\sum_j e_{ij}\right)^2.$$

High values of Q indicate good partitions. The range of Q is $[-\frac{1}{2}, 1)$. Even if the modularity has some problems by design (e.g., the resolution limit [36], unbalanced cluster sizes [37], multiple equivalent, but unstable solutions generated by automorphisms [38]), maximization of Q is the de facto standard formal optimization criterion for graph clustering algorithms.

Appendix B. Measures for Comparing Partitions

We classify the measures that are used in the literature to compare object partitions as three categories [39]:

1. Pair-counting measures.
2. Set-based comparison measures.
3. Information theory based measures.

All these measures come from a general context and, therefore, may be used to compare any object partitions, not only graph partitions. The flip side of the coin is that they do not consider any adjacency information from the underlying graph at all.

The column Abbr. of Tables A1–A4 denotes the Abbreviations used throughout this paper; the column $\mathcal{P} = \mathcal{P}$ denotes the value resulting when identical partitions are compared (max stands for some maximum value depending on the partition).

Appendix B.1. Pair-Counting Measures

All the measures within the first class are based on the four coefficients N_{xy} that count pairs of objects (nodes in our context). Let $\mathcal{P}, \mathcal{Q}$ be partitions of the node set V of a graph G. C and C' denote clusters (subsets of vertices $C, C' \subseteq V$). The coefficients are defined as:

$$N_{11} := \left|\{\{u,v\} \subseteq V \mid (\exists C \in \mathcal{P} : \{u,v\} \subseteq C) \wedge (\exists C' \in \mathcal{Q} : \{u,v\} \subseteq C')\}\right|,$$
$$N_{10} := \left|\{\{u,v\} \subseteq V \mid (\exists C \in \mathcal{P} : \{u,v\} \subseteq C) \wedge (\forall C' \in \mathcal{Q} : \{u,v\} \not\subseteq C')\}\right|,$$
$$N_{01} := \left|\{\{u,v\} \subseteq V \mid (\forall C \in \mathcal{P} : \{u,v\} \not\subseteq C) \wedge (\exists C' \in \mathcal{Q} : \{u,v\} \subseteq C')\}\right|,$$
$$N_{00} := \left|\{\{u,v\} \subseteq V \mid (\forall C \in \mathcal{P} : \{u,v\} \not\subseteq C) \wedge (\forall C' \in \mathcal{Q} : \{u,v\} \not\subseteq C')\}\right|.$$

Please note that $N_{11} + N_{10} + N_{01} + N_{00} = \binom{n}{2} = \frac{n(n-1)}{2}$. One easily can see that for identical partitions $N_{10} = N_{01} = 0$, because two nodes either occur in a cluster together or not. Completely different partitions result in $N_{11} = 0$. All the measures we examined are given in Tables A1 and A2. The RV coefficient is used by Youness and Saporta [40] for partition comparison, and p and q are the cluster counts (e.g., $p = |\mathcal{P}|$) for the two partitions. For a detailed definition of the Lerman index (especially the definitions of the expectation and standard deviation), see Denœud and Guénoche [41].

Table A1. The pair counting measures used in Table 3 [42]. The above measures are similarity measures. Distance measures and non-normalized measures are listed in Table A2. For brevity: $N_{21} = N_{11} + N_{10}$, $N_{12} = N_{11} + N_{01}$, $N'_{01} = N_{00} + N_{01}$ and $N'_{10} = N_{00} + N_{10}$. Abbr., Abbreviation.

Abbr.	Measure	Formula	$\mathcal{P} = \mathcal{P}$
RI	Rand [43]	$\frac{N_{11}+N_{00}}{\binom{n}{2}}$	1.0
ARI	Hubert and Arabie [44]	$\frac{2(N_{00}N_{11}-N_{10}N_{01})}{N'_{01}N_{12}+N'_{10}N_{21}}$	1.0
H	Hamann [45]	$\frac{(N_{11}+N_{00})-(N_{10}+N_{01})}{\binom{n}{2}}$	1.0
CZ	Czekanowski [46]	$\frac{2N_{11}}{2N_{11}+N_{10}+N_{01}}$	1.0
K	Kulczynski [47]	$\frac{1}{2}\left(\frac{N_{11}}{N_{21}}+\frac{N_{11}}{N_{12}}\right)$	1.0
MC	McConnaughey [48]	$\frac{N_{11}^2-N_{10}N_{01}}{N_{21}N_{12}}$	1.0
P	Peirce [49]	$\frac{N_{11}N_{00}-N_{10}N_{01}}{N_{21}N'_{01}}$	1.0
W_I	Wallace [50]	$\frac{N_{11}}{N_{21}}$	1.0
W_{II}	Wallace [50]	$\frac{N_{11}}{N_{12}}$	1.0
FM	Fowlkes and Mallows [51]	$\sqrt{\frac{N_{11}}{N_{21}}\frac{N_{11}}{N_{12}}}$	1.0
Γ	Yule [52]	$\frac{N_{11}N_{00}-N_{10}N_{01}}{\sqrt{N_{21}N_{12}N'_{10}N'_{01}}}$	1.0
SS1	Sokal and Sneath [53]	$\frac{1}{4}\left(\frac{N_{11}}{N_{21}}+\frac{N_{11}}{N_{12}}+\frac{N_{00}}{N'_{10}}+\frac{N_{00}}{N'_{01}}\right)$	1.0
B1	Baulieu [54]	$\frac{\binom{n}{2}^2-\binom{n}{2}(N_{10}+N_{01})+(N_{10}-N_{01})^2}{\binom{n}{2}^2}$	1.0
GL	Gower and Legendre [55]	$\frac{N_{11}+N_{00}}{N_{11}+\frac{1}{2}(N_{10}+N_{01})+N_{00}}$	1.0
SS2	Sokal and Sneath [53]	$\frac{N_{11}}{N_{11}+2(N_{10}+N_{01})}$	1.0
SS3	Sokal and Sneath [53]	$\frac{N_{11}N_{00}}{\sqrt{N_{21}N_{12}N'_{01}N'_{10}}}$	1.0
RT	Rogers and Tanimoto [56]	$\frac{N_{11}+N_{00}}{N_{11}+2(N_{10}+N_{01})+N_{00}}$	1.0
GK	Goodman and Kruskal [57]	$\frac{N_{11}N_{00}-N_{10}N_{01}}{N_{11}N_{00}+N_{10}N_{01}}$	1.0
J	Jaccard [3]	$\frac{N_{11}}{N_{11}+N_{10}+N_{01}}$	1.0
RV	Robert and Escoufier [58]	$\left(N_{11}-\frac{1}{q}N_{21}-\frac{1}{p}N_{12}+\binom{n}{2}\frac{1}{pq}\right)\left[\left(\frac{p-2}{p}N_{21}+\binom{n}{2}\frac{1}{p^2}\right)\left(\frac{q-2}{q}N_{12}+\binom{n}{2}\frac{1}{q^2}\right)\right]^{-\frac{1}{2}}$	1.0

Table A2. Pair counting measures that are not similarity measures. For brevity: $N_{21} = N_{11} + N_{10}$, $N_{12} = N_{11} + N_{01}$, $N'_{01} = N_{00} + N_{01}$ and $N'_{10} = N_{00} + N_{10}$.

Abbr.	Measure	Formula	$\mathcal{P} = \mathcal{P}$
RR	Russel and Rao [59]	$\frac{N_{11}}{\binom{n}{2}}$	max
M	Mirkin and Chernyi [60]	$2(N_{01} + N_{10})$	0.0
Mi	Hilbert [61]	$\sqrt{\frac{N_{10}+N_{01}}{N_{11}+N_{10}}}$	0.0
Pe	Pearson [62]	$\frac{N_{11}N_{00}-N_{10}N_{01}}{N_{21}N_{12}N'_{01}N'_{10}}$	max
B2	Baulieu [54]	$\frac{N_{11}N_{00}-N_{10}N_{01}}{\binom{n}{2}^2}$	max
LI	Lerman [63]	$\frac{N_{11}-E(N_{11})}{\sqrt{\sigma^2(N_{11})}}$	max
NLI	Lerman [63] (normalized)	$\frac{LI(P_1,P_2)}{LI(P_1,P_1)LI(P_2,P_2)}$	1.0
FMG	Fager and McGowan [64]	$\frac{N_{11}}{\sqrt{N_{21}N_{12}}} - \frac{1}{2\sqrt{N_{21}}}$	max

Appendix B.2. Set-Based Comparison Measures

The second class is based on plain set comparison. We investigate three measures (see Table A3), namely the measure of Larsen and Aone [65], the so-called classification error distance [66] and Dongen's metric [67].

Table A3. References and formulas for the three set-based comparison measures used in Table 3. σ is the result of a maximum weighted matching of a bipartite graph. The bipartite graph is constructed from the partitions that shall be compared: the two node sets are derived from the two partitions, and each cluster is represented by a node. By definition, the two node sets are disjoint. The node sets are connected by edges of weight $w_{ij} = \left|\{C_i \cap C'_j \mid C_i \in \mathcal{P}, C'_j \in \mathcal{Q}\}\right|$. As in our context $|\mathcal{P}| = |\mathcal{Q}|$, the found σ is assured to be a perfect (bijective) matching. n is the number of nodes $|V|$.

Abbr.	Measure	Formula	$\mathcal{P} = \mathcal{P}$
LA	Larsen and Aone [65]	$\frac{1}{\lvert\mathcal{P}\rvert}\sum_{C\in\mathcal{P}} \max_{C'\in\mathcal{Q}} \frac{2\lvert C\cap C'\rvert}{\lvert C\rvert+\lvert C'\rvert}$	1.0
d_{CE}	Meilă and Heckerman [66]	$1 - \frac{1}{n}\max_\sigma \sum_{C\in\mathcal{P}} \lvert C \cap \sigma(C)\rvert$	0.0
D	van Dongen [67]	$2n - \sum_{C\in\mathcal{P}} \max_{C'\in\mathcal{Q}} \lvert C\cap C'\rvert - \sum_{C'\in\mathcal{Q}} \max_{C\in\mathcal{P}} \lvert C\cap C'\rvert$	0.0

Appendix B.3. Information Theory-Based Measures

The last class of measures contains those that are rooted in information theory. We show the measures in Table A4, and we recap the fundamentals briefly: the entropy of a random variable X is defined as:

$$H(X) = -\sum_{i=1}^{k} p_i \log p_i$$

with p_i being the probability of a specific incidence. The entropy of a partition can analogously be defined as:

$$H(\mathcal{P}) = -\sum_{C\in\mathcal{P}} \frac{|C|}{n} \log \frac{|C|}{n}.$$

The mutual information of two random variables is:

$$I(X,Y) = \sum_{i=0}^{k}\sum_{j=0}^{l} p_{ij} \log \frac{p_{ij}}{p_i p_j}$$

and again, analogously:

$$MI(\mathcal{P},\mathcal{Q}) = \sum_{C\in\mathcal{P}}\sum_{C'\in\mathcal{Q}} \frac{|C\cap C'|}{n} \log n \frac{|C\cap C'|}{|C||C'|}$$

is the mutual information of two partitions [68]. Meilă [69] introduced the Variation of Information as $VI = H(\mathcal{P}) + H(\mathcal{Q}) - 2MI$.

Table A4. Information theory-based measures used in Table 3. All measures are based on Shannon's definition of entropy. Again, $n = |V|$.

Abbr.	Measure	Formula	$\mathcal{P} = \mathcal{P}$
MI	e.g., Vinh et al. [68]	$\sum_{C\in\mathcal{P}}\sum_{C'\in\mathcal{Q}} \frac{\lvert C\cap C'\rvert}{n} \log n \frac{\lvert C\cap C'\rvert}{\lvert C\rvert\lvert C'\rvert}$	max
NMI_{φ}	Danon et al. [70]	$\frac{\text{MI}}{\varphi(H(\mathcal{P}),H(\mathcal{Q}))}$, $\varphi \in \{\min, \max\}$	1.0
NMI_{Σ}	Danon et al. [70]	$\frac{2\cdot\text{MI}}{H(\mathcal{P})+H(\mathcal{Q})}$	1.0
VI	Meilă [69]	$H(\mathcal{P}) + H(\mathcal{Q}) - 2MI$	0.0

Appendix B.4. Summary

As one can see, all three classes of measures rely mainly on set matching between node sets (clusters), as an alternative definition of $N_{11} = \sum_{C\in\mathcal{P}}\sum_{C'\in\mathcal{Q}} \binom{|C\cap C'|}{2}$ shows [42]. The adjacency information of the graph is completely ignored.

References

1. Melnykov, V.; Maitra, R. CARP: Software for fishing out good clustering algorithms. *J. Mach. Learn. Res.* **2011**, *12*, 69–73.
2. Bader, D.A.; Meyerhenke, H.; Sanders, P.; Wagner, D. (Eds.) *10th DIMACS Implementation Challenge—Graph Partitioning and Graph Clustering*; Rutgers University, DIMACS (Center for Discrete Mathematics and Theoretical Computer Science): Piscataway, NJ, USA, 2012.
3. Jaccard, P. Nouvelles recherches sur la distribution florale. *Bull. Soc. Vaud. Sci. Nat.* **1908**, *44*, 223–270.
4. Horta, D.; Campello, R.J.G.B. Comparing hard and overlapping clusterings. *J. Mach. Learn. Res.* **2015**, *16*, 2949–2997.
5. Romano, S.; Vinh, N.X.; Bailey, J.; Verspoor, K. Adjusting for chance clustering comparison measures. *J. Mach. Learn. Res.* **2016**, *17*, 1–32.
6. Von Luxburg, U.; Williamson, R.C.; Guyon, I. Clustering: Science or art? *JMLR Workshop Conf. Proc.* **2011**, *27*, 65–79.
7. Hennig, C. What are the true clusters? *Pattern Recognit. Lett.* **2015**, *64*, 53–62. [CrossRef]
8. Van Craenendonck, T.; Blockeel, H. *Using Internal Validity Measures to Compare Clustering Algorithms*; Benelearn 2015 Poster Presentations (Online); Benelearn: Delft, The Netherlands, 2015; pp. 1–8.
9. Filchenkov, A.; Muravyov, S.; Parfenov, V. Towards cluster validity index evaluation and selection. In Proceedings of the 2016 IEEE Artificial Intelligence and Natural Language Conference, St. Petersburg, Russia, 10–12 November 2016; pp. 1–8.
10. MacArthur, B.D.; Sánchez-García, R.J.; Anderson, J.W. Symmetry in complex networks. *Discret. Appl. Math.* **2008**, *156*, 3525–3531. [CrossRef]

11. Darga, P.T.; Sakallah, K.A.; Markov, I.L. Faster Symmetry Discovery Using Sparsity of Symmetries. In Proceedings of the 2008 45th ACM/IEEE Design Automation Conference, Anaheim, CA, USA, 8–13 June 2008; pp. 149–154.
12. Katebi, H.; Sakallah, K.A.; Markov, I.L. Graph Symmetry Detection and Canonical Labeling: Differences and Synergies. In *Turing-100. The Alan Turing Centenary*; EPiC Series in Computing; Voronkov, A., Ed.; EasyChair: Manchester, UK, 2012; Volume 10, pp. 181–195.
13. Ball, F.; Geyer-Schulz, A. How symmetric are real-world graphs? A large-scale study. *Symmetry* **2018**, *10*, 29. [CrossRef]
14. Newman, M.E.J.; Girvan, M. Finding and evaluating community structure in networks. *Phys. Rev. E* **2004**, *69*, 026113. [CrossRef] [PubMed]
15. Ovelgönne, M.; Geyer-Schulz, A. An Ensemble Learning Strategy for Graph Clustering. In *Graph Partitioning and Graph Clustering*; Bader, D.A., Meyerhenke, H., Sanders, P., Wagner, D., Eds.; American Mathematical Society: Providence, RI, USA, 2013; Volume 588, pp. 187–205.
16. Wielandt, H. *Finite Permutation Groups*; Academic Press: New York, NY, USA, 1964.
17. James, G.; Kerber, A. The Representation Theory of the Symmetric Group. In *Encyclopedia of Mathematics and Its Applications*; Addison-Wesley: Reading, MA, USA, 1981; Volume 16.
18. Coxeter, H.; Moser, W. Generators and Relations for Discrete Groups. In *Ergebnisse der Mathematik und ihrer Grenzgebiete*; Springer: Berlin, Germany, 1965; Volume 14.
19. Dixon, J.D.; Mortimer, B. Permutation Groups. In *Graduate Texts in Mathematics*; Springer: New York, NY, USA, 1996; Volume 163.
20. Beth, T.; Jungnickel, D.; Lenz, H. *Design Theory*; Cambridge University Press: Cambridge, UK, 1993.
21. Erdős, P.; Rényi, A.; Sós, V.T. On a problem of graph theory. *Stud. Sci. Math. Hung.* **1966**, *1*, 215–235.
22. Burr, S.A.; Erdős, P.; Spencer, J.H. Ramsey theorems for multiple copies of graphs. *Trans. Am. Math. Soc.* **1975**, *209*, 87–99. [CrossRef]
23. Ball, F.; Geyer-Schulz, A. *R Package Partition Comparison*; Technical Report 1-2017, Information Services and Electronic Markets, Institute of Information Systems and Marketing; KIT: Karlsruhe, Germany, 2017.
24. Doob, J.L. Measure Theory. In *Graduate Texts in Mathematics*; Springer: New York, NY, USA, 1994.
25. Hausdorff, F. *Set Theory*, 2nd ed.; Chelsea Publishing Company: New York, NY, USA, 1962.
26. Kuratowski, K. *Topology Volume I*; Academic Press: New York, NY, USA, 1966; Volume 1.
27. Von Neumann, J. Construction of Haar's invariant measure in groups by approximately equidistributed finite point sets and explicit evaluations of approximations. In *Invariant Measures*; American Mathematical Society: Providence, RI, USA, 1999; Chapter 6, pp. 87–134.
28. Ball, F.; Geyer-Schulz, A. Weak invariants of actions of the automorphism group of a graph. *Arch. Data Sci. Ser. A* **2017**, *2*, 1–22.
29. Zachary, W.W. An information flow model for conflict and fission in small groups. *J. Anthropol. Res.* **1977**, *33*, 452–473. [CrossRef]
30. Bock, H.H. *Automatische Klassifikation: Theoretische und praktische Methoden zur Gruppierung und Strukturierung von Daten*; Vandenhoeck und Ruprecht: Göttingen, Germany, 1974.
31. Rossi, R.; Fahmy, S.; Talukder, N. A Multi-level Approach for Evaluating Internet Topology Generators. In Proceedings of the 2013 IFIP Networking Conference, Trondheim, Norway, 2–4 June 2013, pp. 1–9.
32. Rossi, R.A.; Ahmed, N.K. The Network Data Repository with Interactive Graph Analytics and Visualization. In Proceedings of the Twenty-Ninth AAAI Conference on Artificial Intelligence, Austin, TX, USA, 25–30 January 2015.
33. Furst, M.; Hopcroft, J.; Luks, E. Polynomial-time Algorithms for Permutation Groups. In Proceedings of the 21st Annual Symposium on Foundations of Computer Science, Syracuse, NY, USA, 13–15 October 1980; pp. 36–41.
34. McKay, B.D.; Piperno, A. Practical graph isomorphism, II. *J. Symb. Comput.* **2014**, *60*, 94–112. [CrossRef]
35. Babai, L. Graph isomorphism in quasipolynomial time. *arXiv* **2015**, arXiv:1512.03547.
36. Fortunato, S.; Barthélemy, M. Resolution limit in community detection. *Proc. Natl. Acad. Sci. USA* **2007**, *104*, 36–41. [CrossRef] [PubMed]
37. Lancichinetti, A.; Fortunato, S. Limits of modularity maximization in community detection. *Phys. Rev. E* **2011**, *84*, 66122. [CrossRef] [PubMed]

38. Geyer-Schulz, A.; Ovelgönne, M.; Stein, M. Modified randomized modularity clustering: Adapting the resolution limit. In *Algorithms from and for Nature and Life*; Lausen, B., Van den Poel, D., Ultsch, A., Eds.; Studies in Classification, Data Analysis, and Knowledge Organization; Springer International Publishing: Heidelberg, Germany, 2013; pp. 355–363.
39. Meilă, M. Comparing clusterings—An information based distance. *J. Multivar. Anal.* **2007**, *98*, 873–895. [CrossRef]
40. Youness, G.; Saporta, G. Some measures of agreement between close partitions. *Student* **2004**, *51*, 1–12.
41. Denœud, L.; Guénoche, A. Comparison of distance indices between partitions. In *Data Science and Classification*; Batagelj, V., Bock, H.H., Ferligoj, A., Žiberna, A., Eds.; Studies in Classification, Data Analysis, and Knowledge Organization; Springer: Berlin/Heidelberg, Germany, 2006; pp. 21–28.
42. Albatineh, A.N.; Niewiadomska-Bugaj, M.; Mihalko, D. On similarity indices and correction for chance agreement. *J. Classif.* **2006**, *23*, 301–313. [CrossRef]
43. Rand, W.M. Objective criteria for the evaluation of clustering algorithms. *J. Am. Stat. Assoc.* **1971**, *66*, 846–850. [CrossRef]
44. Hubert, L.; Arabie, P. Comparing partitions. *J. Classif.* **1985**, *2*, 193–218. [CrossRef]
45. Hamann, U. Merkmalsbestand und Verwandtschaftsbeziehungen der Farinosae: Ein Beitrag zum System der Monokotyledonen. *Willdenowia* **1961**, *2*, 639–768.
46. Czekanowski, J. "Coefficient of Racial Likeness" und "Durchschnittliche Differenz". *Anthropol. Anz.* **1932**, *9*, 227–249.
47. Kulczynski, S. Zespoly roslin w Pieninach. *Bull. Int. Acad. Pol. Sci. Lett.* **1927**, *2*, 57–203.
48. McConnaughey, B.H. The determination and analysis of plankton communities. *Mar. Res.* **1964**, *1*, 1–40.
49. Peirce, C.S. The numerical measure of the success of predictions. *Science* **1884**, *4*, 453–454. [CrossRef] [PubMed]
50. Wallace, D.L. A method for comparing two hierarchical clusterings: Comment. *J. Am. Stat. Assoc.* **1983**, *78*, 569–576. [CrossRef]
51. Fowlkes, E.B.; Mallows, C.L. A method for comparing two hierarchical clusterings. *J. Am. Stat. Assoc.* **1983**, *78*, 553–569. [CrossRef]
52. Yule, G.U. On the association of attributes in statistics: With illustrations from the material of the childhood society. *Philos. Trans. R. Soc. A* **1900**, *194*, 257–319. [CrossRef]
53. Sokal, R.R.; Sneath, P.H.A. *Principles of Numerical Taxonomy*; W. H. Freeman: San Francisco, CA, USA; London, UK, 1963.
54. Baulieu, F.B. A classification of presence/absence based dissimilarity coefficients. *J. Classif.* **1989**, *6*, 233–246. [CrossRef]
55. Gower, J.C.; Legendre, P. Metric and euclidean properties of dissimilarity coefficients. *J. Classif.* **1986**, *3*, 5–48. [CrossRef]
56. Rogers, D.J.; Tanimoto, T.T. A computer program for classifying plants. *Science* **1960**, *132*, 1115–1118. [CrossRef] [PubMed]
57. Goodman, L.A.; Kruskal, W.H. Measures of association for cross classifications. *J. Am. Stat. Assoc.* **1954**, *49*, 732–764.
58. Robert, P.; Escoufier, Y. A unifying tool for linear multivariate statistical methods: The RV-coefficient. *J. R. Stat. Soc. Ser. C* **1976**, *25*, 257–265. [CrossRef]
59. Russel, P.F.; Rao, T.R. On habitat and association of species of anopheline larvae in south-eastern madras. *J. Malar. Inst. India* **1940**, *3*, 153–178.
60. Mirkin, B.G.; Chernyi, L.B. Measurement of the distance between partitions of a finite set of objects. *Autom. Remote Control* **1970**, *31*, 786–792.
61. Hilbert, D. *Gesammelte Abhandlungen von Hermann Minkowski, Zweiter Band*; Number 2; B. G. Teubner: Leipzig, UK; Berlin, Germany, 1911.
62. Pearson, K. On the coefficient of racial likeness. *Biometrika* **1926**, *18*, 105–117. [CrossRef]
63. Lerman, I.C. Comparing Partitions (Mathematical and Statistical Aspects). In *Classification and Related Methods of Data Analysis*; Bock, H.H., Ed.; North-Holland: Amsterdam, The Netherlands, 1988; pp. 121–132.
64. Fager, E.W.; McGowan, J.A. Zooplankton species groups in the north pacific co-occurrences of species can be used to derive groups whose members react similarly to water-mass types. *Science* **1963**, *140*, 453–460. [CrossRef] [PubMed]

65. Larsen, B.; Aone, C. Fast and Effective Text Mining Using Linear-time Document Clustering. In Proceedings of the Fifth ACM SIGKDD International Conference on Knowledge Discovery and Data Mining, San Diego, CA, USA, 15–18 August 1999; ACM: New York, NY, USA, 1999; pp. 16–22.
66. Meilă, M.; Heckerman, D. An experimental comparison of model-based clustering methods. *Mach. Learn.* **2001**, *42*, 9–29. [CrossRef]
67. Van Dongen, S. *Performance Criteria for Graph Clustering and Markov Cluster Experiments*; Technical Report INS-R 0012; CWI (Centre for Mathematics and Computer Science): Amsterdam, The Netherlands, 2000.
68. Vinh, N.X.; Epps, J.; Bailey, J. Information theoretic measures for clusterings comparison: Variants, properties, normalization and correction for chance. *J. Mach. Learn. Res.* **2010**, *11*, 2837–2854.
69. Meilă, M. Comparing clusterings by the variation of information. In *Learning Theory and Kernel Machines*; Schölkopf, B., Warmuth, M.K., Eds.; Number 2777 in Lecture Notes in Computer Science; Springer: Berlin/Heidelberg, Germany, 2003; pp. 173–187.
70. Danon, L.; Díaz-Guilera, A.; Duch, J.; Arenas, A. Comparing community structure identification. *J. Stat. Mech. Theory Exp.* **2005**, *2005*, P09008. [CrossRef]

symmetry

MDPI

Article

A Multi-Granularity 2-Tuple QFD Method and Application to Emergency Routes Evaluation

Yanlan Mei, Yingying Liang * and Yan Tu *

School of Management, Wuhan University of Technology, Wuhan 430070, China; myl2014@whut.edu.cn
* Correspondence: liangyingying1993@whut.edu.cn (Y.L.); tuyan1988@whut.edu.cn (Y.T.)

Received: 23 September 2018; Accepted: 8 October 2018; Published: 11 October 2018

Abstract: Quality function deployment (QFD) is an effective approach to satisfy the customer requirements (*CRs*). Furthermore, accurately prioritizing the engineering characteristics (*ECs*) as the core of QFD is considered as a group decision making (GDM) problem. In order to availably deal with various preferences and the vague information of different experts on a QFD team, multi-granularity 2-tuple linguistic representation is applied to elucidate the relationship and correlation between *CRs* and *ECs* without loss of information. In addition, the importance of *CRs* is determined using the best worst method (BWM), which is more applicable and has good consistency. Furthermore, we propose considering the relationship matrix and correlation matrix method to prioritize *ECs*. Finally, an example about evaluating emergency routes of metro station is proposed to illustrate the validity of the proposed methodology.

Keywords: quality function deployment; engineering characteristics; group decision making; 2-tuple; metro station; emergency routes

1. Introduction

In order to cope with intense global competitions, enterprises must design the highest quality products that satisfy the voice of customers (VOCs). Quality function deployment (QFD) is an effective method to map customer requirements (*CRs*) into engineering characteristics (*ECs*) in the area of product development [1] and construction industry [2]. The core of QFD is requirements conversion, moreover, the first phase in house of quality (HOQ) mapping *CRs* to *ECs* becomes an essential procedure of implementing QFD [3].

Aiming at implementing QFD successfully, plenty of *CRs* should be acquired, and group decision making (GDM) should be adopted [4]. QFD consists of two major steps: collecting the *CRs* and mapping it to *ECs*, both of which are performed [5,6]. This paper focuses on how the *ECs* in QFD can be prioritized.

There are plenty of methods to prioritize the *ECs*. Fuzzy set theory was widely employed to calculate the rankings of *ECs* under the circumstance of vagueness and impreciseness. Fuzzy multiple objective programming [7], fuzzy goal programming [8], fuzzy relationship and correlations [9], and expected value-based method [10] are proposed to prioritize *ECs*. In addition, Geng et al. [11] integrated the analytic network process to QFD to reflect the initial importance weights of *ECs*. However, the problem is that they paid little attention to the GDM method, which can aggregate different experts' preferences. For the purpose of reaching collective decisions, we combine GDM with QFD.

Kwong et al. [6] put forward the fuzzy GDM method integrated with a fuzzy weighted average to rank *ECs*. Wang [12] adopted the method of aggregating technical importance rather than *CRs* to prioritize *ECs*. With respect to consensus, modified fuzzy clustering was presented so as to reach the consensus of the QFD team [13]. A two-stage GDM was proposed to simultaneously solve the two types of uncertainties (i.e., human assessment on qualitative attributes as well as input information) underlying QFD [4]. However, due to varying personal experience and knowledge, the input information of experts presented with multi-format or multi-granular linguistic preferences makes prioritizing *ECs* more difficult. Therefore, some scholars have focused on the GDM approach based on multi-granularity linguistic environments [14–17]. Xu [18,19] analyzed multiple formats' preferences and provided an approach integrating information in the context of GDM. It is noteworthy that multi-granularity evaluation should be analyzed.

The correlation between *CRs* and *ECs* influencing on the relationship becomes ignored and simplified in the current study. In addition, the linguistic accuracy remains to be discussed. Considering that the 2-tuple linguistic representation can increase the information of precision [20,21]. In order to fill the gap, it is necessary that the QFD methodology is extended with a 2-tuple linguistic environment so as to lessen the loss of information and obtain accurate value of *ECs*. In addition, decision makers may have different knowledge and experience in the process of group decision making, and they may then adopt different linguistic labels to describe the same decision-making problems. This process is denoted as multi-granular linguistic information, which conforms to the actual decision-making process. Therefore, we allow decision makers to employ multi-granular linguistic information, i.e., the linguistic term set has different granularities.

A majority of methods deal with multi-granular linguistic information. Herrera et al. proposed the definition of a basic linguistic term set, and then different linguistic labels can be unified based on a basic linguistic term set [22]. In addition, some transformation methods based on the linguistic hierarchy and extended linguistic hierarchy were presented and applied to a plenty of decision-making problems [23,24]. Among these approaches, the method considering linguistic hierarchy is more flexible and convenient to carry out. In this paper, we adopt this method to deal with the problem of multi-granular linguistic evaluation. For determining the weight of *CRs*, we adopt the best–worst method (BWM) in this paper. This method has good consistency and is easier to implement [25,26]. Our contributions lie in using the BWM to determine the importance of *CRs* and integrate the correlations matrix with the relationship matrix based on a compromise idea, where experts can express their thoughts in different granularities.

In this paper, a GDM approach is integrated with QFD to solve different preferences and prioritize *ECs*. The multi-granularity 2-tuple linguistic information to reflect the attitudes of different experts is employed. This paper is organized as follows: In Section 2, a 2-tuple multi-granularity linguistic representation model, linguistic hierarchies, and a 2-tuple linguistic weighted geometric Bonferroni mean (2TLWGBM) operator are presented. In Section 3, the BWM is applied to compute the weight of *CRs*, and a novel GDM approach to prioritize *ECs* is proposed. An illustrated example about metro stations is provided in Section 4 to demonstrate the applicability of this method. Ultimately, conclusions and future research are marked in Section 5.

2. Preliminaries

In this section, we introduce some basic knowledge about QFD, 2-tuple representation and the 2TLWGBM Operator.

2.1. The Basic Knowledge on QFD

A four-phase QFD model is employed to translate the VOCs to *ECs*, which consists of Product Planning, Part Deployment, Process Planning, and Process and Quality control [3]. The first phase is to collect customer requirements for the product called WHATs and then to transform these needs into *ECs* called HOWs. This phase is so fundamental in product development that the corresponding QFD transformation matrix referred to the HOQ (Figure 1). The HOQ links customer needs to the development team's technical responses, so we focus on this phase in order to translate different preference of customers and experts to prioritize *ECs*. In this paper, we first take the relationship between *CRs* and *ECs* into consideration. In order to transform the importance of *CRs* into *ECs*, the correlation of *CRs* and *ECs* is introduced to modify the initial relationship afterward.

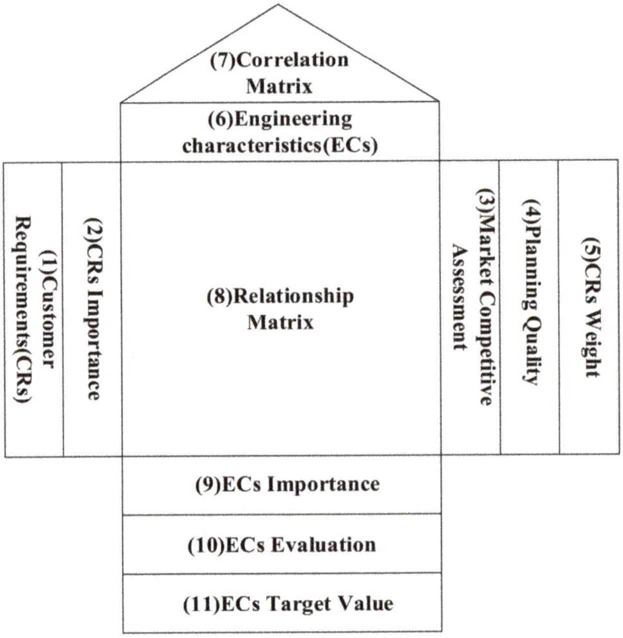

Figure 1. House of Quality (HOQ).

2.2. The 2-Tuple Linguistic Representation

There are numerous formats for representing preference such as linguistic, numerical and 2-tuple linguistic representation. Compared with other forms, 2-tuple linguistic representation makes the assessment more precise and without a loss of information [20]. Next, we will introduce some basic knowledge about 2-tuple representation.

Definition 1 [27]. *Assuming* $S = \{s_1, s_2, \cdots, s_g\}$ *is a linguistic term set and* $\beta \in [0, g]$ represents the consequence of a symbolic aggregation operation. Afterwards, the 2-tuple is expressed as the equivalence to β as follows:

$$\Delta : [0, g] \to S \times [-0.5, 0.5) \tag{1}$$

$$\Delta(\beta) = (s_i, \alpha), \text{ with} \begin{cases} s_i, & i = \text{round}(\beta) \\ \alpha = \beta - i, & \alpha = [-0.5, 0.5) \end{cases} \tag{2}$$

where round $(\cdot)$ *represents the usual round function,* s_i *has the closest index label to* β*, and* α is the value of the symbolic translation.

Definition 2 [27]. *Let* $S = \{s_1, s_2, \cdots, s_g\}$ *be a linguistic term set and* (s_i, α_i) *be a 2-tuple. There is always a function* Δ^{-1} *that can be defined, such that, from a 2-tuple* (s_i, α_i)*, its equivalent numerical value* $\beta \in [0, g] \subset R$ can be obtained, which is described as follows:

$$\Delta^{-1} :\to S \times [-0.5, 0.5) \to [0, g] \tag{3}$$

$$\Delta^{-1}(s_i, \alpha_i) = i + \alpha_i = \beta \tag{4}$$

Definition 3 [28]. *There are 2-tuples* $x = \{(s_1, \alpha_1), (s_2, \alpha_2), \cdots, (s_n, \alpha_n)\}$. Their arithmetic mean is expressed as:

$$(\bar{s}, \bar{\alpha}) = \Delta\left(\frac{1}{n}\sum_{i=1}^{n} \Delta^{-1}(r_i, \alpha_i)\right), \quad \bar{s} \in S, \quad \bar{\alpha} \in [-0.5, 0.5) \tag{5}$$

Definition 4 [28]. *Let* (s_i, α_i) *and* (s_j, α_j) *be two 2-tuple linguistic variables. Their granularities are both* g, and the distance between them is described as follows:

$$d((s_i, \alpha_i), (s_j, \alpha_j)) = \frac{\left|\Delta^{-1}(i + \alpha_i) - \Delta^{-1}(j + \alpha_j)\right|}{g} \tag{6}$$

Definition 5 [28]. *Let* $x = \{(s_1, \alpha_1), (s_2, \alpha_2), \cdots, (s_n, \alpha_n)\}$ *be a set of 2-tuples and* $(\bar{s}, \bar{\alpha})$ be the arithmetic mean of these 2-tuples. The degree of similarity is expressed as

$$\mathrm{sim}((s_{\pi(j)}, \alpha_{\pi(j)}), (\bar{s}, \bar{\alpha})) = 1 - \frac{d((s_{\pi(j)}, \alpha_{\pi(j)}), (\bar{s}, \bar{\alpha}))}{\sum\limits_{j=1}^{n} d((s_{\pi(j)}, \alpha_{\pi(j)}), (\bar{s}, \bar{\alpha}))}, \quad j = 1, 2, \cdots, n \tag{7}$$

Definition 6 [23]. *Let* $LH = \bigcup\limits_{t} l(t, n(t))$, *which is the union of all level* t, *a linguistic hierarchy whose linguistic term set is* $S^{n(t)} = \left\{s_0^{n(t)}, s_1^{n(t)}, \cdots, s_{n(t)-1}^{n(t)}\right\}$. Furthermore, different granularities reflect different preferences under the circumstance of evaluating. The transformation function (TF) between level t *and level* $t\prime$ is defined as

$$\begin{gathered} TF_{t\prime}^{t} : l(t, n(t)) \rightarrow l(t\prime, n(t\prime)) \\ TF_{t\prime}^{t}(s_i^{n(t)}, \alpha^{n(t)}) = \Delta\left(\frac{\Delta^{-1}(s_i^{n(t)}, \alpha^{n(t)}) \cdot (n(t\prime) - 1)}{n(t) - 1}\right) \end{gathered} \tag{8}$$

where t and $t\prime$ represent different levels of linguistic hierarchy.

Note 1. The TF can implement the transformation between different granularities and further achieve a unified linguistic label. Without loss of generality, the transformation usually is carried out from the lower granularity to higher granularity in the process of transformation, i.e., the level $t\prime$ usually corresponds to the maximum granularity.

2.3. The 2TLWGBM Operator

There are numerous operators to aggregate information in different linguistic environments, such as hesitant fuzzy Maclaurin symmetric mean Operators [29], 2-tuple linguistic Muirhead mean operators [30], 2-tuple linguistic Neutrosophic number Bonferroni mean operators [31], and hesitant 2-tuple linguistic prioritized weighted averaging aggregation operator [32] in the context of the 2-tuple environment. In view of the Bonferroni mean (BM) operator capturing the interrelationship between input information and ranking *ECs* under a 2-tuple environment, so the 2TLWGBM operator [33] will be applied to prioritize the sequence of *ECs*. BM is defined as follows:

Definition 7 [33]. *Let* $p, q \geq 0$ *and* $a_i (i = 1, 2, \cdots, n)$ *be a series of non-negative numbers. Then the BM operator is defined as*

$$BM^{p,q}(a_1, a_2, \cdots, a_n) = \left(\frac{1}{n(n-1)} \sum_{\substack{i,j=1 \\ i \neq j}}^{n} a_i^p a_j^q\right)^{\frac{1}{p+q}} \tag{9}$$

Definition 8 [33]. *Let* $\mathbf{x} = \{(r_1, a_1), (r_2, a_2), \cdots, (r_n, a_n)\}$ *be a set of 2-tuple and* $p, q \geq 0$. *In addition,* $\mathbf{w} = (w_1, w_2, \cdots, w_n)^T$ *is the weight vector of* $\mathbf{x}$, *where* $w_i > 0\ (i = 1, 2, \cdots, n)$ *represents the importance degree of* $(r_i, a_i)\ (i = 1, 2, \cdots, n)$, *and* $\sum_{i=1}^{n} w_i = 1$. The 2TLWGBM operator is then expressed as

$$2TLWGBM_w^{p,q}((r_1,a_1),(r_2,a_2),\cdots,(r_n,a_n)) = \Delta\left(\frac{1}{p+q}\left(\prod_{\substack{i,j=1\\ i\neq j}}^{n}\left(p\left(\Delta^{-1}(r_i,a_i)\right)^{w_i}+q\left(\Delta^{-1}(r_j,a_j)\right)^{w_j}\right)\right)^{\frac{1}{n(n-1)}}\right) \tag{10}$$

For the sake of simplicity, it can be set $p = q = 1$, the aggregation operator is indicated as

$$2TLWGBM_w^{1,1}((r_1,a_1),(r_2,a_2),\cdots,(r_n,a_n)) = \Delta\left(\frac{1}{2}\left(\prod_{\substack{i,j=1\\ i\neq j}}^{n}\left(\left(\Delta^{-1}(r_i,a_i)\right)^{w_i}+\left(\Delta^{-1}(r_j,a_j)\right)^{w_j}\right)\right)^{\frac{1}{n(n-1)}}\right) \tag{11}$$

Note 2. Although a majority of aggregation operators have been proposed in recent years, the 2TLWGBM operator has some merits in prioritizing *ECs*. On the one hand, this operator considers the relevance, which accords with the relationship and correlation between *CRs* and *ECs*. On the other hand, it is more flexible owing to the parameter p and q, which makes it more suitable for different decision makers.

3. A Group Decision-Making Approach to Prioritize *ECs*

3.1. Determine the Importance of CRs Based on BWM

Best worst method (BWM) is a MCDM method possessing the advantages in aspects of reaching the consistency and simplifying the calculation with respect to AHP. The core idea of BWM is constructing comparisons relationships between the best attribute (and the worst attribute) to the other attributes. Additionally, an optimization model established ground on consistency is solved to obtain the optical weights. Owing to simple operation and calculation, the BWM is synthesized to determine the importance of *CRs*. The steps are listed as follows:

Step 1. *CRs* $\{CR_1, CR_2, \cdots, CR_n\}$ are chosen, as are the best and the worst CR. The best CR is then compared with the other *CRs* using Number 1–9 is constructed. The best-to-others (BO) vector $A_B = (\alpha_{B1}, \alpha_{B2}, \cdots, \alpha_{Bn})$ is represented where α_{Bj} describes the preference of the best CR over CR_j. Similarly, the Others-to-worst (OW) vector $A_W = (\alpha_{1W}, \alpha_{2W}, \cdots, \alpha_{nW})^T$ is represented where α_{jW} describes the preference of CR_j over the worst CR.

Step 2. The optimal weights of *CRs* are obtained. The optimization model is established to minimize the maximum the difference $\{|\, w_B - \alpha_{Bj}\, w_j \,|\}$ and $\{|\, w_j - \alpha_{jW}\, \omega_W \,|\}$.

$$\begin{aligned} &\min \max_j \left\{\left|w_B - a_{Bj} w_j\right|, \left|w_j - a_{jW} w_W\right|\right\} \\ &s.t. \sum_j w_j = 1 \\ &w_j \geq 0,\ j = 1, 2, \cdots, n \end{aligned} \tag{Model 1}$$

Model 1 can be transformed into a linear programming model as follows:

$$\begin{aligned} &\min \xi \\ &s.t. \left| w_B - a_{Bj} w_j \right| \le \xi,\ j = 1,2,\cdots,n \\ &\left| w_j - a_{jW} w_W \right| \le \xi,\ j = 1,2,\cdots,n \quad \text{(Model 2)} \\ &\sum_j w_j = 1 \\ &w_j \ge 0,\ j = 1,2,\cdots,n \end{aligned}$$

Model 2 is solved to obtain the optimal importance of *CRs* $(w_1^*, w_2^*, \cdots, w_n^*)$ and ξ^*. Alternatively, the bigger ξ^* demonstrates the higher consistency ratio provided by customers. The consistency ratio can be calculated by the proportion between ξ^* and max ξ (Consistency Index).

$$\text{Consistency Ratio} = \frac{\xi^*}{\max \xi} = \frac{\xi^*}{\text{Consistency Index}} \tag{12}$$

where the max ξ is determined according to $(\alpha_{BW} - \xi) \times (\alpha_{BW} - \xi) = (\alpha_{BW} + \xi)$ and $\alpha_{BW} \in \{1, 2, \cdots, 9\}$. The consistency index is listed in Table 1.

Table 1. Consistency index.

α_{BW}	1	2	3	4	5	6	7	8	9
Consistency index	0.00	0.44	1.00	1.63	2.30	3.00	3.73	4.47	5.23

3.2. A Group Decision-Making Approach to Prioritize ECs

In this section, the steps of GDM for multi-granularity 2-tuple linguistic preference to prioritize *ECs* in QFD are given as follows:

Step 3. Different multi-granularity linguistic preferences are obtained.

Suppose the experts EP_k $(k = 1, 2, \cdots, t)$ in QFD product research or design team give the relationship between $CR_i(i = 1, 2, \cdots, n)$ and $EC_j(j = 1, 2, \cdots, s)$ based on different multi-granularities. The kth expert's linguistic term set and evaluation matrix respectively denoted as $S^{n(t)_k} = \left\{ s_i^{n(t)_k} | i = 0, 1, \cdots, n(t) - 1 \right\}$ and $R^k = \left(r_{ij} \right)_{n \times s} r_{ij} \in S^{n(t)_k}$, which is transformed into 2-tuple linguistic evaluation matrix $\widetilde{R}^k = \left(r_{ij}^{n(t)_k}, 0 \right)_{n \times s}, r_{ij}^{n(t)_k} \in S^{n(t)_k}$.

Step 4. Different multi-granularity linguistic preferences are unified.

To begin with, a basic linguistic term set $S^{n(t)_u} = \left\{ s_i^{n(t)_u} | i = 0, 1, \cdots, n(t) - 1 \right\}$ can be chosen, and the relationship matrix can then be transformed applying Equation (8) so as to make 2-tuple linguistic representation reach the same granularity. For instance, the kth expert's judgement matrix is transformed as $\widetilde{R}^{k\prime} = \left(r_{ij}^{n(t)_u}, \alpha_{ij}^{n(t)_u} \right)_{n \times s}, r_{ij}^{n(t)_u} \in S^{n(t)_u}$.

Step 5. All the evaluation matrices are aggregated.

All the evaluation matrices uniformed are aggregated with 2TLWGBM operator in virtue of Equation (10) into $R_{ij}(i = 1, 2, \cdots, n; j = 1, 2, \cdots, s)$. Furthermore, the new matrix represents ultimate relationship matrix between *CRs* and *ECs* in essence.

Step 6. The relationship between *CRs* and *ECs* is modified based on a compromise idea.

After establishing the aggregation matrix, experts give the correlations among *CRs* and *ECs* and the initial HOQ can be obtained, which reflects the relationship R_{ij} between CR_i and

$EC_j, i = 1,2,\cdots,n; j = 1,2,\cdots,s$.It is indispensable that the QFD team estimates the correlations between *CRs* (i.e., $L_{i\xi}(\xi \neq i, \xi = 1,2,\cdots,n)$) and *ECs* (i.e., $T_{i\theta}(\theta \neq i, \theta = 1,2,\cdots,s)$) using 2-tuple based on the basic linguistic set $S_i^{n(t)_u}$. Considering that the assessment result of $L_{i\xi}(\xi \neq i, \xi = 1,2,\cdots,n)$ and $T_{i\theta}(\theta \neq i, \theta = 1,2,\cdots,s)$ has an effect on the initial aggregation matrix of relationship R_{ij} with respect to *CRs* and *ECs*, a higher $L_{i\xi}(\xi \neq i, \xi = 1,2,\cdots,n)$ or $T_{i\theta}(\theta \neq i, \theta = 1,2,\cdots,s)$ implies a benefit to R_{ij}. Consequently, the correlations are taken into account when modifying the relationships between *CRs* and *ECs*. In the process of adjustment, Equation (13) is applied to integrate $L_{i\xi}(\xi \neq i, \xi = 1,2,\cdots,n)$ and $T_{i\theta}(\theta \neq i, \theta = 1,2,\cdots,s)$ into R_{ij}. Furthermore, the modified relationship is computed using the formula as follows:

$$R'_{ij} = \Delta\left(\prod_{v=1}^{V} \Delta^{-1}(s_m, \alpha_m)^{\gamma_v}\right), i = 1,2,\cdots,n; j = 1,2,\cdots,s \tag{13}$$

where $\Delta^{-1}(r_m, a_m)$ is stemming from the set $S = \left\{\Delta^{-1}R_{ij}(r_{ij}^{n(t)_u}, \alpha_{ij}^{n(t)_u}), \Delta^{-1}L_{i\xi}\left(r_{i\xi}^{n(t)_u}, \alpha_{i\xi}^{n(t)_u}\right), \Delta^{-1}T_{j\theta}(r_{j\theta}^{n(t)_u}, \alpha_{j\theta}^{n(t)_u})\right\} \xi \neq i, \theta \neq j$. In addition, the weight of γ_v is corresponding to the proportion of $\Delta^{-1}(r_m, a_m)$. For the sake of reducing the impact from subjectivity, unduly high or unduly low preference values in the correlation matrices are supposed to possess a low weight under the circumstances. That means only moderated assessment giving a higher weight has a small deviation from the true value, which might be advocated in the process of evaluation. Therefore, the weight can be determined by Equation (14).

$$\gamma_v = \frac{\mathrm{sim}((s_m, \alpha_m), (\bar{s}, \bar{\alpha}))}{\sum\limits_{v=1}^{V} sim((s_m, \alpha_m), (\bar{s}, \bar{\alpha}))} \tag{14}$$

Step 7. Integrated *ECs* priorities are determined.

On account of the inconformity of representation, the 2-tuple linguistic form of the relationship matrix, and the numerical value of *CRs* importance, the integrated *ECs* priority $S_{TC_j}(j = 1,2,\cdots,s)$ is calculated by Equation (11).

Step 8. Basic priority of ECs is confirmed.

The linguistic distance $d((s_{TC_j}^{n(t)}, \alpha^{n(t)})(s_{\min}^{n(t)}, \alpha^{n(t)}))$ can be adopted to measure the importance degree, where $(s_{\min}^{n(t)}, \alpha^{n(t)})$ is the minimum value of linguistic term set. Furthermore, the measurement of 2-tuple linguistic distance decides the importance of $EC_j(j = 1,2,\cdots,s)$. Therefore, the normative value of basic priority bpr_j is computed as follows:

$$bpr_j = \frac{d((s_{EC_j}^{n(t)}, \alpha^{n(t)})(s_{\min}^{n(t)}, \alpha^{n(t)}))}{\sum\limits_{j=1}^{s} d((s_{EC_j}^{n(t)}, \alpha^{n(t)})(s_{\min}^{n(t)}, \alpha^{n(t)}))} \tag{13}$$

Step 9. End.

The flow chart of the whole procedures is shown in Figure 2.

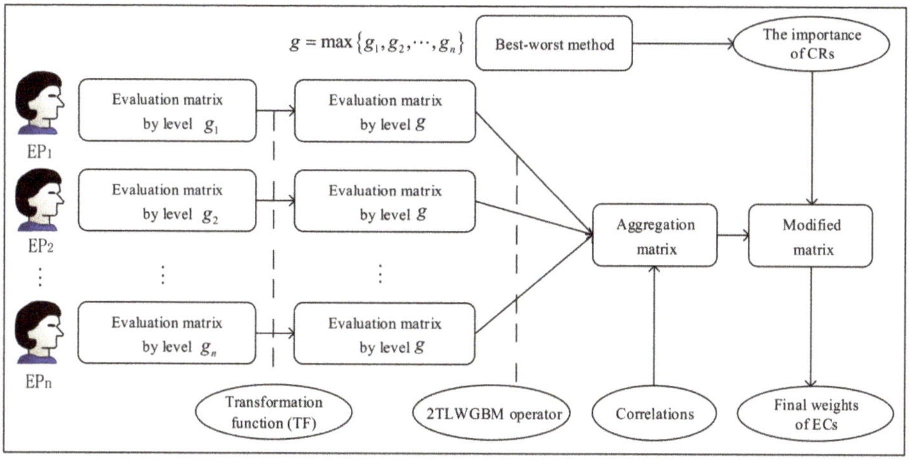

Figure 2. Group decision making for multi-granularity 2-tuple linguistic preference to prioritize engineering characteristics in quality function deployment.

4. Case Study

4.1. Background

The Wuhan metro station is the most common two-floor island structure, which consists mainly of a platform and a station hall. The underground floor is the station hall floor. As shown in Figure 3, the metro station has four main exits and one reserved outlet for docking with the shopping mall and fire curtains are installed at each exit. Therefore, when a crowd passes through the fire curtain in the emergency evacuation process, they have reached the safe area. The station hall floor has four automatic ticket checkers and two emergency dedicated channels. In emergency situations, an automatic ticket checking machine and emergency dedicated channels are in open state. The second underground floor is the platform layer. When an emergency occurs on the platform layer, the crowd must first ascend to the station hall layer and then evacuate through the safety exit.

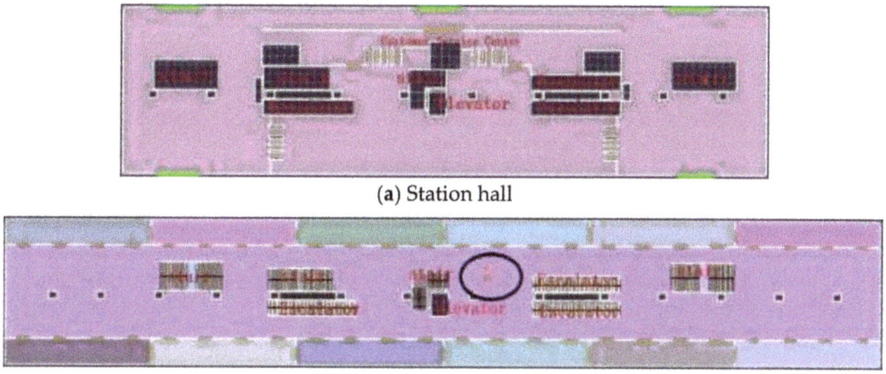

(**a**) Station hall

(**b**) Platform

Figure 3. The structure of metro station in Wuhan.

Taking regional *S* as an example, we analyze the influence factors that have an effect on the evacuation route planning in this area. Five *CRs* and *ECs* are selected in order to determine the weight degree of *ECs*, which can be a basic of evaluating emergency routes.

CR_1 : Expected evacuation time	EC_1 : Number of evacuees per unit time
CR_2 : Crowd density	EC_2 : Managerial capability
CR_3 : Risk level in the region	EC_3 : Risk level of disaster
CR_4 : Possibility of congestion	EC_4 : Organizational situation
CR_5 : Evacuation capability	EC_5 : Evacuation equipment

4.2. Implementation

Step 1. The evaluation *CRs* relationships by passengers are shown in Tables 2 and 3.

Table 2. Best-to-others (BO) vector for passengers.

Passengers	Best	CR_1	CR_2	CR_3	CR_4	CR_5
1	CR_2	3	1	5	9	7
2	CR_2	5	1	4	9	8
3	CR_1	1	2	7	5	9
4	CR_3	4	3	1	7	9
5	CR_1	1	3	9	6	8

Table 3. Others-to-worst (OW) vector for passengers.

Passengers	1	2	3	4	5
Worst	CR_4	CR_4	CR_5	CR_5	CR_3
CR_1	6	4	9	4	9
CR_2	9	9	8	7	7
CR_3	5	7	2	9	1
CR_4	1	1	5	2	6
CR_5	4	3	1	1	3

Step 2. The importance of *CRs* is respectively computed as 0.302, 0.359, 0.187, 0.082 and 0.070, which is determined by the average value by passengers. For example, the model by first passenger is established as follows:

$$\begin{aligned} &\min \xi \\ &s.t. |w_2 - 3w_1| \leq \xi, |w_2 - 5w_3| \leq \xi, \\ &|w_2 - 9w_4| \leq \xi, |w_2 - 7w_5| \leq \xi, \\ &|w_1 - 6w_4| \leq \xi, |w_3 - 5w_4| \leq \xi, \\ &|w_5 - 4w_4| \leq \xi, \sum_{5} w_j = 1, \\ &w_j \geq 0, j = 1, 2, \cdots, 5. \end{aligned}$$

The parameter ξ is obtained as 0.12, and the consistency ratio can be then computed using Equation (12) as 0.023, which indicates it has good consistency.

Step 3. In order to determine the basic priority of these *ECs*, three experts EP_1, EP_2, EP_3 evaluate the importance of *ECs* according to *CRs* given as below (Tables 4–6). They represent preference by using the different linguistic term sets $S_i^{7_1} = \{s_0^7, s_1^7, s_2^7, s_3^7, s_4^7, s_5^7, s_6^7\}$ $S_i^{5_2} = \{s_0^5, s_1^5, s_2^5, s_3^5, s_4^5\}$ $S_i^{9_3} = \{s_0^9, s_1^9, s_2^9, s_3^9, s_4^9, s_5^9, s_6^9, s_7^9, s_8^9\}$.

Table 4. Evaluation matrix R^1 for EP_1.

	EC_1	EC_2	EC_3	EC_4	EC_5
CR_1	$s_5^{7_1}$	$s_6^{7_1}$	$s_0^{7_1}$	$s_4^{7_1}$	$s_0^{7_1}$
CR_2	$s_1^{7_1}$	$s_5^{7_1}$	$s_1^{7_1}$	$s_3^{7_1}$	$s_4^{7_1}$
CR_3	$s_2^{7_1}$	$s_4^{7_1}$	$s_4^{7_1}$	$s_1^{7_1}$	$s_1^{7_1}$
CR_4	$s_3^{7_1}$	$s_5^{7_1}$	$s_1^{7_1}$	$s_6^{7_1}$	$s_5^{7_1}$
CR_5	$s_1^{7_1}$	$s_4^{7_1}$	$s_1^{7_1}$	$s_4^{7_1}$	$s_5^{7_1}$

Table 5. Evaluation matrix R^2 for EP_2.

	EC_1	EC_2	EC_3	EC_4	EC_5
CR_1	$s_3^{5_2}$	$s_4^{5_2}$	$s_0^{5_2}$	$s_3^{5_2}$	$s_1^{5_2}$
CR_2	$s_1^{5_2}$	$s_3^{5_2}$	$s_2^{5_2}$	$s_2^{5_2}$	$s_3^{5_2}$
CR_3	$s_1^{5_2}$	$s_2^{5_2}$	$s_3^{5_2}$	$s_1^{5_2}$	$s_1^{5_2}$
CR_4	$s_2^{5_2}$	$s_3^{5_2}$	$s_1^{5_2}$	$s_4^{5_2}$	$s_3^{5_2}$
CR_5	$s_0^{5_2}$	$s_3^{5_2}$	$s_1^{5_2}$	$s_2^{5_2}$	$s_3^{5_2}$

Table 6. Evaluation matrix R^3 for EP_3.

	EC_1	EC_2	EC_3	EC_4	EC_5
CR_1	$s_6^{9_3}$	$s_8^{9_3}$	$s_1^{9_3}$	$s_5^{9_3}$	$s_0^{9_3}$
CR_2	$s_2^{9_3}$	$s_7^{9_3}$	$s_2^{9_3}$	$s_6^{9_3}$	$s_6^{9_3}$
CR_3	$s_0^{9_3}$	$s_6^{9_3}$	$s_6^{9_3}$	$s_2^{9_3}$	$s_4^{9_3}$
CR_4	$s_3^{9_3}$	$s_7^{9_3}$	$s_2^{9_3}$	$s_8^{9_3}$	$s_6^{9_3}$
CR_5	$s_1^{9_3}$	$s_7^{9_3}$	$s_1^{9_3}$	$s_7^{9_3}$	$s_7^{9_3}$

Step 4. Three evaluation matrices are transformed into 2-tuple representation in Tables 7–9.

Table 7. 2-tuple linguistic evaluation matrix $\widetilde{R}^1$ for EP_1.

	EC_1	EC_2	EC_3	EC_4	EC_5
CR_1	$\left(s_5^{7_1}, 0\right)$	$\left(s_6^{7_1}, 0\right)$	$\left(s_0^{7_1}, 0\right)$	$\left(s_4^{7_1}, 0\right)$	$\left(s_0^{7_1}, 0\right)$
CR_2	$\left(s_1^{7_1}, 0\right)$	$\left(s_5^{7_1}, 0\right)$	$\left(s_1^{7_1}, 0\right)$	$\left(s_3^{7_1}, 0\right)$	$\left(s_4^{7_1}, 0\right)$
CR_3	$\left(s_2^{7_1}, 0\right)$	$\left(s_4^{7_1}, 0\right)$	$\left(s_4^{7_1}, 0\right)$	$\left(s_1^{7_1}, 0\right)$	$\left(s_1^{7_1}, 0\right)$
CR_4	$\left(s_3^{7_1}, 0\right)$	$\left(s_5^{7_1}, 0\right)$	$\left(s_1^{7_1}, 0\right)$	$\left(s_6^{7_1}, 0\right)$	$\left(s_5^{7_1}, 0\right)$
CR_5	$\left(s_1^{7_1}, 0\right)$	$\left(s_4^{7_1}, 0\right)$	$\left(s_1^{7_1}, 0\right)$	$\left(s_4^{7_1}, 0\right)$	$\left(s_5^{7_1}, 0\right)$

Table 8. 2-tuple linguistic evaluation matrix $\widetilde{R}^2$ for EP_2.

	EC_1	EC_2	EC_3	EC_4	EC_5
CR_1	$\left(s_3^{5_2}, 0\right)$	$\left(s_4^{5_2}, 0\right)$	$\left(s_0^{5_2}, 0\right)$	$\left(s_3^{5_2}, 0\right)$	$\left(s_1^{5_2}, 0\right)$
CR_2	$\left(s_1^{5_2}, 0\right)$	$\left(s_3^{5_2}, 0\right)$	$\left(s_2^{5_2}, 0\right)$	$\left(s_2^{5_2}, 0\right)$	$\left(s_3^{5_2}, 0\right)$
CR_3	$\left(s_1^{5_2}, 0\right)$	$\left(s_2^{5_2}, 0\right)$	$\left(s_3^{5_2}, 0\right)$	$\left(s_1^{5_2}, 0\right)$	$\left(s_1^{5_2}, 0\right)$
CR_4	$\left(s_2^{5_2}, 0\right)$	$\left(s_3^{5_2}, 0\right)$	$\left(s_1^{5_2}, 0\right)$	$\left(s_4^{5_2}, 0\right)$	$\left(s_3^{5_2}, 0\right)$
CR_5	$\left(s_0^{5_2}, 0\right)$	$\left(s_3^{5_2}, 0\right)$	$\left(s_1^{5_2}, 0\right)$	$\left(s_2^{5_2}, 0\right)$	$\left(s_3^{5_2}, 0\right)$

Table 9. 2-tuple linguistic evaluation matrix $\widetilde{R}^3$ for EP_3.

	EC_1	EC_2	EC_3	EC_4	EC_5
CR_1	$\left(s_6^{9_3},0\right)$	$\left(s_8^{9_3},0\right)$	$\left(s_1^{9_3},0\right)$	$\left(s_5^{9_3},0\right)$	$\left(s_0^{9_3},0\right)$
CR_2	$\left(s_2^{9_3},0\right)$	$\left(s_7^{9_3},0\right)$	$\left(s_2^{9_3},0\right)$	$\left(s_6^{9_3},0\right)$	$\left(s_6^{9_3},0\right)$
CR_3	$\left(s_0^{9_3},0\right)$	$\left(s_6^{9_3},0\right)$	$\left(s_6^{9_3},0\right)$	$\left(s_2^{9_3},0\right)$	$\left(s_4^{9_3},0\right)$
CR_4	$s_3^{9_3}$	$\left(s_7^{9_3},0\right)$	$\left(s_2^{9_3},0\right)$	$\left(s_8^{9_3},0\right)$	$\left(s_6^{9_3},0\right)$
CR_5	$\left(s_1^{9_3},0\right)$	$\left(s_7^{9_3},0\right)$	$\left(s_1^{9_3},0\right)$	$\left(s_7^{9_3},0\right)$	$\left(s_7^{9_3},0\right)$

Step 5. The aggregation of all the evaluation matrices in Tables 9–11 applying 2TLWGBM operator in Equation (11) into R_{ij} is shown in Table 12.

Table 10. The transformed 2-tuple linguistic evaluation matrix $\widetilde{R}^{1'}$ for EP_1.

	EC_1	EC_2	EC_3	EC_4	EC_5
CR_1	$(s_7^{9_1},-0.33)$	$(s_8^{9_1},0)$	$(s_0^{9_1},0)$	$(s_5^{9_1},0.33)$	$(s_0^{9_1},0)$
CR_2	$(s_1^{9_1},0.33)$	$(s_7^{9_1},-0.33)$	$(s_1^{9_1},0.33)$	$(s_4^{9_1},0)$	$(s_5^{9_1},0.33)$
CR_3	$(s_3^{9_1},-0.33)$	$(s_5^{9_1},0.33)$	$(s_5^{9_1},0.33)$	$(s_1^{9_1},0.33)$	$(s_1^{9_1},0.33)$
CR_4	$(s_4^{9_1},0)$	$(s_7^{9_1},-0.33)$	$(s_1^{9_1},0.33)$	$\left(s_8^{9_3},0\right)$	$(s_7^{9_1},-0.33)$
CR_5	$(s_1^{9_1},0.33)$	$(s_5^{9_1},0.33)$	$(s_1^{9_1},0.33)$	$(s_5^{9_1},0.33)$	$(s_7^{9_1},-0.33)$

Table 11. The transformed 2-tuple linguistic evaluation matrix $\widetilde{R}^{2'}$ for EP_2.

	EC_1	EC_2	EC_3	EC_4	EC_5
CR_1	$(s_6^{9_2},0)$	$(s_8^{9_2},0)$	$(s_0^{9_1},0)$	$(s_6^{9_2},0)$	$(s_2^{9_2},0)$
CR_2	$(s_2^{9_2},0)$	$(s_6^{9_2},0)$	$(s_4^{9_2},0)$	$(s_4^{9_1},0)$	$(s_6^{9_2},0)$
CR_3	$(s_2^{9_2},0)$	$(s_4^{9_2},0)$	$(s_6^{9_2},0)$	$(s_2^{9_2},0)$	$(s_2^{9_2},0)$
CR_4	$(s_4^{9_2},0)$	$(s_6^{9_2},0)$	$(s_2^{9_2},0)$	$(s_8^{9_2},0)$	$(s_6^{9_2},0)$
CR_5	$(s_0^{9_2},0)$	$(s_6^{9_2},0)$	$(s_2^{9_2},0)$	$(s_4^{9_2},0)$	$(s_6^{9_2},0)$

Table 12. The aggregation of all the evaluation matrices.

	EC_1	EC_2	EC_3	EC_4	EC_5
CR_1	$(s_2^9,-0.14)$	$(s_2^9,0.05)$	$(s_0^9,0)$	$(s_2^9,-0.23)$	$(s_0^9,0)$
CR_2	$(s_1^9,0.23)$	$(s_2^9,-0.08)$	$(s_1^9,0.33)$	$(s_2^9,-0.26)$	$(s_2^9,-0.16)$
CR_3	$(s_1^9,-0.23)$	$(s_2^9,-0.23)$	$(s_2^9,-0.16)$	$(s_1^9,0.23)$	$(s_1^9,0.41)$
CR_4	$(s_2^9,-0.48)$	$(s_2^9,-0.08)$	$(s_1^9,0.23)$	$(s_2^9,0.05)$	$(s_2^9,-0.14)$
CR_5	$(s_1^9,-0.35)$	$(s_2^9,-0.1)$	$(s_1^9,0.1)$	$(s_2^9,-0.17)$	$(s_2^9,-0.08)$

Step 6. On the basic of different knowledge and experience, three experts adopt their own linguistic representations to evaluate correlations between *CRs* and *ECs*. These matrices are then aggregated in the same way as the fourth step. Consequently, the initial HOQ is shown in Figure 4.

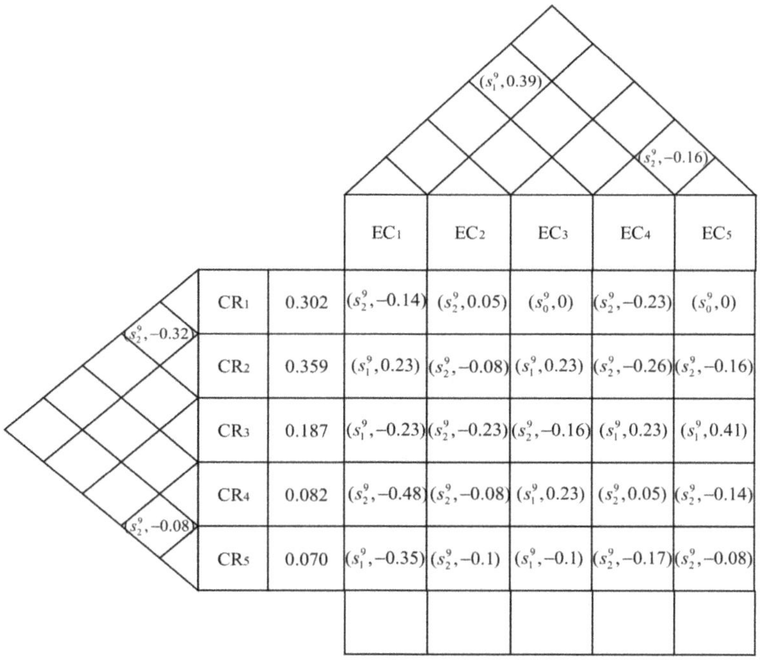

Figure 4. The 2-tuple initial HOQ.

In Figure 4, the correlations between *CRs* and *ECs* are computed in the same way as the relationships between *CRs* and *ECs* are treated. Apparently, an appropriate relationship matrix should take correlations into account, so the modified relationship in virtue of Equations (5)–(7), (12), and (13) is obtained. The result is illustrated in Figure 5.

We take the relationship between CR_1 and EC_1 for an example, the process of calculation is demonstrated as follow

$$(\bar{s},\bar{\alpha}) = \triangle(\tfrac{1}{3}(\triangle^{-1}(s_2^9,-0.14) + (\triangle^{-1}(s_2^9,-0.32) + (\triangle^{-1}(s_1^9,0.39)) \\ = (s_2^9,-0.36)$$

$$d((s_2^9,-0.14),(s_2^9,-0.36)) = \frac{|\triangle^{-1}(2-0.14) - \triangle^{-1}(2-0.36)|}{9} = 0.024$$

Similarly, $d((s_2^9,-0.32),(s_2^9,-0.36)) = 0.004, d((s_1^9,0.39),(s_2^9,-0.36)) = 0.028$

$$sim((s_2^9,-0.14),(s_2^9,-0.36)) = 1 - \frac{0.024}{0.024+0.004+0.028} = 0.571$$

Similarly, $sim((s_2^9,-0.32),(s_2^9,-0.36)) = 0.928, sim((s_1^9,0.39),(s_2^9,-0.36)) = 0.5$
We then compute the weight γ_v by Equation (13)

$$\gamma(s_2^9,-0.14) = \frac{0.571}{0.571+0.928+0.5} = 0.286$$

In the same way, $\gamma(s_2^9,-0.32) = 0.464, \gamma(s_1^9,0.39) = 0.25$

The modified relationship is expressed

$$R'_{11} = \triangle(\triangle^{-1}(s_2^9,-0.14)^{0.286} * \triangle^{-1}(s_2^9,-0.32)^{0.464} * \triangle^{-1}(s_1^9,0.39)^{0.25}) = (s_2^9,-0.35)$$

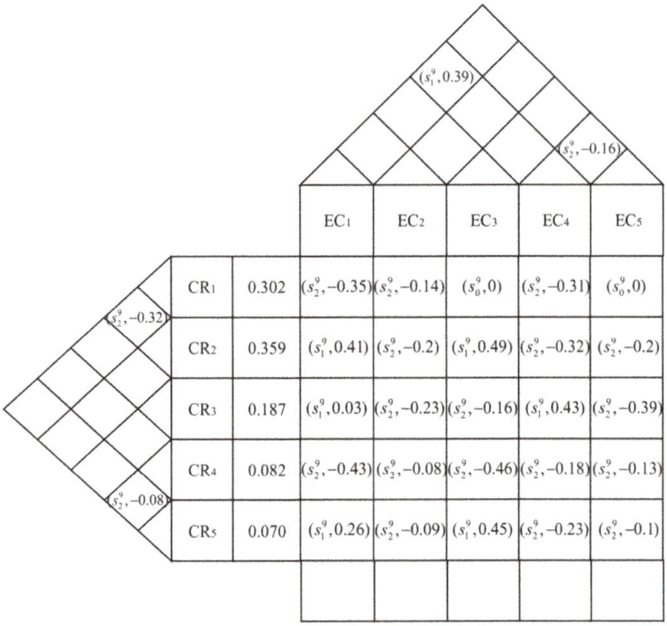

Figure 5. The 2-tuple modified HOQ.

Step 7. After obtaining the modified matrix, the importance of *CRs* should be integrated to reach the final relationships between *CRs* and *ECs*. The result is presented in Table 13. Therefore, the rank of integrated *ECs* priority is $EC_2 \succ EC_4 \succ EC_1 \succ EC_5 \succ EC_3$.

Table 13. The integrated *ECs* priority.

	EC_1	EC_2	EC_3	EC_4	EC_5
Priority	$(s_1^9, 0.07)$	$(s_1^9, 0.13)$	$(s_1^9, -0.18)$	$(s_1^9, 0.11)$	$(s_1^9, -0.16)$

Step 8. The basic priority of *ECs* is computed according to Equations (7) and (14) and Table 13. The minimum value of linguistic term set is $(s_{\min}^{n(t)}, \alpha^{n(t)}) = (s_0^9, 0)$. The ultimate weights of *ECs* are $(s_0^9, 0.215)$ $(s_0^9, 0.228)$ $(s_0^9, 0.165)$ $(s_0^9, 0.223)$ $(s_0^9, 0.169)$.

Step 9. End.

4.3. Managerial Tips

The outcomes of this study are beneficial to planning and selecting the appropriate emergency routes. Moreover, the ranking result can be outlined that decision makers should be paid more attention to management ability. The result indicates that crowd density has a significant influence on emergency route evaluation. Subsequently decision makers should concentrate on these two aspects in order to design and select emergency routes.

In addition, the proposed model is sufficient robust and could be easily implemented in practices for GDM problems. DMs can choose their linguistic preference to evaluate the correlation and relationship between *CRs* and *ECs*. Furthermore, the importance of *ECs* can be adjusted appropriately according to the actual circumstance.

5. Conclusions and Future Research

A systematic GDM approach for prioritizing *ECs* in QFD under the multi-granularity 2-tuple linguistic environment is proposed in this paper. The provided method allows experts from QFD team to evaluate the relationship and correlations between *CRs* and *ECs* in accordance with their experience and preference. For the sake of guaranteeing accurate information, the 2-tuple linguistic representation addressing the vague and imprecise information is utilized. Based on the linguistic hierarchy, different granularities originating from different experts are translated into a basic linguistic term we set in advance. The BWM is applied to determine the importance of *CRs*, which is simple and quick to represent customers' advice.

BM can capture inter-relationships among the aggregated information by taking the conjunction among each pairs of aggregated arguments, for instance, correlations among *CRs*. Therefore, the 2TLWGBM operator is applied to aggregate the evaluation matrix and the importance of *CRs*. In addition, correlations could have an impact on relationship between *CRs* and *ECs*. A modified matrix reflecting the influence is determined in this paper. Compared with other approaches in terms of calculating weight, a method that can lessen the subjectivity of assessment is put forward. Finally, a case study has been calculated and is presented to verify the effectiveness of the proposed method.

In this study, prioritizing *ECs* in QFD is extended to 2-tuple linguistic environment, in which all evaluation matrices from experts are represented by 2-tuple. For one thing, an appropriate and applicable BM operator is employed to deal with the aggregation problem, which should be suitable for accurately prioritizing *ECs* in QFD. Moreover, the degree of similarity is introduced to determine the weight that responds to the effect of correlations, which could obtain a more objective modified matrix.

In future research, the proposed method can be applied to supplier selection, green buildings and new product development. In addition, other GDM approaches can be integrated into QFD to rank the *ECs*, and consensus can be considered. A more reasonable aggregation operator should be developed and applied to QFD. In real life, plenty of problems might be complex and changeful. Establishing a dynamic HOQ is necessary.

Author Contributions: Y.M. drafted the initial manuscript and conceived the model framework. Y.L. provided the relevant literature review and the illustrated example. Y.T. revised the manuscript and analyzed the data.

Funding: This research is supported by the National Social Science Foundation of China (Project No. 15AGL021), the National Natural Science Foundation of China (NSFC) under Project 71801177, and the Project of Humanities and Social Sciences (18YJC63016).

Acknowledgments: The authors would like to thank the anonymous referees and academic editor for their very valuable comments.

Conflicts of Interest: The author declares no conflict of interest.

References

1. Yan, H.B.; Ma, T.J.; Huynh, V.N. Coping with group behaviors in uncertain quality function deployment. *Decis. Sci.* **2014**, *456*, 1025–1052. [CrossRef]
2. Ignatius, J.; Rahman, A.; Yazdani, M.; Aparauskas, J.Å.; Haron, S.H. An integrated fuzzy ANP–QFD approach for green building assessment. *J. Civ. Eng. Manag.* **2016**, *224*, 551–563. [CrossRef]
3. Franceschini, F.; Galetto, M.; Maisano, D.; Mastrogiacomo, L. Prioritisation of engineering characteristics in QFD in the case of customer requirements orderings. *Int. J. Prod. Res.* **2015**, *5313*, 3975–3988. [CrossRef]
4. Yan, H.B.; Ma, T.J. A group decision-making approach to uncertain quality function deployment based on fuzzy preference relation and fuzzy majority. *Eur. J. Oper. Res.* **2015**, *2413*, 815–829. [CrossRef]
5. Liu, C.H.; Wu, H.H. A fuzzy group decision-making approach in quality function deployment. *Qual. Quant.* **2008**, *424*, 527–540. [CrossRef]
6. Kwong, C.K.; Ye, Y.; Chen, Y.; Choy, K.L. A novel fuzzy group decision-making approach to prioritising engineering characteristics in QFD under uncertainties. *Int. J. Prod. Res.* **2011**, *4919*, 5801–5820. [CrossRef]

7. Karsak, E.E. Fuzzy multiple objective programming framework to prioritize design requirements in quality function deployment. *Comput. Ind. Eng.* **2004**, *472*, 149–163. [CrossRef]
8. Chen, L.H.; Weng, M.C. An evaluation approach to engineering design in QFD processes using fuzzy goal programming models. *Eur. J. Oper. Res.* **2006**, *1721*, 230–248. [CrossRef]
9. Kwong, C.K.; Chen, Y.; Bai, H.; Chan, D.S.K. A methodology of determining aggregated importance of engineering characteristics in QFD. *Comput. Ind. Eng.* **2007**, *534*, 667–679. [CrossRef]
10. Liu, J.; Chen, Y.Z.; Zhou, J.; Yi, X.J. An exact expected value-based method to prioritize engineering characteristics in fuzzy quality function deployment. *Int. J. Fuzzy Syst.* **2016**, *184*, 630–646. [CrossRef]
11. Geng, X.L.; Chu, X.N.; Xue, D.Y.; Zhang, Z.F. An integrated approach for rating engineering characteristics' final importance in product-service system development. *Comput. Ind. Eng.* **2010**, *594*, 585–594. [CrossRef]
12. Wang, Y.M. A fuzzy-normalisation-based group decision-making approach for prioritising engineering design requirements in QFD under uncertainty. *Int. J. Prod. Res.* **2012**, *5023*, 6963–6977. [CrossRef]
13. Chen, L.H.; Ko, W.C.; Tseng, C.Y. Fuzzy approaches for constructing house of quality in QFD and its applications: A group decision-making method. *IEEE Trans. Eng. Manag.* **2013**, *601*, 77–87. [CrossRef]
14. Zhang, Z.F.; Chu, X.N. Fuzzy group decision-making for multi-format and multi-granularity linguistic judgments in quality function deployment. *Expert Syst. Appl.* **2009**, *365*, 9150–9158. [CrossRef]
15. Wang, X.T.; Xiong, W. An integrated linguistic-based group decision-making approach for quality function deployment. *Expert Syst. Appl.* **2011**, *3812*, 14428–14438. [CrossRef]
16. Wang, Z.Q.; Fung, R.Y.K.; Li, Y.L.; Pu, Y. A group multi-granularity linguistic-based methodology for prioritizing engineering characteristics under uncertainties. *Comput. Ind. Eng.* **2016**, *91*, 178–187. [CrossRef]
17. Bo, C.; Zhang, X.; Shao, S.; Smarandache, F. Multi-granulation neutrosophic rough sets on a single domain and dual domains with applications. *Symmetry* **2018**, *10*, 296. [CrossRef]
18. Xu, Z.S. Multiple-attribute group decision making with different formats of preference information on attributes. *IEEE Trans. Syst. Man Cybern. Part B* **2007**, *376*, 1500–1511.
19. Xu, Z.S. Group decision making based on multiple types of linguistic preference relations. *Inf. Sci.* **2008**, *1782*, 452–467. [CrossRef]
20. Martínez, L.; Herrera, F. An overview on the 2-tuple linguistic model for computing with words in decision making: Extensions, applications and challenges. *Inf. Sci.* **2012**, *207*, 1–18. [CrossRef]
21. Ju, Y.B.; Liu, X.Y.; Wang, A.H. Some new Shapley 2-tuple linguistic Choquet aggregation operators and their applications to multiple attribute group decision making. *Soft Comput.* **2010**, *20*, 4037–4053. [CrossRef]
22. Herrera, F.; Herrera-Viedma, E.; Martínez, L. A fusion approach for managing multi-granularity linguistic term sets in decision making. *Fuzzy Set. Syst.* **2000**, *114*, 43–58. [CrossRef]
23. Herrera, F.; Martínez, L. A model based on linguistic 2-tuples for dealing with multigranular hierarchical linguistic contexts in multi-expert decision-making. *IEEE Trans. Syst. Man Cybern. Part B* **2001**, *312*, 227–234. [CrossRef] [PubMed]
24. Espinilla, M.; Liu, J. Martinez, L. An extended hierarchical linguistic model for decision-making problems. *Comput. Intell.* **2011**, *27*, 489–512. [CrossRef]
25. Rezaei, J. Best-worst multi-criteria decision-making method. *Omega* **2015**, *53*, 49–57. [CrossRef]
26. Stević, Ž.; Pamučar, D.; Kazimieras Zavadskas, E.; Ćirović, G.; Prentkovskis, O. The selection of wagons for the internal transport of a logistics company: A novel approach based on rough BWM and rough SAW methods. *Symmetry* **2017**, *9*, 264. [CrossRef]
27. Herrera, F.; Martinez, L. A 2-tuple fuzzy linguistic representation model for computing with words. *IEEE Trans. Fuzzy Syst.* **2000**, *86*, 746–752.
28. Wei, G.W.; Zhao, X.F. Some dependent aggregation operators with 2-tuple linguistic information and their application to multiple attribute group decision making. *Expert Syst. Appl.* **2012**, *395*, 5881–5886. [CrossRef]
29. Qin, J.D.; Liu, X.W.; Pedrycz, W. Hesitant fuzzy Maclaurin symmetric mean operators and its application to multiple-attribute decision making. *Int. J. Fuzzy Syst.* **2015**, *174*, 509–520. [CrossRef]
30. Qin, J.D.; Liu, X.W. 2-tuple linguistic Muirhead mean operators for multiple attribute group decision making and its application to supplier selection. *Kybernetes* **2016**, *451*, 2–29. [CrossRef]
31. Wang, J.; Wei, G.; Wei, Y. Models for green supplier selection with some 2-tuple linguistic neutrosophic number Bonferroni mean operators. *Symmetry* **2018**, *10*, 131. [CrossRef]

32. Wang, L.; Wang, Y.; Liu, X. Prioritized aggregation operators and correlated aggregation operators for hesitant 2-tuple linguistic variables. *Symmetry* **2018**, *10*, 39. [CrossRef]
33. Jiang, X.P.; Wei, G.W. Some Bonferroni mean operators with 2-tuple linguistic information and their application to multiple attribute decision making. *J. Intell. Fuzzy Syst.* **2014**, *275*, 2153–2162.

symmetry

MDPI

Article

The Structure Theorems of Pseudo-BCI Algebras in Which Every Element is Quasi-Maximal

Xiaoying Wu [1] and Xiaohong Zhang [1,2,*]

1 Department of Mathematics, Shaanxi University of Science & Technology, Xi'an 710021, China; 46018@sust.edu.cn
2 Department of Mathematics, Shanghai Maritime University, Shanghai 201306, China
* Correspondence: zhangxh@shmtu.edu.cn or zhangxiaohong@sust.edu.cn

Received: 19 September 2018; Accepted: 29 September 2018; Published: 8 October 2018

Abstract: For mathematical fuzzy logic systems, the study of corresponding algebraic structures plays an important role. Pseudo-BCI algebra is a class of non-classical logic algebras, which is closely related to various non-commutative fuzzy logic systems. The aim of this paper is focus on the structure of a special class of pseudo-BCI algebras in which every element is quasi-maximal (call it QM-pseudo-BCI algebras in this paper). First, the new notions of quasi-maximal element and quasi-left unit element in pseudo-BCK algebras and pseudo-BCI algebras are proposed and some properties are discussed. Second, the following structure theorem of QM-pseudo-BCI algebra is proved: every QM-pseudo-BCI algebra is a KG-union of a quasi-alternating BCK-algebra and an anti-group pseudo-BCI algebra. Third, the new notion of weak associative pseudo-BCI algebra (WA-pseudo-BCI algebra) is introduced and the following result is proved: every WA-pseudo-BCI algebra is a KG-union of a quasi-alternating BCK-algebra and an Abel group.

Keywords: fuzzy logic; pseudo-BCI algebra; quasi-maximal element; KG-union; quasi-alternating BCK-algebra

1. Introduction

In the study of t-norm based fuzzy logic systems [1–9], algebraic systems (such as residuated lattices, BL-algebras, MTL-algebras, pseudo-BL algebras, pseudo-MTL algebras, et al.) play an important role. In this paper, we discuss pseudo-BCI/BCK algebras which are connected with non-commutative fuzzy logic systems (such that non-commutative residuared lattices, pseudo-BL/pseudo-MTL algebras).

BCK-algebras and BCI-algebras were introduced by Iséki [10] as algebras induced by Meredith's implicational logics BCK and BCI. The name of BCK-algebra and BCI-algebra originates from the combinatories B, C, K, I in combinatory logic. The notion of pseudo-BCK algebra was introduced by G. Georgescu and A. Iorgulescu in [11] as a non-commutative extension of BCK-algebras. Then, as common generalization of pseudo-BCK algebras and BCI-algebras, W.A. Dudek and Y.B. Jun introduced the concept of pseudo-BCI algebra in [12]. In fact, there are many other non-classical logic algebraic systems related to BCK- and BCI-algebras, such as BCC-algebra, BZ-algebra and so forth, some monographs and papers on these topics can be found in [7–9,13–18].

Pseudo-BCI-algebras are algebraic models of some extension of a noncommutative version of the BCI-logic, the corresponding logic is called pseudo-BCI logic [19]. P. Emanovský and J. Kühr studied some properties of pseudo-BCI algebras, X.L. Xin et al. [20] investigated monadic pseudo BCI-algebras and corresponding logics and some authors discussed the filter (ideal) theory of pseudo-BCI algebras [21–28]. Moreover, some notions of period, state and soft set are applied to pseudo-BCI algebras [29–31].

In this paper, we further study the structure characterizations of pseudo-BCI algebras. By using the notions of quasi-maximal element, quasi-left unit element, KG-union and direct product, we give the structure theorem of the class of pseudo-BCI algebras in which every element is quasi-maximal (call they QM-pseudo-BCI algebras). Moreover, we introduce weak associative property in pseudo-BCI algebras, discuss basic properties of weak associative pseudo-BCI algebra (WA-pseudo-BCI algebra) and establish the structure theorem of WA-pseudo-BCI algebra.

It should be noted that the original definition of pseudo-BCI/BCK algebra is different from the definition used in this paper. They are dual. We think that the logical semantics of this algebraic structure can be better represented by using the present definition.

2. Preliminaries

Definition 1 ([10,16]). *An algebra $(A; \rightarrow, 1)$ of type (2,0) is called a BCI-algebra if the following conditions are satisfied for all x, y, z from A:*

(1) $x \rightarrow y \leq (y \rightarrow z) \rightarrow (x \rightarrow z)$,
(2) $x \leq (x \rightarrow y) \rightarrow y$,
(3) $x \leq x$,
(4) $x \leq y$, $y \leq x$ *imply* $x = y$, *where* $x \leq y$ *means* $x \rightarrow y = 1$. *An algebra $(A; \rightarrow, 1)$ of type (2,0) is called a BCK-algebra if it is a BCI-algebra and satisfies:*
(5) $x \rightarrow 1 = 1,\ \forall x \in A$.

Definition 2 ([10,16]). *A BCK-algebra $(A; \rightarrow, 1)$ is called bounded if there exists unique element 0 such that $0 \rightarrow x = 1$ for any $x \in A$.*

Definition 3 ([13,14]). *A BCK-algebra $(A; \rightarrow, 1)$ is called quasi-alternating BCK-algebra if it satisfies the following axiom: $\forall\, x, y \in X$, $x \neq y$ implies $x \rightarrow y = y$.*

Definition 4 ([9,11]). *A pseudo-BCK algebra is a structure $(A; \leq, \rightarrow, \rightsquigarrow, 1)$, where "$\leq$" is a binary relation on A, "$\rightarrow$" and "$\rightsquigarrow$" are binary operations on A and "1" is an element of A, verifying the axioms: for all $x, y, z \in A$,*

(1) $x \rightarrow y \leq (y \rightarrow z) \rightsquigarrow (x \rightarrow z)$, $x \rightsquigarrow y \leq (y \rightsquigarrow z) \rightarrow (x \rightsquigarrow z)$,
(2) $x \leq (x \rightarrow y) \rightsquigarrow y$, $x \leq (x \rightsquigarrow y) \rightarrow y$
(3) $x \leq x$,
(4) $x \leq 1$,
(5) $x \leq y$, $y \leq x \Rightarrow x = y$,
(6) $x \leq y \rightsquigarrow x \rightarrow y = 1 \Leftrightarrow x \rightsquigarrow y = 1$.

If $(A; \leq, \rightarrow, \rightsquigarrow, 1)$ is a pseudo-BCK algebra satisfying $x \rightarrow y = x \rightsquigarrow y$ for all $x, y \in A$, then $(A; \rightarrow, 1)$ is a BCK-algebra.

Proposition 1 ([9,11]). *Let $(A; \leq, \rightarrow, \rightsquigarrow, 1)$ be a pseudo-BCK algebra, then A satisfy the following properties ($\forall x, y, z \in A$):*

(1) $x \leq y \Rightarrow y \rightarrow z \leq x \rightarrow z$, $y \rightsquigarrow z \leq x \rightsquigarrow z$
(2) $x \leq y, y \leq z \Rightarrow x \leq z$,
(3) $x \rightsquigarrow (y \rightarrow z) = y \rightarrow (x \rightsquigarrow z)$,
(4) $x \leq y \rightarrow z \Leftrightarrow y \leq x \rightsquigarrow z$,
(5) $x \rightarrow y \leq (z \rightarrow x) \rightarrow (z \rightarrow y)$, $x \rightsquigarrow y \leq (z \rightsquigarrow x) \rightsquigarrow (z \rightsquigarrow y)$,
(6) $x \leq y \rightarrow x$, $x \leq y \rightsquigarrow x$,
(7) $1 \rightarrow x = x$, $1 \rightsquigarrow x = x$,
(8) $x \leq y \Rightarrow z \rightarrow x \leq z \rightarrow y$, $z \rightsquigarrow x \leq z \rightsquigarrow y$,
(9) $((y \rightarrow x) \rightsquigarrow x) \rightarrow x = y \rightarrow x$, $((y \rightsquigarrow x) \rightarrow x) \rightsquigarrow x = y \rightsquigarrow x$.

Definition 5 ([[12]). *A pseudo-BCI algebra is a structure (A; $\leq$, $\rightarrow$, $\rightsquigarrow$, 1), where "$\leq$" is a binary relation on A, "$\rightarrow$" and "$\rightsquigarrow$" are binary operations on A and "1" is an element of A, verifying the axioms: for all $x, y, z \in A$,*

(1) $x \rightarrow y \leq (y \rightarrow z) \rightsquigarrow (x \rightarrow z),\ x \rightsquigarrow y \leq (y \rightsquigarrow z) \rightarrow (x \rightsquigarrow z)$,
(2) $x \leq (x \rightarrow y) \rightsquigarrow y,\ x \leq (x \rightsquigarrow y) \rightarrow y$,
(3) $x \leq x$,
(4) *if* $x \leq y$ *and* $y \leq x$, *then* $x = y$,
(5) $x \leq y$ *iff* $x \rightarrow y = 1$ *iff* $x \rightsquigarrow y = 1$.

Note that, every pseudo-BCI algebra satisfying $x \rightarrow y = x \rightsquigarrow y$ for all $x, y \in A$ is a BCI-algebra.

Proposition 2 ([12,22,24]). *Let (A; $\leq$, $\rightarrow$, $\rightsquigarrow$, 1) be a pseudo-BCI algebra, then A satisfy the following properties ($\forall x, y, z \in A$):*

(1) *if* $1 \leq x$, *then* $x = 1$,
(2) *if* $x \leq y$, *then* $y \rightarrow z \leq x \rightarrow z$ *and* $y \rightsquigarrow z \leq x \rightsquigarrow z$,
(3) *if* $x \leq y$ *and* $y \leq z$, *then* $x \leq z$,
(4) $x \rightsquigarrow (y \rightarrow z) = y \rightarrow (x \rightsquigarrow z)$,
(5) $x \leq y \rightarrow z$, *iff* $y \leq x \rightsquigarrow z$
(6) $x \rightarrow y \leq (z \rightarrow x) \rightarrow (z \rightarrow y),\ x \rightsquigarrow y \leq (z \rightsquigarrow x) \rightsquigarrow (z \rightsquigarrow y)$,
(7) *if* $x \leq y$, *then* $z \rightarrow x \leq z \rightarrow y$ *and* $z \rightsquigarrow x \leq z \rightsquigarrow y$,
(8) $1 \rightarrow x = x,\ 1 \rightsquigarrow x = x$,
(9) $((y \rightarrow x) \rightsquigarrow x) \rightarrow x = y \rightarrow x,\ ((y \rightsquigarrow x) \rightarrow x) \rightsquigarrow x = y \rightsquigarrow x$,
(10) $x \rightarrow y \leq (y \rightarrow x) \rightsquigarrow 1,\ x \rightsquigarrow y \leq (y \rightsquigarrow x) \rightarrow 1$,
(11) $(x \rightarrow y) \rightarrow 1 = (x \rightarrow 1) \rightsquigarrow (y \rightarrow 1), (x \rightsquigarrow y) \rightsquigarrow 1 = (x \rightsquigarrow 1) \rightarrow (y \rightarrow 1)$
(12) $x \rightarrow 1 = x \rightsquigarrow 1$.

Definition 6 ([10,24]). *A pseudo-BCI algebra A is said to be an anti-grouped pseudo-BCI algebra if it satisfies the following identities:*

$$\textit{for any } x \in A,\ (x \rightarrow 1) \rightarrow 1 = x \textit{ or } (x \rightsquigarrow 1) \rightsquigarrow 1 = x.$$

Proposition 3 ([24]). *A pseudo-BCI algebra A is anti-grouped if and only if it satisfies:*

(G1) for all $x, y, z \in A$, $(x \rightarrow y) \rightarrow (x \rightarrow z) = y \rightarrow z$ *and*
(G2) for all $x, y, z \in A$, $(x \rightsquigarrow y) \rightsquigarrow (x \rightsquigarrow z) = y \rightsquigarrow z$.

Proposition 4 ([24]). *Let* **A** *= (A; $\leq$, $\rightarrow$, $\rightsquigarrow$, 1) be an anti-grouped pseudo-BCI algebra. Define* $\boldsymbol{\Phi}$**(A)** *= (A; +, −, 1) by*

$$x + y = (x \rightarrow 1) \rightarrow y = (y \rightsquigarrow 1) \rightsquigarrow x,\ \forall x, y \in A;$$

$$-x = x \rightarrow 1 = x \rightsquigarrow 1,\ \forall x \in A.$$

Then $\boldsymbol{\Phi}$**(A)** *is a group. Conversely, let* **G** *= (G; +, −, 1) be a group. Define* $\boldsymbol{\Psi}$**(G)** *= (G; $\leq$, $\rightarrow$, $\rightsquigarrow$, 1), where*

$$x \rightarrow y = (-x) + y,\ x \rightsquigarrow y = y + (-x),\ \forall x, y \in G;$$

$$x \leq y \textit{ if and only if } (-x) + y = 1 \ (\textit{or } y + (-x) = 1),\ \forall x, y \in G.$$

Then, $\boldsymbol{\Psi}$**(G)** *is an anti-grouped pseudo-BCI algebra. Moreover, the mapping* $\boldsymbol{\Phi}$ *and* $\boldsymbol{\Psi}$ *are mutually inverse.*

Definition 7 ([27]). *Let (A; ≤, →, ⇝, 1) be a pseudo-BCI algebra. Denote*

$$K(A) = \{x \in A | x \leq 1\};$$

$$AG(A) = \{x \in A | (x \to 1) \to 1 = x\}.$$

We say that K(A) is the pseudo-BCK part of A and AG(A) is the anti-grouped part of A.

Definition 8 ([28]). *A pseudo-BCI algebra A is said to be a T-type if it satisfies the following identities:*
(T1) for all $x \in A$, $(x \to 1) \to 1 = x \to 1$, or $(x \rightsquigarrow 1) \rightsquigarrow 1 = x \rightsquigarrow 1$.

Proposition 5 ([28]). *A pseudo-BCI algebra A is T-type if and only if it satisfies:*
(T2) for all $x \in A$, $x \to (x \to 1) = 1$, or $x \rightsquigarrow (x \rightsquigarrow 1) = 1$.

3. Some New Concepts and Results

By the definition of pseudo-BCI/BCK algebra, we know that the direct product of two pseudo-BCI/BCK algebras is a pseudo-BCI/BCK algebra. That is, we have the following lemma.

Lemma 1 ([20]). *Let $(X; \to_X, \rightsquigarrow_X, 1_X)$ and $(Y; \to_Y, \rightsquigarrow_Y, 1_Y)$ be two pseudo-BCI algebras. Define two binary operators →, ⇝ on X × Y as follwos: for any $(x_1, y_1), (x_2, y_2) \in X \times Y$,*

$$(x_1, y_1) \to (x_2, y_2) = (x_1 \to_X x_2, y_1 \to_Y y_2);$$

$$(x_1, y_1) \rightsquigarrow (x_2, y_2) = (x_1 \rightsquigarrow_X x_2 y_1 \rightsquigarrow_Y y_2);$$

and denote $1 = (1_X, 1_Y)$. Then $(X \times Y; \to, \rightsquigarrow, 1)$ is a pseudo-BCI algebra.

By the results in [18,20], we can easy to verify that the following lemma (the proof is omitted).

Lemma 2. *Let (K; →, ⇝, 1) be a pseudo-BCK algebra, (G; →, ⇝, 1) an anti-grouped pseudo-BCI algebra and K∩G = {1}. Denote A = KG b and define the operations →, ⇝ on A as follows:*

$$x \to y = \begin{cases} x \to y & if\ x, y \in K\ or\ x, y \in G \\ y & if\ x \in K, \quad y \in G \\ x \to 1 & if\ y \in K\{1\}, \quad x \in G \end{cases}$$

$$x \rightsquigarrow y = \begin{cases} x \rightsquigarrow y & if\ x, y \in K\ or\ x, y \in G \\ y & if\ x \in K, \quad y \in G \\ x \rightsquigarrow 1 & if\ y \in K\{1\}, \quad x \in G \end{cases}$$

Then (A; →, ⇝, 1) is a pseudo-BCI algebra.

Definition 9. *Let K be a pseudo-BCK algebra and G be an anti-grouped pseudo-BCI algebra, K∩G = {1}. If the operators →, ⇝ are defined on A = K∪G according to Lemma 2, then (A; →, ⇝, 1) is a pseudo-BCI algebra, we call A to be a KG-union of K and G and denote by $A = K \oplus_{KG} G$.*

Definition 10. *Let (X, ≤) is a partial ordered set with 1 as a constant element. For x in X, we call x a quasi-maximal element of X, if for any $a \in X$, $x \leq a \Rightarrow x = a$ or $a = 1$.*

Definition 11. *Let (G,*) be a grouoid, $x \in G$. Then x is called a quasi-left unit element of G, if it satisfies:*

$$\forall y \in G,\ x * y = y \ when\ x \neq y.$$

Theorem 1. *Let $(A; \leq, \rightarrow, \rightsquigarrow, 1)$ be pseudo-BCK algebra. Then the following conditions are equivalent:*

(a1) $\forall x \in A$, *x is a quasi-maximal element;*
(a2) $\forall x \in A, y \in A - \{1\}$, $x \leq y$ *implies* $x = y$;
(a3) $\forall x \in A$, *x is a quasi-left unit elemen w.r.t* $\rightarrow, \rightsquigarrow$, *that is,* $x \neq y$ *implies* $x \rightarrow y = y$ *and* $x \rightsquigarrow y = y$;
(a4) $\forall x, y \in A$, $x \neq y$ *implies* $x \rightarrow y = y$;
(a5) $\forall x, y \in A$, $x \neq y$ *implies* $x \rightsquigarrow y = y$.

Proof. (a1) $\Rightarrow$ (a2): Suppose that $x \in A, y \in A - \{1\}$ *and* $x \leq y$.

Case 1: If $x = 1$, it is follows that $1 = x \leq y \leq 1$, that is, $x = y = 1$.

Case 2: If $x \neq 1$, by (a1) and Definition 10, from $x \leq y$ and $y \neq 1$, we have $x = y$. Therefore, (a2) hold.

(a2) $\Rightarrow$ (a3): For any x, y in A, by Proposition 1 (6) and Definition 4 (2), we have $x \leq y \rightarrow x$, $y \leq x \rightarrow y, x \leq (x \rightarrow y) \rightsquigarrow y$. Assume $x \neq y$. If $y \rightarrow x = 1$, then $x \rightarrow y \neq 1$ (since, if $x \rightarrow y = 1$, then form $y \rightarrow x = 1$ and $x \rightarrow y = 1$ we get $x = y$, this is contradictory to the hypothesis $x \neq y$). Thus, from $y \leq x \rightarrow y$ and $x \rightarrow y \neq 1$, using (a2) we have $y = x \rightarrow y$.

If $y \rightarrow x \neq 1$, from this and $x \leq y \rightarrow x$ *and* applying (a2), we have $x = y \rightarrow x$. Thus,

(i) when $(x \rightarrow y) \rightsquigarrow y = 1$, we can get $x \rightarrow y \leq y \leq x \rightarrow y$, that is, $y = x \rightarrow y$;
(ii) when $(x \rightarrow y) \rightsquigarrow y \neq 1$, from this and $x \leq (x \rightarrow y) \rightsquigarrow y$, using (a2) we have $x = (x \rightarrow y) \rightsquigarrow y$. Combine the aforementioned conclusion $x = y \rightarrow x$, we can get

$$x = y \rightarrow x = y \rightarrow ((x \rightarrow y) \rightsquigarrow y) = (x \rightarrow y) \rightsquigarrow (y \rightarrow y) = (x \rightarrow y) \rightsquigarrow 1 = 1,$$

It follows that $y = 1 \rightarrow y = x \rightarrow y$.

Therefore, based on the above cases we know that $x \neq y$ implies $y = x \rightarrow y$.

Similarly, we can prove that $x \neq y$ implies $y = x \rightsquigarrow y$.

(a3) $\Rightarrow$ (a4): Obviously.

(a4) $\Rightarrow$ (a5): Suppose $x \neq y$. Applying (a4), $x \rightarrow y = y$. Also, by Definition 4 (2), $x \leq (x \rightsquigarrow y) \rightarrow y$, thus $x \rightarrow [(x \rightsquigarrow y) \rightarrow y] = 1$.

Case 1: If $x \neq (x \rightsquigarrow y) \rightarrow y$, using (a4), $x \rightarrow [(x \rightsquigarrow y) \rightarrow y] = [(x \rightsquigarrow y) \rightarrow y]$. Hence, $(x \rightsquigarrow y) \rightarrow y = 1$. Moreover,

$$y \rightarrow (x \rightsquigarrow y) = x \rightsquigarrow (y \rightarrow y) = x \rightsquigarrow 1 = 1.$$

Therefore, $y = x \rightsquigarrow y$. Case 2: If $x = (x \rightsquigarrow y) \rightarrow y$, then $x \rightsquigarrow y = y$. In fact, if $x \rightsquigarrow y \neq y$, using (a4), $(x \rightsquigarrow y) \rightarrow y = y$, it follows that $x = y$, this is a contradiction with $x \neq y$.

By above results we know that (a5) hold.

(a5) $\Rightarrow$ (a1): Assume that $x \in X, a \in X$ and $x \leq a$. Then $x \rightsquigarrow a = 1$. If $x \neq a$, by (a5), $x \rightsquigarrow a = a$, then $a = x \rightsquigarrow a = 1$. This means that $x \leq a$ implies $x = a$ or $a = 1$. □

By Theorem 1 and Definition 3 we get

Corollary 1. *Let $(A; \leq, \rightarrow, \rightsquigarrow, 1)$ be a pseudo-BCK algebra. Then every element of A is quasi-maximal if and only if A is a quasi-alternating BCK-algebra.*

4. The Class of Pseudo-BCI Algebras in Which Every Element is Quasi-Maximal

Example 1. *Let A = {a, b, c, d, e, f, g, 1}. Define operations $\rightarrow$ and $\rightsquigarrow$ on A as following Cayley Tables 1 and 2. Then A is pseudo-BCI algebra in which every element is quasi-maximal.*

Table 1. The Cayley table of operation $\rightarrow$.

$\rightarrow$	**a**	**b**	**c**	**d**	**e**	**f**	**g**	**1**
a	1	b	c	d	e	f	g	1
b	a	1	c	d	e	f	g	1
c	c	c	1	f	g	d	e	c
d	d	d	e	1	c	g	f	d
e	f	f	g	c	1	e	d	f
f	e	e	d	g	f	1	c	e
g	g	g	c	d	e	f	1	g
1	a	b	c	d	e	f	g	1

Table 2. The Cayley table of operation $\rightsquigarrow$.

$\rightsquigarrow$	**a**	**b**	**c**	**d**	**e**	**f**	**g**	**1**
a	1	b	c	d	e	f	g	1
b	a	1	c	d	e	f	g	1
c	c	c	1	f	g	d	e	c
d	d	d	e	1	c	g	f	d
e	f	f	g	c	1	e	d	f
f	e	e	d	g	f	1	c	e
g	g	g	c	d	e	f	1	g
1	a	b	c	d	e	f	g	1

Definition 12. *A pseudo-BCI/BCK algebra A is said to be a QM-pseudo-BCI/BCK algebra if every element of A is quasi-maximal.*

Theorem 2. *Let $(A; \leq, \rightarrow, \rightsquigarrow, 1)$ be a pseudo-BCI algebra. Then A is a QM-pseudo-BCI algebra if and only if it satisfies:*

$$for\ any\ x,\ y \in A - \{1\},\ x \leq y \Rightarrow x = y.$$

Proof. If A is a QM-pseudo-BCI algebra, by Definitions 10 and 12, the above condition is satisfied.

Conversely, assume that $x,\ y \in A,\ x \leq y$. If $x = 1$, then $1 = x \leq y$, it follows that $x = y = 1$, by Proposition 2 (1). If $x \neq 1, y \neq 1$, then $x = y$ by the condition. This means that x is a quasi- maximal element in A, hence, A is a QM-pseudo-BCI algebra. □

By Theorem 1 we know that a pseudo-BCK algebra is a QM-pseudo-BCK algebra if and only if it is a quasi-alternating BCK-algebra. It will be proved that any QM-pseudo-BCI algebra is constructed by the combination of a quasi-alternating BCK-algebra and an anti-grouped pseudo- BCI algebra (a group-like algebra).

Lemma 3 ([27]). *Let A be a pseudo-BCI algebra, K(A) the pseudo-BCK part of A. If AG(A) = (A − K(A))∪{1} is subalgebra of A, then $(\forall x, y \in A)$*

(1) *If $x \in K(A)$ and $y \in A - K(A)$, then $x \rightarrow y = x \rightsquigarrow y = y$.*
(2) *If $x \in A - K(A)$ and $y \in K(A)$, then $x \rightarrow y = x \rightsquigarrow y = x \rightarrow 1$.*

Applying the results in [24,27] we can easy to verify that the following lemma is true (the proof is omitted).

Lemma 4. *Let A be an anti-grouped pseudo-BCI algebra. Then*

(1) *for any x, y in A, $x \leq y$ implies $x = y$;*

(2) *for any* x, y *in* A, $x = (x \to y) \rightsquigarrow y = (x \rightsquigarrow y) \to y$.

Theorem 3. *Let A be a pseudo-BCI algebra, K(A) the pseudo-BCK part of A and AG(A) the anti-grouped part of A. The following statements are equivalent:*

(1) A is a QM-pseudo-BCI algebra;
(2) *K(A) is quasi-alternating BCK-algebras and AG(A) = (A − K(A)) ∪ {1};*
(3) $\forall x, y \in A, x \neq y$ *implies* $(x \to y) \rightsquigarrow y = (x \to 1) \to 1$;
(4) $\forall x, y \in A, x \neq y$ *implies* $(x \rightsquigarrow y) \to y = (x \to 1) \to 1$;
(5) $\forall x, y \in A, x \neq y$ *implies* $(x \to y) \to y = (x \to 1) \to 1$;
(6) $\forall x, y \in A, x \neq y$ *implies* $(x \rightsquigarrow y) \rightsquigarrow y = (x \to 1) \to 1$.

Proof. (1) ⇒ (2): Suppose that A is a QM-pseudo-BCI algebra. Then, for all $x, y \in K(A)$, by Corollary 1, we can know $K(A)$ is quasi-alternating BCK-algebras. If $x \in A - K(A)$, then $x \to 1 \neq 1$ and $(x \to 1) \to 1 \neq 1$. Since $x \leq (x \to 1) \to 1$, by Definition 12 we have $x = (x \to 1) \to 1$. Thus, $(A - K(A)) \cup \{1\} \subseteq AG(A)$. On the other hand, obviously, $AG(A) \subseteq (A - K(A)) \cup \{1\}$. Hence $AG(A) = (A - K(A)) \cup \{1\}$.

(2) ⇒ (3): Assume that (2) hold. For any x, y in A, $x \neq y$,

Case 1: $x, y \in K(A)$. Then $x \to 1 = y \to 1 = 1$. Because $K(A)$ is quasi-alternating BCK-algebra, using Theorem 1, $x \to y = y$. Thus

$$(x \to y) \rightsquigarrow y = y \rightsquigarrow y = 1 = 1 \to 1 = (x \to 1) \to 1.$$

Case 2: $x, y \in AG(A)$. Since $AG(A)$ is an anti-grouped pseudo-BCI subalgebra of A, then by Lemma 4 we get

$$(x \to y) \rightsquigarrow y = x = (x \to 1) \to 1.$$

Case 3: $x \in K(A), y \in AG(A)$. Then $x \to 1 = 1$. Applying Lemma 3 (1), $x \to y = y$. Then

$$(x \to y) \rightsquigarrow y = y \rightsquigarrow y = 1 = 1 \to 1 = (x \to 1) \to 1.$$

Case 4: $x \in AG(A)$, $y \in K(A)$. Then $x = (x \to 1) \to 1$, $y \to 1 = 1$. Applying Lemma 3 (2), $x \to y = x \to 1$. When $x = 1$, then $(x \to y) \rightsquigarrow y = (x \to 1) \to 1$; when $x \neq 1$, then $x \to 1 \in A - K(A)$, using Lemma 3 (2),

$$(x \to 1) \rightsquigarrow y = (x \to 1) \to 1$$

Hence,

$$(x \to y) \rightsquigarrow y = (x \to 1) \rightsquigarrow y = (x \to 1) \to 1.$$

(3) ⇒ (1): Assume that $x \leq y$ and $x \neq y$. We will prove that $y = 1$. By (3), we have

$$y = 1 \rightsquigarrow y = (x \to y) \rightsquigarrow y = (x \to 1) \to 1.$$

Case 1: when $x \in K(A)$, then $x \to 1 = 1$, so $y = 1$. Case 2: when $x \in X - K(A)$, then $(x \to 1) \to 1 = x$, so $y = x$, this is a contradiction with $x \neq y$.

Therefore, for all $x \in A$, x is a quasi-maximal element of A.

(4) ⇒ (2): Suppose (4) hold. For any x, y in A.

If $x, y \in K(A)$, $x \neq y$, by (4),

$$(x \rightsquigarrow y) \to y = (x \to 1) \to 1 = 1.$$

Then, $x \rightsquigarrow y \leq y$. Since $K(A)$ is a pseudo-BCK subalgebra of A, using Proposition 1 (6), $y \leq x \rightsquigarrow y$. It follows that $y \leq x \rightsquigarrow y \leq y$, that is, $x \rightsquigarrow y = y$. Thus, applying Theorem 1, $K(A)$ is a quasi-alternating BCK-algebra.

If $x \in A - K(A)$, we prove that $(x \to 1) \to 1 = x$. Assume $(x \to 1) \to 1 \neq x$, by (4), we have

$$\{[(x \to 1) \to 1] \rightsquigarrow x\} \to x = \{[(x \to 1) \to 1] \to 1\} \to 1.$$

Using Proposition 2 (9) and (12),

$$\{[(x \to 1) \to 1] \to 1\} \to 1 = (x \to 1) \to 1.$$

Thus

$$\{[(x \to 1) \to 1] \rightsquigarrow x\} \to x = (x \to 1) \to 1.$$

Moreover, applying Proposition 2 (9), (11) and (12) we have

$$\begin{aligned} & \{[(x \to 1) \to 1] \rightsquigarrow x\} \to 1 \\ = \; & \{[(x \to 1) \to 1] \rightsquigarrow 1\} \to (x \rightsquigarrow 1) \\ = \; & \{[(x \to 1) \rightsquigarrow 1] \to 1\} \to (x \rightsquigarrow 1) \\ = \; & (x \to 1) \to (x \rightsquigarrow 1) \\ = \; & 1. \end{aligned}$$

This means that $((x \to 1) \to 1) \to x \in K(A)$. By Lemma 3 (1),

$$\{[(x \to 1) \to 1] \rightsquigarrow x\} \to x = x.$$

Hence, $(x \to 1) \to 1 = x$. This is contraction with $(x \to 1) \to 1 \neq x$. Therefore, $(x \to 1) \to 1 = x$ and $x \in AG(A)$. It follows that $(A - K(A)) \cup \{1\} \subseteq AG(A)$. Obviously, $AG(A) \subseteq (A - K(A)) \cup \{1\}$. So $AG(A) = (A - K(A)) \cup \{1\}$.

(2) $\Rightarrow$ (4): It is similar to (2) $\Rightarrow$ (3). It follows that (4) $\Leftrightarrow$ (2).

Similarly, we can prove (5) $\Leftrightarrow$ (2), (6) $\Leftrightarrow$ (2). □

Theorem 4. *Let $(A; \leq, \to, \rightsquigarrow, 1)$ be a pseudo-BCI algebra, $AG(A)$ the anti-grouped part of A, $K(A)$ the pseudo-BCK part of A. Then A is a QM-pseudo-BCI algebra if and only if $K(A)$ is a quasi-alternating BCK-algebra and $A = K(A) \oplus_{KG} AG(A)$.*

Proof. If A is a QM-pseudo-BCI algebra, then $K(A)$ is a quasi-alternating BCK-algebra and $A = K(A) \oplus_{KG} AG(A)$, by Lemma 3 and Theorem 3.

Conversely, if $K(A)$ is a quasi-alternating BCK-algebra, then every element in $K(A)$ is quasi-maximal; if $A = K(A) \oplus_{KG} AG(A)$, then $AG(A) = (A - K(A)) \cup \{1\}$, it follows that every element in $A - K(A)$ is quasi-maximal. By Definition 12, we know that A is a QM-pseudo-BCI algebra. □

5. Weak Associative Pseudo-BCI Algebras

Definition 13. *A pseudo-BCI/BCK algebra A is said to be weak associative, if it satisfies:*

$$for\ any,\ y, z \in A,\ (x \to y) \to z = x \to (y \to z)\ when\ (x \neq y,\ x \neq z).$$

Example 2 *Let $A = \{a, b, c, d, e, f, 1\}$. Define operation $\to$ on A as following Cayley Table 3. Then A is a weak associative pseudo-BCI algebra, where $\rightsquigarrow = \to$.*

Table 3. The Cayley table of the operation $\rightarrow$.

$\rightarrow$	a	b	c	d	e	f	1
a	1	b	c	d	e	f	1
b	a	1	c	d	e	f	1
c	a	b	1	d	e	f	1
d	d	d	d	1	f	e	d
e	e	e	e	f	1	d	e
f	f	f	f	e	d	1	f
1	a	b	c	d	e	f	1

Theorem 5. *Let $(A; \leq, \rightarrow, \rightsquigarrow, 1)$ be a weak associative pseudo-BCI algebra. Then A is a QM-pseudo-BCI algebra and a T-type pseudo-BCI algebra.*

Proof. For any x, y in A, $x \neq y$, then (by Definition 13)

$$(x \rightarrow y) \rightarrow y = x \rightarrow (y \rightarrow y) = x \rightarrow 1.$$

Thus, if $x \neq 1$, then $(x \rightarrow 1) \rightarrow 1 = x \rightarrow 1$. Obviously, when $x = 1$, $(x \rightarrow 1) \rightarrow 1 = x \rightarrow 1$. Hence, from Definition 13 we get that for any x, y in A, $x \neq y \Rightarrow (x \rightarrow y) \rightarrow y = (x \rightarrow 1) \rightarrow 1$. Applying Theorem 3 (5) we know that A is a QM-pseudo-BCI algebra.

Moreover, we already prove that $(x \rightarrow 1) \rightarrow 1 = x \rightarrow 1$ for any x in A, by Definition 8 we know that A is a T-type pseudo-BCI algebra. □

The inverse of Theorem 5 is not true. Since $(d \rightarrow c) \rightarrow c \neq d \rightarrow 1$, so the QM-pseudo-BCI algebra in Example 1 is not weak associative. The following example shows that a T-type pseudo-BCI algebra may be not a QM-pseudo-BCI algebra.

Example 3. *Let $A = \{a, b, c, d, 1\}$. Define operations $\rightarrow$ and $\rightsquigarrow$ on A as following Cayley Tables 4 and 5. Then A is a T-type pseudo-BCI algebra but it is not a QM-pseudo-BCI algebra, since*

$$(b \rightarrow c) \rightarrow a = a \neq 1 = b \rightarrow (c \rightarrow a).$$

Table 4. The operation $\rightarrow$ in the T-type pseudo-BCI algebra.

$\rightarrow$	a	b	c	d	1
a	1	1	1	d	1
b	b	1	1	d	1
c	b	b	1	d	1
d	d	d	d	1	d
1	a	b	c	d	1

Table 5. The operation $\rightsquigarrow$ in the T-type pseudo-BCI algebra.

$\rightsquigarrow$	a	b	c	d	1
a	1	1	1	d	1
b	c	1	1	d	1
c	a	b	1	d	1
d	d	d	d	1	d
1	a	b	c	d	1

Lemma 5 ([16,24]). *Let $(A; \rightarrow, 1)$ be a BCI-algebra. Then the following statements are equivalent:*

(1) *A is associative, that is,* $(x \to y) \to z = x \to (y \to z)$ *for any* x, y, z *in* A;
(2) *for any* x *in* A, $x \to 1 = x$;
(3) *for all* x, y *in* A, $x \to y = y \to x$.

Theorem 6. *Let* $(A; \leq, \to, \rightsquigarrow, 1)$ *be a weak associative pseudo-BCI algebra, AG(A) the anti-grouped part of A, K(A) the pseudo-BCK part of A. Then*

(1) *K(A) is quasi-alternating BCK-algebra and AG(A) = (A − K(A))∪{1};*
(2) *For any* x *in* $AG(A)$, $x \to 1 = x \rightsquigarrow 1 = x$;
(3) *For any* x, y *in* A, $x \to y = x \rightsquigarrow y$, that is, A is a BCI-algebra;
(4) *AG(A) is an Abel group, that is, AG(A) is associative BCI-algebra.*

Proof. (1) It follows from Theorems 5 and 3.

(2) For any x in $AG(A)$, then $(x \to 1) \to 1 = x$. We will prove that $x \to 1 = x$.
If $x = 1$, obviously, $x \to 1 = x$.
If $x \neq 1$, then $(x \to 1) \to 1 = x \to 1$ by Definition 13. Thus,

$$x \to 1 = (x \to 1) \to 1 = x.$$

Applying Proposition 2 (12) we have

$$x \rightsquigarrow 1 = x \to 1 = x.$$

(3) For any x, y in A,

(i) when x, y in $K(A)$, by (1), $K(A)$ is a BCK-algebra, so $x \to y = x \rightsquigarrow y$;
(ii) when x, y in $(A - K(A))$, by (1) and (2), applying Proposition 2 (11),

$$x \to y = (x \to y) \to 1 = (x \to 1) \rightsquigarrow (y \to 1) = x \rightsquigarrow y;$$

(iii) when x in $K(A)$, y in $(A - K(A))$, using Lemma 3 (1), $x \to y = x \rightsquigarrow y$;
(iv) when y in $K(A)$, x in $(A - K(A))$, using Lemma 3 (2), $x \to y = x \rightsquigarrow y$;

Therefore, for all x, y in A, $x \to y = x \rightsquigarrow y$. It follows that A is a BCI-algebra.

(4) Applying (2), by Lemma 5 we know that $AG(A)$ is an Abel group, that is, AG(A) is associative BCI-algebra. □

From Theorems 6 and 4 we immediately get

Theorem 7. *Let* $(A; \leq, \to, \rightsquigarrow, 1)$ *be a pseudo-BCI algebra, AG(A) the anti-grouped part of A, K(A) the pseudo-BCK part of A. Then A is a weak associative pseudo-BCI algebra if and only if K(A) is a quasi-alternating BCK-algebra, AG(A) is an Abelian group and* $A = K(A) \oplus_{KG} AG(A)$.

Theorem 8. *Let* $(A; \leq, \to, \rightsquigarrow, 1)$ *be a pseudo-BCI algebra. Then the following conditions are equivalent:*

(1) *for any* $x, y, z \in A$, $(x \to y) \to z = x \to (y \to z)$ *when* $(x \neq y, x \neq z)$;
(2) *for any* $x, y, z \in A$, $(x \rightsquigarrow y) \rightsquigarrow z = x \rightsquigarrow (y \rightsquigarrow z)$ *when* $(x \neq y, x \neq z)$;
(3) *for any* $x, y, z \in A$, $(x \to y) \rightsquigarrow z = x \to (y \rightsquigarrow z)$ *when* $(x \neq y, x \neq z)$;
(4) *for any* $x, y, z \in A$, $(x \rightsquigarrow y) \to z = x \rightsquigarrow (y \to z)$ *when* $(x \neq y, x \neq z)$.

Proof. $(1) \Rightarrow (2)$: It follows from Definition 13 and Theorem 6.

$(2) \Rightarrow (1)$: Similar to the discussion process from Definition 13 to Theorem 6, we can obtain a result similar to Theorem 6. That is, from (2) we can get that A is a BCI-algebra. Hence, (2) implies (1).

Similarly, (3) $\Leftrightarrow$ (1) and (4) $\Leftrightarrow$ (1). □

Finally, we discuss the relationships among general pseudo-BCI algebras, QM-pseudo-BCI algebras and weak associative pseudo-BCI algebras (WA-pseudo-BCI algebras).

In fact, in every T-type pseudo-BCI algebra, there is a maximal WA-pseudo-BCI subalgebra. That is, if $(A; \leq, \rightarrow, \rightsquigarrow, 1)$ is a T-type pseudo-BCI algebra, $AG(A)$ the anti-grouped part of A, $K(A)$ the pseudo-BCK part of A, then $K_{qm}(A) \cup AG(A)$ is a WA-pseudo-BCI subalgebra of A, where $K_{qm}(A)$ is the set of all quasi-maximal element in $K(A)$. For example, $\{c, d, 1\}$ is a WA-pseudo-BCI subalgebra of the pseudo-BCI algebra A in Example 3.

In general, in every pseudo-BCI algebra, there is a maximal QM-pseudo-BCI subalgebra. That is, if $(A; \leq, \rightarrow, \rightsquigarrow, 1)$ is a pseudo-BCI algebra, $AG(A)$ the anti-grouped part of A, $K(A)$ the pseudo-BCK part of A, then $K_{qm}(A) \cup AG(A)$ is a QM-pseudo-BCI subalgebra of A, where $K_{qm}(A)$ is the set of all quasi-maximal element in $K(A)$.

6. Conclusions

In the study of pseudo-BCI algebras, the structures of various special pseudo-BCI algebras are naturally an important problem. At present, the structures of several subclasses such as quasi-alternating pseudo-BCI algebras and anti-grouped pseudo-BCI algebras are clear. In this paper, we have studied an important subclass of pseudo-BCI algebras, that is, QM-pseudo-BCI algebras in which every element is quasi-maximal. We obtain a very clear structure theorem of this subclass. At the same time, we have studied a class of more special pseudo-BCI algebras, that is, weak associative (WA) pseudo-BCI algebras in which every element is weak associative and obtained the structure theorem of this subclass. These results enrich the research content of pseudo-BCI algebras and clearly presented the relationships between various subclasses, which can be illustrated as Figure 1. Finally, we show that the two types of pseudo-BCI algebras are very important, since (1) every pseudo-BCI algebra contains a subalgebra which is QM-pseudo-BCI algebra, (2) every T-type pseudo-BCI algebra contains a subalgebra which is WA-pseudo-BCI algebra. As a further study direction, we will discuss the integration of related topics in the light of some new research findings in [32–34].

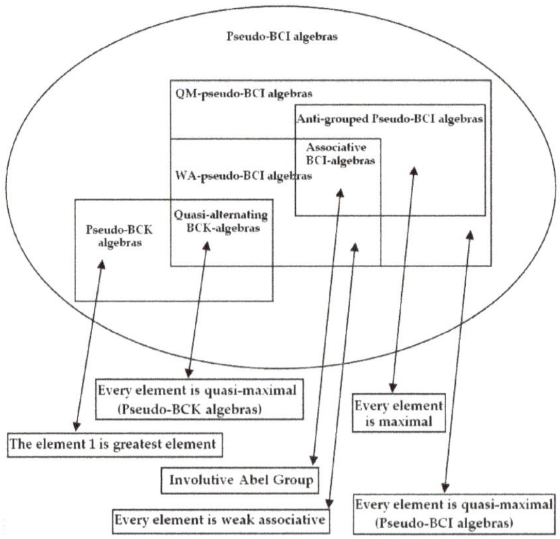

Figure 1. Main results in this paper.

Author Contributions: The contributions of the authors are roughly equal. X.Z. and X.W. initiated the research; X.W. wrote the draft and X.Z. completed final version.

Funding: This research was funded by National Natural Science Foundation of China grant number 61573240.

Conflicts of Interest: The authors declare no conflict of interest.

References

1. Klement, E.P.; Mesiar, R.; Pap, E. *Triangular Norms*; Springer: Dordrecht, The Netherlands, 2000.
2. Flaminio, T. Strong non-standard completeness for fuzzy logics. *Soft Comput.* **2008**, *12*, 321–333. [CrossRef]
3. Aguzzoli, S.; Flaminio, T.; Ugolini, S. Equivalences between subcategories of MTL-algebras via Boolean algebras and prelinear semihoops. *J. Logic Comput.* **2017**, *27*, 2525–2549. [CrossRef]
4. Hájek, P. *Metamathematics of Fuzzy Logic*; Springer: Dordrecht, The Netherlands, 1998.
5. Esteva, F.; Godo, L. Monoidal t-norm based logic: Towards a logic for left-continuous t-norms. *Fuzzy Sets Syst.* **2001**, *124*, 271–288. [CrossRef]
6. Wang, G.J.; Zhou, H.J. *Introduction to Mathematical Logic and Resolution Principle*; Alpha International Science Ltd.: Oxford, UK, 2009.
7. Zhang, X.H. *Fuzzy Logics and Algebraic Analysis*; Science Press: Beijing, China, 2008.
8. Zhang, X.H.; Dudek, W.A. BIK+-logic and non-commutative fuzzy logics. *Fuzzy Syst. Math.* **2009**, *23*, 8–20.
9. Iorgulescu, A. *Implicative-Groups vs. Groups and Generalizations*; Bucharest Academy of Economic Studies: Bucharest, Romania, 2018.
10. Iséki, K. An algebra related with a propositional calculus. *Proc. Jpn. Acad.* **1966**, *42*, 26–29. [CrossRef]
11. Georgescu, G.; Iorgulescu, A. Pseudo-BCK algebras: An extension of BCK algebras. In *Combinatorics, Computability and Logic*; Springer: London, UK, 2001; pp. 97–114.
12. Dudek, W.A.; Jun, Y.B. Pseudo-BCI algebras. *East Asian Math. J.* **2008**, *24*, 187–190.
13. Pu, Y.S. Quasi-alternating BCI-algebra and others. *J. Hanzhong Teach. Coll.* **1985**, *3*, 1–8.
14. Zhang, X.H.; Yang, Y.B. A kind of BCI-algebra and its structure. *J. Northwest Norm. Univ. (Nat. Sci.)* **1992**, *28*, 27–31.
15. Dudek, W.A.; Zhang, X.H. On atoms in BCC-algebras, Discuss. *Math. Algebra Stoch. Methods* **1995**, *15*, 81–85.
16. Huang, Y.S. *BCI-Algebra*; Science Press: Beijing, China, 2006.
17. Dudek, W.A.; Zhang, X.H.; Wang, Y.Q. Ideals and atoms of BZ-algebras. *Math. Slov.* **2009**, *59*, 387–404.
18. Zhang, X.H. BCC-algebras and residuated partially-ordered groupoid. *Math. Slov.* **2013**, *63*, 397–410.
19. Dymek, G.; Kozanecka-Dymek, A. Pseudo-BCI-logic. *Bull. Sect. Logic* **2013**, *42*, 33–41.
20. Emanovský, P.; Kühr, J. Some properties of pseudo-BCK- and pseudo-BCI-algebras. *Fuzzy Sets Syst.* **2018**, *339*, 1–16. [CrossRef]
21. Xin, X.L.; Fu, Y.L.; Lai, Y.Y.; Wang, J.T. Monadic pseudo BCI-algebras and corresponding logics. *Soft Comput.* **2018**, 1–12. [CrossRef]
22. Jun, Y.B.; Kim, H. S.; Neggers, J. On pseudo-BCI ideals of pseudo-BCI algebras. *Mat. Vesnik* **2006**, *58*, 39–46.
23. Dymek, G. Atoms and ideals of pseudo-BCI-algebras. *Comment. Math.* **2012**, *52*, 73–90.
24. Zhang, X.H.; Jun, Y.B. Anti-grouped pseudo-BCI algebras and anti-grouped filters. *Fuzzy Syst. Math.* **2014**, *28*, 21–33.
25. Zhang, X.H. Fuzzy anti-grouped filters and fuzzy normal filters in pseudo-BCI algebras. *J. Intell. Fuzzy Syst.* **2017**, *33*, 1767–1774. [CrossRef]
26. Zhang, X.H.; Gong, H.J. Implicative pseudo-BCK algebras and implicative pseudo-filters of pseudo-BCK algebras. In Proceedings of the 2010 IEEE International Conference on Granular Computing, San Jose, CA, USA, 14–16 August 2010; Volume 62, pp. 615–619.
27. Zhang, X.H. Pseudo-BCK part and anti-grouped part of pseudo-BCI algebras. In Proceedings of the IEEE International Conference on Progress in Informatics and Computing, Shanghai, China, 10–12 December 2010; Volume 1, pp. 127–131.
28. Zhang, X.H.; Lu, Y.F.; Mao, X.Y. T-type pseudo-BCI algebras and T-type pseudo-BCI filters. In Proceedings of the 2010 IEEE International Conference on Granular Computing, San Jose, CA, USA, 14–16 August 2010; pp. 839–844.
29. Dymek, G. On a period of elements of pseudo-BCI algebras. *Discuss. Math. Gener. Algebra Appl.* **2015**, *35*, 21–31. [CrossRef]

30. Xin, X.L.; Li, Y.J.; Fu, Y.L. States on pseudo-BCI algebras. *Eur. J. Pure Appl. Math.* **2017**, *10*, 455–472.
31. Zhang, X.H.; Park, C.; Wu, S.P. Soft set theoretical approach to pseudo-BCI algebras. *J. Intell. Fuzzy Syst.* **2018**, *34*, 559–568.
32. Zhang, X.H.; Smarandache, F.; Liang, X.L. Neutrosophic duplet semi-group and cancellable neutrosophic triplet groups. *Symmetry* **2017**, *9*, 275. [CrossRef]
33. Zhang, X.H.; Bo, C.X.; Smarandache, F.; Park, C. New operations of totally dependent- neutrosophic sets and totally dependent-neutrosophic soft sets. *Symmetry* **2018**, *10*, 187. [CrossRef]
34. Zhang, X.H.; Bo, C.X.; Smarandache, F.; Dai, J.H. New inclusion relation of neutrosophic sets with applications and related lattice structure. *Int. J. Mach. Learn. Cyber.* **2018**, *9*, 1753–1763. [CrossRef]

symmetry

MDPI

Article

A Note on the Minimum Size of a Point Set Containing Three Nonintersecting Empty Convex Polygons

Qing Yang [1], Zengtai You [2] and Xinshang You [3,*]

1 Faculty of Accounting, Shanxi University of Finance and Economics, Taiyuan 030006, China; yangqing@tju.edu.cn
2 College of Computer Science and Engineering, Dalian Minzu University, Dalian 116600, China; zengtaiyou@163.com
3 College of Economics and Management, Shandong University of Science and Technology, Qingdao 266590, China
* Correspondence: youxinshang@tju.edu.cn

Received: 6 September 2018; Accepted: 26 September 2018; Published: 29 September 2018

Abstract: Let P be a planar point set with no three points collinear, k points of P be a *k-hole* of P if the k points are the vertices of a convex polygon without points of P. This article proves 13 is the smallest integer such that any planar points set containing at least 13 points with no three points collinear, contains a 3-hole, a 4-hole and a 5-hole which are pairwise disjoint.

Keywords: planar point set; convex polygon; disjoint holes

1. Introduction

In this paper, we deal with the finite planar point set P in *general position*, that is to say, no three points in P are collinear. In 1935, Erdős and Szekeres [1], posed a famous combinational geometry question: Whether for every positive integer $m \geq 3$, there exists a smallest integer $ES(m)$, such that any set of n points ($n \geq ES(m)$), contains a subset of m points which are the vertices of a convex polygon. It is a long standing open problem to evaluate the exact value of $ES(m)$. Erdős and Szekeres [2] showed that $ES(m) \geq 2^{m-2}+1$, which is also conjectured to be sharp. We have known that $ES(4) = 5$ and $ES(5) = 9$. Then by using computer, Szekeres and Peters [3] proved that $ES(6) = 17$. The value of $ES(m)$ for all $m > 6$ is unknown.

For a planar point set P, let k points of P be a *k-hole* of P if the k points are the vertices of a convex polygon whose interior contains no points of P. Erdős posed another famous question in 1978. He asked whether for every positive integer k, there exists a smallest integer $H(k)$, such that any set of at least $H(k)$ points in the plane, contains a *k-hole*. It is obvious that $H(3) = 3$. Esther Klein showed $H(4) = 5$. Harborth [4] determined $H(5) = 10$, and also gave the configuration of nine points with no empty convex pentagons. Horton [5] showed that it was possible to construct arbitrarily large set of points without a 7-hole, That is to say $H(k)$ does not exist for $k \geq 7$. The existence of $H(6)$ had been proved by Gerken [6] and Nicolás [7], independently. In [8], Urabe first studied the disjoint holes problems when hewas considering the question about partitioning of planar point sets. Let $Ch(P)$ stand for the *convex hull* of a point set P. A family of holes $\{H_i\}_{i\in I}$ is called pairwise disjoint if $Ch(H_i) \cap Ch(H_j) = \emptyset, i \neq j; i \in I, j \in I$. These holes are disjoint with each other. Determine the smallest integer $n(k_1, ..., k_l)$, $k_1 \leq k_2 \leq ... \leq k_l$, such that any set of at least $n(k_1, ..., k_l)$ points of the plane, contains a k_i*-hole* for every i, $1 \leq i \leq l$, where the holes are disjoint. From [9], we know $n(2,4) = 6$, $n(3,3) = 6$. Urabe [8] showed that $n(3,4) = 7$, while Hosono and Urabe [10] showed that $n(4,4) = 9$. In [11], Hosono and Urabe also gave $n(3,5) = 10$, $12 \leq n(4,5) \leq 14$ and $16 \leq n(5,5) \leq 20$.

The result $n(3,4) = 7$ and $n(4,5) \leq 14$ were re-authentication by Wu and Ding [12]. Hosono and Urabe [9] proved $n(4,5) \leq 13$. $n(4,5) = 12$ by Bhattacharya and Das was published in [13], who also discussed the convex polygons and pseudo-triangles [14]. Hosono and Urabe also changed the lower bound on $H(5,5)$ to 17 [9], and Bhattacharya and Das showed the upper bound on $n(5,5)$ to 19 [15]. Recently, more detailed discussions about two holes are published in [16]. Hosono and Urabe in [9] showed $n(2,3,4) = 9$, $n(2,3,5) = 11$, $n(4,4,4) = 16$. We showed $n(3,3,5) = 12$ in [17]. We have proved that $n(3,3,5) = 12$ [17], $n(4,4,5) \leq 16$ [18] and also discuss a disjoint holes problem in preference [19]. In this paper, we will continue discussing this problem and prove that $n(3,4,5) = 13$.

2. Definitions

The *vertices* are on convex hull of the given points,from the remaining *interior points*. Let $V(P)$ denote a set of the vertices and $I(P)$ be a set of the interior points of P. $|P|$ stands for the number of points contained in P. Let $p_1, p_2, ..., p_k$ be k points of P, we know that $p_1, p_2, ..., p_k$ be a k-hole H when the k points are the vertices of a convex polygon whose interior does not contain any point of P. And we simply say $H = (p_1 p_2 ... p_k)_k$. As in [9], let $l(a,b)$ be the line passing points a and b. Determine the closed half-plane with $l(a,b)$, who contains c or does not contain c by $H(c;ab)$ or $H(\bar{c};ab)$, respectively. R is a region in the plane. An interior point of R is an element of a given point set P in its interior, and we say R is empty when R contains no interior points, and simply $R = \emptyset$. The interior region of the angular domain determined by the points a, b and c is a convex cone. It is denoted by $\gamma(a;b,c)$. a is the apex. b and c are on the boundary of the angular domain. If $\gamma(a;b,c)$ is not empty, we define an interior point of $\gamma(a;b,c)$ be attack point $\alpha(a;b,c)$, such that $\gamma(a;b,\alpha(a;b,c))$ is empty, as shown in Figure 1.

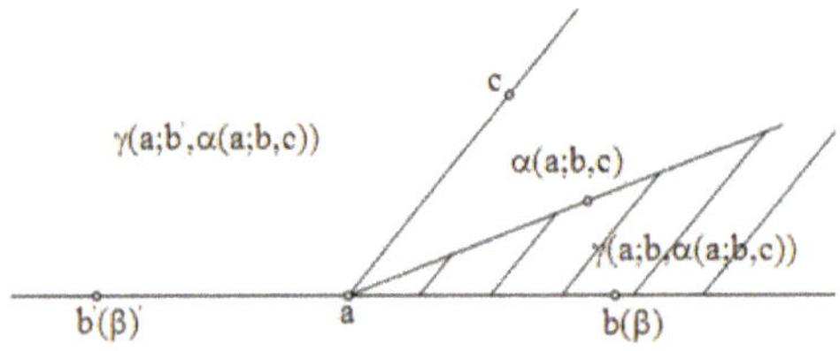

Figure 1. Figure of attack point.

For $\beta = b$ or $\beta = c$ of $\gamma(a;b,c)$, let β' be a point such that a is on the line segment $\overline{\beta\beta'}$. $\gamma(a;b',c)$ means that a lies on the segment $\overline{bb'}$. Let $v_1, v_2, v_3, v_4 \in P$ and $(v_1v_2v_3v_4)_4$ be a 4-hole, as shown in Figure 2. We name $l(v_3,v_4)$ a separating line, denoted by $SL(v_3,v_4)$ or SL_4 for simple, when all of the remaining points of P locate in $H(\overline{v_1};v_3v_4)$.

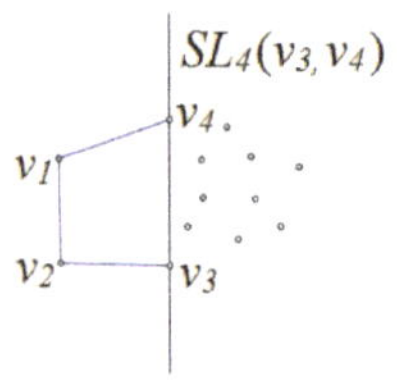

Figure 2. Figure of separating line.

We identify indices modulo t, when indexing a set of t points.

3. Main Result and Proof

Theorem 1. *[9] For any planar point set with at least 13 points in general position, if there exists a separating line SL_4, which separates a 4-hole from all of the remaining points, we always can find a 3-hole, a 4-hole and a 5-hole which are pairwise disjoint.*

From [20], we know that $13 \leq n(3,4,5) \leq 14$. In this note we will give the exact value of $n(3,4,5)$, that is the following theorem.

Theorem 2. *$n(3,4,5) = 13$, that is to say, 13 is the smallest integer such that any planar point set with at least 13 points in general position, we always can find a 3-hole, a 4-hole and a 5-hole which are pairwise disjoint.*

Proof. Let P be a 13 points set. $CH(P) = \{v_1, v_2, ..., v_l\}$. If we can find a 5-hole and a disjoint convex region with at least 7 points remained, we are done by $n(3,4) = 7$ [8]. That is to say, if we find a straight line which separates a 5-hole from at least 7 points remained, the result is correct. We call such a line a cutting line through two points u and v in P, denoted by $L_5(u,v)$. If we can find a 4-hole and the vertices number of the remaining points is more than 4, we are done by Theorem 1, where the two parts are disjoint. That is to say, if we can find such a cutting line through two points m and n in P, denoted by $L_4(m,n)$, our conclusion is correct. Therefore, in the following proof, if we can find a cutting line $L_5(u,v)$ or $L_4(m,n)$, our conclusion must be true.

In the following, we will assume there does not exist a separating line SL_4. Then there must exist a point p_i, such that $\gamma(p_i; v_i, v'_{i-1})$ and $\gamma(p_i; v_{i-1}, v'_i)$ are empty, as shown in Figure 3. Considering the 13 points, it is easy to know the conclusion is obvious right when $|V(P)| \geq 7$. Next, we discuss the considerations that $3 \leq |V(P)| \leq 6$.

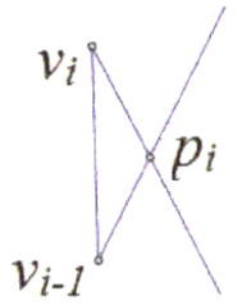

Figure 3. Figure of point determined by two separating lines.

Case 1 $|V(P)| = 6$.

Let $v_i \in V(P)$ for $i = 1,2,...6$. As shown in Figure 4, we have the points p_i for $i = 1,2,...6$, such that the shaded region is empty and we have 1 point p_7 remained.

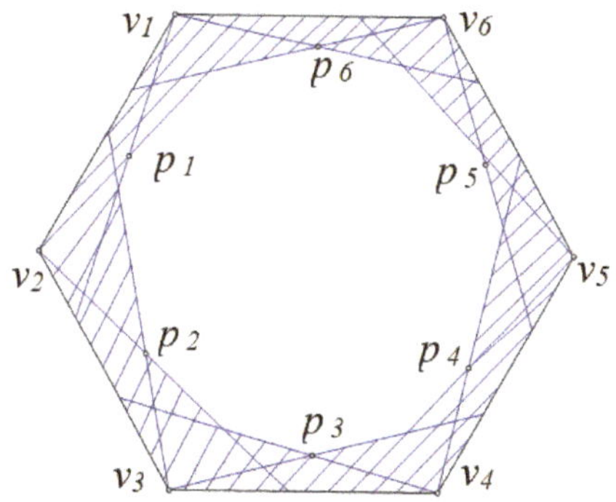

Figure 4. Figure of $|V(P)| = 6$

As the isomorphism of geometry from Figure 4, we only discuss one case. And the rest could be obtained in the same way.

Assume $\gamma(v_1; p_1, v_3) \cap \gamma(v_3; v_1, p_2) = \emptyset$. We have a cutting line $L_5(v_1, \alpha(v_1; v_3, v_6))$.

Assume $\gamma(v_1; p_1, v_3) \cap \gamma(v_3; v_1, p_2) \neq \emptyset$. We have a cutting line $L_5(v_1, p_4)$.

Case 2 $|V(P)| = 5$.

Let $v_i \in V(P)$ for $i = 1,2,3,4,5$. We have 5 friend points p_i for $i = 1,2,3,4,5$ as shown in Figure 5. Then we have 3 points r_1, r_2, r_3 remained.

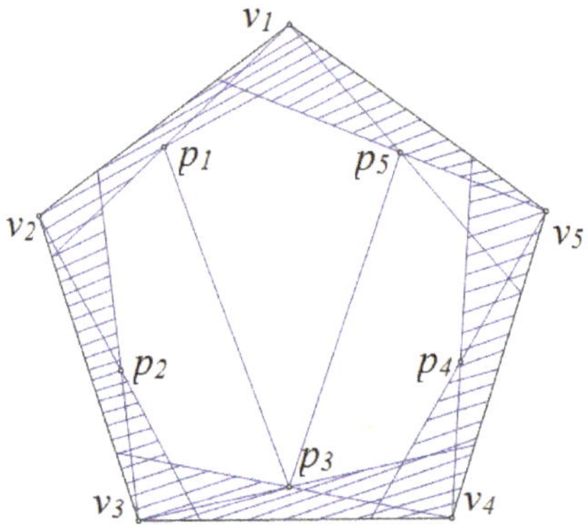

Figure 5. Figure of $|V(P)| = 5$.

Assume $\gamma(p_1; v_1', p_3) \cap \gamma(p_2; v_3', v_2') = \emptyset$. We have a cutting line $L_5(p_1, \alpha(p_1; p_3, v_2'))$.

Assume $\gamma(p_3; v_3', p_5) \cap \gamma(p_4; v_4', v_5') = \emptyset$. We have a cutting line $L_5(p_3, \alpha(p_3; p_5, p_1))$.

Assume $\gamma(p_1; v_1', p_3) \cap \gamma(p_2; v_3', v_2') \neq \emptyset$ and $\gamma(p_3; v_3', p_5) \cap \gamma(p_4; v_4', v_5') \neq \emptyset$. Suppose $\gamma(p_1; v_2', p_3) \cap \gamma(p_5; v_5', p_3) = \emptyset$. If $\gamma(p_1; v_1', p_3) \cap \gamma(p_2; v_3', v_2')$ has two of the remaining points say $r_1, r_2, r_3 \in \gamma(p_5; p_3, v_5)$, let $r_1 = \alpha(p_3; p_1, v_4')$: and if $r_2 \in \gamma(r_1; p_2, p_3') \neq \emptyset$, we have a cutting line $L_5(r_1, p_3)$; and if $r_2 \in \gamma(r_1; p_1', p_3)$, we have $(v_2 v_3 p_2)_3$, $(p_1 r_1 r_2 p_3 v_1)_5$ and a 4-hole from the remaining points; and if $r_2 \in \gamma(r_1; p_2, p_1')$, we have a cutting line $L_5(p_1, r_1)$. If $\gamma(p_5; p_3, v_5) \cap \gamma(v_4; p_3, p_4)$ has two of the remaining points, symmetrically, the conclusion is also right. Suppose $\gamma(p_1; v_2', p_3) \cap \gamma(p_5; v_5', p_3) \neq \emptyset$. We may suppose $r_1 \in \gamma(p_1; v_1', p_3) \cap \gamma(p_2; v_3', v_2')$, $r_2 \in \gamma(p_1; v_2', p_3) \cap \gamma(p_5; v_5', p_3)$, $r_3 \in \gamma(p_3; v_4, p_5) \cap \gamma(p_4; v_4', v_5')$. If $\gamma(r_2; p_1, p_3') \neq \emptyset$, we have $(v_2 v_3 p_2)_3$, $(p_1 r_1 p_3 r_2 v_1)_5$ and a 4-hole from the remaining points. If $\gamma(r_2; p_3, p_1') \neq \emptyset$, we have $(v_2 v_3 p_2)_3$, $(r_2 p_1 r_1 p_3 \alpha(r_2; p_3, p_1'))_5$ and a 4-hole from the remaining points. If $\gamma(r_2; p_1, p_3') = \emptyset$ and $\gamma(r_2; p_3, p_1') = \emptyset$, we have $(v_4 v_5 p_4)_3$, $(r_3 p_5 v_1 r_2 p_3)_5$ and a 4-hole from the remaining points.

Case 3 $|V(P)| = 4$.

Let $v_i \in V(P)$ for $i = 1,2,3,4$. We have 4 friend points p_i for $i = 1,2,3,4$. Then we have 5 points r_1, r_2, r_3, r_4, r_5 remained as shown in Figure 6.

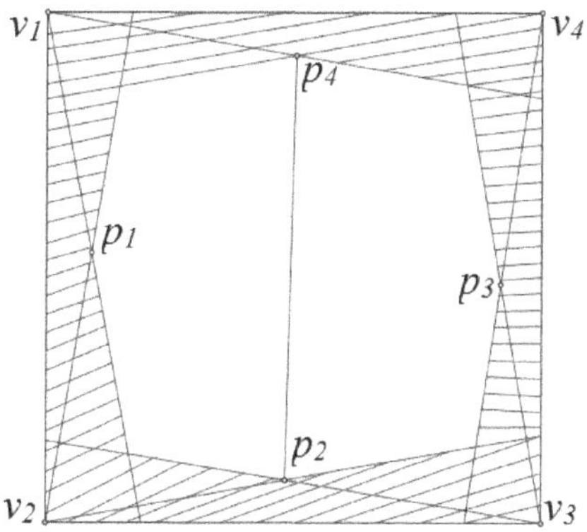

Figure 6. Figure of $|V(P)| = 4$.

If $\gamma(p_1;v_1',v_2') \cap H(p_1;p_2p_4) = \emptyset$ or $\gamma(p_3;v_3',v_4') \cap H(p_3;p_2p_4) = \emptyset$, we have a cutting line $L_5(p_4,\alpha(p_4;p_2,v_1'))$ or $L_5(p_4,\alpha(p_4;p_2,v_4'))$. Then we will consider that $\gamma(p_1;v_1',v_2') \cap H(p_1;p_2p_4) \neq \emptyset$ and $\gamma(p_3;v_3',v_4') \cap H(p_3;p_2p_4) \neq \emptyset$.

Assume one of the five points say $r_1 \in \gamma(p_1;v_1',v_2') \cap H(p_1;p_2p_4)$ and the remaining four say $r_i \in \gamma(p_3;v_3',v_4') \cap H(p_3;p_2p_4)$, $i = 2,3,4,5$. (If $\gamma(p_1;v_1',v_2') \cap H(p_1;p_2p_4)$ has four points and $\gamma(p_1;v_1',v_2') \cap H(p_1;p_2p_4)$ has one point, symmetrically, the conclusion is also right). Let $r_2 = \alpha(p_4;p_2,v_1')$.

Suppose $r_1 \in \gamma(p_1;v_1',p_2)$ or $r_1 \in \gamma(p_1;v_2',p_4)$. We always have a cutting line $L_5(p_2,p_4)$. Suppose $r_1 \in \gamma(p_1;p_4,r_2) \cap H(p_1;p_2p_4))$. We have $(v_1v_4p_4)_3$, $(p_1v_2p_2r_2r_1)_5$ and a 4-hole from the remaining points. Suppose $r_1 \in \gamma(p_1;p_2,r_2) \cap H(p_1;p_2p_4)$. We have $(v_2v_3p_2)_3$, $(p_1v_1p_4r_2r_1)_5$ and a 4-hole from the remaining points.

Assume two of the five points, say $r_1,r_2 \in \gamma(p_1;v_1',v_2') \cap H(p_1;p_2p_4)$ and the remaining three say $r_i \in \gamma(p_3;v_3',v_4') \cap H(p_3;p_2p_4)$, $i = 3,4,5$. (If $\gamma(p_1;v_1',v_2') \cap H(p_1;p_2p_4)$ has three points and $\gamma(p_1;v_1',v_2') \cap H(p_1;p_2p_4)$ has two points, symmetrically, our conclusion is also right.)

Suppose $\gamma(p_2;v_1,p_4) = \emptyset$. If $\gamma(p_2;v_1,p_1) \neq \emptyset$, let $r_1 = \alpha(p_2;v_1,p_1)$, we have $(r_2p_1v_2)_3$, $(p_4v_1r_1p_2\alpha(p_2;p_4,v_2'))_5$ and a 4-hole from the remaining points. If $\gamma(p_2;v_1,p_1) = \emptyset$, we have $(r_1r_2v_2)_3$, $(p_4v_1p_1p_2\alpha(p_2;p_4,v_2'))_5$ and a 4-hole from the remaining points. Suppose $\gamma(p_2;v_1,p_4) \neq \emptyset$. Let $r_1 = \alpha(p_2;p_4,v_1)$. If $r_2 \in \gamma(r_1;p_1,p_2')$, we have $(v_1v_4p_4)_3$, $(r_1r_2p_1v_2p_2)_5$ and a 4-hole from the remaining points. If $r_2 \in \gamma(r_1;p_1,p_4')$, we have $(v_2p_2v_3)_3$, $(v_1p_1r_2r_1p_4)_5$ and a 4-hole from the remaining points. If $r_2 \in \gamma(r_1;p_2,p_4')$, we have $(v_1v_2p_1)_3$, $(p_4r_1r_2p_2\alpha(p_2;p_4,v_2')_5$ and a 4-hole from the remaining points.

Case 4 $|V(P)| = 3$.

Let $v_1,v_2,v_3 \in V(P)$. We have 3 friend points p_1,p_2,p_3 and 7 points remained. As shown in Figure 7, denote $\gamma(p_1;v_2',p_3) \cap \gamma(p_3;v_3',p_1) = T_1$, $\gamma(p_1;v_1',p_2) \cap \gamma(p_2;v_3',p_1) = T_2$, $\gamma(p_2;v_2',p_3) \cap \gamma(p_3;v_1',p_1) = T_3$.

Without loss of generality, we assume $|T_3| \geq |T_1| \geq |T_2|$.

(1) $|T_3| = 7$.

We have a cutting line $L_5(p_2,\alpha(p_2;p_3,v_2')))$.

(2) $|T_3| = 6$.

Name the remaining one r_1. If $r_1 \in \gamma(p_3;v_3',p_1)$ or $r_1 \in \gamma(p_2;v_3',p_1)$, we have a cutting line $L_5(p_2,p_3)$. If $r_1 \in \gamma(p_3;p_1,p_2) \cap \gamma(p_1;p_2,p_3)$: and if $\gamma(r_1;p_3,p_1') \neq \emptyset$, we have a cutting line

$L_5(r_1, \alpha(r_1; p_3, p_1'))$; and if $\gamma(r_1; p_3, p_1') = \emptyset$, we have $(v_1v_3p_3)_3$, $(r_1p_1v_2p_2\alpha(r_1; p_2, p_1'))_5$ and a 4-hole from the remaining points.

(3) $|T_3| = 5$.

Name the remaining two points r_1, r_2. Then we will discuss the region $\gamma(p_3; v_1, p_1)$, as shown in Figure 8.

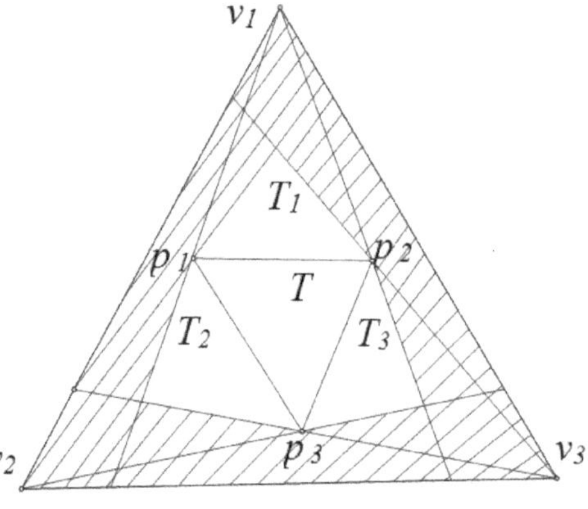

Figure 7. Figure of $|V(P)| = 5$.

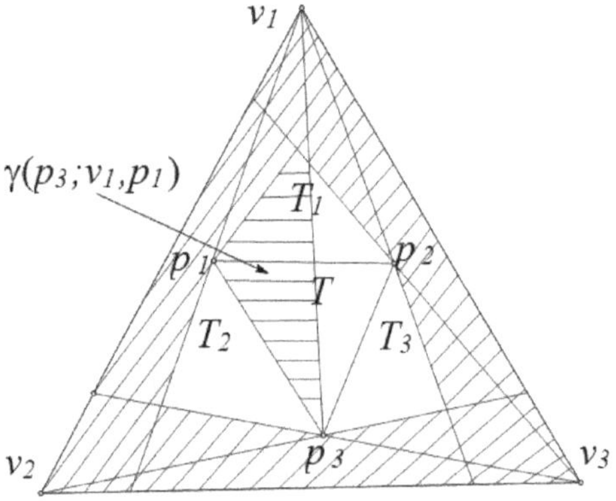

Figure 8. Figure of $|T_3| = 5$

Assume $\gamma(p_3; v_1, p_1) = \emptyset$. (If $\gamma(p_1; p_2, v_2) = \emptyset$, by the similar reason our conclusion is also right.) Let $r_1 = \alpha(p_3; p_1, p_2)$. Suppose $r_1 \in \gamma(p_2; p_1, p_3)$.

If $r_2 \in \gamma(r_1; p_3, p_1')$, we have a cutting line $L_5(p_3, r_2)$. If $r_2 \in \gamma(p_2; r_1, p_1)$: and if $\gamma(r_1; p_3, p_1') \neq \emptyset$, we have $(r_2p_2v_2)_3$, $(p_3v_1p_1r_1\alpha(r_1; p_3, p_1'))_5$ and a 4-hole from the remaining points; and if $\gamma(r_1; p_3, p_1') = \emptyset$, we have $(v_1v_2p_1)_3$, $(p_3r_1r_2p_2\alpha(p_2; p_3, v_3))_5$ and a 4-hole from the remaining points. Suppose $r_1 \in \gamma(p_2; p_1, v_3')$. If $r_2 \in \gamma(r_1; p_3, p_1')$, we have a cutting line $L_5(p_3, r_2)$. If $r_2 \in \gamma(r_1; p_1', p_3')$, we have $(r_1v_2p_2)_3$, $(p_3v_1p_1r_1\alpha(r_1; p_3, p_2))_5$ and a 4-hole from the remaining points.

Assume $\gamma(p_3; v_1, p_1) \neq \emptyset$ and $\gamma(p_1; p_2, v_2) \neq \emptyset$. Then we suppose $\gamma(p_3; v_1, p_1)$ has one point say r_1 and $\gamma(p_1; p_2, v_2)$ has one point say r_2. If $\gamma(r_1; p_1, p_2') \neq \emptyset$, we have a cutting line $L_5(p_2, r_1)$. If $\gamma(r_1; p_1, p_2') = \emptyset$, we have a cutting line $L_5(r_1, \alpha(r_1; p_2, p_3)$.

(4) $|T_3| = 4$.

Name the remaining three points r_1, r_2, r_3. Then we will discuss the region $\gamma(p_3; p_1, v_3')$, as shown in Figure 9.

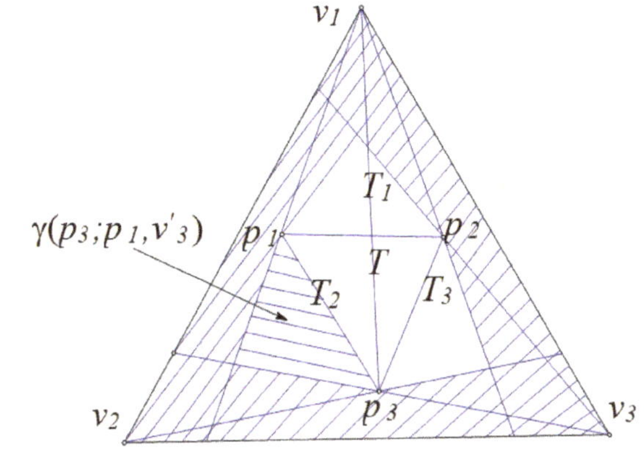

Figure 9. Figure of $|T_3| = 4$

(a) Assume $r_1, r_2, r_3 \in \gamma(p_3; p_1, v_3')$. Let $r_1 = \alpha(p_3; p_1, v_3')$. We have $(v_1p_2p_3)_3$, $(r_1p_1v_2p_2p_3)_5$ and a 4-hole from the remaining points.

(b) Assume two of $r_i, i = 1, 2, 3$, say $r_1, r_2 \in \gamma(p_3; p_1, v_3')$. Suppose $r_3 \in \gamma(p_2; p_1, p_3) \cap \gamma(p_3; p_1, p_2)$. If $\gamma(r_3; p_1, p_2') \neq \emptyset$: we have a 4-hole from $\{r_4, r_5, r_6, r_7, v_3\}$, $(p_1v_2p_2r_3\alpha(r_3; p_1, p_2'))_5$ and a 3-hole from the remaining points. If $\gamma(r_3; p_1, p_2') \neq \emptyset$, we have$(r_3p_1v_2p_2\alpha(r_3; p_2, p_1'))_5$, $(r_1r_2v_1)_3$ and a 4-hole from the remaining points. If $\gamma(r_3; p_1, p_2') = \emptyset$ and $\gamma(r_3; p_1, p_2') = \emptyset$, we have a cutting line $L_4(p_2, r_3)$.

(c) Assume one of $r_i, i = 1, 2, 3$, say $r_1 \in \gamma(p_3; p_1, v_3')$.

Suppose $\gamma(r_3; p_1, v_2) = \emptyset$. We have a cutting line $L_5(p_3, r_2)$.

Suppose $\gamma(p_3; p_1, v_2) \neq \emptyset$. Let $r_2 = \alpha(p_3; p_1, v_2)$. If $r_2 \in \gamma(p_1; v_1', p_2)$, we have a cutting line $L_5(r_2, \alpha(r_2; p_3, p_2))$. Then we suppose $r_2 \in \gamma(p_1; p_2, p_3)$. If $r_1 \in \gamma(r_2; p_2', p_1)$: and if $r_3 \in \gamma(r_2; p_3, p_1')$, we have a cutting line $L_5(r_2, r_3)$; and if $r_3 \in \gamma(r_2; p_2, p_1')$, we have $(v_1r_1p_3)_3$, $(p_1v_2p_2r_3r_2)_5$ and a 4-hole from the remaining points; and if $r_3 \in \gamma(r_2; p_2, v_2)$, we have $(v_1v_3p_3)_3$, $(r_1p_1v_2r_3r_2)_5$ and a 4-hole from the remaining points; and if $r_3 \in \gamma(r_2; v_2, p_3')$, we have a cutting line $L_5(v_2, p_3)$. If $r_1 \in \gamma(r_2; p_2', p_3)$: and if $r_3 \in \gamma(r_2; p_3, p_1')$, we have a cutting line $L_5(r_2, \alpha(r_2; p_3, p_1'))$; and if $r_3 \in \gamma(r_2; p_2, p_1')$, we have $(v_1r_1p_3)_3$, $(r_2p_1v_2p_2r_3)_5$ and a 4-hole from the remaining points; and if $r_3 \in \gamma(r_2; r_1', p_2)$, we have $(v_1v_2p_1)_3$, $(p_3r_1r_2r_3p_2)_5$ and a 4-hole from the remaining points; and if $r_3 \in \gamma(r_2; v_2, r_1')$, we have $(v_1v_3p_3)_3$, $(r_1p_1v_2r_3r_2)_5$ and a 4-hole from the remaining points; and if $r_3 \in \gamma(r_2; p_3', v_2)$, we have $(v_1r_1p_1)_3$, $(p_3r_2r_3v_2p_2)_5$ and a 4-hole from the remaining points.

(d) Assume $\gamma(p_3; p_1, v_3') = \emptyset$. By the same reason, we also assume $\gamma(p_1; p_2, v_1') = \emptyset$. Then we will discuss the region $\gamma(v_1; p_1, p_2)$, as shown in Figure 10.

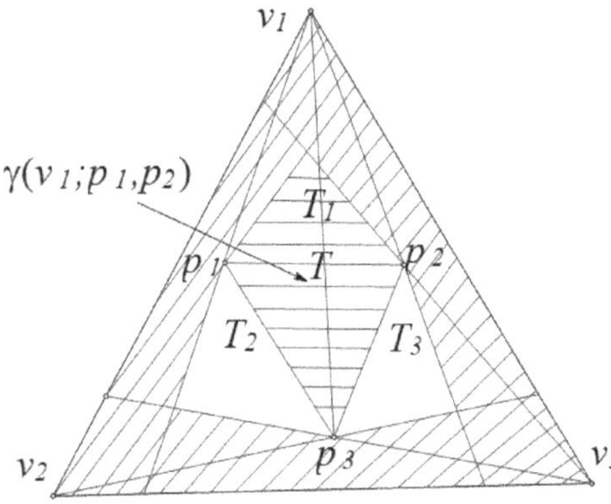

Figure 10. Figure of $|T_3| = 4$ with shaded region nonempty.

(d1) Suppose $\gamma(v_1; p_1, p_2) = \emptyset$. Let $r_1 = \alpha(p_1; p_3, p_2)$ within $(p_1p_2p_3)$.

If $\gamma(r_1; p_1, p_3') \neq \emptyset$, we have $(v_2v_3p_2)_3$, $(r_1p_3v_1p_1\alpha(r_1; p_1, p_3'))_5$ and a 4-hole from the remaining points.

If $\gamma(r_1; p_1, p_3') = \emptyset$: and if $\gamma(r_1; p_3, p_1') \neq \emptyset$, we have $(v_2v_3p_2)_3$, $(p_3v_1p_1r_1\alpha(r_1; p_3, p_1))_5$ and a 4-hole from the remaining points; and if $\gamma(r_1; p_3, p_1') = \emptyset$, let $r_2 = \alpha(r_1; p_3', p_1')$ within $(p_1p_2p_3)$, we have $(v_1v_3p_3)_3$, $(r_1p_1v_2r_2r_3)_5$ and a 4-hole from the remaining points when $r_3 \in \gamma(r_2; r_1, v_2') \cap \gamma(r_1; r_2, p_1')$, we have $(v_1p_1r_1p_3)_4$, $(r_3r_2v_2p_2\alpha(r_3; p_2, r_2))_5$ and a 3-hole from the remaining points when $r_3 \in \gamma(r_2; p_1', v_2')$ and $\gamma(p_3; p_2, v_1') \cap \gamma(r_3; p_2, r_2') \neq \emptyset$, we have $(v_1v_2p_1)_3$, $(p_3r_1r_2r_3\alpha(r_3; p_3, r_2'))_5$ and a 4-hole from the remaining points when $r_3 \in \gamma(r_2; p_1', v_2')$ and $\gamma(p_3; p_2, v_1') \cap \gamma(r_3; p_2, r_2') = \emptyset$, we have $(v_1r_1p_3)_3$, $(p_1v_2p_2r_3r_2)_5$ and a 4-hole from the remaining points when $r_3 \in \gamma(r_2; p_1', p_2)$, we have $(v_1v_2p_1)_3$, $(p_3r_1r_2r_3p_2)_5$ and a 4-hole from the remaining points when $r_3 \in \gamma(r_2; r_1', p_2)$.

(d2) Suppose $\gamma(v_1; p_1, p_2)$ has one of the r_1, r_2, r_3, say $r_1 \in \gamma(r_1; p_1, p_2)$. Let $r_2 = \alpha(p_2; p_1, p_3)$.

If $r_2 \in \gamma(r_1; p_2, p_3)$, we have $(v_2v_3p_2)_3$, $(r_1p_1v_1p_3r_2)_5$ and a 4-hole from the remaining points.

If $r_2 \in \gamma(r_1; p_1, p_3)$: and if $r_3 \in \gamma(r_2; r_1, p_3)$, we have $(v_1v_2p_1)_3$, $(r_3r_2r_1p_2p_3)_5$ and a 4-hole from the remaining points; and if $r_3 \in \gamma(r_2; p_3, p_1')$, we have $(v_2v_3p_2)_3$, $(v_1p_1r_2r_3p_3)_5$ and a 4-hole from the remaining points; and if $r_3 \in \gamma(r_2; p_1', v_1')$, we have a cutting line $L_5(r_1, \alpha(r_1; p_2, p_1'))$ when $\gamma(r_1; p_2, p_1') \neq \emptyset$, we have $(v_2v_3p_2)_3$, $(r_3r_2p_1r_1\alpha(r_3; r_1, r_2'))_5$ and a 4-hole from the remaining points when $\gamma(r_1; p_2, p_1') = \emptyset$ and $\gamma(r_3; r_1, r_2') \neq \emptyset$, we have $(r_1p_1v_2p_2)_4$, $(p_3v_1r_2r_3\alpha(r_3; p_3, r_2'))_5$ and a 3-hole from the remaining points when $\gamma(r_1; p_2, p_1') = \emptyset$ and $\gamma(r_3; r_1, r_2') = \emptyset$.

If $r_2 \in \gamma(r_1; p_2, p_3)$, we have $(v_2v_3p_2)_3$, $(p_3v_1p_1r_1r_2)_5$ and a 4-hole from the remaining points.

(d3) Suppose $\gamma(v_1; p_1, p_2)$ has two of the points r_1, r_2, r_3, say $r_1, r_2 \in \gamma(r_1; p_1, p_2)$. Let $r_1 = \alpha(p_2; p_1, p_3)$.

If $\gamma(r_1; p_2, p_1') \neq \emptyset$, we have a cutting line $L_5(r_1, \alpha(r_1; p_2, p_1'))$.

If $\gamma(r_1; p_2, p_1') = \emptyset$, let $r_2 = \alpha(p_1; p_2, p_2)$: and if $r_2 \in \gamma(v_1; p_2, p_3)$, we have a cutting line $L_5(r_2, r_3)$ when $r_3 \in \gamma(r_2; p_1, p_3)$, we have $(v_1p_1v_2)_3$, $(p_3r_2r_3r_1p_2)_5$ and a 4-hole from the remaining points when $r_3 \in \gamma(r_2; p_3, r_1)$, we have $(v_2v_3p_2)_3$, $(r_2v_1p_1r_1p_3)_5$

and a 4-hole from the remaining points when $r_3 \in \gamma(r_1; r_2, p_1^{'})$, we have $(v_1p_3v_3)_3$, $(r_1p_1v_2p_2r_3)_5$ and a 4-hole from the remaining points when $r_3 \in \gamma(r_1; p_2, p_1^{'})$; and if $r_2 \in \gamma(v_1; p_1, p_2)$, we have $(p_1r_1p_2v_2)_4$, $(p_1v_1r_1r_3\alpha(r_3; p_3, r_2^{'}))_5$ and a 3-hole from the remaining points when $\gamma(r_2; r_3, p_1^{'}) \cap \gamma(p_2; p_3, v_2^{'}) \neq \emptyset$, we have a cutting line $L_5(r_1, \alpha(r_1; p_2, p_1^{'}))$, when $\gamma(r_1; p_2, p_1^{'}) \neq \emptyset$, we have $(v_1v_3p_3)_3$, $(r_3r_2p_1r_1\alpha(r_1; r_3, p_1^{'}))_5$ and a 4-hole from the remaining points when $\gamma(r_2; r_3, p_1^{'}) \cap \gamma(p_2; p_3, v_2^{'}) = \emptyset$ and $\gamma(r_1; p_2, p_1^{'}) = \emptyset$.

(d4) Suppose $\gamma(v_1; p_1, p_2)$ has all of the three points r_1, r_2, r_3. Let $r_1 = \alpha(p_1; p_3, p_2)$, $r_2 = \alpha(p_1; p_2, p_3)$.

If $\gamma(r_1; p_3, p_1^{'}) \neq \emptyset$ or $\gamma(r_2; p_2, p_1^{'}) \neq \emptyset$, we always have a cutting line L_5.

If $\gamma(r_1; p_3, p_1^{'}) = \emptyset$ and $\gamma(r_2; p_2, p_1^{'}) = \emptyset$: and if $r_3 \in \gamma(r_1; p_1, p_3^{'})$, we have a cutting line $L_5(p_3, r_1)$; and if $r_3 \in \gamma(r_1; p_3^{'}, p_2) \cap \gamma(r_2; p_2^{'}, r_1)$, we have $(v_1v_2p_1)_3$, $(p_3r_1r_3r_2p_2)_5$ and a 4-hole from the remaining points; and if $r_3 \in \gamma(r_2; p_1, p_2^{'})$, we have a cutting line $L_5(p_2, r_2)$; and if $r_3 \in \gamma(r_2; p_1^{'}, v_2^{'})$, we have $(v_1p_1r_1p_3)_4$, $(r_3r_2v_2p_2\alpha(r_3; p_2, r_2^{'}))_5$ and a 3-hole from the remaining points when $\gamma(r_3; p_2, r_2^{'}) \cap \gamma(v_1; p_2, p_3) \neq \emptyset$, we have $(v_2v_3p_2)_3$, $(r_1p_1r_2r_3\alpha(r_3; p_3, r_2^{'}))_5$ and a 4-hole from the remaining points when $\gamma(r_3; p_2, r_2^{'}) \cap \gamma(v_1; p_2, p_3) \neq \emptyset$.

(5) $|T_3| = 3$. Let $r_1, r_2, r_3 \in T_3$.

(a) $|T_1| = 3$.

Let $r_4, r_5, r_6 \in T_1$. Name the remaining one point r_7. Assume $r_7 \in \gamma(v_2; p_3, p_2)$, as shown in Figure 11.

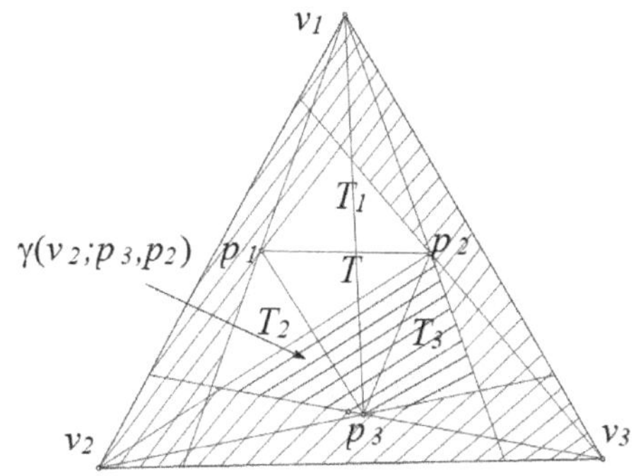

Figure 11. Figure of $|T_1| = 3$.

Symmetrically, our conclusion is also right when $r_7 \in \gamma(v_2; p_3, p_1)$. Let $r_4 = \alpha(p_3; p_1, v_3^{'})$. We have $(r_5r_6v_1)_3$, $(r_4p_1v_2r_7p_3)_5$ and a 4-hole from the remaining points.

(b) $|T_1| = 2$.

Let $r_4, r_5 \in T_1$. Name the remaining two points r_6, r_7.

(b1) $|T_2| = 2$. Let $r_6, r_7 \in T_2$.

Assume $\gamma(v_1; p_1, p_2) = \emptyset$. Let $r_4 = \alpha(p_2; v_1, p_3)$. Suppose $r_5 \in \gamma(r_4; p_2^{'}, p_3)$. We have a cutting line $L_5(p_1, p_3)$. Suppose $r_5 \in \gamma(r_4; v_1^{'}, p_3)$. If $\gamma(r_5; p_3, r_4^{'}) \neq \emptyset$, we have a cutting line $L_5(r_5, \alpha(r_5; p_3, v_4^{'}))$. If $\gamma(r_5; p_3, r_4^{'}) = \emptyset$, we have a cutting line $L_5(r_1, \alpha(r_1; p_1, p_2)$ where $r_1 = \alpha(p_1; p_3, p_2)$. Suppose $r_5 \in \gamma(r_4; p_2^{'}, v_1^{'})$. We have $(r_6r_7v_2)_3$, $(r_4v_1p_1p_2r_5)_5$ and a 4-hole from the remaining points.

Assume $\gamma(v_1; p_1, p_2)$ has one of r_4, r_5. Let $r_4 \in \alpha(v_1; p_1, p_2)$. Suppose $r_5 \in \alpha(r_4; p_1', v_1)$. If $\gamma(r_4; p_2, v_1') = \emptyset$, we have $(r_2r_3v_3)_3$, $(r_5r_4p_2r_1p_3)_5$ and a 4-hole from the remaining points where $r_1 = \alpha(p_2; p_3, v_3)$. If $\gamma(r_4; p_2, v_1') \neq \emptyset$, we have $(p_1v_2r_7)_3$, $(v_1r_4r_6p_2r_5)_5$ and a 4-hole from the remaining points where $r_6 = \alpha(r_4; p_2, v_1')$.

Assume $\gamma(v_1; p_1, p_2)$ has r_4, r_5. Let $r_4 \in \alpha(p_2; v_1, p_1)$, $r_1 = \alpha(p_2; p_3, v_3)$. we have $(r_2r_3v_3)_3$, $(p_2r_1p_3v_1r_4)_5$ and a 4-hole from the remaining points.

(b2) $|T_2| = 1$.

Let $r_6 \in T_2$ and $r_7 \in (p_1p_2p_3)$, as shown in Figure 12.

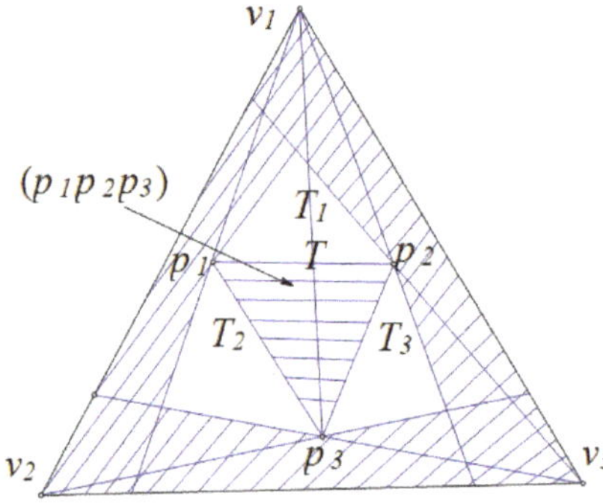

Figure 12. Figure of $|T_2| = 1$.

Assume $r_6 \in \gamma(r_7; p_3', p_2)$. We have $(r_2r_3v_3)_3$, $(p_3r_7r_6p_2r_1)_5$ and a 4-hole from the remaining points where $r_1 = \alpha(p_2; p_3, v_3)$. Assume $r_6 \in \gamma(r_7; p_3', v_2)$. We have $(r_4r_5v_1)_3$, $(p_1v_2r_6r_7p_3)_5$ and a 4-hole from the remaining points. Assume $r_6 \in \gamma(r_7; p_1, v_2)$. If $\gamma(r_7; r_6', p_2) \neq \emptyset$, we have $(r_2r_3v_3)_3$, $(p_6r_7r_1p_2v_2)_5$ and a 4-hole from the remaining points where $r_1 = \alpha(r_7; p_2, r_6')$. If $\gamma(r_7; r_6', p_2) = \emptyset$, we have $(v_2v_3p_2)_3$, $(p_1r_6r_7r_1p_3)_5$ and a 4-hole from the remaining points where $r_1 = \alpha(r_7; p_3, p_1')$.

(b3) $|T_2| = 0$.

Let $r_6, r_7 \in (p_1p_2p_3)$. Then we will discuss the region $\gamma(p_3; p_1, v_1)$, as shown in Figure 13.

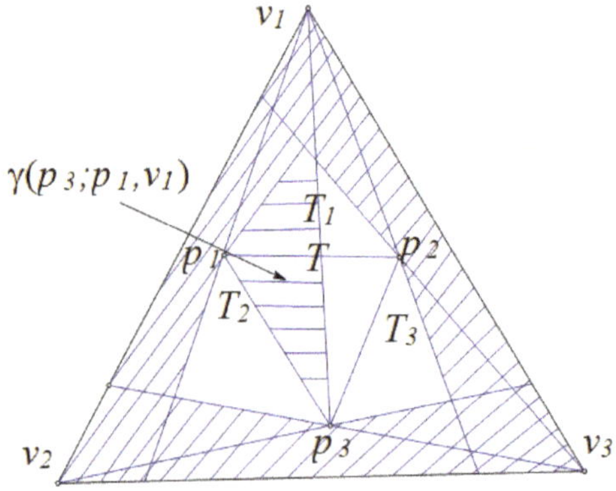

Figure 13. Figure of $|T_2| = 0$.

Assume $\gamma(p_3;p_1,v_1)=\emptyset$. Suppose $\gamma(r_6;p_3,v_2^{'})\neq\emptyset$. We have $(p_3p_1v_2r_6\alpha(r_6;p_3,v_2^{'}))_5$, $(r_4r_5v_1)_3$ and a 4-hole from the remaining points. Suppose $\gamma(r_6;p_3,v_2^{'})=\emptyset$. If $\gamma(r_7;r_6^{'},p_2)\cap\gamma(p_2;r_7,v_2^{'})\neq\emptyset$, we have $(r_2r_3v_3)_3$, $(r_6v_2p_2r_1r_7)_5$ and a 4-hole from the remaining points where $r_1=\alpha(r_7;p_2,r_6^{'})$. If $\gamma(r_7;r_6^{'},p_2)\cap\gamma(p_2;r_7,v_2^{'})=\emptyset$, we have $(v_2v_3p_2)_3$, $(r_1r_7r_6p_1\alpha(r_1;p_1,p_3))_5$ and a 4-hole from the remaining points where $r_1=\alpha(p_1;r_7,p_3)$ within $\gamma(p_3;p_2,v_3)$.

Assume $\gamma(p_3;p_2,v_1)=\emptyset$. We have $(r_2r_3v_3)_3$, $(r_1p_3r_6v_2p_2)_5$ and a 4-hole from the remaining points where $r_1=\alpha(p_3;p_2,v_3)$ and $r_6=\alpha(p_3;v_2,p_1)$.

Assume $\gamma(p_3;p_1,v_2)\neq\emptyset$ and $\gamma(p_3;p_1,v_2)\neq\emptyset$. We may assume $r_6\in\gamma(p_3;p_1,v_2)$ and $r_7\in\gamma(p_3;p_1,v_2)$. Suppose $r_7\in\gamma(r_6;p_2,p_1^{'})$. We have $(r_4r_5v_1)_3$, $(r_6r_7p_2v_2p_1)_5$ and a 4-hole from the remaining points. Suppose $r_7\in\gamma(r_6;p_3,p_1^{'})$. If $\gamma(r_7;r_6^{'},p_2)\neq\emptyset$, we have $(r_2r_3v_3)_3$, $(r_7r_6v_2p_2r_1)_5$ and a 4-hole from the remaining points where $r_1=\alpha(p_2;r_7,v_2^{'})$. If $\gamma(r_7;r_6^{'},p_2)=\emptyset$: and if $\gamma(r_7;p_1,v_1)=\emptyset$, we have $(r_2v_3p_2)_3$, $(v_1p_1r_6r_7r_4)_5$ and a 4-hole from the remaining points where $r_4=\alpha(r_7;p_1,p_3)$ within $\gamma(p_3;p_1,v_1)$; and if $\gamma(r_7;p_1,v_1)\neq\emptyset$, we have $(r_2v_3p_2)_3$, $(r_4p_1r_6r_7r_1)_5$ and a 4-hole from the remaining points where $r_4=\alpha(r_7;p_1,v_1)$.

(c) $|T_1|=1$. Let $r_4\in T_1$.

(c1) $|T_2|=1$. Let $r_5\in T_2$ and $r_6,r_7\in(p_1p_2p_3)$.

Firstly, consider $r_4\in\gamma(v_1;p_1,p_2)$, then we will discuss the region $\gamma(v_1;p_1,p_2)\cap(p_1p_2p_3)=\emptyset$, as shown in Figure 14.

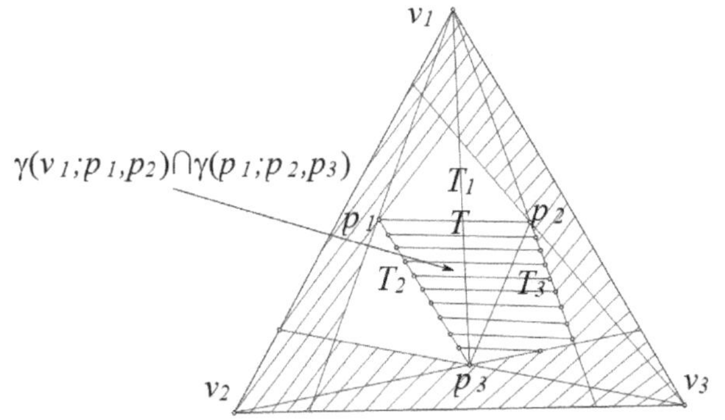

Figure 14. Figure of $|T_1|=1$ and $|T_2|=1$.

Assume $\gamma(v_1;p_1,p_2)\cap(p_1p_2p_3)=\emptyset$. We have a cutting line $L_5(r_4,\alpha(r_4;p_2,p_1^{'}))$. Assume $\gamma(v_1;p_1,p_2)\cap(p_1p_2p_3)\neq\emptyset$. Let $r_6=\alpha(p_2;p_1,v_1)$. If $\gamma(r_6;p_2,p_1^{'})\neq\emptyset$, we have a cutting line $L_5(r_4,\alpha(r_6;p_2,p_1^{'}))$. Then we may assume $\gamma(r_6;p_2,p_1^{'})=\emptyset$.

Suppose $r_5\in\gamma(r_6;v_2,r_4^{'})$. If $r_7\in\gamma(r_6;p_2^{'},r_4)$, we have $(p_1v_2r_5)_3$, $(r_4r_7r_6p_2\alpha(r_4;p_2,p_3))_5$ and a 4-hole from the remaining points. If $r_7\in\gamma(r_6;p_1^{'},r_4)$, we have a cutting line $L_5(r_4,r_6)$.

Suppose $r_5\in\gamma(r_6;v_2,p_1)$. If $\gamma(r_6;r_5^{'},p_1^{'})\neq\emptyset$, we have $(v_1r_4p_1)_3$, $(r_6r_5v_2p_2\alpha(r_6;p_2,r_5^{'}))_5$ and a 4-hole from the remaining points. If $\gamma(r_6;r_5^{'},p_1^{'})=\emptyset$: and if $r_7\in\gamma(r_6;r_4,r_5^{'})$, we have $(v_2v_3p_2)_3$, $(r_4p_1r_5r_6r_7)_5$ and a 4-hole from the remaining points; and if $r_7\in\gamma(r_6;r_4,p_2^{'})$, we have $(p_1v_2r_5)_3$, $(r_4r_7r_6p_2\alpha(r_4;p_2,p_3))_5$ and a 4-hole from the remaining points. Suppose $r_5\in\gamma(r_6;p_2,r_4^{'})$. If $r_7\in\gamma(r_6;p_2^{'},r_4)$, we have $(p_1v_2r_5)_3$, $(p_2r_6r_7r_4\alpha(r_4;p_2,p_3))_5$ and a 4-hole from the remaining points. If $r_7 \in \gamma(r_6;r_4,p_2) \cap H(r_6;r_4p_2)$, we have $(p_1v_2r_6r_5)_3$, $(p_3r_4r_7p_2\alpha(p_3;p_2,v_1^{'}))_5$ and a 3-hole from the remaining points. If $r_7 \in \gamma(p_2;r_4,v_1)$, we have $(v_1v_2p_1)_3$, $(r_1r_6r_5p_2r_7)_5$ and a 4-hole from the remaining points. If $r_7 \in \gamma(p_2;v_1,p_3)$, we have $(p_1v_2r_5)_3$, $(v_1r_4r_6p_2r_7)_5$ and a 4-hole from the remaining points.

Secondly, consider $r_4 \in \gamma(v_1;p_2,p_3)$, then we will discuss the region $\gamma(r_4;p_2,p_3) \cap (p_1p_2p_3) = \emptyset$, as shown in Figure 15.

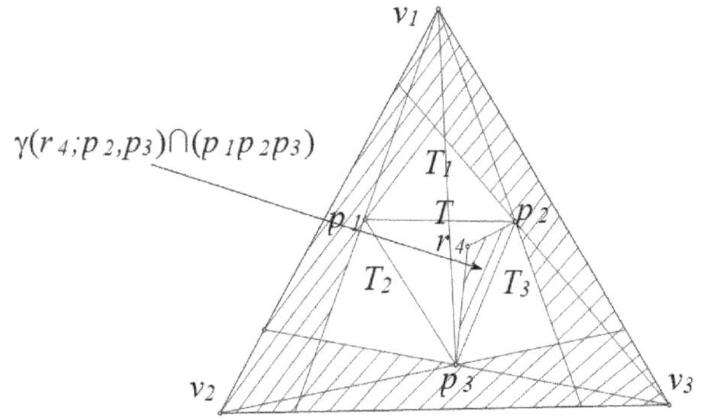

Figure 15. Figure of $|T_1| = 1$ and $|T_2| = 1$ with shaded region nonempty.

Assume $\gamma(r_4;p_2,p_3) \cap (p_1p_2p_3) = \emptyset$. We have $(r_2r_3v_3)_3$, $(r_1p_3r_4r_6p_2)_5$ and a 4-hole from the remaining points where $r_1 = \alpha(p_3;p_2,v_3)$, $r_6 = \alpha(r_4;p_2,p_1)$. Assume $\gamma(r_4;p_1,p_2) \cap (p_1p_2p_3) = \emptyset$. We have $L_5(p_2,r_4)$. Assume $\gamma(r_4;p_2,p_3) \cap (p_1p_2p_3) \neq \emptyset$ and $\gamma(r_4;p_1,p_2) \cap (p_1p_2p_3) \neq \emptyset$. Then we may assume $r_6 \in \gamma(r_4;p_2,p_3)$, $r_7 \in \gamma(r_4;p_1,p_2)$. Suppose $r_6 \in \gamma(r_4;v_1^{'},p_3) \cap (p_1p_2p_3)$. If $\gamma(r_6;r_4^{'},p_3) \neq \emptyset$, we have $(p_1r_5p_2r_7)_4$, $(v_1r_4r_6r_1p_3)_5$ and $(r_2r_3v_3)_3$ where $r_1 = \alpha(r_6;p_3,r_4^{'})$. If $\gamma(r_6;r_4^{'},p_3) = \emptyset$: and if $r_7 \in \gamma(r_4;v_1,p_2) \cap \gamma(v_1;p_2,r_4)$, we have $L_5(p_2;r_4)$; and if $r_7 \in \gamma(r_4;r_5,p_2) \cap \gamma(p_2;p_1,v_1)$, we have $L_5(r_4;r_7)$; and if $r_7 \in \gamma(r_4;p_1,r_5)$, we have $(v_1v_2p_1)_3$, $(r_4r_7r_5p_2r_6)_5$ and a 4-hole from the remaining points. Suppose $r_6 \in \gamma(r_4;v_1^{'},p_2) \cap (p_1p_2p_3)$. If $r_7 \in \gamma(v_1;p_1,p_2) \cap (p_1p_2p_3)$, we have $(r_5v_2p_1)_3$, $(v_1r_7p_2r_6r_4)_5$ and a 4-hole from the remaining points. If $r_7 \in \gamma(v_1;p_2,r_4) \cap \gamma(r_4;p_1,p_2)$: and if $r_7 \in \gamma(r_7;r_4^{'},p_1)$, we have $L_5(r_4,r_7)$; and if $r_5 \in \gamma(r_7;r_4^{'},p_2)$, we have $(v_1v_2p_1)_3$, $(r_4r_7r_5p_2r_6)_5$ and a 4-hole from the remaining points.

(c2) $|T_2| = 0$.

Denote $r_1,r_2,r_3 \in T_3$, $r_4 \in T_2$, $r_5,r_6,r_7 \in (p_1p_2p_3)$. Let $r_5 = \alpha(p_3;p_1,p_2)$ within $(p_1p_2p_3)$. If $\gamma(r_5;p_1^{'},p_3) \neq \emptyset$, we have $L_5(r_5;\alpha(r_5;p_3,p_1^{'}))$. Then we assume $\gamma(r_5;p_1^{'},p_3) = \emptyset$, and we will discuss the region $\gamma(r_5;p_1,p_2) \cap (p_1p_2p_3) = \emptyset$, as hown in Figure 16.

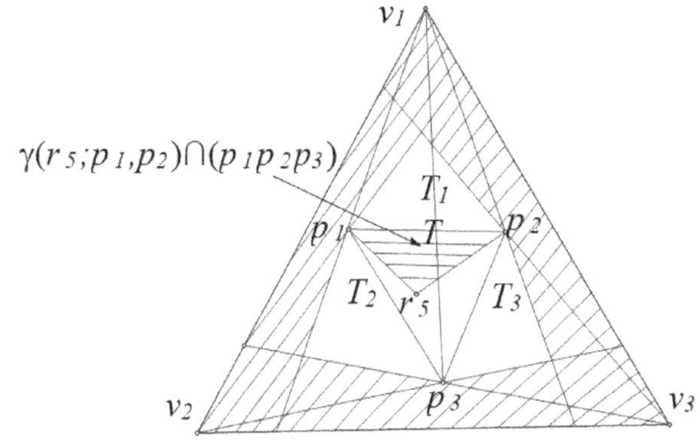

Figure 16. Figure of $|T_1| = 1$ and $|T_2| = 0$.

Assume $\gamma(r_5; p_1, p_2) \cap (p_1p_2p_3) = \emptyset$, we have $(v_1r_4p_3)_3$, $(r_5p_1v_2p_2\alpha(p_2; r_5, p_3))_5$ and a 4-hole from the remaining points.

Assume $\gamma(r_5; p_2, p_3) \cap (p_1p_2p_3) = \emptyset$. Let $p_6 = \alpha(r_5; p_2, p_3')$. Suppose $r_4 \in \gamma(r_5; p_3, r_6')$. We have $(r_2r_3v_3)_3$, $(p_2r_1p_3r_4\alpha(r_4; p_2, p_3'))_5$ and a 4-hole from the remaining points where $r_1 = \alpha(p_2; p_3, v_2')$. Suppose $r_4 \in \gamma(r_5; p_1, r_6')$. We have $(r_2r_3r_4)_3$, $(p_2r_1p_3r_5r_6)_5$ and a 4-hole from the remaining points where $r_1 = \alpha(p_2; p_3, v_2')$.

Assume $\gamma(r_5; p_1, p_2) \cap (p_1p_2p_3) \neq \emptyset$ and $\gamma(r_5; p_2, p_3) \cap (p_1p_2p_3) \neq \emptyset$. Without loss of generality, we suppose $r_6 \in \gamma(r_5; p_1, p_2)$, $r_7 \in \gamma(r_5; p_2, p_3)$.

Firstly, we may assume $r_6 \in \gamma(r_5; v_2, p_2)$. Suppose $r_4 \in \gamma(r_6; p_7, p_2')$. We have $L_5(p_2, r_6)$. Suppose $r_4 \in \gamma(r_5; r_6', p_1) \cap H(r_5; r_6p_2)$. We have a cutting line $L_5(r_5, r_6)$. Suppose $r_4 \in \gamma(r_6; p_6', p_1')$. If $r_7 \in \gamma(r_4; p_2, p_3)$, we have $(v_1v_2p_1)_3$, $(v_4r_5r_6p_2r_7)_5$ and a 4-hole from the remaining points. If $r_7 \in \gamma(p_2; r_4, p_5)$, we have $(v_3r_2r_3)_3$, $(v_1p_1r_6r_5)_4$ and $(p_3r_4r_7p_2r_1)_5$ where $r_1 = \alpha(p_2; p_3, v_2')$.

Secondly, we have may assume $r_6 \in \gamma(r_5; v_2, p_3')$, we have $(v_1p_1r_4)_3$, $(r_5r_6v_2p_2r_7)_5$ and a 4-hole from the remaining points.

(d) $|T_1| = 0$. $|T_2| = 0$.

Let $r_4, r_5, r_6, r_7 \in (p_1p_2p_3)$. And $r_1 = \alpha(p_3; p_2, v_3)$, $r_4 = \alpha(p_3; p_2, p_1)$, $r_5 = \alpha(p_2; p_1, r_4)$. If $\gamma(r_4; p_2, p_3') \neq \emptyset$, we have $(r_2r_3v_3)_3$, $(p_2r_1p_3r_4\alpha(r_4; p_2, p_3'))_5$ and a 4-hole from the remaining points. Assume $r_5 \in \gamma(r_4; p_1, p_3)$. If $\gamma(r_5; p_2, p_1') \neq \emptyset$, we have a cutting line $L_5(r_5; \alpha(r_5; p_2, p_1'))$. Then we will discuss the region $\gamma(v_4; p_1, p_2') \cap (p_1p_2p_3)$ and $\gamma(r_4; p_1, p_3') \cap \gamma(p_1; p_5, r_4)$, as shown in Figure 17.

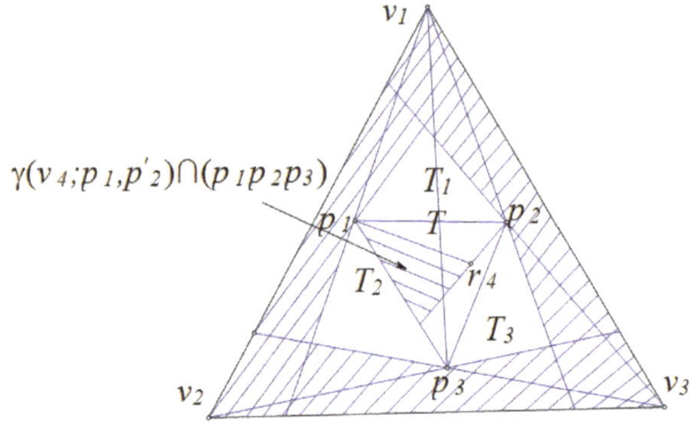

Figure 17. Figure of $|T_1| = 1$ and $|T_2| = 0$ with shaded region nonempty.

Assume $\gamma(v_4; p_1, p_2') \cap (p_1p_2p_3) = \emptyset$. We have $(r_7r_5v_2)_3$, $(r_4p_3v_1p_1r_6)_5$ and a 4-hole from the remaining points where $r_6 = \alpha(r_4; p_1, p_3')$.

Assume $\gamma(r_4; p_1, p_3') \cap \gamma(p_1; p_5, r_4) = \emptyset$. Let $r_6 = \alpha(p_1; p_3, r_4)$. Suppose $r_7 \in \gamma(r_6; r_4, p_1')$. We have $(v_1v_2p_1)_3$, $(p_2r_4r_7r_6r_5)_5$ and a 4-hole from the remaining points. Suppose $r_7 \in \gamma(r_6; r_4, v_1') \cap \gamma(r_4; p_1, p_2')$. We have $(r_1r_2r_3)_3$, $(p_1v_2p_2r_5)_4$ and $(r_4p_3v_1r_6r_7)_5$. Suppose $r_7 \in \gamma(r_6; r_5, v_1') \cap \gamma(r_4; p_1, p_2')$. We have $(v_3r_2r_3)_3$, $(r_4p_2r_1p_3)_4$ and $(r_6v_1p_2r_5r_7)_5$. Suppose $r_7 \in \gamma(r_6; r_5, p_2')$. We have $(v_1v_2p_2)_3$, $(r_4p_3r_6r_7r_5)_5$ and a 4-hole from the remaining points. Suppose $r_7 \in \gamma(r_6; p_1, r_3')$. We have a cutting line $L_5(p_3, r_6)$.

Assume $\gamma(v_4; p_1, p_2') \cap (p_1p_2p_3) \neq \emptyset$ and $\gamma(r_4; p_1, p_3') \cap \gamma(p_1; p_5, r_4) \neq \emptyset$. Without loss of generality, assume $r_6 \in \gamma(r_4; p_1, p_2') \cap (p_1p_2p_3)$, $r_7 \in \gamma(r_4; p_1, p_3') \cap \gamma(p_1; p_5, r_4')$.

Suppose $r_6 \in \gamma(r_5; p_3, p_1')$. We have a cutting line $L_5(r_6, \alpha(r_6; p_1, p_3'))$.

Suppose $r_6 \in \gamma(r_5; p_3, p_1) \cap \gamma(v_1; r_4, p_3)$. If $r_7 \in \gamma(r_5; p_3, p_2') \cap \gamma(p_1; r_5, r_4)$, we have $(v_2p_2r_5)_3$, $(v_1p_1r_7r_4r_6)_5$ and a 4-hole from the remaining points. If $r_7 \in \gamma(r_5; p_3', p_1') \cap \gamma(r_4; p_1, p_3')$, we have $(v_2p_2p_3)_3$, $(v_1p_1r_5r_7r_6)_5$ and a 4-hole from the remaining points.

Suppose $r_6 \in \gamma(r_5; p_3, p_1) \cap \gamma(v_1; r_1, p_1)$. If $r_7 \in \gamma(r_6; r_4, v_1')$, we have $(v_1r_6r_7r_4p_3)_5$, $(p_1v_2p_2r_5)_4$ and $(r_1r_2r_3)_3$. If $r_7 \in \gamma(r_6; r_5, v_1')$, we have $(v_2v_3p_2)_3$, $(v_1p_1r_5r_7r_6)_5$ and a 4-hole from the remaining points.

(6) $|T_3| = 2$.

Let $r_1, r_2 \in T_3$ and $r_1 = \alpha(p_2; p_3, v_1')$. Assume $r_2 \in \gamma(r_1; p_2, v_3)$. We have $(p_2r_1r_2v_3)_4$ and the remaining 9 points are in $H(\overline{v_3}; p_2p_3)$, as shown in Figure 18.

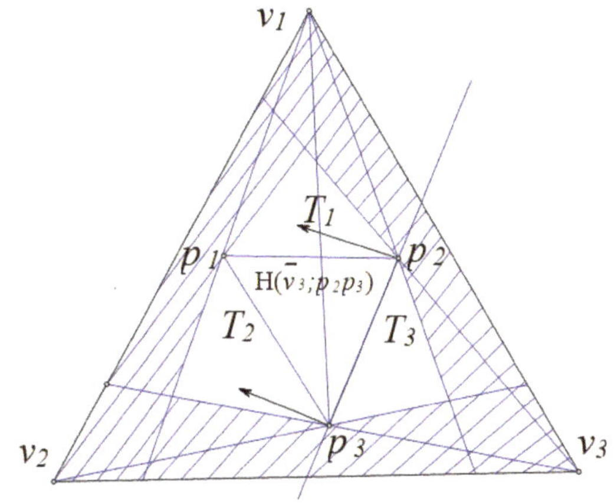

Figure 18. Figure of $|T_3| = 2$.

By the discussion of Part One, we know our conclusion is right. Assume $r_2 \in \gamma(r_1; p_3^{'}, v_2)$. We have $(p_3r_1r_2v_2)_4$. By the discussion of Part One, we know our conclusion is also right. Assume $r_2 \in \gamma(r_1; p_3^{'}, p_2)$. We have a cutting line $L_5(p_2, \alpha(p_2; p_3, p_1))$.

(7) $|T_3| = 1$.

Let $r_1 \in T_3$, $r_2 \in T_1$, $r_3 \in T_2$ and $r_4, r_5, r_6, r_7 \in (p_1p_2p_3)$. Let $r_4 = \alpha(p_3; p_2, p_1)$ within $(p_1p_2p_3)$. Assume $r_4 \in \gamma(p_3; p_1, v_1)$, as shown in Figure 19.

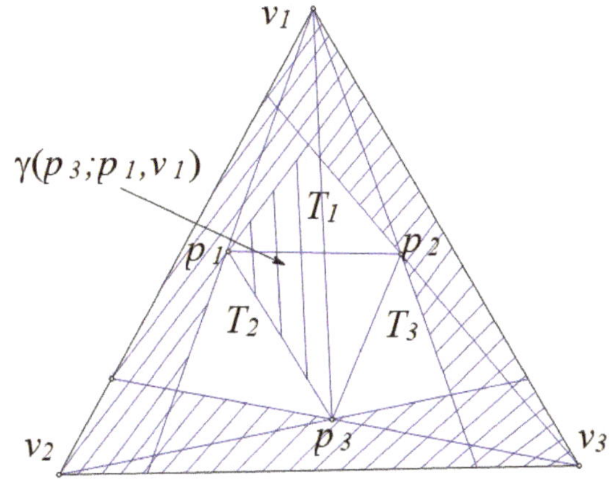

Figure 19. Figure of $|T_3| = 1$.

If $r_2 \in \gamma(v_1; p_2, p_3)$, we have a cutting line $L_5(r_2, \alpha(r_2; p_2, p_1))$. If $r_2 \in \gamma(v_1; p_2, p_1)$, we have a cutting line $L_5(v_1, \alpha(v_1; p_2, p_1))$. Assume $r_4 \in \gamma(p_3; p_2, v_1)$. If $\gamma(r_4; p_3, p_2^{'}) \neq \emptyset$, we have a cutting line $L_5(r_4, \alpha(r_4; p_3, p_2^{'}))$. If $\gamma(r_4; p_2, p_3^{'}) \neq \emptyset$, we have a cutting line $L_5(r_4; \alpha(r_4; p_2, p_3^{'}))$. If $\gamma(r_4; p_3, p_2^{'}) = \emptyset$ and $\gamma(r_4; p_2, p_3^{'}) = \emptyset$: and if $r_1 \in \gamma(r_4; p_2, v_3)$, we have $(r_4p_3v_3r_1)_4$; and if $r_1 \in \gamma(r_4; p_3, v_3)$, we have $(p_2r_4r_1v_3)_4$. Then the remaining 9 points are all in $H(\overline{v_3}; p_2p_3)$. By the discussion of Part One, our conclusion is right.

(8) $|T_3| = 0$.

Then $|T_2| = 0$, $|T_1| = 0$ and $r_i \in (p_1p_2p_3)$ for $i = 1, ..., 7$. Let $r_1 = \alpha(p_1; p_3, p_2)$. If $r_1 \in \gamma(p_1; p_3, v_3)$, as shown in Figure 20.

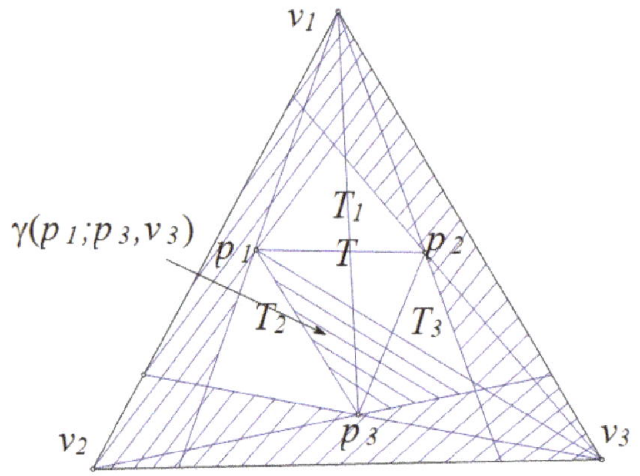

Figure 20. Figure of $|T_1| = 0$ and $|T_2| = 0$.

We have $(v_1p_1r_1p_3)_4$ and the remaining 9 points are all in $H(\overline{p_3}; p_1r_1)$. By the discussion of Part One, our conclusion is right. If $r_1 \in \gamma(p_1; p_3, v_3)$: and if $\gamma(r_1; p_1, p_3^{'}) = \emptyset$, we have $(v_1p_3r_1p_1)_4$ and the remaining 9 points are all in $H(\overline{v_1}; p_3r_1)$; and if $\gamma(r_1; p_1, p_3^{'}) \neq \emptyset$, we have a cutting line $L_5(r_1, \alpha(r_1; p_1, p_3^{'}))$. □

4. Conclusions

In this paper, we discuss a classical discrete geometry problem. After detailed proof, conclusion shows that a general planar point set contains a 3-hole, a 4-hole and a 5-hole, with at least 13 points. As $30 \leq n(6) \leq 463$ [16,21] and $n(7)$ does not exist, the proposed theorem will contribute to the theoretical research to some degree. Discrete geometry is a meaningful tool to study social networks. Therefore, our conclusion could be used to deal with some complex network problems. For example, under the environment of competition social structure, the structural holes which have been studied by many economists, are part of an important research branch of discrete geometry.

Author Contributions: Conceptualization, Z.Y.; Funding acquisition, Q.Y.; Methodology, Q.Y. and X.Y.

Funding: National Social Science Fund of China (18CGL018).

Acknowledgments: National Social Science Fund of China (18CGL018). This fund covers the costs to publish in open access.

Conflicts of Interest: The author declares no conflict of interest.

References

1. Erdős, P.; Szekers, G. A combinatorial problem in geometry. *Compos. Math.* **1935**, *2*, 463–470.
2. Erdős, P.; Szekers, G. On some extremum problem in elementary geometry. *Ann. Univ. Sci. Budapest* **1960**, *3*, 53–62.
3. Szekeres, G.; Peters, L. Computer solution to the 17-point Erdős-Szekeres problem. *ANZIAM J.* **2006**, *48*, 151–164. [CrossRef]
4. Harborth, H. Konvexe Funfeck in ebenen Punktmengen. *Elem. Math.* **1978**, *33*, 116–118.
5. Horton, J. Sets with no empty convex 7-gons. *Can. Math. Bull.* **1983**, *26*, 482–484. [CrossRef]
6. Gerken, T. Empty convex hexagons in planar point sets. *Discret. Comput. Geom.* **2008**, *39*, 239–272. [CrossRef]
7. Nicolás, C. The empty hexagon theorem. *Discret. Comput. Geom.* **2007**, *38*, 389–397. [CrossRef]
8. Urabe, M. On a partition into convex polygons. *Discret. Appl. Math.* **1996**, *64*, 179–191. [CrossRef]

9. Hosono, K.; Urabe, M. A minimal planar point set with specified disjoint empty convex subsets. *Lect. Notes Comput. Sci.* **2008**, *4535*, 90–100.
10. Hosono, K.; Urabe, M. On the number of disjoint convex quadrilaterals for a planar point set. *Comput. Geom. Theory Appl.* **2001**, *20*, 97–104. [CrossRef]
11. Hosono, K.; Urabe, M. On the minimum size of a point set containing two non-intersecting empty convex polygons. *Lect. Notes Comput. Sci.* **2005**, *3742*, 117–122.
12. Wu, L.; Ding, R. Reconfirmation of two results on disjoint empty convex polygons. *Lect. Notes Comput. Sci.* **2007**, *4381*, 216–220.
13. Bhattacharya, B.; Das, S. On the minimum size of a point set containing a 5-hole and a disjoint 4-hole. *Stud. Sci. Math. Hung.* **2011**, *48*, 445–457. [CrossRef]
14. Bhattacharya, B.B. Sandip Das, On pseudo-convex partitions of a planar point set. *Discret. Math.* **2013**, *313*, 2401–2408. [CrossRef]
15. Bhattacharya, B.; Das, S. Disjoint Empty Convex Pentagons in Planar Point Sets. *Period. Math. Hung.* **2013**, *66*, 73–86. [CrossRef]
16. Hosono, K.; Urabe, M. Specifed holes with pairwise disjoint interiors in planar point sets. *AKCE Int. J. Graph. Comb.* **2018**. [CrossRef]
17. You, X.; Wei, X. On the minimum size of a point set containing a 5-hole and double disjoint 3-holes. *Math. Notes* **2013**, *97*, 951–960. [CrossRef]
18. You, X.; Wei, X. A note on the upper bound for disjoint convex partitions. *Math. Notes* **2014**, *96*, 268–274. [CrossRef]
19. You, X.; Wei, X. A note on the value about a disjoint convex partition problem. *Ars Comb.* **2014**, *115*, 459–465.
20. You, X.; Chen, T. A note on the value about a disjoint convex partition problem. *Math. Notes*, **2018**, *104*, 135–149.
21. Koshelev, V.A. On Erdős-Szekeres problem for empty hexagons in the plane. *Modelirovanie i Analiz Informatsionnykh Sistem* **2009**, *16*, 22–74.

symmetry

MDPI

Article

Multi-Granulation Graded Rough Intuitionistic Fuzzy Sets Models Based on Dominance Relation

Zhan-ao Xue [1,2,*], Min-jie Lv [1,2], Dan-jie Han [1,2] and Xian-wei Xin [1,2]

1 College of Computer and Information Engineering, Henan Normal University, Xinxiang 453007, China; lmj2921419592@163.com (M.-j.L.); handanjie2017@163.com (D.-j.H.); Xin_XianWei@163.com (X.-w.X.)
2 Engineering Lab of Henan Province for Intelligence Business & Internet of Things, Henan Normal University, Xinxiang 453007, China
* Correspondence: 121017@htu.edu.cn

Received: 7 August 2018; Accepted: 26 September 2018; Published: 28 September 2018

Abstract: From the perspective of the degrees of classification error, we proposed graded rough intuitionistic fuzzy sets as the extension of classic rough intuitionistic fuzzy sets. Firstly, combining dominance relation of graded rough sets with dominance relation in intuitionistic fuzzy ordered information systems, we designed type-I dominance relation and type-II dominance relation. Type-I dominance relation reduces the errors caused by single theory and improves the precision of ordering. Type-II dominance relation decreases the limitation of ordering by single theory. After that, we proposed graded rough intuitionistic fuzzy sets based on type-I dominance relation and type-II dominance relation. Furthermore, from the viewpoint of multi-granulation, we further established multi-granulation graded rough intuitionistic fuzzy sets models based on type-I dominance relation and type-II dominance relation. Meanwhile, some properties of these models were discussed. Finally, the validity of these models was verified by an algorithm and some relative examples.

Keywords: graded rough sets; rough intuitionistic fuzzy sets; dominance relation; logical conjunction operation; logical disjunction operation; multi-granulation

1. Introduction

Pawlak proposed a rough set model in 1982, which is a significant method in dealing with uncertain, incomplete, and inaccurate information [1]. Its key strategy is to consider the lower and upper approximations based on precise classification.

As a tool, the classic rough set is based on precise classification. It is too restrictive for some problems in the real world. Considering this defect of classic rough sets, Yao proposed the graded rough sets (GRS) model [2]. Then researchers paid more attention to it and relative literatures began to accumulate on its theory and application. GRS can be defined as the lower approximation being $\underline{R}_k(X) = \{x \mid |[x]_R| - |[x]_R \cap X| \leq k, x \in U\}$ and the upper approximation being $\overline{R}_k(X) = \{x \mid |[x]_R \cap X| > k, x \in U\}$. $|[x]_R \cap X|$ is the absolute number of the elements of $|[x]_R|$ inside X and should be called internal grade, $|[x]_R| - |[x]_R \cap X|$ is the absolute number of the elements of $|[x]_R|$ outside X and should be called external grade. $\overline{R}_k(X)$ means union of the elements whose equivalence class' internal-grade about X is greater than k, $\underline{R}_k(X)$ means union of the elements whose equivalence class' external grade about X is at most k [3].

In the view of granular computing [4], the classic rough set is a single-granulation rough set. However, in the real world, we need multiple granularities to analyze and solve problems and Qian et al. proposed multi-granulation rough sets solving this issue [5]. Subsequently, multi-granulation rough sets were extended in References [6–9]. In addition, in the viewpoint of the degrees of classification error, Hu et al. and Wang et al. established a novel model of multi-granulation graded covering rough sets [10,11]. Simultaneously, Wu et al. constructed graded multi-granulation rough sets [12].

References [13–17] discussed GRS in a multi-granulation environment. Moreover, for GRS, it has been studied that the equivalence relation has been extended to the dominance relation [13,14], the limited tolerance relation [17] and so forth [10,11]. In general, all these aforementioned studies have naturally contributed to the development of GRS.

Inspired by the research reported in References [5,13–17], intuitionistic fuzzy sets (IFS) are also a theory which describe uncertainty [18]. IFS consisting of a membership function and a non-membership function are commonly encountered in uncertainty, imprecision, and vagueness [18]. The notion of IFS, proposed by Atanassov, was initially developed in the framework of fuzzy sets [19]. Furthermore, it can describe the "fuzzy concept" of "not this and not that", that is to say, neutral state or neutral degree, thus it is more precise to portray the ambiguous nature of the objective world. IFS theory is applicable in decision-making, logical planning, medical diagnosis, machine learning, and market forecasting, etc. Applications of IFS have attracted people's attention and achieved fruitful results [20–27].

In recent years, IFS have been a hot research topic in uncertain information systems [6,28–30]. For example, in the development of IFS theory, Ai et al. proposed intuitionistic fuzzy line integrals and gave their concrete values in Reference [31]. Zhang et al. researched the fuzzy logic algebraic system and neutrosophic sets as generalizations of IFS in References [23,26,27]. Furthermore, Guo et al. provided the dominance relation of intuitionistic fuzzy information systems [30].

Both rough sets and IFS not only describe uncertain information but also have strong complementarity in practical problems. As such many researchers have studied the combination of rough sets and IFS, namely, rough intuitionistic fuzzy sets (RIFS) and intuitionistic fuzzy rough sets (IFRS) [32]. For example, Huang et al., Gong et al., Zhang et al., He et al., and Tiwari et al. effectively developed IFRS respectively from uncertainty measures, variable precision rough sets, dominance–based rough sets, interval-valued IFS, and attribute selection [29,30,33–35]. Additionally, Zhang and Chen, Zhang and Yang, Huang et al. studied dominance relation of IFRS [19–21]. With respect to RIFS, Xue et al. provided a multi-granulation covering the RIFS model [9].

The above models did not consider the classification of some degrees of error [6–9,29,30,33,36] in dominance relation on GRS and dominance relation in intuitionistic fuzzy ordered information systems [37]. Therefore, in this paper, firstly, we introduce GRS into RIFS to get graded rough intuitionistic fuzzy sets (GRIFS) solving this problem. Then, considering the need for more precise sequence information in the real world, based on dominance relation of GRS and an intuitionistic fuzzy ordered information system, we respectively perform logical conjunction and disjunction operation to gain type-I dominance relation and type-II dominance relation. After that, we use type-I dominance relation and type-II dominance relation thereby replacing equivalence relation to generalize GRIFS. We design two novel models of GRIFS based on type-I dominance relation and type-II dominance relation. In addition, to accommodate a complex environment, we further extend GRIFS models based on type-I dominance relation and type-II dominance relation, respectively, to multi-granulation GRIFS models based on type-I dominance relation and type-II dominance relation. These models present a new path to extract more flexible and accurate information.

The rest of this paper is organized as follows. In Section 2, some basic concepts of IFS and GRS, RIFS are briefly reviewed, at the same time, we give the definition of GRS based on dominance relation. In Section 3, we respectively propose two novel models of GRIFS models based on type-I dominance relation and type-II dominance relation and verify the validity of these two models. In Section 4, the basic concepts of multi-granulation RIFS are given. Then, we propose multi-granulation GRIFS models based on type-I dominance relation and type-II dominance relation, and provide the concepts of optimistic and pessimistic multi-granulation GRIFS models based on type-I dominance relation and type-II dominance relation, respectively. In Section 5, we use an algorithm and example to study and illustrate the multi-granulation GRIFS models based on type-I dominance relation and type-II dominance relation, respectively. In Section 6, we conclude the paper and illuminate on future research.

2. Preliminaries

Definition 1 ([22]). *Let U be a non-empty classic universe of discourse. U is denoted by:*

$$A = \{< x,\ \mu_A(x), \nu_A(x) > |x \in U\},$$

A can be viewed as IFS on U, where $\mu_A(x) : U \to [0,1]$ and $\nu_A(x) : U \to [0,1]$. $\mu_A(x)$ and $\nu_A(x)$ are denoted as membership and non-membership degrees of the element x in A, satisfying $0 \leq \mu_A(x) + \nu_A(x) \leq 1$. For $\forall\, x \in U$, the hesitancy degree is $\pi_A(x) = 1 - \mu_A(x) - \nu_A(x)$, noticeably, $\pi_A(x) :\ U \to [0,1]$. $\forall\, A,\ B \in IFS(U)$, the basic operations of A and B are given as follows:

(1) $A \subseteq B \Leftrightarrow \mu_A(x) \leq \mu_B(x),\ \nu_A(x) \geq \nu_B(x), \forall x \in U,$
(2) $A = B \Leftrightarrow \mu_A(x) = \mu_B(x),\ \nu_A(x) = \nu_B(x),\ \forall x \in U,$
(3) $A \cup B = \{< x,\ \max\{\mu_A(x),\ \mu_B(x)\},\ \min\{\nu_A(x),\ \nu_B(x)\} > |x \in U\},$
(4) $A \cap B = \{< x,\ \min\{\mu_A(x),\ \mu_B(x)\},\ \max\{\nu_A(x),\ \nu_B(x)\} > |x \in U\},$
(5) $\sim A = \{< x,\ \nu_A(x),\ \mu_A(x) > |x \in U\}.$

Definition 2 ([2]). *Let (U, R) be an approximation space, assume $k \in N$, where N is the natural number set. Then GRS can be defined as follows:*

$$\underline{R}_k(X) = \{x||[x]_R| - |[x]_R \cap X| \leq k,\ x \in U\},$$
$$\overline{R}_k(X) = \{x||[x]_R \cap X| > k,\ x \in U\}.$$

$\underline{R}_k(X)$ and $\overline{R}_k(X)$ can be considered as the lower and upper approximations of X with respect to the graded k. Then we call the pair $(\underline{R}_k(X),\ \overline{R}_k(X))$ GRS. When $k = 0$, $\underline{R}_0(X) = \underline{R}(X)$, $\overline{R}_0(X) = \overline{R}(X)$. However, in general, $\underline{R}_k(X)\overline{R}_k(X)$, $\overline{R}_k(X)\underline{R}_k(X)$.

In Reference [4], the positive and negative domains of X are given as follows:

$$POS(X) = \underline{R}_k(X) \cap \overline{R}_k(X),\ NEG(X) = \neg(\underline{R}_k(X) \cup \overline{R}_k(X)).$$

Definition 3 ([36]). *If we denote $R_{\overline{a}}^{\geq} = \{(x_i, x_j) \in U \times U : f(x_i) \geq f(x_j), \forall a \in A\}$ where A is a subset of the attributes set and $f(x)$ is the value of attribute a, then $[x]_{\overline{a}}^{\geq}$ is referred to as the dominance class of dominance relation $R_{\overline{a}}^{\geq}$. Moreover, we denote approximation space based on dominance relations by $S = (U, R_{\overline{a}}^{\geq})$.*

Definition 4. *Let $(U, R_{\overline{a}}^{\geq})$ be an information approximation. $U/R_{\overline{a}}^{\geq}$ is the set of dominance classes induced by a dominance relation $R_{\overline{a}}^{\geq}$, and $[x]_{\overline{a}}^{\geq}$ is called the dominance class containing x. Assume $k \in N$, where N is the natural number set. GRS based on dominance relation can be defined:*

$$\underline{R_{\overline{k}}^{\geq}}(X) = \{x||[x]_{\overline{a}}^{\geq}| - |[x]_{\overline{a}}^{\geq} \cap X| \leq k,\ x \in U\},$$
$$\overline{R_{\overline{k}}^{\geq}}(X) = \{x||[x]_{\overline{a}}^{\geq} \cap X| > k,\ x \in U\}.$$

When $k = 0$, $(\overline{R_{\overline{0}}^{\geq}}(X),\ \underline{R_{\overline{0}}^{\geq}}(X))$ will be rough sets based on dominance relation.

Example 1. *Suppose there are nine patients $U = \{x_1, x_2, x_3, x_4, x_5, x_6, x_7, x_8, x_9\}$, they may suffer from a cold. According to their fever, we get $U/R_{\overline{a}}^{\geq} = \{\{x_1, x_2, x_4\}, \{x_3, x_8\}, \{x_6, x_8\}, \{x_5, x_7, x_8, x_9\}\}$, $X \subseteq U$. Then suppose $X = \{x_1, x_2, x_4, x_7, x_9\}$, we can obtain GRS based on dominance relation.*

The demonstration process is given as follows:

Suppose $k = 1$, then we can get,

$$[x_1]_{\overline{a}}^{\geq} = [x_2]_{\overline{a}}^{\geq} = [x_4]_{\overline{a}}^{\geq} = \{x_1, x_2, x_4\},\ [x_3]_{\overline{a}}^{\geq} = [x_8]_{\overline{a}}^{\geq} = \{x_3, x_8\},\ [x_6]_{\overline{a}}^{\geq} = [x_8]_{\overline{a}}^{\geq} = \{x_6, x_8\},$$
$$[x_5]_{\overline{a}}^{\geq} = [x_7]_{\overline{a}}^{\geq} = [x_8]_{\overline{a}}^{\geq} = [x_9]_{\overline{a}}^{\geq} = \{x_5, x_7, x_8, x_9\}.$$

Then, we can calculate $\underline{R}_1^{\geq}(X)$, $\overline{R}_1^{\geq}(X)$ *and* $POS(X)$, $NEG(X)$.

$$\underline{R}_1^{\geq}(X) = \{x_1, x_2, x_4\},\ \overline{R}_1^{\geq}(X) = \{x_1, x_2, x_4, x_5, x_7, x_8, x_9\}.$$

$$POS(X) = \underline{R}_1^{\geq}(X) \cap \overline{R}_1^{\geq}(X) = \{x_1, x_2, x_4\} \cap \{x_1, x_2, x_4, x_5, x_7, x_8, x_9\} = \{x_1, x_2, x_4\},$$
$$NEG(X) = \neg(\underline{R}_1^{\geq}(X) \cup \overline{R}_1^{\geq}(X)) = \neg(\{x_1, x_2, x_4\} \cup \{x_1, x_2, x_4, x_5, x_7, x_8, x_9\}) = \{x_3, x_6\}.$$

Through the above analysis, we can see x_1, x_2, *and* x_4 *patients suffering from a cold disease and* x_3 *and* x_6 *patients not having a cold disease.*

When $k = 0$, $(\overline{R}_0^{\geq}(X), \underline{R}_0^{\geq}(X))$ *will be rough sets based on dominance relation.*

Definition 5 ([8,32,35]). *Let X be a non-empty set and R be an equivalence relation on X. Let B be IFS in X with the membership function* $\mu_B(x)$ *and non-membership function* $\nu_B(x)$. *The lower and upper approximations, respectively, of B are IFS of the quotient set X/R with*

(1) Membership function defined by

$$\mu_{\underline{R}(B)}(X_i) = \inf\{\mu_B(x) |\ x \in X_i\},\ \mu_{\overline{R}(B)}(X_i) = \sup\{\mu_B(x) |\ x \in X_i\}.$$

(2) Non-membership function defined by

$$\nu_{\underline{R}(B)}(X_i) = \sup\{\nu_B(x) |\ x \in X_i\},\ \nu_{\overline{R}(B)}(X_i) = \inf\{\nu_B(x) |\ x \in X_i\}.$$

In this way, we can prove $\underline{R}(B)$ *and* $\overline{R}(B)$ *are IFS.*

For $\forall x \in X_i$, *we can obtain,*

$$\mu_B(x) + \nu_B(x) \leq 1, \mu_B(x) \leq 1 - \nu_B(x), \sup\{\mu_B(x) |\ x \in X_i\} \leq \sup\{1 - \nu_B(x) |\ x \in X_i\}, \sup\{\mu_B(x) |\ x \in X_i\} \leq 1 - \inf\{\nu_B$$

Hence $\overline{R}(B)$ *is IFS. Similarly, we can prove that* $\underline{R}(B)$ *is IFS. The RIFS of* $\underline{R}(B)$ *and* $\overline{R}(B)$ *are given as ollows:*

$$\underline{R}(B) = \{< x, \inf_{y \in [x]_i} \mu_B(y), \sup_{y \in [x]_i} \nu_B(y) > | x \in U\},$$

$$\overline{R}(B) = \{< x, \sup_{y \in [x]_i} \mu_B(y), \inf_{y \in [x]_i} \nu_B(y) > | x \in U\}.$$

3. GRIFS Model Based on Dominance Relation

In this section, we propose a GRIFS model based on dominance relation. Moreover, this model contains a GRIFS model based on type-I dominance relation and GRIFS model based on type-II dominance relation, respectively. Then we employ an example to demonstrate the validity of these two models, and finish by discussing some basic properties of these two models.

Definition 6 ([37]). *If* (U, A, V, f) *is an intuitionistic fuzzy ordered information system, so* $(R')_a^{\geq} = \{(x, y) \in U \times U | f_a(y) \geq f_a(x), \forall a \in A\}$ *can be called dominance relation in the intuitionistic fuzzy ordered information system.*

$$\begin{aligned} [x]_a^{\geq'} &= \{y | (x, y) \in (R')_a^{\geq}, \forall a \in A, y \in U\} \\ &= \{y | \mu_a(y) \geq \mu_a(x),\ \nu_a(y) \leq \nu_a(x), \forall a \in A, y \in U\} \end{aligned}$$

$[x]_a^{\geq'}$ *is dominance class of x in terms of dominance relation* $(R')_a^{\geq}$.

3.1. GRIFS Model Based on Type-I Dominance Relation

Definition 7. *Let $IS^{\geq^{\mathrm{I}}} = (U, A, V, f)$ be an intuitionistic fuzzy ordered information system and $R_{\overline{a}}^{\geq}$ be a dominance relation of the attribute set A. Suppose X is the GRS of $R_{\overline{a}}^{\geq}$ on U, $a \in A$, and IFS B on U about attribute a satisfies dominance relation $(R')_{\overline{a}}^{\geq}$. The lower approximation $\underline{R_{\overline{k}}}^{\geq^{\mathrm{I}}}(B)$ and the upper approximation $\overline{R_{\overline{k}}}^{\geq^{\mathrm{I}}}(B)$ with respect to the graded k are given as follows:*

When $k \geq 1$, we can gain,

$$\underline{R_{\overline{k}}}^{\geq^{\mathrm{I}}}(B) = \{< x, \inf_{y \in (\overset{j}{\underset{s=1}{\wedge}}(([x]_a^{\geq})_s \wedge [x]_a^{\geq'}))} (\mu_B(y) \wedge \mu'_B(y)), \sup_{y \in (\overset{j}{\underset{s=1}{\wedge}}(([x]_a^{\geq})_s \wedge [x]_a^{\geq'}))} (\nu_B(y) \vee \nu'_B(y)) > | x \in U\},$$

$$\overline{R_{\overline{k}}}^{\geq^{\mathrm{I}}}(B) = \{< x, \sup_{y \in (\overset{j}{\underset{s=1}{\wedge}}(([x]_a^{\geq})_s \wedge [x]_a^{\geq'}))} (\mu_B(y) \vee \mu'_B(y)), \inf_{y \in (\overset{j}{\underset{s=1}{\wedge}}(([x]_a^{\geq})_s \wedge [x]_a^{\geq'}))} (\nu_B(y) \wedge \nu'_B(y)) > | x \in U\}.$$

$$\mu'_B(y) = \frac{|\overline{R_{\overline{k}}^{\geq}}(X) \cap \underline{R_{\overline{k}}^{\geq}}(X)|}{|U|}, \ \nu'_B(y) = \frac{|\neg(\overline{R_{\overline{k}}^{\geq}}(X) \cup \underline{R_{\overline{k}}^{\geq}}(X))|}{|U|}.$$

Obviously, $0 \leq \mu'_B(y) \leq 1$, $0 \leq \nu'_B(y) \leq 1$, $j = 1, 2, \cdots, n$.

When $k = 0$, $\mu'_B(y)$ and $\nu'_B(y)$ degenerate to be calculated by the classical rough set. However, under these circumstances, the model is still valid, we call this model RIFS based on type-I dominance relation.

Note that, in GRIFS model based on type-I dominance relation, we let $[x]_a^{\geq}$ and $[x]_a^{\geq'}$ perform a conjunction operation $\wedge$, this is to say $\geq^{\mathrm{I}}$ means $\overset{j}{\underset{s=1}{\wedge}}(([x]_a^{\geq})_s \wedge [x]_a^{\geq'})$.

Note that, $\overset{j}{\underset{s=1}{\wedge}}(([x]_a^{\geq})_s \wedge [x]_a^{\geq'})$ in GRIFS model based on type-I dominance relation, if x have j dominance classes $[x]_a^{\geq}$ of dominance relation $R_{\overline{a}}^{\geq}$on GRS, we perform a conjunction operation $\wedge$ of j dominance classes $[x]_a^{\geq}$ and $[x]_a^{\geq'}$.

According to Definition 7, the following theorem can be obtained.

Theorem 1. *Let $IS^{\geq^{\mathrm{I}}} =< U, A, V, f >$ be an intuitionistic fuzzy ordered information system, and B be IFS on U. Then a GRIFS model based on type-I dominance relation has these following properties:*

$$(1)\ \underline{R_{\overline{k}}}^{\geq^{\mathrm{I}}}(B) \subseteq B \subseteq \overline{R_{\overline{k}}}^{\geq^{\mathrm{I}}}(B),$$
$$(2)\ \mathrm{A} \subseteq \mathrm{B},\ \underline{R_{\overline{k}}}^{\geq^{\mathrm{I}}}(A) \subseteq \underline{R_{\overline{k}}}^{\geq^{\mathrm{I}}}(B),\ \overline{R_{\overline{k}}}^{\geq^{\mathrm{I}}}(A) \subseteq \overline{R_{\overline{k}}}^{\geq^{\mathrm{I}}}(B),$$
$$(3)\ \underline{R_{\overline{k}}}^{\geq^{\mathrm{I}}}(A \cap B) = \underline{R_{\overline{k}}}^{\geq^{\mathrm{I}}}(A) \cap \underline{R_{\overline{k}}}^{\geq^{\mathrm{I}}}(B),\ \overline{R_{\overline{k}}}^{\geq^{\mathrm{I}}}(A \cup B) = \overline{R_{\overline{k}}}^{\geq^{\mathrm{I}}}(A) \cup \overline{R_{\overline{k}}}^{\geq^{\mathrm{I}}}(B).$$

Proof. (1) From Definition 7, we can get,

$$\inf_{y \in (\overset{j}{\underset{s=1}{\wedge}}(([x]_a^{\geq})_s \wedge [x]_a^{\geq'}))} (\mu_B(y) \wedge \mu'_B(y)) \leq \mu_B(x) \leq \sup_{y \in (\overset{j}{\underset{s=1}{\wedge}}(([x]_a^{\geq})_s \wedge [x]_a^{\geq'}))} (\mu_B(y) \vee \mu'_B(y)) \Leftrightarrow \mu_{\underline{R_{\overline{k}}}^{\geq^{\mathrm{I}}}(B)}(x) \leq \mu_B(x) \leq \mu_{\overline{R_{\overline{k}}}^{\geq^{\mathrm{I}}}(B)}(x),$$
$$\sup_{y \in (\overset{j}{\underset{s=1}{\wedge}}(([x]_a^{\geq})_s \wedge [x]_a^{\geq'}))} (\nu_B(y) \vee \nu'_B(y)) \geq \nu_B(x) \geq \inf_{y \in (\overset{j}{\underset{s=1}{\wedge}}(([x]_a^{\geq})_s \wedge [x]_a^{\geq'}))} (\nu_B(y) \wedge \nu'_B(y)) \Leftrightarrow \nu_{\underline{R_{\overline{k}}}^{\geq^{\mathrm{I}}}(B)}(x) \geq \nu_B(x) \geq \nu_{\overline{R_{\overline{k}}}^{\geq^{\mathrm{I}}}(B)}(x),$$

Hence, $\underline{R_{\overline{k}}}^{\geq^{\mathrm{I}}}(B) \subseteq B \subseteq \overline{R_{\overline{k}}}^{\geq^{\mathrm{I}}}(B)$.

(2) Based on Definition 1 and $A \subseteq B$,

Thus we can get, $\mu_A(x) \leq \mu_B(x)$, $\nu_A(x) \geq \nu_B(x)$.
From Definition 7, we can get, $\mu'_A(y) = \mu'_B(y)$, $\nu'_A(y) = \nu'_B(y)$.

Then, in the GRIFS model based on type-I dominance relation, we can get,

$$\inf_{y\in(\bigwedge_{s=1}^{j}(([x]_a^{\geq})_s\wedge[x]_a^{\geq'}))}(\mu_A(y)\wedge\mu'_A(y))\leq\inf_{y\in(\bigwedge_{s=1}^{j}(([x]_a^{\geq})_s\wedge[x]_a^{\geq'}))}(\mu_B(y)\wedge\mu'_B(y))\Leftrightarrow\mu_{\underline{R_k^{\geq}}^{\mathrm{I}}(A)}(x)\leq\mu_{\underline{R_k^{\geq}}^{\mathrm{I}}(B)}(x),$$

$$\sup_{y\in(\bigwedge_{s=1}^{j}(([x]_a^{\geq})_s\wedge[x]_a^{\geq'}))}(\nu_A(y)\vee\nu'_A(y))\geq\sup_{y\in(\bigwedge_{s=1}^{j}(([x]_a^{\geq})_s\wedge[x]_a^{\geq'}))}(\nu_B(y)\vee\nu'_B(y))\Leftrightarrow\nu_{\underline{R_k^{\geq}}^{\mathrm{I}}(A)}(x)\geq\nu_{\underline{R_k^{\geq}}^{\mathrm{I}}(B)}(x).$$

Thus we can get, $\underline{R_k^{\geq}}^{\mathrm{I}}(A)\subseteq\underline{R_k^{\geq}}^{\mathrm{I}}(B)$.

In the same way, we can get, $\overline{R_k^{\geq}}^{\mathrm{I}}(A)\subseteq\overline{R_k^{\geq}}^{\mathrm{I}}(B)$.

(3) From Definition 7, we can get,

$$\begin{aligned}\mu_{\underline{R_k^{\geq}}^{\mathrm{I}}(A\cap B)}(x)&=\inf_{y\in(\bigwedge_{s=1}^{j}(([x]_a^{\geq})_s\wedge[x]_a^{\geq'}))}(\mu_{A\cap B}(y)\wedge\mu'_{A\cap B}(y))=(\inf_{y\in(\bigwedge_{s=1}^{j}(([x]_a^{\geq})_s\wedge[x]_a^{\geq'}))}(\mu_A(y)\wedge\mu'_A(y)))\wedge(\inf_{y\in(\bigwedge_{s=1}^{j}(([x]_a^{\geq})_s\wedge[x]_a^{\geq'}))}(\mu_B(y)\wedge\mu'_B(y)))\\&=\mu_{\underline{R_k^{\geq}}^{\mathrm{I}}(A)}(x)\wedge\mu_{\underline{R_k^{\geq}}^{\mathrm{I}}(B)}(x),\\\nu_{\underline{R_k^{\geq}}^{\mathrm{I}}(A\cap B)}(x)&=\sup_{y\in(\bigwedge_{s=1}^{j}(([x]_a^{\geq})_s\wedge[x]_a^{\geq'}))}(\nu_{A\cap B}(y)\vee\nu'_{A\cap B}(y))=(\sup_{y\in(\bigwedge_{s=1}^{j}(([x]_a^{\geq})_s\wedge[x]_a^{\geq'}))}(\nu_A(y)\vee\nu'_A(y)))\wedge(\sup_{y\in(\bigwedge_{s=1}^{j}(([x]_a^{\geq})_s\wedge[x]_a^{\geq'}))}(\nu_B(y)\vee\nu'_B(y)))\\&=\nu_{\underline{R_k^{\geq}}^{\mathrm{I}}(A)}(x)\wedge\nu_{\underline{R_k^{\geq}}^{\mathrm{I}}(B)}(x),\end{aligned}$$

Thus we can get, $\underline{R_k^{\geq}}^{\mathrm{I}}(A\cap B)=\underline{R_k^{\geq}}^{\mathrm{I}}(A)\cap\underline{R_k^{\geq}}^{\mathrm{I}}(B)$.

In the same way, we can get $\overline{R_k^{\geq}}^{\mathrm{I}}(A\cup B)=\overline{R_k^{\geq}}^{\mathrm{I}}(A)\cup\overline{R_k^{\geq}}^{\mathrm{I}}(B)$. □

Example 2. *In a city, the court administration needs to recruit 3 staff. Applicants who pass the application, preliminary examination of qualifications, written examination, interview, qualification review, political review, and physical examination can be employed. In order to facilitate the calculation, we simplify the enrollment process to qualification review, written test, interview. At present, 12 people have passed the preliminary examination of qualifications, and 9 of them have passed the written examination (administrative professional ability test and application).* $U=\{x_1,x_2,x_3,x_4,x_5,x_6,x_7,x_8,x_9\}$ *is the domain. We can get* $U/R_a^{\geq}=\{\{x_1,x_2,x_4\},\{x_3,x_8\},\{x_7\},\{x_4,x_5,x_6,x_9\}\}$ *according to the "excellent" and "pass" of the two results. In addition, through the interview of 9 people, the following IFS can be obtained, and we suppose* $X=\{x_1,x_4,x_5,x_6,x_9\}$, $X\subseteq U$.

$$B=\left\{\frac{[0.9,0]}{x_1},\frac{[0.8,0.1]}{x_2},\frac{[0.65,0.3]}{x_3},\frac{[0.85,0.1]}{x_4},\frac{[0.95,0.05]}{x_5},\frac{[0.7,0.3]}{x_6},\frac{[0.5,0.2]}{x_7},\frac{[0.87,0.1]}{x_8},\frac{[0.75,0.2]}{x_9}\right\}.$$

To solve the above problems, we can use the model described in References [38,39], which are rough sets based on dominance relation.

First, according to $U/R_a^{\geq}$, we can get,

$$\underline{R}^{\geq}(X)=\{x_4,x_5,x_6,x_9\},\ \overline{R}^{\geq}(X)=\{x_1,x_2,x_4,x_5,x_6,x_9\},$$

Through rough sets based on dominance relation, we can get some applicants with better written test scores. However, regarding IFS B, we cannot use rough sets based on dominance relation to handle the data. Therefore, we are even less able to get the final result with the model. To process the interview data, we need to use another model, described in Reference [40]. Through data processing, we can obtain the dominance classes as follows:

$$[x_1]_a^{\geq'}=\{x_1\},\ [x_2]_a^{\geq'}=\{x_2,x_4,x_5,x_8\},\ [x_3]_a^{\geq'}=\{x_3,x_4,x_5,x_6,x_8,x_9\},\ [x_4]_a^{\geq'}=\{x_4,x_5\},$$
$$[x_5]_a^{\geq'}=\{x_5\},\ [x_6]_a^{\geq'}=\{x_6,x_8,x_9\},\ [x_7]_a^{\geq'}=\{x_7,x_8,x_9\},\ [x_8]_a^{\geq'}=\{x_8\},\ [x_9]_a^{\geq'}=\{x_9\}.$$

From the above analysis, we can get,

$$x_5 \geq x_1 \geq x_8 \geq x_4 \geq x_2 \geq x_9 \geq x_6 \geq x_3 \geq x_7$$

Through dominance relation in the intuitionistic fuzzy ordered information system, we can get some applicants with better interview results, but we still cannot get the final results. To get this result, we need to analyze the applicants who have better written test scores and better written test scores. Based on the above conclusions, we can determine that only x_5 and x_4 applicants meet the requirements. However, the performance of others is not certain. If they only need one or two staff, then this analysis can help us to choose the applicant. However, we need 3 applicants, so we cannot get the result in this way. However, there is a model in Definition 6 that can help us get the results. The calculation process is as follows:

According to Example 1, when $k = 1$, we can get

$$\underline{R}_{\overline{1}}^{\geq}(X) = \{x_1, x_2, x_4, x_5, x_6, x_7, x_9\},\ \overline{R}_{\overline{1}}^{\geq}(X) = \{x_1, x_2, x_4, x_5, x_6, x_7, x_9\},$$

According to Definitions 7 and 8, we can then get,

$$\mu'_B(y) = \frac{|\overline{R}_{\overline{1}}^{\geq}(X) \cap \underline{R}_{\overline{1}}^{\geq}(X)|}{|U|} = \frac{7}{9} \approx 0.78,\ \nu'_B(y) = \frac{|\neg(\overline{R}_{\overline{1}}^{\geq}(X) \cup \underline{R}_{\overline{1}}^{\geq}(X))|}{|U|} = \frac{2}{9} \approx 0.22.$$

So, according to Definition 6 and Example 1, we can compute the conjunction operation of $[x]_{\overline{a}}^{\geq}$ and $[x]_{\overline{a}}^{\geq'}$, and the results are as Table 1.

Table 1. The conjunction operation of $[x]_{\overline{a}}^{\geq}$ and $[x]_{\overline{a}}^{\geq'}$.

x	$[x]_{\overline{a}}^{\geq}$	$[x]_{\overline{a}}^{\geq'}$	$[x]_{\overline{a}}^{\geq} \wedge [x]_{\overline{a}}^{\geq'}$
x_1	$\{x_1, x_2, x_4\}$	$\{x_1\}$	$\{x_1\}$
x_2	$\{x_1, x_2, x_4\}$	$\{x_2, x_4, x_5, x_8\}$	$\{x_2, x_4\}$
x_3	$\{x_3, x_8\}$	$\{x_3, x_4, x_5, x_6, x_8, x_9\}$	$\{x_3, x_8\}$
x_4	$\{x_1, x_2, x_4\}, \{x_4, x_5, x_6, x_9\}$	$\{x_4, x_5\}$	$\{x_4\}$
x_5	$\{x_4, x_5, x_6, x_9\}$	$\{x_5\}$	$\{x_5\}$
x_6	$\{x_4, x_5, x_6, x_9\}$	$\{x_6, x_8, x_9\}$	$\{x_6, x_9\}$
x_7	$\{x_7\}$	$\{x_7, x_8, x_9\}$	$\{x_7\}$
x_8	$\{x_3, x_8\}$	$\{x_8\}$	$\{x_8\}$
x_9	$\{x_4, x_5, x_6, x_9\}$	$\{x_9\}$	$\{x_9\}$

GRIFS model based on type-I dominance relation can be obtained as follows:

$$\underline{R}_{\overline{1}}^{\geq^{\mathrm{I}}}(B) = \left\{\frac{[0.78, 0.22]}{x_1}, \frac{[0.78, 0.22]}{x_2}, \frac{[0.65, 0.3]}{x_3}, \frac{[0.78, 0.22]}{x_4}, \frac{[0.78, 0.22]}{x_5}, \frac{[0.7, 0.3]}{x_6}, \frac{[0.5, 0.22]}{x_7}, \frac{[0.78, 0.22]}{x_8}, \frac{[0.75, 0.22]}{x_9}\right\},$$
$$\overline{R}_{\overline{1}}^{\geq^{\mathrm{I}}}(B) = \left\{\frac{[0.9, 0]}{x_1}, \frac{[0.85, 0.1]}{x_2}, \frac{[0.78, 0.1]}{x_3}, \frac{[0.85, 0.1]}{x_4}, \frac{[0.95, 0.05]}{x_5}, \frac{[0.87, 0.1]}{x_6}, \frac{[0.78, 0.1]}{x_7}, \frac{[0.87, 0.1]}{x_8}, \frac{[0.78, 0.2]}{x_9}\right\}.$$

Comprehensive analysis $\underline{R}_{\overline{1}}^{\geq^{\mathrm{I}}}(B)$ and $\overline{R}_{\overline{1}}^{\geq^{\mathrm{I}}}(B)$, we can conclude that x_5, x_1, x_8, x_2 and x_4 applicants are more suitable for the position in the pessimistic situation. From this example we can see that our model is able to handle more complicated situations than the previous theories, and it can help us get more accurate results.

3.2. GRIFS Model Based on Type-II Dominance Relation

Definition 8. *Let U be a non-empty set and A be the attribute set on U, and $a \in A$, $R_{\overline{a}}^{\geq}$ is a dominance relation of attribute A. Let X be GRS of $R_{\overline{a}}^{\geq}$ on U, and IFS B on U about attribute a satisfies dominance relation $(R')_{\overline{a}}^{\geq}$. The lower and upper approximations of B with respect to the graded k are given as follows:*

When $k \geq 1$, we can get,

$$\underline{R_k^{\geq}}^{\Pi}(B) = \{< x, \inf_{y \in (\overset{j}{\underset{s=1}{\vee}}(([x]_a^{\geq})_s \vee [x]_a^{\geq'}))} (\mu_B(y) \wedge \mu'_B(y)), \sup_{y \in (\overset{j}{\underset{s=1}{\vee}}(([x]_a^{\geq})_s \vee [x]_a^{\geq'}))} (\nu_B(y) \vee \nu'_B(y)) > | x \in U\},$$

$$\overline{R_k^{\geq}}^{\Pi}(B) = \{< x, \sup_{y \in (\overset{j}{\underset{s=1}{\vee}}(([x]_a^{\geq})_s \vee [x]_a^{\geq'}))} (\mu_B(y) \vee \mu'_B(y)), \inf_{y \in (\overset{j}{\underset{s=1}{\vee}}(([x]_a^{\geq})_s \vee [x]_a^{\geq'}))} (\nu_B(y) \wedge \nu'_B(y)) > | x \in U\}.$$

$$\mu'_B(y) = \frac{|\overline{R_k^{\geq}}(X) \cap \underline{R_k^{\geq}}(X)|}{|U|}, \nu'_B(y) = \frac{|\neg(\overline{R_k^{\geq}}(X) \cup \underline{R_k^{\geq}}(X))|}{|U|}.$$

Obviously, $0 \leq \mu'_B(y) \leq 1$, $0 \leq \nu'_B(y) \leq 1$, $j = 1, 2, \cdots, n$.

When $k = 0$, $\mu'_B(y)$ and $\nu'_B(y)$ are calculated from the classical rough set. However, under these circumstances the model is still valid and we call this model RIFS based on type-II dominance relation.

Note that in the GRIFS model based on type-II dominance relation, we perform a disjunction operation $\vee$ on $[x]_a^{\geq}$ and $[x]_a^{\geq'}$, this is to say $\geq^{\Pi}$ means $\overset{j}{\underset{s=1}{\vee}}(([x]_a^{\geq})_s \vee [x]_a^{\geq'})$.

Note that, $\overset{j}{\underset{s=1}{\vee}}(([x]_a^{\geq})_s \vee [x]_a^{\geq'})$ in the GRIFS model based on type-II dominance relation. If x have j dominance classes $[x]_a^{\geq}$ of dominance relation $R_a^{\geq}$ on GRS, we perform a disjunction operation $\vee$ of j dominance classes $[x]_a^{\geq}$ and $[x]_a^{\geq'}$, respectively.

According to Definition 8, the following theorem can be obtained.

Theorem 2. *Let $IS^{\geq^{\Pi}} =< U, A, V, f >$ be an intuitionistic fuzzy ordered information system, and B be IFS on U. Then GRIFS model based on type-II dominance relation will have the following properties:*

$$(1)\ \underline{R_k^{\geq}}^{\Pi}(B) \subseteq B \subseteq \overline{R_k^{\geq}}^{\Pi}(B),$$
$$(2)\ A \subseteq B, \underline{R_k^{\geq}}^{\Pi}(A) \subseteq \underline{R_k^{\geq}}^{\Pi}(B),\ \overline{R_k^{\geq}}^{\Pi}(A) \subseteq \overline{R_k^{\geq}}^{\Pi}(B),$$
$$(3)\ \underline{R_k^{\geq}}^{\Pi}(A \cap B) = \underline{R_k^{\geq}}^{\Pi}(A) \cap \underline{R_k^{\geq}}^{\Pi}(B), \overline{R_k^{\geq}}^{\Pi}(A \cup B) = \overline{R_k^{\geq}}^{\Pi}(A) \cup \overline{R_k^{\geq}}^{\Pi}(B).$$

Proof. The proving process of Theorem 2 is similar to Theorem 1. □

Example 3. *Nine senior university students are going to graduate from a computer department and they want to work for a famous internet company. Let $U = \{x_1, x_2, x_3, x_4, x_5, x_6, x_7, x_8, x_9\}$ be the domain. The company has a campus recruitment at this university. Based on their confidence in programming skills, we get the following IFS B whether they succeed in the campus recruitment or not. At the same time, according to programming skills grades in school, $U/R_a^{\geq} = \{\{x_1, x_2, x_4\}, \{x_4, x_5, x_6, x_9\}, \{x_3, x_8\}, \{x_7\}\}$ can be obtained. We suppose $X = \{x_1, x_4, x_5, x_6, x_9\}$, $X \subseteq U$.*

$$B = \left\{ \frac{[0.9, 0]}{x_1}, \frac{[0.8, 0.1]}{x_2}, \frac{[0.65, 0.3]}{x_3}, \frac{[0.85, 0.1]}{x_4}, \frac{[0.95, 0.05]}{x_5}, \frac{[0.7, 0.3]}{x_6}, \frac{[0.5, 0.2]}{x_7}, \frac{[0.87, 0.1]}{x_8}, \frac{[0.75, 0.2]}{x_9} \right\}.$$

We can try to use rough sets based on dominance relation to solve the above problems, as described in Reference [38].

First, according to $U/R_a^{\geq}$, we can get the result as follows.

$$\underline{R}^{\geq}(X) = \{x_4, x_5, x_6, x_9\}, \overline{R}^{\geq}(X) = \{x_1, x_2, x_4, x_5, x_6, x_9\},$$

From the upper and lower approximations, we can get that x_4, x_5, x_6 and x_9 students may pass the campus interview. However, we cannot use the rough set based on dominance relation to deal with the data of the test scores of their programming skills. In order to process B, we need to use another model, outlined in Reference [40]. The result is as follows:

$$x_5 \geq x_1 \geq x_8 \geq x_4 \geq x_2 \geq x_9 \geq x_6 \geq x_3 \geq x_7$$

Through IFS, we can get that x_4, x_2, x_1 and x_7 students are better than other students. From the above analysis, we can get student x_4 who can be successful in the interview. However, we are not sure about other students. At the same time, from the process of analysis, we find that different models are built for the examples, and the predicted results will have deviation. Our model is based on GRS based on dominance relation and the dominance relation in intuitionistic fuzzy ordered information system. Thus, we can use the model to predict the campus interview.

Consequently, according to Definition 8 and Example 1, we can compute the disjunction operation of $[x]_a^{\geq}$ and $[x]_a^{\geq'}$, the results are as Table 2.

Table 2. The disjunction operation of $[x]_a^{\geq}$ and $[x]_a^{\geq'}$.

x	$[x]_a^{\geq}$	$[x]_a^{\geq'}$	$[x]_a^{\geq} \vee [x]_a^{\geq'}$
x_1	$\{x_1, x_2, x_4\}$	$\{x_1\}$	$\{x_1, x_2, x_4\}$
x_2	$\{x_1, x_2, x_4\}$	$\{x_2, x_4, x_5, x_8\}$	$\{x_1, x_2, x_4, x_5, x_8\}$
x_3	$\{x_3, x_8\}$	$\{x_3, x_4, x_5, x_6, x_8, x_9\}$	$\{x_3, x_4, x_5, x_6, x_8, x_9\}$
x_4	$\{x_1, x_2, x_4\}, \{x_4, x_5, x_6, x_9\}$	$\{x_4, x_5\}$	$\{x_1, x_2, x_4, x_5, x_6, x_9\}$
x_5	$\{x_4, x_5, x_6, x_9\}$	$\{x_5\}$	$\{x_4, x_5, x_6, x_9\}$
x_6	$\{x_4, x_5, x_6, x_9\}$	$\{x_6, x_8, x_9\}$	$\{x_4, x_5, x_6, x_8, x_9\}$
x_7	$\{x_7\}$	$\{x_7, x_8, x_9\}$	$\{x_7, x_8, x_9\}$
x_8	$\{x_3, x_8\}$	$\{x_8\}$	$\{x_3, x_8\}$
x_9	$\{x_4, x_5, x_6, x_9\}$	$\{x_9\}$	$\{x_4, x_5, x_6, x_9\}$

GRIFS model based on type-II dominance relation can be obtained as follows:

$$\underline{R}_{\tilde{1}}^{\geq \Pi}(B) = \left\{ \frac{[0.78,0.22]}{x_1}, \frac{[0.78,0.22]}{x_2}, \frac{[0.65,0.3]}{x_3}, \frac{[0.7,0.22]}{x_4}, \frac{[0.7,0.22]}{x_5}, \frac{[0.7,0.3]}{x_6}, \frac{[0.5,0.22]}{x_7}, \frac{[0.65,0.3]}{x_8}, \frac{[0.7,0.3]}{x_9} \right\},$$
$$\overline{R}_{\tilde{1}}^{\geq \Pi}(B) = \left\{ \frac{[0.9,0]}{x_1}, \frac{[0.95,0]}{x_2}, \frac{[0.95,0.05]}{x_3}, \frac{[0.95,0]}{x_4}, \frac{[0.95,0.05]}{x_5}, \frac{[0.95,0.05]}{x_6}, \frac{[0.87,0.1]}{x_7}, \frac{[0.87,0.1]}{x_8}, \frac{[0.95,0.05]}{x_9} \right\}.$$

Through the above analysis, the students' interviews prediction can be obtained. x_4, x_2 and x_1 students are better than others. From this example, the model can help us to analyze the same situation though two kinds of dominance relations. Therefore, this example can be analyzed more comprehensively

4. Multi-Granulation GRIFS Models Based on Dominance Relation

In this section, we give the multi-granulation RIFS conception, and then propose optimistic and pessimistic multi-granulation GRIFS models based on type-I dominance relation and type-II dominance relation, respectively. These four models are constructed by multiple granularities GRIFS models based on type-I and type-II dominance relation. Finally, we discuss some properties of these models.

Definition 9 ([39]). *Let $IS =< U, A, V, f >$ be an information system, $A_1, A_2, \cdots, A_m \subseteq A$, and R_{A_i} is an equivalence relation of x in terms of attribute set A. $[x]_{A_i}$ is the equivalence class of R_{A_i}, $\forall B \subseteq U$, B is IFS. Then the optimistic multi-granulation lower and upper approximations of A_i can be defined as follows:*

$$\underline{\sum_{i=1}^{m} R^{O}_{A_i}}(B) = \{< x, \mu_{\underline{\sum_{i=1}^{m} R^{O}_{A_i}}(B)}(x), \nu_{\underline{\sum_{i=1}^{m} R^{O}_{A_i}}(B)}(x) > | x \in U\},$$
$$\overline{\sum_{i=1}^{m} R^{O}_{A_i}}(B) = \{< x, \mu_{\overline{\sum_{i=1}^{m} R^{O}_{A_i}}(B)}(x), \nu_{\overline{\sum_{i=1}^{m} R^{O}_{A_i}}(B)}(x) > | x \in U\},$$
$$\mu_{\underline{\sum_{i=1}^{m} R^{O}_{A_i}}(B)}(x) = \bigvee_{i=1}^{m} \inf_{y \in [x]_{A_i}} \mu_B(y), \ \nu_{\underline{\sum_{i=1}^{m} R^{O}_{A_i}}(B)}(x) = \bigwedge_{i=1}^{m} \sup_{y \in [x]_{A_i}} \nu_B(y),$$
$$\mu_{\overline{\sum_{i=1}^{m} R^{O}_{A_i}}(B)}(x) = \bigwedge_{i=1}^{m} \sup_{y \in [x]_{A_i}} \mu_B(y), \ \nu_{\overline{\sum_{i=1}^{m} R^{O}_{A_i}}(B)}(x) = \bigvee_{i=1}^{m} \inf_{y \in [x]_{A_i}} \nu_B(y).$$

where $[x]_{A_i}$ is the equivalence class of x in terms of the equivalence relation A_i. $[x]_{A_1}, [x]_{A_2}, \cdots, [x]_{A_m}$ are m equivalence classes, and $\vee$ is a disjunction operation.

Definition 10 ([39]). *Let $IS =< U, A, V, f >$ be an information system, $A_1, A_2, \cdots, A_m \subseteq A$, and R_{A_i} is an equivalence relation of x in terms of attribute set A. $[x]_{A_i}$ is the equivalence class of R_{A_i}, $\forall B \subseteq U$, B is IFS. Then the pessimistic multi-granulation lower and upper approximations of A_i can be easily obtained by:*

$$\underline{\sum_{i=1}^{m} R^{P}_{A_i}}(B) = \{< x, \mu_{\underline{\sum_{i=1}^{m} R^{P}_{A_i}}(B)}(x), \nu_{\underline{\sum_{i=1}^{m} R^{P}_{A_i}}(B)}(x) > | x \in U\},$$
$$\overline{\sum_{i=1}^{m} R^{P}_{A_i}}(B) = \{< x, \mu_{\overline{\sum_{i=1}^{m} R^{P}_{A_i}}(B)}(x), \nu_{\overline{\sum_{i=1}^{m} R^{P}_{A_i}}(B)}(x) > | x \in U\},$$

$$\mu_{\underline{\sum_{i=1}^{m} R^{P}_{A_i}}(B)}(x) = \bigwedge_{i=1}^{m} \inf_{y \in [x]_{A_i}} \mu_B(y), \ \nu_{\underline{\sum_{i=1}^{m} R^{P}_{A_i}}(B)}(x) = \bigvee_{i=1}^{m} \sup_{y \in [x]_{A_i}} \nu_B(y),$$
$$\mu_{\overline{\sum_{i=1}^{m} R^{P}_{A_i}}(B)}(x) = \bigvee_{i=1}^{m} \sup_{y \in [x]_{A_i}} \mu_B(y), \ \nu_{\overline{\sum_{i=1}^{m} R^{P}_{A_i}}(B)}(x) = \bigwedge_{i=1}^{m} \inf_{y \in [x]_{A_i}} \nu_B(y).$$

where $[x]_{A_i}$ is the equivalence class of x in terms of the equivalence relation A_i. $[x]_{A_1}, [x]_{A_2}, \cdots, [x]_{A_m}$ are m equivalence classes, and $\wedge$ is a conjunction operation.

4.1. GRIFS Model Based on Type-I Dominance Relation

Definition 11. *Let $IS^{\geq I} =< U, A, V, f >$ be an intuitionistic fuzzy ordered information system, $A_1, A_2, \cdots, A_m \subseteq A$. $(R^{\geq}_{\overline{a}})_i$ is a dominance relation of x in terms of attribute $A_i, a \in A_i$, where $([x]^{\geq}_{\overline{a}})_i$ is the dominance class of $(R^{\geq}_{\overline{a}})_i$. Suppose X is GRS of $(R^{\geq}_{\overline{a}})_i$ and B is IFS on U. IFS B with respect to attribute a satisfies dominance relation $((R')^{\geq}_{\overline{a}})_i$. Therefore, the lower and upper approximations of B with respect to the graded k are given as follows:*

When $k \geq 1$, we can get,

$$\underline{\sum_{i=1}^{m} R^{O}_{A_i}}^{\geq I}_{(k)}(B) = \{< x, \mu_{\underline{\sum_{i=1}^{m} R^{O}_{A_i}}^{\geq I}_{(k)}(B)}(x), \nu_{\underline{\sum_{i=1}^{m} R^{O}_{A_i}}^{\geq I}_{(k)}(B)}(x) > | x \in U\},$$
$$\overline{\sum_{i=1}^{m} R^{O}_{A_i}}^{\geq I}_{(k)}(B) = \{< x, \mu_{\overline{\sum_{i=1}^{m} R^{O}_{A_i}}^{\geq I}_{(k)}(B)}(x), \nu_{\overline{\sum_{i=1}^{m} R^{O}_{A_i}}^{\geq I}_{(k)}(B)}(x) > | x \in U\},$$
$$\mu'_{B_i}(y) = \frac{|\overline{R}^{\geq}_{k}(X) \cap \underline{R}^{\geq}_{k}(X)|}{|U|}, \ \nu'_{B_i}(y) = \frac{|\neg(\overline{R}^{\geq}_{k}(X) \cup \underline{R}^{\geq}_{k}(X))|}{|U|}.$$

We can get GRS in $A_1, A_2, \cdots, A_m$, then there will be $\mu'_{B_1}(y), \mu'_{B_2}(y), \mu'_{B_3}(y), \cdots, \mu'_{B_m}(y)$ and $\nu'_{B_1}(y), \nu'_{B_2}(y), \nu'_{B_3}(y), \cdots, \nu'_{B_m}(y)$. Subsequently, we can obtain,

$$\mu_{\underline{\sum_{i=1}^{m} R^{O}_{A_i}}^{\geq^{I}}_{(k)}(B)}(x) = \bigvee_{i=1}^{m} \inf_{y \in (\bigwedge_{s=1}^{j}(([x]^{\geq}_{a})_s \wedge [x]^{\geq'}_{a}))_i} (\mu_B(y) \wedge \mu'_{B_i}(y)), \quad \nu_{\underline{\sum_{i=1}^{m} R^{O}_{A_i}}^{\geq^{I}}_{(k)}(B)}(x) = \bigwedge_{i=1}^{m} \sup_{y \in (\bigwedge_{s=1}^{j}(([x]^{\geq}_{a})_s \wedge [x]^{\geq'}_{a}))_i} (\nu_B(y) \vee \nu'_{B_i}(y)),$$

$$\mu_{\overline{\sum_{i=1}^{m} R^{O}_{A_i}}^{\geq^{I}}_{(k)}(B)}(x) = \bigwedge_{i=1}^{m} \sup_{y \in (\bigwedge_{s=1}^{j}(([x]^{\geq}_{a})_s \wedge [x]^{\geq'}_{a}))_i} (\mu_B(y) \vee \mu'_{B_i}(y)), \quad \nu_{\overline{\sum_{i=1}^{m} R^{O}_{A_i}}^{\geq^{I}}_{(k)}(B)}(x) = \bigvee_{i=1}^{m} \inf_{y \in (\bigwedge_{s=1}^{j}(([x]^{\geq}_{a})_s \wedge [x]^{\geq'}_{a}))_i} (\nu_B(y) \wedge \nu'_{B_i}(y)).$$

Obviously, $0 \leq \mu'_B(y) \leq 1$, $0 \leq \nu'_B(y) \leq 1$, $j = 1, 2, \cdots, n$.

When $\underline{\sum_{i=1}^{m} R^{O}_{A_i}}^{\geq^{I}}_{(k)}(B) \neq \overline{\sum_{i=1}^{m} R^{O}_{A_i}}^{\geq^{I}}_{(k)}(B)$, *B is an optimistic multi-granulation GRIFS model based on type-I dominance relation.*

When $k = 0$, $\mu'_{B_i}(y)$ and $\nu'_{B_i}(y)$ are calculated through the classical rough set. However, under these circumstances the model is still valid and we call this model an optimistic multi-granulation RIFS based on type-I dominance relation.

Definition 12. *Let $IS^{\geq^{I}} =< U, A, V, f >$ be an intuitionistic fuzzy ordered information system, $A_1, A_2, \cdots, A_m \subseteq A$. $(R^{\geq}_{\tilde{a}})_i$ is a dominance relation of x in terms of attribute A_i, where $([x]^{\geq}_{\tilde{a}})_i$ is the dominance class of $(R^{\geq}_{\tilde{a}})_i$. Suppose X is GRS of $(R^{\geq}_{\tilde{a}})_i$ and B is IFS on U. IFS B about attribute a satisfies dominance relation $((R')^{\geq}_{\tilde{a}})_i$, $a \in A_i$. Then the lower and upper approximations of B with respect to the graded k are given as follows:*

When $k \geq 1$, we can get,

$$\underline{\sum_{i=1}^{m} R^{p}_{A_i}}^{\geq^{I}}_{(k)}(B) = \{< x, \mu_{\underline{\sum_{i=1}^{m} R^{p}_{A_i}}^{\geq^{I}}_{(k)}(B)}(x), \nu_{\underline{\sum_{i=1}^{m} R^{p}_{A_i}}^{\geq^{I}}_{(k)}(B)}(x) > | x \in U\},$$

$$\overline{\sum_{i=1}^{m} R^{p}_{A_i}}^{\geq^{I}}_{(k)}(B) = \{< x, \mu_{\overline{\sum_{i=1}^{m} R^{p}_{A_i}}^{\geq^{I}}_{(k)}(B)}(x), \nu_{\overline{\sum_{i=1}^{m} R^{p}_{A_i}}^{\geq^{I}}_{(k)}(B)}(x) > | x \in U\},$$

$$\mu'_{B_i}(y) = \frac{|\overline{R}^{\geq}_{k}(X) \cap \underline{R}^{\geq}_{k}(X)|}{|U|}, \quad \nu'_{B_i}(y) = \frac{|\neg(\overline{R}^{\geq}_{k}(X) \cup \underline{R}^{\geq}_{k}(X))|}{|U|}.$$

We can obtain GRS in $A_1, A_2, \cdots, A_m$, then there will be $\mu'_{B_1}(y), \mu'_{B_2}(y), \mu'_{B_3}(y), \cdots, \mu'_{B_m}(y)$ and $\nu'_{B_1}(y), \nu'_{B_2}(y), \nu'_{B_3}(y), \cdots, \nu'_{B_m}(y)$. Subsequently, we can obtain,

$$\mu_{\underline{\sum_{i=1}^{m} R^{p}_{A_i}}^{\geq^{I}}_{(k)}(B)}(x) = \bigwedge_{i=1}^{m} \inf_{y \in (\bigwedge_{s=1}^{j}(([x]^{\geq}_{a})_s \wedge [x]^{\geq'}_{a}))_i} (\mu_B(y) \wedge \mu'_{B_i}(y)), \quad \nu_{\underline{\sum_{i=1}^{m} R^{p}_{A_i}}^{\geq^{I}}_{(k)}(B)}(x) = \bigvee_{i=1}^{m} \sup_{y \in (\bigwedge_{s=1}^{j}(([x]^{\geq}_{a})_s \wedge [x]^{\geq'}_{a}))_i} (\nu_B(y) \vee \nu'_{B_i}(y)),$$

$$\mu_{\overline{\sum_{i=1}^{m} R^{p}_{A_i}}^{\geq^{I}}_{(k)}(B)}(x) = \bigvee_{i=1}^{m} \sup_{y \in (\bigwedge_{s=1}^{j}(([x]^{\geq}_{a})_s \wedge [x]^{\geq'}_{a}))_i} (\mu_B(y) \vee \mu'_{B_i}(y)), \quad \nu_{\overline{\sum_{i=1}^{m} R^{p}_{A_i}}^{\geq^{I}}_{(k)}(B)}(x) = \bigwedge_{i=1}^{m} \inf_{y \in (\bigwedge_{s=1}^{j}(([x]^{\geq}_{a})_s \wedge [x]^{\geq'}_{a}))_i} (\nu_B(y) \wedge \nu'_{B_i}(y)).$$

Obviously, $0 \leq \mu'_B(y) \leq 1$, $0 \leq \nu'_B(y) \leq 1$, $j = 1, 2, \cdots, n$.

When $\underline{\sum_{i=1}^{m} R^{p}_{A_i}}^{\geq^{I}}_{(k)}(B) \neq \overline{\sum_{i=1}^{m} R^{p}_{A_i}}^{\geq^{I}}_{(k)}(B)$, *B is a pessimistic multi-granulation GRIFS model based on type-I dominance relation.*

When $k = 0$, $\mu'_{B_i}(y)$ and $\nu'_{B_i}(y)$ are calculated through the classical rough set. However, under these circumstances the model is still valid and we call this model a pessimistic multi-granulation RIFS based on type-I dominance relation.

Note that, $(\bigwedge_{s=1}^{j}(([x]_{\bar{a}}^{\geq})_s \wedge [x]_{\bar{a}}^{\geq'}))_i$ *in multi-granulation GRIFS models based on type-I dominance relation. If x have j dominance classes* $[x]_{\bar{a}}^{\geq}$ *of dominance relation* $R_{\bar{a}}^{\geq}$ *on GRS, we perform a conjunction operation* $\wedge$ *of j dominance classes* $[x]_{\bar{a}}^{\geq}$ *and* $[x]_{\bar{a}}^{\geq'}$*, respectively.*

Note that multi-granulation GRIFS models based on type-I dominance relation are formed by combining multiple granularities GRIFS models based on type-I dominance relation.

According to Definitions 11 and 12, the following theorem can be obtained.

Theorem 3. *Let* $IS^{\geq^{I}} = < U, A, V, f >$ *be an intuitionistic fuzzy ordered information system,* $A_1, A_2, \cdots, A_m \subseteq A$*, and B be IFS on U. Then the optimistic and pessimistic multi-granulation GRIFS models based on type-I dominance relation have the following properties:*

$$\underline{\sum_{i=1}^{m} R^{O}_{A_i}}^{\geq^{I}}_{(k)}(B) = \bigcup_{i=1}^{m} \underline{R_{A_i}}^{\geq^{I}}_{(k)}(B), \quad \overline{\sum_{i=1}^{m} R^{O}_{A_i}}^{\geq^{I}}_{(k)}(B) = \bigcap_{i=1}^{m} \overline{R_{A_i}}^{\geq^{I}}_{(k)}(B).$$

$$\underline{\sum_{i=1}^{m} R^{p}_{A_i}}^{\geq^{I}}_{(k)}(B) = \bigcap_{i=1}^{m} \underline{R_{A_i}}^{\geq^{I}}_{(k)}(B), \quad \overline{\sum_{i=1}^{m} R^{p}_{A_i}}^{\geq^{I}}_{(k)}(B) = \bigcup_{i=1}^{m} \overline{R_{A_i}}^{\geq^{I}}_{(k)}(B).$$

Proof. One can derive them from Definitions 7, 11, and 12. □

4.2. GRIFS Model Based on Type-II Dominance Relation

Definition 13. *Let* $IS^{\geq^{II}} = < U, A, V, f >$ *be an intuitionistic fuzzy ordered information system,* $A_1, A_2, \cdots, A_m \subseteq A$*, and U be the universe of discourse.* $(R_{\bar{a}}^{\geq})_i$ *is a dominance relation of x in terms of attribute* A_i*,* $a \in A_i$*, where* $([x]_{\bar{a}}^{\geq})_i$ *is the dominance class of* $(R_{\bar{a}}^{\geq})_i$*. Suppose X is GRS of* $(R_{\bar{a}}^{\geq})_i$ *and B is IFS on U. IFS B about attribute a satisfies dominance relation* $((R')_{\bar{a}}^{\geq})_i$*. So the lower and upper approximations of B with respect to the graded k are given as follows:*

When $k \geq 1$*, we can get,*

$$\underline{\sum_{i=1}^{m} R^{O}_{A_i}}^{\geq^{II}}_{(k)}(B) = \{< x, \mu_{\underline{\sum_{i=1}^{m} R^{O}_{A_i}}^{\geq^{II}}_{(k)}(B)}(x), \nu_{\underline{\sum_{i=1}^{m} R^{O}_{A_i}}^{\geq^{II}}_{(k)}(B)}(x) > | x \in U\},$$

$$\overline{\sum_{i=1}^{m} R^{O}_{A_i}}^{\geq^{II}}_{(k)}(B) = \{< x, \mu_{\overline{\sum_{i=1}^{m} R^{O}_{A_i}}^{\geq^{II}}_{(k)}(B)}(x), \nu_{\overline{\sum_{i=1}^{m} R^{O}_{A_i}}^{\geq^{II}}_{(k)}(B)}(x) > | x \in U\},$$

$$\mu'_{B_i}(y) = \frac{|\overline{R}_k^{\geq}(X) \cap \underline{R}_k^{\geq}(X)|}{|U|}, \quad \nu'_{B_i}(y) = \frac{|\neg(\overline{R}_k^{\geq}(X) \cup \underline{R}_k^{\geq}(X))|}{|U|}.$$

We can obtain GRS in $A_1, A_2, \cdots, A_m$*, then there will be* $\mu'_{B_1}(y), \mu'_{B_2}(y), \mu'_{B_3}(y), \cdots, \mu'_{B_m}(y)$ *and* $\nu'_{B_1}(y), \nu'_{B_2}(y), \nu'_{B_3}(y), \cdots, \nu'_{B_m}(y)$*. Subsequently, we can obtain,*

$$\mu_{\underline{\sum_{i=1}^{m} R^{O}_{A_i}}^{\geq^{II}}_{(k)}(B)}(x) = \bigvee_{i=1}^{m} \inf_{y \in (\bigvee_{s=1}^{j}(([x]_{\bar{a}}^{\geq})_s \vee [x]_{\bar{a}}^{\geq'}))_i} (\mu_B(y) \wedge \mu'_{B_i}(y)), \quad \nu_{\underline{\sum_{i=1}^{m} R^{O}_{A_i}}^{\geq^{II}}_{(k)}(B)}(x) = \bigwedge_{i=1}^{m} \sup_{y \in (\bigvee_{s=1}^{j}(([x]_{\bar{a}}^{\geq})_s \vee [x]_{\bar{a}}^{\geq'}))_i} (\nu_B(y) \vee \nu'_{B_i}(y)),$$

$$\mu_{\overline{\sum_{i=1}^{m} R^{O}_{A_i}}^{\geq^{II}}_{(k)}(B)}(x) = \bigwedge_{i=1}^{m} \sup_{y \in (\bigvee_{s=1}^{j}(([x]_{\bar{a}}^{\geq})_s \vee [x]_{\bar{a}}^{\geq'}))_i} (\mu_B(y) \vee \mu'_{B_i}(y)), \quad \nu_{\overline{\sum_{i=1}^{m} R^{O}_{A_i}}^{\geq^{II}}_{(k)}(B)}(x) = \bigvee_{i=1}^{m} \inf_{y \in (\bigvee_{s=1}^{j}(([x]_{\bar{a}}^{\geq})_s \vee [x]_{\bar{a}}^{\geq'}))_i} (\nu_B(y) \wedge \nu'_{B_i}(y)).$$

Obviously, $0 \leq \mu'_B(y) \leq 1,\ 0 \leq \nu'_B(y) \leq 1,\ j = 1, 2, \cdots, n.$

When $\underline{\sum_{i=1}^{m} R^{O}_{A_i}}^{\geq \Pi}_{(k)}(B) \neq \overline{\sum_{i=1}^{m} R^{O}_{A_i}}^{\geq \Pi}_{(k)}(B)$*, B is an optimistic multi-granulation GRIFS model based on type-II dominance relation.*

When $k = 0$*,* $\mu'_{B_i}(y)$ and $\nu'_{B_i}(y)$ *are calculated from the classical rough set. Under these circumstances, the model is still valid.*

Definition 14. *Let* $IS^{\geq \Pi} =< U, A, V, f >$ *be an intuitionistic fuzzy ordered information system,* $A_1, A_2, \cdots, A_m \subseteq A$*.* $(R^{\geq}_{\overline{a}})_i$ *is a dominance relation of x in terms of attribute* $A_i, a \in A_i$*, where* $([x]^{\geq}_{\overline{a}})_i$ *is the dominance class of* $(R^{\geq}_{\overline{a}})_i$*. Suppose X is GRS of* $(R^{\geq}_{\overline{a}})_i$ *on U and B is IFS on U. IFS B with respect to attribute a satisfies dominance relation* $((R')^{\geq}_{\overline{a}})_i$*. Then lower and upper approximations of B with respect to the graded k are as follows:*

When $k \geq 1$*, we can get,*

$$\underline{\sum_{i=1}^{m} R^{p}_{A_i}}^{\geq \Pi}_{(k)}(B) = \{< x, \mu_{\underline{\sum_{i=1}^{m} R^{p}_{A_i}}^{\geq \Pi}_{(k)}(B)}(x), \nu_{\underline{\sum_{i=1}^{m} R^{p}_{A_i}}^{\geq \Pi}_{(k)}(B)}(x) > | x \in U\},$$

$$\overline{\sum_{i=1}^{m} R^{p}_{A_i}}^{\geq \Pi}_{(k)}(B) = \{< x, \mu_{\overline{\sum_{i=1}^{m} R^{p}_{A_i}}^{\geq \Pi}_{(k)}(B)}(x), \nu_{\overline{\sum_{i=1}^{m} R^{p}_{A_i}}^{\geq \Pi}_{(k)}(B)}(x) > | x \in U\},$$

$$\mu'_{B_i}(y) = \frac{|\overline{R}^{\geq}_{\overline{k}}(X) \cap \underline{R}^{\geq}_{\overline{k}}(X)|}{|U|},\ \nu'_{B_i}(y) = \frac{|\neg(\overline{R}^{\geq}_{\overline{k}}(X) \cup \underline{R}^{\geq}_{\overline{k}}(X))|}{|U|}.$$

We can obtain GRS in $A_1, A_2, \cdots, A_m$*, then there will be* $\mu'_{B_1}(y), \mu'_{B_2}(y), \mu'_{B_3}(y), \cdots, \mu'_{B_m}(y)$ *and* $\nu'_{B_1}(y), \nu'_{B_2}(y), \nu'_{B_3}(y), \cdots, \nu'_{B_m}(y)$*. Subsequently, we can obtain,*

$$\mu_{\underline{\sum_{i=1}^{m} R^{p}_{A_i}}^{\geq \Pi}_{(k)}(B)}(x) = \bigwedge_{i=1}^{m} \inf_{y \in (\bigvee_{s=1}^{j}(([x]^{\geq}_{\overline{a}})_s \vee [x]^{\geq'}_{\overline{a}}))_i} (\mu_B(y) \wedge \mu'_{B_i}(y)),\ \nu_{\underline{\sum_{i=1}^{m} R^{p}_{A_i}}^{\geq \Pi}_{(k)}(B)}(x) = \bigvee_{i=1}^{m} \sup_{y \in (\bigvee_{s=1}^{j}(([x]^{\geq}_{\overline{a}})_s \vee [x]^{\geq'}_{\overline{a}}))_i} (\nu_B(y) \vee \nu'_{B_i}(y)),$$

$$\mu_{\overline{\sum_{i=1}^{m} R^{p}_{A_i}}^{\geq \Pi}_{(k)}(B)}(x) = \bigvee_{i=1}^{m} \sup_{y \in (\bigvee_{s=1}^{j}(([x]^{\geq}_{\overline{a}})_s \vee [x]^{\geq'}_{\overline{a}}))_i} (\mu_B(y) \vee \mu'_{B_i}(y)),\ \nu_{\overline{\sum_{i=1}^{m} R^{p}_{A_i}}^{\geq \Pi}_{(k)}(B)}(x) = \bigwedge_{i=1}^{m} \inf_{y \in (\bigvee_{s=1}^{j}(([x]^{\geq}_{\overline{a}})_s \vee [x]^{\geq'}_{\overline{a}}))_i} (\nu_B(y) \wedge \nu'_{B_i}(y)).$$

Obviously, $0 \leq \mu'_B(y) \leq 1,\ 0 \leq \nu'_B(y) \leq 1,\ j = 1, 2, \cdots, n$*.*

When $\underline{\sum_{i=1}^{m} R^{p}_{A_i}}^{\geq \Pi}_{(k)}(B) \neq \overline{\sum_{i=1}^{m} R^{p}_{A_i}}^{\geq \Pi}_{(k)}(B)$*, B is a pessimistic multi-granulation GRIFS model based on type-II dominance relation.*

When $k = 0$*,* $\mu'_{B_i}(y)$ and $\nu'_{B_i}(y)$ *are calculated from the classical rough set. Under these circumstances, the model is still valid.*

Note that, in $(\bigvee_{s=1}^{j}(([x]^{\geq}_{\overline{a}})_s \vee [x]^{\geq'}_{\overline{a}}))_i$*, if x have j dominance classes* $[x]^{\geq}_{\overline{a}}$ *of dominance relation* $R^{\geq}_{\overline{a}}$ *on GRS, we perform a disjunction operation* $\vee$ *of j dominance classes* $[x]^{\geq}_{\overline{a}}$ *and* $[x]^{\geq'}_{\overline{a}}$*, respectively.*

Note that multi-granulation GRIFS models based on type-II dominance relation are formed by combining multiple granularities GRIFS models based on type-II dominance relation.

According to Definitions 13 and 14, the following theorem can be obtained.

Theorem 4. *Let $IS^{\geq^{\Pi}} = < U, A, V, f >$ be an intuitionistic fuzzy ordered information system, $A_1, A_2, \cdots, A_m \subseteq A$, and $IFS\ B \subseteq U$. Then optimistic and pessimistic multi-granulation GRIFS models based on type-II dominance relation have the following properties:*

$$\underline{\sum_{i=1}^{m} R^{O}_{A_i}}^{\geq^{\Pi}}_{(k)}(B) = \bigcup_{i=1}^{m} \underline{R_{A_i}}^{\geq^{\Pi}}_{(k)}(B), \quad \overline{\sum_{i=1}^{m} R^{O}_{A_i}}^{\geq^{\Pi}}_{(k)}(B) = \bigcap_{i=1}^{m} \overline{R_{A_i}}^{\geq^{\Pi}}_{(k)}(B),$$

$$\underline{\sum_{i=1}^{m} R^{P}_{A_i}}^{\geq^{\Pi}}_{(k)}(B) = \bigcap_{i=1}^{m} \underline{R_{A_i}}^{\geq^{\Pi}}_{(k)}(B), \quad \overline{\sum_{i=1}^{m} R^{P}_{A_i}}^{\geq^{\Pi}}_{(k)}(B) = \bigcup_{i=1}^{m} \overline{R_{A_i}}^{\geq^{\Pi}}_{(k)}(B).$$

Proof. One can derive them from Definitions 7, 13 and 14. □

5. Algorithm and Example Analysis

5.1. Algorithm

Through Examples 1–3, we can conclude that the GRIFS model is effective, and now we use multi-granulation GRIFS models based on dominance relation to predict results under the same situations again as Algorithm 1.

Algorithm 1. Computing multi-granulation GRIFS models based on dominance relation.

Input: $IS = < U, A, V, f >, X \subseteq U$, IFS $B \subseteq U$, k is a natural number
Output: Multi-granulation GRIFS models based on dominance relation
1:**if** ($U \neq \phi$ **and** $A \neq \phi$)
2: **if** can build up GRS
3: **if** ($k \geq 1$ && i = 1 to m && $a \in A_i$)
4: compute $\mu'_B(y)$ and $\nu'_B(y)$, $[x]^{\geq}_{a}$ and $[x]^{\geq'}_{a}$, for each $A_i \subseteq A$;
5: then compute $\wedge$ and $\vee$ of $\mu_B(y)$ and $\mu'_{B_i}(y)$, $\nu'_{B_i}(y)$ and $\nu_B(y)$ and compute $\wedge$ and $\vee$ of $[x]^{\geq}_{a}$ and $[x]^{\geq'}_{a}$;
6: **if** ($x \in U$ && $\forall y \in (\overset{j}{\underset{s=1}{*}}(([x]^{\geq}_{a})_s * [x]^{\geq'}_{a}))_i$)
7: **for** (i = 1 to m && $1 \leq j \leq n$)
8: compute $\mu_{\underline{\sum_{i=1}^{m} R^{\Delta}_{A_i}}^{\geq^{\bullet}}_{(k)}(B)}(x)$, $\nu_{\underline{\sum_{i=1}^{m} R^{\Delta}_{A_i}}^{\geq^{\bullet}}_{(k)}(B)}(x)$, $\mu_{\overline{\sum_{i=1}^{m} R^{\Delta}_{A_i}}^{\geq^{\bullet}}_{(k)}(B)}(x)$, $\nu_{\overline{\sum_{i=1}^{m} R^{\Delta}_{A_i}}^{\geq^{\bullet}}_{(k)}(B)}(x)$;
9: **end**
10: compute $\underline{\sum_{i=1}^{m} R^{\Delta}_{A_i}}^{\geq^{\bullet}}_{(k)}(B)$, $\overline{\sum_{i=1}^{m} R^{\Delta}_{A_i}}^{\geq^{\bullet}}_{(k)}(B)$.
11: **end**
12: **end**
13: **end**
14:**else**
return NULL
15:**end**

Note that Δ represents optimistic and pessimistic and $*$ means $\wedge$ or $\vee$ operation, and in $(\overset{j}{\underset{s=1}{*}}(([x]^{\geq}_{a})_s * [x]^{\geq'}_{a}))_i$, $(\overset{j}{\underset{s=1}{*}}(([x]^{\geq}_{a})_s * [x]^{\geq'}_{a}))_i$ represent $(\overset{j}{\underset{s=1}{\wedge}}(([x]^{\geq}_{a})_s \wedge [x]^{\geq'}_{a}))_i$ or $(\overset{j}{\underset{s=1}{\vee}}(([x]^{\geq}_{a})_s \vee [x]^{\geq'}_{a}))_i$ in this algorithm. $\bullet$ represents I or II.

Through this algorithm, we next illustrate these models by example again.

5.2. An Illustrative Example

We use this example to illustrate Algorithm 1 of multi-granulation GRIFS models based on type-I and type-II dominance relation. According to Algorithm 1, we will not discuss this case where k is 0. There are 9 patients. Let $U = \{x_1, x_2, x_3, x_4, x_5, x_6, x_7, x_8, x_9\}$ be the domain. Next, we analyzed these 9 patients from these symptoms of fever and salivation. The set of condition attributes are A = {fever, salivation, streaming nose}. For fever, we can get $U/R^{\geq} = \{\{x_1, x_2, x_4\}, \{x_3, x_8\}, \{x_7\}, \{x_4, x_5, x_6, x_9\}\}$, for salivation there is $U/R^{\geq} = \{\{x_1\}, \{x_1, x_4\}, \{x_3, x_5, x_6\}, \{x_5, x_6\}, \{x_6, x_9\}, \{x_2, x_7, x_8\}\}$, and for streaming nose $U/R^{\geq} = \{\{x_1\}, \{x_1, x_2, x_4\}, \{x_3, x_5, x_6\}, \{x_4, x_6, x_7, x_9\}, \{x_2, x_7, x_8\}\}$. According to the cold disease, these patients have the have the following IFS

$$B = \left\{ \frac{[0.9, 0]}{x_1}, \frac{[0.8, 0.1]}{x_2}, \frac{[0.65, 0.3]}{x_3}, \frac{[0.85, 0.1]}{x_4}, \frac{[0.95, 0.05]}{x_5}, \frac{[0.7, 0.3]}{x_6}, \frac{[0.5, 0.2]}{x_7}, \frac{[0.87, 0.1]}{x_8}, \frac{[0.75, 0.2]}{x_9} \right\}.$$

Suppose $X = \{x_1, x_4, x_5, x_6, x_9\}$, $k = 1$. Then we can obtain multi-granulation GRIFS models based on type-I and type-II dominance relation through Definitions 11–14. Results are as follows.

For fever, according to $U/R^{\geq}$, we can get,

$$\underline{R}_1^{\geq}(X) = \{x_1, x_2, x_4, x_5, x_6, x_7, x_9\},\ \overline{R}_1^{\geq}(X) = \{x_1, x_2, x_4, x_5, x_6, x_7, x_9\},$$
$$\mu'_{B_1}(y) = \frac{|\overline{R}_1^{\geq}(X) \cap \underline{R}_1^{\geq}(X)|}{|U|} = \frac{7}{9} \approx 0.78,\ \nu'_{B_1}(y) = \frac{|\neg(\overline{R}_1^{\geq}(X) \cup \underline{R}_1^{\geq}(X))|}{|U|} = \frac{2}{9} \approx 0.22.$$

Similarly, for salivation and streaming nose, the results are as follows:

$$\mu'_{B_2}(y) = \frac{6}{9} \approx 0.67,\ \nu'_{B_2}(y) = \frac{3}{9} \approx 0.33.\ \mu'_{B_2}(y) = \frac{8}{9} \approx 0.89,\ \nu'_{B_2}(y) = \frac{1}{9} \approx 0.11.$$

According to Definitions 11–14, we can obtain multi-granulation GRIFS models based on type-I dominance relation and type-II dominance relation.

For $\mu'_{B_1}(y)$ and $\nu'_{B_1}(y)$, $\mu'_{B_2}(y)$ and $\nu'_{B_2}(y)$ and $\mu'_{B_3}(y)$ and $\nu'_{B_3}(y)$, the results are the followings as Table 3.

Table 3. The conjunction and disjunction operation of $\mu_B(y)$ and $\mu'_{B_1}(y)$.

x	x_1	x_2	x_3	x_4	x_5	x_6	x_7	x_8	x_9
$\mu_B(y)$	0.9	0.8	0.65	0.85	0.95	0.7	0.5	0.87	0.75
$\mu_B(y)$	0	0.1	0.3	0.1	0.05	0.3	0.2	0.1	0.2
$\mu_B(y) \wedge \mu'_{B_1}(y)$	0.78	0.78	0.65	0.78	0.78	0.7	0.5	0.72	0.75
$\nu_B(y) \wedge \nu'_{B_1}(y)$	0.22	0.22	0.3	0.22	0.22	0.3	0.22	0.22	0.22
$\mu_B(y) \vee \mu'_{B_1}(y)$	0.9	0.8	0.78	0.85	0.95	0.78	0.78	0.78	0.78
$\nu_B(y) \wedge \nu'_{B_1}(y)$	0	0.1	0.22	0.1	0.05	0.22	0.2	0.1	0.2
$\mu_B(y) \wedge \mu'_{B_2}(y)$	0.67	0.67	0.65	0.67	0.67	0.67	0.5	0.6	0.67
$\nu_B(y) \vee \nu'_{B_2}(y)$	0.33	0.33	0.33	0.33	0.33	0.33	0.33	0.33	0.33
$\mu_B(y) \vee \mu'_{B_2}(y)$	0.89	0.8	0.67	0.85	0.95	0.7	0.67	0.67	0.75
$\nu_B(y) \wedge \nu'_{B_2}(y)$	0.11	0.1	0.3	0.1	0.05	0.3	0.2	0.1	0.2
$\mu_B(y) \wedge \mu'_{B_3}(y)$	0.9	0.8	0.65	0.85	0.89	0.7	0.5	0.87	0.75
$\nu_B(y) \vee \nu'_{B_3}(y)$	0	0.11	0.3	0.11	0.11	0.3	0.2	0.11	0.2
$\mu_B(y) \vee \mu'_{B_3}(y)$	0.9	0.89	0.89	0.89	0.95	0.89	0.89	0.89	0.89
$\nu_B(y) \wedge \nu'_{B_3}(y)$	0	0.1	0.11	0.1	0.05	0.11	0.11	0.1	0.11

Then, according to Definition 6, for B, we can get $[x]_a^{\geq'}$. Then, the conjunction operation of $[x]_a^{\geq}$ and $[x]_a^{\geq'}$ can be computed as Table 1.

For fever, we can get GRIFS based on type-I dominance relation as follows:

$$\underline{R}_1^{\geq I}(B) = \left\{ \frac{[0.78,0.22]}{x_1}, \frac{[0.78,0.22]}{x_2}, \frac{[0.65,0.3]}{x_3}, \frac{[0.78,0.22]}{x_4}, \frac{[0.78,0.22]}{x_5}, \frac{[0.7,0.3]}{x_6}, \frac{[0.5,0.22]}{x_7}, \frac{[0.78,0.22]}{x_8}, \frac{[0.75,0.22]}{x_9} \right\},$$

$$\overline{R}_1^{\geq I}(B) = \left\{ \frac{[0.9,0]}{x_1}, \frac{[0.85,0.1]}{x_2}, \frac{[0.78,0.1]}{x_3}, \frac{[0.85,0.1]}{x_4}, \frac{[0.95,0.05]}{x_5}, \frac{[0.87,0.1]}{x_6}, \frac{[0.78,0.1]}{x_7}, \frac{[0.87,0.1]}{x_8}, \frac{[0.78,0.2]}{x_9} \right\}.$$

For streaming nose, similar to Table 1, we can obtain $[x]_a^{\geq} \wedge [x]_a^{\geq'}$ as Table 4.

Table 4. The conjunction operation of $[x]_a^{\geq}$ and $[x]_a^{\geq'}$.

x	$[x]_a^{\geq}$	$[x]_a^{\geq'}$	$[x]_a^{\geq} \wedge [x]_a^{\geq'}$
x_1	$\{x_1\}, \{x_1, x_2, x_4\}$	$\{x_1\}$	$\{x_1\}$
x_2	$\{x_1, x_2, x_4\}, \{x_2, x_7, x_8\}$	$\{x_2, x_4, x_5, x_8\}$	$\{x_2\}$
x_3	$\{x_3, x_5, x_6\}$	$\{x_3, x_4, x_5, x_6, x_8, x_9\}$	$\{x_3, x_5, x_6\}$
x_4	$\{x_1, x_2, x_4\}, \{x_4, x_6, x_7, x_9\}$	$\{x_4, x_5\}$	$\{x_4\}$
x_5	$\{x_3, x_5, x_6\}$	$\{x_5\}$	$\{x_5\}$
x_6	$\{x_3, x_5, x_6\}, \{x_4, x_6, x_7, x_9\}$	$\{x_6, x_8, x_9\}$	$\{x_6\}$
x_7	$\{x_2, x_7, x_8\}, \{x_4, x_6, x_7, x_9\}$	$\{x_7, x_8, x_9\}$	$\{x_7\}$
x_8	$\{x_2, x_7, x_8\}$	$\{x_8\}$	$\{x_8\}$
x_9	$\{x_4, x_6, x_7, x_9\}$	$\{x_9\}$	$\{x_9\}$

For salivation, similar to Table 1, we can obtain $[x]_a^{\geq} \wedge [x]_a^{\geq'}$ as Table 5.

Table 5. The conjunction operation of $[x]_a^{\geq}$ and $[x]_a^{\geq'}$.

x	$[x]_a^{\geq}$	$[x]_a^{\geq'}$	$[x]_a^{\geq} \wedge [x]_a^{\geq'}$
x_1	$\{x_1, x_2\}$	$\{x_1\}$	$\{x_1\}$
x_2	$\{x_1, x_2\}, \{x_2, x_4\}$	$\{x_2, x_4, x_5, x_8\}$	$\{x_2\}$
x_3	$\{x_3, x_8\}$	$\{x_3, x_4, x_5, x_6, x_8, x_9\}$	$\{x_3, x_8\}$
x_4	$\{x_2, x_4\}$	$\{x_4, x_5\}$	$\{x_4\}$
x_5	$\{x_5, x_6\}$	$\{x_5\}$	$\{x_5\}$
x_6	$\{x_5, x_6\}$	$\{x_6, x_8, x_9\}$	$\{x_6\}$
x_7	$\{x_7, x_9\}, \{x_7, x_8, x_9\}$	$\{x_7, x_8, x_9\}$	$\{x_7, x_9\}$
x_8	$\{x_3, x_8\}, \{x_7, x_8, x_9\}$	$\{x_8\}$	$\{x_8\}$
x_9	$\{x_7, x_9\}, \{x_7, x_8, x_9\}$	$\{x_9\}$	$\{x_9\}$

For streaming nose, we can get GRIFS based on type-I dominance relation as follows:

$$\underline{R}_1^{\geq I}(B) = \left\{ \frac{[0.9,0.11]}{x_1}, \frac{[0.8,0.11]}{x_2}, \frac{[0.65,0.3]}{x_3}, \frac{[0.85,0.11]}{x_4}, \frac{[0.89,0.11]}{x_5}, \frac{[0.7,0.3]}{x_6}, \frac{[0.5,0.2]}{x_7}, \frac{[0.87,0.11]}{x_8}, \frac{[0.75,0.2]}{x_9} \right\},$$

$$\overline{R}_1^{\geq I}(B) = \left\{ \frac{[0.9,0]}{x_1}, \frac{[0.89,0.1]}{x_2}, \frac{[0.95,0.05]}{x_3}, \frac{[0.89,0.1]}{x_4}, \frac{[0.95,0.05]}{x_5}, \frac{[0.89,0.11]}{x_6}, \frac{[0.89,0.11]}{x_7}, \frac{[0.89,0.1]}{x_8}, \frac{[0.89,0.11]}{x_9} \right\}.$$

For salivation, we can get GRIFS based on type-I dominance relation as follows:

$$\underline{R}_1^{\geq I}(B) = \left\{ \frac{[0.67,0.33]}{x_1}, \frac{[0.67,0.33]}{x_2}, \frac{[0.65,0.33]}{x_3}, \frac{[0.67,0.33]}{x_4}, \frac{[0.67,0.33]}{x_5}, \frac{[0.67,0.33]}{x_6}, \frac{[0.67,0.33]}{x_7}, \frac{[0.67,0.33]}{x_8}, \frac{[0.67,0.33]}{x_9} \right\},$$

$$\overline{R}_1^{\geq I}(B) = \left\{ \frac{[0.9,0]}{x_1}, \frac{[0.8,0.1]}{x_2}, \frac{[0.87,0.1]}{x_3}, \frac{[0.85,0.1]}{x_4}, \frac{[0.95,0.05]}{x_5}, \frac{[0.7,0.3]}{x_6}, \frac{[0.75,0.2]}{x_7}, \frac{[0.87,0.1]}{x_8}, \frac{[0.75,0.2]}{x_9} \right\}.$$

For $[x_2]_{a}^{\geq} = \{x_1, x_2\}$, $[x_2]_{a}^{\geq} = \{x_2, x_4\}$ and $[x_2]_{a}^{\geq'} = \{x_2, x_4, x_5, x_8\}$, based on Definitions 11–14, we should perform the conjunction operation of them, respectively.

$$([x_1]_{a}^{\geq} \wedge [x_1]_{a}^{\geq'}) \wedge ([x_1]_{a}^{\geq} \wedge [x_1]_{a}^{\geq'}) = (\{x_1\} \wedge \{x_1, x_1, x_4\}) \wedge (\{x_1, x_2\} \wedge \{x_1, x_2, x_4\}) = \{x_1\} \wedge \{x_1, x_2\} = \{x_1\}.$$

Similarly, for x_2, x_4, x_6 and x_7, we can get the results as Tables 4 and 5.

Therefore, according to the Definitions 11 and 12 and the above calculations, we can get multi-granulation GRIFS models based on a type-I dominance relation as follows:

$$\underline{\sum_{i=1}^{3} R_{A_i}^{O}}_{1}^{\geq^{\mathrm{I}}}(B) = \left\{ \frac{[0.89,0.11]}{x_1}, \frac{[0.8,0.11]}{x_2}, \frac{[0.65,0.3]}{x_3}, \frac{[0.85,0.11]}{x_4}, \frac{[0.89,0.11]}{x_5}, \frac{[0.7,0.3]}{x_6}, \frac{[0.67,0.2]}{x_7}, \frac{[0.87,0.11]}{x_8}, \frac{[0.75,0.2]}{x_9} \right\},$$

$$\overline{\sum_{i=1}^{3} R_{A_i}^{O}}_{1}^{\geq^{\mathrm{I}}}(B) = \left\{ \frac{[0.9,0]}{x_1}, \frac{[0.8,0.1]}{x_2}, \frac{[0.78,0.1]}{x_3}, \frac{[0.85,0.1]}{x_4}, \frac{[0.95,0.05]}{x_5}, \frac{[0.7,0.3]}{x_6}, \frac{[0.75,0.2]}{x_7}, \frac{[0.87,0.1]}{x_8}, \frac{[0.75,0.2]}{x_9} \right\}.$$

$$\underline{\sum_{i=1}^{3} R_{A_i}^{P}}_{1}^{\geq^{\mathrm{I}}}(B) = \left\{ \frac{[0.67,0.33]}{x_1}, \frac{[0.67,0.33]}{x_2}, \frac{[0.65,0.33]}{x_3}, \frac{[0.67,0.33]}{x_4}, \frac{[0.67,0.33]}{x_5}, \frac{[0.67,0.33]}{x_6}, \frac{[0.5,0.33]}{x_7}, \frac{[0.67,0.33]}{x_8}, \frac{[0.67,0.33]}{x_9} \right\},$$

$$\overline{\sum_{i=1}^{3} R_{A_i}^{P}}_{1}^{\geq^{\mathrm{I}}}(B) = \left\{ \frac{[0.89,0]}{x_1}, \frac{[0.85,0.1]}{x_2}, \frac{[0.95,0.05]}{x_3}, \frac{[0.85,0.1]}{x_4}, \frac{[0.95,0.05]}{x_5}, \frac{[0.87,0.05]}{x_6}, \frac{[0.78,0.1]}{x_7}, \frac{[0.87,0.1]}{x_8}, \frac{[0.78,0.1]}{x_9} \right\}.$$

From the above results, Figures 1 and 2 can be drawn as follows:

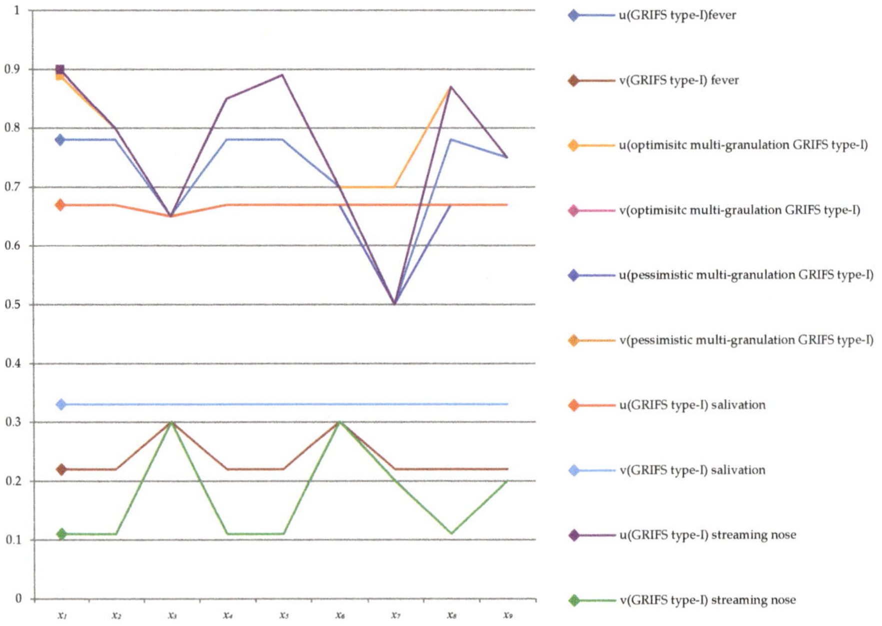

Figure 1. The lower approximation of GRIFS based on type-I dominance relation, as well as optimistic and pessimistic multi-granulation GRIFS based on type-I dominance relation.

For Figure 1, we can obtain,

$$\mu(y)_{OI1} \geq \mu(y)_{GIn1} \Theta\, \mu(y)_{GIf1} \Theta(y)_{GIs1} \geq \mu(y)_{PI1}, \nu(y)_{GIs1} \geq \nu(y)_{PI1} \geq \nu(y)_{OI1} = \nu(y)_{GIf1} \geq \nu(y)_{GIn1};$$

Note:

Θ represents $\leq$ or $\geq$;
$\mu(y)_{GIf1}$ and $\nu(y)_{GIf1}$ represent GRIFS type-I dominance relation (fever);
$\mu(y)_{GIs1}$ and $\nu(y)_{GIs1}$ represent GRIFS type-I dominance relation (salivation);
$\mu(y)_{GIn1}$ and $\nu(y)_{GIn1}$ represent GRIFS type-I dominance relation (streaming nose);
$\mu(y)_{OI1}$ and $\nu(y)_{OI1}$ represent optimistic multi-granulation GRIFS type-I dominance relation;
$\mu(y)_{PI1}$ and $\nu(y)_{PI1}$ represent pessimistic multi-granulation GRIFS type-I dominance relation;

From Figure 1, we can get that x_1, x_2, x_4, x_5 and x_8 patients have the disease, and x_7 patients do not have the disease.

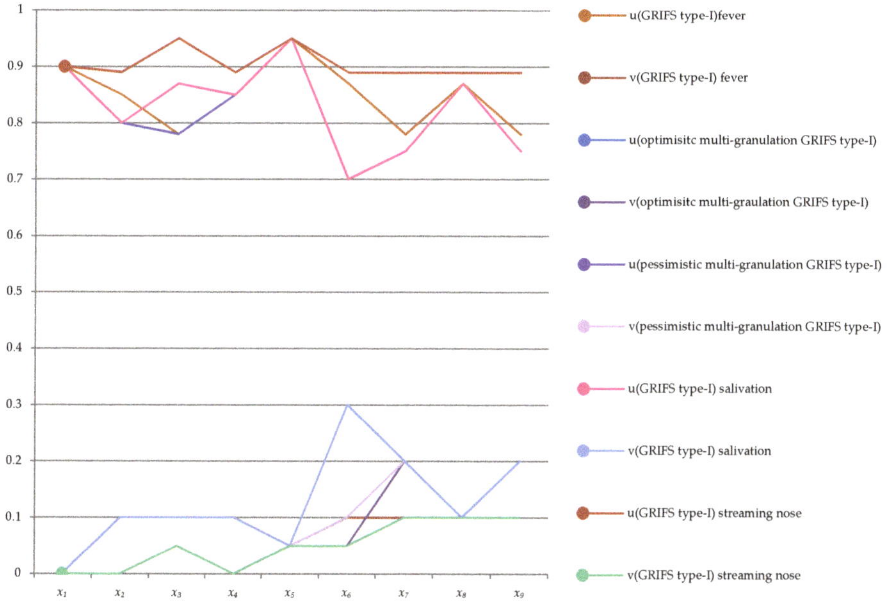

Figure 2. The upper approximation of GRIFS based on type-I dominance relation, as well as optimistic and pessimistic multi-granulation GRIFS based on type-I dominance relation.

Then, from Figure 2, we can obtain,

$$\mu(y)_{OI2} = \mu(y)_{GIn2} \geq \mu(y)_{PI2}\Theta\, \mu(y)_{GIs2}\Theta\, \mu(y)_{GIf2}, \nu(y)_{GIs2} \geq \nu(y)_{OI2}\Theta\, \nu(y)_{PI2}\Theta\, \nu(y)_{GIf2} \geq \nu(y)_{GIn2};$$

Note:

Θ represents $\leq$ or $\geq$;
$\mu(y)_{GIf2}$ and $\nu(y)_{GIf2}$ represent GRIFS type-I dominance relation (fever);
$\mu(y)_{GIs2}$ and $\nu(y)_{GIs2}$ represent GRIFS type-I dominance relation (salivation);
$\mu(y)_{GIn2}$ and $\nu(y)_{GIn2}$ represent GRIFS type-I dominance relation (streaming nose);
$\mu(y)_{OI2}$ and $\nu(y)_{OI2}$ represent optimistic multi-granulation GRIFS type-I dominance relation;
$\mu(y)_{PI2}$ and $\nu(y)_{PI2}$ represent pessimistic multi-granulation GRIFS type-I dominance relation;

From Figure 2, we can get that x_1, x_2, x_3, x_4, x_5, x_6, x_8, and x_9 patients have the disease, and x_6 patients do not have the disease.

For multi-granulation GRIFS models based on type-II dominance relation, the calculations for this model are similar to multi-granulation GRIFS models based on type-I dominance relation.

Firstly, for streaming nose, we can compute the disjunction operation of $[x]_a^{\geq}$ and $[x]_a^{\geq'}$, and the results are as Table 6.

Table 6. The disjunction operation of $[x]_a^{\geq}$ and $[x]_a^{\geq'}$.

x	$[x]_a^{\geq}$	$[x]_a^{\geq'}$	$[x]_a^{\geq} \vee [x]_a^{\geq'}$
x_1	$\{x_1\}, \{x_1, x_2, x_4\}$	$\{x_1\}$	$\{x_1, x_2, x_4\}$
x_2	$\{x_1, x_2, x_4\}, \{x_2, x_7, x_8\}$	$\{x_2, x_4, x_5, x_8\}$	$\{x_1, x_2, x_4, x_5, x_7, x_8\}$
x_3	$\{x_3, x_5, x_6\}$	$\{x_3, x_4, x_5, x_6, x_8, x_9\}$	$\{x_3, x_4, x_5, x_6, x_8, x_9\}$
x_4	$\{x_1, x_2, x_4\}, \{x_4, x_6, x_7, x_9\}$	$\{x_4, x_5\}$	$\{x_1, x_2, x_4, x_5, x_6, x_7, x_9\}$
x_5	$\{x_3, x_5, x_6\}$	$\{x_5\}$	$\{x_3, x_5, x_6\}$
x_6	$\{x_3, x_5, x_6\}, \{x_4, x_6, x_7, x_9\}$	$\{x_6, x_8, x_9\}$	$\{x_3, x_4, x_5, x_6, x_7, x_8, x_9\}$
x_7	$\{x_2, x_7, x_8\}, \{x_4, x_6, x_7, x_9\}$	$\{x_7, x_8, x_9\}$	$\{x_2, x_4, x_6, x_7, x_8, x_9\}$
x_8	$\{x_2, x_7, x_8\}$	$\{x_8\}$	$\{x_2, x_7, x_8\}$
x_9	$\{x_4, x_6, x_7, x_9\}$	$\{x_9\}$	$\{x_4, x_6, x_7, x_9\}$

Next, for salivation, we can compute the disjunction operation of $[x]_a^{\geq}$ and $[x]_a^{\geq'}$, and the results are as Table 7.

Table 7. The disjunction operation of $[x]_a^{\geq}$ and $[x]_a^{\geq'}$.

x	$[x]_a^{\geq}$	$[x]_a^{\geq'}$	$[x]_a^{\geq} \vee [x]_a^{\geq'}$
x_1	$\{x_1, x_2\}$	$\{x_1\}$	$\{x_1, x_2\}$
x_2	$\{x_1, x_2\}, \{x_2, x_4\}$	$\{x_2, x_4, x_5, x_8\}$	$\{x_1, x_2, x_4, x_5, x_8\}$
x_3	$\{x_3, x_8\}$	$\{x_3, x_4, x_5, x_6, x_8, x_9\}$	$\{x_3, x_4, x_5, x_6, x_8, x_9\}$
x_4	$\{x_2, x_4\}$	$\{x_4, x_5\}$	$\{x_2, x_4, x_5\}$
x_5	$\{x_5, x_6\}$	$\{x_5\}$	$\{x_5, x_6\}$
x_6	$\{x_5, x_6\}$	$\{x_6, x_8, x_9\}$	$\{x_5, x_6, x_8, x_9\}$
x_7	$\{x_7, x_9\}, \{x_7, x_8, x_9\}$	$\{x_7, x_8, x_9\}$	$\{x_7, x_8, x_9\}$
x_8	$\{x_3, x_8\}, \{x_7, x_8, x_9\}$	$\{x_8\}$	$\{x_3, x_7, x_8, x_9\}$
x_9	$\{x_7, x_9\}, \{x_7, x_8, x_9\}$	$\{x_9\}$	$\{x_7, x_8, x_9\}$

Then, for fever, we compute the disjunction operation of $[x]_a^{\geq}$ and $[x]_a^{\geq'}$, and these results are shown as Table 8.

Table 8. The disjunction operation of $[x]_a^{\geq}$ and $[x]_a^{\geq'}$.

x	$[x]_a^{\geq}$	$[x]_a^{\geq'}$	$[x]_a^{\geq} \vee [x]_a^{\geq'}$
x_1	$\{x_1, x_2, x_4\}$	$\{x_1\}$	$\{x_1, x_2, x_4\}$
x_2	$\{x_1, x_2, x_4\}$	$\{x_2, x_4, x_5, x_8\}$	$\{x_1, x_2, x_4, x_5, x_8\}$
x_3	$\{x_3, x_8\}$	$\{x_3, x_4, x_5, x_6, x_8, x_9\}$	$\{x_3, x_4, x_5, x_6, x_8, x_9\}$
x_4	$\{x_1, x_2, x_4\}, \{x_4, x_5, x_6, x_9\}$	$\{x_4, x_5\}$	$\{x_1, x_2, x_4, x_5, x_6, x_9\}$
x_5	$\{x_4, x_5, x_6, x_9\}$	$\{x_5\}$	$\{x_4, x_5, x_6, x_9\}$
x_6	$\{x_4, x_5, x_6, x_9\}$	$\{x_6, x_8, x_9\}$	$\{x_4, x_5, x_6, x_8, x_9\}$
x_7	$\{x_7\}$	$\{x_7, x_8, x_9\}$	$\{x_7, x_8, x_9\}$
x_8	$\{x_3, x_8\}$	$\{x_8\}$	$\{x_3, x_8\}$
x_9	$\{x_4, x_5, x_6, x_9\}$	$\{x_9\}$	$\{x_4, x_5, x_6, x_9\}$

For streaming nose, we can get GRIFS based on type-II dominance relation,

$$\underline{R}_1^{\geq \Pi}(B) = \left\{ \frac{[0.8, 0.11]}{x_1}, \frac{[0.5, 0.2]}{x_2}, \frac{[0.65, 0.3]}{x_3}, \frac{[0.5, 0.3]}{x_4}, \frac{[0.65, 0.3]}{x_5}, \frac{[0.5, 0.3]}{x_6}, \frac{[0.5, 0.3]}{x_7}, \frac{[0.5, 0.2]}{x_8}, \frac{[0.5, 0.3]}{x_9} \right\},$$

$$\overline{R}_1^{\geq \Pi}(B) = \left\{ \frac{[0.9, 0]}{x_1}, \frac{[0.95, 0]}{x_2}, \frac{[0.95, 0.05]}{x_3}, \frac{[0.95, 0]}{x_4}, \frac{[0.95, 0.05]}{x_5}, \frac{[0.95, 0.05]}{x_6}, \frac{[0.89, 0.1]}{x_7}, \frac{[0.89, 0.1]}{x_8}, \frac{[0.89, 0.1]}{x_9} \right\}.$$

For salivation, we can get GRIFS based on type-II dominance relation,

$$\underline{R}_1^{\geq\Pi}(B) = \left\{ \frac{[0.67,0.33]}{x_1}, \frac{[0.67,0.33]}{x_2}, \frac{[0.65,0.33]}{x_3}, \frac{[0.67,0.33]}{x_4}, \frac{[0.67,0.33]}{x_5}, \frac{[0.67,0.33]}{x_6}, \frac{[0.5,0.33]}{x_7}, \frac{[0.5,0.33]}{x_8}, \frac{[0.5,0.33]}{x_9} \right\},$$
$$\overline{R}_1^{\geq\Pi}(B) = \left\{ \frac{[0.9,0]}{x_1}, \frac{[0.95,0]}{x_2}, \frac{[0.95,0.05]}{x_3}, \frac{[0.95,0.05]}{x_4}, \frac{[0.95,0.05]}{x_5}, \frac{[0.95,0.05]}{x_6}, \frac{[0.87,0.1]}{x_7}, \frac{[0.87,0.1]}{x_8}, \frac{[0.87,0.1]}{x_9} \right\}.$$

For fever, GRIFS type-II dominance relation can be calculated as follows:

$$\underline{R}_1^{\geq\Pi}(B) = \left\{ \frac{[0.78,0.22]}{x_1}, \frac{[0.78,0.22]}{x_2}, \frac{[0.65,0.3]}{x_3}, \frac{[0.7,0.22]}{x_4}, \frac{[0.7,0.22]}{x_5}, \frac{[0.7,0.3]}{x_6}, \frac{[0.5,0.22]}{x_7}, \frac{[0.65,0.3]}{x_8}, \frac{[0.7,0.3]}{x_9} \right\},$$
$$\overline{R}_1^{\geq\Pi}(B) = \left\{ \frac{[0.9,0]}{x_1}, \frac{[0.95,0]}{x_2}, \frac{[0.95,0.05]}{x_3}, \frac{[0.95,0]}{x_4}, \frac{[0.95,0.05]}{x_5}, \frac{[0.95,0.05]}{x_6}, \frac{[0.87,0.1]}{x_7}, \frac{[0.87,0.1]}{x_8}, \frac{[0.95,0.05]}{x_9} \right\}.$$

Based on Definitions 13 and 14, the condition of these patients based on multi-granulation GRIFS type-II dominance relation can be obtained as follows:

$$\underline{\sum_{i=1}^{3} R_{A_i}^{O}}_{1}^{\geq\Pi}(B) = \left\{ \frac{[0.8,0.11]}{x_1}, \frac{[0.78,0.2]}{x_2}, \frac{[0.65,0.3]}{x_3}, \frac{[0.7,0.22]}{x_4}, \frac{[0.7,0.3]}{x_5}, \frac{[0.7,0.3]}{x_6}, \frac{[0.5,0.22]}{x_7}, \frac{[0.65,0.2]}{x_8}, \frac{[0.7,0.3]}{x_9} \right\},$$
$$\overline{\sum_{i=1}^{3} R_{A_i}^{O}}_{1}^{\geq\Pi}(B) = \left\{ \frac{[0.9,0]}{x_1}, \frac{[0.95,0]}{x_2}, \frac{[0.95,0.05]}{x_3}, \frac{[0.95,0.05]}{x_4}, \frac{[0.95,0.05]}{x_5}, \frac{[0.95,0.05]}{x_6}, \frac{[0.87,0.1]}{x_7}, \frac{[0.87,0.1]}{x_8}, \frac{[0.87,0.1]}{x_9} \right\}.$$
$$\underline{\sum_{i=1}^{3} R_{A_i}^{P}}_{1}^{\geq\Pi}(B) = \left\{ \frac{[0.67,0.33]}{x_1}, \frac{[0.5,0.33]}{x_2}, \frac{[0.65,0.33]}{x_3}, \frac{[0.5,0.4]}{x_4}, \frac{[0.65,0.33]}{x_5}, \frac{[0.5,0.33]}{x_6}, \frac{[0.5,0.33]}{x_7}, \frac{[0.5,0.33]}{x_8}, \frac{[0.5,0.33]}{x_9} \right\},$$
$$\overline{\sum_{i=1}^{3} R_{A_i}^{P}}_{1}^{\geq\Pi}(B) = \left\{ \frac{[0.9,0]}{x_1}, \frac{[0.95,0]}{x_2}, \frac{[0.95,0.05]}{x_3}, \frac{[0.95,0]}{x_4}, \frac{[0.95,0.05]}{x_5}, \frac{[0.95,0.05]}{x_6}, \frac{[0.89,0.1]}{x_7}, \frac{[0.89,0.1]}{x_8}, \frac{[0.95,0.05]}{x_9} \right\}.$$

Then, from Figure 3, we can obtain,

$$\mu(y)_{O\Pi3} \geq \mu(y)_{G\Pi f3}\Theta\, \mu(y)_{G\Pi n3}\Theta\, \mu(y)_{G\Pi s3} \geq \mu(y)_{P\Pi3}, \nu(y)_{P\Pi3} \geq \nu(y)_{G\Pi s3}\Theta\, \nu(y)_{O\Pi3}\Theta\, \nu(y)_{G\Pi f3}\Theta\, \nu(y)_{G\Pi n3};$$

Note:

Θ represents $\leq$ or $\geq$;
$\mu(y)_{G\Pi f3}$ and $\nu(y)_{G\Pi f3}$ represent GRIFS type-II dominance relation (fever);
$\mu(y)_{G\Pi s3}$ and $\nu(y)_{G\Pi s3}$ represent GRIFS type-II dominance relation (salivation);
$\mu(y)_{G\Pi n3}$ and $\nu(y)_{G\Pi n3}$ represent GRIFS type-II dominance relation (streaming nose);
$\mu(y)_{O\Pi3}$ and $\nu(y)_{O\Pi3}$ represent optimistic multi-granulation GRIFS type-II dominance relation;
$\mu(y)_{P\Pi3}$ and $\nu(y)_{P\Pi3}$ represent pessimistic multi-granulation GRIFS type-II dominance relation;

From Figure 3, we can see that x_1, x_2, x_4 patients have the disease, and $x_3, x_5, x_6, x_7, x_8, x_9$ patients do not have the disease.

Then, from Figure 4, we can obtain,

$$\mu(y)_{O\Pi4} \geq \mu(y)_{G\Pi n4}\Theta\, \mu(y)_{G\Pi f4}\Theta\, \mu(y)_{G\Pi s4} \geq \mu(y)_{P\Pi4}, \nu(y)_{P\Pi4}\Theta\, \nu(y)_{G\Pi s4}\Theta\, \nu(y)_{G\Pi f4}\Theta\, \nu(y)_{G\Pi n4}\Theta\, \nu(y)_{O\Pi4};$$

Note:

Θ represents $\leq$ or $\geq$;
$\mu(y)_{G\Pi f4}$ and $\nu(y)_{G\Pi f4}$ represent GRIFS type-II dominance relation (fever);
$\mu(y)_{G\Pi s4}$ and $\nu(y)_{G\Pi s4}$ represent GRIFS type-II dominance relation (salivation);
$\mu(y)_{G\Pi n4}$ and $\nu(y)_{G\Pi n4}$ represent GRIFS type-II dominance relation (streaming nose);
$\mu(y)_{O\Pi4}$ and $\nu(y)_{O\Pi4}$ represent optimistic multi-granulation GRIFS type-II dominance relation;
$\mu(y)_{P\Pi4}$ and $\nu(y)_{P\Pi4}$ represent pessimistic multi-granulation GRIFS type-II dominance relation;

From Figure 4, we can see that $x_1, x_2, x_3, x_4, x_5, x_6, x_7, x_8$ and x_9 patients have the disease.

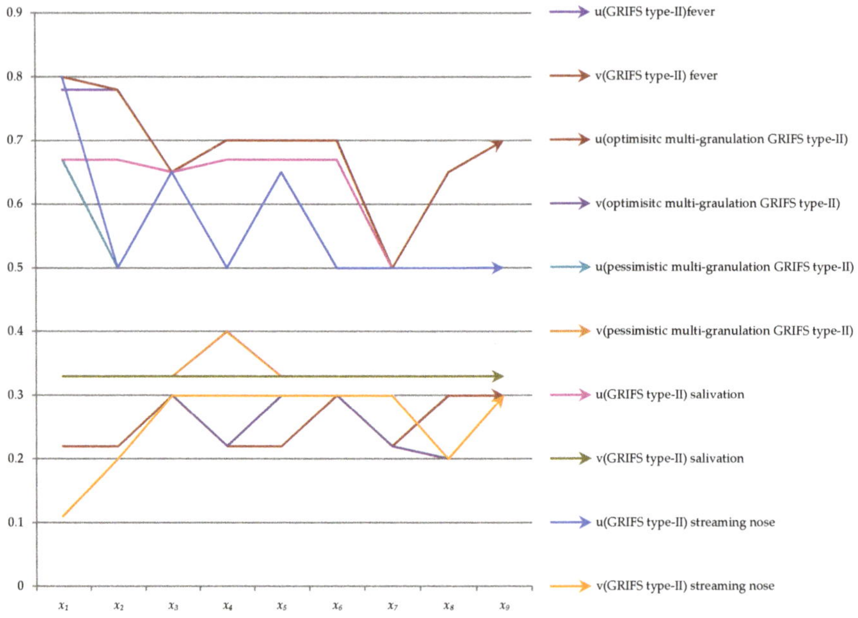

Figure 3. The lower approximation of GRIFS based on type-II dominance relation, as well as optimistic and pessimistic multi-granulation GRIFS based on type-II dominance relation.

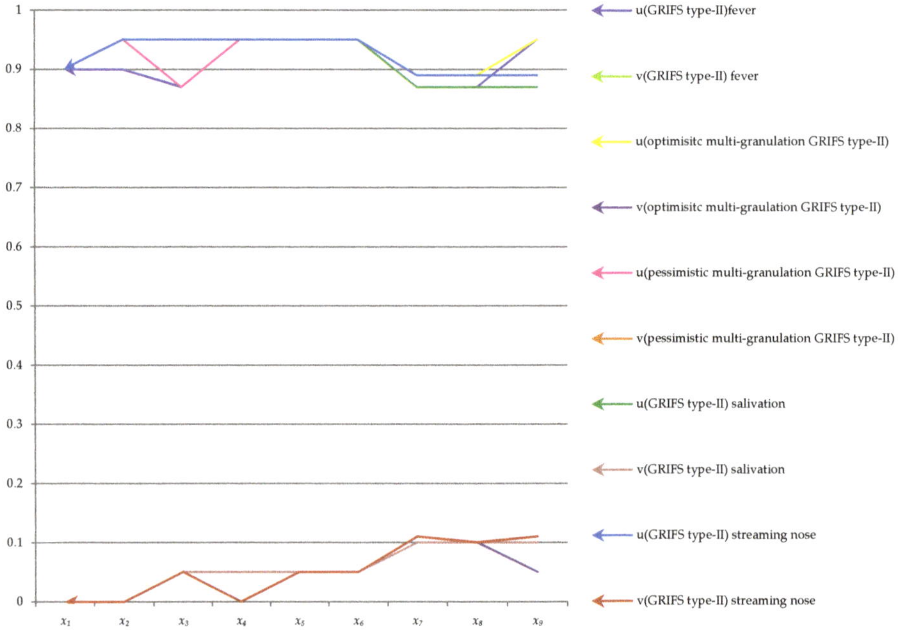

Figure 4. The upper approximation of GRIFS based on type-II dominance relation, as well as optimistic and pessimistic multi-granulation GRIFS based on type-II dominance relation.

From Figures 1 and 2, x_1, x_2, x_4 and x_8 patients have the disease, x_6 and x_7 patient do not have the disease. From Figures 3 and 4, x_1, x_2 and x_4 patients have the disease, x_3, x_5, x_6, x_7, x_8 and x_9 patients do not have the disease. Furthermore, this example proves the accuracy of Algorithm 1.

This example analyzes and discusses multi-granulation GRIFS models based on dominance relation. From conjunction and disjunction operations of two kinds of dominance classes perspective, we analyzed GRIFS models based on type-I dominance relation and type-II dominance relation and also optimistic and pessimistic multi-granulation GRIFS models based on type-I dominance relation and type-II dominance relation, respectively. Through the analysis of this example, the validity of these multi-granulation GRIFS models based on type-I dominance relation and type-II dominance relation models can be obtained.

6. Conclusions

These theories of GRS and RIFS are extensions of the classical rough set theory. In this paper, we proposed a series of models on GRIFS based on dominance relation, which were based on the combination of GRS, RIFS, and dominance relations. Moreover, these models of multi-granulation GRIFS models based on dominance relation were established on GRIFS models based on dominance relation using multiple dominance relations on the universe. The validity of these models was demonstrated by giving examples. Compared with GRS based on dominance relation, GRIFS models based on dominance relation can be more precise. Compared with GRIFS models based on dominance relation, multi-granulation GRIFS models based on dominance relation can be more accurate. It can be demonstrated using the algorithm, and our methods provide a way to combine GRS and RIFS. Our next work is to study the combination of GRS and variable precision rough sets on the basis of our proposed methods.

Author Contributions: Z.-a.X. and M.-j.L. initiated the research and wrote the paper, D.-j.H. participated in some of the search work, and X.-w.X. supervised the research work and provided helpful suggestions.

Funding: This research received no external funding.

Acknowledgments: This work is supported by the national natural science foundation of China under Grant Nos. 61772176, 61402153, and the scientific and technological project of Henan Province of China under Grant Nos. 182102210078, 182102210362, and the Plan for Scientific Innovation of Henan Province of China under Grant No. 18410051003, and the key scientific and technological project of Xinxiang City of China under Grant No. CXGG17002.

Conflicts of Interest: The authors declare no conflicts of interest.

References

1. Pawlak, Z. Rough sets. *Int. J. Comput. Inf. Sci.* **1982**, *11*, 341–356. [CrossRef]
2. Yao, Y.Y.; Lin, T.Y. Generalization of rough sets using modal logics. *Intell. Autom. Soft Comput.* **1996**, *2*, 103–119. [CrossRef]
3. Zhang, X.Y.; Mo, Z.W.; Xiong, F.; Cheng, W. Comparative study of variable precision rough set model and graded rough set model. *Int. J. Approx. Reason.* **2012**, *53*, 104–116. [CrossRef]
4. Zadeh, L.A. *Fuzzy Sets, Fuzzy Logic, and Fuzzy Systems*; World Scientific Publishing Corporation: Singapore, 1996; pp. 433–448.
5. Qian, Y.H.; Liang, J.Y.; Yao, Y.Y.; Dang, C.Y. MGRS: A multi-granulation rough set. *Inf. Sci.* **2010**, *180*, 949–970. [CrossRef]
6. Xu, W.H.; Wang, Q.R.; Zhang, X.T. Multi-granulation fuzzy rough sets in a fuzzy tolerance approximation space. *Int. J. Fuzzy Syst.* **2011**, *13*, 246–259.
7. Lin, G.P.; Liang, J.Y.; Qian, Y.H. Multigranulation rough sets: From partition to covering. *Inf. Sci.* **2013**, *241*, 101–118. [CrossRef]
8. Liu, C.H.; Pedrycz, W. Covering-based multi-granulation fuzzy rough sets. *J. Intell. Fuzzy Syst.* **2015**, *30*, 303–318. [CrossRef]

9. Xue, Z.A.; Si, X.M.; Xue, T.Y.; Xin, X.W.; Yuan, Y.L. Multi-granulation covering rough intuitionistic fuzzy sets. *J. Intell. Fuzzy Syst.* **2017**, *32*, 899–911.
10. Hu, Q.J. Extended Graded Rough Sets Models Based on Covering. Master's Thesis, Shanxi Normal University, Linfen, China, 21 March 2016. (In Chinese)
11. Wang, H.; Hu, Q.J. Multi-granulation graded covering rough sets. In Proceedings of the International Conference on Machine Learning and Cybernetics, Guangzhou, China, 12–15 July 2015; Institute of Electrical and Electronics Engineers Computer Society: New York, NY, USA, 2015.
12. Wu, Z.Y.; Zhong, P.H.; Hu, J.G. Graded multi-granulation rough sets. *Fuzzy Syst. Math.* **2014**, *28*, 165–172. (In Chinese)
13. Yu, J.H.; Zhang, X.Y.; Zhao, Z.H.; Xu, W.H. Uncertainty measures in multi-granulation with different grades rough set based on dominance relation. *J. Intell. Fuzzy Syst.* **2016**, *31*, 1133–1144. [CrossRef]
14. Yu, J.H.; Xu, W.H. Multigranulation with different grades rough set in ordered information system. In Proceedings of the International Conference on Fuzzy Systems and Knowledge Discovery, Zhangjiajie, China, 15–17 August 2015; Institute of Electrical and Electronics Engineers Incorporated: New York, NY, USA, 2016.
15. Wang, X.Y.; Shen, J.Y.; Shen, J.L.; Shen, Y.X. Graded multi-granulation rough set based on weighting granulations and dominance relation. *J. Shandong Univ.* **2017**, *52*, 97–104. (In Chinese)
16. Zheng, Y. Graded multi-granularity rough sets based on covering. *J. Shanxi Normal Univ.* **2017**, *1*, 5–9. (In Chinese)
17. Shen, J.R.; Wang, X.Y.; Shen, Y.X. Variable grade multi-granulation rough set. *J. Chin. Comput. Syst.* **2016**, *37*, 1012–1016. (In Chinese)
18. Atanassov, K.T.; Rangasamy, P. Intuitionistic fuzzy sets. *Fuzzy Sets Syst.* **1986**, *20*, 87–96. [CrossRef]
19. Zadeh, L.A. Fuzzy sets. *Inf. Control* **1965**, *8*, 338–353. [CrossRef]
20. Huang, B.; Zhuang, Y.L.; Li, H.X. Information granulation and uncertainty measures in interval-valued intuitionistic fuzzy information systems. *Eur. J. Oper. Res.* **2013**, *231*, 162–170. [CrossRef]
21. Slowinski, R.; Vanderpooten, D. A generalized definition of rough approximations based on similarity. *IEEE Trans. Knowl. Data Eng.* **1996**, *12*, 331–336. [CrossRef]
22. Chang, K.H.; Cheng, C.H. A risk assessment methodology using intuitionistic fuzzy set in FMEA. *Int. J. Syst. Sci.* **2010**, *41*, 1457–1471. [CrossRef]
23. Zhang, X.H. Fuzzy anti-grouped filters and fuzzy normal filters in pseudo-BCI algebras. *J. Intell. Fuzzy Syst.* **2017**, *33*, 1767–1774. [CrossRef]
24. Gong, Z.T.; Zhang, X.X. The further investigation of variable precision intuitionistic fuzzy rough set model. *Int. J. Mach. Learn. Cybern.* **2016**, *8*, 1565–1584. [CrossRef]
25. He, Y.P.; Xiong, L.L. Generalized inter-valued intuitionistic fuzzy soft rough set and its application. *J. Comput. Anal. Appl.* **2017**, *23*, 1070–1088.
26. Zhang, X.H.; Bo, C.X.; Smarandache, F.; Dai, J.H. New inclusion relation of neutrosophic sets with applications and related lattice structure. *Int. J. Mach. Learn. Cybern.* **2018**, *9*, 1753–1763. [CrossRef]
27. Zhang, X.H.; Smarandache, F.; Liang, X.L. Neutrosophic duplet semi-group and cancellable neutrosophic triplet groups. *Symmetry* **2017**, *9*, 275. [CrossRef]
28. Huang, B.; Guo, C.X.; Li, H.X.; Feng, G.F.; Zhou, X.Z. An intuitionistic fuzzy graded covering rough set. *Knowl.-Based Syst.* **2016**, *107*, 155–178. [CrossRef]
29. Tiwari, A.K.; Shreevastava, S.; Som, T. Tolerance-based intuitionistic fuzzy-rough set approach for attribute reduction. *Expert Syst. Appl.* **2018**, *101*, 205–212. [CrossRef]
30. Guo, Q.; Ming, Y.; Wu, L. Dominance relation and reduction in intuitionistic fuzzy information system. *Syst. Eng. Electron.* **2014**, *36*, 2239–2243.
31. Ai, A.H.; Xu, Z.S. Line integral of intuitionistic fuzzy calculus and their properties. *IEEE Trans. Fuzzy Syst.* **2018**, *26*, 1435–1446. [CrossRef]
32. Rizvi, S.; Naqvi, H.J.; Nadeem, D. Rough intuitionistic fuzzy sets. In Proceedings of the 6th Joint Conference on Information Sciences, Research Triangle Park, NC, USA, 8–13 March 2002; Duke University/Association for Intelligent Machinery: Durham, NC, USA, 2002.
33. Zhang, X.X.; Chen, D.G.; Tsang, E.C.C. Generalized dominance rough set models for the dominance intuitionistic fuzzy information systems. *Inf. Sci.* **2017**, *378*, 1339–1351. [CrossRef]

34. Zhang, Y.Q.; Yang, X.B. An intuitionistic fuzzy dominance-based rough set. In Proceedings of the 7th International Conference on Intelligent Computing, Zhengzhou, China, 11–14 August 2011; Springer: Berlin, Germany, 2011.
35. Huang, B.; Zhuang, Y.L.; Li, H.X.; Wei, D.K. A dominance intuitionistic fuzzy-rough set approach and its applications. *Appl. Math. Model.* **2013**, *37*, 7128–7141. [CrossRef]
36. Zhang, W.X.; Wu, W.Z.; Liang, J.Y.; Li, D.Y. *Theory and Method of Rough Sets*; Science Press: Beijing, China, 2001. (In Chinese)
37. Wen, X.J. Uncertainty measurement for intuitionistic fuzzy ordered information system. Master's Thesis, Shanxi Normal University, Linfen, China, 21 March 2015. (In Chinese)
38. Lezanski, P.; Pilacinska, M. The dominance-based rough set approach to cylindrical plunge grinding process diagnosis. *J. Intell. Manuf.* **2018**, *29*, 989–1004. [CrossRef]
39. Huang, B.; Guo, C.X.; Zhang, Y.L.; Li, H.X.; Zhou, X.Z. Intuitionistic fuzzy multi-granulation rough sets. *Inf. Sci.* **2014**, *277*, 299–320. [CrossRef]
40. Greco, S.; Matarazzo, B.; Slowinski, R. An algorithm for induction decision rules consistent with the dominance principle. In Proceedings of the 2nd International Conference on Rough Sets and Current Trends in Computing, Banff, AB, Canada, 16–19 October 2000; Springer: Berlin, Germany, 2011.

symmetry

MDPI

Article

Four Operators of Rough Sets Generalized to Matroids and a Matroidal Method for Attribute Reduction

Jingqian Wang [1] and Xiaohong Zhang [2,*]

[1] College of Electrical & Information Engineering, Shaanxi University of Science & Technology, Xi'an 710021, China; wangjingqianw@163.com
[2] School of Arts and Sciences, Shaanxi University of Science & Technology, Xi'an 710021, China
* Correspondence: zhangxiaohong@sust.edu.cn or zxhonghz@263.net

Received: 8 August 2018; Accepted: 14 September 2018; Published: 19 September 2018

Abstract: Rough sets provide a useful tool for data preprocessing during data mining. However, many algorithms related to some problems in rough sets, such as attribute reduction, are greedy ones. Matroids propose a good platform for greedy algorithms. Therefore, it is important to study the combination between rough sets and matroids. In this paper, we investigate rough sets and matroids through their operators, and provide a matroidal method for attribute reduction in information systems. Firstly, we generalize four operators of rough sets to four operators of matroids through the interior, closure, exterior and boundary axioms, respectively. Thus, there are four matroids induced by these four operators of rough sets. Then, we find that these four matroids are the same one, which implies the relationship about operators between rough sets and matroids. Secondly, a relationship about operations between matroids and rough sets is presented according to the induced matroid. Finally, the girth function of matroids is used to compute attribute reduction in information systems.

Keywords: rough set; matroid; operator; attribute reduction

1. Introduction

Rough set theory was proposed by Pawlak [1,2] in 1982 as a mathematical tool to deal with various types of data in data mining. There are many practical problems have been solved by it, such as rule extraction [3,4], attribute reduction [5–7], feature selection [8–10] and knowledge discovery [11]. In Pawlak's rough sets, the relationships of objects are equivalence relations. However, it is well known that this requirement is excessive in practice [12,13]. Hence, Pawlak's rough sets have been extended by relations [14,15], coverings [16–18] and neighborhoods [6,19]. They have been combined with other theories including topology [20], lattice theory [21,22], graph theory [23,24] and fuzzy set theory [25,26].

However, many optimization issues related to rough sets, including attribute reduction, are NP-hard. Therefore, the algorithms to deal with them are often greedy ones [27]. Matroid theory [28–30] is a generalization of graph and linear algebra theories. It has been used in information coding [31] and cryptology [32]. Recently, the combination between rough sets and matroids has attracted many interesting research. For example, Zhu and Wang [33] established a matroidal structure through the upper approximation number and studied generalized rough sets with matroidal approaches. Liu and Zhu [34] established a parametric matroid through the lower approximation operator of rough sets. Li et al. [35,36] used matroidal approaches to investigate rough sets through closure operators. Su and Zhu [37] presented three types of matroidal structures of covering-based rough sets. Wang et al. [38] induced a matroid named 2-circuit matroid by equivalence relations, and equivalently formulated attribute reduction with matroidal approaches. Wang and Zhu used matrix approaches to study the 2-circuit matroid [39], and used contraction operation in matroids to study some relationships between a subset and the upper approximation of this subset in rough sets [40]. Unfortunately, all of these papers never study matroids and rough sets through the positive, negative and boundary operators of rough sets. Thus, it is necessary to further study rough sets and matroids by these operators in this paper. In addition, only Wang et al. [38] presented two equivalent descriptions of attribute reduction by closure operators and rank functions of matroids, respectively. We consider presenting a novel approach to attribute reduction through the girth function of matroids in this paper.

In this paper, we mainly use the positive operator, the negative operator and the boundary operator to study matroids and rough sets, and propose a method to compute attribute reduction in information systems through the girth function of matroids. Firstly, we generalize the positive (the lower approximation operator), upper approximation, negative and boundary operators of rough sets to the interior, closure, exterior and boundary operators of matroids respectively. Among them, the upper and lower approximation operators have been studied in [35]. Thus, there are four matroids induced by these four operators of rough sets. Then, the relationship between these four matroids is studied, which implies the relationship about operators between rough sets and matroids. In fact, these four matroids are the same one. Secondly, a relationship about the restriction operation both in matroids and rough sets is proposed. Finally, a matroidal approach is proposed to compute attribute reduction in information systems through the girth function of matroids, and an example about attribute reduction is solved. Using this matroidal approach, we can compute attribute reduction through their results "2" and "∞".

The rest of this paper is organized as follows. Section 2 recalls some basic notions about rough sets, information systems and matroids. In Section 3, we generalize four operators of rough sets to four operators of matroids, respectively. In addition, we study the relationship between four matroids induced by these four operators of rough sets. Moreover, a relationship about operations between matroids and rough sets is presented. In Section 4, an equivalent formulation of attribute reduction through the girth function is presented. Based on the equivalent formulation, a novel method is proposed to compute attribute reduction in information systems. Finally, Section 5 concludes this paper and indicates further works.

2. Basic Definitions

In this section, we review some notions in Pawlak's rough sets, information systems and matroids.

2.1. Pawlak's Rough Sets and Information Systems

The definition of approximation operators is presented in [1,41].

Let R an equivalence relation on U. For any $X \subseteq U$, a pair of approximation $\overline{R}(X)$ and $\underline{R}(X)$ of X are defined by

$$\overline{R}(X) = \{x \in U : RN(x) \bigcap X \neq \emptyset\},$$
$$\underline{R}(X) = \{x \in U : RN(x) \subseteq X\},$$

where $RN(x) = \{y \in U : xRy\}$. $\overline{R}$ and $\underline{R}$ are called the upper and lower approximation operators with respect to R, respectively.

In this paper, U is a nonempty and finite set called universe. Let $-X$ be the complement of X in U and $\emptyset$ be the empty set. We have the following conclusions about $\overline{R}$ and $\underline{R}$.

Proposition 1. *Refs. [1,41] Let R be an equivalence relation on U. For any $X, Y \subseteq U$,*

(1L) $\underline{R}(U) = U$, (1H) $\overline{R}(U) = U$,
(2L) $\underline{R}(\phi) = \phi$, (2H) $\overline{R}(\phi) = \phi$,
(3L) $\underline{R}(X) \subseteq X$, (3H) $X \subseteq \overline{R}(X)$,
(4L) $\underline{R}(X \bigcap Y) = \underline{R}(X) \bigcap \underline{R}(Y)$, (4H) $\overline{R}(X \bigcup Y) = \overline{R}(X) \bigcup \overline{R}(Y)$,
(5L) $\underline{R}(\underline{R}(X)) = \underline{R}(X)$, (5H) $\overline{R}(\overline{R}(X)) = \overline{R}(X)$,
(6L) $X \subseteq Y \Rightarrow \underline{R}(X) \subseteq \underline{R}(Y)$, (6H) $X \subseteq Y \Rightarrow \overline{R}(X) \subseteq \overline{R}(Y)$,
(7L) $\underline{R}(-\underline{R}(X)) = -\underline{R}(X)$, (7H) $\overline{R}(-\overline{R}(X)) = -\overline{R}(X)$,
(8LH) $\underline{R}(-X) = -\overline{R}(X)$, (9LH) $\underline{R}(X) \subseteq \overline{R}(X)$.

On the basis of the upper and lower approximation operators with respect to R, one can define three operators to divide the universe, namely, the negative operator NEG_R, the positive operator POS_R and the boundary operator BND_R:

$$NEG_R(X) = U - \overline{R}(X),$$
$$POS_R(X) = \underline{R}(X),$$
$$BND_R(X) = \overline{R}(X) - \underline{R}(X).$$

An information system [38] is an ordered pair $IS = (U, A)$, where U is a nonempty finite set of objects and A is a nonempty finite set of attributes such that $a : U \rightarrow V_a$ for any $a \in A$, where V_a is called the value set of a. For all $B \subseteq A$, the indiscernibility relation induced by B is defined as follows:

$$IND(B) = \{(x, y) \in U \times U : \forall b \in B, b(x) = b(y)\}.$$

Definition 1. *(Reduct [38]) Let $IS = (U, A)$ be an information system. For all $B \subseteq A$, B is called a reduct of IS, if the following two conditions hold:*

(1) $IND(B) \neq IND(B - b)$ for any $b \in B$,
(2) $IND(B) = IND(A)$.

2.2. Matroids

Definition 2. *(Matroid [29,30]) Let U is a finite set, and $\mathbf{I}$ is a nonempty subset of 2^U (the set of all subsets of U). $(U, \mathbf{I})$ is called a matroid, if the following conditions hold:*
(I1) If $I \in \mathbf{I}$ and $I' \subseteq I$, then $I' \in \mathbf{I}$.
(I2) If $I_1, I_2 \in \mathbf{I}$ and $|I_1| < |I_2|$, then there exists $e \in I_2 - I_1$ such that $I_1 \bigcup \{e\} \in \mathbf{I}$, where $|I|$ denotes the cardinality of I.

Let $M = (U, \mathbf{I})$ be a matroid. We shall often write $U(M)$ for U and $\mathbf{I}(M)$ for $\mathbf{I}$, particularly when several matroids are being considered. The members of $\mathbf{I}$ are the independent sets of M.

Example 1. *Let $U = \{a_1, a_2, a_3, a_4, a_5\}$ and $\mathbf{I} = \{\emptyset, \{a_1\}, \{a_2\}, \{a_3\}, \{a_4\}, \{a_5\}, \{a_1, a_3\}, \{a_1, a_4\}, \{a_1, a_5\}, \{a_2, a_3\}, \{a_2, a_4\}, \{a_2, a_5\}, \{a_3, a_4\}, \{a_3, a_5\}, \{a_4, a_5\}, \{a_1, a_3, a_4\}, \{a_1, a_3, a_5\}, \{a_1, a_4, a_5\}, \{a_2, a_3, a_4\}, \{a_2, a_3, a_5\}, \{a_2, a_4, a_5\}\}$. Then, $M = (U, \mathbf{I})$ is a matroid.*

In order to make some expressions brief, some denotations are presented. Let $\mathbf{A} \subseteq 2^U$. Then,

$$Min(\mathbf{A}) = \{X \in \mathbf{A} : \forall Y \in \mathbf{A}, Y \subseteq X \Rightarrow X = Y\},$$
$$Max(\mathbf{A}) = \{X \in \mathbf{A} : \forall Y \in \mathbf{A}, X \subseteq Y \Rightarrow X = Y\},$$
$$Opp(\mathbf{A}) = \{X \subseteq U : X \notin \mathbf{A}\}.$$

The set of all circuits of M is defined as $\mathbf{C}(M) = Min(Opp(\mathbf{I}))$. The rank function r_M of M is denoted by $r_M(X) = max\{|I| : I \subseteq X, I \in \mathbf{I}\}$ for any $X \subseteq U$. $r_M(X)$ is called the rank of X in M. The closure operator cl_M of M is defined as

$$cl_M(X) = \{u \in U : r_M(X) = r_M(X \bigcup \{u\})\} \text{ for all } X \subseteq U.$$

We call $cl_M(X)$ the closure of X in M. X is called a closed set if $cl_M(X) = X$, and we denote the family of all closed sets of M by $\mathbf{F}(M)$. The closure axiom of a matroid is introduced in the following proposition.

Proposition 2. *(Closure axiom [29,30]) Let cl be an operator of U. Then, there exists one and only one matroid M such that $cl = cl_M$ iff cl satisfies the following four conditions:*
$(CL1)$ $X \subseteq cl(X)$ for any $X \subseteq U$;
$(CL2)$ If $X \subseteq Y \subseteq U$, then $cl(X) \subseteq cl(Y)$;
$(CL3)$ $cl(cl(X)) = cl(X)$ for any $X \subseteq U$;
$(CL4)$ For any $x, y \in U$, if $y \in cl(X \bigcup \{x\}) - cl(X)$, then $x \in cl(X \bigcup \{y\})$.

Example 2. *(Continued from Example 1) Let $X = \{a_3, a_4\}$. Then,*
$C(M) = Min(Opp(I)) = \{\{a_1, a_2\}, \{a_3, a_4, a_5\}\}$,
$r_M(X) = max\{|I| : I \subseteq X, I \in I\} = 2$,
$cl_M(X) = \{u \in U : r_M(X) = r_M(X \bigcup \{u\})\} = \{a_3, a_4, a_5\}$,
$F(M) = \{\emptyset, \{a_3\}, \{a_4\}, \{a_5\}, \{a_1, a_2\}, \{a_1, a_2, a_3\}, \{a_1, a_2, a_4\}, \{a_1, a_2, a_5\}, \{a_3, a_4, a_5\}, \{a_1, a_2, a_3, a_4, a_5\}\}$.

Based on $\mathbf{F}(M)$, the interior operator int_M of M is defined as

$$int_M(X) = \bigcup\{Y \subseteq X : U - Y \in \mathbf{F}(M)\} \text{ for any } X \subseteq U.$$

$int_M(X)$ is called the interior of X in M. X is called a open set if $int_M(X) = X$. The following proposition shows the interior axiom of a matroid.

Proposition 3. *(Interior axiom [29,30]) Let int be an operator of U. Then, there exists one and only one matroid M such that $int = int_M$ iff int satisfies the following four conditions:*
$(INT1)$ $int(X) \subseteq X$ for any $X \subseteq U$,
$(INT2)$ If $X \subseteq Y \subseteq U$, then $int(X) \subseteq int(Y)$,
$(INT3)$ $int(int(X)) = int(X)$ for any $X \subseteq U$,
$(INT4)$ For any $x, y \in U$, if $y \in int(X) - int(X - \{x\})$, then $x \notin int(X - \{y\})$.

Example 3. *(Continued from Example 2) $int_M(X) = \bigcup\{Y \subseteq X : U - Y \in F(M)\} = \{a_3, a_4\}$.*

Based on the closure operator cl_M, the exterior operator ex_M and the boundary operator bo_M of M are defined as

$$ex_M(X) = -cl_M(X) \text{ and } bo_M(X) = cl_M(X) \bigcap cl_M(-X) \text{ for all } X \subseteq U.$$

$ex_M(X)$ is called the exterior of X in M, and $bo_M(X)$ is called the boundary of X in M. The following two propositions present the exterior and boundary axioms, respectively.

Proposition 4. *(Exterior axiom [42]) Let ex be an operator of U. Then, there exists one and only one matroid M such that $ex = ex_M$ iff ex_M satisfies the following four conditions:*
$(EX1)$ $X \bigcap ex(X) = \emptyset$ for any $X \subseteq U$;
$(EX2)$ If $X \subseteq Y \subseteq U$, then $ex(Y) \subseteq ex(X)$;
$(EX3)$ $ex(-ex(X)) = ex(X)$ for any $X \subseteq U$;
$(EX4)$ For any $x, y \in U$, if $y \in ex(X) - ex(X \bigcup \{x\})$, then $x \notin ex(X \bigcup \{y\})$.

Proposition 5. *(Boundary axiom [42]) Let bo be an operator of U. Then, there exists one and only one matroid M such that $bo = bo_M$ iff bo satisfies the following five conditions:*
$(BO1)$ $bo(X) = bo(-X)$ *for any* $X \subseteq U$;
$(BO2)$ $bo(bo(X)) \subseteq bo(X)$ *for any* $X \subseteq U$;
$(BO3)$ $X \bigcap Y \bigcap (bo(X) \bigcup bo(Y)) \subseteq X \bigcap Y \bigcap bo(X \bigcap Y)$ *for any* $X, Y \subseteq U$;
$(BO4)$ *For any* $x, y \in U$, *if* $y \in bo(X \bigcup \{x\}) - bo(X)$, *then* $x \in bo(X \bigcup \{y\})$;
$(BO5)$ $bo(X \bigcup bo(X)) \subseteq bo(X)$ *for any* $X \subseteq U$.

Example 4. *(Continued from Example 2)* $ex_M(X) = U - \{a_3, a_4, a_5\} = \{a_1, a_2\}$, $bo_M(X) = cl_M(X) \bigcap cl_M(-X) = \{a_3, a_4, a_5\} \bigcap \{a_1, a_2, a_5\} = \{a_5\}$.

The following proposition shows some relationships between these above four operators, namely cl_M, int_M, ex_M and bo_M.

Proposition 6. *Ref. [42] Let $M = (U, \mathbf{I})$ be a matroid. For all $X \subseteq U$, the following statements hold:*
(1) $int_M(X) = -cl_M(-X)$ *and* $cl_M(X) = -int_M(-X)$;
(2) $cl_M(bo_M(X)) = bo_M(X)$;
(3) $bo_M(ex_M(X)) = bo_M(-X)$.

3. The Relationship about Operators between Rough Sets and Matroids

In this section, four matroids are induced by four operators of rough sets. These four matroids are induced by the lower approximation operator $\underline{R}$ (because $\underline{R} = POS_R$, we only consider $\underline{R}$), the upper approximation operator $\overline{R}$, the negative operator NEG_R and the boundary operator BND_R through the interior axiom, the closure axiom, the exterior axiom and the boundary axiom, respectively. Among them, the upper approximation operator $\overline{R}$ has been studied in [35]. Then, the relationship between these four matroids are studied, and we find that these four are the same one. According to this work, we present the relationship about operators between rough sets and matroids.

3.1. Four Matroids Induced by Four Operators of Rough Sets

In this subsection, we generalize the positive operator (the lower approximation operator), the upper approximation operator, the negative operator and the boundary operator of rough sets to the interior operator, the closure operator, the exterior operator and the boundary operator of matroids, respectively. Firstly, the following lemma is proposed.

Lemma 1. *Refs. [1,41] Let R be an equivalence relation on U. For any $x, y \in U$, if $x \in RN(y)$, then $y \in RN(x)$.*

The following proposition shows that the lower approximation operator $\underline{R}$ satisfies the interior axiom of matroids.

Proposition 7. *Let R be an equivalence relation on U. Then, $\underline{R}$ satisfies $(INT1)$, $(INT2)$, $(INT3)$ and $(INT4)$ of Proposition 3.*

Proof. By $(1L)$, $(6L)$ and $(5L)$ of Proposition 1, $\underline{R}$ satisfies $(INT1)$, $(INT2)$ and $(INT3)$, respectively. $(INT4)$: For any $x, y \in U$, if $y \in \underline{R}(X) - \underline{R}(X - \{x\})$, then $y \in \underline{R}(X)$ but $y \notin \underline{R}(X - \{x\})$. Hence, $RN(y) \subseteq X$ but $RN(y) \nsubseteq X - \{x\}$. Therefore, $x \in RN(y)$. According to Lemma 1, $y \in RN(x)$. Hence, $RN(x) \nsubseteq X - \{y\}$, i.e., $x \notin \underline{R}(X - \{y\})$. □

Inspired by Proposition 7, there is a matroid such that $\underline{R}$ is its interior operator.

Definition 3. *Let R be an equivalence relation on U. The matroid whose interior operator is* $\underline{R}$ *is denoted by* $M(\underline{R})$. *We say* $M(\underline{R})$ *is the matroid induced by* $\underline{R}$.

Corollary 1. *Let R be an equivalence relation on U. Then,* $int_{M(\underline{R})} = POS_R$.

Proof. According to Definition 3, $int_{M(\underline{R})} = \underline{R}$. Since $POS_R = \underline{R}$, so $int_{M(\underline{R})} = POS_R$. □

The upper approximation operator $\overline{R}$ satisfies the closure axiom in [35,38].

Proposition 8. *Refs. [35,38] Let R be an equivalence relation on U. Then,* $\overline{R}$ *satisfies* $(CL1)$, $(CL2)$, $(CL3)$ *and* $(CL4)$ *of Proposition 2.*

Proposition 8 determines the second matroid induced by $\overline{R}$.

Definition 4. *Let R be an equivalence relation on U. The matroid whose closure operator is* $\overline{R}$ *is denoted by* $M(\overline{R})$. *We say that* $M(\overline{R})$ *is the matroid induced by* $\overline{R}$.

The negative operator NEG_R satisfies the exterior axiom.

Proposition 9. *Let R be an equivalence relation on U. Then,* NEG_R *satisfies* $(EX1)$, $(EX2)$, $(EX3)$ *and* $(EX4)$ *of Proposition 4.*

Proof. $(EX1)$: For any $X \subseteq U$, $NEG_R(X) = U - \overline{R}(X)$. According to $(3H)$ of Proposition 1, $X \subseteq \overline{R}(X)$. Therefore, $X \bigcap NEG_R(X) = \emptyset$;
$(EX2)$: According to $(6H)$ of Proposition 1, if $X \subseteq Y \subseteq U$, then $\overline{R}(X) \subseteq \overline{R}(Y)$. Therefore, $U - \overline{R}(Y) \subseteq U - \overline{R}(Y)$, i.e., $NEG_R(Y) \subseteq NEG_R(X)$;
$(EX3)$: For any $X \subseteq U$, $NEG_R(X) = U - \overline{R}(X)$. Hence, $-NEG_R(X) = U - NEG_R(X) = U - (U - \overline{R}(X)) = \overline{R}(X)$. Therefore, $NEG_R(-NEG_R(X)) = NEG_R(\overline{R}(X)) = U - \overline{R}(\overline{R}(X))$. According to $(5H)$ of Proposition 1, $\overline{R}(\overline{R}(X)) = \overline{R}(X)$. Hence, $NEG_R(-NEG_R(X)) = U - \overline{R}(X) = NEG_R(X)$;
$(EX4)$: For any $x, y \in U$, if $y \in NEG_R(X) - NEG_R(X \bigcup \{x\})$, then $y \in NEG_R(X)$ but $y \notin NEG_R(X \bigcup \{x\})$, i.e., $y \in U - \overline{R}(X)$ but $y \notin U - \overline{R}(X \bigcup \{x\})$. Since $\overline{R}(X) \subseteq U$ and $\overline{R}(X \bigcup \{x\}) \subseteq U$, so $y \in \overline{R}(X \bigcup \{x\})$ but $y \notin \overline{R}(X)$. Hence, $RN(y) \bigcap (X \bigcup \{x\}) \neq \emptyset$ but $RN(y) \bigcap X = \emptyset$. Therefore, $RN(y) \bigcap \{x\} \neq \emptyset$, i.e., $x \in RN(y)$. According to Lemma 1, $y \in RN(x)$. Hence, $RN(x) \bigcap (X \bigcup \{y\}) \neq \emptyset$, i.e., $x \in \overline{R}(X \bigcup \{y\})$. Therefore, $x \notin U - \overline{R}(X \bigcup \{y\})$, i.e., $x \notin NEG_R(X \bigcup \{y\})$. □

Proposition 9 determines the third matroid such that NEG_R is its exterior operator.

Definition 5. *Let R be an equivalence relation on U. The matroid whose exterior operator is* NEG_R *is denoted by* $M(NEG_R)$. *We say* $M(NEG_R)$ *is the matroid induced by* NEG_R.

In order to certify the boundary operator BND_R satisfies the boundary axiom, the following two lemmas are proposed.

Lemma 2. *Refs. [1,41] Let R be an equivalence relation on U. For all* $X, Y \subseteq U$, $\overline{R}(X \bigcap Y) \subseteq \overline{R}(X) \bigcap \overline{R}(Y)$.

Lemma 3. *Let R be an equivalence relation on U. If* $X \subseteq Y \subseteq U$, *then* $X \bigcap BND_R(Y) \subseteq BND_R(X)$.

Proof. For any $x \in X \bigcap BND_R(Y)$, $X \bigcap BND_R(Y) = X \bigcap (\overline{R}(X) - \underline{R}(X)) = X \bigcap \overline{R}(X) \bigcap \overline{R}(-X))$. Since $X \subseteq Y \subseteq U$, so $-Y \subseteq -X \subseteq U$. According to $(6H)$ of Proposition 1, $X \bigcap \overline{R}(X) \bigcap \overline{R}(-X)) = X \bigcap \overline{R}(-X) \subseteq \overline{R}(X) \bigcap \overline{R}(-X) = BND_R(X)$. Hence, $x \in BND_R(X)$, i.e., $X \bigcap BND_R(Y) \subseteq BND_R(X)$. □

The boundary operator BND_R satisfies the boundary axiom.

Proposition 10. *Let R be an equivalence relation on U. Then, BND_R satisfies $(BO1)$, $(BO2)$, $(BO3)$, $(BO4)$ and $(BO5)$ of Proposition 5.*

Proof. $(BO1)$: According to $(8LH)$ of Proposition 1, $\underline{R}(-X) = -\overline{R}(X)$. For any $X \subseteq U$,

$$\begin{aligned} BND_R(-X) &= \overline{R}(-X) - \underline{R}(-X) \\ &= \overline{R}(-X) \bigcap (U - \underline{R}(-X)) \\ &= (-\underline{R}(X)) \bigcap \overline{R}(X) \\ &= \overline{R}(X) \bigcap (-\underline{R}(X)) \\ &= \overline{R}(X) - \underline{R}(X) \\ &= BND_R(X). \end{aligned}$$

$(BO2)$: For any $X \subseteq U$,

$$\begin{aligned} BND_R(BND_R(X)) &= \overline{R}(BND_R(X)) - \underline{R}(BND_R(X)) \\ &= \overline{R}(BND_R(X)) \bigcap (U - \underline{R}(BND_R(X))) \\ &= \overline{R}(BND_R(X)) \bigcap (-\underline{R}(BND_R(X))) \\ &= \overline{R}(BND_R(X)) \bigcap (\overline{R}(-BND_R(X))) \\ &\subseteq \overline{R}(BND_R(X)) \\ &= \overline{R}(\overline{R}(X) - \underline{R}(X)) \\ &= \overline{R}(\overline{R}(X) \bigcap (-\underline{R}(X))). \end{aligned}$$

According to Lemma 1, we know

$$\begin{aligned} \overline{R}(\overline{R}(X) \bigcap (-\underline{R}(X))) &\subseteq \overline{R}(\overline{R}(X)) \bigcap \overline{R}(-\underline{R}(X)) \\ &= \overline{R}(X) \bigcap \overline{R}(-\underline{R}(X)) \\ &= \overline{R}(X) - \underline{R}(X) \\ &= BND_R(X). \end{aligned}$$

Hence, $BND_R(BND_R(X)) \subseteq BND_R(X)$;
$(BO3)$: For any $X, Y \subseteq U$, $X \bigcap Y \bigcap (BND_R(X) \bigcup BND_R(Y)) = X \bigcap Y \bigcap ((\overline{R}(X) - \underline{R}(X)) \bigcup (\overline{R}(Y) - \underline{R}(Y))) = X \bigcap Y \bigcap ((\overline{R}(X) \bigcap \overline{R}(-X)) \bigcup (\overline{R}(Y) \bigcap \overline{R}(-Y))) \subseteq X \bigcap Y \bigcap (\overline{R}(-X) \bigcup \overline{R}(-Y))$. According to $(4H)$ of Proposition 1, we know $X \bigcap Y \bigcap (\overline{R}(-X) \bigcup \overline{R}(-Y)) = X \bigcap Y \bigcap \overline{R}((-X) \bigcup (-Y)) = X \bigcap Y \bigcap \overline{R}(-(X \bigcap Y))$. According to $(6H)$ of Proposition 1, we know $X \bigcap Y \subseteq \overline{R}(X \bigcap Y)$. Therefore, $X \bigcap Y \bigcap \overline{R}(-(X \bigcap Y)) = X \bigcap Y \bigcap \overline{R}(-(X \bigcap Y)) \bigcap \overline{R}(X \bigcap Y) = X \bigcap Y \bigcap BND_R(X \bigcap Y)$.
$(BO4)$: When $x = y$ or $x \in X$, it is straightforward. When $y \in X$, it does not hold. (In fact, we suppose $y \in X$. If $y \in BND_R(X \bigcup \{x\})$, according to Lemma 3, we know $y \in X \bigcap BND_R(X \bigcup \{x\}) \subseteq BND_R(X)$,

which is contradictory with $y \in BND_R(X\bigcup\{x\}) - BND_R(X)$. Hence, $y \notin X$.) We only need to prove it for $x \neq y$ and $x, y \notin X$. If $y \in BND_R(X\bigcup\{x\}) - BND_R(X)$, since

$$\begin{aligned}
& BND_R(X\bigcup\{x\}) - BND_R(X) \\
= \; & (\overline{R}(X\bigcup\{x\}) - \underline{R}(X\bigcup\{x\})) - (\overline{R}(X) - \underline{R}(X)) \\
= \; & (\overline{R}(X\bigcup\{x\}) \bigcap \overline{R}(-(X\bigcup\{x\}))) - (\overline{R}(X) \bigcap \overline{R}(-X)) \\
= \; & (\overline{R}(X\bigcup\{x\}) \bigcap \overline{R}(-(X\bigcup\{x\}))) \bigcap ((-\overline{R}(X)) \bigcup (-\overline{R}(-X))) \\
= \; & (\overline{R}(X\bigcup\{x\}) \bigcap \overline{R}(-(X\bigcup\{x\})) \bigcap (-\overline{R}(X))) \bigcup (\overline{R}(X\bigcup\{x\}) \\
& \bigcap (\overline{R}(-(X\bigcup\{x\}))) \bigcap (-\overline{R}(-X))) \\
= \; & \overline{R}(X\bigcup\{x\}) \bigcap \overline{R}(-(X\bigcup\{x\})) \bigcap (-\overline{R}(X)) \\
= \; & \overline{R}(X\bigcup\{x\}) \bigcap (-\overline{R}(X)) \bigcap \overline{R}(-(X\bigcup\{x\})) \\
= \; & (\overline{R}(X\bigcup\{x\}) - \overline{R}(X)) \bigcap \overline{R}(-(X\bigcup\{x\})),
\end{aligned}$$

then $y \in \overline{R}(X\bigcup\{x\}) - \overline{R}(X)$ and $y \in \overline{R}(-(X\bigcup\{x\}))$. According to Proposition 8, we have $x \in \overline{R}(X\bigcup\{y\})$. Since $y \in \overline{R}(-(X\bigcup\{x\}))$, so $x \in \overline{R}(-(X\bigcup\{y\}))$. Hence, $y \in \overline{R}(X\bigcup\{y\}) \bigcap \overline{R}(-(X\bigcup\{y\}))$, i.e., $y \in BND_R(X\bigcup\{y\})$.
$(BO5)$: For any $X, Y \subseteq U$,

$$\begin{aligned}
BND_R(X\bigcup BND_R(X)) & = \overline{R}(X\bigcup BND_R(X)) - \underline{R}(X\bigcup BND_R(X)) \\
& = \overline{R}(X\bigcup BND_R(X)) \bigcap \overline{R}(-(X\bigcup BND_R(X))) \\
& = \overline{R}(X\bigcup BND_R(X)) \bigcap \overline{R}((-X) \bigcap (BND_R(-X))) \\
& \subseteq \overline{R}(X\bigcup BND_R(X)) \bigcap \overline{R}(-X) \\
& = \overline{R}(X\bigcup(\overline{R}(X) \bigcap \overline{R}(-X))) \bigcap \overline{R}(-X) \\
& = \overline{R}(\overline{R}(X) \bigcap U) \bigcap \overline{R}(-X) \\
& = \overline{R}(\overline{R}(X)) \bigcap \overline{R}(-X).
\end{aligned}$$

According to $(5H)$ and $(8LH)$ of Proposition 1, $\overline{R}(\overline{R}(X)) \bigcap \overline{R}(-X) = \overline{R}(X) \bigcap \overline{R}(-X) = \overline{R}(X) - \underline{R}(X) = BND_R(X)$. Therefore, $BND_R(X\bigcup BND_R(X)) \subseteq BND_R(X)$. □

Proposition 8 determines the fourth matroid such that BND_R is its boundary operator.

Definition 6. *Let R be an equivalence relation on U. The matroid whose boundary operator is BND_R is denoted by $M(BND_R)$. We say that $M(BND_R)$ is the matroid induced by BND_R.*

3.2. The Relationship between These Four Matroids

This subsection studies the relationship between these four matroids in the above subsection. In fact, these four matroids are the same one.

Theorem 1. *Let R be an equivalence relation on U. Then,*

$$M(\underline{R}) = M(\overline{R}) = M(NEG_R) = M(BND_R).$$

Proof. (1) On one hand, $M(\underline{R})$ and $M(\overline{R})$ have the same grand U. On the other hand, according to Definition 3, we know $int_{M(\underline{R})}(X) = \underline{R}(X)$ for any $X \subseteq U$. By Proposition 6, $cl_{M(\underline{R})}(X) = -int_{M(\underline{R})}(-X) = -\underline{R}(-X)$. According to $(8LH)$ of Proposition 1, $-\underline{R}(-X) = \overline{R}(X)$. Hence, $cl_{M(\underline{R})}(X) = \overline{R}(X)$. According to Definition 4, $cl_{M(\overline{R})}(X) = \overline{R}(X)$. Therefore, $cl_{M(\underline{R})}(X) = cl_{M(\overline{R})}(X)$, i.e., $M(\underline{R}) = M(\overline{R})$.
(2) On one hand, $M(\overline{R})$ and $M(NEG_R)$ have the same grand U. On the other hand, according to Definition 4, we know $cl_{M(\overline{R})} = \overline{R}$. For any $X \subseteq U$, $ex_{M(\overline{R})}(X) = -cl_{M(\overline{R})}(X) = -\overline{R}(X) = U -$

$\overline{R}(X) = NEG_R(X)$. By Definition 5, $ex_{M(NEG_R)}(X) = NEG_R(X)$. Hence, $ex_{M(\overline{R})}(X) = ex_{M(NEG_R)}(X)$, i.e., $M(\overline{R}) = M(NEG_R)$.
(3) On one hand, $M(\overline{R})$ and $M(NEG_R)$ have the same grand U. On the other hand, according to Definition 4, we have $cl_{M(\overline{R})} = \overline{R}$. For all $X \subseteq U$, $bo_{M(\overline{R})}(X) = cl_{M(\overline{R})}(X) \bigcap cl_{M(\overline{R})}(-X) = \overline{R}(X) \bigcap \overline{R}(-X) = \overline{R}(X) \bigcap \overline{R}(-X) = \overline{R}(X) - \underline{R}(X) = BND_R(X)$. According to Definition 6, $bo_{M(NEG_R)}(X) = BND_R(X)$. Therefore, $bo_{M(\overline{R})}(X) = bo_{M(NEG_R)}(X)$, i.e., $M(\overline{R}) = M(NEG_R)$. □

Definition 7. *Let R be an equivalence relation on U. The matroid whose interior operator, closure operator, exterior operator and boundary operator are $\underline{R}$, $\overline{R}$, NEG_R and BND_R is defined as $M(R)$. We say that $M(R)$ is the matroid induced by R.*

According to the above definition, we have the relationship about operators between rough sets and matroids as Table 1:

Table 1. The relationship about operators between rough sets and matroids.

$M(R)$ Is the Matroid Induced by R
$int_{M(R)} = \underline{R} = POS_R$
$cl_{M(R)} = \overline{R}$
$ex_{M(R)} = NEG_R$
$bo_{M(R)} = BND_R$

3.3. The Relationship about Operations between Matroids and Rough Sets

In this subection, a relationship about the restriction operation both in matroids and rough sets is proposed. First of all, two definitions of these two operations are presented in the following two definitions.

Definition 8. *(Restriction [29,30]) Let $M = (U, \mathbf{I})$ be a matroid. For $X \subseteq U$, the restriction of M to X is defined as $M|X = (X, \mathbf{I}_X)$, where $\mathbf{I}_X = \{I \subseteq X : I \in \mathbf{I}\}$.*

Not that $\mathbf{C}(M|X) = \{C \subseteq X : C \in \mathbf{C}(M)\}$. For an equivalence relation R on U, there is also a definition of restriction of R. For any $X \subseteq U$, $R|X$ is an equivalence relation called the restriction of R to X, where $R|X = \{(x, y) \in X \times X : (x, y) \in R\}$, $X \times X$ is the product set of X and X. According to Definition 7, $M(R|X)$ is a matroid on X.

In [38], the set of independent sets of $M(R)$ is proposed in the following lemma.

Lemma 4. *Ref. [38] Let R be an equivalence relation on U. Then,*

$$\mathbf{I}(M(R)) = \{X \subseteq U : \forall x, y \in X, x \neq y \Rightarrow (x, y) \notin R\}.$$

Example 5. *Let R be an equivalence relation on U with $U = \{a, b, c, d, e\}$, and $U/R = \{\{a, b\}, \{c, d, e\}\}$. According to Lemma 4, $\mathbf{I}(M(R)) = \{\emptyset, \{a\}, \{b\}, \{c\}, \{d\}, \{e\}, \{a, c\}, \{b, c\}, \{a, d\}, \{b, d\}, \{a, e\}, \{b, e\}\}$.*

Proposition 11. *Let R be an equivalence relation on U and $X \subseteq U$. Then, $M(R|X) = M(R)|X$.*

Proof. For any $X \subseteq U$, $R|X$ is an equivalence relation on X. Thus, $M(R|X)$ is a matroid on X. By Definition 8, $M(R)|X$ is a matroid on X. Therefore, we need to prove only $\mathbf{I}(M(R|X)) = \mathbf{I}(M(R)|X)$. According to Lemma 4, $\mathbf{I}(M(R|X)) = \{Y \subseteq X : \forall x, y \in Y, x \neq y \Rightarrow (x, y) \notin R|X\}$, $\mathbf{I}(M(R)|X) = \{Y \subseteq X : \forall x, y \in Y, x \neq y \Rightarrow (x, y) \notin R\}$. On one hand, since $R|X \subseteq R$, $\mathbf{I}(M(R)|X) \subseteq \mathbf{I}(M(R|X))$. On the other hand, suppose $Y \in \mathbf{I}(M(R|X)) - \mathbf{I}(M(R)|X)$. For any $x, y \in Y$, if $x \neq y$, then $(x, y) \notin R|X$ but $(x, y) \in R$. Therefore, $x, y \notin X$ but $x, y \in U$, i.e., $x, y \in U - X$. Hence, $Y \subseteq U - X$, which is contradictory with $Y \subseteq X$, i.e., $Y \in \mathbf{I}(M(R|X)) - \mathbf{I}(M(R)|X)$. Thus, $\mathbf{I}(M(R|X)) \subseteq \mathbf{I}(M(R)|X)$. □

Example 6. *(Continued from Example 5) Let* $X = \{a,b,c\}$. *According to Definition 8,* $I(M(R)|X) = \{\emptyset,\{a\},\{b\},\{c\},\{a,c\},\{b,c\}\}$, *and* $M(R)|X = (X, I(M(R)|X))$. *Since* $R|X = \{(a,a),(b,b),(c,c),(a,b),(b,a)\}$, *so* $X/(R|X) = \{\{a,b\},\{c\}\}$. *According to Lemma 4,* $I(M(R|X)) = \{\emptyset,\{a\},\{b\},\{c\},\{a,c\},\{b,c\}\}$, *and* $M(R|X) = (X, I(M(R|X))$. *Therefore,* $M(R|X) = M(R)|X$.

4. A Matroidal Approach to Attribute Reduction through the Girth Function

In this section, a matroidal approach is proposed to compute attribute reduction in information systems through the girth function of matroids.

4.1. An Equivalent Formulation of Attribute Reduction through the Girth Function

Lemma 5. *Ref. [15] Let* R_1 *and* R_2 *be two equivalence relations on* U*, respectively. Then,* $\underline{R_1} = \underline{R_2}$ *if and only if* $R_1 = R_2$.

Based on Lemma 5, we propose a necessary and sufficient condition for two equivalence relations induce the same matroids.

Proposition 12. *Let* R_1 *and* R_2 *be two equivalence relations on* U*, respectively. Then,* $M(R_1) = M(R_2)$ *if and only if* $R_1 = R_2$.

Proof. According to Definition 7, $M(R_1)$ and $M(R_2)$ have the same grand U. According to Proposition 3, Proposition 7 and Lemma 5,

$$\begin{aligned} M(R_1) = M(R_2) &\Leftrightarrow int_{M(R_1)} = int_{M(R_2)} \\ &\Leftrightarrow \underline{R_1} = \underline{R_2} \\ &\Leftrightarrow R_1 = R_2. \end{aligned}$$

□

An equivalent formulation of attribute reduction in information systems is presented from the viewpoint of matroids.

Proposition 13. *Let* $IS = (U, A)$ *be an information system. For all* $B \subseteq A$*,* B *is a reduct of* IS *if and only if it satisfies the following two conditions:*

(1) For all $b \in B$, $M(IND(B)) \neq M(IND(B-b))$;
(2) $M(IND(B)) = M(IND(A))$.

Proof. Since $IND(A)$, $IND(B)$ and $IND(B-b)$ are equivalence relations on U, $M(IND(A))$, $M(IND(B))$ and $M(IND(B-b))$ are matroids on U. According to Proposition 12,
(1) For all $b \in B$, $M(IND(B)) \neq M(IND(B-b)) \Leftrightarrow IND(B) \neq IND(B-b)$;
(2) $M(IND(B)) = M(IND(A)) \Leftrightarrow IND(B) = IND(A)$.
According to Definition 1, it is immediate. □

In Proposition 13, the equivalent formulation of attribute reduction is not convenient for us to compute the attribute reduction. We consider to use the girth function of matroids to compute it.

Definition 9. *(Girth function [29,30]) Let* $M = (U, I)$ *be a matroid. The girth* $g(M)$ *of* M *is defined as:*

$$g(M) = \begin{cases} min\{|C| : C \in C(M)\}, & C(M) \neq \emptyset; \\ \infty, & C(M) = \emptyset. \end{cases}$$

For all $X \subseteq U$*, the girth function* g_M *is defined as* $g_M(X) = g(M|X)$. $g_M(X)$ *is called the girth of* X *in* M.

According to Definition 9, the girth function is related to circuits. Thus, the following lemma presents the family of all circuits of $M(R)$.

Lemma 6. *Ref. [38] Let R be an equivalence relation on U. Then,*

$$\mathbf{C}(M(R)) = \{\{x,y\} \subseteq U : x \neq y \wedge (x,y) \in R\}.$$

Example 7. *(Continued from Example 5)* $\mathbf{C}(M(R)) = \{\{a,b\},\{c,d\},\{c,e\},\{d,e\}\}$.

Based on the characteristics of the matroid induced by an equivalence relation, a type of matroids is abstracted, which is called a 2-circuit matroid. M is called a 2-circuit matroid if $|C| = 2$ for all $C \in \mathbf{C}(M)$. Note that, if $\mathbf{C}(M) = \emptyset$, then M is also a 2-circuit matroid. In this section, we don't consider this case. The matroid $M(R)$ is a 2-circuit matroid.

Proposition 14. *Let R be an equivalence relation on U and $X \subseteq U$. Then,*

$$g(M(R)) = \begin{cases} 2, & \mathbf{C}(M(R)) \neq \emptyset; \\ \infty, & \mathbf{C}(M(R)) = \emptyset; \end{cases}$$

$$g_{M(R)}(X) = \begin{cases} 2, & \mathbf{C}(M(R)|X) \neq \emptyset; \\ \infty, & \mathbf{C}(M(R)|X) = \emptyset. \end{cases}$$

Proof. Since $M(R)$ is a 2-circuit matroid, $|C| = 2$ for all $C \in \mathbf{C}(M(R))$. According to Definition 9, it is immediate. □

Corollary 2. *Let R be an equivalence relation on U and $X \subseteq U$. Then,*

$$g(M(R)) = \begin{cases} 2, & \exists x \in U, s.t., |RN(x)| \geq 2; \\ \infty, & otherwise, \end{cases}$$

$$g_{M(R)}(X) = \begin{cases} 2, & \exists x \in X, s.t., |RN(x) \bigcap X| \geq 2; \\ \infty, & otherwise. \end{cases}$$

Proof. According to Lemma 6,

$$\begin{aligned} \mathbf{C}(M(R)) \neq \emptyset & \Leftrightarrow \exists x,y \subseteq U, s.t., x \neq y \wedge (x,y) \in R \\ & \Leftrightarrow \exists x \in U, s.t., |RN(x)| \geq 2. \end{aligned}$$

Hence,

$$g(M(R)) = \begin{cases} 2, & \exists x \in U, s.t., |RN(x)| \geq 2; \\ \infty, & otherwise. \end{cases}$$

Since $\mathbf{C}(M(R)|X) = \{C \subseteq X : C \in \mathbf{C}(M(R))\} = \{\{x,y\} \subseteq X : x \neq y \wedge (x,y) \in R\}$,

$$\begin{aligned} \mathbf{C}(M(R)|X) \neq \emptyset & \Leftrightarrow \exists x,y \subseteq X, s.t., x \neq y \wedge (x,y) \in R \\ & \Leftrightarrow \exists x \in U, s.t., |RN(x) \bigcap X| \geq 2. \end{aligned}$$

Hence,

$$g_{M(R)}(X) = \begin{cases} 2, & \exists x \in X, s.t., |RN(x) \bigcap X| \geq 2; \\ \infty, & otherwise. \end{cases}$$

□

Lemma 7. *Refs. [1,41] Let R_1 and R_2 be two equivalence relations on U, respectively. Then, for any $x \in U$,*

$$(R_1 \bigcap R_2)N(x) = R_1N(x) \bigcap R_2N(x).$$

According to Corollary 2, the girth function of the matroid induced by attribute subsets is presented in the following proposition.

Proposition 15. *Let $IS = (U, A)$ be an information system and $X \subseteq U$. Then, for all $B \subseteq A$,*

$$g(M(IND(B))) = \begin{cases} 2, & \exists x \in U, s.t., |\bigcap_{R_i \in B} R_iN(x)| \geq 2; \\ \infty, & otherwise, \end{cases}$$

$$g_{M(IND(B))}(X) = \begin{cases} 2, & \exists x \in X, s.t., |(\bigcap_{R_i \in B} R_iN(x)) \bigcap X| \geq 2; \\ \infty, & otherwise. \end{cases}$$

Proof. According to Lemma 7 and Corollary 2, it is immediate. □

Note that R_i in R_iN denotes the equivalence relation induced by attribute $R_i \in A$. According to the girth axiom, we know that a matroid is corresponding to one and only one girth function.

Proposition 16. *(Girth axiom [29,30]) Let $g : 2^U \rightarrow \mathbb{Z}^+ \bigcup \{0, \infty\}$ be a function. Then, there exists one and only one matroid M such that $g = g_M$ iff g satisfies the following three conditions:*
(G1) If $X \subseteq U$ and $g(X) < \infty$, then X has a subset Y such that $g(X) = g(Y) = |Y|$.
(G2) If $X \subseteq Y \subseteq U$, then $g(X) \geq g(Y)$.
(G3) If X and Y are distinct subsets of U with $g(X) = |X|$, $g(Y) = |Y|$, then $g((X \bigcup Y) - \{e\}) < \infty$ for any $e \in X \bigcap Y$.

Inspired by Propositions 13 and 16, we can use the girth function in matroids to compute attribute reduction.

Theorem 2. *Let $IS = (U, A)$ be an information system. For all $B \subseteq A$, B is a reduct of IS if and only if it satisfies the following two conditions:*

(1) For all $b \in B$, there exists $X \subseteq U$ such that $g_{M(IND(B))}(X) \neq g_{M(IND(B-b))}(X)$.
(2) For all $X \subseteq U$, $g_{M(IND(B))}(X) = g_{M(IND(A))}(X)$.

Proof. According to Propositions 13 and 16, it is immediate. □

4.2. The Process of the Matroidal Methodology

In this subsection, we give the process of the matroidal approach to compute attribute reduction in information systems according to the equivalent description in Section 4.1.

In order to obtain all results of an information system $IS = (U, A)$, we need to compute $g_{M(IND(B))}(X)$ for all $B \subseteq A$ and $X \subseteq U$ based on Theorem 2. According to Definition 1, we know a reduct of IS will not be $\emptyset$. Hence, we only consider $B \subseteq A$ and $B \neq \emptyset$. On the other hand, for all $X \subseteq U$ and $B \subseteq A$, if $|X| \leq 1$, then $g_{M(IND(B))}(X) = g_{M(IND(A))}(X)$. According to Theorem 2, we only consider X whose $|X| \geq 2$. Therefore, the process is shown as follows:

- **Input**: An information system $IS = (U, A)$, where $U = \{u_1, u_2, \cdots, u_n\}$ and $A = \{a_1, a_2, \cdots, a_m\}$.
- **Output**: All results of IS.
- **Step 1**: Suppose $B_i \subset A$ ($B_i \neq \emptyset$ and $i = 1, 2, \cdots, 2^m - 2$), we compute all $IND(B_i)$ and $IND(A)$.

- **Step 2**: For any $i = 1, 2, \cdots, 2^m - 2$, we compute $g_{M(IND(B_i))}(X)$ and $g_{M(IND(A))}(X)$ for any $X \subseteq U$ and $|X| \geq 2$.
- **Step 3**: Obtain all results of IS according to Theorem 2.

4.3. An Applied Example

Example 8. *Let us consider the following information system $IS = (U, A)$ as is shown in Table 2.*

Table 2. An information system.

	a_1	a_2	a_3
u_1	0	1	0
u_2	1	2	2
u_3	1	0	0
u_4	2	1	1
u_5	1	1	2

Let $B_1 = \{a_1\}$, $B_2 = \{a_2\}$, $B_3 = \{a_3\}$, $B_4 = \{a_1, a_2\}$, $B_5 = \{a_1, a_3\}$, $B_6 = \{a_2, a_3\}$, $A = \{a_1, a_1, a_3\}$. g_{B_i} denotes $g_{M(IND(B_i))}$ for $1 \leq i \leq 6$ and g_A denotes $g_{M(IND(A))}$. All girth functions induced by attribute subsets as is shown in Table 3.

Table 3. Girth functions induced by attribute subsets.

	g_{B_1}	g_{B_2}	g_{B_3}	g_{B_4}	g_{B_5}	g_{B_6}	g_A
u_1, u_2	∞	∞	∞	∞	∞	∞	∞
u_1, u_3	∞	∞	2	∞	∞	∞	∞
u_1, u_4	∞	2	∞	∞	∞	∞	∞
u_1, u_5	∞	2	∞	∞	∞	∞	∞
u_2, u_3	2	∞	∞	∞	∞	∞	∞
u_2, u_4	∞	∞	∞	∞	∞	∞	∞
u_2, u_5	∞	∞	2	∞	2	∞	∞
u_3, u_4	∞	∞	∞	∞	∞	∞	∞
u_3, u_5	∞	∞	∞	∞	∞	∞	∞
u_4, u_5	∞	2	∞	∞	∞	∞	∞
u_1, u_2, u_3	2	∞	2	∞	∞	∞	∞
u_1, u_2, u_4	∞	2	∞	∞	∞	∞	∞
u_1, u_2, u_5	∞	2	2	∞	2	∞	∞
u_1, u_3, u_4	∞	2	2	∞	∞	∞	∞
u_1, u_3, u_5	∞	2	2	∞	∞	∞	∞
u_1, u_4, u_5	∞	2	∞	∞	∞	∞	∞
u_2, u_3, u_4	2	∞	∞	∞	∞	∞	∞
u_2, u_3, u_5	2	∞	∞	∞	2	∞	∞
u_2, u_4, u_5	∞	2	2	∞	2	∞	∞
u_3, u_4, u_5	∞	2	∞	∞	∞	∞	∞
u_1, u_2, u_3, u_4	2	2	2	∞	∞	∞	∞
u_1, u_2, u_3, u_5	2	2	2	∞	2	∞	∞
u_1, u_2, u_4, u_5	∞	2	2	∞	2	∞	∞
u_1, u_3, u_4, u_5	∞	2	2	∞	∞	∞	∞
u_2, u_3, u_4, u_5	2	2	2	∞	2	∞	∞
u_1, u_2, u_3, u_4, u_5	2	2	2	∞	2	∞	∞

Accordingly, there are two reducts of IS: $B_4 = \{a_1, a_2\}$ and $B_6 = \{a_2, a_3\}$.

5. Conclusions

In this paper, we generalize four operators of rough sets to four operators of matroids through the interior axiom, the closure axiom, the exterior axiom and the boundary axiom, respectively. Moreover, we present a matroidal approach to compute attribute reduction in information systems. The main conclusions in this paper and the continuous work to do are listed as follows:

1. There are four matroids induced by these four operators of rough sets. In fact, these four matroids are the same one, which implies the relationship about operators between rough sets and matroids. In this work, we assume an equivalence relation. However, there are other structures have been used in rough set theory, among them, tolerance relations [43], similarity relations [44], and binary relations [15,45]. Hence, they can suggest as a future research, the possibility of extending their ideas to these types of settings.
2. The girth function of matroids is used to compute attribute reduction in information systems. This work can be viewed as a bridge linking matroids and information systems in the theoretical impact. In the practical impact, it is a novel method by which calculations will become algorithmic and can be implemented by a computer. Based on this work, we can use the girth function of matroids for attribute reduction in decision systems in the future.
3. In the future, we will further expand the research content of this paper based on some new studies on neutrosophic sets and related algebraic structures [46–50].

Author Contributions: This paper is written through contributions of all authors. The individual contributions and responsibilities of all authors can be described as follows: the idea of this whole thesis was put forward by X.Z., and he also completed the preparatory work of the paper. J.W. analyzed the existing work of rough sets and matroids, and wrote the paper.

Funding: This research is funded by the National Natural Science Foundation of China under Grant Nos. 61573240 and 61473239.

Conflicts of Interest: The authors declare no conflicts of interest.

References

1. Pawlak, Z. *Rough Sets: Theoretical Aspects of Reasoning about Data*; Kluwer Academic Publishers: Boston, MA, USA, 1991.
2. Pawlak, Z. Rough sets. *Int. J. Comput. Inf. Sci.* **1982**, *11*, 341–356. [CrossRef]
3. Du, Y.; Hu, Q.; Zhu, P.; Ma, P. Rule learning for classification based on neighborhood covering reduction. *Inf. Sci.* **2011**, *181*, 5457–5467. [CrossRef]
4. Wang, X.; Tsang, E.C.; Zhao, S.; Chen, D.; Yeung, D.S. Learning fuzzy rules from fuzzy samples based on rough set technique. *Inf. Sci.* **2007**, *177*, 4493–4514. [CrossRef]
5. Chen, J.; Lin, Y.; Lin, G.; Li, J.; Ma, Z. The relationship between attribute reducts in rough sets and minimal vertex covers of graphs. *Inf. Sci.* **2015**, *325*, 87–97. [CrossRef]
6. Dai, J.; Hu, Q.; Hu, H. Neighbor inconsistent pair selection for attribute reduction by rough set approach. *IEEE Trans. Fuzzy Syst.* **2018**, *26*, 937–950. [CrossRef]
7. Yang, Y.; Chen, D.; Wang, H. Active sample selection based incremental algorithm for attribute reduction with rough sets. *IEEE Trans. Fuzzy Syst.* **2017**, *25*, 825–838. [CrossRef]
8. Chen, Y.; Miao, D.; Wang, R.; Wu, K. A rough set approach to feature selection based on power set tree. *Knowl.-Based Syst.* **2011**, *24*, 275–281. [CrossRef]
9. Javidi, M.; Eskandari, S. Streamwise feature selection: A rough set method. *Int. J. Mach. Learn. Cybern.* **2018**, *9*, 667–676. [CrossRef]
10. Cai, Z.; Zhu, W. Multi-label feature selection via feature manifold learning and sparsity regularization. *Int. J. Mach. Learn. Cybern.* **2018**, 9, 1321–1334. [CrossRef]
11. Johnson, J.A.; Liu, M.; Chen, H. Unification of knowledge discovery and data mining using rough sets approach in a real-world application. *Rough Sets Curr. Trends Comput.* **2001**, *2005*, 330–337.
12. Alcantud, J.C.R. Revealed indifference and models of choice behavior. *J. Math. Psychol.* **2002**, *46*, 418–430. [CrossRef]
13. Luce, R.D. Semiorders and a theory of utility discrimination. *Econometrica* **1956**, *24*, 178–191. [CrossRef]
14. Wang, C.; He, Q.; Shao, M.; Xu, Y.; Hu, Q. A unified information measure for general binary relations. *Knowl.-Based Syst.* **2017**, *135*, 18–28. [CrossRef]

15. Zhu, W. Generalized rough sets based on relations. *Inf. Sci.* **12007**, *177*, 4997–5001. [CrossRef]
16. Zhu, W.; Wang, F. Reduction and axiomatization of covering generalized rough sets. *Inf. Sci.* **2003**, *152*, 217–230. [CrossRef]
17. Zhu, W. Relationship among basic concepts in covering-based rough sets. *Inf. Sci.* **2009**, *179*, 2478–2486. [CrossRef]
18. Yang, B.; Hu, B. On some types of fuzzy covering-based rough sets. *Fuzzy Sets Syst.* **2017**, 312, 36–65. [CrossRef]
19. Wang, C.; He, Q.; Wang, X.; Chen, D.; Qian, Y. Feature selection based on neighborhood discrimination index. *IEEE Trans. Neural Netw. Learn. Syst.* **2018**, *29*, 2986–2999. [CrossRef] [PubMed]
20. Zhao, Z. On some types of covering rough sets from topological points of view. *Int. J. Approx. Reason.* **2016**, *68*, 1–14. [CrossRef]
21. Gao, N.; Li, Q.; Han, H.; Li, Z. Axiomatic approaches to rough approximation operators via ideal on a complete completely distributive lattice. *Soft Comput.* **2018**, *22*, 2329–2339. [CrossRef]
22. Zhang, X.; Miao, D.; Liu, C.; Le, M. Constructive methods of rough approximation operators and multigranulation rough sets. *Knowl.-Based Syst.* **2016**, *91*, 114–125. [CrossRef]
23. Chiaselotti, G.; Ciucci, D.; Gentile, T. Rough set theory and digraphs. *Fundam. Inf.* **2017**, *153*, 291–325. [CrossRef]
24. Wang, J.; Zhu, W. Applications of bipartite graphs and their adjacency matrices to covering-based rough sets. *Fundam. Inf.* **2017**, *156*, 237–254. [CrossRef]
25. Liang, D.; Xu, Z.; Liu, D. Three-way decisions with intuitionistic fuzzy decision-theoretic rough sets based on point operators. *Inf. Sci.* **2017**, *375*, 183–201. [CrossRef]
26. Zhang, X.; Dai, J.; Yu, Y. On the union and intersection operations of rough sets based on various approximation spaces. *Inf. Sci.* **2015**, *292*, 214–229. [CrossRef]
27. Qian, Y.; Liang, J.; Pedrycz, W.; Dang, C. Positive approximation: An accelerator for attribute reduction in rough set theory. *Artif. Intell.* **2010**, *174*, 597–618. [CrossRef]
28. Edmonds, J. Matroids and the greedy algorithm. *Math. Program.* **1971**, *1*, 127–136. [CrossRef]
29. Lai, H. *Matroid Theory*; Higher Education Press: Beijing, China, 2001.
30. Oxley, J.G. *Matroid Theory*; Oxford University Press: Oxford, UK, 1993.
31. Liu, S.; Manganiello, F.; Kschischang, F.R. Matroidal structure of skew polynomial rings with application to network coding. *Finite Fields Their Appl.* **2017**, *46*, 236–246. [CrossRef]
32. Dougherty, R.; Freiling, C.; Zeger, K. Networks, matroids, and non-Shannon information inequalities. *IEEE Trans. Inf. Theory* **2007**, *53*, 1949–1969. [CrossRef]
33. Zhu, W.; Wang, S. Matroidal approaches to generalized rough sets based on relations. *Int. J. Mach. Learn. Cybern.* **2011**, *2*, 273–279. [CrossRef]
34. Liu, Y.; Zhu, W. Parametric matroid of rough set. *Int. J. Uncertain. Fuzziness Knowl.-Based Syst.* **2015**, *23*, 893–908. [CrossRef]
35. Li, X.; Liu, S. Matroidal approaches to rough set theory via closure operators. *Int. J. Approx. Reason.* **2012**, *53*, 513–527. [CrossRef]
36. Li, X.; Yi, H.; Liu, S. Rough sets and matroids from a lattice-theoretic viewpoint. *Inf. Sci.* **2016**, *342*, 37–52. [CrossRef]
37. Su, L.; Zhu, W. Closed-set lattice and modular matroid induced by covering-based rough sets. *Int. J. Mach. Learn. Cybern.* **2017**, *8*, 191–201. [CrossRef]
38. Wang, S.; Zhu, Q.; Zhu, W.; Min, F. Matroidal structure of rough sets and its characterization to attribute reduction. *Knowl.-Based Syst.* **2012**, 36, 155–161. [CrossRef]
39. Wang, J.; Zhu, W. Applications of matrices to a matroidal structure of rough sets. *J. Appl. Math.* **2013**, *2013*. [CrossRef]
40. Wang, J.; Zhu, W. Contraction to matroidal structure of rough sets. In *Rough Sets and Knowledge Technology*; Springer: Berlin/Heidelberg, Germany, 2013; Volume 8171, pp. 75–86.
41. Yao, Y. Constructive and algebraic methods of theory of rough sets. *Inf. Sci.* **1998**, *109*, 21–47. [CrossRef]
42. Guo, J.; Li, S. The property of operator in matroid. *J. Shaanxi Normal Univ.* **2007**, *35*, 13–16.
43. Skowron, A.; Stepaniuk, J. Tolerance approximation spaces. *Fundam. Inf.* **1996**, *27*, 245–253.
44. Slowinski, R.; Vanderpooten, D. A generalized definition of rough approximations based on similarity. *IEEE Trans. Knowl. Data Eng.* **2000**, *12*, 331–336. [CrossRef]

45. Liu, G.; Zhu, W. The algebraic structures of generalized rough set theory. *Inf. Sci.* **2008**, *178*, 4105–4113. [CrossRef]
46. Zhang, X.; Smarandache, F.; Liang, X. Neutrosophic duplet semi-group and cancellable neutrosophic triplet groups. *Symmetry* **2017**, *9*, 275. [CrossRef]
47. Zhang, X. Fuzzy anti-grouped filters and fuzzy normal filters in pseudo-BCI algebras. *J. Intell. Fuzzy Syst.* **2017**, *33*, 1767–1774. [CrossRef]
48. Zhang, X.; Bo, C.; Smarandache, F.; Park, C. New operations of totally dependent-neutrosophic sets and totally dependent-neutrosophic soft sets. *Symmetry* **2018**, *10*, 187. [CrossRef]
49. Zhang, X.; Bo, C.; Smarandache, F.; Dai, J. New inclusion relation of neutrosophic sets with applications and related lattice structure. *Int. J. Mach. Learn. Cybern.* **2018**, *9*, 1753–1763. [CrossRef]
50. Zhang, X.; Park, C.; Wu, S. Soft set theoretical approach to pseudo-BCI algebras. *J. Intell. Fuzzy Syst.* **2018**, *34*, 559–568. [CrossRef]

symmetry

MDPI

Article

Some Results on Multigranulation Neutrosophic Rough Sets on a Single Domain

Hu Zhao [1,*] and Hong-Ying Zhang [2]

1 School of Science, Xi'an Polytechnic University, Xi'an 710048, China
2 School of Mathematics and Statistics, Xi'an Jiaotong University, Xi'an 710049, China; Zhyemily@mail.xjtu.edu.cn
* Correspondence: zhaohu@xpu.edu.cn or zhaohu2007@yeah.net; Tel.: +86-187-0294-2326

Received: 11 August 2018; Accepted: 13 September 2018; Published: 19 September 2018

Abstract: As a generalization of single value neutrosophic rough sets, the concept of multi-granulation neutrosophic rough sets was proposed by Bo et al., and some basic properties of the pessimistic (optimistic) multigranulation neutrosophic rough approximation operators were studied. However, they did not do a comprehensive study on the algebraic structure of the pessimistic (optimistic) multigranulation neutrosophic rough approximation operators. In the present paper, we will provide the lattice structure of the pessimistic multigranulation neutrosophic rough approximation operators. In particular, in the one-dimensional case, for special neutrosophic relations, the completely lattice isomorphic relationship between upper neutrosophic rough approximation operators and lower neutrosophic rough approximation operators is proved.

Keywords: neutrosophic set; neutrosophic rough set; pessimistic (optimistic) multigranulation neutrosophic approximation operators; complete lattice

1. Introduction

In order to deal with imprecise information and inconsistent knowledge, Smarandache [1,2] first introduced the notion of neutrosophic set by fusing a tri-component set and the non-standard analysis. A neutrosophic set consists of three membership functions, where every function value is a real standard or non-standard subset of the nonstandard unit interval $]0^-, 1^+[$. Since then, many authors have studied various aspects of neutrosophic sets from different points of view, for example, in order to apply the neutrosophic idea to logics, Rivieccio [3] proposed neutrosophic logics which is a generalization of fuzzy logics and studied some basic properties. Guo and Cheng [4] and Guo and Sengur [5] obtained good applications in cluster analysis and image processing by using neutrosophic sets. Salama and Broumi [6] and Broumi and Smarandache [7] first introduced the concept of rough neutrosophic sets, handled incomplete and indeterminate information, and studied some operations and their properties.

In order to apply neutrosophic sets conveniently, Wang et al. [8] proposed single valued neutrosophic sets by simplifying neutrosophic sets. Single valued neutrosophic sets can also be viewed as a generalization of intuitionistic fuzzy sets (Atanassov [9]). Single valued neutrosophic sets have become a new majorly research issue. Ye [10–12] proposed decision making based on correlation coefficients and weighted correlation coefficient of single valued neutrosophic sets, and gave an application of proposed methods. Majumdar and Samant [13] studied similarity, distance and entropy of single valued neutrosophic sets from a theoretical aspect.

Şahin and Küçük [14] gave a subsethood measure of single valued neutrosophic sets based on distance and showed its effectiveness through an example. We know that there's a certain connection among fuzzy rough approximation operators and fuzzy relations (resp., fuzzy topologies, information systems [15–17]). Hence, Yang et al. [18] firstly proposed neutrosophic relations and studied some kinds of kernels and closures of neutrosophic relations. Subsequently they proposed single valued neutrosophic rough sets [19] by fusing single valued neutrosophic sets and rough sets (Pawlak, [20]), and they studied some properties of single value neutrosophic upper and lower approximation operators. As a generalization of single value neutrosophic rough sets, Bao and Yang [21] introduced p-dimension single valued neutrosophic refined rough sets, and they also gave some properties of p-dimension single valued neutrosophic upper and lower approximation operators.

As another generalization of single value neutrosophic rough sets, Bo et al. [22] proposed the concept of multi-granulation neutrosophic rough sets and obtained some basic properties of the pessimistic (optimistic) multigranulation neutrosophic rough approximation operators. However, the lattice structures of those rough approximation operators in references [19,21,22], were not well studied. Following this idea, Zhao and Zhang [23] gave the supremum and infimum of the p-dimension neutrosophic upper and lower approximation operators, but they did not study the relationship between the p-dimension neutrosophic upper approximation operators and the p-dimension neutrosophic lower approximation operators, especially in the one-dimensional case. Inspired by paper [23], a natural problem is: Can the lattice structure of pessimistic (optimistic) multigranulation neutrosophic approximation operators be given?

In the present paper, we study the algebraic structure of optimistic (pessimistic) multigranulation single valued neutrosophic approximation operators.

The structure of the paper is organized as follows. The next section reviews some basic definitions of neutrosophic sets and one-dimensional multi-granulation rough sets. In Section 3, the lattice structure of the pessimistic multigranulation neutrosophic rough approximation operators are studied. In Section 4, for special neutrosophic relations, a one-to-one correspondence relationship between neutrosophic upper approximation operators and lower approximation operators is given. Finally, Section 5 concludes this article and points out the deficiencies of the current research.

2. Preliminaries

In this section, we briefly recall several definitions of neutrosophic set (here "neutrosophic set" refers exclusively to "single value neutrosophic set") and one-dimensional multi-granulation rough set.

Definition 1 ([8]). *A neutrosophic set B in X is defined as follows: $\forall a \in X$,*

$$B = (T_A(a), I_A(a), F_A(a)),$$

where $T_A(a)$, $I_A(a)$, $F_A(a) \in [0,1]$, $0 \leq \sup T_A(a) + \sup I_A(a) + \sup F_A(a) \leq 3$. The set of all neutrosophic sets on X will be denoted by $\mathbb{SVNS}(X)$.

Definition 2 ([11]). *Let C and D be two neutrosophic sets in X, if*

$$T_C(a) \leq T_D(a),\ I_C(a) \geq I_D(a) \text{ and } F_C(a) \geq F_D(a)$$

for each $a \in X$, then we called C is contained in D, i.e., $C \Subset D$. If $C \Subset D$ and $D \Subset C$, then we called C is equal to D, denoted by $C = D$.

Definition 3 ([18]). *Let A and B be two neutrosophic sets in X,*

(1) The union of A and B is a s neutrosophic set C, denoted by $A \Cup B$, where $\forall x \in X$,

$$T_C(a) = max\{T_A(a), T_B(a)\},\ I_C(a) = min\{I_A(a), I_B(a)\}, \text{ and}$$
$$F_C(a) = min\{F_A(a), F_B(a)\}.$$

(2) The intersection of A and B is a neutrosophic set D, denoted by $A \Cap B$, where $\forall x \in X$,

$$T_D(a) = min\{T_A(a), T_B(a)\},\ I_D(a) = max\{I_A(a), I_B(a)\},\ and$$
$$F_D(a) = max\{F_A(a), F_B(a)\}.$$

Definition 4 ([18]). *A neutrosophic relation R in X is defined as follows:*

$$R = \{< (a,b), T_R(a,b), I_R(a,b), F_R(a,b) >| \ (a,b) \in X \times X\},$$

where $T_R : X \times X \to [0,1], I_R : X \times X \to [0,1], F_R : X \times X \to [0,1]$, and

$$0 \leq \sup T_R(a,b) + \sup I_R(a,b) + \sup F_R(a,b) \leq 3.$$

The family of all neutrosophic relations in X will be denoted by $\mathbf{SVNR}(X)$, and the pair (X,R) is called a neutrosophic approximation space.

Definition 5 ([19]). *Let (X,R) be a neutrosophic approximation space, $\forall A \in \mathbb{SVNS}(X)$, the lower and upper approximations of A with respect to (X,R), denoted by $\underline{R}(A)$ and $\overline{R}(A)$, are two neutrosophic sets whose membership functions are defined as: $\forall a \in X$,*

$$T_{\underline{R}(A)}(a) = \underset{b\in X}{\wedge}[F_R(a,b) \vee T_A(b)],\ I_{\underline{R}(A)}(a) = \underset{b\in X}{\vee}[(1 - I_R(a,b)) \wedge I_A(b)],$$
$$F_{\underline{R}(A)}(a) = \underset{b\in X}{\vee}[T_R(a,b) \wedge F_A(b)],\ T_{\overline{R}(A)}(a) = \underset{b\in X}{\vee}[T_R(a,b) \wedge T_A(b)],$$
$$I_{\overline{R}(A)}(a) = \underset{b\in X}{\wedge}[I_R(a,b) \vee I_A(b)],\ F_{\overline{R}(A)}(a) = \underset{b\in X}{\wedge}[F_R(a,b) \vee F_A(b)].$$

The pair $(\underline{R}(A), \overline{R}(A))$ is called the one-dimensional multi-granulation rough set (also called single value neutrosophic rough set or one-dimension single valued neutrosophic refined rough set) of A with respect to (X,R). $\underline{R}$ and $\overline{R}$ are referred to as the neutrosophic lower and upper approximation operators,respectively.

Lemma 1 ([19]). *Let R_1 and R_2 be two neutrosophic relations in X, $\forall A \in \mathbb{SVNS}(X)$, we have*

(1) $\underline{R_1 \Cup R_2}(A) = \underline{R_1}(A) \Cap \underline{R_2}(A)$;
(2) $\overline{R_1 \Cup R_2}(A) = \overline{R_1}(A) \Cup \overline{R_2}(A)$;
(3) $\underline{R_1 \Cap R_2}(A) \Supset \underline{R_1}(A) \Cup \underline{R_2}(A) \Supset \underline{R_1}(A) \Cap \underline{R_2}(A)$;
(4) $\overline{R_1 \Cap R_2}(A) \Subset \overline{R_1}(A) \Cap \overline{R_2}(A)$.

3. The Lattice Structure of the Pessimistic Multigranulation Neutrosophic Rough Approximation Operators

In this section, set $M = \{R_1, R_2, \cdots, R_n\} = \{R_i\}_{i=\overline{1,n}}$ is called a multigranulation neutrosophic relations set on X if each R_i is a neutrosophic relation on X. In this case, the pair (X, M) will be called an n-dimensional multigranulation neutrosophic apptoximation space.

Definition 6 ([22]). *Let (X,M) be an n-dimensional multigranulation neutrosophic apptoximation space. We define two pairs of approximation operators as follows, for all $\forall A \in \mathbb{SVNS}(X)$ and $a \in X$,*

$$M^O(A) = (\underline{M}^O(A), \overline{M}^O(A)),\ M^P(A) = (\underline{M}^P(A), \overline{M}^P(A)),$$

where

$$T_{\underline{M}^O(A)}(a) = \vee_{i=1}^n T_{\underline{R_i}(A)}(a),\ I_{\underline{M}^O(A)}(a) = \wedge_{i=1}^n I_{\underline{R_i}(A)}(a),\ F_{\underline{M}^O(A)}(a) = \wedge_{i=1}^n F_{\underline{R_i}(A)}(a).$$
$$T_{\overline{M}^O(A)}(a) = \wedge_{i=1}^n T_{\overline{R_i}(A)}(a),\ I_{\overline{M}^O(A)}(a) = \vee_{i=1}^n I_{\overline{R_i}(A)}(a),\ F_{\overline{M}^O(A)}(a) = \vee_{i=1}^n F_{\overline{R_i}(A)}(a).$$

$$T_{\underline{M}^P(A)}(a) = \wedge_{i=1}^n T_{\underline{R_i}(A)}(a),\ I_{\underline{M}^P(A)}(a) = \vee_{i=1}^n I_{\underline{R_i}(A)}(a),\ F_{\underline{M}^P(A)}(a) = \vee_{i=1}^n F_{\underline{R_i}(A)}(a).$$
$$T_{\overline{M}^P(A)}(a) = \vee_{i=1}^n T_{\overline{R_i}(A)}(a),\ I_{\overline{M}^P(A)}(a) = \wedge_{i=1}^n I_{\overline{R_i}(A)}(a),\ F_{\overline{M}^P(A)}(a) = \wedge_{i=1}^n F_{\overline{R_i}(A)}(a).$$

Then the pair $M^O(A) = (\underline{M}^O(A), \overline{M}^O(A))$ *is called an optsimistic multigranulation neutrosophic rough set, and the pair* $M^P(A) = (\underline{M}^P(A), \overline{M}^P(A))$ *is called an pessimistic multigranulation neutrosophic rough set* $\overline{M}^O$ *and* $\overline{M}^P$ *are referred to as the optimistic and pessimistic multigranulation neutrosophic upper approximation operators, respectively. Similarly,* $\underline{M}^O$ *and* $\underline{M}^P$ *are referred to as the optimistic and pessimistic multigranulation neutrosophic lower approximation operators, respectively.*

Remark 1. *If* $n = 1$*, then the multigranulation neutrosophic rough set will degenerated to a one-dimensional multi-granulation rough set (see Definition 5). In the following, the family of all multigranulation neutrosophic relations set on X will be denoted by* $n - \mathbf{SVNR}(X)$*. Defined a relation* $\sqsubseteq$ *on* $n - \mathbf{SVNR}(X)$ *as follows:* $M \sqsubseteq N$ *if and only if* $M_i \Subset N_i$*, then* $(n - \mathbf{SVNR}(X), \sqsubseteq)$ *is a poset, where* $M = \{M_i\}_{i=\overline{1,n}}$ *and* $N = \{N_i\}_{i=\overline{1,n}}$*.*

$\forall \{M^j\}_{j\in\Lambda} \subseteq n - \mathbf{SVNR}(X)$*, where* $M^j = \left\{M_i^j\right\}_{i=\overline{1,n}}$ *and* Λ *be a index set, we can define union and intersection of* M^j *as follows:*

$$\underset{j\in\Lambda}{\vee} M^j = \left\{\Cup_{j\in\Lambda} M_i^j\right\}_{i=\overline{1,n}},\ \underset{j\in\Lambda}{\wedge} M^j = \left\{\Cap_{j\in\Lambda} M_i^j\right\}_{i=\overline{1,n}},$$

where

$$T_{\Cup_{j\in\Lambda} M_i^j}(a,b) = \underset{j\in\Lambda}{\vee} T_{M_i^j}(a,b),\ I_{\Cup_{j\in\Lambda} M_i^j}(a,b) = \underset{j\in\Lambda}{\wedge} I_{M_i^j}(a,b),$$
$$F_{\Cup_{j\in\Lambda} M_i^j}(a,b) = \underset{j\in\Lambda}{\wedge} F_{M_i^j}(a,b), T_{\Cap_{j\in\Lambda} M_i^j}(a,b) = \underset{j\in\Lambda}{\wedge} T_{M_i^j}(a,b),$$
$$I_{\Cap_{j\in\Lambda} M_i^j}(a,b) = \underset{j\in\Lambda}{\vee} I_{M_i^j}(a,b),\ F_{\Cap_{j\in\Lambda} M_i^j}(a,b) = \underset{j\in\Lambda}{\vee} F_{M_i^j}(a,b).$$

Then $\underset{j\in\Lambda}{\vee} M^j$ *and* $\underset{j\in\Lambda}{\wedge} M^j$ *are two multigranulation neutrosophic relations on X, and we easily show that* $\underset{j\in\Lambda}{\vee} M^j$ *and* $\underset{j\in\Lambda}{\wedge} M^j$ *are infimum and supremum of* $\{M^j\}_{j\in\Lambda}$*, respectively. Hence we can easily obtain the following theorem:*

Theorem 1. $(n - \mathbf{SVNR}(X), \sqsubseteq, \wedge, \vee)$ *is a complete lattice,* $\widetilde{X_n} = \{\underbrace{X_n, X_n, \cdots, X_n}_{n}\}$ *and* $\widetilde{\varnothing_N} = \{\underbrace{\varnothing_N, \varnothing_N, \cdots, \varnothing_N}_{n}\}$ *are its top element and bottom element, respectively, where* X_n *and* $\varnothing_N$ *are two neutrosophic relations in X and defined as follows:* $\forall (a,b) \in X \times X$, $T_{X_N}(a,b) = 1$, $I_{X_N}(a,b) = 0$, $F_{X_N}(a,b) = 0$ *and* $T_{\varnothing_N}(a,b) = 0$, $I_{\varnothing_N}(a,b) = 1$, $F_{\varnothing_N}(a,b) = 1$*. In particular,* $(\mathbf{SVNR}(X), \Subset, \Cup, \Cap)$ *is a complete lattice.*

Theorem 2. *Let* $M = \{R_i\}_{i=\overline{1,n}}$ *and* $N = \{Q_i\}_{i=\overline{1,n}}$ *be two multigranulation neutrosophic relations set on X,* $\forall A \in \mathbb{SVNS}(X)$*, we have*

(1) $\underline{M \vee N}^O(A) \Subset \underline{M}^O(A) \Cap \underline{N}^O(A)$, $\underline{M \vee N}^P(A) = \underline{M}^P(A) \Cap \underline{N}^P(A)$;
(2) $\overline{M \vee N}^O(A) \Supset \overline{M}^O(A) \Cup \overline{N}^O(A)$, $\overline{M \vee N}^P(A) = \overline{M}^P(A) \Cup \overline{N}^P(A)$;
(3) $\underline{M \wedge N}^O(A) \Supset \underline{M}^O(A) \Cup \underline{N}^O(A) \Supset \underline{M}^O(A) \Cap \underline{N}^O(A)$, $\underline{M \wedge N}^P(A) \Supset \underline{M}^P(A) \Cup \underline{N}^P(A) \Supset \underline{M}^P(A) \Cap \underline{N}^P(A)$;
(4) $\overline{M \wedge N}^O(A) \Subset \overline{M}^O(A) \Cap \overline{N}^O(A)$, $\overline{M \wedge N}^P(A) \Subset \overline{M}^P(A) \Cap \overline{N}^P(A)$.

Proof. We only show that the case of the optimistic multigranulation neutrosophic approximation operators.

(1) $\forall a \in X$, by Lemma 1 and Definition 6, we have the following:

$$\begin{aligned}
&T_{\underline{M \vee N}^O(A)}(a) \\
&= \vee_{i=1}^{n} T_{\underline{R_i \Cup Q_i}(A)}(a) = \vee_{i=1}^{n} T_{\underline{R_i}(A) \Cap \underline{Q_i}(A)}(a) \\
&= \vee_{i=1}^{n} \left[T_{\underline{R_i}(A)}(a) \wedge T_{\underline{Q_i}(A)}(a) \right] \\
&\le \left[\vee_{i=1}^{n} T_{\underline{R_i}(A)}(a) \right] \wedge \left[\vee_{i=1}^{n} T_{\underline{Q_i}(A)}(a) \right] \\
&= T_{\underline{M}^O(A)}(a) \wedge T_{\underline{N}^O(A)}(a) \\
&= T_{\underline{M}^O(A) \Cap \underline{N}^O(A)}(a),
\end{aligned}$$

$$\begin{aligned}
&I_{\underline{M \vee N}^O(A)}(a) \\
&= \wedge_{i=1}^{n} I_{\underline{R_i \Cup Q_i}(A)}(a) = \wedge_{i=1}^{n} I_{\underline{R_i}(A) \Cap \underline{Q_i}(A)}(a) \\
&= \wedge_{i=1}^{n} \left[I_{\underline{R_i}(A)}(a) \vee I_{\underline{Q_i}(A)}(a) \right] \\
&\ge \left[\wedge_{i=1}^{n} I_{\underline{R_i}(A)}(a) \right] \vee \left[\wedge_{i=1}^{n} I_{\underline{Q_i}(A)}(a) \right] \\
&= I_{\underline{M}^O(A)}(a) \vee I_{\underline{N}^O(A)}(a) \\
&= I_{\underline{M}^O(A) \Cap \underline{N}^O(A)}(a),
\end{aligned}$$

$$\begin{aligned}
&F_{\underline{M \vee N}^O(A)}(a) \\
&= \wedge_{i=1}^{n} F_{\underline{R_i \Cup Q_i}(A)}(a) = \wedge_{i=1}^{n} F_{\underline{R_i}(A) \Cap \underline{Q_i}(A)}(a) \\
&= \wedge_{i=1}^{n} \left[F_{\underline{R_i}(A)}(a) \vee F_{\underline{Q_i}(A)}(a) \right] \\
&\ge \left[\wedge_{i=1}^{n} F_{\underline{R_i}(A)}(a) \right] \vee \left[\wedge_{i=1}^{n} F_{\underline{Q_i}(A)}(a) \right] \\
&= F_{\underline{M}^O(A)}(a) \vee F_{\underline{N}^O(A)}(a) \\
&= F_{\underline{M}^O(A) \Cap \underline{N}^O(A)}(a).
\end{aligned}$$

Hence, $\underline{M \vee N}^O(A) \Subset \underline{M}^O(A) \Cap \underline{N}^O(A)$.

(2) $\forall a \in X$, by Lemma 1 and Definition 6, we have the following:

$$\begin{aligned}
&T_{\overline{M \vee N}^O(A)}(a) \\
&= \wedge_{i=1}^{n} T_{\overline{R_i \Cup Q_i}(A)}(a) = \wedge_{i=1}^{n} T_{\overline{R_i}(A) \Cup \overline{Q_i}(A)}(a) \\
&= \wedge_{i=1}^{n} \left[T_{\overline{R_i}(A)}(a) \vee T_{\overline{Q_i}(A)}(a) \right] \\
&\ge \left[\wedge_{i=1}^{n} T_{\overline{R_i}(A)}(a) \right] \vee \left[\wedge_{i=1}^{n} T_{\overline{Q_i}(A)}(a) \right] \\
&= T_{\overline{M}^O(A)}(a) \vee T_{\overline{N}^O(A)}(a) = T_{\overline{M}^O(A) \Cup \overline{N}^O(A)}(a),
\end{aligned}$$

$$\begin{aligned}
&I_{\overline{M \vee N}^O(A)}(a) \\
&= \vee_{i=1}^{n} I_{\overline{R_i \Cup Q_i}(A)}(a) = \vee_{i=1}^{n} I_{\overline{R_i}(A) \Cup \overline{Q_i}(A)}(a) \\
&= \vee_{i=1}^{n} \left[I_{\overline{R_i}(A)}(a) \wedge I_{\overline{Q_i}(A)}(a) \right] \\
&\le \left[\vee_{i=1}^{n} I_{\overline{R_i}(A)}(a) \right] \wedge \left[\vee_{i=1}^{n} I_{\overline{Q_i}(A)}(a) \right] \\
&= I_{\overline{M}^O(A)}(a) \wedge I_{\overline{N}^O(A)}(a) = I_{\overline{M}^O(A) \Cup \overline{N}^O(A)}(a),
\end{aligned}$$

$$\begin{aligned}
&F_{\overline{M \vee N}^O(A)}(a) \\
&= \vee_{i=1}^{n} F_{\overline{R_i \Cup Q_i}(A)}(a) = \vee_{i=1}^{n} F_{\overline{R_i}(A) \Cup \overline{Q_i}(A)}(a) \\
&= \vee_{i=1}^{n} \left[F_{\overline{R_i}(A)}(a) \wedge F_{\overline{Q_i}(A)}(a) \right] \\
&\le \left[\vee_{i=1}^{n} F_{\overline{R_i}(A)}(a) \right] \wedge \left[\vee_{i=1}^{n} F_{\overline{Q_i}(A)}(a) \right] \\
&= F_{\overline{M}^O(A)}(a) \wedge F_{\overline{N}^O(A)}(a) = F_{\overline{M}^O(A) \Cup \overline{N}^O(A)}(a).
\end{aligned}$$

Hence, $\overline{M \vee N}^O(A) \Supset \overline{M}^O(A) \Cup \overline{N}^O(A)$.

(3) $\forall a \in X$, by Lemma 1 and Definition 6, we have the following:

$$
\begin{aligned}
&T_{\underline{M \wedge N}^O(A)}(a) \\
&= \vee_{i=1}^{n} T_{\underline{R_i \Cap Q_i}(A)}(a) \geq \vee_{i=1}^{n} T_{\underline{R_i}(A) \Cup \underline{Q_i}(A)}(a) \\
&= \vee_{i=1}^{n} \left[T_{\underline{R_i}(A)}(a) \vee T_{\underline{Q_i}(A)}(a) \right] \\
&= \left[\vee_{i=1}^{n} T_{\underline{R_i}(A)}(a) \right] \vee \left[\vee_{i=1}^{n} T_{\underline{Q_i}(A)}(a) \right] \\
&= T_{\underline{M}^O(A)}(a) \vee T_{\underline{N}^O(A)}(a) \geq T_{\underline{M}^O(A)}(a) \wedge T_{\underline{N}^O(A)}(a),
\end{aligned}
$$

$$
\begin{aligned}
&I_{\underline{M \wedge N}^O(A)}(a) \\
&= \wedge_{i=1}^{n} I_{\underline{R_i \Cap Q_i}(A)}(a) \leq \wedge_{i=1}^{n} I_{\underline{R_i}(A) \Cup \underline{Q_i}(A)}(a) \\
&= \wedge_{i=1}^{n} \left[I_{\underline{R_i}(A)}(a) \wedge I_{\underline{Q_i}(A)}(a) \right] \\
&= \left[\wedge_{i=1}^{n} I_{\underline{R_i}(A)}(a) \right] \wedge \left[\wedge_{i=1}^{n} I_{\underline{Q_i}(A)}(a) \right] \\
&= I_{\underline{M}^O(A)}(a) \wedge I_{\underline{N}^O(A)}(a) \leq I_{\underline{M}^O(A)}(a) \vee I_{\underline{N}^O(A)}(a),
\end{aligned}
$$

$$
\begin{aligned}
&F_{\underline{M \wedge N}^O(A)}(a) \\
&= \wedge_{i=1}^{n} F_{\underline{R_i \Cap Q_i}(A)}(a) \leq \wedge_{i=1}^{n} F_{\underline{R_i}(A) \Cup \underline{Q_i}(A)}(a) \\
&= \wedge_{i=1}^{n} \left[F_{\underline{R_i}(A)}(a) \wedge F_{\underline{Q_i}(A)}(a) \right] \\
&= \left[\wedge_{i=1}^{n} F_{\underline{R_i}(A)}(a) \right] \wedge \left[\wedge_{i=1}^{n} F_{\underline{Q_i}(A)}(a) \right] \\
&= F_{\underline{M}^O(A)}(a) \wedge F_{\underline{N}^O(A)}(a) \leq F_{\underline{M}^O(A)}(a) \vee F_{\underline{N}^O(A)}(a).
\end{aligned}
$$

Hence, $\underline{M \wedge N}^o(A) \Supset \underline{M}^o(A) \Cup \underline{N}^o(A) \Supset \underline{M}^o(A) \Cap \underline{N}^o(A)$.

(4) $\forall a \in X$, by Lemma 1 and Definition 6, we have the following:

$$
\begin{aligned}
&T_{\overline{M \wedge N}^O(A)}(a) \\
&= \wedge_{i=1}^{n} T_{\overline{R_i \Cap Q_i}(A)}(a) \leq \wedge_{i=1}^{n} T_{\overline{R_i}(A) \Cap \overline{Q_i}(A)}(a) \\
&= \wedge_{i=1}^{n} \left[T_{\overline{R_i}(A)}(a) \wedge T_{\overline{Q_i}(A)}(a) \right] \\
&= \left[\wedge_{i=1}^{n} T_{\overline{R_i}(A)}(a) \right] \wedge \left[\wedge_{i=1}^{n} T_{\overline{Q_i}(A)}(a) \right] \\
&= T_{\overline{M}^O(A)}(a) \wedge T_{\overline{N}^O(A)}(a) = T_{\overline{M}^O(A) \Cap \overline{N}^O(A)}(a),
\end{aligned}
$$

$$
\begin{aligned}
&I_{\overline{M \wedge N}^O(A)}(a) \\
&= \vee_{i=1}^{n} I_{\overline{R_i \Cap Q_i}(A)}(a) \geq \vee_{i=1}^{n} I_{\overline{R_i}(A) \Cap \overline{Q_i}(A)}(a) \\
&= \vee_{i=1}^{n} \left[I_{\overline{R_i}(A)}(a) \vee I_{\overline{Q_i}(A)}(a) \right] \\
&= \left[\vee_{i=1}^{n} I_{\overline{R_i}(A)}(a) \right] \vee \left[\vee_{i=1}^{n} I_{\overline{Q_i}(A)}(a) \right] \\
&= I_{\overline{M}^O(A)}(a) \vee T_{\overline{N}^O(A)}(a) = I_{\overline{M}^O(A) \Cap \overline{N}^O(A)}(a),
\end{aligned}
$$

$$
\begin{aligned}
&F_{\overline{M \wedge N}^O(A)}(a) \\
&= \vee_{i=1}^{n} F_{\overline{R_i \Cap Q_i}(A)}(a) \geq \vee_{i=1}^{n} F_{\overline{R_i}(A) \Cap \overline{Q_i}(A)}(a) \\
&= \vee_{i=1}^{n} \left[F_{\overline{R_i}(A)}(a) \vee F_{\overline{Q_i}(A)}(a) \right] \\
&= \left[\vee_{i=1}^{n} F_{\overline{R_i}(A)}(a) \right] \vee \left[\vee_{i=1}^{n} F_{\overline{Q_i}(A)}(a) \right] \\
&= F_{\overline{M}^O(A)}(a) \vee F_{\overline{N}^O(A)}(a) = F_{\overline{M}^O(A) \Cap \overline{N}^O(A)}(a).
\end{aligned}
$$

Hence, $\overline{M \wedge N}^O(A) \Subset \overline{M}^O(A) \Cap \overline{N}^O(A)$. □

From Theorem 2, we can easily obtain the following corollary:

Corollary 1. *Let* $M = \{R_i\}_{i=\overline{1,n}}$ *and* $N = \{Q_i\}_{i=\overline{1,n}}$ *be two multigranulation neutrosophic relations set on* X. *If* $M \sqsubseteq N$, *then* $\forall A \in \mathbb{SVNS}(X)$,

$$\underline{N}^O(A) \Subset \underline{M}^O(A),\ \underline{N}^P(A) \Subset \underline{M}^P(A)),\ \overline{M}^O(A) \Subset \overline{N}^O(A),\ \overline{M}^P(A) \Subset \overline{N}^P(A).$$

Let $H_n^P = \left\{\overline{M}^P \mid M \in n - \mathbf{SVNR}(X)\right\}$ and $L_n^P = \left\{\underline{M}^P \mid M \in n - \mathbf{SVNR}(X)\right\}$ be the set of pessimistic multigranulation neutrosophic upper and lower approximation operators in X, respectively.

- Defined a relation $\hat{\lesssim}$ on H_n^P as follows: $\overline{M}^P \hat{\lesssim} \overline{N}^P$ if and only if $\overline{M}^P(A) \Subset \overline{N}^P(A)$ for each $A \in \mathbb{SVNS}(X)$. Then $(H_n^P, \hat{\lesssim})$ is a poset.
- Defined a relation $\hat{\lesssim}$ on L_n^P as follows: $\underline{M}^P \hat{\lesssim} \underline{N}^P$ if and only if $\underline{N}^P(A) \Subset \underline{M}^P(A)$ for each $A \in \mathbb{SVNS}(X)$. Then $(L_n^P, \hat{\lesssim})$ is a poset.

Let $H_n^O = \left\{\overline{M}^O \mid M \in n - \mathbf{SVNR}(X)\right\}$ and $L_n^O = \left\{\underline{M}^O \mid M \in n - \mathbf{SVNR}(X)\right\}$ be the set of optimistic multigranulation neutrosophic upper and lower approximation operators in X, respectively.

- Defined a relation $\hat{\lesssim}$ on H_n^O as follows: $\overline{M}^O \hat{\lesssim} \overline{N}^O$ if and only if $\overline{M}^O(A) \Subset \overline{N}^O(A)$ for each $A \in \mathbb{SVNS}(X)$. Then $(H_n^O, \hat{\lesssim})$ is a poset.
- Defined a relation $\hat{\lesssim}$ on L_n^O as follows: $\underline{M}^O \hat{\lesssim} \underline{N}^O$ if and only if $\underline{N}^O(A)) \Subset \underline{M}^O(A)$ for each $A \in \mathbb{SVNS}(X)$. Then $(L_n^O, \hat{\lesssim})$ is a poset.

Theorem 3. *(1)* $\forall \left\{\overline{M_i}^P\right\}_{i \in I} \subseteq (H_n^P, \hat{\lesssim})$ *and I be a index set, we can define union and intersection of* $\overline{M_i}^P$ *as follows:*

$$\hat{\vee}_{i \in I} \overline{M_i}^P = \overline{\vee_{i \in I} M_i}^P,\ \hat{\wedge}_{i \in I} \overline{M_i}^P = \overline{[\wedge_{i \in I} M_i]}^P,$$

where $[\wedge_{i \in I} M_i] = \vee\left\{M \in n - \mathbf{SVNR}(X) \mid \forall A \in \mathbb{SVNS}(X), \overline{M}^P(A) \Subset \Cap_{i \in I} \overline{M_i}^P(A)\right\}$. *Then* $\hat{\vee}_{i \in I} \overline{M_i}^P$ *and* $\hat{\wedge}_{i \in I} \overline{M_i}^P$ *are supremum and infimum of* $\left\{\overline{M_i}^P\right\}_{i \in I}$, *respectively.*

(2) $\forall \left\{\underline{M_i}^P\right\}_{i \in I} \subseteq (L_n^P, \hat{\lesssim})$ *and I be a index set, we can define union and intersection of* $\underline{M_i}^P$ *as follows:*

$$\hat{\vee}_{i \in I} \underline{M_i}^P = \underline{\vee_{i \in I} M_i}^P,\ \hat{\wedge}_{i \in I} \underline{M_i}^P = \underline{[\vee_{i \in I} M_i]}^P,$$

where $[\vee_{i \in I} M_i] = \vee\left\{M \in n - \mathbf{SVNR}(X) \mid \forall A \in \mathbb{SVNS}(X), \Cup_{i \in I} \underline{M_i}^P(A) \Subset \underline{M}^P(A)\right\}$. *Then* $\hat{\vee}_{i \in I} \underline{M_i}^P$ *and* $\hat{\wedge}_{i \in I} \underline{M_i}^P$ *are supremum and infimum of* $\left\{\underline{M_i}^P\right\}_{i \in I}$, *respectively.*

Proof. We only show (1).

Let $M = \vee_{i \in I} M_i$, then $M_i \sqsubseteq M$ for each $i \in I$. By Corollary 1, we have $\overline{M_i}^P(A) \Subset \overline{M}^P(A)$ for any $A \in \mathbb{SVNS}(X)$. Thus $\overline{M_i}^P \hat{\lesssim} \overline{M}^P$. If $M^\star$ is a multigranulation neutrosophic relations set such that $\overline{M_i}^P \hat{\lesssim} \overline{M^\star}^P$ for each $i \in I$, then $A \in \mathbb{SVNS}(X)$, $\overline{M_i}^P(A) \Subset \overline{M^\star}^P(A)$. Hence,

$$\overline{M}^P(A) = \overline{\vee_{i \in I} M_i}^P(A) = \Cup_{i \in I} \overline{M_i}^P(A) \Subset \overline{M^\star}^P(A).$$

Thus $\overline{M}^P \hat{\lesssim} \overline{M^\star}^P$. So $\hat{\vee}_{i \in I} \overline{M_i}^P = \overline{\vee_{i \in I} M_i}^P$ is the supremum of $\left\{\overline{M_i}^P\right\}_{i \in I}$.

Let $Q = [\wedge_{i \in I} M_i]$, then $\forall B \in \mathbb{SVNS}(X)$, we have

$$\overline{Q}^P(B) = \overline{[\wedge_{i \in I} M_i]}^P(B) \Subset \Cap_{i \in I} \overline{M_i}^P(B) \Subset \overline{M_i}^P(B).$$

Thus $\overline{Q}^P \hat{\lesssim} \overline{M_i}^P$ for each $i \in I$. If M^* is a multigranulation neutrosophic relations set such that $\overline{M^*}^P \hat{\lesssim} \overline{M_i}^P$ for each $i \in I$, then

$$\overline{M^*}^P(A) \Subset \Cap_{i\in I}\overline{M_i}^P(A).$$

By the construction of $[\underset{i\in I}{\wedge} M_i]$, we can easily obtain $M^* \sqsubseteq [\underset{i\in I}{\wedge} M_i] = Q$. Hence,

$$\overline{M^*}^P \hat{\lesssim} \overline{[\underset{i\in I}{\wedge} M_i]}^P = \overline{Q}^P,$$

So $\underset{i\in I}{\hat{\wedge}} \overline{M_i}^P = \overline{[\underset{i\in I}{\wedge} M_i]}^P$ is the infimum of $\left\{\overline{M_i}^P\right\}_{i\in I}$. $\square$

Remark 2. *(1) $\forall A \in \mathbb{SVNS}(X)$, $\forall a \in X$, we can calculate that the following formula holds.*

$$T_{\overline{\widetilde{\varnothing_N}}^P(A)}(a) = 0,\ I_{\overline{\widetilde{\varnothing_N}}^P(A)}(a) = 1,\ F_{\overline{\widetilde{\varnothing_N}}^P(A)}(a) = 1,$$

$$T_{\underline{\widetilde{\varnothing_N}}^P(A)}(a) = 1,\ I_{\underline{\widetilde{\varnothing_N}}^P(A)}(a) = 0,\ F_{\underline{\widetilde{\varnothing_N}}^P(A)}(a) = 0.$$

Hence, $\forall M \in n-\mathbf{SVNR}(X)$, $\overline{\widetilde{\varnothing_N}}^P(A) \Subset \overline{M}^P(A)$ and $\underline{M}^P(A) \Subset \underline{\widetilde{\varnothing_N}}^P(A)$. It shows that $\overline{\widetilde{\varnothing_N}}^P \hat{\lesssim} \overline{M}^P$ and $\underline{\widetilde{\varnothing_N}}^P \hat{\lesssim} \underline{M}^P$, i.e., $\overline{\widetilde{\varnothing_N}}^P$ is the bottom element of $(H_n^P, \hat{\lesssim})$ and $\underline{\widetilde{\varnothing_N}}^P$ is the bottom element of $(L_n^P, \hat{\lesssim})$. By Theorem 3, we have the following result: Both $(H_n^P, \hat{\lesssim}, \hat{\wedge}, \hat{\vee})$ and $(L_n^P, \hat{\lesssim}, \hat{\wedge}, \hat{\vee})$ are complete lattices.

(2) Similarly, we can prove that both $(H_n^O, \hat{\lesssim}, \hat{\wedge}, \hat{\vee})$ and $(L_n^O, \hat{\lesssim}, \hat{\wedge}, \hat{\vee})$ are complete lattices if we can use the generalization formula of

$$\overline{M \vee N}^O(A) \Subset \overline{M}^O(A) \Cup \overline{N}^O(A) \text{ and } \underline{M \vee N}^O(A) \Supset \underline{M}^O(A) \Cap \underline{N}^O(A),$$

However, by Theorem 2, we known that

$$\overline{M \vee N}^O(A) \Supset \overline{M}^O(A) \Cup \overline{N}^O(A) \text{ and } \underline{M \vee N}^O(A) \Subset \underline{M}^O(A) \Cap \underline{N}^O(A).$$

So, naturally, there is the following problem:

How to give the supremum and infimum of the optimistic multigranulation neutrosophic rough approximation operators?

In the one-dimensional case, for convenience, we will use $H = \{\overline{R} \mid R \in \mathbf{SVNR}(X)\}$ and $L = \{\underline{R} \mid R \in \mathbf{SVNR}(X)\}$ to denote the set of neutrosophic upper and lower approximation operators in X, respectively. According to Lemma 1, Remark 2 and Theorem 3, we have the following result: both $(H, \leq, \wedge, \vee)$ and $(L, \leq, \wedge, \vee)$ are complete lattices (it is also the one-dimensional case of Reference [23]).

4. The Relationship between Complete Lattices $(H,\leq,\wedge,\vee)$ and $(L,\leq,\wedge,\vee)$

In this section, we will study the relationship between complete lattices $(H, \leq, \wedge, \vee)$ and $(L, \leq, \wedge, \vee)$. Set

$$\mathcal{A} = \left\{\mathbf{SVNR}(X) \mid \forall R_1, R_2 \in \mathbf{SVNR}(X), \overline{R_1} \leq \overline{R_2} \Leftrightarrow R_1 \Subset R_2 \Leftrightarrow \underline{R_1} \leq \underline{R_2}\right\}.$$

Firstly, we will give an example to illustrate that $\mathcal{A}$ is not an empty family.

Example 1. *Let $X = \{a\}$ be a single point set, R_1 and R_2 are two single valued neutrosophic relations in X.*

(1) *If $\overline{R_1} \leq \overline{R_2}$, then $R_1 \Subset R_2$. In fact, if $\overline{R_1} \leq \overline{R_2}$, then $\overline{R_1}(A) \Subset \overline{R_2}(A)$ for each $A \in \mathbb{SVNS}(\{a\})$.*

Thus, $\forall a \in X$,

$$T_{\overline{R_1}(A)}(a) \leq T_{\overline{R_2}(A)}(a),\ I_{\overline{R_1}(A)}(a) \geq I_{\overline{R_2}(A)}(a), \text{and } F_{\overline{R_1}(A)}(a) \geq F_{\overline{R_2}(A)}(a).$$

Moreover, $T_{R_1}(a,a) \wedge T_A(a) \leq T_{R_2}(a,a) \wedge T_A(a)$, $I_{R_1}(a,a) \vee I_A(a) \geq I_{R_2}(a,a) \vee I_A(a)$, *and* $F_{R_1}(a,a) \vee F_A(a) \geq F_{R_2}(a,a) \vee F_A(a)$. *Considering the arbitrariness of* A, *in particular, take* $A = \{< a,(1,0,0) >\}$, *we have* $T_{R_1}(a,a) \leq T_{R_2}(a,a)$, $I_{R_1}(a,a) \geq I_{R_2}(a,a)$ *and* $F_{R_1}(a,a) \geq F_{R_2}(a,a)$.

Hence, $R_1 \Subset R_2$.

Similarly, we also can show that the following result:

(2) *If* $\underline{R_1} \leq \underline{R_2}$, *then* $R_1 \Subset R_2$. *So, by (1), (2) and Corollary 1, we have* $\mathbf{SVNR}(\{a\}) \in \mathcal{A}$, *i.e.,* $\mathcal{A}$ *is not an empty family.*

Now, we will give the relationship between complete lattices $(H, \leq, \wedge, \vee)$ and $(L, \leq, \wedge, \vee)$.

Proposition 1. *If* $\mathbf{SVNR}(X) \in \mathcal{A}$, *then* $[\Cap_{i\in I} R_i] = \Cap_{i\in I} R_i = [\Cup_{i\in I} R_i]$, *where* I *is a index set, and* $R_i \in \mathbf{SVNR}(X)$ *for each* $i \in I$.

Proof. We first show that $[\Cap_{i\in I} R_i] = \Cap_{i\in I} R_i$. Let R be a neutrosophic relation in X such that $\Cap_{i\in I}\overline{R}_i(A) \Supset \overline{R}(A)$ for each $A \in \mathbb{SVNS}(X)$, then $\overline{R}_i \geq \overline{R}$, this is equivalent to $R_i \Supset R$ since $\mathbf{SVNR}(X) \in \mathcal{A}$. Thus $\Cap_{i\in I} R_i \Supset R$. Moreover, by the construction of $[\Cap_{i\in I} R_i]$, we have $\Cap_{i\in I} R_i \Supset [\Cap_{i\in I} R_i]$. On the other hand, we can show that $\Cap_{i\in I}\overline{R}_i(A) \Supset \overline{\Cap_{i\in I} R_i}(A)$ for each $A \in \mathbb{SVNS}(X)$. So

$$[\Cap_{i\in I} R_i] = \Cup\{R \in \mathbf{SVNR}(X) \mid \forall A \in \mathbb{SVNS}(X), \Cap_{i\in I}\overline{R}_i(A) \Supset \overline{R}(A)\} \Supset \Cap_{i\in I} R_i.$$

Hence $[\Cap_{i\in I} R_i] = \Cap_{i\in I} R_i$.

Now, we show that $\Cap_{i\in I} R_i = [\Cup_{i\in I} R_i]$. Let R be a single valued neutrosophic relation in such that $\Cup_{i\in I}\underline{R}_i(A) \Subset \underline{R}(A)$ for each $A \in \mathbb{SVNS}(X)$, then $\underline{R}_i \geq \underline{R}$, this is equivalent to $R_i \Supset R$ since $\mathbf{SVNR}(X) \in \mathcal{A}$. Thus $\Cap_{i\in I} R_i \Supset R$. Moreover, by the construction of $[\Cup_{i\in I} R_i]$. We have $\Cap_{i\in I} R_i \Supset [\Cup_{i\in I} R_i]$.

On the other hand, we can show that $\Cup_{i\in I}\underline{R}_i(A) \Subset \underline{\Cap_{i\in I} R_i}(A)$ for each $A \in \mathbb{SVNS}(X)$. So

$$[\Cup_{i\in I} R_i] = \Cup\{R \in \mathbf{SVNR}(X) \mid \forall A \in \mathbb{SVNS}(X), \Cup_{i\in I}\underline{R}_i(A) \Subset \underline{R}(A)\} \Supset \Cap_{i\in I} R_i.$$

Hence, $[\Cup_{i\in I} R_i] = \Cap_{i\in I} R_i$.

From above proved, we know that $[\Cap_{i\in I} R_i] = \Cap_{i\in I} R_i = [\Cup_{j\in J} R_j]$. □

Theorem 4. *If* $\mathbf{SVNR}(X) \in \mathcal{A}$, *then* $(\mathbf{SVNR}(X), \Subset, \Cup, \Cap)$ *and* $(H, \leq, \wedge, \vee)$ *are complete lattice isomorphism.*

Proof. Define a mapping $\phi_{12} : \mathbf{SVNR}(X) \to H$ as follows: $\forall R \in \mathbf{SVNR}(X)$, $\phi_{12}(R) = \overline{R}$. Obviously, ϕ_{12} is surjective. If $\overline{R_1} = \overline{R_2}$, notice that $\mathbf{SVNR}(X) \in \mathcal{A}$, we know that $R_1 = R_2$. So ϕ_{12} is one-one. $\forall \{R_i\}_{i\in I} \subseteq \mathbf{SVNR}(X)$ and I be a index set. By Theorem 3 and Proposition 1, we have

$$\phi_{12}(\Cup_{i\in I} R_i) = \overline{\Cup_{i\in I} R_i} = \bigvee_{i\in I} \overline{R_i} = \bigvee_{i\in I} \phi_{12}(R_i),$$

and

$$\phi_{12}(\Cap_{i\in I} R_i) = \overline{\Cap_{i\in I} R_i} = \overline{[\Cap_{i\in I} R_i]} = \bigwedge_{i\in I} \overline{R_i} = \bigwedge_{i\in I} \phi_{12}(R_i).$$

Hence, ϕ_{12} preserves arbitrary union and arbitrary intersection. □

From above proved, we know that $(\mathbf{SVNR}(X), \Subset, \Cup, \Cap)$ and $(H, \leq, \wedge, \vee)$ are complete lattice isomorphism.

Theorem 5. *If* $\mathbf{SVNR}(X) \in \mathcal{A}$*, then* $(\mathbf{SVNR}(X), \Subset, \Cup, \Cap)$ *and* $(L, \leq, \wedge, \vee)$ *are complete lattice isomorphism.*

Proof. Define a mapping $\phi_{13} : \mathbf{SVNR}(X) \to L$ as follows:$\forall R \in \mathbf{SVNR}(X)$, $\phi_{12}(R) = \underline{R}$. Obviously, ϕ_{13} is surjective. If $\underline{R_1} = \underline{R_2}$, notice that $\mathbf{SVNR}(X) \in \mathcal{A}$, we know that $R_1 = R_2$. So ϕ_{13} is one-one. $\forall \{R_i\}_{i\in I} \subseteq \mathbf{SVNR}(X)$ and I be an index set. By Theorem 3 and Proposition 1, we have

$$\phi_{13}(\Cup_{i\in I} R_i) = \underline{\Cup_{i\in I} R_i} = \vee_{i\in I} \underline{R_i} = \vee_{i\in I} \phi_{13}(R_i),$$

and

$$\phi_{13}(\Cap_{i\in I} R_i) = \underline{\Cap_{i\in I} R_i} = \underline{[\Cup_{i\in I} R_i]} = \wedge_{i\in I} \underline{R_i} = \wedge_{i\in I} \phi_{13}(R_i).$$

Hence, ϕ_{13} preserves arbitrary union and arbitrary intersection. □

From the above proof, we know that $(\mathbf{SVNR}(X), \Subset, \Cup, \Cap)$ and $(L, \leq, \wedge, \vee)$ are complete lattice isomorphism.

Theorem 6. *If* $\mathbf{SVNR}(X) \in \mathcal{A}$*, then* $(H, \leq, \wedge, \vee)$ *and* $(L, \leq, \wedge, \vee)$ *are complete lattice isomorphism.*

Proof. Through Theorems 4 and 5, we immediately know that the conclusion holds. We can also prove it by the following way:

Define a mapping $\phi_{23} : H \to L$ as follows: $\forall \overline{R} \in H$, $\phi_{23}(\overline{R}) = \underline{R}$. Through Theorems 4 and 5, there must be one and only one $R \in \mathbf{SVNR}(X)$ such that $\phi_{23}(\overline{R}) = \underline{R}$ for each $\underline{R} \in L$. This shows ϕ_{23} is surjective. If $\underline{R_1} = \underline{R_2}$, notice that $\mathbf{SVNR}(X) \in \mathcal{A}$, we know that $\overline{R_1} = \overline{R_2}$. So ϕ_{23} is one-one. $\forall \{\overline{R_i}\}_{i\in I} \subseteq H$ and I be a index set. Through Theorem 3 and Proposition 1, we have

$$\phi_{23}(\vee_{i\in I} \overline{R_i}) = \phi_{23}(\overline{\Cup_{i\in I} R_i}) = \underline{\Cup_{i\in I} R_i} = \vee_{i\in I} \underline{R_i} = \vee_{i\in I} \phi_{13}(\overline{R_i}),$$

and

$$\phi_{13}(\wedge_{i\in I} \overline{R_i}) = \phi_{13}(\overline{[\Cap_{i\in I} R_i]}) = \underline{[\Cap_{i\in I} R_i]} = \underline{[\Cup_{i\in I} R_i]} = \wedge_{i\in I} \underline{R_i} = \wedge_{i\in I} \phi_{23}(\overline{R_i}).$$

Hence, ϕ_{23} preserves arbitrary union and arbitrary intersection. So, $(H, \leq, \wedge, \vee)$ and $(L, \leq, \wedge, \vee)$ are complete lattice isomorphism. □

Remark 3. *Through Theorems 4–6, we can ascertain that* ϕ_{12}*,*ϕ_{13} *and* ϕ_{23} *are isomorphic mappings among complete lattices. Moreover, the following diagram can commute, i.e.,* $\phi_{23} \circ \phi_{12} = \phi_{13}$ *(see Figure 1).*

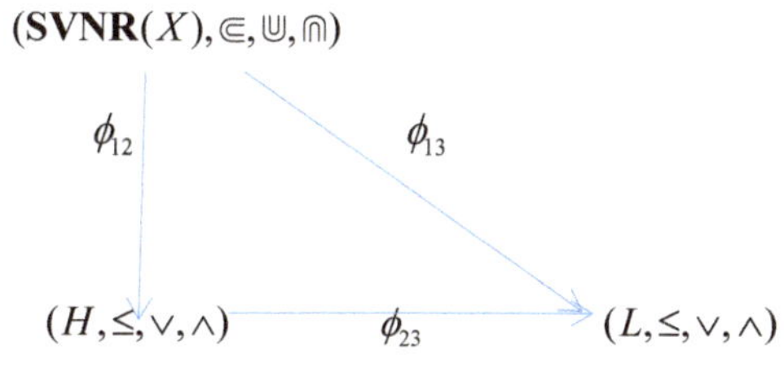

Figure 1. Correspondence relationship among three complete lattices.

5. Conclusions

Following the idea of multigranulation neutrosophic rough sets on a single domain as introduced by Bo et al. (2018), we gave the lattice structure of the pessimistic multigranulation neutrosophic rough approximation operators. In the one-dimensional case, for each special $\mathbf{SVNR}(X)$, we gave a one-to-one correspondence relationship between complete lattices $(H, \leq)$ and $(L, \leq)$. Unfortunately, at the moment, we haven't solved the following problems:

(1) Can the supremum and infimum of the optimistic multigranulation neutrosophic rough approximation operators be given?
(2) For any set , are $(H, \leq)$ and $(L, \leq)$ isomorphic between complete lattices?

Author Contributions: H.Z. provided the idea of the paper and proved the theorems. H.-Y.Z. provided helpful suggestions.

Funding: This research received no external funding.

Acknowledgments: The work is partly supported by the National Natural Science Foundation of China (Grant No. 61473181, 11771263 and 11671007), the Doctoral Scientific Research Foundation of Xi'an Polytechnic University (Grant No. BS1426), the Construction Funding of Xi'an Polytechnic University for Mathematics (Grant No. 107090701), and the Scientific Research Program Funded by Shaanxi Provincial Education Department (2018).

Conflicts of Interest: The authors declare no conflict of interest.

References

1. Smarandache, F. *Neutrosophy: Neutrosophic Probability, Set, and Logic*; American Research Press: Rehoboth, MA, USA, 1998.
2. Smarandache, F. Neutrosophic set—-A generialization of the intuitionistics fuzzy sets. *Int. J. Pure Appl. Math.* **2005**, *24*, 287–297.
3. Rivieccio, U. Neutrosophic logics: Prospects and problems. *Fuzzy Sets Syst.* **2008**, *159*, 1860–1868. [CrossRef]
4. Guo, Y.; Cheng, H.D. A new neutrosophic approach to image segmentation. *Pattern Recogn.* **2009**, *42*, 587–595. [CrossRef]
5. Guo, Y.; Sengur, A. NCM: Neutrosophic c-means clustering algorithm. *Pattern Recogn.* **2015**, *48*, 2710–2724. [CrossRef]
6. Salama, A.A.; Broumi, S. Roughness of neutrosophic sets. *Elixir Appl. Math.* **2014**, *74*, 26833–26837.
7. Broumi, S.; Smarandache, F. Rough neutrosophic sets. *Ital. J. Pure Appl. Math.* **2014**, *32*, 493–502.
8. Wang, H.; Smarandache, F.; Zhang, Y.Q.; Sunderraman, R. Single valued neutrosophic sets, Multispace Multistruct. *Google Sch.* **2010**, *4*, 410–413.
9. Atanassov, K. Intuitionistic fuzzy sets. *Fuzzy Sets Syst.* **1986**, *20*, 87–96. [CrossRef]
10. Ye, J. Multicriteria decision-making method using the correlation coefficient under single-valued neutrosophic environment. *Int. J. Gen. Syst.* **2013**, *42*, 386–394. [CrossRef]
11. Ye, J. Improved correlation coefficients of single valued neutrosophic sets and interval neutrosophic sets for multiple attribute decision making. *J. Intell. Fuzzy Syst.* **2014**, *27*, 2453–2462.
12. Ye, S.; Ye, J. Dice similarity measure among single valued neutrosophic multisets and its applcation in medical diagnosis. *Neutrosophic Sets Syst.* **2014**, *6*, 48–53.
13. Majumdar, P.; Samant, S.K. On similarity and entropy of neutrosophic sets. *J. Intell. Fuzzy Syst.* **2014**, *26*, 1245–1252.
14. Şahin, R.; Küçük, A. Subsethood measure for single valued neutrosophic sets. *J. Intell. Fuzzy Syst.* **2015**, *29*, 525–530. [CrossRef]
15. Li, Z.W.; Cui, R.C. *T*-similarity of fuzzy relations and related algebraic structures. *Fuzzy Sets Syst.* **2015**, *275*, 130–143. [CrossRef]
16. Li, Z.W.; Cui, R.C. Similarity of fuzzy relations based on fuzzy topologies induced by fuzzy rough approximation operators. *Inf. Sci.* **2015**, *305*, 219–233. [CrossRef]
17. Li, Z.W.; Liu, X.F.; Zhang, G.Q.; Xie, N.X.; Wang, S.C. A multi-granulation decision-theoretic rough set method for distributed fc-decision information systems: An application in medical diagnosis. *Appl. Soft Comput.* **2017**, *56*, 233–244. [CrossRef]
18. Yang, H.L.; Guo, Z.L.; She, Y.H.; Liao, X.W. On single valued neutrosophic relations. *J. Intell. Fuzzy Syst.* **2016**, *30*, 1045–1056. [CrossRef]

19. Yang, H.L.; Zhang, C.L.; Guo, Z.L.; Liu, Y.L.; Liao, X.W. A hybrid model of single valued neutrosophic sets and rough sets: Single valued neutrosophic rough set model. *Soft Comput.* **2017**, *21*, 6253–6267. [CrossRef]
20. Pawlak, Z. Rough sets. *Int. J. Comput. Inf. Sci.* **1982**, *11*, 341–356. [CrossRef]
21. Bao, Y.L.; Yang, H.L. On single valued neutrosophic refined rough set model and its applition. *J. Intell. Fuzzy Syst.* **2017**, *33*, 1235–1248. [CrossRef]
22. Bo, C.X.; Zhang, X.H.; Shao, S.T.; Smarandache, F. Multi-Granulation Neutrosophic Rough Sets on a Single Domain and Dual Domains with Applications. *Symmetry* **2018**, *10*, 296. [CrossRef]
23. Zhao, H.; Zhang, H.Y. A result on single valued neutrosophic refined rough approximation operators. *J. Intell. Fuzzy Syst.* **2018**, 1–8. [CrossRef]

symmetry

MDPI

Article

Fixed Points Results in Algebras of Split Quaternion and Octonion

Mobeen Munir [1,*], Asim Naseem [2], Akhtar Rasool [1], Muhammad Shoaib Saleem [3] and Shin Min Kang [4,5,*]

[1] Division of Science and Technology, University of Education, Lahore 54000, Pakistan; akhtar.ue@gmail.com
[2] Department of Mathematics, Government College University, Lahore 54000, Pakistan; asimroz@gmail.com
[3] Department of Mathematics, University of Okara, Okara 56300, Pakistan; shaby45@yahoo.com
[4] Department of Mathematics and RINS, Gyeongsang National University, Jinju 52828, Korea
[5] Center for General Education, China Medical University, Taichung 40402, Taiwan
[*] Correspondence: mmunir@ue.edu.pk (M.M.); smkang@gnu.ac.kr (S.M.K.)

Received: 3 August 2018; Accepted: 14 September 2018; Published: 17 September 2018

Abstract: Fixed points of functions have applications in game theory, mathematics, physics, economics and computer science. The purpose of this article is to compute fixed points of a general quadratic polynomial in finite algebras of split quaternion and octonion over prime fields $\mathbb{Z}_p$. Some characterizations of fixed points in terms of the coefficients of these polynomials are also given. Particularly, cardinalities of these fixed points have been determined depending upon the characteristics of the underlying field.

Keywords: fixed point; split-quaternion; quadratic polynomial; split-octonion

Subject Classification (2010): 30C35; 05C31

1. Introduction

Geometry of space-time can be understood by the choice of convenient algebra which reveals hidden properties of the physical system. These properties are best describable by the reflections of symmetries of physical signals that we receive and of the algebra using in the measurement process [1–3]. Thus, we need normed division algebras with a unit element for the better understanding of these systems. For these reasons, higher dimension algebras have been an immense source of inspiration for mathematicians and physicists as their representations pave the way towards easy understanding of universal phenomenons. These algebras present nice understandings towards general rotations and describe some easy ways to consider geometric problems in mechanics and dynamical systems [4,5].

Quaternion algebra have been playing a central role in many fields of sciences such as differential geometry, human imaging, control theory, quantum physics, theory of relativity, simulation of particle motion, 3D geo-phones, multi-spectral images, signal processing including seismic velocity analysis, seismic waveform de-convolution, statistical signal processing and probability distributions (see [6–8] and references therein). It is known that rotations of 3D-Minkowski spaces can be represented by the algebra of split quaternions [5]. Applications of these algebras can be traced in the study of Graphenes, Black holes, quantum gravity and Gauge theory. A classical application of split quaternion is given in [1] where Pavsic discussed spin gauge theory. Quantum gravity of $2 + 1$ dimension has been described by Carlip in [2] using split quaternions. A great deal of research is in progress where authors are focused on considering matrices of quaternions and split-quaternions [9–12]. The authors in [13] gave a fast structure-preserving method to compute singular value decomposition of quaternion matrices. Split quaternions play a vital role in geometry and physical models in four-dimensional spaces as the elements of split quaternion are used to express Lorentzian rotations [14]. Particularly, the geometric and physical applications of split quaternions require solving split quaternionic equations [15,16]. Similarly, octonion and split octonion algebras play important role in mathematical physics. In [8], authors discussed ten dimensional space-time with help of these eight dimensional algebras. In [16], authors gave comprehensive applications of split octonions in geometry. Anastasiou developed M-theory algebra with the help of octonions [3].

This article mainly covers finite algebras of split quaternion and split octonion over prime fields $\mathbb{Z}_p$. Split quaternion algebra over $\mathbb{R}$ was in fact introduced by James Cockle in 1849 on already established quaternions by Hamilton in 1843. Both of these algebras are actually associative, but non-commutative, non-division ring generated by four basic elements. Like quaternion, it also forms a four dimensional real vector space equipped with a multiplicative operation. However, unlike the quaternion algebra, the split quaternion algebra contains zero divisors, nilpotent and nontrivial idempotents. For a detailed description of quaternion and its generalization (octonions), please follow [15–18]. As mathematical structures, both are algebras over the real numbers which are isomorphic to the algebra of 2×2 real matrices. The name split quaternion is used due to the division into positive and negative terms in the modulus function. The set $(1, \hat{i}, \hat{j}, \hat{k})$ forms a basis. The product of these elements are $\hat{i}^2 = -1$, $\hat{j}^2 = 1 = \hat{k}^2$, $\hat{i}\hat{j} = \hat{k} = -\hat{j}\hat{i}$, $\hat{j}\hat{k} = -\hat{i} = -\hat{k}\hat{j}$, $\hat{k}\hat{i} = \hat{j} = -\hat{i}\hat{k}$, $\hat{i}\hat{j}\hat{k} = 1$. It follows from the defining relations that the set $(\pm 1, \pm i, \pm j, \pm k)$ is a group under split quaternion multiplication which is isomorphic to the dihedral group of a square. Following Table 1 encodes the multiplication of basis split quaternions.

Table 1. Split quaternion multiplication table.

.	**1**	$\hat{i}$	$\hat{j}$	$\hat{k}$
1	1	$\hat{i}$	$\hat{j}$	$\hat{k}$
$\hat{i}$	$\hat{i}$	-1	$\hat{k}$	$-\hat{j}$
$\hat{j}$	$\hat{j}$	$-\hat{k}$	1	$-\hat{i}$
$\hat{k}$	$\hat{k}$	$\hat{j}$	$\hat{i}$	1

The split octonion is an eight-dimensional algebraic structure, which is non-associative algebra over some field with basis 1, $\acute{t}_1$, $\acute{t}_2$, $\acute{t}_3$, $\acute{t}_4$, $\acute{t}_5$, $\acute{t}_6$ and $\acute{t}_7$. The subtraction and addition in split octonions is computed by subtracting and adding corresponding terms and their coefficients. Their multiplication is given in this table. The product of each term can be given by multiplication of the coefficients and a multiplication table of the unit split octonions is given following Table 2.

Table 2. Split octonions' multiplication table.

.	$\acute{t}_1$	$\acute{t}_2$	$\acute{t}_3$	$\acute{t}_4$	$\acute{t}_5$	$\acute{t}_6$	$\acute{t}_7$
$\acute{t}_1$	-1	$\acute{t}_3$	$-\acute{t}_2$	$-\acute{t}_7$	$\acute{t}_6$	$-\acute{t}_5$	$\acute{t}_4$
$\acute{t}_2$	$-\acute{t}_3$	-1	$\acute{t}_1$	$-\acute{t}_6$	$-\acute{t}_7$	$\acute{t}_4$	$\acute{t}_5$
$\acute{t}_3$	$\acute{t}_2$	$-\acute{t}_1$	-1	$\acute{t}_5$	$-\acute{t}_4$	$-\acute{t}_7$	$\acute{t}_6$
$\acute{t}_4$	$\acute{t}_7$	$\acute{t}_6$	$-\acute{t}_5$	1	$-\acute{t}_3$	$\acute{t}_2$	$\acute{t}_1$
$\acute{t}_5$	$-\acute{t}_6$	$\acute{t}_7$	$\acute{t}_4$	$\acute{t}_3$	1	$-\acute{t}_1$	$\acute{t}_2$
$\acute{t}_6$	$\acute{t}_5$	$-\acute{t}_4$	$\acute{t}_7$	$-\acute{t}_2$	$\acute{t}_1$	1	$\acute{t}_3$
$\acute{t}_7$	$-\acute{t}_4$	$-\acute{t}_5$	$-\acute{t}_6$	$-\acute{t}_1$	$-\acute{t}_2$	$-\acute{t}_3$	1

From the table, we get very useful results:

$$\acute{t}_i^2 = -1, \forall i = 1, ..., 3,$$

$$\acute{t}_i^2 = 1, \forall i = 4, ..., 7,$$

$$\acute{t}_i \acute{t}_j = -\acute{t}_j \acute{t}_i, \forall i \neq j.$$

Brand in [19] computed the roots of a quaternion over $\mathbb{R}$. Strictly speaking, he proved mainly De Moivres theorem and then used it to find nth roots of a quaternion. His approach paved way for finding roots of a quaternion in an efficient and intelligent way. Ozdemir in [20] computed the roots of a split quaternion. In [21], authors discussed Euler's formula and De Moivres formula for quaternions. In [15], authors gave some geometrical applications of the split quaternion. It is important to mention that these two algebras can also be constructed for $\mathbb{Z}_p$ over prime finite fields of characteristic P. In this way, we obtain finite algebras with entirely different properties. Recently, the ring of quaternion over $\mathbb{Z}_p$ was studied by Michael Aristidou in [22,23], where they computed the idempotents and nilpotents in $\mathbb{H}/\mathbb{Z}_p$. In [18], authors computed the roots of a general quadratic polynomial in algebra of split quaternion over $\mathbb{R}$. They also computed fixed points of general quadratic polynomials in the same sittings. A natural question arises as to what happens with the same situations over $\mathbb{Z}_p$. Authors in [24] discussed split-quaternion over Z_p in algebraic settings.

In the present article, we first obtain the roots of a general quadratic polynomial in the algebra of split quaternion over $\mathbb{Z}_p$. Some characterizations of fixed points in terms of the coefficients of these polynomials are also given. As a consequence, we give some computations about algebraic properties of particular classes of elements in this settings. We also give examples as well as the codes that create these examples with ease. For a computer program, we refer to Appendix A at the end of the article. We hope that our results will be helpful in understanding the communication in machine language and cryptography.

Definition 1. *Let $x \in \mathbb{H}_s$, $x = a_0 + a_1\hat{i} + a_2\hat{j} + a_3\hat{k}$ where $a_i \in \mathbb{R}$. The* conjugate *of x is defined as $\bar{x} = a_0 - a_1\hat{i} - a_2\hat{j} - a_3\hat{k}$. The square of pseudo-norm of x is given by*

$$N(x) = x\bar{x} = a_0^2 + a_1^2 - a_3^2 - a_4^2. \tag{1}$$

Definition 2. *Let $x = a_0 + \sum_{i=1}^{7} a_i t_i \in \mathbb{O}_s/\mathbb{Z}_p$. The* conjugate *of x is defined as*

$$\begin{aligned}
\overline{x} &= \overline{a_0 + \sum_{i=1}^{7} a_i t_i} \\
&= a_0 + \sum_{i=1}^{7} a_i \overline{t_i} \\
&= a_0 - \sum_{i=1}^{7} a_i t_i \\
&= a_0 + \sum_{i=1}^{7} a_i' t_i,
\end{aligned}$$

where $a_i' = -a_i$ where $i = 1, 2, ..., 7$. The square of pseudo-norm of x is given by

$$N(x) = x\overline{x} = \sum_{i=0}^{3} a_i^2 - \sum_{i=4}^{7} a_i^2.$$

2. Main Results

In this section, we formulate our main results. At first, we give these results for split quaternions and then we move towards split octonions.

2.1. Some Fixed Points Results of Quadratic Functions in Split Quaternions over the Prime Field

We first solve a general quadratic polynomial in algebra of split quaternion. As a consequence, we find fixed points of associated functions in this algebra.

Theorem 1. *The quadratic equation $ax^2 + bx + c = 0$ $a, b, c \in \mathbb{Z}_p$, where p is an odd prime and $p \nmid a$, has root $x = a_0 + a_1\hat{i} + a_2\hat{j} + a_3\hat{k} \in \mathbb{H}_s/\mathbb{Z}_p$ if and only if $a_0 = \frac{p-b}{2a}$ and $a_1^2 - a_2^2 - a_3^2 = (\frac{p^2-b^2}{4a^2}) + \frac{c}{a}$.*

Proof.

$$x = a_0 + a_1\hat{i} + a_2\hat{j} + a_3\hat{k}, \tag{2}$$
$$x^2 = (a_0 + a_1\hat{i} + a_2\hat{j} + a_3\hat{k})^2, \tag{3}$$

$$\begin{aligned}
&= a_0^2 - a_1^2 + a_2^2 + a_3^2 + 2a_0a_1\hat{i} + 2a_0a_2\hat{j} + 2a_0a_3\hat{k} \\
&= a_0^2 + a_0^2 - \|x\| + 2a_0a_1\hat{i} + 2a_0a_2\hat{j} + 2a_0a_3\hat{k} \\
&= 2a_0^2 - \|x\| + 2a_0a_1\hat{i} + 2a_0a_2\hat{j} + 2a_0a_3\hat{k} \\
&= 2a_0(a_0 + a_1\hat{i} + a_2\hat{j} + a_3\hat{k}) - \|x\| \\
&= 2a_0x - \|x\|.
\end{aligned}$$

Putting x and x^2 into $ax^2 + bx + c = 0$, we have

$$\begin{aligned}
2aa_0x - a\|x\| + bx + c &= 0, \\
(2aa_0 + b)x - a\|x\| + c &= 0, \\
(2aa_0 + b)(a_0 + a_1\hat{i} + a_2\hat{j} + a_3\hat{k}) - a(a_0^2 + a_1^2 - a_2^2 - a_3^2) + c &= 0, \\
(2aa_0 + b)a_0 + (2aa_0 + b)(a_1\hat{i} + a_2\hat{j} + a_3\hat{k}) - a(a_0^2 + a_1^2 - a_2^2 - a_3^2) + c &= 0.
\end{aligned}$$

Comparing vector terms in the above equation, we get

$$2aa_0 + b = 0, \tag{4}$$
$$a_0 = \frac{-b}{2a} = \frac{p-b}{2a}. \tag{5}$$

Comparing constant terms, we get

$$(2aa_0 + b)a_0 - a(a_0^2 + a_1^2 - a_2^2 - a_3^2) + c = 0, \tag{6}$$
$$(2aa_0 + b)a_0 - aa_0^2 + c = a(a_1^2 - a_2^2 - a_3^2), \tag{7}$$
$$aa_0^2 + ba_0 + c = a(a_1^2 - a_2^2 - a_3^2), \tag{8}$$
$$(aa_0 + b)a_0 + c = a(a_1^2 - a_2^2 - a_3^2), \tag{9}$$
$$(a(\frac{p-b}{2a}) + b)\frac{p-b}{2a} + c = a(a_1^2 - a_2^2 - a_3^2), \tag{10}$$
$$\frac{p^2 - b^2}{4a^2} + \frac{c}{a} = a_1^2 - a_2^2 - a_3^2. \tag{11}$$

□

On the basis of the above results 2.1, we arrive at a new result given as

Theorem 2. *The fixed point of function* $f(x) = x^2 + (b+1)x + c$ *where* $a, b, c \in \mathbb{Z}_p$, p *is an odd prime and* $p \nmid a$ *is* $x = a_0 + a_1\hat{i} + a_2\hat{j} + a_3\hat{k} \in \mathbb{H}_s/\mathbb{Z}_p$ *if and only if* $a_0 = \frac{p-b}{2a}$ *and* $a_1^2 - a_2^2 - a_3^2 = (\frac{p^2-b^2}{4a^2}) + \frac{c}{a}$.

Proof. It is enough to give a new relation $f(x) = g(x) + x$, where $g(x) = x^2 + bx + c$. Then, existence of fixed points for $f(x)$ is equivalent to the solutions of $g(x)$. Then, the required result is immediate from the above theorem. □

Theorem 3. *Let* p *be an odd prime,* $p \nmid a$, *if* $x = a_0 + a_1\hat{i} + a_2\hat{j} + a_3\hat{k} \in \mathbb{H}_s/\mathbb{Z}_p$ *is a root of quadratic equation* $x^2 + bx + c = 0$, *where* $a, b, c \in \mathbb{Z}_p$. *Then, conjugate of* x *i.e.,* $\bar{x} = a_0 - a_1\hat{i} - a_2\hat{j} - a_3\hat{k} \in \mathbb{H}_s/\mathbb{Z}_p$ *is also the root of quadratic equation* $x^2 + bx + c = 0$.

Proof. The proof follows simply by using condition of Theorem 1 applied on the conjugate of x. □

Theorem 4. *Let* p *be an odd prime,* $p \nmid a$, *if* $x = a_0 + a_1\hat{i} + a_2\hat{j} + a_3\hat{k} \in \mathbb{H}_s/\mathbb{Z}_p$ *be the fixed point of function* $f(x) = x^2 + (b+1)x + c$, *where* $a, b, c \in Z_p$. *Then, the conjugate of* x *i.e.,* $\bar{x} = a_0 - a_1\hat{i} - a_2\hat{j} - a_3\hat{k} \in \mathbb{H}_s/\mathbb{Z}_p$ *also be the fix point of function* $f(x) = x^2 + (b+1)x + c$.

Proof. Again, it is enough to use relation $f(x) = g(x) + x$ where $g(x) = x^2 + bx + c$. Then, the existence of fixed points for $f(x)$ is equivalent to the solutions of $g(x)$. Then, the required result is immediate from the above theorem. □

The following two theorems are new results about the number of fixed points of $f(x) = x^2 + (b+1)x + c$.

Theorem 5. $|Fix(f)| = \begin{cases} p^2, & b = 0, c = 0, \\ p^2 + p + 2, & c = 0, b \neq 0. \end{cases}$

Proof. We split the proof in cases.

Case 1: For $c = 0$ and $b = 0$, we obtain two $\mathbb{H}_s/\mathbb{Z}_p \cong \mathbb{M}_2(\mathbb{Z}_p)$,where p is prime. It is easy to see that $\mathbb{H}_s/\mathbb{Z}_p$ and $\mathbb{M}_2(\mathbb{Z}_p)$ are isomorphic as algebras, the map $\varphi : \mathbb{H}_s/\mathbb{Z}_p \longmapsto \mathbb{M}_2(\mathbb{Z}_p)$ is defined as $\varphi(a_0 + a_1\hat{i} + a_2\hat{j} + a_3\hat{k}) = a_0\binom{10}{01} + a_1\binom{0p-1}{10} + a_2\binom{0p-1}{p-10} + a_3\binom{p-10}{01}$. As $\mathbb{H}_s/\mathbb{Z}_p \cong \mathbb{M}_2(\mathbb{Z}_p)$, so we find the number of nilpotent elements in $\mathbb{M}_2(\mathbb{Z}_p)$. It is well-known by Fine and Herstein that the probability that $n \times n$ matrix over a Galois field having p^α elements have $p^{\alpha.n}$ nilpotent elements. As in our case, $\alpha = 1$ and $n = 2$, thus the probability that the 2×2 matrix over $\mathbb{Z}_p$ has p^{-2} nilpotent elements:

$$\frac{|nil(\mathbb{M}_2(\mathbb{Z}_p))|}{|(\mathbb{M}_2(\mathbb{Z}_p|))} = p^{-2}, \tag{12}$$

$$\frac{|nil(\mathbb{M}_2(\mathbb{Z}_p))|}{p^4} = p^{-2}, \tag{13}$$

$$|nil(\mathbb{M}_2(\mathbb{Z}_p))| = p^2. \tag{14}$$

Case 2: For $c = 0$ and $b \neq 0$, we obtain as many points as there are matrices $M_2(\mathbb{Z}_p)$ because of the above isomorphism, and, using the argument given in 2, we arrive at the result. □

Theorem 6. *Let* $b \neq 0$ *and* $c \neq 0$. *Then,* $|Fix(f)| = \begin{cases} p^2 - p, & p \equiv 1(mod3), \\ p^2 + p, & p \equiv 2(mod3), \\ 3, & p = 3. \end{cases}$

Proof. Case 1: For $p = 3$, there is nothing to prove.
Case 2: For $p \equiv 1(mod3)$, we have two further cases:

case I: If $p \equiv 3(mod4)$,
$x^2 + y^2 = z$ has a unique solution for z = 0.
$x^2 + y^2 = z$ has $(p+1)$ options for $z \neq 0$, thus $(p+1)(p-1)$ options in all.
Thus, we get that $x^2 + y^2 = z$ has total number of solutions $(p+1)(p-1) + 1 = p^2 - 1 + 1 = p^2$.
Now, when $z = 0$, we get no solution for a_1:

$$\begin{aligned} \sharp &= 1(p+1) + 2(\frac{p-1}{2})(p+1) && (15)\\ &= p + 1 + p^2 - 1 && (16)\\ &= p^2 + p. && (17) \end{aligned}$$

case II: If $p \equiv 1(mod4)$,
$x^2 + y^2 = z$ has $(2p-1)$ solutions for z = 0.
$x^2 + y^2 = z$ has $(p-1)$ options for $z \neq 0$, thus $(p-1)(p-1)$ options in all.
Thus, we get that $x^2 + y^2 = z$ has total number of solutions
$(p-1)(p-1) + (2p-1) = p^2 - p - p + 1 + 2p - 1 = p^2$.
Now, when $z = 0$, we get two solutions for a_1:

$$\begin{aligned} \sharp &= 2(2p-1) + 2(\frac{p-3}{2})(p-1) + 1(p-1) && (18)\\ &= 4p - 2 + p^2 - p - 3p + 3 + p - 1 && (19)\\ &= p^2 + p. && (20) \end{aligned}$$

Case 3: For $p \equiv 2(mod3)$, we have two further cases:
case I: If $p \equiv 3(mod4)$
$x^2 + y^2 = z$ has a unique solution for z = 0.
$x^2 + y^2 = z$ has $(p+1)$ options for $z \neq 0$. So $(p+1)(p-1)$ options in all.
Thus we get, $x^2 + y^2 = z$ has total number of solutions $(p+1)(p-1) + 1 = p^2 - 1 + 1 = p^2$
Now, when $z = 0$, we get no solution for a_1:

$$\begin{aligned} \sharp &= 1(2) + 2(\frac{p-3}{2})(p+1) + 1(p+1) && (21)\\ &= 2 + p^2 + p - 3p - 3 + p + 1 && (22)\\ &= p^2 - p. && (23) \end{aligned}$$

case II: If $p \equiv 1(mod4)$,
$x^2 + y^2 = z$ has $(2p-1)$ solutions for z = 0.
$x^2 + y^2 = z$ has $(p-1)$ options for $z \neq 0$. So $(p-1)(p-1)$ options in all.
Thus we get, $x^2 + y^2 = z$ has total number of solutions
$(p-1)(p-1) + (2p-1) = p^2 - p - p + 1 + 2p - 1 = p^2$.
Now, when $z = 0$, we get two solutions for a_1.

$$\begin{aligned} \sharp &= 1(p-1) + 2(\frac{p-1}{2})(p-1) && (24)\\ &= p - 1 + p^2 - p - p + 1 && (25)\\ &= p^2 - p. && (26) \end{aligned}$$

□

2.2. Some Algebraic Consequences about $\mathbb{H}_s/\mathbb{Z}_p$

We can understand the algebraic structure of $\mathbb{H}_s/\mathbb{Z}_p$ with ease. The following results are simple facts obtained from the previous section.

Corollary 1. *Let p be an odd prime, an element*

$$x = a_0 + a_1\hat{i} + a_2\hat{j} + a_3\hat{k} \in \mathbb{H}_s/\mathbb{Z}_p \tag{27}$$

is idempotent $\Leftrightarrow a_0 = \frac{p+1}{2}$ *and* $a_1^2 - a_2^2 - a_3^2 = \frac{p^2-1}{4}$.

Proof. Taking $a = 1, b = p - 1$ and $c = p$ in the above theorem, we have

$$\begin{aligned} x^2 + (p-1)x + p &= 0, \\ x^2 - x &= 0, \\ x^2 &= x \end{aligned}$$

has root

$$x = a_0 + a_1\hat{i} + a_2\hat{j} + a_3\hat{k},$$

where

$$\begin{aligned} a_0 &= \frac{p-b}{2a} \\ &= \frac{p+1}{2}, \end{aligned}$$

and

$$\begin{aligned} a_1^2 - a_2^2 - a_3^2 &= \frac{p^2-b^2}{4a^2} + \frac{c}{a} = \frac{p^2-(-1)^2}{4(1)^2} + \frac{0}{1} \\ &= \frac{p^2-1}{4}. \end{aligned}$$

In other words, we can say x is idempotent. □

We also present similar results but without proof as they can be derived similarly.

Corollary 2. *Let p be an odd prime an element and*

$$x = a_0 + a_1\hat{i} + a_2\hat{j} + a_3\hat{k} \in \mathbb{H}_s/\mathbb{Z}_p \tag{28}$$

is idempotent if and only if $a_0 = \frac{p+1}{2}$ *and* $\|x\| = 0$.

Corollary 3. *Let p be an odd prime and* $x \in \mathbb{H}_s/\mathbb{Z}_p$*. If x is an idempotent, then* $\|x\| = 0$.

Corollary 4. *Let p be an odd prime. If* $x \in \mathbb{H}_s/\mathbb{Z}_p$ *is idempotent, then* $\bar{x}$ *is also an idempotent.*

Corollary 5. *Let p be an odd prime. If* $x \in \mathbb{H}_s/\mathbb{Z}_p$ *and x is of the form* $x = a_0$*. If x is idempotent, then it is either 0 or 1.*

Corollary 6. *Let p be an odd prime and $x \in \mathbb{H}_s/\mathbb{Z}_p$ of the form*

$$x = a_0 + a_1\hat{\imath} + a_2\hat{\jmath} + a_3\hat{k}, \tag{29}$$

where at least one $a_i \neq 0$. Then, x is not an idempotent.

Corollary 7. *Let p be an odd prime, and the quadratic equation $x^2 = 0$ has root $x = a_0 + a_1\hat{\imath} + a_2\hat{\jmath} + a_3\hat{k} \in \mathbb{H}_s/\mathbb{Z}_p$, where $a_0 = \frac{p}{2}$ and $a_1^2 - a_2^2 - a_3^2 = \frac{p^2}{4}$.*

Proof. Taking $a = 1$, $b = 0$ and $c = 0$ in the above theorem, we have that

$$\begin{aligned} x^2 + (p)x + o &= 0, \\ x^2 &= 0 \end{aligned}$$

has root

$$x = a_0 + a_1\hat{\imath} + a_2\hat{\jmath} + a_3\hat{k},$$

where

$$\begin{aligned} a_0 &= \frac{p-b}{2a} = \frac{p-0}{2} \\ &= \frac{p}{2}, \end{aligned}$$

and

$$\begin{aligned} a_1^2 - a_2^2 - a_3^2 &= \frac{p^2 - b^2}{4a^2} + \frac{c}{a} = \frac{p^2 - (0)^2}{4(1)^2} + \frac{0}{1} \\ a_1^2 - a_2^2 - a_3^2 &= \frac{p^2}{4}. \end{aligned}$$

In other words, we can say x is nilpotent. □

2.3. *Some Fixed Points Results of Quadratic Functions in Split Octonions over the Prime Field*

Theorem 7. *The quadratic equation $ax^2 + bx + c = 0$ where $a, b, c \in \mathbb{Z}_p$, p is an odd prime and $p \nmid a$ has root $x = a_0 + \sum_{i=1}^{7} a_i \acute{t}_i \in \mathbb{O}_s/\mathbb{Z}_p$ if and only if $a_0 = \frac{p-b}{2a}$ and $\sum_{i=1}^{3} a_i^2 - \sum_{i=4}^{7} a_i^2 = (\frac{p^2-b^2}{4a^2}) + \frac{c}{a}$.*

Proof.

$$ax^2 + bx + c = 0.$$

Take $x = a_0 + \sum_{i=1}^{7} a_i f_i$, we have

$$\begin{aligned}
x^2 &= (a_0 + \sum_{i=1}^{7} a_i f_i)^2 \\
&= (a_0)^2 + (\sum_{i=1}^{7} a_i f_i)^2 + 2a_0 \sum_{i=1}^{7} a_i f_i \\
&= (a_0)^2 - \sum_{i=1}^{7} a_i^2 + 2a_0 \sum_{i=1}^{7} a_i f_i where (\sum_{i=1}^{7} a_i f_i)^2 = -\sum_{i=1}^{7} a_i^2 \\
&= (a_0)^2 + a_0^2 - ||x|| + 2a_0 \sum_{i=1}^{7} a_i f_i where ||x|| = a_0 + \sum_{i=1}^{7} a_i^2 \\
&= 2(a_0)^2 - ||x|| + 2a_0 \sum_{i=1}^{7} a_i f_i \\
&= 2(a_0)^2 + 2a_0 \sum_{i=1}^{7} a_i f_i - ||x||, \\
&= 2a_0 x - ||x||.
\end{aligned}$$

Putting it in the above equation, we get

$$a(2a_0 x - ||x||) + bx + c = 0, \tag{30}$$

$$2aa_0 x - a||x|| + bx + c = 0, \tag{31}$$

$$(2aa_0 + b)x - a||x|| + c = 0. \tag{32}$$

Here, $x = a_0 + \sum_{i=1}^{7} a_i f_i$ and $||x|| = a_0^2 + \sum_{i=1}^{3} a_i^2 - \sum_{i=3}^{7} a_i^2$, we have

$$\begin{aligned}
(2aa_0 + b)(a_0 + \sum_{i=1}^{7} a_i f_i) - a[a_0^2 + \sum_{i=1}^{3} a_i^2 - \sum_{i=3}^{7} a_i^2] + c &= o, \\
(2aa_0 + b)a_0 + (2aa_0 + b) \sum_{i=1}^{7} a_i f_i - aa_0^2 - a \sum_{i=1}^{3} a_i^2 + a \sum_{i=3}^{7} a_i^2 + c &= o,
\end{aligned}$$

Comparing vector terms on both sides, we have

$$\begin{aligned}
(2aa_0 + b)a_i &= 0, \\
2aa_0 + b &= 0, \\
a_0 &= \frac{-b}{2a}, \\
a_0 &= \frac{p - b}{2a}.
\end{aligned}$$

Comparing constant terms on both sides, we have

$$\begin{aligned}
(2aa_0 + b)a_0 - aa_0^2 - a \sum_{i=1}^{3} a_i^2 + a \sum_{i=4}^{7} a_i^2 + c &= o, \\
2aa_0^2 + ba_0 - aa_0^2 + c &= a \sum_{i=1}^{3} a_i^2 - a \sum_{i=4}^{7} a_i^2, \\
a_0(aa_0 + b) + c &= a[\sum_{i=1}^{3} a_i^2 - \sum_{i=4}^{7} a_i^2],
\end{aligned}$$

where $a_0 = \frac{p-b}{2a}$.

$$\begin{aligned}
(\frac{p-b}{2a})(a(\frac{p-b}{2a})+b)+c &= a[\sum_{i=1}^{3} a_i^2 - \sum_{i=4}^{7} a_i^2], \\
(\frac{p-b}{2a})(\frac{p+b}{2})+c &= a[\sum_{i=1}^{3} a_i^2 - \sum_{i=4}^{7} a_i^2], \\
(\frac{p^2-b^2}{4a})+c &= a[\sum_{i=1}^{3} a_i^2 - \sum_{i=4}^{7} a_i^2], \\
(\frac{p^2-b^2}{4a^2})+\frac{c}{a} &= \sum_{i=1}^{3} a_i^2 - \sum_{i=4}^{7} a_i^2.
\end{aligned}$$

□

Theorem 8. *The fixed points of function $f(x) = ax^2 + (b+1)x + c$ are $x = a_0 + \sum_{i=1}^{7} a_i \acute{t}_i \in \mathbb{O}_s/\mathbb{Z}_p$, where $a_0 = \frac{p-b}{2a}$ and $\sum_{i=1}^{3} a_i^2 - \sum_{i=4}^{7} a_i^2 = (\frac{p^2-b^2}{4a^2}) + \frac{c}{a}$.*

Proof. It is enough to use relation $f(x) = g(x) + x$ where $g(x) = ax^2 + bx + c$. Then, the existence of fixed points for $f(x)$ is equivalent to the solutions of $g(x)$. Then, the required result is immediate from the above theorem. □

Corollary 8. *The fixed point of function $f(x) = x^2 + x$ are $x = a_0 + \sum_{i=1}^{7} a_i \acute{t}_i \in \mathbb{O}_s/\mathbb{Z}_p$ where $a_0 = \frac{p}{2}$ and $\sum_{i=1}^{3} a_i^2 - \sum_{i=4}^{7} a_i^2 = \frac{p^2}{4}$.*

Proof. It is obvious from the above theorem, only by taking $a = 1$, $b = 0$ and $c = 0$. □

Theorem 9. *Let p be an odd prime. If $x = a_0 + \sum_{i=1}^{7} a_i \acute{t}_i \in \mathbb{O}_s/\mathbb{Z}_p$ is the root of the quadratic equation $ax^2 + bx + c = 0$ $a, b, c \in Z_p$, then $\overline{x} = a_0 + \sum_{i=1}^{7} \acute{a}_i \acute{t}_i \in \mathbb{O}_s/\mathbb{Z}_p$ is also the root of the quadratic equation $ax^2 + bx + c = 0$ $a, b, c \in Z_p$.*

Proof.

$$\begin{aligned}
\overline{x} &= \overline{a_0 + \sum_{i=1}^{7} a_i \acute{t}_i} = a_0 + \sum_{i=1}^{7} \acute{a}_i \overline{\acute{t}_i} = a_0 - \sum_{i=1}^{7} a_i \acute{t}_i && (33) \\
&= a_0 + \sum_{i=1}^{7} \acute{a}_i \acute{t}_i, && (34)
\end{aligned}$$

where $\acute{a}_i = -a_i$ where $i = 1, 2, ..., 7$ as

$$a_0 = \frac{p-b}{2a}$$

and

$$\begin{aligned}
\sum_{i=1}^{3} \acute{a}_i^2 - \sum_{i=4}^{7} \acute{a}_i^2 &= \sum_{i=1}^{3} (-a_i)^2 - \sum_{i=4}^{7} (-a_i)^2 && (35) \\
&= \sum_{i=1}^{3} (a_i)^2 - \sum_{i=4}^{7} (a_i)^2 && (36) \\
&= \frac{p^2-b^2}{4a^2} + \frac{c}{a}. && (37)
\end{aligned}$$

It implies that $\overline{x}$ is the root of the quadratic equation $ax^2 + bx + c = 0$ $a, b, c \in Z_p$. □

Theorem 10. *If the function $f(x) = ax^2 + (b+1)x + c$ has fixed point $x = a_0 + \sum_{i=1}^{7} a_i t_i \in \mathbb{O}_s/\mathbb{Z}_p$, then $\overline{x} = a_0 + \sum_{i=1}^{7} \acute{a}_i \acute{t}_i \in \mathbb{O}_s/\mathbb{Z}_p$ also is the fixed point of function $f(x) = ax^2 + (b+1)x + c$.*

Proof. It is enough to use relation $f(x) = g(x) + x$, where $g(x) = ax^2 + bx + c$. Then, the existence of fixed points for $f(x)$ is equivalent to the solutions of $g(x)$. Then, the required result is immediate from the above theorem. □

3. Some Algebraic Consequences about $\mathbb{O}_s/\mathbb{Z}_p$

Proposition 1. *Let p be an odd prime and an element*

$$x = a_0 + \sum_{i=1}^{7} a_i \acute{t}_i \in \mathbb{O}_s/\mathbb{Z}_p$$

is idempotent $\Leftrightarrow a_0 = \frac{p+1}{2}$ and

$$\sum_{i=1}^{3} a_i^2 - \sum_{i=4}^{7} a_i^2 = \frac{p^2-1}{4}.$$

Proof. Taking $a = 1, b = p-1$ and $c = p$ in the above theorem, we have

$$x^2 + (p-1)x + p = 0, \tag{38}$$
$$x^2 - x = 0, \tag{39}$$
$$x^2 = x \tag{40}$$

has root

$$x = a_0 + \sum_{i=1}^{7} a_i \acute{t}_i, \tag{41}$$

where

$$a_0 = \frac{p-b}{2a} = \frac{p-p+1}{2} \tag{42}$$
$$= \frac{1}{2}, \tag{43}$$

and

$$\sum_{i=1}^{3} a_i^2 - \sum_{i=4}^{7} a_i^2 = \frac{p^2-b^2}{4a^2} + \frac{c}{a} = \frac{p^2-(-1)^2}{4(1)^2} + \frac{0}{1} \tag{44}$$
$$= \frac{p^2-1}{4}. \tag{45}$$

In other words, we can say that x is idempotent. □

Proposition 2. *Let p be an odd prime and element*

$$x = a_0 + \sum_{i=1}^{7} a_i \acute{t}_i \in \mathbb{O}_s/\mathbb{Z}_p \tag{46}$$

is idempotent if and only if $a_0 = \frac{p+1}{2}$ and $||x|| = 0$.

Proposition 3. *Let p be an odd prime and $x \in \mathbb{O}_s/\mathbb{Z}_p$. If x is an idempotent, then $||x|| = 0$.*

Proposition 4. *Let p be an odd prime. If* $x \in \mathbb{O}_s/\mathbb{Z}_p$ *is idempotent, then* $\bar{x}$ *is also an idempotent.*

Proposition 5. *Let p be an odd prime. If* $x = a_0 \in \mathbb{O}_s/\mathbb{Z}_p$ *is idempotent, then it is either* 0 *or* 1*.*

Proposition 6. *Let p be an odd prime and* $x \in \mathbb{O}_s/\mathbb{Z}_p$ *be of the form*

$$x = \sum_{i=1}^{7} a_i \acute{t}_i, \tag{47}$$

where at least one $a_i \neq 0$*. Then, x is not an idempotent.*

Proposition 7. *Let p be an odd prime and the quadratic equation* $x^2 = 0$ *has root* $x = a_0 + \sum_{i=1}^{7} a_i \acute{t}_i \in \mathbb{O}_s/\mathbb{Z}_p$*, where* $a_0 = \frac{p}{2}$ *and* $\sum_{i=1}^{3} a_i^2 - \sum_{i=4}^{7} a_i^2 = \frac{p^2}{4}$*.*

Proof. Taking $a = 1, b = 0$ and $c = 0$ in the above theorem, we have

$$x^2 + (p)x + o = 0, \tag{48}$$

$$x^2 = 0 \tag{49}$$

has root

$$x = a_0 + \sum_{i=1}^{7} a_i \acute{t}_i, \tag{50}$$

where

$$a_0 = \frac{p-b}{2a} = \frac{p-0}{2} \tag{51}$$

$$= \frac{p}{2} \tag{52}$$

and

$$\sum_{i=1}^{3} a_i^2 - \sum_{i=4}^{7} a_i^2 = \frac{p^2 - b^2}{4a^2} + \frac{c}{a} = \frac{p^2 - (0)^2}{4(1)^2} + \frac{0}{1} \tag{53}$$

$$= \frac{p^2}{4}. \tag{54}$$

In other words, we can say that x is nilpotent. □

Using results of the previous section and programs mentioned in the Appendix A, we can give many examples.

4. Examples

In this section, we add examples relating to the previous section. These results are generated by the codes given in Appendix A. These along with other examples can be created using codes, and results can be applied to crypto systems and communication channel systems.

Example 1. *We find all solutions of* $x^2 - x = 0$ *over* $\mathbb{H}_s/\mathbb{Z}_7$*. As above, we see that, if* $x \in \mathbb{H}_s/\mathbb{Z}_7$*, then* $a_0 = \frac{7-(-1)}{2} = 4$ *and following are the values of* a_1*,* a_2 *and* a_3*, respectively, satisfying the equation* $-a_1^2 + a_2^2 + a_3^2 = -5$ *or* $a_1^2 - a_2^2 - a_3^2 = 5$.

$$\begin{array}{cccccccc}
(0,1,1) & (0,1,6) & (0,3,0) & (0,4,0) & (0,1,1) & (0,6,1) & (0,6,6) & (1,1,3) \\
(1,1,4) & (1,3,1) & (1,3,6) & (1,4,1) & (1,4,6) & (1,6,4) & (2,2,3) & (2,2,4) \\
(2,3,2) & (2,3,3) & (2,3,5) & (2,4,2) & (2,4,3) & (2,4,5) & (2,5,3) & (2,5,4) \\
(3,0,2) & (3,0,5) & (3,2,0) & (3,3,3) & (3,3,4) & (3,4,3) & (3,4,4) & (3,5,0) \\
(4,0,2) & (4,0,5) & (4,2,0) & (4,3,3) & (4,3,4) & (4,4,3) & (4,4,4) & (4,5,0) \\
(5,0,6) & (5,2,3) & (5,2,4) & (5,3,2) & (5,3,5) & (5,4,2) & (5,4,5) & (5,5,3) \\
(5,5,4) & (5,6,0) & (6,1,4) & (6,3,6) & (6,4,1) & (6,4,6) & (6,6,3) & (6,6,4)
\end{array}.$$

Example 2. *We compute all solutions of* $2x^2 + x = 0$ *over* $\mathbb{H}_s/\mathbb{Z}_5$*. As above, we see that, if* $x \in \mathbb{H}_s/\mathbb{Z}_5$*, then* $a_0 = 1$ *and following are the values of* a_1*,* a_2 *and* a_3*, respectively, satisfying the equation* $-a_1^2 + a_2^2 + a_3^2 = -4$ *or* $a_1^2 - a_2^2 - a_3^2 = 4$.

$$\begin{array}{ccccc}
(0,0,1) & (1,4,1) & (2,2,4) & (3,1,2) & (3,4,2) \\
(0,0,4) & (1,4,4) & (2,3,1) & (3,1,3) & (3,4,3) \\
(0,1,0) & (2,0,0) & (2,3,4) & (3,2,1) & (4,1,1) \\
(0,4,0) & (2,1,2) & (2,4,2) & (3,2,4) & (4,1,4) \\
(1,1,1) & (2,1,3) & (2,4,3) & (3,3,1) & (4,4,1) \\
(1,1,4) & (2,2,1) & (3,0,0) & (3,3,4) & (4,4,4)
\end{array}.$$

Example 3. *We compute all solutions of* $x^2 + x + 1 = 0$ *over* $\mathbb{H}_s/\mathbb{Z}_7$*. As above, we see that, if* $x \in \mathbb{H}_s/\mathbb{Z}_7$*, then* $a_0 = 3$ *and following are the values of* a_1*,* a_2 *and* a_3*, respectively satisfying the equation* $-a_1^2 + a_2^2 + a_3^2 = -6$ *or* $a_1^2 - a_2^2 - a_3^2 = 6$.

$$\begin{array}{ccccccc}
(0,0,1) & (1,0,3) & (6,0,3) & (2,1,2) & (5,1,2) & (3,1,3) & (4,1,3) \\
(0,0,6) & (1,0,4) & (6,0,4) & (2,1,5) & (5,1,5) & (3,1,4) & (4,1,4) \\
(0,1,0) & (1,1,6) & (6,1,6) & (2,2,1) & (5,2,1) & (3,3,1) & (4,3,1) \\
(0,2,2) & (1,3,0) & (6,3,0) & (2,2,6) & (5,2,6) & (3,3,6) & (4,3,6) \\
(0,2,5) & (1,4,0) & (6,4,0) & (2,5,1) & (5,5,1) & (3,4,1) & (4,4,1) \\
(0,5,2) & (1,6,1) & (6,6,1) & (2,5,6) & (5,5,6) & (3,4,6) & (4,4,6) \\
(0,5,5) & (1,1,1) & (6,1,1) & (2,6,2) & (5,6,2) & (3,6,3) & (4,6,2) \\
(0,6,0) & (1,6,6) & (6,6,6) & (2,6,5) & (5,6,5) & (3,6,4) & (4,6,5)
\end{array}.$$

Example 4. *We compute all solutions of* $x^2 = 0$ *over* $\mathbb{H}_s/\mathbb{Z}_5$*. As above, we see that, if* $x \in \mathbb{H}_s/\mathbb{Z}_5$*, then* $a_0 = 0$ *and following are the values of* a_1*,* a_2 *and* a_3*, respectively, satisfying the equation* $-a_1^2 + a_2^2 + a_3^2 = 0$ *or* $a_1^2 - a_2^2 - a_3^2 = 0$.

$$\begin{array}{ccccc}
(0,0,0) & (0,3,4) & (1,4,0) & (2,0,3) & (3,3,0) \\
(0,1,2) & (0,4,2) & (4,0,1) & (2,2,0) & (0,3,1) \\
(0,1,3) & (0,4,3) & (4,0,4) & (2,3,0) & (1,1,0) \\
(0,2,1) & (1,0,1) & (4,1,0) & (3,0,2) & (2,0,2) \\
(0,2,4) & (1,0,4) & (4,4,0) & (3,0,3) & (3,2,0)
\end{array}.$$

Example 5. *We compute all solutions of* $x^2 - x = 0$ *over* $\mathbb{O}_s/\mathbb{Z}_3$ *(idempotents in the split octonion algebra). As above we see that* $x = a_0 + \sum_{i=1}^{7} a_i t_i \in \mathbb{O}_s/\mathbb{Z}_3$ *where* $a_0 = \frac{3-(-1)}{2} = 2$ *and following is the values of* a_1*,* a_2*,* a_3*,* a_4*,* a_5*,* a_6 *and* a_7 *respectively satisfying the equation* $\sum_{i=1}^{3} a_i^2 - \sum_{i=4}^{7} a_i^2 = (\frac{p^2-b^2}{4a^2}) + \frac{c}{a} = 2$. *We do so by putting values for* $p = 3$*,* $a = 1$*,* $b = -1$*,* $c = 0$ *in above given code.*

$$(2,1,0,1,1,0,2); (2,1,0,1,1,1,0); (2,1,0,1,1,2,0); (2,1,0,1,2,0,1);$$
$$(2,1,0,1,2,0,2); (2,1,0,1,2,1,0); (2,1,0,1,2,2,0); (2,1,0,2,0,1,1);$$
$$(2,1,0,2,0,1,2); (2,1,0,2,0,2,1); (2,1,0,2,0,2,2); (2,1,0,2,1,0,1);$$
$$(2,1,0,2,1,0,2); (2,1,0,2,1,1,0); (2,1,0,2,1,2,0); (2,1,0,2,2,0,1);$$
$$(2,1,0,2,2,0,2); (2,1,0,2,2,1,0); (2,1,0,2,2,2,0); (2,1,1,0,0,0,1);$$
$$(2,1,1,0,0,0,2); (2,1,1,0,0,1,0); (2,1,1,0,0,2,0); (2,1,1,0,1,0,0);$$
$$(2,1,1,0,2,0,0); (2,1,1,1,0,0,0); (2,1,1,1,1,1,1); (2,1,1,1,1,1,2);$$
$$(2,1,1,1,1,2,1); (2,1,1,1,1,1,1); (2,1,1,1,1,1,2); (2,1,1,1,1,2,1);$$
$$(2,1,1,1,1,2,2); (2,1,1,1,2,1,1); (2,1,1,1,2,1,2); (2,1,1,1,2,2,1);$$
$$(2,1,1,1,2,2,2); (2,1,1,2,0,0,0); (2,1,1,2,1,1,1); (2,1,1,2,1,1,2);$$
$$(2,1,1,2,1,2,1); (2,1,1,2,1,2,2); (2,1,1,2,2,1,1); (2,1,1,2,2,1,2);$$
$$(2,1,1,2,2,2,1); (2,1,1,2,2,2,2); (2,1,2,0,0,0,1); (2,1,2,0,0,0,2);$$
$$(2,1,2,0,0,1,0); (2,1,2,0,0,2,0); (2,1,2,0,1,0,0); (2,1,2,0,2,0,0);$$
$$(2,1,2,1,0,0,0); (2,1,2,1,1,1,1); (2,1,2,1,1,1,2); (2,1,2,1,1,2,1);$$
$$(2,1,2,1,1,2,2); (2,1,2,1,2,1,1); (2,1,2,1,2,1,2); (2,1,2,1,2,2,1);$$
$$(2,1,2,1,2,2,2); (2,1,2,2,0,0,0); (2,1,2,2,1,1,1); (2,1,2,2,1,1,2);$$
$$(2,1,2,2,1,2,1); (2,1,2,2,1,2,2); (2,1,2,2,2,1,1); (2,1,2,2,2,1,2);$$
$$(2,1,2,2,2,2,1); (2,1,2,2,2,2,2); (2,2,0,0,0,0,0); (2,2,0,0,1,1,1);$$
$$(2,2,0,0,1,1,2); (2,2,0,0,1,2,1); (2,2,0,0,1,2,2); (2,2,0,0,1,2,2);$$
$$(2,2,0,0,2,1,1); (2,2,0,0,2,1,2); (2,2,0,0,2,2,1); (2,2,0,0,2,2,2);$$
$$(2,2,0,1,0,1,1); (2,2,0,1,0,1,2); (2,2,0,1,0,2,1); (2,2,0,1,0,2,2);$$
$$(2,2,0,1,1,0,1); (2,2,0,1,1,0,2); (2,2,0,1,1,1,0); (2,2,0,1,1,2,0);$$
$$(2,2,0,1,2,0,1); (2,2,0,1,2,0,2); (2,2,0,1,2,1,0); (2,2,0,1,2,2,0);$$
$$(2,2,0,2,0,1,1); (2,2,0,2,0,1,1); (2,2,0,2,0,1,2); (2,2,0,2,0,2,1);$$
$$(2,2,0,2,0,2,2); (2,2,0,2,1,0,1); (2,2,0,2,1,0,2); (2,2,0,2,1,1,0);$$
$$(2,2,0,2,1,2,0); (2,2,0,2,2,0,1); (2,2,0,2,2,0,2); (2,2,0,2,2,1,0);$$
$$(2,2,0,2,2,2,0); (2,2,1,0,0,0,1); (2,2,1,0,0,0,2); (2,2,1,0,0,1,0);$$
$$(2,2,1,0,0,2,0); (2,2,1,0,1,0,0); (2,2,1,0,2,0,0); (2,2,1,1,0,0,0);$$
$$(2,2,1,1,1,1,1); (2,2,1,1,1,1,1); (2,2,1,1,1,1,2); (2,2,1,1,1,2,1);$$
$$(2,2,1,1,1,2,2); (2,2,1,1,2,1,1); (2,2,1,1,2,1,2); (2,2,1,1,2,2,1);$$
$$(2,2,1,1,2,2,2); (2,2,1,2,0,0,0); (2,2,1,2,1,1,1); (2,2,1,2,1,1,2);$$
$$(2,2,1,2,1,2,1); (2,2,1,2,1,2,2); (2,2,1,2,2,1,1); (2,2,1,2,2,1,2);$$
$$(2,2,1,2,2,2,1); (2,2,1,2,2,2,2); (2,2,2,0,0,0,1); (2,2,2,0,0,0,2);$$
$$(2,2,2,0,0,1,0); (2,2,2,0,0,2,0); (2,2,2,0,1,0,0); (2,2,2,0,2,0,0);$$
$$(2,2,2,1,0,0,0); (2,2,2,1,1,1,1); (2,2,2,1,1,1,2); (2,2,2,1,1,2,1);$$
$$(2,2,2,1,1,2,2); (2,2,2,1,2,1,2); (2,2,2,1,2,2,1); (2,2,2,1,2,2,2);$$
$$(2,2,2,2,0,0,0); (2,2,2,2,1,1,1); (2,2,2,2,1,1,2); (2,2,2,2,1,2,1);$$
$$(2,2,2,2,1,2,2); (2,2,2,2,2,1,1); (2,2,2,2,2,1,2); (2,2,2,2,2,2,1);$$
$$(2,2,2,2,2,2,2).$$

5. Conclusions and Further Directions

In this article, we produced some general results about fixed points of a general quadratic polynomial in algebras of split quaternion and octonion over $\mathbb{Z}_p$. We not only characterized these points in terms of the coefficients of these polynomials but also gave the cardinality of these points and also the programs that produced fixed points. We arrived at the following new results for a general quadratic function.

Theorem 11. $|Fix(f)| = \begin{cases} p^2, & b = 0, c = 0, \\ p^2 + p + 2, & c = 0, b \neq 0. \end{cases}$

Theorem 12. *Let $b \neq 0$ and $c \neq 0$. Then,* $|Fix(f)| = \begin{cases} p^2 - p, & p \equiv 1(mod3); \\ p^2 + p, & p \equiv 2(mod3); \\ 3, & p = 3. \end{cases}$

We also give the following two new results for the fixed points of a general quadratic quaternionic equation without proofs. Proofs are left as an open problem.

Theorem 13. $|Fix(f)| = \begin{cases} p^6, & b = 0, c = 0; \\ p^6 + p^3, & c = 0, b \neq 0. \end{cases}$

Theorem 14. *Let $b \neq 0$ and $c \neq 0$. Then,* $|Fix(f)| = \begin{cases} p^6 + p^3, & p \equiv 1(mod3); \\ p^6 - p^3, & p \equiv 2(mod3); \\ p^6, & p = 3. \end{cases}$

We like to remark that new results can be obtained for a general cubic polynomials in these algebras.

6. Data Availability Statement

No such data has been used to prove these results.

Author Contributions: M.M. conceived the idea and drafted the manuscript. Computations have been done by A.R., A.N. and M.S.S., S.M.K. edited and made final corrections. All authors read and approved the final manuscript.

Funding: This research received no external funding.

Acknowledgments: The authors are thankful to the reviewers for their valuable comments and suggestions. The authors are thankful to the University of Education Lahore for providing us a platform to present this article at ICE 2018.

Conflicts of Interest: The authors declare that they have no competing interests.

Appendix A. Computer Codes

Here, we put together some programs to compute fixed points and roots easily.

Appendix A.1. Program for Finding Solutions of the Quadratic Equation in $\mathbb{H}_s/\mathbb{Z}_p$

Following codes, count and print the number of solutions of quadratic equation $ax^2 + bx + c = 0$ in $\mathbb{H}_s/\mathbb{Z}_p$. These codes print the string a_1, a_2, a_3 with the understanding that the co-efficient $a_0 = \frac{p-b}{2a}$ is fixed in $\mathbb{H}_s/\mathbb{Z}_p$ and satisfying the relation $a_1^2 - a_2^2 - a_3^2 = (\frac{p^2-b^2}{4a^2}) + \frac{c}{a}$ or $-a_1^2 + a_2^2 + a_3^2 = (\frac{-p^2+b^2}{4a^2}) - \frac{c}{a}$ for $\mathbb{H}_s/\mathbb{Z}_p$.

CODE: This code will give solutions of the quadratic equation only by putting values for p, a, b, c, where p is an odd prime and $a, b, c \varepsilon \mathbb{Z}_p$.

```
#include<iostream>
```

```
#include<conio.h>
using namespace std;

main()
{
int a1, a2, a3, p, n, a, b, c, count;
count=0;
cout<<"Enter value for p: ";
cin>>p;
cout<<"Enter value for a: ";
cin>>a;
cout<<"Enter value for b: ";
cin>>b;
cout<<"Enter value for c: ";
cin>>c;
n=((p*p-b*b)/(4*a*a))+(c/a);
while(n<0)
n=n+p;
for(int i=0; i<p; i++)
{
a3=i;
for(int j=0; j<p; j++)
{
a2=j;
for(int k=0; k<p; k++)
{
a1=k;
int sum=(a1*a1)-(a2*a2)-(a3*a3);
while(sum<0)
sum=sum+p;
if(sum%p==n)
{
count++;
cout<<a1<<" "<<a2<<" "<<a3<<endl;
}
}
}
}
cout<<"\nCount: "<<count;
getch();
}
```

Appendix A.2. Program for Finding Roots of the Quadratic Equation in $\mathbb{O}_s/\mathbb{Z}_p$

Following codes, count and print the number of solutions of quadratic equation $ax^2 + bx + c = 0$ in $\mathbb{O}_s/\mathbb{Z}_p$. These codes print the string $a_1, a_2, a_3, a_4, a_5, a_6, a_7$ with the understanding that the co-efficient $a_0 = \frac{p-b}{2a}$ is fixed in $\mathbb{O}_s/\mathbb{Z}_p$ and satisfying the relation $\sum_{i=1}^{3} a_i^2 - \sum_{i=4}^{7} a_i^2 = (\frac{p^2-b^2}{4a^2}) + \frac{c}{a}$ for $\mathbb{O}_s/\mathbb{Z}_p$.

CODE: This code will give solutions of the quadratic equation only by putting values for p, a, b, c, where p is an odd prime and $a, b, c \in \mathbb{Z}_p$.

```
#include <iostream>
```

```
#include <fstream>
using namespace std;
int main(){
  int a1,a2,a3,a4,a5,a6,a7;
  int sum=0;
  int p;
  int n=0;
  int count=2;
  int totalCount=0;
  cout<<"Enter value of p: ";
  cin>>p;
  n = ((p*p -1)/4)%p;
  for(int i=0;i<p;i++)
  {
   a1 = i;
   for(int j=0; j<p; j++)
   {
    a2 = j;
    for(int k=0; k<p; k++)
    {
     a3 = k;
      for(int l=0; l<p;l++)
      {
       a4 = l;
       for(int m=0; m<p; m++)
       {
        a5 = m;
        for(int q=0; q<p; q++)
        {
         a6 = q;
         for(int r=0; r<p;r++)
         {
          a7 = r;
          totalCount++;
          cout<<a1<<" "<<a2<<" "<<a3<<" "<<a4<<" "<<a5
              <<" "<<a6<<" "<<a7<<endl;
          //dataFile << a1 << endl;
          sum = a1*a1+a2*a2+a3*a3-a4*a4-a5*a5-a6*a6-a7*a7;
          if(sum%p == n)
           count++;
          }
         }
        }
       }
      }
     }
    }
    cout<<"Total Count is: "<<totalCount<<endl;
    cout<<"Count is: "<<count<<endl;
    system("pause");
```

```
return 0;
}
```

References

1. Pavsic, M. Spin gauge theory of gravity in clifford space: A realization of kaluza-klein theory in four-dimensional spacetime. *Int. J. Mod. Phys. A* **2006**, *21*, 5905–5956. [CrossRef]
2. Carlip, S. *Quantum Gravity in* $2 + 1$ *Dimensions*; Cambridge University Press: Cambridge, UK, 2003.
3. Anastasiou, A.; Borsten, L.; Duff, M.J. An octonionic formulation of the M-theory algebra. *J. High Energy Phys.* **2014**, *2014*, 22. [CrossRef]
4. Kula, L.; Yayli, Y. Split Quaternions and Rotations in Semi Euclidean Space. *J. Korean Math. Soc.* **2007**, *44*, 1313–1327. [CrossRef]
5. Ozdemir, M.; Ergin, A. Rotations with Timelike Quaternions in Minkowski 3-Space. *J. Geom. Phys.* **2006**, *56*, 322–336. [CrossRef]
6. Dzhunushaliev, V. Cosmological constant, supersymmetry, nonassociativity, and big numbers. *Eur. Phys. J. C* **2015**, *75*, 86. [CrossRef]
7. Dzhunushaliev, V. A non-associative quantum mechanics. *Found. Phys. Lett.* **2006**, *19*, 157–167. [CrossRef]
8. Pavsic, M. Kaluza-klein theory without extra dimensions, curved Clifford space. *Phys. Lett. B* **2005**, *614*, 85–95. [CrossRef]
9. Alagz, Y.; Oral, K.; Yce, S. Split Quaternion Matrices. *Misk. Mathem. Not.* **2012**, *13*, 223–232.
10. Erdogdu, M.; Ozedemir, M. On Eigenvalues of Split Quaternion Matrics. *Adv. Appl. Clifford Algebras* **2013**, *23*, 615–623. [CrossRef]
11. Zhang, F. Quaternions and matrices of quaternions. *Linear Algebra Appl.* **1997**, *251*, 21–57. [CrossRef]
12. Lin, Y.; Wang, Q. Completing a 2×2 bock matrix of real quaternions with a partial specified inverse. *J. Appl. Math.* **2013**, *2013*, 271–278. [CrossRef]
13. Li, Y.; Wei, M.; Zhang, F.; Zhao, J. A fast structure-preserving method for computing the singular value decomposition of quaternion matrices. *Appl. Math. Comput.* **2014**, *235*, 157–167. [CrossRef]
14. Chung, K.; Sudbery, A. Octonions and the Lorentz and conformal groups of ten-dimensional space-time. *Phys. Lett. B* **1987**, *198*, 161–164. [CrossRef]
15. Gogberashvili, M.; Sakhelashvili, O. Geometrical Applications of Split Octonions. *Adv. Math. Phys.* **2015**, *2015*, 196708. [CrossRef]
16. Conway, J.H.; Smith, D.A. On Quaternions and Octonions: Their Geometry, Arithematic, and Symmetry. *Bull. Am. Math. Soc.* **2005**, *42*, 229–243.
17. John, C.B. The octonions. *Bull. Am. Math. Soc.* **2002**, *39*, 145–205.
18. Kwun, Y.C.; Munir, M.; Nazeer, W.; Kang, S. Some Fixed points Results of Quadratic Functions in Split Quaternions. *J. Funct. Spaces* **2106**, *2016*, 3460257. [CrossRef]
19. Brand, L. The roots of a quaternion. *Am. Math. Mon.* **1942**, *49*, 519–520. [CrossRef]
20. Ozdemir, M. The Roots of a Split Quaternion. *Appl. Math. Lett.* **2009**, *22*, 258–263. [CrossRef]
21. Cho, E. Euler's formula and De Moivre's formula for quaternions. *Mo. J. Math. Sci.* **1999**, *11*, 80–83.
22. Aristiou, M.; Demetre, A. Idempotent Elements in Quaternion Rings over $\mathbb{Z}_p$. *Int. J. Algebraic* **2012**, *6*, 249–254.
23. Miguel, C.J.; Serodio, R. On the Structure of Quaternion Rings over $\mathbb{Z}_p$. *Int. J. Algebra* **2011**, *5*, 1313–1325.
24. Munir, M.; Nizami, A.; Nazeer, W.; Kang, S. On Elements of Split Quaternions over Z_p. *Glob. J. Pure Appl. Math.* **2016**, *12*, 4253–4271.

symmetry

MDPI

Article

Optimizing the High-Level Maintenance Planning Problem of the Electric Multiple Unit Train Using a Modified Particle Swarm Optimization Algorithm

Jianping Wu [1], Boliang Lin [1,*], Hui Wang [2], Xuhui Zhang [3], Zhongkai Wang [2] and Jiaxi Wang [1]

1. School of Traffic and Transportation, Beijing Jiaotong University, Beijing 100044, China; jianpingwu@bjtu.edu.cn (J.W.); wangjiaxi@bjtu.edu.cn (J.W.)
2. Institute of Computing Technology, China Academy of Railway Sciences, Beijing 100081, China; wanghui2215@163.com (H.W.); winter-light@163.com (Z.W.)
3. Department of Vehicles, China Railway Shanghai Bureau Group Co., Ltd., Shanghai 200071, China; xhzhang121@163.com

* Correspondence: bllin@bjtu.edu.cn; Tel.: +86-10-5168-7149

Received: 17 July 2018; Accepted: 17 August 2018; Published: 19 August 2018

Abstract: Electric multiple unit (EMU) trains' high-level maintenance planning is a discrete problem in mathematics. The high-level maintenance process of the EMU trains consumes plenty of time. When the process is undertaken during peak periods of the passenger flow, the transportation demand may not be fully satisfied due to the insufficient supply of trains. In contrast, if the process is undergone in advance, extra costs will be incurred. Based on the practical requirements of high-level maintenance, a 0–1 programming model is proposed. To simplify the description of the model, candidate sets of delivery dates, i.e., time windows, are generated according to the historical data and maintenance regulations. The constraints of the model include maintenance regulations, the passenger transportation demand, and capacities of workshop. The objective function is to minimize the mileage losses of all EMU trains. Moreover, a modified particle swarm algorithm is developed for solving the problem. Finally, a real-world case study of Shanghai Railway is conducted to demonstrate the proposed method. Computational results indicate that the (approximate) optimal solution can be obtained successfully by our method and the proposed method significantly reduces the solution time to 500 s.

Keywords: Electric multiple unit trains; high-level maintenance planning; time window; 0–1 programming model; particle swarm algorithm

1. Introduction

In China, high-speed railway has become a priority option for the long trip due to its convenience and comfortableness, and it account for 60% of rail total passenger traffic. It has long been a difficult problem for the China Railway (CR) that supply enough available Electric Multiple Unit (EMU) trains in tourist rush seasons to fulfill heavy transportation tasks. The EMU trains are the unique vehicles running on the China high-speed railway. In addition, they have a high purchase cost and complicated maintenance regulations. Therefore, the problem becomes much worse.

The gigh-level maintenance planning (HMP) is an important part of the operation and maintenance management for the EMU trains, which covers the third-level maintenance, the fourth-level maintenance and the fifth-level maintenance. Due to the complex regular preventive maintenance, uneven distribution of passenger flow, limited maintenance capacity of workshop and several weeks for maintenance service time, the EMU trains' HMP needs to be scheduled in advance, of which the planning horizon lasts for a natural year or more. The HMP's aim is to provide enough available EMU trains for the long peak period of the passenger flow (such as the Spring Festival and the summer holiday) under the conditions of limited maintenance resources and the regular preventive maintenance policy. The HMP is a prerequisite for the EMU operational plan, the second-level maintenance plan, the job shop scheduling at workshop, and the outsourcing plan for high-level maintenance workloads (see Figure 1).

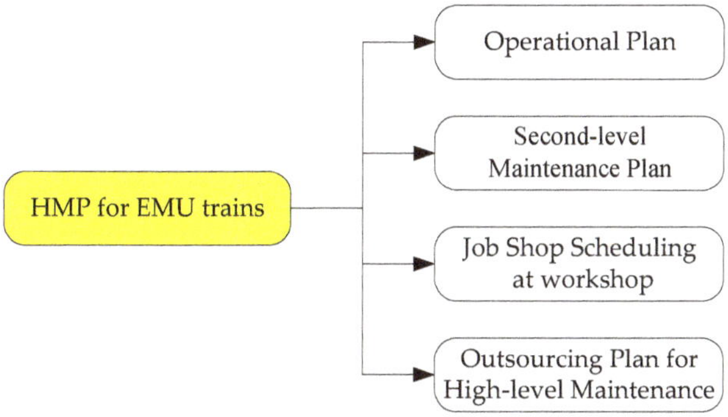

Figure 1. The position of high-level maintenance plan (HMP) in the operation and maintenance management system for the electric multiple unit (EMU) trains.

In the CR system, according to the maintenance regulations [1], the EMU trains that have been put into operation in the first period will undergo the high-level maintenance (HM) procedures together in the next few years. When lots of EMU trains are sent to workshop in tourist rush season, it will lead to the travel needs of passengers cannot be met. Not only the operating income but also the traveler satisfaction level on high-speed railway transportation will decrease. Meanwhile, it will increase the maintenance costs when the HM procedures are undergone in advance. Due to the limited capacities of workshops, the maintenance service time will be prolonged when plenty of EMU trains undergo the maintenance procedures together. Therefore, to scientifically formulate the HMP that meets the travel demands and reduces maintenance cost as much as possible is a complicated combinatorial optimization problem that needs to be solved in the field of high-speed railway operation.

For the maintenance system, Stuchly et al. [2] and Rezvanizaniani et al. [3] developed a condition based maintenance management system, Shimada [4] introduced the accident prevention maintenance system, and Cheng and Tsao [5] proposed a preventive and corrective maintenances system. These maintenance systems can reduce the maintenance costs and improve the utilization efficiency of EMU train.

Many experts and scholars study the first-level and the second-level maintenances. Maróti and Kroon [6,7] proposed that adjust the operation schedule ahead of time to ensure the maintenance procedures can be undertook in time. Tsuji et al. [8] developed a novel approach based on ant colony optimization and Wang et al. [9] designed an algorithm based column generation to solve the EMU train maintenance plan problem. Giacco et al. [10] integrated the rolling stock circulation problem and short-term maintenance planning, and a mixed-integer linear-programming formulation was proposed.

In other fields, scheduled maintenance planning problems also have been researched. Ziarati et al. [11], Lingaya et al. [12], and Wang et al. [13] studied the locomotive operation and maintenance. Moudania and Félix [14] and Mehmet and Bilge [15] developed the aircraft operation and maintenance. Budai et al. [16] researched the long-term planning of railway maintenance works. In addition, Grigoriev et al. [17] tried to find the length of maintenance plan to minimize the total operation costs.

There are a few relevant literatures available for the EMU trains HMP problem. Lin et al. [18] designed a state function to show the state of EMU train on each day during the planning horizon, and a non-linear 0–1 programming model and its solution strategy was proposed. Wu et al. [19] proposed a time-state network to optimize the EMU trains HMP problem. Li et al. [20] presented a forecast method to estimate the maintenance quantity of the EMU trains in future.

Compared to the existing researches, a 0–1 programming model and solution strategy were proposed. In the mathematic model, all necessary regulations and practical constraints were considered.

The remainder of this paper is organized as follows. The problem description of HMP in the CR system is presented in Section 2. In Section 3, a 0–1 programming model is proposed, and then the solution algorithm on the basis of particle swarm algorithm is designed in Section 4. An empirical case is provided to verify the effectiveness of the model and the algorithm in Section 5. The last section gives some conclusions and the possible areas of further research.

2. HMP Problem at CR

The HMP, a typical discrete system, is a tactical plan that determines when the EMU trains to undergo the high-level maintenance. The length of the planning horizon lasts for about one year. Each train undergoes the high-level maintenance at most once during the planning horizon because of the interval between two adjacent HM processes is longer than the span of planning horizon; the order of the maintenance level for a train is the third-level, the fourth-level, the third-level, the fifth level, the third-level and so on until they are scrapped [1]. Therefore, the level of the high-level maintenance can be deduced beforehand according to the records and regulations of maintenance. According to the maintenance regulations, each train has a time window during which the train can be delivered to workshop on any day. The lower bound of the time window is the earliest date on which a train can be delivered to workshop while the upper bound is the latest date. The detailed problem descriptions can be referred to Lin et al. [18] and Wu et al. [19]. Here, we focus on the generation steps of the time window.

The estimated time of arrival (eta) of the HM can be easily calculated [20]. In this process, the average daily operating mileage is used to describe the daily usage of EMU trains before HM procedures. The notations used in the generation process of the time window are listed in Table 1.

Table 1. Notations used in the generation process of the time window.

Notations	Definition
e	The index of EMU trains;
E	The set of all EMU trains;
E'	An empty set;
eta_e	The eta of the HM for the EMU train e;
$e(m)$	The type of the EMU train e;
$e(g)$	The maintenance level of the EMU train e;
R^+_{mg}	The maximum value of the difference between the actual operating mileage before the HM and the eta_e for the train of which the type is m and maintenance level is g;
R^-_{mg}	The maximum value of the difference between the eta_e and the actual operating mileage before the HM for the train of which the type is m and the maintenance level is g;
l_e	The average daily operating mileage for the EMU train e before HM.

The time window of the start time for the HM can be generated as follows.

Step 1. Take an EMU train e from E, and calculate the *eta* of HM [20];

Step 2. According to $e(m)$ and $e(g)$, determine the offset range of the operating mileage for the EMU train e: $[-R^-_{e(m),e(g)}, R^+_{e(m),e(g)}]$;

Step 3. Calculate the offset range of the time window for the EMU train e: $[-R^-_{e(m),e(g)}/l_e, R^+_{e(m),e(g)}/l_e]$;

Step 4. Determine the time window of the EMU train e: $[eta_e - R^-_{e(m),e(g)}/l_e, eta_e + R^+_{e(m),e(g)}/l_e]$, and set $E = E - \{e\}$, $E' = E' + \{e\}$;

Step 5. If $E = \phi$, turn to Step 6, otherwise turn to Step 1;

Step 6. Set $E = E'$, over.

The EMU train's time window is continuous if we set the "day" as the minimum time unit. In this way, the time window can be presented by a time interval.

In addition, the HMP aims to ensure that there are enough well-conditioned trains to meet the passenger transport demand. We set a maximum HM rate to guarantee it. The HM rate is the ratio of the number of trains in HM state to the fleet size.

3. Mathematical Model of the HMP Problem

In this section, we propose a 0–1 programming model for the HMP problem. The constraints of the model include the maintenance interval, the passenger transportation demand, and the capacity of workshop. The objective function is to minimize the mileage losses of all EMU trains.

3.1. Notations

The all notations that used in the model are listed in Table 2.

Table 2. Notations used in the model.

Notations	Definition
Indices	
e	The index of EMU trains;
m	The index of types for the EMU trains;
t	The index of dates during the planning horizon;
g	The index of the maintenance level;
Set	
E	The set of all EMU trains; $e \in E$
M	The set of all types; $m \in M$
T	The set of all dates during the planning horizon; $t \in T$
G	The set of all maintenance levels; $g \in G$
Input parameters	
$e(g)$	The maintenance level of the EMU train e;
$e(m)$	The type of the EMU train e;
c	The unit penalty fee for the unused mileage before the HM;
eta_e	The eta for the EMU train e;
l_e	The average daily operating mileage for the EMU train e before the HM;
WS_e	The first date of the time window for the EMU train e, $WS_e = eta_e - R^-_{e(m),e(g)}/l_e$ (See Section 2);
WE_e	The last date of the time window for the EMU train e, $WE_e = eta_e + R^+_{e(m),e(g)}/l_e$ (See Section 2);
α_m	The conversion coefficient for train of which the type is m, indicates whether the train includes sixteen cars or not, if yes, then α_m = 2; otherwise, α_m = 1;
θ_t	The maximum HM rate on the t-th day;
Inv	The fleet size (the standard set);
d^t_m	The maximum number of an EMU train in the HM state for the m type on the t-th day;
b_g	The maximum number of an EMU train in the g-th level maintenance state;
N_g	The maximum number of an acceptable EMU train in the g-th level maintenance state at the same time;
Q	A sufficiently large positive number;
H^g_m	The maintenance service time for an EMU train with the m-th type and the g-th level;
J^g_m	The minimum interval time for delivering another train after an EMU train with the m-th type and the g-th level enters the workshop;
$\|T\|$	The length of the planning horizon;
$\|E\|$	The number of the EMU trains which need to be maintained during the planning horizon;

Table 2. *Cont.*

Notations	Definition
Decision Variables	
x_e^t	Binary variable, indicates whether the EMU train e selects the t-th day to start the HM procedures during the planning horizon, x_e^t = 1 if yes, x_e^t = 0 otherwise;
y_e^t	Binary variable, indicates whether the EMU train e in the $e(g)$-th level maintenance state on the t-th day during the planning horizon, y_e^t = 1 if yes, y_e^t = 0 otherwise;
z_e^t	Binary variable, indicate whether the EMU train e in the $e(g)$-th level delivery interval on the t-th day during the planning horizon, z_e^t = 1 if yes, z_e^t = 0 otherwise;

3.2. Optimization Objective

The objective function of the mathematical model is to maximize the service efficiency of the EMU trains, i.e., to minimize the unutilized mileage. In this way, the objective function can be presented as follows.

$$\min Z = c \sum_{t\in[WS_e,WE_e]} \sum_{e\in E} (WE_e - t) l_e x_e^t \tag{1}$$

3.3. Constraints Analysis

According to Section 2, each train e can choose one and only one delivery date during the time window. This is the uniqueness constraint.

$$\sum_{t\in[WS_e,WE_e]} x_e^t = 1 \quad \forall e \in E \tag{2}$$

Any time t out of the time window for the train e cannot be selected.

$$x_e^t = 0 \quad \forall e \in E, t \notin [WS_e, WE_e] \text{ and } t \in T \tag{3}$$

The θ_t is a variable according to travel demand which should be guaranteed. Theoretically, the value of θ_t is different for each day during the planning horizon. To describe this requirement, a set of constraints established as follows.

$$\sum_{e\in E} \alpha_{e(m)} y_e^t \le \theta_t \cdot Inv \qquad \forall t \in T \tag{4}$$

Because of the various itineraries, the number of the trains with the specific type in the HM state must be less than the given threshold value on the t-th day. Constraints in this respect can be expressed as follows.

$$\sum_{e\in E|e(m)=m} y_e^t \le d_m^t \qquad \forall m \in M,\ t \in T \tag{5}$$

The number of trains in each level of the HM state should not be exceeded the capacity of workshops. A set of constraints can be listed as follows.

$$\sum_{e\in E|e(g)=g} y_e^t \le b_g \qquad \forall g \in G,\ t \in T \tag{6}$$

Meanwhile, restricted by the limited resources, only a few trains are permitted to enter the workshop over several days. This situation can be described in the form of mathematical inequalities as follows.

$$\sum_{e\in E|e(g)=g} z_e^t \le N_g \qquad \forall g \in G,\ t \in T \tag{7}$$

In addition, the logical relationships between those three sets of decision variables are presented as follows.

$$(x_e^k - 1)Q \le \left(\sum_{t=k}^{k+H_{e(m)}^{e(g)}-1} y_e^t - H_{e(m)}^{e(g)} \right) \le (1 - x_e^k)Q \qquad k \in [WS_e, WE_e],\ \forall e \in E \tag{8}$$

$$\sum_{t=1}^{|T|+H_{e(m)}^{e(g)}} y_e^t = H_{e(m)}^{e(g)} \qquad \forall\, e \in E \tag{9}$$

$$(x_e^k - 1)Q \le \left(\sum_{t=k}^{k+J_{e(m)}^{e(g)}-1} z_e^t - J_{e(m)}^{e(g)} \right) \le (1 - x_e^k)Q \qquad k \in [WS_e, WE_e],\ \forall e \in E \tag{10}$$

$$\sum_{t=1}^{|T|+J_{e(m)}^{e(g)}} z_e^t = J_{e(m)}^{e(g)} \qquad \forall\, e \in E \tag{11}$$

Finally, all of the decision variables are binary variables.

$$x_e^t,\ y_e^t,\ z_e^t \in \{0,1\} \qquad \forall\, e \in E, t \in T \tag{12}$$

3.4. Model Construction

On the basis of above analysis, a 0–1 programming model for the EMU train HMP problem is proposed as follows.

$$\text{HMP model: } \min Z = c \sum_{t \in [WS_e, WE_e]} \sum_{e \in E} (WE_e - t) l_e x_e^t$$

$$\text{s.t.} \quad \sum_{t \in [WS_e, WE_e]} x_e^t = 1 \qquad \forall\, e \in E$$

$$x_e^t = 0 \qquad \forall\, e \in E, t \notin [WS_e, WE_e] \text{ and } t \in T$$

$$\sum_{e \in E} \alpha_{e(m)} y_e^t \le \theta_t \cdot Inv \qquad \forall\, t \in T$$

$$\sum_{e \in E | e(m) = m} y_e^t \le d_m^t \qquad \forall m \in M,\ t \in T$$

$$\sum_{e \in E | e(g) = g} y_e^t \le b_g \qquad \forall\, g \in G,\ t \in T$$

$$\sum_{e \in E | e(g) = g} z_e^t \le N_g \qquad \forall\, g \in G,\ t \in T$$

$$(x_e^k - 1)Q \le \left(\sum_{t=k}^{k+H_{e(m)}^{e(g)}-1} y_e^t - H_{e(m)}^{e(g)} \right) \le (1 - x_e^k)Q \quad k \in [WS_e, WE_e],\ \forall e \in E$$

$$\sum_{t=1}^{|T|+H_{e(m)}^{e(g)}} y_e^t = H_{e(m)}^{e(g)} \qquad \forall\, e \in E$$

$$(x_e^k - 1)Q \le \left(\sum_{t=k}^{k+J_{e(m)}^{e(g)}-1} z_e^t - J_{e(m)}^{e(g)} \right) \le (1 - x_e^k)Q \quad k \in [WS_e, WE_e],\ \forall e \in E$$

$$\sum_{t=1}^{|T|+J_{e(m)}^{e(g)}} z_e^t = J_{e(m)}^{e(g)} \qquad \forall\, e \in E$$

$$x_e^t,\ y_e^t,\ z_e^t \in \{0,1\} \qquad \forall\, e \in E, t \in T$$

We can see from the HMP model that the number of all decision variables equals to $3 \times |T| \times |E|$ that is the product of two factors: the time span of the planning horizon and the number of trains. But the search space will reach $|T|^{|E|}$ according to the model. It is too complicated to be solved by using CPLEX or Gurobi within a reasonable time. Thus, a meta-heuristic solution strategy based on the particle swarm optimization (PSO) algorithm is designed to address this problem.

4. Modified Particle Swarm Optimization Algorithm

PSO algorithm is a population based stochastic optimization technique motivated by social behavior of organisms. PSO algorithm has the advantage of fast convergence speed and high accuracy solution and it is easy to be applied in most areas [21], which makes it attract great attention from researchers. The algorithm can also be used in solving the combinatorial optimization problem [22,23]. In this section, we present a modified particle swarm optimization (MPSO) algorithm to solve the HMP model based on analysis and preprocess.

4.1. Processing of Model Constraints

The constraint conditions of the HMP model need to be processed before applying the MPSO algorithm. Constraints (2), (3), (8)–(12) are the logical relationships, and they can be observed by the specific encoding rules (see the latter section), while the others can be removed by the penalty function method. For the value of the penalty factor, it is necessary to combine the actual application of the EMU trains and the strength of the constraint. Among the Inequations (4)–(7), the Inequation (7) has the strongest constraint and the Inequation (4) has the weakest constraint. The relationship of the penalty coefficient is $\lambda_4 > \lambda_3 > \lambda_2 > \lambda_1$. The optimization model can be presented as follows.

$$\begin{aligned} \min W = & \ c \sum_{t \in [WS_e, WE_e]} \sum_{e \in E} (WE_e - t) l_e x_e^t + \lambda_1 \sum_{t \in T} \max\{0, \sum_{e \in E} \alpha_{e(m)} y_e^t - \theta_t \cdot Inv\} \\ & + \lambda_2 \sum_{t \in T} \sum_{m \in M} \max\{0, \sum_{e \in E | e(m)=m} y_e^t - d_m^t\} + \lambda_3 \sum_{t \in T} \sum_{g \in G} \max\{0, \sum_{e \in E | e(g)=g} y_e^t - b_g\} + \lambda_4 \sum_{t \in T} \sum_{g \in G} \max\{0, \sum_{e \in E | e(g)=g} z_e^t - N_g\} \end{aligned} \tag{13}$$

s.t. (2), (3), (8)–(12).

4.2. General Particle Swarm Optimization Algorithm

In general PSO algorithm, the basic update equations of the velocity and position of the particles are as follows:

$$V_i(r+1) = \omega V_i(r) + c_1 \xi (HB_i(r) - P_i(r)) + c_2 \eta (GB(r) - P_i(r)) \tag{14}$$

$$P_i(r+1) = P_i(r) + V_i(r+1) \tag{15}$$

where $V_i(r)$ and $P_i(r)$ denote the velocity and the position, respectively, for particle i in the $r-$ th iteration. $HB_i(r)$ denotes the best position in the history for particle i by the end of the $r-$ th iteration; $GB(r)$ denotes the best position in the history for all of the particles by the end of the $r-$ th iteration. ω denotes the inertia weight; c_1 denotes the self-learning factor; c_2 denotes the social learning factor; ξ and η are the random numbers in [0, 1].

4.3. MPSO and Solution Strategy

4.3.1. Inertia Weight

In order to make the particles have a better search ability in the early stage of evolution and have a better development ability in the later stage of the evolution, the linear time-varying inertia weight is adopted in this paper. The inertia weight can be calculated as follows.

$$\omega(r) = \omega_{\max} - (\omega_{\max} - \omega_{\min}) \cdot r / MAXR \tag{16}$$

where $\omega(r)$ denotes the inertia weight at the r-th iteration. $\omega_{\max}$ and $\omega_{\min}$ denotes the maximum inertia weight and the minimum inertia weight, respectively. And $MAXR$ denotes the maximum number of the evolution iterations.

4.3.2. Learning Factor

In the same way, in order to make the particles strengthen the global search ability in the early stage and converge to the global optimum in the later period, we decrease the self-learning factor and increase the social learning factor continuously during the process of optimization. The calculation formulae are as follows.

$$c_1(r) = c'_1 + (c''_1 - c'_1) \cdot r / MAXR \tag{17}$$

$$c_2(r) = c'_2 + (c''_2 - c'_2) \cdot r / MAXR \tag{18}$$

where $c_1(r)$ and $c_2(r)$ denote the self-learning factor and the social learning factor in the $r -$ th iteration. c'_1 and c''_1 denote the initial value and the final value for the self-learning factor; c'_2 and c''_2 denote the initial value and the final value for the social learning factor; they are the constants.

4.3.3. Update Equations

The particle continuously updates its position in the search space at an unfixed speed. The velocity represents the variation of position in magnitude and direction like the definition in classical physics, and it has the same dimension as the position. Let $hbest_i(r)$ denote the corresponding fitness value of $HB_i(r)$. Let $gbest(r)$ denote the corresponding fitness value of $GB(r)$. The fitness value can be calculated by the formula (13). The update equations of the position and velocity for the particle i are as follows.

$$V_i(r+1) = \omega(r) \cdot V_i(r) + c_1(r) \cdot \xi(r) \cdot (HB_i(r) - P_i(r)) + c_2(r) \cdot \eta(r) \cdot (GB(r) - P_i(r)) \tag{19}$$

$$P_i(r+1) = P_i(r) + V_i(r+1) \tag{20}$$

where $\xi(r)$ and $\eta(r)$ are the random numbers in [0,1].

Each dimension of the velocity of a particle is limited to an interval $[-V_{\max}, V_{\max}]$, and if it is out of the interval, we set the boundary value of the interval as the actual velocity component. Similarly, the components of the position vector are limited to the time window for each train. Let $pbest_i(r)$ denote the fitness value for $P_i(r)$.

4.3.4. Encoding Rules and Initial Solution

It is a crucial step to make the particle of the MPSO and the solution of a certain problem correspond with each other. We use a particle to represent an overall HMP for all of the EMU trains. According to Equations (2) and (3), we set the dimensionality of a particle to $|E|$. The value of each dimension represents the start time of the HM for the corresponding EMU train. The detailed description is shown in Figure 2 with the help of schematic diagram.

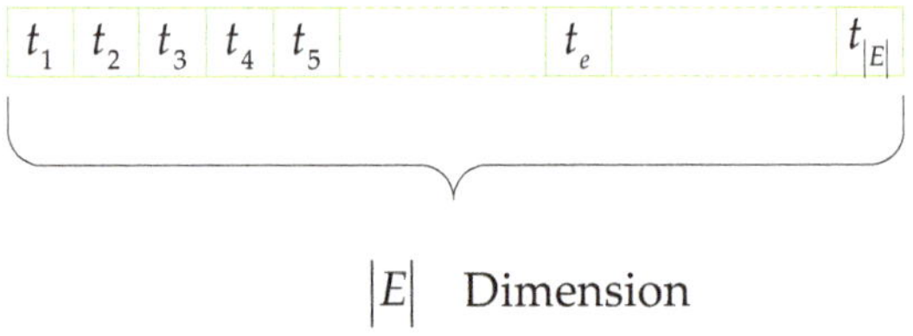

Figure 2. Encoding Rules (a particle).

In Figure 2, t_e denotes the start time of the HM for the e − th EMU train. Therefore, $x_e^t = 1$ when the value of t_e is determined, and the t in x_e^t equals to t_e. Because of the maintenance level and the type of each train can be ensured in advance, the value of $H_{e(m)}^{e(g)}$ and $J_{e(m)}^{e(g)}$ for the train e is known. Then, the value of y_e^t and z_e^t can be determined accordingly, i.e., $y_e^t = 1$ $(t \in [t_e, t_e + H_{e(m)}^{e(g)} - 1])$; $z_e^t = 1$ $(t \in [t_e, t_e + J_{e(m)}^{e(g)} - 1])$. In this way, the Constraints (2), (3), (8)–(12) are well handled.

The initial solution can be generated by selecting the start time of the HM during the time window for each train. The selection is stochastic in this process.

4.4. Algorithm Steps

Step 1. Generate the time window for all of the EMU trains that need to be maintained during the planning horizon, turn to Step 2.

Step 2. Initialization. Assign values to related parameters including the size of particle swarm I, $\omega_{\max}$, $\omega_{\min}$, c'_1, c''_1, c'_2, c''_2, $MAXR$ and $V_{\max}$. Generate the initial solution $P_i(0)$ according to Section 4.3.1, and generate the initial velocity $V_i(0)$ randomly. Set $r = 0$, turn to Step 3.

Step 3. Calculate the fitness value $pbest_i(r)$ of each particle, turn to Step 4.

Step 4. Compare the fitness values of $P_i(r)$ with $HB_i(r)$. If $pbest_i(r) < hbest_i(r)$, then $HB_i(r) = P_i(r)$, $hbest_i(r) = pbest_i(r)$. Turn to Step 5.

Step 5. Compare the fitness values of $HB_i(r)(i \in I)$ with $GB(r)$. If $hbest_i(r) < gbest(r)$, then $GB(r) = HB_i(r)$, $gbest(r) = hbest_i(r)$. Turn to Step 6.

Step 6. Update $\omega(r)$, $c_1(r)$ and $c_2(r)$ according to Formulas (16)–(18), turn to Step 7.

Step 7. Update $V_i(r)$ and $P_i(r)$ according to Formulas (19) and (20). If any dimension in $V_i(r)$ out of $[-V_{\max}, V_{\max}]$, we set the boundary value of the interval as an actual value. If any dimension in $P_i(r)$ out of $[WS_e, WE_e]$, we set the boundary value of the time window as an actual value, turn to Step 8.

Step 8. $r = r + 1$. If $r > MAXR$ or $gbest(r) = gbest(r - 200)$ $(r > 200)$, turn to Step 9; otherwise, turn to Step 3.

Step 9. Make $GB(r)$ feasible according to the HMP model, and output $GB(r)$ and $gbest(r)$. Over.

5. Case Study

In this section, we implement the proposed method to solve a practice problem. The detailed description of the case can be found in the literature [20]. The proposed model is solved by the commercial optimization solver, e.g., Gurobi, as well as the MPSO algorithm. The exact method is coded in Python 2.7 and implemented within Spyder 3.1.4 and the MPSO algorithm is implemented in C++. All the computational experiments are conducted on the computer with Intel Core i5-6200U CPU and 8 GB RAM.

According to the literature [20], some parameters are valued as follows. $Inv = 115$, $|E| = 60$, $|T| = 533$. To protect data confidentiality, we can only generate the time window in advance for each train and use an ID number replace the train. The initial conditions of trains when $|T| = 0$, which include the train ID, type, the average daily operating mileage, the time window, the maintenance level and the maintenance service time, are listed in Table 3.

Table 3. Initial conditions of all trains.

ID	Type	l_e (km)	Time Window	Level	$H_{e(m)}^{e(g)}$ (Day)
1	m1	1600	[64,164]	3	50
2	m1	1600	[64,164]	3	50
3	m1	1600	[90,190]	3	50
4	m1	1600	[124,224]	3	50
5	m1	1600	[144,244]	3	50
6	m1	1600	[188,288]	3	50
7	m1	1600	[200,300]	3	50
8	m1	1600	[274,374]	4	55
9	m1	1600	[370,470]	4	55
10	m1	1600	[379,479]	4	55
11	m1	1600	[429,479]	4	55
12	m2	1600	[72,172]	3	50
13	m2	1600	[158,258]	3	50
14	m2	1600	[216,316]	3	50
15	m2	1600	[264,354]	3	50
16	m2	1600	[387,484]	3	50
17	m2	1600	[396,484]	3	50
18	m2	1600	[409,484]	3	50
19	m2	1600	[443,484]	4	55
21	m3	2000	[134,234]	5	60
21	m3	2000	[149,249]	3	40
22	m3	2000	[150,250]	3	40
23	m3	2000	[153,253]	3	40
24	m3	2000	[158,258]	5	60
25	m3	2000	[158,258]	3	40
26	m3	2000	[159,259]	3	40
27	m3	2000	[167,267]	3	40
28	m3	2000	[172,272]	4	55
29	m3	2000	[172,272]	3	40
30	m3	2000	[182,282]	3	40
31	m3	2000	[184,284]	3	40
32	m3	2000	[185,285]	3	40
33	m3	2000	[190,290]	3	40
34	m3	2000	[190,290]	3	40
35	m3	2000	[191,291]	3	40
36	m3	2000	[192,292]	3	40
37	m3	2000	[193,293]	3	40
38	m3	2000	[206,306]	3	40
39	m3	2000	[209,309]	3	40
40	m3	2000	[211,311]	3	40
41	m3	2000	[211,311]	3	40
42	m3	2000	[249,349]	3	40
43	m3	2000	[276,376]	3	40
44	m3	2000	[280,380]	3	40
45	m3	2000	[281,381]	3	40
46	m3	2000	[289,389]	3	40
47	m3	2000	[290,390]	3	40
48	m3	2000	[299,399]	3	40
49	m3	2000	[300,400]	3	40
50	m3	2000	[309,409]	3	40
51	m3	2000	[309,409]	3	40
52	m3	2000	[320,420]	3	40
53	m3	2000	[321,421]	3	40
54	m3	2000	[325,425]	3	40
55	m3	2000	[395,494]	3	40
56	m3	2000	[398,494]	3	40
57	m3	2000	[425,494]	3	40
58	m3	2000	[429,494]	3	40
59	m3	2000	[448,494]	3	40
60	m3	2000	[459,494]	3	40

From Table 3, it can be seen that there are 60 EMU trains will undergo HM procedures during the planning period. Among them, there are 101 candidate dates at most, and the minimum values are only 36 candidate dates.

During the planning horizon, the HM rate is valued as follows:

$$\theta_t = \begin{cases} 0 & if\ t \in [149, 189) \\ 7\% & if\ t \in [189, 318) \\ 6\% & if\ t \in [318, 380) \\ 0 & if\ t = 533 \\ 9\% & otherwise \end{cases} \tag{21}$$

All of these trains undergo the HM procedures in the factory. The maximal capacity is ten, i.e., $b_1 + b_2 + b_3 = 10$. Meanwhile, the receiving capability per day for a factory is limited, e.g., N_3+N_4+N_5 = 1; J^3_{m1} = J^4_{m1} = J^5_{m1} = J^3_{m2} = J^4_{m2} = J^5_{m2} = J^3_{m3} = J^4_{m3} = J^5_{m3} = 1. According to the demand, d^t_{m1} = d^t_{m2} = 4, d^t_{m3} = 10 ($t \in T$). Therefore, the Formula (13) can be converted to the following form.

$$\begin{aligned} \min W = c \sum_{t\in[WS_e, WE_e]} \sum_{e\in E} (WE_e - t) l_e x^t_e + \lambda_1 \sum_{t\in T} \max\{0, \sum_{e\in E} \alpha_{e(m)} y^t_e - \theta_t Inv\} \\ + \lambda_2 \sum_{t\in T} \sum_{m\in M} \max\{0, \sum_{e\in E|e(m)=m} y^t_e - d^t_m\} + \lambda_3 \sum_{t\in T} \max\{0, -10 + \sum_{g\in G} \sum_{e\in E|e(g)=g} y^t_e\} + \lambda_4 \sum_{t\in T} \max\{0, -1 + \sum_{g\in G} \sum_{e\in E|e(g)=g} z^t_e\} \end{aligned} \tag{22}$$

In addition, the values of other parameters are as follows. c = 0.001, λ_1 = 100, λ_2 = 500, λ_3 = 800, λ_4 = 1000; I = 1000, ω_{max} = 1.2, ω_{min} = 0.8, c'_1 = 2.5, c''_1 = 0.5, c'_2 = 0.5, c''_2 = 2.5, V_{max} = (WE_e-WS_e+1)/2 $MAXR$ = 1000.

Based on the data given above, the HMP model is solved by the proposed algorithm. The program runs for 500 s. The returned optimal fitness value is 3,213,121. The curve of the optimal fitness value in the iterative process is depicted in Figure 3.

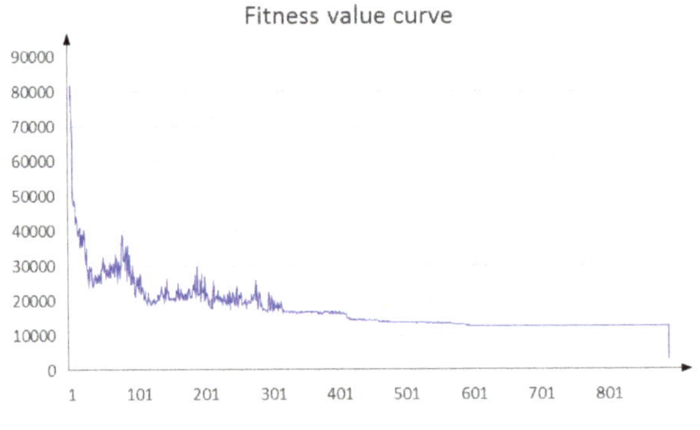

Figure 3. The optimal fitness value curve.

As can be seen in Figure 3, in the first 300 iterations, the algorithm has a strong search capability, which can effectively avoid the occurrence of the premature phenomena; and from about the 300th to the 690th iteration, the development ability of the algorithm is strengthened, which is helpful to search the optimal solution; the fitness value remains the same in the last 200 iterations, indicating that the (approximate) optimal solution for the HMP problem has been generated. The (approximate) optimal solution is listed in Table 4, and the first column is the train's ID; the second column is the start date of HM procedures denoted by t_e.

Table 4. Optimal solution.

e	t_e	e	t_e	e	t_e	e	t_e	e	t_e	e	t_e
1	108	11	481	21	182	31	280	41	310	51	409
2	103	12	142	22	220	32	252	42	349	52	419
3	153	13	212	23	250	33	278	43	376	53	421
4	193	14	333	24	240	34	273	44	379	54	425
5	198	15	360	25	248	35	282	45	381	55	472
6	300	16	460	26	188	36	276	46	388	56	498
7	312	17	492	27	218	37	238	47	390	57	500
8	387	18	458	28	268	38	303	48	398	58	502
9	482	19	490	29	270	39	243	49	400	59	496
10	480	20	190	30	246	40	308	50	407	60	494

We present the daily HM rates from the (approximate) optimal solution, and compare those with the predefined maintenance rate thresholds (see Figure 4).

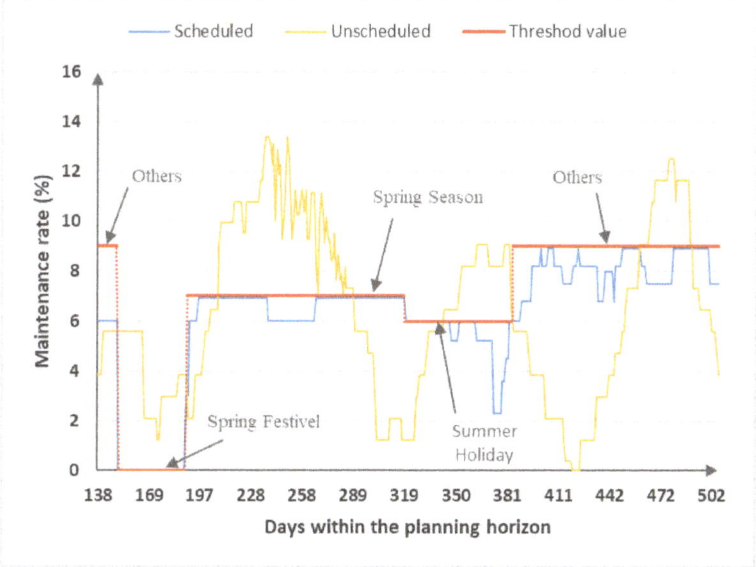

Figure 4. The distribution of the high-level maintenance rate in 2017.

From Figure 4, we can see that the HM rate of the optimal solution remains below the threshold in each period. Therefore, the proposed HMP model and the algorithm can meet the travel demands. However, the HM rate from the unscheduled solution fluctuates sharply.

In addition, to compare the performance of the proposed algorithm, we solve the HMP model by the Gurobi solver because the HMP model is linear. The detailed numerical comparison results of two solution methods, i.e., the Gurobi and the MPSO algorithm, are shown in Table 5.

Table 5. Numerical comparison results.

Method	z (km)	Time Consumption (s)
Gurobi	3,213,121	3186 s
MPSO	3,213,121	500 s

We can conclude that our proposed approach is efficient and effective from Table 5. This result demonstrates that the proposed algorithm is more efficient than the commercial optimization solver with reducing 84.31% in the solution time consumption.

6. Conclusions

This paper researches the electric multiple unit train's HMP problem. A 0–1 integer programming model and a modified particle swarm optimization algorithm are proposed. The objective function of the model is to minimize the unutilized mileage for all trains, and the model considers the necessary regulations and practical constraints, including passenger transport demand, workshop maintenance capacity, and maintenance regulations. A real-world instance demonstrates that the proposed method can efficaciously obtain the (approximate) optimal solution (see Table 4), and the solution strategy significantly reduces the solution time to 500 s (see Table 5). This result also demonstrates that the proposed algorithm is more efficient than the commercial optimization solver with reducing 84.31% in the solution time consumption. Optimize the workshop's overhaul process to shorten the maintenance service time is needed in future research.

Author Contributions: The authors contributed equally to this work.

Acknowledgments: This work was supported by the National Railway Administration of the People's Republic of China under the Grant Number KF2017-015, and the China Railway (formerly Ministry of Railways) under the Grant Number 2015X004-C. We would like to thank Lirong Diao of State Intellectual Property Office of the P.R.C. and Yinan Zhao of Beijing Jiaotong University for her valuable comments and suggestions.

Conflicts of Interest: The authors declare no conflicts of interest.

References

1. China Railway. *User Manual for EMU Operation and Maintenance Procedures*, 1st ed.; Railway Publishing House: Beijing, China, 2013.
2. Stuchly, V.; Grencik, J.; Poprocky, R. Railway vehicle maintenance and information systems. *WIT Trans. Built Environ.* **2000**, *50*, 885–894.
3. Rezvanizanianil, S.M.; Valibeiglool, M.; Asgharil, M.; Barabady, J.; Kumar, U. Reliability centered maintenance for rolling stock: A case study in coaches' wheel sets of passenger trains of Iranian railway. In Proceedings of the IEEE International Conference on Industrial Engineering and Engineering Management, Singapore, 8–11 December 2008; pp. 516–520.
4. Shimada, N. Rolling stock maintenance for safe and stable transport. *Jpn. Railw. Eng.* **2006**, *46*, 4–7.
5. Cheng, Y.H.; Tsao, H.L. Rolling stock maintenance strategy selection, spares parts' estimation, and replacements' interval calculation. *Int. J. Prod. Econ.* **2010**, *128*, 404–412. [CrossRef]
6. Maróti, G.; Kroon, L. Maintenance routing for train units: The transition model. *Transp. Sci.* **2005**, *39*, 518–525. [CrossRef]
7. Maróti, G.; Kroon, L. Maintenance routing for train units: The interchange model. *Comput. Oper. Res.* **2007**, *34*, 1121–1140. [CrossRef]
8. Tsuji, Y.; Kuroda, M.; Imoto, Y. Rolling stock planning for passenger trains based on ant colony optimization. *Trans. Jpn. Soc. Mech. Eng. Part C* **2010**, *76*, 397–406. [CrossRef]
9. Wang, Y.; Liu, J.; Miao, J.R. Column generation algorithms based optimization method for maintenance scheduling of multiple units. *China Railw. Sci.* **2010**, *31*, 115–120.
10. Giacco, G.L.; D'Ariano, A.; Pacciarelli, D. Rolling stock rostering optimization under maintenance constraints. *J. Intell. Transp. Syst.* **2014**, *18*, 95–105. [CrossRef]
11. Ziaratia, K.; Soumisa, F.; Desrosiers, J.; Gélinas, S.; Saintonge, A. Locomotive assignment with heterogeneous consists at CN North America. *Eur. J. Oper. Res.* **1997**, *97*, 281–292. [CrossRef]
12. Lingaya, N.; Cordeau, J.F.; Desaulniers, G.; Desrosiers, J.; Soumis, F. Operational car assignment at VIA Rail Canada. *Transp. Res. B Meth.* **2002**, *36*, 755–778. [CrossRef]
13. Wang, L.; Ma, J.J.; Lin, B.L.; Chen, L.; Wen, X.H. Method for optimization of freight locomotive scheduling and routing problem. *J. China Railw. Soc.* **2014**, *36*, 7–15. [CrossRef]

14. Moudania, W.E.; Félix, M.C. A dynamic approach for aircraft assignment and maintenance scheduling by airlines. *J. Air Transp. Manag.* **2000**, *6*, 233–237. [CrossRef]
15. Mehmet, B.; Bilge, Ü. Operational aircraft maintenance routing problem with remaining time consideration. *Eur. J. Oper. Res.* **2014**, *235*, 315–328. [CrossRef]
16. Budai, G.; Huisman, D.; Dekker, R. Scheduling preventive railway maintenance activities. *J. Oper. Res. Soc.* **2006**, *57*, 1035–1044. [CrossRef]
17. Grigoriev, A.; Klundert, J.V.D.; Spieksma, F.C.R. Modeling and solving the periodic maintenance problem. *Eur. J. Oper. Res.* **2006**, *172*, 783–797. [CrossRef]
18. Lin, B.L.; Lin, R.X. An Approach to the high-level maintenance planning for EMU trains based on simulated annealing. *arXiv* **2017**, arXiv:1704.02752v1. Available online: https://arxiv.org/abs/1704.02752 (accessed on 10 April 2017).
19. Wu, J.P.; Lin, B.L.; Wang, J.X.; Liu, S.Q. A network-based method for the EMU train high-level maintenance planning problem. *Appl. Sci.* **2018**, *8*, 2. [CrossRef]
20. Li, Y.; Zhang, W.J.; Jia, Z.K. Forecast method of annual senior over haul amount for EMU. *J. Transp. Eng.* **2013**, *13*, 102–107.
21. Wang, D.W.; Wang, J.W.; Wang, H.F.; Zhang, R.Y; Guo, Z. *Intelligent Optimization Method*; High Education Press: Beijing, China, 2007; pp. 217–259.
22. Li, J.; Lin, B.L.; Wang, Z.K.; Chen, L.; Wang, J.X. A pragmatic optimization method for motor train set assignment and maintenance scheduling problem. *Discrete Dyn. Nat. Soc.* **2016**, *3*, 1–13. [CrossRef]
23. Wang, J.X.; Lin, B.L.; Jin, J.C. Optimizing the shunting schedule of electric multiple units depot using an enhanced particle swarm optimization algorithm. *Comput. Intell. Neurosci.* **2016**, *1*, 1–11. [CrossRef] [PubMed]

symmetry

MDPI

Article

Binary Icosahedral Group and 600-Cell

Jihyun Choi and Jae-Hyouk Lee *

Department of Mathematics, Ewha Womans University 52, Ewhayeodae-gil, Seodaemun-gu, Seoul 03760, Korea; katherinejh.choi@gmail.com

* Correspondence: jaehyoukl@ewha.ac.kr; Tel.: +82-2-3277-3346

Received: 10 July 2018; Accepted: 26 July 2018; Published: 7 August 2018

Abstract: In this article, we have an explicit description of the binary isosahedral group as a 600-cell. We introduce a method to construct binary polyhedral groups as a subset of quaternions $\mathbb{H}$ via spin map of $SO(3)$. In addition, we show that the binary icosahedral group in $\mathbb{H}$ is the set of vertices of a 600-cell by applying the Coxeter–Dynkin diagram of H_4.

Keywords: binary polyhedral group; icosahedron; dodecahedron; 600-cell

MSC: 52B10; 52B11; 52B15

---

1. Introduction

The classification of finite subgroups in $SL_n(\mathbb{C})$ derives attention from various research areas in mathematics. Especially when $n = 2$, it is related to McKay correspondence and ADE singularity theory [1].

The list of finite subgroups of $SL_2(\mathbb{C})$ consists of cyclic groups ($\mathbb{Z}_n$), binary dihedral groups corresponded to the symmetry group of regular $2n$-gons, and binary polyhedral groups related to regular polyhedra. These are related to the classification of regular polyhedrons known as Platonic solids. There are five platonic solids (tetrahedron, cubic, octahedron, dodecahedron, icosahedron), but, as a regular polyhedron and its dual polyhedron are associated with the same symmetry groups, there are only three binary polyhedral groups (binary tetrahedral group $2T$, binary octahedral group $2O$, binary icosahedral group $2I$) related to regular polyhedrons. Moreover, it is a well-known fact that there is a correspondence between binary polyhedral groups and vertices of 4-polytopes as follows:

$$\begin{aligned} 2T &\leftrightarrow \text{vertices of 24-cell,} \\ 2O &\leftrightarrow \text{vertices of dual compound of 24-cell,} \\ 2I &\leftrightarrow \text{vertices of 600-cell,} \end{aligned}$$

where the dual compound of 24-cell means by the compound polytopes obtained from 24-cell and its dual polytope, which is also a 24-cell [2–5].

As the symmetries of polyhedrons are isometries, the related finite subgroups are also considered as the subgroups of $SU(2)$. As $SU(2) = Sp(1)$ is a spin group of $SO(3)$, we can regard $2T$, $2O$, and $2I$ as subgroups of quaternions $\mathbb{H}$. From this point of view, it is also well known that the vertices of 24 cell correspond to roots of D_4, and the set of vertices of the dual compound of 24-cell, which is the union of a 24-cell and a dual 24-cell forms a roots of F_4. The 600-cell is a complicated case of a reflection group of H_4-type [3,6].

The aim of this article is to provide explicit description of a binary icosahedron group $2I$ as a 600-cell. By applying spin covering map from $Sp(1)$ to $SO(3)$, we introduce a method to construct the binary polyhedral groups in terms of quaternions from the symmetries of regular polyhedrons. Then, by applying the theory of reflection groups along the Coxeter–Dynkin diagram, we show that the subgroup $2I$ in $\mathbb{H}$ is indeed the set of vertices of a 600-cell. We also discuss $2T$ related to 24-cell, but, because the dual compound of a 24-cell is not regular, its relation to $2O$ will be discussed in another article.

2. Binary Polyhedral Groups in Quaternions

Every finite subgroup of $SL_2(\mathbb{C})$ is conjugate to a finite subgroup of $SU(2)$ so that the classification of the finite subgroup of $SL_2(\mathbb{C})$ including binary polyhedral groups corresponds to the classification of the finite subgroup of $SU(2)$. As $SU(2) \simeq Sp(1)$, we can identify the binary polyhedral groups as certain subsets in quaternions $\mathbb{H}$. In fact, $Sp(1)$ is not only a unit sphere in $\mathbb{H}$ but also the spin group $Spin(3)$, which is a 2-covering map of $SO(3)$. In this section, we explain how an element in $SO(3)$ lifts to quaternions in $Sp(1)$.

The algebra of quaternions $\mathbb{H}$ is the four-dimensional vector space over $\mathbb{R}$ defined by

$$\mathbb{H} := \{a + bi + cj + dk \mid a, b, c, d \in \mathbb{R}\}$$

satisfying relations $i^2 = j^2 = k^2 = ijk = -1$. The quaternionic conjugate of $q = a + bi + cj + dk$ is defined by

$$\bar{q} := a - bi - cj - dk$$

and the corresponding norm $|q|$ is also defined by $|q| := \sqrt{q\bar{q}} = \sqrt{a^2 + b^2 + c^2 + d^2}$. Along this norm, quaternions satisfy $|pg| = |p|\,|g|$, which implies that it is one of the normed algebras whose classification consists of real numbers $\mathbb{R}$, complex numbers $\mathbb{C}$, quaternions $\mathbb{H}$, and octonions $\mathbb{O}$. A quaternion q is called *real* if $\bar{q} = q$ and is called *imaginary* if $\bar{q} = -q$. According to these facts, we can divide $\mathbb{H}$ into a real part and an imaginary part:

$$\mathbb{H} \simeq \mathbb{R}^4 \simeq Re(\mathbb{H}) \oplus Im(\mathbb{H}) = \mathbb{R} \oplus \mathbb{R}^3.$$

It is well known that the set of unit sphere $S^3 = \{q \in \mathbb{H} \mid |q| = 1\}$ in $\mathbb{H}$ is a Lie group $Sp(1)$, which is also isomorphic to $SU(2)$ as follows:

$$\begin{array}{ccc} Sp(1) & \simeq & SU(2) = \left\{ \begin{pmatrix} a & b \\ -\bar{b} & \bar{a} \end{pmatrix} \middle| \; a, b \in \mathbb{C}, |a|^2 + |b|^2 = 1 \right\}, \\ q = a + bj & \longleftrightarrow & \begin{pmatrix} a & b \\ -\bar{b} & \bar{a} \end{pmatrix}. \end{array}$$

Below, we use the identification between $\mathbb{R}^3$ and $Im(\mathbb{H})$. Along this, a vector $v = (v_1, v_2, v_3)$ in $\mathbb{R}^3$ (resp. a quaternion $q = a_1 i + a_2 j + a_3 k$ in $Im(\mathbb{H})$) is corresponded to a quaternion $(v)^{\#} = v_1 i + v_2 j + v_3 k$ in $Im(\mathbb{H})$ (resp. a vector $\vec{q} = (a_1, a_2, a_3)$ in $\mathbb{R}^3$).

Now, we define a map Φ,which is given by an action of $Sp(1)$ on $\mathrm{Im}(\mathbb{H}) \cong \mathbb{R}^3$

$$\begin{array}{llllll} \Phi : Sp(1) & \to & SO(3), & & & \\ x & \mapsto & \Phi(x) := \rho_x & : \mathrm{Im}(\mathbb{H}) & \to & \mathrm{Im}(\mathbb{H}), \\ & & & v & \mapsto & xv\bar{x}. \end{array}$$

As a matter of fact, the map ρ_x must be defined as

$$\begin{array}{llll} \rho_x & : \mathbb{R}^3 & \to & \mathbb{R}^3 \\ & v & \mapsto & \overrightarrow{(xv^{\#}\bar{x})} \end{array}$$

so that ρ_x is in $SO(3)$. However, we use a simpler definition instead. It is well known that Φ is a 2-covering map, which is also a group homomorphism.

In the next section, we will consider the preimage of Φ to define the lifting of symmetry groups of polyhedrons in $\mathbb{R}^3$, which are subgroups of $SO(3)$. For this purpose, we consider ρ_x further in below.

We observe that the multiplication of two pure quaternions p, q in $Im(\mathbb{H})$ can be written by the cross product $\times$ and the standard inner product $\cdot$ on $\mathbb{R}^3$:

$$pq = -\vec{p}\cdot\vec{q} + (\vec{p}\times\vec{q})^{\#}.$$

After we denote $x = x_0 + x_+$ ($x_0 \in \mathrm{Re}(\mathbb{H})$, $x_+ \in \mathrm{Im}(\mathbb{H})$), $\rho_x(v)$ can be written as

$$\begin{aligned}\rho_x(v) &= xv\bar{x} = (x_0 + x_+)v(x_0 - x_+)\\ &= \left(x_0 v - \vec{v}\cdot\overrightarrow{x_+} + (\overrightarrow{x_+}\times\vec{v})^{\#}\right)(x_0 - x_+)\\ &= (x_0^2 - |\overrightarrow{x_+}|^2)v + 2\left(\overrightarrow{x_+}\cdot\overrightarrow{v}\right)x_+ + 2x_0\left(\overrightarrow{x_+}\times\vec{v}\right)^{\#}.\end{aligned}$$

Here, since $|x|^2 = x_0^2 + |\overrightarrow{x_+}|^2$, we denote $x = \cos\theta + \sin\theta\,\frac{x_+}{|x_+|}$ where $\cos\theta = x_0$ and $\sin\theta = |\overrightarrow{x_+}|^2$ for some $\theta \in [0, \pi)$.

Now, to understand the meaning of $\rho_x(v)$, we consider two cases for $\overrightarrow{v}$ case (1) $\overrightarrow{v} \perp \overrightarrow{x_+}$ and case (2) $\overrightarrow{v}\,//\,\overrightarrow{x_+}$.

(1) (Case $\overrightarrow{v} \perp \overrightarrow{x_+}$) Since $\overrightarrow{v}\cdot\overrightarrow{x_+} = 0$, we have

$$\begin{aligned}\rho_x(v) &= (x_0^2 - |\overrightarrow{x_+}|^2)v + 2x_0\left(\overrightarrow{x_+}\times\vec{v}\right)^{\#}\\ &= (x_0^2 - |\overrightarrow{x_+}|^2)v + 2x_0|\overrightarrow{x_+}|\left(\frac{\overrightarrow{x_+}}{|\overrightarrow{x_+}|}\times\overrightarrow{v}\right)^{\#}\\ &= \cos 2\theta\ v + \sin 2\theta\left(\frac{\overrightarrow{x_+}}{|\overrightarrow{x_+}|}\times\overrightarrow{v}\right)^{\#}.\end{aligned}$$

(2) (Case $\overrightarrow{v}\,//\,\overrightarrow{x_+}$) After we denote $\overrightarrow{v} = t\,\overrightarrow{x_+}$ for some $t \in \mathbb{R}$,

$$\begin{aligned}\rho_x(v) &= (x_0^2 - |\overrightarrow{x_+}|^2)tx_+ + 2\left(\overrightarrow{x_+}\cdot t\,\overrightarrow{x_+}\right)x_+\\ &= \left(t\,(x_0^2 - |\overrightarrow{x_+}|^2) + 2\left(\overrightarrow{x_+}\cdot t\,\overrightarrow{x_+}\right)\right)x_+\\ &= t\,(x_0^2 + |\overrightarrow{x_+}|^2)x_+ = tx_+ = v.\end{aligned}$$

By the two cases above, we conclude $\rho_x(v)$ presents the rotation of vector $\overrightarrow{v}$ in $\mathbb{R}^3$ with respect to the axis $\overrightarrow{x_+}$ by $2\cos^{-1}\ x_0 \in [0, 2\pi)$.

By applying the above, we have the following lemma.

Lemma 1. *For each element A in $SO(3)$ presenting a rotation with respect to a unit vector $\vec{a}$ for angle $\alpha \in [0, 2\pi)$, the preimage of $\Phi : Sp(1) \to SO(3)$ is given as*

$$\Phi^{-1}(A) = \left\{\pm\left(\cos\frac{\alpha}{2} + \sin\frac{\alpha}{2}\,(\vec{a})^{\#}\right)\right\} \subset \mathbb{H}.$$

Note if we choose unit vector $-\vec{a}$ instead $\vec{a}$ in the lemma, then the rotation performed for angle $-\alpha$. Therefore, the related lifting is given by $\cos\left(-\frac{\alpha}{2}\right) + \sin\left(-\frac{\alpha}{2}\right)(-\vec{a})^{\#} = \cos\frac{\alpha}{2} + \sin\frac{\alpha}{2}\,(\vec{a})^{\#}$. Hence $\Phi^{-1}(A)$ is well defined.

By applying Lemma 1, we can consider the preimage of any subset G in $SO(3)$. We call the preimage $\Phi^{-1}(G)$ the *lift* of G in $Sp(1) \subset \mathbb{H}$. When G is one of the symmetry groups of regular polyhedrons, the lift $\Phi^{-1}(G)$ is called *binary polyhedral group*. In particular, we consider *binary tetrahedral group* $2T$, *binary octahedral group* $2O$, and *binary icosahedral group* $2I$, which are lifts of symmetry groups of tetrahedron, octahedron and icosahedron with order 24, 48, 120, respectively.

2.1. Symmetry Groups of Regular Polyhedrons

A polyhedron considered here is convex and regular. According to convention, we denote a regular polyhedron by $\{p, q\}$, which means that the polyhedron has only one type of face which is a p-gon, and each vertex is contained in q faces. It is well known that there are only five regular polyhedrons, which are also called *Platonic solids*. Up to duality, we consider three classes of regular polyhedrons as follows:

$$\begin{array}{ccc} \text{Tetrahedron} & \leftrightarrow & \text{Tetrahedron(self-dual)}, \\ \text{Octahedron} & \leftrightarrow & \text{Cube}, \\ \text{Icosahedron} & \leftrightarrow & \text{Dodecahedron}. \end{array}$$

As the special linear group $SO(3)$ is generated by rotations on $\mathbb{R}^3$, we consider rotations of $\mathbb{R}^3$ preserving a regular polyhedron to study the symmetry group of it. When the axis of the rotation crosses a vertex (the barycenter of an edge, the barycenter or a face resp.), we call the rotation *vertex symmetry* (*edge symmetry*, *face symmetry* resp.). For instance, the tetrahedron has two different types of axes of rotations. One is the line passing through a vertex and the barycenter of the opposite face, and the other is the line connecting barycenters of the edges at the opposite position. We also say that a symmetry has *order n* if the order of the corresponding rotation is n. Note that the order of each edge symmetry is 2. One can figure all the possible orders of each type of symmetry for regular polyhedrons, as shown in Table 1.

Table 1. Order of symmetries.

Polyhedron	Tetrahedron	Octahedron	Icosahedron
Point Symmetry	3	4, 2	5
Edge Symmetry	2	2	2
Face Symmetry	3	3	3

Construction of binary polyhedral groups

Now, we will introduce a construction which provides a way to find the elements of binary polyhedral groups related to regular polyhedrons when the set of vertices of regular polyhedrons are given.

Assume we have a regular polyhedron $\{p, q\}$ whose barycenter is the origin of $\mathbb{R}^3$ and let $\{P_i\}$ be the set of vertices of the regular polyhedron:

(1) Find the barycenters of vertices, edges and faces (The barycenter of each vertex is itself).
(2) For each barycenter, derive all the related symmetries in $SO(3)$ by identifying corresponding axis of rotations and its order.
(3) For each symmetry obtained from step 2, we get related lifts in $\mathbb{H}$ by applying Lemma 1. It is useful to observe that we obtain the axis of rotation and its order instead of related angle where there can be more than one related angle.
(4) The union of lifts is a subset of binary polyhedral groups. In fact, its union with $\{\pm 1\}$ is the binary polyhedral group by counting elements.

Note: From the above, it is clear that two regular polyhedrons which are dual to each other are associated with the same binary polyhedral groups.

For example, let B be a barycenter of order 3. Then, there are two related angles $\frac{\pi}{3}$ and $\frac{2\pi}{3}$. Thus, the corresponding lift is

$$\left\{ \pm \left(\cos\frac{\pi}{3} + \sin\frac{\pi}{3}\frac{B}{|B|} \right), \pm \left(\cos\frac{2\pi}{3} + \sin\frac{2\pi}{3}\frac{B}{|B|} \right) \right\}.$$

Here, we also observe that, if we begin with $-B$ instead of B, the corresponding lifts are still the same.

Binary tetrahedral group

We consider a tetrahedron consisting of vertices $\{P_1, P_2, P_3, P_4\}$.

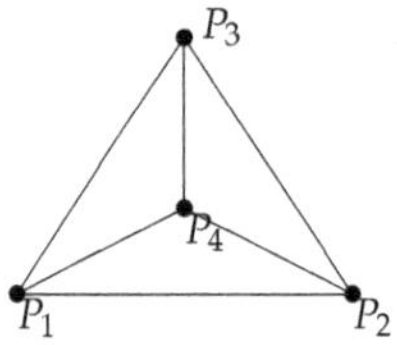

Tetrahedron

(1) Since vertex symmetry has order 3, each vertex symmetry has two angles $\frac{\pi}{3}$ and $\frac{2\pi}{3}$ and the union of lifts of vertex symmetries is

$$V_T := \bigcup_i \left\{ \pm \left(\cos\frac{\pi}{3} + \sin\frac{\pi}{3}\frac{P_i}{|P_i|} \right), \pm \left(\cos\frac{2\pi}{3} + \sin\frac{2\pi}{3}\frac{P_i}{|P_i|} \right) \right\}.$$

Thus, we have

$$|V_T| = |\text{vertices of a tetrahedron}| \times 4 = 16.$$

(2) As the edge symmetry has order 2, each edge symmetry has only one angle $\frac{\pi}{2}$ so that the related lift is

$$\left\{ \pm \left(\cos\frac{\pi}{2} + \sin\frac{\pi}{2} \frac{\frac{P_i + P_j}{2}}{\left|\frac{P_i + P_j}{2}\right|} \right) \right\}$$

and the union is given as

$$E_T := \bigcup_{i \neq j} \left\{ \pm \left(\frac{P_i + P_j}{|P_i + P_j|} \right) \right\}.$$

Since $P_1 + P_2 + P_3 + P_4 = 0$, two barycenters $\frac{1}{2}(P_1 + P_2)$ and $\frac{1}{2}(P_3 + P_4)$ of edges have the same lifts of edge symmetries. Similarly, the pairs of edges produce the same lifts of edge symmetries, and we get

$$|E_T| = \frac{|\text{edges of a tetrahedron}|}{2} \times 2 = 6.$$

(3) For a barycenter of face consisting of $\{P_1, P_2, P_3\}$, we have a relation

$$\frac{P_1 + P_2 + P_3}{3} = -\frac{P_4}{3}$$

since $P_1 + P_2 + P_3 + P_4 = 0$. Thus, the related lift of face symmetry is the same as the lift of vertex lift for a vertex P_4. Similarly, each lift of face symmetry corresponds to the lift of vertex symmetry.

Finally, the union $V_T \cup E_T$ of lifts of symmetries of a tetrahedron is a subset binary tetrahedral group $2T$ in $Sp(1)$. Since

$$|V_T \cup E_T \cup \{\pm 1\}| = 16 + 6 + 2 = 24 = |2T|,$$

the union $V_T \cup E_T \cup \{\pm 1\}$ is a binary tetrahedral group, namely

$$2T = V_T \cup E_T \cup \{\pm 1\}.$$

If we choose vertices $\{P_1, P_2, P_3, P_4\}$ of a tetrahedron as

$$\left\{P_1 = \frac{i+j+k}{2}, P_2 = \frac{i-j-k}{2}, P_3 = \frac{-i+j-k}{2}, P_4 = \frac{-i-j+k}{2}\right\},$$

the corresponding binary tetrahedral group is obtained as

$$2\hat{T} =: \left\{\pm 1,\ \pm i,\ \pm j,\ \pm k,\ \frac{1}{2}(\pm 1 \pm i \pm j \pm k)\right\}.$$

Remark: The subset $2\hat{T}$ is the unit integral quaternions which is also known as *Hurwitz integral quaternions*. ([7,8])

Binary Octahedral Group

We consider an octahedron consisting of vertices $\{P_i\ i = 1, ..., 8\}$ as below.

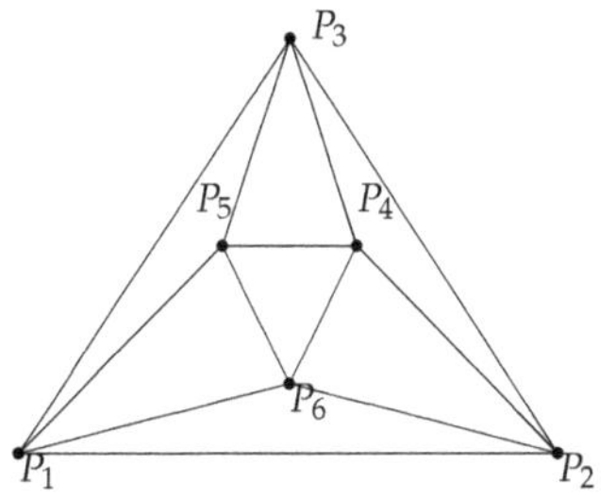

Octahedron

(1) The possible orders of vertex symmetry are 2 and 4. The vertex symmetry with order 4 has two angles $\frac{\pi}{4}$ and $\frac{3\pi}{4}$ and one with order 2 has one angle $\frac{\pi}{2}$. Thus, the union of lifts of vertex symmetries is

$$V_O := \bigcup_i \left\{\pm\left(\cos\frac{\pi}{4} + \sin\frac{\pi}{4}\frac{P_i}{|P_i|}\right),\ \pm\left(\cos\frac{3\pi}{4} + \sin\frac{3\pi}{4}\frac{P_i}{|P_i|}\right),\ \pm\left(\cos\frac{\pi}{2} + \sin\frac{\pi}{2}\frac{P_i}{|P_i|}\right)\right\}.$$

Since two antipodal vertices produce the same lifts of vertex symmetries, we obtain that

$$|V_O| = \frac{|\text{vertices of an octahedron}|}{2} \times 6 = 18.$$

(2) As the edge symmetry has order 2, each edge symmetry has only one angle $\frac{\pi}{2}$ so that the union of the related lift is

$$E_O := \bigcup \left\{\pm\left(\frac{P_i + P_j}{|P_i + P_j|}\right)\right\},$$

where the union is performed for all the pairs of P_i and P_j form an edge. For the barycenter of an edge given by $\frac{1}{2}(P_i + P_j)$, there is exactly one edge whose barycenter is antipodal to $\frac{1}{2}(P_i + P_j)$. Moreover, the pair of edges produce the same lifts of edge symmetries. Therefore, we get

$$|E_O| = \frac{|\text{edges of an octahedron}|}{2} \times 2 = 12.$$

(3) For a barycenter of face given as $\frac{P_i+P_j+P_k}{3}$, the face symmetry has order 3 and it is related to two angles $\frac{\pi}{3}$ and $\frac{2\pi}{3}$. Thus, the lifts of a face symmetry is

$$\left\{ \pm\left(\cos\frac{\pi}{3}+\sin\frac{\pi}{3}\frac{\frac{P_i+P_j+P_k}{3}}{\left|\frac{P_i+P_j+P_k}{3}\right|}\right), \pm\left(\cos\frac{2\pi}{3}+\sin\frac{2\pi}{3}\frac{\frac{P_i+P_j+P_k}{3}}{\left|\frac{P_i+P_j+P_k}{3}\right|}\right)\right\}$$

and the union of lifting of face symmetries is

$$F_O := \bigcup\left\{ \pm\left(\cos\frac{\pi}{3}+\sin\frac{\pi}{3}\frac{P_i+P_j+P_k}{|P_i+P_j+P_k|}\right), \pm\left(\cos\frac{2\pi}{3}+\sin\frac{2\pi}{3}\frac{P_i+P_j+P_k}{|P_i+P_j+P_k|}\right)\right\}.$$

Since the octahedron is symmetric for origin, for the barycenter of a face given by $\frac{1}{3}(P_i+P_j+P_k)$, there is exactly one face whose barycenter is antipodal to $\frac{1}{3}(P_i+P_j+P_k)$, and the pair of faces produce the same lifts of face symmetries. Therefore, we deduce

$$|F_O| = \frac{|\text{faces of an octahedron}|}{2}\times 4 = 16.$$

Finally, the union $V_O \cup E_O \cup F_O$ of lifts of symmetries of an octahedron is a subset of the binary octahedral group $2O$ in $Sp(1)$. Since

$$|V_O \cup E_O \cup F_O \cup \{\pm 1\}| = 18+12+16+2 = 48 = |2O|,$$

the union $V_O \cup E_O \cup F_O \cup \{\pm 1\}$ is a binary octahedral group, namely

$$2O = V_O \cup E_O \cup F_O \cup \{\pm 1\}.$$

One can take P_is as follows:

$$P_1 = i, P_2 = j, P_3 = k, P_4 = -i, P_5 = -j, P_6 = -k$$

so as to obtain

$$2\hat{O} := \left\{ \begin{array}{c} \pm 1,\ \pm i,\ \pm j,\ \pm k,\ \frac{1}{2}(\pm 1 \pm i \pm j \pm k), \\ \frac{1}{\sqrt{2}}(\pm 1 \pm i), \frac{1}{\sqrt{2}}(\pm 1 \pm j), \frac{1}{\sqrt{2}}(\pm 1 \pm k), \frac{1}{\sqrt{2}}(\pm i \pm j), \frac{1}{\sqrt{2}}(\pm i \pm k), \frac{1}{\sqrt{2}}(\pm j \pm k) \end{array} \right\}.$$

Binary Icosahedral Group

Since both the regular icosahedron and its dual regular dodecahedron produce the binary icosahedral group, we consider a regular dodecahedron in $\mathbb{R}^3$ instead of a regular icosahedron. Moreover, for the sake of convenience, one can choose specific coordinates of vertices of a dodecahedron in $\mathbb{R}^3$ such as

$$\left\{(\pm 1, \pm 1, \pm 1),\ (\pm\tau, \pm\frac{1}{\tau}, 0),\ (0, \pm\tau, \pm\frac{1}{\tau}),\ (\pm\frac{1}{\tau}, 0, \pm\tau)\right\},$$

where $\tau = \frac{\sqrt{5}+1}{2} = 2\cos\frac{\pi}{5}$ and $\frac{1}{\tau} = \frac{\sqrt{5}-1}{2} = -2\cos\frac{2\pi}{5}$. It is also useful to know $\sin\frac{\pi}{5} = \frac{\sqrt{10-2\sqrt{5}}}{4}$ and $\sin\frac{2\pi}{5} = \frac{\sqrt{10+2\sqrt{5}}}{4}$.

In the following diagram, we consider the given set of vertices as a subset in $\operatorname{Im}\mathbb{H} = \mathbb{R}^3$ and depict the configuration among the vertices.

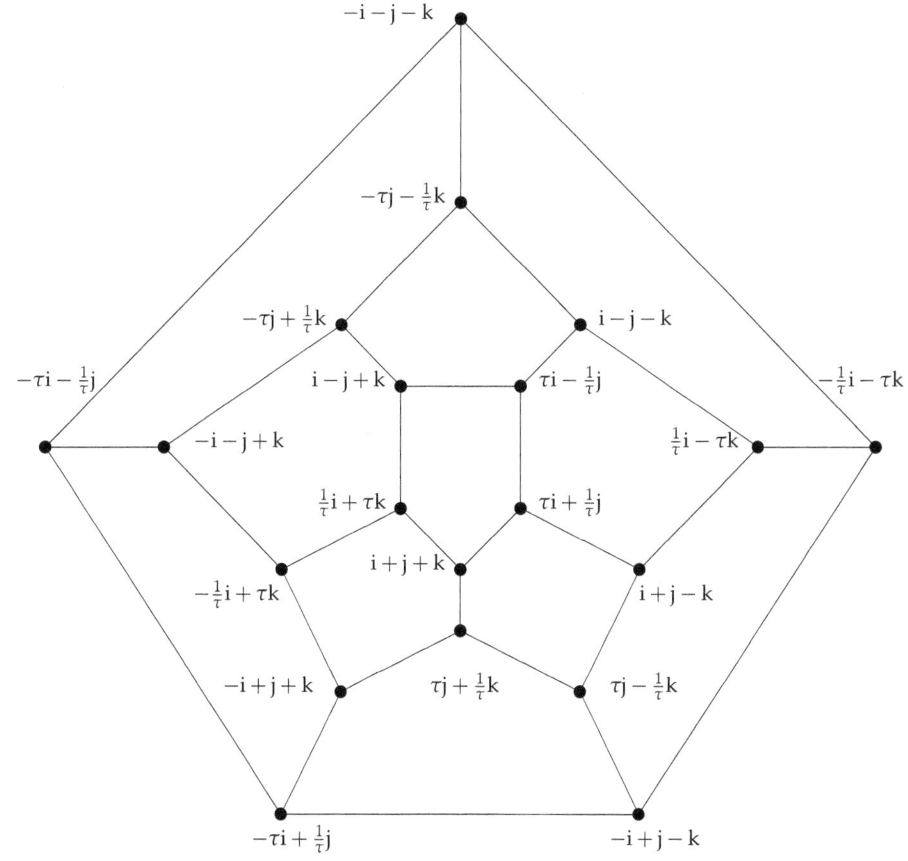

dodecahedron

For the above dodecahedron, we denote the set of vertices as $\{P_i\ i = 1, \ldots, 20\}$ without a specific choice of order.

(1) Since the possible order of each vertex symmetry is 3, the vertex symmetry has two angles $\frac{\pi}{3}$ and $\frac{2\pi}{3}$. Thus, the union of lifts of vertex symmetries is

$$V_I := \bigcup_i \left\{ \pm\left(\cos\frac{\pi}{3} + \sin\frac{\pi}{3}\frac{P_i}{|P_i|}\right), \pm\left(\cos\frac{2\pi}{3} + \sin\frac{2\pi}{3}\frac{P_i}{|P_i|}\right) \right\}.$$

Since the dodecahedron is symmetric for origin, each vertex and its antipodal vertex produce the same lifts of vertex symmetries. Thus, we obtain that

$$|V_I| = \frac{|\text{vertices of a dodecahedron}|}{2} \times 4 = 40.$$

(2) As before, the edge symmetry has order 2, and each edge symmetry has only one angle $\frac{\pi}{2}$ so that the union of the related lifts is

$$E_I := \bigcup \left\{ \pm \left(\frac{P_i + P_j}{|P_i + P_j|} \right) \right\},$$

where the union is performed for all the pairs of P_i and P_j form an edge. Just like the lifts of edge symmetries for an octahedron, the pair of antipodal edges produce the same lifts of edge symmetries. Therefore, we get

$$|E_I| = \frac{|\text{edges of a dodecahedron}|}{2} \times 2 = 30.$$

(3) For a barycenter of face given as $\frac{1}{5}\left(P_{i_1} + P_{i_2} + P_{i_3} + P_{i_4} + P_{i_5}\right)$ where P_{i_l} $(l = 1,2,3,4,5)$ forms a face, the face symmetry has order 5 and it is related to four angles $\frac{a\pi}{5}$ $(a = 1,2,3,4)$. Thus, the lifts of a face symmetry are

$$\left\{ \pm \left(\cos\frac{a\pi}{5} + \sin\frac{a\pi}{5} \frac{\frac{1}{5}\left(P_{i_1} + P_{i_2} + P_{i_3} + P_{i_4} + P_{i_5}\right)}{\left|\frac{1}{5}\left(P_{i_1} + P_{i_2} + P_{i_3} + P_{i_4} + P_{i_5}\right)\right|} \right) (a = 1,2,3,4) \right\}$$

and the union of lifts of face symmetries is

$$F_I := \bigcup \left\{ \pm \left(\cos\frac{a\pi}{5} + \sin\frac{a\pi}{5} \frac{P_{i_1} + P_{i_2} + P_{i_3} + P_{i_4} + P_{i_5}}{|P_{i_1} + P_{i_2} + P_{i_3} + P_{i_4} + P_{i_5}|} \right) (a = 1,2,3,4) \right\}.$$

Since a pair of antipodal faces produce the same lifts of face symmetries, we deduce

$$|F_I| = \frac{|\text{faces of a dodecahedron}|}{2} \times 8 = 48.$$

Finally, the union $V_I \cup E_I \cup F_I$ of lifts of symmetries of a dodecahedron is a subset of a binary icosahedral group $2I$ in $Sp(1)$. Since

$$|V_I \cup E_I \cup F_I \cup \{\pm 1\}| = 40 + 30 + 48 + 2 = 120 = |2I|,$$

the union $V_I \cup E_I \cup F_I \cup \{\pm 1\}$ is a binary icosahedral group, namely

$$2I = V_I \cup E_I \cup F_I \cup \{\pm 1\}.$$

For the given vertices we have, we can obtain

$$V_I = \left\{ \frac{1}{2}(\pm 1 \pm i \pm j \pm k), \frac{1}{2}\left(\pm 1 \pm \tau j \pm \frac{1}{\tau}k\right), \frac{1}{2}\left(\pm 1 \pm \frac{1}{\tau}i \pm \tau k\right), \frac{1}{2}\left(\pm 1 \pm \tau i \pm \frac{1}{\tau}j\right) \right\}$$

$$E_I = \left\{ \pm i, \pm j, \pm k, \frac{1}{2}\left(\pm i \pm \frac{1}{\tau}j \pm \tau k\right), \frac{1}{2}\left(\pm \frac{1}{\tau}i \pm \tau j \pm k\right), \frac{1}{2}\left(\pm \tau i \pm j \pm \frac{1}{\tau}k\right) \right\}$$

$$F_I = \left\{ \begin{array}{l} \frac{1}{2}\left(\pm \frac{1}{\tau} \pm \tau i \pm k\right), \frac{1}{2}\left(\pm \tau \pm \frac{1}{\tau}i \pm j\right), \frac{1}{2}\left(\pm \tau \pm \frac{1}{\tau}j \pm k\right), \\ \frac{1}{2}\left(\pm \tau \pm i \pm \frac{1}{\tau}k\right), \frac{1}{2}\left(\pm \frac{1}{\tau} \pm i \pm \tau j\right), \frac{1}{2}\left(\pm \frac{1}{\tau} \pm j \pm \tau k\right) \end{array} \right\}.$$

As a result, we can identify all the elements of the binary icosahedral group as

$$2\hat{I} := \left\{ \begin{array}{c} \pm 1,\ \pm i,\ \pm j,\ \pm k,\ \frac{1}{2}\left(\pm 1 \pm i \pm j \pm k\right) \\ \frac{1}{2}\left(\pm i \pm \frac{1}{\tau} j \pm \tau k\right), \frac{1}{2}\left(\pm 1 \pm \tau j \pm \frac{1}{\tau} k\right), \frac{1}{2}\left(\pm \frac{1}{\tau} \pm \tau i \pm k\right), \frac{1}{2}\left(\pm \tau \pm \frac{1}{\tau} i \pm j\right), \\ \frac{1}{2}\left(\pm \frac{1}{\tau} i \pm \tau j \pm k\right), \frac{1}{2}\left(\pm \tau \pm \frac{1}{\tau} j \pm k\right), \frac{1}{2}\left(\pm \tau \pm i \pm \frac{1}{\tau} k\right), \frac{1}{2}\left(\pm \frac{1}{\tau} \pm i \pm \tau j\right), \\ \frac{1}{2}\left(\pm \tau i \pm j \pm \frac{1}{\tau} k\right), \frac{1}{2}\left(\pm \frac{1}{\tau} \pm j \pm \tau k\right), \frac{1}{2}\left(\pm 1 \pm \frac{1}{\tau} i \pm \tau k\right), \frac{1}{2}\left(\pm 1 \pm \tau i \pm \frac{1}{\tau} j\right) \end{array} \right\}.$$

Theorem 1. *The finite subsets $2\hat{T}$, $2\hat{O}$ and $2\hat{I}$ in $\mathbb{H}$ defined as above are a binary tetrahedral group, a binary octahedral group, and a binary icosahedral group, respectively.*

Note that it is well known that the a subset $\left\{\pm 1, \pm i, \pm j, \pm k, \frac{1}{2}(\pm 1 \pm i \pm j \pm k)\right\}$ in $2I$ is the vertices of 24-cell and the complementary subset in $2I$ is the vertices of a snub 24-cell.

3. 600-Cell

The Coxeter–Dynkin diagrams are the way of describing the group generated by reflections. For each graph, *node* represents a mirror (or a reflection hypersurface) and the label m attached to a branch between nodes marks the dihedral angle $\frac{\pi}{m}$ between two mirrors. By convention, no label is attached to a branch if the corresponding dihedral angle is $\frac{\pi}{3}$. When all the dihedral angles are $\frac{\pi}{3}$, the diagram is called *simply laced*. Ringed nodes present so called active mirrors where there is a point P not to sit in the hyperplanes of reflections corresponded to the mirrors. By successive applying the reflections in the diagram to the point P, we obtain a polytope whose symmetry group is the Weyl group generated by the Coxeter–Dynkin diagram. Moreover, the combinatorics of subpolytopes can also be decoded by the Coxeter–Dynkin diagram when it is simply laced with one ringed node (see [7,9,10]). In fact, a similar method can be applied for the diagram, which is not simply laced or has more than one ringed node.

The Coxeter–Dynkin diagram of 24-cell is an example of simply laced with one ringed node.

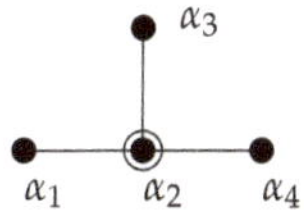

Coxeter–Dynkin diagram of 24-cell

The Weyl group associated with this diagram is D_4-type. In [7], the subpolytopes of 24-cell as shown in Table 2 are described by using the Coxeter–Dynkin diagram.

Table 2. Subpolytopes of 24-cell.

Subpolytope	Vertices	Edges	Faces	Cells
total number	24	96	96 ({3})	24 ({3,3})

The Coxeter–Dynkin diagram of 600-cell is given by

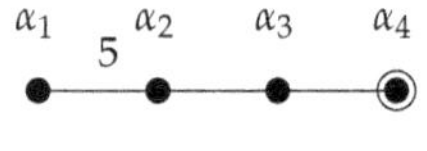

Coxeter–Dynkin diagram of 600-cell

whose Weyl group is H_4-type. Thus, the diagram is not simply laced and has one ringed node.

(1) Vertices

By removing a ringed node, we obtain the isotropy subgroup in the Weyl group of H_4 which fixed a vertex in the 600-cell. Here, the corresponding isotropy group is H_3 and we can compute the total number of vertices as

$$\frac{|H_4|}{|H_3|} = \frac{14400}{120} = 120.$$

For the remaining diagram above, we ring a node connected to the removed node. Then, we obtain the Coxeter–Dynkin diagram of an Icosahedron, which implies that the vertex figure of 600 cell is an icosahedron.

(2) Edges

For edges, we consider the ringed node that performs one reflection corresponding to an edge.

For the isotropy subgroup of the edge, we remove the unringed node connected to the ringed node. In addition, the remaining diagram generates the isotropy subgroup $H_2 \times A_1$. Thus, we compute the total number of edges as

$$\frac{|H_4|}{|H_2||A_1|} = \frac{14400}{10 \cdot 2} = 720.$$

(3) Faces

For faces, we consider the ringed node and extend the diagram to unringed nodes so as to obtain a subdiagram of A_2-type. The subdiagram of A_2 with one ringed node generate $\{3\}$, namely a triangle. Thus, the faces of 600-cell are all triangles.

For the isotropy subgroup of a face, we remove any unringed node connected to the subdiagram of a face. The remaining subdiagram generates the isotropy subgroup of a face, which is $A_1 \times A_2$. Thus, we compute the total number of faces as

$$\frac{|H_4|}{|A_1||A_2|} = \frac{14400}{2!3!} = 1200.$$

(4) Cells

To obtain a cell in a 600-cell, we consider an extended diagram from the ringed nodes to unringed nodes so as to obtain a subdiagram of A_3. The diagram of type A_3 with one ringed node on one side represents a tetrahedron.

For the isotropy subgroup of a cell, we consider that any unringed node connected to the subdiagram of A_3, and the subdiagram given by removing the node generates the isotropy subgroup, which is A_3. Thus, we compute the total number of cells, Table 3 shows the subpolytopes of 600:

$$\frac{|H_4|}{|A_3|} = \frac{14400}{4!} = 600.$$

Table 3. Subpolytopes of 600-cell.

Subpolytopes	Vertices	Edges	Faces	Cells
total number	120	720	1200 ($\{3\}$)	600 ($\{3,3\}$)

4. Binary Polyhedral Groups as Polytopes

In this section, we show that the binary icosahedral group $2I$ in $\mathbb{H}$ is the set of vertices of a 600-cell. Thus, the convex hull of $2I$ in $\mathbb{H}$ is a 600-cell.

For each α in $\mathbb{H}$ with $|\alpha| = 1$, we define a reflection on $\mathbb{H}$ as

$$\begin{array}{rccc} \sigma_\alpha : & \mathbb{H} & \longrightarrow & \mathbb{H}, \\ & x & \mapsto & \sigma_\alpha(x) := x - 2\,(\vec{x} \cdot \vec{\alpha})\,\alpha. \end{array}$$

Since $\mathbb{H}$ is a normed division algebra, $\sigma_\alpha(x)$ is also written as $\sigma_a(x) = -a\,\bar{x}\,a$ via quaternionic multiplication (see Ref. [7]). Since σ_a is a reflection for a vector α, σ_a has eigenvalues ± 1 where α is an eigenvector of -1 and the hyperplane perpendicular to α is the eigenspace of 1. Moreover, it is not an element in $SO(3)$.

Binary tetrahedral group $2\hat{T}$ in $\mathbb{H}$ and 24-cell

For 24-cell, we consider the Coxeter–Dynkin diagram of type D_4 given in Section 3, where

$$\alpha_1 = i, \alpha_2 = \frac{1}{2}(1 + i + j + k), \alpha_3 = j, \alpha_4 = k.$$

In Ref. [7], the Weyl group generated by the Coxeter–Dynkin diagram acts on the binary tetrahedral group $2\hat{T}$. Moreover, it is shown that $2\hat{T}$ is the set of vertices of a 24-cell. In fact, it is also the subset of unit integral quaternions.

Binary icosahedral group $2\hat{I}$ in $\mathbb{H}$ and 600-cell

Similarly, for 600-cell, we consider the Coxeter–Dynkin diagram of Type H_4 given in Section 3, where

$$\alpha_1 = i, \alpha_2 = \frac{1}{2}\left(\tau i + j - \frac{1}{\tau}k\right), \alpha_3 = j, \alpha_4 = \frac{1}{2}\left(-\frac{1}{\tau} + j + \tau k\right).$$

The Weyl group generated by the reflections $\{\sigma_{\alpha_i}\ i = 1,2,3,4\}$ is denoted by W_H. In below, we want to show that **(1)** the Weyl group W_H acts on the binary icosahedral group $2\hat{I}$, and **(2)** $2\hat{I}$ is a single orbit, where it corresponds to the set of vertices of 600-cell.

Lemma 2. *The Weyl group W_H acts on the binary icosahedral group $2\hat{I}$ in $\mathbb{H}$.*

Proof. Since the Weyl group W_H is generated by the reflections σ_{α_i} $(i = 1,2,3,4)$, we show that each σ_{α_i} acts on $2I$. For an arbitrary element $a + bi + cj + dk \in \mathbb{H}$, the reflections are written as follows:

$$\begin{aligned}
\sigma_{\alpha_1}(a+bi+cj+dk) &= a - bi + cj + dk,\\
\sigma_{\alpha_2}(a+bi+cj+dk) &= a + \left(-\frac{1}{2\tau}b - \frac{\tau}{2}c + \frac{1}{2}d\right) i\\
&\quad + \left(-\frac{\tau}{2}b + \frac{1}{2}c + \frac{1}{2\tau}d\right) j + \left(\frac{1}{2}b + \frac{1}{2\tau}c + \frac{\tau}{2}d\right) k,\\
\sigma_{\alpha_3}(a+bi+cj+dk) &= a + bi - cj + dk,\\
\sigma_{\alpha_4}(a+bi+cj+dk) &= \left(\frac{\tau}{2}a + \frac{1}{2\tau}c + \frac{1}{2}d\right) + bi\\
&\quad + \left(\frac{1}{2\tau}a + \frac{1}{2}c - \frac{\tau}{2}d\right) j + \left(\frac{1}{2}a - \frac{\tau}{2}c - \frac{1}{2\tau}d\right) k.
\end{aligned}$$

It is easy to see that σ_{α_1} and σ_{α_3} act on $2\hat{I}$. By choosing $\{1, i, j, k\}$ as an ordered orthonormal basis of $\mathbb{H}$, σ_{α_2} and σ_{α_4} can be written as

$$\sigma_{\alpha_2} = \begin{pmatrix} 1 & 0 & 0 & 0 \\ 0 & -\frac{1}{2\tau} & -\frac{\tau}{2} & \frac{1}{2} \\ 0 & -\frac{\tau}{2} & \frac{1}{2} & \frac{1}{2\tau} \\ 0 & \frac{1}{2} & \frac{1}{2\tau} & \frac{\tau}{2} \end{pmatrix} \text{ and } \sigma_{\alpha_4} = \begin{pmatrix} \frac{\tau}{2} & 0 & \frac{1}{2\tau} & \frac{1}{2} \\ 0 & 1 & 0 & 0 \\ \frac{1}{2\tau} & 0 & \frac{1}{2} & -\frac{\tau}{2} \\ \frac{1}{2} & 0 & -\frac{\tau}{2} & -\frac{1}{2\tau} \end{pmatrix}.$$

In addition, these are similar because $\sigma_{\alpha_4} = S^t \sigma_{\alpha_2} S,$ where S is an orthogonal matrix

$$\begin{pmatrix} 0 & 1 & 0 & 0 \\ 0 & 0 & 0 & 1 \\ 0 & 0 & 1 & 0 \\ 1 & 0 & 0 & 0 \end{pmatrix}.$$

In fact, S is an element in $SO(4)$ defined by

$$1 \to k,\ i \to 1,\ j \to j,\ k \to i$$

and one can check that S acts on $2\hat{I}$ by simple calculation. Thus, it suffices to show that σ_{α_2} acts on $2I$ to check σ_{α_2} and σ_{α_4} act on $2\hat{I}$. For σ_{α_2}, we consider 3×3 submatrix A of σ_{α_2} defined as

$$A := \begin{pmatrix} -\frac{1}{2\tau} & -\frac{\tau}{2} & \frac{1}{2} \\ -\frac{\tau}{2} & \frac{1}{2} & \frac{1}{2\tau} \\ \frac{1}{2} & \frac{1}{2\tau} & \frac{\tau}{2} \end{pmatrix}.$$

This is a automorphism of $\operatorname{Im}\mathbb{H}$ which satisfies $A^t A = Id$ and $\det A = -1$. Moreover, A also acts on

$$\left\{(\pm 1, \pm 1, \pm 1),\ (\pm\tau, \pm\frac{1}{\tau}, 0),\ (0, \pm\tau, \pm\frac{1}{\tau}),\ (\pm\frac{1}{\tau}, 0, \pm\tau)\right\},$$

which is our choice of the vertices of a dodecahedron. Since A is a reflection, it is also a symmetry of the dodecahedron so that it also acts on the set of edges and the set of faces. According to the construction of binary icosahedral group $2\hat{I}$ in Section 2.1, the action of A on the icosahedron induces the action of σ_{α_2} on $2\hat{I}$. For example, an edge symmetry given by an edge $P_i + P_j$ is sent to another given by $AP_i + AP_j$ because

$$\sigma_{\alpha_2}\left(\frac{P_i + P_j}{|P_i + P_j|}\right) = \frac{AP_i + AP_j}{|P_i + P_j|} = \frac{AP_i + AP_j}{|AP_i + AP_j|}.$$

Similarly, we conclude that σ_{α_2} acts on $2\hat{I}$. □

By applying the above lemma, we obtain the following theorem.

Theorem 2. *The set $2\hat{I}$ of a binary icosahedral group is an orbit of the Weyl group W_H, and it is the set of vertices of a 600-cell.*

Proof. By Lemma 2, the Weyl group W_H acts on $2I$. Now, we consider an element 1 in $2\hat{I}$ and its orbit $W_H\{1\} \subset 2\hat{I}$. Since 1 is perpendicular to $\alpha_1 = i, \alpha_2 = \frac{1}{2}\left(\tau i + j - \frac{1}{\tau}k\right)$, $\alpha_3 = j$ and $1 \cdot \alpha_4 = (1,0,0,0) \cdot \left(-\frac{1}{2\tau}, 0, \frac{1}{2}, \frac{\tau}{2}\right) \neq 0$, the orbit $W_H\{1\}$ is given by the following Coxeter–Dynkin diagram 3 of 600-cell. By Section 3, we have $|W_H\{1\}| = 120 = |2\hat{I}|$. Therefore, we conclude $W_H\{1\} = 2\hat{I}$ and this gives the theorem. □

Author Contributions: This article is written by both authors.

Funding: The second author was supported by the Basic Science Research Program through the National Research Foundation of Korea (NRF) funded by the Ministry of Education, Science and Technology (No.2016R1D1A1B03931673).

Acknowledgments: The authors gratefully thank the Referees for the constructive comments and recommendations, which definitely helped to improve the readability and quality of the paper.

Conflicts of Interest: The authors declare no conflicts of interest.

References

1. Van Hoboken, J. Platonic Solids, Binary Polyhedral Groups Kleinian Singularities and Lie Algebras of Type ADE. Master's Thesis, University of Amsterdam, Amsterdam, The Netherlands, March 2002.
2. Coxeter, H.S.M. The binary polyhedral groups, and other generalizations of the quaternion group. *Duke Math. J.* **1940**, *7*, 367–379. [CrossRef]
3. Koca, M.; Al-Ajmi, M.; Koç, R. Group theoretical analysis of 600-cell and 120-cell 4D polytopes with quaternions. *J. Phys. A* **2007**, *40*, 7633–7642. [CrossRef]
4. Koca, M.; Al-Ajmi, M.; Koç, R. Quaternionic representation of the Coxeter group $W(H_4)$ and the polyhedra. *J. Phys. A* **2006**, *39*, 14047–14054. [CrossRef]
5. Koca, M.; Al-Ajmi, M.; Koca, N.O. Quaternionic representation of snub 24-cell and its dual polytope derived from E_8 root system. *Linear Algebra Appl.* **2011**, *434*, 977–989. [CrossRef]
6. Dechant, P.-P. Clifford algebra unveils a surprising geometric significance of quaternionic root systems of Coxeter groups. *Adv. Appl. Clifford Algebra* **2013**, *23*, 301–321. [CrossRef]
7. Chang, W.N.; Lee, J.H.; Lee, S.H.; Lee, Y.J. Gosset polytopes in integral octonions. *Czechoslov. Math. J.* **2014**, *64*, 683–702. [CrossRef]
8. Conway, J.H.; Smith, D.A. *On Quaternions and Octonions: Their Geometry, Arithmetic, and Symmetry*; AK Peters: Natick, MA, USA, 2003.
9. Coxeter, H.S.M. *Regular Polytopes*, 3rd ed.; Dover Publications, Inc.: New York, NY, USA, 1973.
10. Lee, J.-H. Gosset polytopes in Picard groups of del Pezzo surfaces. *Can. J. Math.* **2012**, *64*, 123–150. [CrossRef]

symmetry

MDPI

Article

Enumeration of Strongly Regular Graphs on up to 50 Vertices Having S_3 as an Automorphism Group

Marija Maksimović

Department of Mathematics, University of Rijeka, Rijeka 51000, Croatia; mmaksimovic@math.uniri.hr; Tel.: +38-5051-584-665

Received: 17 May 2018; Accepted: 08 June 2018; Published: 11 June 2018

Abstract: One of the main problems in the theory of strongly regular graphs (SRGs) is constructing and classifying SRGs with given parameters. Strongly regular graphs with parameters $(37, 18, 8, 9)$, $(41, 20, 9, 10)$, $(45, 22, 10, 11)$, $(49, 24, 11, 12)$, $(49, 18, 7, 6)$ and $(50, 21, 8, 9)$ are the only strongly regular graphs on up to 50 vertices that still have to be classified. In this paper, we give the enumeration of SRGs with these parameters having S_3 as an automorphism group. The construction of SRGs in this paper is a step in the classification of SRGs on up to 50 vertices.

Keywords: strongly regular graph; automorphism group; orbit matrix

1. Introduction

We assume that the reader is familiar with the basic notions of the theory of finite groups. For basic definitions and properties of strongly regular graphs, we refer the reader to [1–3].

A graph is regular if all its vertices have the same valency; a simple regular graph $\Gamma = (\mathcal{V}, \mathcal{E})$ is strongly regular with parameters (v, k, λ, μ) if it has $|\mathcal{V}| = v$ vertices, valency k, and if any two adjacent vertices are together adjacent to λ vertices, while any two nonadjacent vertices are together adjacent to μ vertices. A strongly regular graph with parameters (v, k, λ, μ) is usually denoted by $\mathrm{SRG}(v, k, \lambda, \mu)$. An automorphism of a strongly regular graph Γ is a permutation of vertices of Γ, such that every two vertices are adjacent if and only if their images are adjacent.

By $S(V)$, we denote the symmetric group on the nonempty set V. If $G \leq S(V)$ and $x \in V$, then the set $xG = \{xg | g \in G\}$ is called a G-orbit of x. The set $G_x = \{g \in G | xg = x\}$ is called a stabilizer of x in G. If G is finite, then $|xG| = \frac{|G|}{|G_x|}$. By G_x^g, we denote a conjugate subgroup $g^{-1}G_x g$ of G_x.

One of the main problems in the theory of strongly regular graphs (SRGs) is constructing and classifying SRGs with given parameters. A frequently-used method of constructing combinatorial structures is the construction of combinatorial structures with a prescribed automorphism group. Orbit matrices of block designs have been used for such a construction of combinatorial designs since the 1980s. However, orbit matrices of strongly regular graphs had not been introduced until 2011. Namely, Majid Behbahani and Clement Lam introduced the concept of orbit matrices of strongly regular graphs in [4]. They developed an algorithm for the construction of orbit matrices of strongly regular graphs with an automorphism group of prime order and the construction of corresponding strongly regular graphs.

A method of constructing strongly regular graphs admitting an automorphism group of composite order using orbit matrices is introduced and presented in [5]. Using this method, we classify strongly regular graphs with parameters $(37, 18, 8, 9)$, $(41, 20, 9, 10)$, $(45, 22, 10, 11)$, $(49, 18, 7, 6)$, $(49, 24, 11, 12)$ and $(50, 21, 8, 9)$ having S_3 as an automorphism group. These graphs are the only strongly regular graphs with up to 50 vertices that still have to be classified. Enumeration of SRGs with these parameters having a non-abelian automorphism group of order six, i.e., the construction of SRGs with these parameters in this paper, is a step in that classification. Using this construction, we show that there is no SRG$(37,18,8,9)$ having S_3 as an automorphism group. Furthermore, we show that there are 80 SRGs$(41,20,9,10)$, 288 SRGs$(45,22,10,11)$, 72 SRGs$(49,24,11,12)$, 34 SRGs$(49,18,7,6)$ and 45 SRGs$(50,21,8,9)$ having a non-abelian automorphism group of order six.

The paper is organized as follows: After a brief description of the terminology and some background results, in Section 2, we describe the concept of orbit matrices, based on the work of Behbahani and Lam [4]. In Section 3, we explain the method of construction of strongly regular graphs from their orbit matrices presented in [5]. In Section 4, we apply this method to construct strongly regular graphs with parameters $(37,18,8,9)$, $(41,20,9,10)$, $(45,22,10,11)$, $(49,18,7,6)$, $(49,24,11,12)$ and $(50,21,8,9)$ having a non-abelian automorphism group of order six.

For the construction of orbit matrices and graphs, we have used our own computer programs written for GAP [6]. Isomorphism testing for the obtained graphs and the analysis of their full automorphism groups are conducted using the Grape package for GAP [7].

2. Orbit Matrices of Strongly Regular Graphs

Orbit matrices of block designs have been frequently used for the construction of block designs with a presumed automorphism group, see, e.g., [8–11]. In 2011, Behbahani and Lam introduced the concept of orbit matrices of SRGs (see [4]). While Behbahani and Lam were mostly focused on orbit matrices of strongly regular graphs admitting an automorphism of prime order, a general definition of an orbit matrix of a strongly regular graph is given in [12].

Let Γ be an SRG(v,k,λ,μ) and A be its adjacency matrix. Suppose an automorphism group G of Γ partitions the set of vertices V into b orbits $O_1,\ldots,O_b$, with sizes $n_1,\ldots,n_b$, respectively. The orbits divide A into submatrices $[A_{ij}]$, where A_{ij} is the adjacency matrix of vertices in O_i versus those in O_j. We define matrices $C=[c_{ij}]$ and $R=[r_{ij}]$, $1\le i,j\le b$, such that c_{ij} is the column sum of A_{ij} and r_{ij} is the row sum of A_{ij}. The matrix R is related to C by:

$$r_{ij}n_i = c_{ij}n_j. \tag{1}$$

Since the adjacency matrix is symmetric, it follows that:

$$R = C^T. \tag{2}$$

The matrix R is the row orbit matrix of the graph Γ with respect to G, and the matrix C is the column orbit matrix of the graph Γ with respect to G.

Behbahani and Lam showed that orbit matrices $R=[r_{ij}]$ and $R^T=C=[c_{ij}]$ satisfy the condition:

$$\sum_{s=1}^{b} c_{is}r_{sj}n_s = \delta_{ij}(k-\mu)n_j + \mu n_i n_j + (\lambda-\mu)c_{ij}n_j.$$

Since $R=C^T$, it follows that:

$$\sum_{s=1}^{b} \frac{n_s}{n_j} c_{is}c_{js} = \delta_{ij}(k-\mu) + \mu n_i + (\lambda-\mu)c_{ij} \tag{3}$$

and:

$$\sum_{s=1}^{b} \frac{n_s}{n_j} r_{si}r_{sj} = \delta_{ij}(k-\mu) + \mu n_i + (\lambda-\mu)r_{ji}.$$

Therefore, in [12], we introduced the following definition of orbit matrices of strongly regular graphs.

Definition 1. *A $(b \times b)$-matrix $R = [r_{ij}]$ with entries satisfying conditions:*

$$\sum_{j=1}^{b} r_{ij} = \sum_{i=1}^{b} \frac{n_i}{n_j} r_{ij} = k \tag{4}$$

$$\sum_{s=1}^{b} \frac{n_s}{n_j} r_{si} r_{sj} = \delta_{ij}(k-\mu) + \mu n_i + (\lambda - \mu) r_{ji} \tag{5}$$

where $0 \leq r_{ij} \leq n_j$, $0 \leq r_{ii} \leq n_i - 1$ and $\sum_{i=1}^{b} n_i = v$, is called a row orbit matrix for a strongly regular graph with parameters (v, k, λ, μ) and the orbit length distribution $(n_1, \ldots, n_b)$.

Definition 2. *A $(b \times b)$-matrix $C = [c_{ij}]$ with entries satisfying conditions:*

$$\sum_{i=1}^{b} c_{ij} = \sum_{j=1}^{b} \frac{n_j}{n_i} c_{ij} = k \tag{6}$$

$$\sum_{s=1}^{b} \frac{n_s}{n_j} c_{is} c_{js} = \delta_{ij}(k-\mu) + \mu n_i + (\lambda - \mu) c_{ij} \tag{7}$$

where $0 \leq c_{ij} \leq n_i$, $0 \leq c_{ii} \leq n_i - 1$ and $\sum_{i=1}^{b} n_i = v$, is called a column orbit matrix for a strongly regular graph with parameters (v, k, λ, μ) and the orbit length distribution $(n_1, \ldots, n_b)$.

3. The Method of Construction

A method of constructing strongly regular graphs admitting an automorphism group of composite order using orbit matrices is introduced and presented in [5]. In this section, we will give a brief overview of this method.

For the construction of strongly regular graphs with parameters (v, k, λ, μ), we first check whether these parameters are feasible (see [2]). Then, we select the group G and assume that it acts as an automorphism group of an SRG(v, k, λ, μ). The construction of strongly regular graphs admitting an action of a presumed automorphism group, using orbit matrices, consists of the following two basic steps:

- Construction of orbit matrices for the presumed automorphism group
- Construction of strongly regular graphs from the obtained orbit matrices (indexing of orbit matrices)

We could use row or column orbit matrices, but since we are constructing matrices row by row, it is more convenient for us to use column orbit matrices. For the construction of orbit matrices for the presumed automorphism group, we need to determine all possible orbit length distributions $(n_1, n_2, \ldots, n_b)$ for an action of the group G. Suppose an automorphism group G of Γ partitions the set of vertices V into b orbits $O_1, \ldots, O_b$, with sizes $n_1, \ldots, n_b$. Obviously, n_i is a divisor of $|G|$, $i = 1, \ldots, b$, and:

$$\sum_{i=1}^{b} n_i = v.$$

When determining the orbit length distribution, we also use the following result that can be found in [13].

Theorem 1. *Let $s < r < k$ be the eigenvalues of an SRG(v, k, λ, μ), then:*

$$\phi \leq \frac{\max(\lambda, \mu)}{k - r} v,$$

where ϕ is the number of fixed points for a nontrivial automorphism group G.

For each orbit length distribution we construct column orbit matrices. For the construction of orbit matrices, we first need to find prototypes.

3.1. Prototypes for a Row of a Column Orbit Matrix

A prototype for a row of a column orbit matrix C gives us information about the number of occurrences of each integer as an entry of a particular row of C. Behbahani and Lam [4,13] introduced the concept of a prototype for a row of a column orbit matrix C of a strongly regular graph with a presumed automorphism group of prime order. We will generalize this concept and describe a prototype for a row of a column orbit matrix C of a strongly regular graph under a presumed automorphism group of composite order. Prototypes are useful in the first step of the construction of strongly regular graphs, namely the construction of column orbit matrices.

Suppose an automorphism group G of a strongly regular graph Γ with parameters (v, k, λ, μ) partitions the set of vertices V into b orbits $O_1, \ldots, O_b$, of sizes $n_1, \ldots, n_b$. With $l_i, i = 1, \ldots, \rho$, we denote all divisors of $|G|$ in ascending order ($l_1 = 1, \ldots, l_\rho = |G|$).

3.1.1. Prototypes for a Fixed Row

Consider the row r of a column orbit matrix C. We say that it is a fixed row of a matrix C if $n_r = 1$, i.e., if it corresponds to an orbit of length one. The entries in this row are either zero or one. Let d_{l_i} denote the number of orbits whose length are $l_i, i = 1, \ldots, \rho$.

Let x_e denote the number of occurrences of an element $e \in \{0, 1\}$ at the positions of the row r that correspond to the orbits of length one. It follows that:

$$x_0 + x_1 = d_1, \tag{8}$$

where d_1 is the number of orbits of length one. Since the diagonal elements of the adjacency matrix of a strongly regular graph are equal to zero, it follows that $x_0 \geq 1$.

Let $y_e^{(l_i)}$ denote the number of occurrences of an element $e \in \{0, 1\}$ at the positions of the row r that correspond to the orbits of length l_i $(i = 2, \ldots, \rho)$. We have:

$$y_0^{(l_i)} + y_1^{(l_i)} = d_{l_i}, \quad i = 2, \ldots, \rho \tag{9}$$

Because the row sum of an adjacency matrix of Γ is equal to k, it follows that:

$$x_1 + \sum_{i=2}^{\rho} l_i \cdot y_1^{(l_i)} = k. \tag{10}$$

The vector:

$$p_1 = (x_0, x_1; y_0^{(l_2)}, y_1^{(l_2)}; \ldots; y_0^{(l_\rho)}, y_1^{(l_\rho)})$$

whose components are nonnegative integer solutions of the equalities (8), (9) and (10) is called a prototype for a fixed row.

3.1.2. Prototypes for a Non-Fixed Row

Let us consider the row r of a column orbit matrix C, where $n_r \neq 1$. Let d_{l_i} denote the number of orbits whose length is $l_i, i = 1, \ldots, \rho$.

If a fixed vertex is adjacent to a vertex from an orbit $O_i, 1 \leq i \leq b$, then it is adjacent to all vertices from the orbit O_i. Therefore, the entries at the positions corresponding to fixed columns are either zero or n_r. Let x_e denote the number of occurrences of an element $e \in \{0, n_r\}$ at those positions of the row r, which correspond to the orbits of length one. We have:

$$x_0 + x_{n_r} = d_1. \tag{11}$$

The entries at the positions corresponding to the orbits whose lengths are greater than one are $0, 1, \ldots, n_r - 1$ or n_r. The entry at the position (r, r) is $0 \le c_{rr} \le n_r - 1$, since the diagonal elements of the adjacency matrix of strongly regular graphs are zero.

Let $y_e^{(l_i)}$ denote the number of occurrences of an element $e \in \{0, \ldots, n_r\}$ of row r at the positions that correspond to the orbits of length l_i $(i = 2, \ldots, \rho)$. From (1) and (2), we conclude that:

$$c_{ri}n_i = c_{ir}n_r, \tag{12}$$

where $c_{ir} \in \{0, \ldots, n_i\}$. If $c_{ri} \cdot \frac{n_i}{n_r} \notin \{0, \ldots, n_i\}$, then $y_{c_{ri}}^{(n_i)} = 0$. It follows that:

$$\sum_{e=0}^{n_r} y_e^{(l_i)} = d_{l_i}, \quad i = 2, \ldots, \rho. \tag{13}$$

Since the row sum of an adjacency matrix is equal to k, we have that:

$$x_{n_r} + \sum_{i=2}^{\rho} \sum_{h=1}^{n_r} y_h^{(l_i)} \cdot h \cdot \frac{n_{l_i}}{n_r} = k, \tag{14}$$

From (3), we conclude that:

$$\sum_{s=1}^{b} c_{rs}c_{rs}n_s = (k - \mu)n_r + \mu n_r^2 + (\lambda - \mu)c_{rr}n_r,$$

where $c_{rr} \in \{0, \ldots, n_r - 1\}$. It follows that:

$$n_r^2 x_{n_r} + \sum_{i=2}^{\rho} \sum_{h=1}^{n_r} y_h^{(l_i)} \cdot h^2 \cdot n_{l_i} = (k - \mu)n_r + \mu n_r^2 + (\lambda - \mu)c_{rr}n_r, \tag{15}$$

The vector:

$$p_{n_r} = (x_0, x_{n_r}; y_0^{l_2}, \ldots, y_{n_r}^{l_2}; \ldots; y_0^{l_\rho}, \ldots, y_{n_r}^{l_\rho}),$$

whose components are nonnegative integer solutions of Equalities (11), (13), (14) and (15) is called a prototype for a row corresponding to the orbit of length n_r.

Using prototypes, we construct an orbit matrix row by row.

Not every orbit matrix gives rise to a strongly regular graph, while, on the other hand, a single orbit matrix may produce several nonisomorphic strongly regular graphs. Further, nonisomorphic orbit matrices may produce isomorphic graphs. Therefore, the constructed graphs need to be checked for isomorphism.

Theorem 2. *Let $\Gamma = (V, E)$ be a strongly regular graph, $G \le Aut(\Gamma)$, and let $(b \times b)$-matrix C be a column orbit matrix of the graph Γ with respect to the group G. Further, let α be an element of $S(V)$ with the following property: if $\alpha(i) = j$, then the stabilizer G_{x_i} is conjugate to G_{x_j}, where $x_i, x_j \in V$ and $O_i = x_iG, O_j = x_jG$. Then, there exists permutation $g^* \in C_{S(V)}(G)$ such that $\alpha(i) = j \iff g^*(O_i) = O_j$.*

Definition 3. *Let $A = (a_{ij})$ be an $(b \times b)$-matrix and $\alpha \in S_b$. The matrix $B = A\alpha$ is the $(b \times b)$-matrix $B = (b_{ij})$, where $b_{\alpha(i)\alpha(j)} = a_{ij}$. If $A\alpha = A$, then α is called an automorphism of the matrix A.*

Definition 4. *Let an $(b \times b)$-matrix $A = (a_{ij})$ be the orbit matrix of a strongly regular graph Γ with respect to the group $G \le Aut(\Gamma)$. A mapping $\alpha \in S_b$ is called an isomorphism from A to $B = A\alpha$ if the following condition holds: if $\alpha(i) = j$, then the stabilizer G_{x_i} is conjugate to G_{x_j}. We say that the orbit matrices A and B are isomorphic. If $A\alpha = A$, then α is called an automorphism of the orbit matrix A. All automorphisms of an orbit matrix A form the full automorphism group of A, denoted by $Aut(A)$.*

During the construction of orbit matrices, for the elimination of isomorphic structures, we use permutations that satisfy the conditions from Theorem 2, i.e., isomorphisms from Definition 4.

The next big step of the construction of graphs, called indexing, often cannot be performed in a reasonable amount of time. To make such a construction possible, for a refinement of constructed orbit matrices, we use the composition series:

$$\{1\} = H_0 \trianglelefteq H_1 \trianglelefteq \cdots \trianglelefteq H_n = G,$$

of a solvable automorphism group G of a strongly regular graph. Let Γ be a strongly regular graph and $H \trianglelefteq G \leq \mathrm{Aut}(\Gamma)$. Each G-orbit of Γ decomposes to H-orbits of the same size (see [9]). Therefore, each orbit matrix for the group G decomposes to orbit matrices for the group H, and the following theorem holds [5].

Theorem 3. *Let Ω be a finite nonempty set, $H \lhd G \leq S(\Omega)$, $x \in \Omega$ and $xG = \bigsqcup_{i=1}^{h} x_i H$. Then, a group G/H acts transitively on the set $\{x_i H \mid i = 1, 2, \ldots, h\}$.*

Therefore, after we have constructed corresponding orbit matrices for the group G, we continue until we find all refinements for the normal subgroup $H_{n-1} \trianglelefteq G$. In the next step, we obtain orbit matrices for the group H_{n-2}, H_{n-3}, and so on. Our last step is the construction of the corresponding orbit matrices for the subgroup $H_0 = \{1\}$, i.e., construction of adjacency matrices of the strongly regular graphs. The concept of the G-isomorphism of two-block designs was introduced in [14]. For the elimination of mutually-isomorphic structures, we use the concept of G-isomorphism.

Definition 5. *Let $\Gamma_1 = (V, E_1)$ and $\Gamma_2 = (V, E_2)$ be strongly regular graphs, and let $G \leq Aut(\Gamma_1) \cap Aut(\Gamma_2) \leq S(V)$. An isomorphism $\alpha : \Gamma_1 \to \Gamma_2$ is called a G-isomorphism from Γ_1 onto Γ_2 if there is an automorphism $\tau : G \to G$ such that for each $x, y \in V$ and each $g \in G$, the following holds:*

$$(\alpha x)(\tau g) = \alpha y \Leftrightarrow xg = y.$$

If α is a $G-$isomorphism from Γ_1 to Γ_2, then the vertices x_i and x_j are in the same G-orbit if and only if the vertices $\alpha(x_i)$ and $\alpha(x_j)$ are in the same G-orbit.

Lemma 1. *Let $\Gamma_1 = (V, E_1)$ and $\Gamma_2 = (V, E_2)$ be strongly regular graphs, and let $G \leq Aut(\Gamma_1) \cap Aut(\Gamma_2) \leq S = S(V)$. A permutation $\alpha \in S$ is a G-isomorphism from Γ_1 onto Γ_2 if and only if α is an isomorphism from Γ_1 to Γ_2 and $\alpha \in N_S(G)$, where $N_S(G)$ is the normalizer of G in S.*

In each step of refinement of an orbit matrix A, we eliminate isomorphic orbit matrices using the automorphisms from $Aut(A)$, because each automorphism of an orbit matrix determines an G-isomorphism.

4. SRGs with up to 50 Vertices Having S_3 as an Automorphism Group

SRGs with parameters $(37, 18, 8, 9)$, $(41, 20, 9, 10)$, $(45, 22, 10, 11)$, $(49, 24, 11, 12)$, $(49, 18, 7, 6)$ and $(50, 21, 8, 9)$ are the only strongly regular graphs on up to 50 vertices that still have to be classified [2,15]. According to [2], it is known that strongly regular graphs with these parameters exist, but their final enumeration result is not known. In this section, we present the results of the constructed strongly regular graphs with parameters $(37, 18, 8, 9)$, $(41, 20, 9, 10)$, $(45, 22, 10, 11)$, $(49, 18, 7, 6)$, $(49, 24, 11, 12)$ and $(50, 21, 8, 9)$ having $S_3 \cong Z_3 : Z_2 \cong \langle \rho\phi | \rho^3 = 1, \phi^2 = 1, \phi\rho\phi = \rho^{-1} \rangle$ as an automorphism group. In each case, we construct strongly regular graphs by using the algorithm described in Section 3. The orbit lengths for an action of the group G at the set of points of a graph can get values from the set $\{1, 2, 3, 6\}$. Using the program Mathematica [16], we get all possible orbit length distributions

(d_1, d_2, d_3, d_6) for the action of S_3 on a particular SRG that satisfy Theorem 1. For each orbit length distribution, we find the corresponding prototypes using Mathematica. Using our own programs, which are written for GAP [6], we construct all orbit matrices for a given orbit length distribution. Having in mind the action of the whole group, we refine the constructed orbit matrices. For the refinement, we use the composition series

$$\{1\} \trianglelefteq \langle \rho \rangle \trianglelefteq S_3$$

and obtain orbit matrices for the action of the subgroup $Z_3 \triangleleft S_3$. In this step, each orbit of length two and six decomposes to two orbits of length one and three, respectively. In the final step of the construction, we obtain adjacency matrices of the strongly regular graphs with particular parameters admitting a non-abelian automorphism group of order six. Finally, we check isomorphisms of strongly regular graphs and determine orders of the full automorphism groups using the Grape package for GAP [7].

4.1. SRGs(37,18,8,9)

In this section, we present the results of SRGs(37,18,8,9) having S_3 as an automorphism group. According to [17], there are at least 6760 SRGs(37,18,8,9), and none of them have S_3 as an automorphism group. We show that there are no strongly regular graphs with parameters (37,18,8,9) having a non-abelian automorphism group of order six.

We get 176 possibilities for orbit length distributions, but only three give rise to orbit matrices. In Table 1, we present the number of mutually-nonisomorphic orbit matrices for each orbit length distribution, the number of orbit matrices for Z_3 (obtained by the refinement) and the number of constructed SRGs with parameters (37,18,8,9). These calculations prove Theorem 4.

Table 1. Number of orbit matrices and SRGs(37,18,8,9) for the automorphism group S_3.

Distribution	#OM-S_3	#OM-Z_3	#SRGs
(1,0,0,6)	3	6	0
(1,0,4,4)	3	3	0
(1,0,8,2)	3	3	0

Theorem 4. *There are no strongly regular graphs with parameters* $(37, 18, 8, 9)$ *having an automorphism group isomorphic to the symmetric group* S_3.

4.2. SRGs(41,20,9,10)

In this section, we present the results of SRGs$(41, 20, 9, 10)$ having S_3 as an automorphism group. We show that there are exactly 80 strongly regular graphs with parameters $(41, 20, 9, 10)$ having a non-abelian automorphism group of order six.

We get 216 possibilities for orbit length distributions, but only one gives rise to any orbit matrices. In Table 2, we present the number of mutually-nonisomorphic orbit matrices for each orbit length distribution, the number of orbit matrices for Z_3 (obtained by the refinement) and the number of constructed SRGs with parameters $(41, 20, 9, 10)$. These calculations prove Theorem 5. Information about the orders of the full automorphism groups is presented in Table 3.

Table 2. Number of orbit matrices and SRGs(41,20,9,10) for the automorphism group S_3.

Distribution	#OM-S_3	#OM-Z_3	#SRGs
(1,2,4,4)	10	10	80

Theorem 5. *Up to isomorphism, there are exactly 80 strongly regular graphs with parameters* $(41, 20, 9, 10)$ *having an automorphism group isomorphic to the symmetric group* S_3.

Table 3. SRGs with parameters (41,20,9,10) having S_3 as an automorphism group.

$\|Aut(\Gamma)\|$	#SRGs
6	80

The adjacency matrices of the constructed SRGs can be found at [18].

4.3. SRGs(45,22,10,11)

In this section, we present the results of SRGs$(45, 22, 10, 11)$ having S_3 as an automorphism group. We show that there are exactly 288 strongly regular graphs with parameters $(45, 22, 10, 11)$ having a non-abelian automorphism group of order six.

We get 309 possibilities for orbit length distributions, but only one gives rise to any orbit matrices. In Table 4, we present the number of mutually-nonisomorphic orbit matrices for each orbit length distribution, the number of orbit matrices for Z_3 (obtained by the refinement) and the number of constructed SRGs with parameters $(45, 22, 10, 11)$. These calculations prove Theorem 6. Information about orders of the full automorphism groups is presented in Table 5.

Table 4. Number of orbit matrices and SRGs(45,22,10,11) for the automorphism group S_3.

Distribution	#OM-S_3	#OM-Z_3	#SRGs
(1,4,4,4)	7	7	288

Table 5. SRGs with parameters (45,22,10,11) having S_3 as an automorphism group

$\|Aut(\Gamma)\|$	#SRGs
6	288

Theorem 6. *Up to isomorphism, there are exactly 288 strongly regular graphs with parameters* $(45, 22, 10, 11)$ *having an automorphism group isomorphic to the symmetric group* S_3.

The adjacency matrices of the constructed SRGs can be found at [19].

4.4. SRGs(49,18,7,6)

In the paper [5], we proved the following theorem.

Theorem 7. *Up to isomorphism, there are exactly* 36 *strongly regular graphs with parameters* $(49, 18, 7, 6)$ *having an automorphism group isomorphic to the symmetric group* S_3.

Two of these graphs have not been constructed in [4,20]. The adjacency matrices of the constructed SRGs can be found at [21].

4.5. SRGs(49,24,11,12)

In this section, we present the results of SRGs$(49, 24, 11, 12)$ having S_3 as an automorphism group. We show that there are exactly 72 strongly regular graphs with parameters $(49, 24, 11, 12)$ having a non-abelian automorphism group of order six.

We get 435 possibilities for orbit length distributions, but only a few give rise to orbit matrices. In Table 6, we present the number of mutually-nonisomorphic orbit matrices for each orbit length distribution, the number of orbit matrices for Z_3 (obtained by the refinement) and the number of constructed SRGs with parameters $(49, 24, 11, 12)$. Thus, we prove Theorem 8. Information about orders of the full automorphism groups is presented in Table 7.

Table 6. Number of orbit matrices and SRGs$(49, 24, 11, 12)$ for the automorphism group S_3.

Distribution	#OM-S_3	#OM-Z_3	#SRGs	Distribution	#OM-S_3	#OM-Z_3	#SRGs
(0,2,3,6)	8	16	6	(1,3,0,7)	6	6	0
(0,2,5,5)	4	0	0	(1,3,2,6)	10	2	0
(0,2,7,4)	8	0	0	(1,3,6,4)	2	2	0
(1,0,0,8)	2	15	2	(1,6,0,6)	1	0	0
(1,0,2,7)	20	32	0	(3,2,0,7)	4	10	0
(1,0,8,4)	26	24	12	(3,2,6,4)	6	16	0
(1,0,10,3)	2	0	0	(4,0,9,3)	6	0	0
(1,0,12,2)	16	0	0	(5,1,0,7)	2	4	0
(1,0,14,1)	12	0	0	(5,1,6,4)	2	2	12
				(7,0,0,7)	2	2	40

Table 7. SRGs with parameters $(49, 24, 11, 12)$ having S_3 as an automorphism group.

$\|\mathrm{Aut}(\Gamma)\|$	#SRGs
6	42
18	22
24	4
126	4

Theorem 8. *Up to isomorphism, there are exactly 72 strongly regular graphs with parameters* $(49, 24, 11, 12)$ *having an automorphism group isomorphic to the symmetric group* S_3.

The adjacency matrices of the constructed SRGs can be found at [22].

4.6. SRGs(50,21,8,9)

In this section, we present the results of SRGs$(50, 21, 8, 9)$ having S_3 as an automorphism group. According to [17], there are 18 graphs obtained from the 18 Steiner (2,4,25) systems, and three of them have S_3 as an automorphism group. We show that there are exactly 45 strongly regular graphs with parameters $(50, 21, 8, 9)$ having a non-abelian automorphism group of order six. Hence, to our best knowledge, 42 of the constructed strongly regular graphs are new.

We get 340 possibilities for orbit length distributions, but only a few give rise to orbit matrices. In Table 8, we present the number of mutually-nonisomorphic orbit matrices for each orbit length distribution, the number of orbit matrices for Z_3 (obtained by the refinement) and the number of constructed SRGs with parameters $(50, 21, 8, 9)$. Thus, we prove Theorem 9. Information about the orders of the full automorphism groups is presented in Table 9.

Theorem 9. *Up to isomorphism, there are exactly 45 strongly regular graphs with parameters* $(50, 21, 8, 9)$ *having an automorphism group isomorphic to the symmetric group* S_3.

The adjacency matrices of the constructed SRGs can be found at [23].

In Table 10, we summarize the obtained results, i.e., give a list of all the obtained strongly regular graphs and orders of their full automorphism groups.

Table 8. Number of orbit matrices and SRGs(50, 21, 8, 9) for the automorphism group S_3.

Distribution	#OM-S_3	#OM-Z_3	#SRGs	Distribution	#OM-S_3	#OM-Z_3	#SRGs
(0,1,2,7)	10	3	2	(2,0,2,7)	10	16	0
(0,1,4,6)	10	4	6	(2,0,8,4)	22	24	12
(0,1,6,5)	12	21	6	(2,0,10,3)	2	0	0
(0,1,8,4)	8	8	1	(2,0,12,2)	27	0	0
(0,4,2,6)	2	2	0	(2,0,14,1)	14	0	0
(0,4,4,5)	4	3	16	(2,3,0,7)	2	3	0
(0,4,6,4)	3	4	0	(2,3,2,6)	6	1	0
(0,4,8,2)	4	6	0	(2,3,6,4)	2	4	0
(1,2,3,6)	10	20	5	(4,2,6,4)	6	12	0
(1,2,5,5)	2	0	0	(5,0,9,3)	2	0	0
(1,2,7,4)	4	0	0	(6,1,6,4)	1	1	4

Table 9. SRGs with parameters (50, 21, 8, 9) having S_3 as an automorphism group.

$\|Aut(\Gamma)\|$	#SRGs
6	35
18	6
72	1
150	1
336	1
504	1

Table 10. SRGs on up to 50 vertices having S_3 as an automorphism group.

(v, k, λ, μ)	$\|Aut(\Gamma)\|$	#SRGs
(41, 20, 9, 10)	6	80
(45, 22, 10, 11)	6	288
(49, 18, 7, 6)	6	18
(49, 18, 7, 6)	12	2
(49, 18, 7, 6)	18	2
(49, 18, 7, 6)	24	4
(49, 18, 7, 6)	48	1
(49, 18, 7, 6)	72	4
(49, 18, 7, 6)	126	1
(49, 18, 7, 6)	144	2
(49, 18, 7, 6)	1008	1
(49, 18, 7, 6)	1764	1
(49, 24, 11, 12)	6	42
(49, 24, 11, 12)	18	22
(49, 24, 11, 12)	24	4
(49, 24, 11, 12)	126	4
(50, 21, 8, 9)	6	35
(50, 21, 8, 9)	18	6
(50, 21, 8, 9)	72	1
(50, 21, 8, 9)	150	1
(50, 21, 8, 9)	336	1
(50, 21, 8, 9)	504	1

Funding: This research was funded by [Croatian Science Foundation] Grant number [1637].

Acknowledgments: Special thanks to Dean Crnković.

Conflicts of Interest: The founding sponsors had no role in the design of the study; in the collection, analyses or interpretation of data; in the writing of the manuscript; nor in the decision to publish the results.

Abbreviations

The following abbreviations are used in this manuscript:

SRG Strongly regular graph

References

1. Beth, T.; Jungnickel, D.; Lenz, H. *Design Theory Volume I*, 2nd ed.; Cambridge University Press: Cambridge, UK, 1999.
2. Brouwer, A.E. Strongly Regular Graphs. In *Handbook of Combinatorial Designs*, 2nd ed.; Colbourn, C.J., Dinitz, J.H., Eds.; Chapman & Hall/CRC: Boca Raton, FL, USA, 2007; pp. 852–868.
3. Tonchev, V.D. *Combinatorial Configurations: Designs, Codes, Graphs*; Longman Scientific & Technical: New York, NY, USA, 1988.
4. Behbahani, M.; Lam, C. Strongly regular graphs with non-trivial automorphisms. *Discret. Math.* **2011**, *311*, 132–144. [CrossRef]
5. Crnković, D.; Maksimović, M. Construction of strongly regular graphs having an automorphism group of composite order. under review.
6. GAP Groups. Algorithms, Programming—A System for Computational Discrete Algebra, Version 4.7.2. 2013. Available online: http://www.gap-system.org (accessed on 20 January 2017).
7. Soicher, L.H. The GRAPE Package for GAP, Version 4.6.1. 2012. Available online: http://www.maths.qmul.ac.uk/~leonard/grape/ (accessed on 4 April 2018).
8. Janko, Z. Coset enumeration in groups and constructions of symmetric designs. *Ann. Discret. Math.* **1992**, *52*, 275–277.
9. Crnković, D.; Rukavina, S. Construction of block designs admitting an abelian automorphism groups. *Metrika* **2005**, *62*, 175–183. [CrossRef]
10. Crnković, D.; Rukavina, S.; Schmidt, M. A Classification of all Symmetric Block Designs of Order Nine with an Automorphism of Order Six. *J. Combin. Des.* **2006**, *14*, 301–312. [CrossRef]
11. Crnković, D.; Pavčević, M.O. Some new symmetric designs with parameters (64,28,12). *Discret. Math.* **2001**, *237*, 109–118. [CrossRef]
12. Crnković, D.; Maksimović, M.; Rodrigues, B.G.; Rukavina, S. Self-orthogonal codes from the strongly regular graphs on up to 40 vertices. *Adv. Math. Commun.* **2016**, *10*, 555–582. [CrossRef]
13. Behbahani, M. On Strongly Regular Graphs. Ph.D. Thesis, Concordia University, Montreal, QC, Canada, May 2009.
14. Ćepulić, V. On Symmetric Block Designs (40,13,4) with Automorphisms of Order 5. *Discret. Math.* **1994**, *128*, 45–60. [CrossRef]
15. Brouwer, A.E. Parameters of Strongly Regular Graphs. Available online: http://www.win.tue.nl/~aeb/graphs/srg/srgtab1-50.html (accessed on 1 April 2018).
16. Wolfram Mathematica, Version 7.0.0. 2008. Available online: http://www.wolfram.com/mathematica/ (accessed on 4 April 2018).
17. Spence, E. Strongly Regular Graphs on at Most 64 Vertices. Available online: http://www.maths.gla.ac.uk/~es/srgraphs.php (accessed on 10 April 2018).
18. Maksimović, M. SRGs(41,20,9,10) having S_3 as an automorphism group. Available online: http://www.math.uniri.hr/~mmaksimovic/srg41.txt (accessed on 10 June 2018).
19. Maksimović, M. SRGs(45,22,10,11) having S_3 as an automorphism group. Available online: http://www.math.uniri.hr/~mmaksimovic/srg45.txt (accessed on 10 June 2018).
20. Behbahani, M.; Lam, C.; Östergård, P.R.J. On triple systems and strongly regular graphs. *J. Combin. Theory Ser. A* **2012**, *119*, 1414–1426. [CrossRef]
21. Maksimović, M. SRGs(49,18,7,6) having S_3 as an automorphism group. Available online: http://www.math.uniri.hr/~mmaksimovic/srg49s3.txt (accessed on 10 June 2018).
22. Maksimović, M. SRGs(49,24,11,12) having S_3 as an automorphism group. Available online: http://www.math.uniri.hr/~mmaksimovic/srgs49.txt (accessed on 10 June 2018).
23. Maksimović, M. SRGs(50,21,8,9) having S_3 as an automorphism group. Available online: http://www.math.uniri.hr/~mmaksimovic/srg50.txt (accessed on 10 June 2018).

MDPI
St. Alban-Anlage 66
4052 Basel
Switzerland
Tel. +41 61 683 77 34
Fax +41 61 302 89 18
www.mdpi.com

Symmetry Editorial Office
E-mail: symmetry@mdpi.com
www.mdpi.com/journal/symmetry

www.ingramcontent.com/pod-product-compliance
Lightning Source LLC
LaVergne TN
LVHW071934080526
838202LV00064B/6608
* 9 7 8 3 0 3 9 2 8 1 9 0 9 *